Biology Today

An Issues Approach

Biology Today

An Issues Approach

Eli C. Minkoff
Pamela J. Baker

*Department of Biology
Bates College,
Lewiston, Maine*

The McGraw-Hill Companies, Inc.

New York   St. Louis   San Francisco   Auckland   Bogotá   Caracas
Lisbon   London   Madrid   Mexico City   Milan   Montreal   New Delhi
San Juan   Singapore   Sydney   Tokyo   Toronto

McGraw-Hill

*A Division of The **McGraw·Hill** Companies*

Biology Today: An Issues Approach
Copyright © 1996 by The McGraw-Hill Companies, Inc. All rights reserved.
Printed in the United States of America. Except as permitted under the
United States Copyright Act of 1976, no part of this publication may be
reproduced or distributed in any form or by any means, or stored in a data
base or retrieval system, without the prior written permission of the
publisher.

1 2 3 4 5 6 7 8 9 DOW DOW 9 0 9 8 7 6 5

ISBN 0-07-042629-5

This book was set in New Aster by GTS Graphics.
The editors were Denise Schanck and Ty McConnell;
the design was done by Keithley and Associates, Inc.;
the production supervisor was Elizabeth J. Strange.
R. R. Donnelley & Sons Company was printer and binder.

Library of Congress Cataloging-in-Publication Data 95-081972

About the Authors

Eli C. Minkoff is Professor of Biology at Bates College. He received his bachelor's degree, Magna Cum Laude, from Columbia University and his M.A. and Ph.D. from Harvard University. He has also taught at Harvard University, Northeastern University, and the University of Southern Maine. A member of Phi Beta Kappa and the Society for the Study of Evolution, Dr. Minkoff has published numerous articles in scholarly journals and encyclopedic reference works. He is also the author of several texts and laboratory manuals, including the textbook *Evolutionary Biology*.

Pamela J. Baker is Associate Professor of Biology at Bates College. She holds two bachelor of science degrees (from the University of Wales at Swansea, Magna Cum Laude, and from Bates College, Cum Laude). She also holds an M.A. and Ph.D. from the State University of New York at Buffalo. Her research focuses on the function of the immune system in periodontal disease and on the investigation of new antimicrobial agents. She has published many articles in research journals and in reference works and is the author of a children's sign language dictionary, *My First Book of Sign*.

Contents

Preface

Those who have been teaching introductory biology for the last two or three decades have been burdened with the supposition that *absolutely everything* needs to be covered in a single course. Textbooks written in this tradition are weighty, encyclopedic works with a thousand or more pages and hefty pricetags. Students are exposed to the *results* of biology without gaining understanding of biology as a *process of discovery*. Students soon forget much of the information because it lacks meaning to them. Studies have shown that many students are discouraged from further biology courses by their experiences with introductory courses that use an encyclopedic approach.

Our book represents an attempt to get away from this "tyranny of coverage" by using an issues-oriented approach to the teaching of biology, one that emphasizes coherent understanding of selected issues rather than an attempt to cover everything. The issues we have chosen are current topics that students are likely to see in the news or to read about in books of general interest. It is our feeling that students (especially those not majoring in biology) are more likely to remember this material if it is meaningfully related to issues of concern to them. Our approach accordingly helps students to experience the connections among the fields of biology, the interdisciplinary nature of today's biology, and the intimate connections between biological and social issues. We also hope to instill in students the feeling that biology is both interesting and relevant to their lives, and that a further understanding of biology can be a delight rather than a burden.

We are also committed as teachers to fostering understanding of biology, what some now call "biological literacy." Thus, we have not overlooked the teaching of "facts," but we have chosen to teach these facts in a context that emphasizes how these facts are produced, organized, and used to solve problems. Thus the issues we have selected are ones that are not only of current importance, but ones that lend themselves as vehicles for teaching the major concepts of biology. One such list of major concepts is contained in the pamphlet *Developing Biological Literacy*, issued by the Biological Sciences Curriculum Study (BSCS). As the table following this Preface shows, we have covered all of these concepts, some of them in several places. We have also endeavored to cover them in a way that enables students to see the connections between them, and we have provided a list of further connections at the end of each chapter.

Biology as a discipline has become fragmented to the extent that different perspectives on the same problem, for example, molecular perspectives and environmental perspectives, are often taught in separate courses with no reference to each other. We aim for a more comprehensive view of each issue. The current understanding of each issue is covered from different perspectives, which often include cellular and molecular perspectives, organismal or individual perspectives, and global or population perspectives, combined as appropriate. Coverage of each issue also includes its social context, both historical and contemporary. Phrasing ideas as "our current understanding" will help students to realize the ongoing nature of discovery and to identify the processes that are necessary for new ideas to be accepted. In no case should students assume that we have covered all there is to say on any issue.

One of the aims of our approach is to educate good citizens, biologists and nonbiologists alike, with an understanding that will enable them to evaluate scientific arguments and make appropriate decisions affecting their own lives and the well-being of society. We are committed to teaching science as a human activity that impinges upon other aspects of society and gives rise to social issues that require discussion. Citizens are increasingly called upon to deal with science-based issues throughout their lives, in the foods they choose to eat, the medicines they take, and the very air they breathe. Legislators, juries, and corporate managers need to make important decisions, affecting many lives, based in part on the findings of science. We think that all good citizens need to know science and the way that scientists work, and they also need to know how science can be used and how it can be misused.

We aim to stimulate critical thinking and questioning rather than memorization. Science cannot be learned by memorizing facts. Science is a process rather than a result, and it is an understanding of that process that we hope to instill in students, beginning in Chapter 1. To help students appreciate this process, we have presented multiple interpretations or points of view as much as possible. Societal and ethical issues will be mentioned wherever relevant, and Chapter 2 is devoted to an examination of ethical principles. We encourage teachers to set aside time for class discussions to further stimulate student thought, or for students to set up such discussions among themselves informally. We have included Thought Questions at the end of each section of each chapter with such class discussions in mind. A few thought questions have factual "right" answers; most do not. Some questions require students to do further reading; many questions encourage students to think about the limitations of available data or to think about the applications and implications of science. We encourage students with differing viewpoints to discuss these questions among themselves and to ask, "What further information would help us resolve our differences or reach decisions?" We also invite students and their instructors to write us at **biobook@abacus.bates.edu** with comments, questions, or suggestions.

In addition to the Table of Biological Concepts, this book contains the following study aids:

1. **Chapter outlines and section headings:** The book is divided into sections, each of which begins with a section heading that states its major idea. These section headings and important subheadings also appear in the table of contents as a guide to both previewing and reviewing the contents of each chapter.

2. **Illustrations:** Since many of the concepts of biology can be understood and remembered visually, the book is well illustrated with photographs and drawings. The captions to these illustrations often provide another important avenue to understanding.

3. **Tables:** Some tables usefully summarize important sections of text; other tables provide specific examples of material mentioned in the text.

4. **Boxes:** Boxes provide the background for the material in the text. The material presented in boxes is just as important as the main text but is presented in this "stand alone" fashion so that it does not break the flow of the text.

5. **Thought Questions:** At the end of each major section of a chapter are a series of Thought Questions. Some of these are "check your understanding" questions for which the answers may be found in the text. More, however, are questions that ask students to think about the limits of data or about applications of biological findings in society. These questions can form the basis for discussion either in class or in informal study groups. Many will require further reading beyond what is presented in the text.

6. **Chapter Summaries:** The key ideas from each chapter are reviewed in a summary at the end of the chapter.

7. **Key Terms:** Important vocabulary is highlighted in bold type in the text. All of the Key Terms from a chapter are listed at the end of the chapter with page references. Key Terms are mainly concepts rather than specific examples (that is, neurotransmitter rather than the name of a specific neurotransmitter).

8. **Connections to Other Chapters:** At the end of each chapter, a series of sentences relate the concepts of that chapter to the material in other chapters.

9. **A Classification of Organisms:** A taxonomic summary appears as an appendix at the end of the book.

10. **Bibliography:** Suggestions for further readings are presented at the end of the book, organized by chapter.

11. **Glossary:** Each of the Key Terms is defined in the glossary at the end of the book.

ACKNOWLEDGMENTS

We would like to take this opportunity to thank the many people who reviewed portions of this text and provided us with helpful suggestions. In alphabetical order, they were: Lee Abrahamsen, Bates College; Gregory Anderson, Bates College; Andrew N. Ash, Pembroke State University; David Baker; Bruce Bourque, Bates College; Joe W. Camp, Jr., Purdue University North Central; William J. Campbell, Louisiana Tech University; Mary Colavito-Shepanski, Santa Monica College; Diane Cowan, formerly of Bates College; Martha Crunkleton, Bates College; David Cummiskey, Bates College; Christine P. Curran, University of Cincinnati; Mark Dixon, Bates College; Elizabeth Eames, Bates College; Lynn A. Ebersole, Northern Kentucky University; Edward Goldin, former student at Bates College; Helen Greenwood, University of Southern Maine; David Handley, University of Maine Agricultural Research Station; Pat Hauslein, St. Cloud State University; Alan R. P. Journet, Southeast Missouri State University; Donald E. Keith, Tarleton State University; John Kelsey, Bates College; Sharon Kinsman, Bates College; Virginia G. Latta, Jefferson State Community College; Richard J. Meyer, Humboldt State University; Glendon R. Miller, Wichita State University; Nancy Minkoff; Neil Minkoff, Lahey-Hitchcock Medical Center; Jane Noble-Harvey, University of Delaware; Mark Okrent, Bates College; Lois Ongley, Bates College; Joseph G. Pelliccia, Bates College; Karen Rasmussen, Bowdoin College; Gary Shields, Kirkwood Community College; Barbara Y. Stewart, Swarthmore College; Gregory J. Stewart, West Georgia College; Robert Thomas, Bates College; Robert M. Thornton, University of California–Davis; Robin W. Tyser, University of Wisconsin; James E. Urban, Kansas State University; Aaron Wallack, Cognex Corporation; Linda Wallack; Thomas Wenzel, Bates College; Anne Williams, Bates College; Thomas M. Wolf, Washburn University of Topeka; and H. Elton Woodward, Daytona Beach Community College. We also thank Sylvia Warren for her many helpful suggestions and Laura Malloy for introducing us to the Biological Sciences Curriculum Study pamphlet. Special thanks are also due to Kathi Prancan, Denise Schanck, and Ty McConnell, and the rest of the staff at McGraw-Hill who helped us throughout the process of following this book to completion.

Eli C. Minkoff
Pamela J. Baker

TABLE OF BIOLOGICAL CONCEPTS

The following matrix indicates which biological concepts (grouped by theme) are developed in each of the sixteen chapters. A red arrow (▶) marks coverage of a concept within a chapter.

Chapter	**EVOLUTION: PATTERNS OF CHANGE**						**EVOLUTION: PRODUCTS OF CHANGE**						**INTERACTION AND INTERDEPENDENCE**					**GENETIC CONTINUITY AND REPRODUCTION**							**GROWTH, DEVELOPMENT, AND DIFFERENTIATION**				**ENERGY, MATTER, AND ORGANIZATION**				**MAINTENANCE OF A DYNAMIC EQUILIBRIUM**				
	The dynamic Earth	Forces of evolutionary change	Patterns of evolution	Extinction	Conservation biology	Population genetics	Origin and characteristics of life	Specialization and adaptation	Species and speciation	Phylogenetic classification	Biodiversity	Biogeography	Biosphere	Ecosystems	Community structure	Population ecology	Environmental factors	The gene	DNA, the genetic material	Gene action	Patterns of inheritance	Reproduction	Molecular genetics	Genetics and biotechnology	Patterns of growth	Patterns of development	Differentiation	Form and function	Molecular structure	Hierarchy of organization	Matter	Energy and metabolism	Detection of environmental stimuli	Movement	Homeostasis	Health and disease	Behavior
1. Biology as a Science (**)							▶											▶	▶														▶				
2. Ethics and Social Decision Making (**)																																				▶	
3. Human Genes and Genomes																		▶	▶	▶	▶	▶	▶	▶			▶	▶								▶	
4. Evolution		▶	▶				▶	▶	▶	▶	▶	▶				▶	▶				▶	▶						▶									
5. Variation among Human Populations		▶				▶	▶	▶	▶	▶					▶	▶				▶	▶						▶								▶	▶	
6. The Population Explosion				▶										▶	▶	▶						▶					▶				▶						▶
7. Sociobiology and Reproductive Strategies	▶	▶													▶	▶						▶						▶								▶	▶
8. Nutrition Affects Health															▶														▶	▶	▶	▶			▶	▶	
9. Cancer and Cancer Therapy																			▶	▶			▶			▶	▶					▶			▶	▶	
10. Brain Functions and Malfunctions		▶																												▶	▶	▶		▶	▶	▶	
11. The Use and Abuse of Drugs																												▶		▶		▶			▶	▶	▶
12. The Mind–Body Connection			▶																									▶	▶	▶	▶				▶	▶	▶
13. AIDS and HIV			▶																			▶	▶				▶								▶	▶	▶
14. Bioengineering													▶	▶												▶		▶				▶			▶	▶	
15. Plants and Crop Production	▶	▶	▶					▶					▶	▶	▶	▶	▶						▶		▶	▶	▶	▶				▶			▶	▶	
16. Biodiversity, Extinction, and Endangered Habitats	▶			▶	▶	▶		▶			▶	▶	▶	▶	▶	▶	▶															▶				▶	

(**) NOTE: The process of scientific discovery and the ethical and social implications of science are also covered in all subsequent chapters.

Biology Today

An Issues Approach

Biology as a Science

Biology is the scientific study of living systems. Our gardens, our pets, our trees, and our fellow humans are all examples of living systems. We can look at them, admire them, write poems about them, and enjoy their company. The Nuer, a pastoral people of Africa, care for their cattle and attach great emotional value to each of them. They write poetry about—and occasionally to—their cattle, they name themselves after their favorite cows or bulls, and they move from place to place according to the needs of their cattle for new pastures. They come to know individual cattle very well, almost as members of the family. Like many other people who live close to the land, the Nuer have also acquired a vast store of useful knowledge about the many animal and plant species that occur in their region. Scientific understanding grew out of a thorough familiarity with the environment, supplemented by a tradition of systematic testing. This chapter deals with the methods of science in general, and with the application of those methods to the study of living systems.

Living systems share all or most of the properties listed in Box 1.1, a set of properties on which most biologists agree. Because living systems are complex and continually changing, an understanding of these systems often requires special methods of investigation or ways of formulating thoughts. This chapter deals with those special methods, a type of investigation that has come to be called **science.** Many people think that science is defined by its subject matter, but this is not correct. *Science is defined instead by its methods.*

Science is a distinctly human activity. As such, it cannot be divorced from other aspects of human life. Science has always been affected by the lives of individual scientists, including their lives outside the laboratory. Religious ideas, political ideas, and social prejudices have all found their way into theories that have been labeled scientific by their adherents. Most scientists seek to reduce the effect of such influences; the first step in doing so is to admit that influences of this kind are always present.

A. Science Is Based on Testing Falsifiable Hypotheses

Hypotheses Are Central to Science

The essence of science is the formulation and testing of certain kinds of statements called **hypotheses.** At the moment of its inception, a hypothesis is a tentative explanation of events that occur or of how something works. What makes science distinctive is that its hypotheses are then subjected to rigorous testing. One result of the testing process is that many hypotheses are rejected as false. Eliminating one hypothesis often helps us frame the next hypothesis.

Hypotheses must be statements about the observable universe. These hypotheses must also be

---

BOX 1.1   CHARACTERISTICS OF LIVING SYSTEMS

Anything is considered to be a living system if it exhibits growth, metabolism, homeostasis, and selective response at some time during its existence. Living systems are composed of organisms that can be either single-celled or multicellular. Organisms belong to populations of similar organisms, at least some of which are capable of reproducing.

- Metabolism. All living things take energy-rich materials from their environment and release other materials that, on the average, have a lower energy content. Some of the energy is used to carry on life processes, but some also accumulates and is released only upon death.
- Motion. Most (but not all) living systems convert some of the energy they use into motion of some sort, including internal motion within cells.
- Selective response. All living systems have some capability of responding selectively to certain external stimuli and not to others. Many organisms respond to offensive stimuli by withdrawing. All organisms have some capacity to distinguish needed nutrients from other chemicals and to respond appropriately in most cases.
- Homeostasis. All living systems have at least some capacity to change potentially harmful or threatening conditions into conditions

more favorable to their continuing existence, e.g., by metabolizing certain toxic chemicals into less harmful ones.
- Growth and biosynthesis. All living things go through phases during which they make more of their own material at the expense of some of the materials around them.
- Genetic material. All living organisms contain hereditary information derived from previously existing organisms. This genetic material takes the form of a nucleic acid (either DNA or RNA) in all known cases.
- Reproduction. All living beings have some capacity to make other organisms similar to themselves by transmitting at least some of their genetic material.
- Population structure. All living organisms belong to populations of similar organisms. Populations can be defined retrospectively as organisms related by common descent. Among organisms capable of sexual processes, populations may also be defined prospectively to include all those organisms that can interbreed with one another.

Viruses strain these definitions of living systems. Viruses contain genetic material yet they do not exist as cells, and replicate only inside and with the help of some other organism.

---

formulated in such a way that they can be tested by comparison to the world of experience. To be a hypothesis, a statement must be either **verifiable** (confirmable) or **falsifiable** (capable of being shown to be untrue). This process of testing by comparison to the observable universe is called *empirical testing.* Observations gathered in testing any hypothesis are generally called **data.**

Statements that are not hypotheses.   It follows from the above definition of hypotheses that certain types of statements cannot be used as scientific hypotheses.

Moral judgments and religious concepts are excluded from science because they are not falsifi-

able. For example, the statement "there is a God" cannot be disproven or falsified by any possible demonstration of empirical fact or observation.

Moral or ethical views concerning human conduct have an influence on science, but such moral questions cannot be decided by evidence alone (Chapter 2).

Judgments about what ideas or things are valuable are not subject to falsification by hypothesis testing. Well-designed opinion surveys can gather and summarize data about how many people agree or disagree with a particular value judgment—for example, "Education is important for its own sake"—but survey results cannot be used as evidence that any particular judgment is "true."

Judgments of what things are beautiful or likable are called esthetic judgments. The statements "Jazz is good music," "Salsa is better than ketchup," and "My roommate is a nice person" are examples of esthetic judgments and are not falsifiable hypotheses. Poetry, literature, and art are judged by esthetic criteria unrelated to the scientific principle of falsifiability.

Specific versus general hypotheses. Hypotheses that are easy to verify generally tell us very little. For example, the hypothesis "The sun will rise in the east tomorrow morning" can be tested by awakening early, facing east, and observing what happens. If the sun does rise, then our hypothesis is verified, or confirmed; if the sun does not rise, then our hypothesis is falsified, or disconfirmed. However, the confirmation of this hypothesis is far from an important scientific discovery. It is relatively unimportant because it is too specific, which is exactly what makes it verifiable.

Suppose, now, that we examine the much bolder hypothesis "The sun will rise in the east *every* morning." We can test this second hypothesis in the same way that the first hypothesis was tested, by rising early and facing east, and we can also declare that the hypothesis would be falsified if the sun failed to rise. But what if the sun does rise? Does this verify that the sun will rise *every* morning? Suppose we decide to watch the sunrise five days in a row, or five thousand? A single failure of the sun to rise will absolutely falsify the hypothesis, but no finite number of sunrises would be sufficient to verify the hypothesis for all time. This is the kind of hypothesis that science usually examines: hypotheses that are absolutely falsifiable, but not absolutely verifiable.

Falsified hypotheses are rejected, and new hypotheses (which may in some cases be modifications of the original hypotheses) are suggested in their place. If testing a hypothesis does not reject it, we may want to generalize the hypothesis. For example, if a hypothesis tested using rats has not been falsified, we may want to apply the hypothesis to people as well, or to all animals. However, we can never know how far we can extrapolate (generalize) results unless we continue to try to falsify our premise under different conditions. In this way, the testing of hypotheses allows us to draw conclusions about the observable world, but only to the extent that we have tested many possible circumstances and conditions.

The importance of falsifiable hypotheses in science was first emphasized by the philosopher Karl Popper. In particular, it was Popper who first pointed out that the distinction between science and other disciplines can be based upon the fact that scientists are always seeking to test falsifiable hypotheses. According to Popper, scientists perform the curious exercise of trying to falsify the hypotheses that they believe in, then publishing the results. Scientists convince each other to believe in certain hypotheses according to how rigorously or how often they have tried to falsify these hypotheses and failed.

A definition of science. **Science** may now be defined as a method of investigation based on the testing of falsifiable hypotheses that take the form of universal generalizations that can be falsified but never absolutely verified. Notice that this makes scientific statements *tentative*, or provisional, subject to possible falsification on the next occasion that a test is conducted. Even the most cherished scientific belief can be falsified—for example, the sun may fail to rise tomorrow morning. Repeated exposure of our hypotheses to possible falsification increases our confidence in these hypotheses when they are not falsified, but no amount of testing can guarantee absolute truth.

Any hypothesis that is tested again and again without ever being falsified is considered to be well supported and comes to be generally accepted. It may be used as the basis for formulating further hypotheses, so there is soon a cluster of related hypotheses, supported by the results of many tests, which is then called a *theory*, as described below.

Ways of devising hypotheses. **Deduction** is reasoning from the general to the specific. Deductive logic of the "If . . . then" form is frequently used to set up testable hypotheses: "*If* organisms of type X require oxygen to live, *then* this individual of type X will die if I put it in an atmosphere without oxygen." Often contrasted with deduction is another type of reasoning called **induction** (or, more properly, *inductive generalization*), reasoning from the specific to the general. This type of reasoning is commonly used in day-to-day life: "I like the pizza in restaurants A, B, C, and D; therefore I will like pizza in any other restaurant." Induction never guarantees the truth of any conclusions drawn— "The next restaurant may serve pizza that I don't like." As we have seen above, science also uses inductive methods to generalize from specific hypotheses. If a certain drug slows down the heart rate in this rat and that rat, perhaps it will also have the same effect in other mammals. If it also slows the heart rate in two humans and a turtle, then maybe it will do so in many other animals, perhaps all animals. Induction also produces the "ah hah" moments for scientists, in which a series of seemingly unrelated observations suddenly coalesce into a cohesive picture.

Deduction and induction are only two of the many ways in which scientists go about the business of formulating hypotheses. Other ways include (1) intuition or imagination, (2) esthetic preferences, (3) religious and philosophical ideas, (4) comparison and analogy with other processes, and (5) serendipity, or the discovery of one thing while looking for something else. Moreover, these ways may be mixed or combined. For example, Albert Einstein declared that he arrived at his hypotheses about the physics of the universe by considering esthetic qualities such as beauty or simplicity and by asking, "If I were God, how would I have made the world?" Einstein also said that "imagination is more important than knowledge," a remark that is particularly true for the formulation of hypotheses (Fig. 1.1). Nobel prize–winning physicist Niels Bohr said that his hypothesis of atomic structure (the heavy nucleus in the center, with the electrons circling rapidly around it, "like a miniature solar system") first occurred to him by analogy with our solar system. Alexander Fleming found the first antibiotic as the result of a laboratory accident: On dishes of bacteria that should have been thrown away earlier, he observed clear areas where fungi had overgrown the bacteria. His hypothesis, that a product of the fungi had killed the bacteria, was validated by tests and that fungal product is what we now know as penicillin. As these several examples show, *hypotheses are formed by all kinds of logical and extralogical processes*, which is one more reason why they must be subjected to rigorous testing afterward.

Hypothesis testing in variable systems. Biological systems are complex and variable. No individual animal or plant is exactly like any other animal or plant. At any instant in time, they may differ in their internal conditions, in their external conditions, or in the way these conditions are interacting. Further, the same individual is not exactly the same from one day to the next. Because living systems vary, tests must be repeated. If a hypothesis is tested in one animal, or one cell, and the organism responds in a particular way, the re-

ROSE IS ROSE reprinted by permission of UFS, Inc.

FIGURE 1.1
Imaginative hypotheses may originate from various logical or extralogical processes,
especially from young scientists. Does the idea shown here qualify as a scientific hypothesis?

sult is far less reliable as a means of prediction than it would be if 10 animals, or 100 cells, all responded in the same way. What often happens, however, is that 9 out of 10 animals, or 94 of 100 cells, respond in one way and the remainder in another way. When some individuals respond differently from the rest, it may be the result of a source of variation that has not yet been identified. Scientists who devote their attention to the anomalous cases may sometimes discover new phenomena that were previously ignored.

Interpretation of the results from tests on variable systems usually requires statistical treatment to find out whether the observed differences are "real" or can be explained by random variation.

For example, scientists who suspected that dietary fats were contributing to the risks for heart disease obtained important evidence on this hypothesis by comparing heart attack rates in populations with low fat consumption with heart attack rates in populations with high fat consumption. Once the rates of heart attack in the populations under study had been determined, the results were analyzed statistically to determine whether the preliminary findings—that high fat consumption was associated with increased risk of heart attack—were meaningful or could have arisen by chance or from sampling error. (Error occurs when the people picked for the sample studied are not representative of the general population, either because the sample is too small or because it was not chosen randomly.) Studies of the relationship between high fat diets and heart disease have indeed found significantly higher heart attack rates (and also significantly higher breast cancer rates) in populations with higher levels of dietary fat consumption.

What Is a Theory?

A **theory** is a cluster of related hypotheses that share a common vocabulary and a common subject matter. Theories develop after the results of many tests have accumulated. The language of the theory nearly always refers to certain entities that cannot be seen, and are hence "theoretical." Examples of such entities are *electron, force, gene, enzyme, reflex, fitness, species, ecosystem, extinction, community, aversion,* and *family.* Each unseen entity can be defined in terms of the observable effects that it produces. For example, geneticists observe that the offspring of individual organisms, behave *as if* certain hereditary factors (genes) are present. Physicists have observed that metallic foils behave *as if* they are composed of atoms. We can even distinguish different types of genes or types of atoms, but they remain unseen nevertheless.

The language of science. Scientists sometimes coin words to denote new theoretical concepts and sometimes they give existing words new meaning in the context of the new theory. The intent is to foster accurate communication among scientists, but sometimes the specialized use of familiar words (like *altruism* by population ecologists and sociobiologists or *self* by immunologists) serves as a barrier to communication between scientists and nonscientists and even between scientists from different fields.

In any case, we should recognize that scientific terms mean only what a particular theory says they mean, or what scientists who support that theory understand them to mean. When opposing theories are in conflict, a frequently occurring problem is that the same word is used by different groups of scientists to mean different things, or that the same thing is described in different words by the supporters of different theories.

Productive theories. One of the most important features of a good theory is that it will often suggest new and different hypotheses to test. A theory of this kind is a stimulus to further research and is sometimes called a *productive* theory. Sometimes, two or more theories may offer competing explanations for the same data, a situation that almost always stimulates research from several directions. A theory may be productive for a while and then no longer stimulate new research, in part because research is a peculiarly human endeavor. The theories that last are the ones that remain productive the longest, while the less productive theories are often abandoned without ever being fully disproved. In some cases, it is the falsification of one of its hypotheses (or the failure of a crucial test) that causes a theory to be rejected—remember that the hypotheses that make up a theory are always subject to possible falsification. Even a long-cherished theory may be abandoned (or greatly modified) if it no longer holds predictive or explanatory power.

Theoretical models.   Many theories can be communicated using a simplified mathematical or visual form, called a **model.** Such a model, while not a formal part of the theory, can nevertheless be an important teaching tool in helping communicate the theory to other people. For example, Bohr's conceptualization of the atom in terms of electrons circling around the nucleus like a miniature solar system was the model of atomic structure for generations of students. However, models are analogies. Like other analogies, models are comparable to the phenomena they describe only so far, and no further. Attempts to determine *how far* an analogy holds often suggest new hypotheses to test or lead to new ways of testing old hypotheses. The planetary model of atomic structure is a case in point. With the development of quantum physics, it became clear that the solar system model was inadequate to explain the behavior of atomic particles. Similarly, the model of genes on a chromosome as a linear sequence, "like beads on a string," which was popularized by early twentieth-century geneticists, has been supplanted by the double-helix model. Thus, theories and models are not simply opinions or points of view; they are mechanisms for generating ideas that can be tested and for communicating these ideas.

THOUGHT QUESTIONS

1. In a group, discuss the hypothesis shown in Fig. 1.1. Is it falsifiable? If you believe so, then explain what sorts of observations might falsify it. How could we go about testing such an idea?
2. Which of the following are falsifiable statements? For each statement that you think is falsifiable, explain what sort of observation might falsify it.

    This horse is a cinch to win in the next race.
    Pearl Jam is a better musical group than The Rolling Stones.
    In a maze that they have never seen before, rats will turn right just about as often as they will turn left.

    The hungry cat will eat because hunger awakens the food spirits within the cat.
    The hungry cat will eat because hunger awakens an innate food-seeking drive within the cat.
    The angles of a triangle always add up to 180 degrees.
    If these two plants are crossed, approximately half of the offspring will resemble one parent and half will resemble the other.
    It is wrong to inflict pain on a cat.
    Restaurant A is better than restaurant B.
    The average science major at this school gets better grades than the average humanities major.
    All people should be treated equally under the law.

3. Which of the following are examples of inductive reasoning? Which are deductive?

    All green leaves I have ever tested contain chlorophyll, so the green leaves on that tree contain chlorophyll, too.
    If chlorophyll is soluble in alcohol, and if this leaf contains chlorophyll, then I should be able to dissolve the chlorophyll from this leaf by soaking it in alcohol.
    If all adult female birds lay eggs, then this female chick will lay eggs if raised to maturity.
    If all known species of birds are egg-laying, then the next species to be discovered will be egg-laying, too.
    If chemical X destroys vitamin C, then I should be able to produce the symptoms of vitamin C deficiency by feeding these animals only food that has been treated with chemical X.

B. Testing Hypotheses Varies in Different Branches of Science

Hypotheses are tested by comparing them to the real world, i.e., by making empirical observations. In this, science differs from pure mathematics, which examines only theoretically defined concepts. The sciences differ from one another, however, in the ways in which hypotheses are tested.

Experimental Science

Some scientists test hypotheses by conducting **experiments**—artificially contrived situations set up for the express purpose of testing some hypothesis. Most **experimental sciences** aim, in one way or another, to answer questions of the form "How does X work?" The scientist designs an experiment such that a certain outcome is expected (or not expected); then the results of the experiment are determined *objectively*, which means, in this context, *without bias either for or against the hypothesis being tested*, or *without any bias that would impair the falsifiability of the hypothesis*.

Many experiments involve comparison of an experimental situation or group with a *control* situation or **control group** in which all variables are ideally held the same except for the one being tested. For example, animals given a new drug are compared to a similar group of animals—the control group—that are not given the drug. To make the results strictly comparable, the control group should be given a substance similar to whatever is given to the experimental group, but lacking the one ingredient thought to be essential. In order to make sure that any difference in outcome can be attributed to the presence or absence of the drug, care must be taken to ensure that the two groups are equivalent in every other way: similar animals, similar cages, similar temperatures, similar diets, and so on.

The use of the word *control* in an experimental context is different from the usual use of the word. Scientists try to control (standardize) the possible sources of variation in their experiments. They are not exercising authority over the animals or predetermining the outcome of the experiment. We often see confusion over the term *control* in the caricature of the "mad scientist," a literary character who usually prides himself (nearly always *him*self) on his ability to control everything, often extending into a quest to control the world.

As an example of the experimental approach, consider the following experiment in bacterial genetics that was conducted by Joshua and Esther Lederberg, part of the basis for Joshua's subsequent Nobel prize. Most bacteria are killed by an antibiotic like streptomycin, but the Lederbergs exposed the common intestinal bacterium *Escherichia coli* to this drug and were able to isolate a number of streptomycin-resistant bacteria. They allowed these bacteria to reproduce and were able to show that resistance to streptomycin was inherited by their offspring. In other words, a permanent genetic change had occurred; such changes are called *mutations* (Chapter 3). Now, the Lederbergs had two hypotheses to test. The first hypothesis was that the mutation had been *induced,* or caused, by exposure to the streptomycin. The second was that the mutation had occurred before (and therefore independently of) exposure to the streptomycin. In order to distinguish between these hypotheses, the Lederbergs devised the experiment shown in Fig. 1.2. In this experiment, a copy, or replica, of the original plate of bacteria was made. Only the replica, not the original, was exposed to streptomycin, and the position of each bacterial colony was noted. The induced mutation hypothesis predicted that the mutation would occur whenever a bacterium was exposed to streptomycin. In fact, most of the bacteria died (thus falsifying this hypothesis), but an occasional colony proved to be streptomycin-resistant. The prior mutation hypothesis predicted that the mutation for streptomycin resistance had occurred before the exposure to streptomycin. To test this second hypothesis, the Lederbergs went back and tested the colonies from the original plate. They discovered that the same colonies that were streptomycin-resistant on the replica plate were also streptomycin-resistant on the original plate. This finding was consistent with the prior mutation hypothesis for this particular sample of bacteria.

The prior mutation hypothesis had been tested and not falsified in the case of one mutation for drug resistance in one species of bacteria. How far could the finding be generalized? From this one experiment alone, one cannot tell. However, other investigators repeated the experiment for other mutations and other species of microorganisms. So far, the hypothesis of prior mutation has not been falsified. It is difficult to test the hypothesis in large or long-lived organisms, but most scientists are willing to assume the truth of the hypothesis for *all* organisms. There are many species (and thousands of mutations for each species) that have never been tested in this way, which leaves a good deal of opportunity for the hypothesis to be falsified at some time in the future.

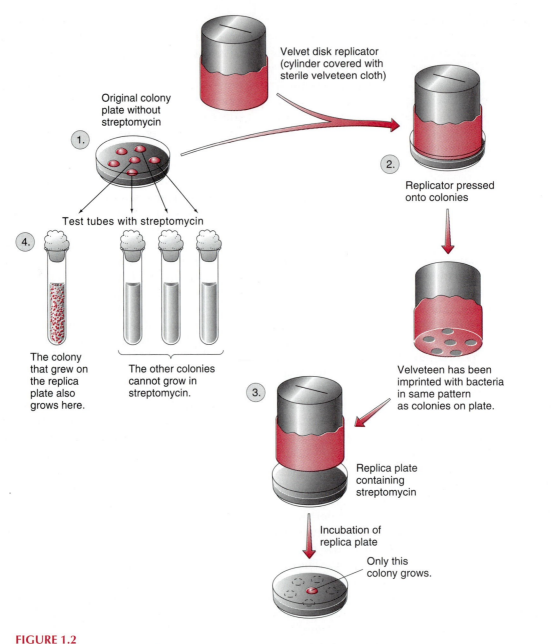

Velvet disk replicator
(cylinder covered with
sterile velveteen cloth)

Original colony
plate without
streptomycin

1.

2.

Replicator pressed
onto colonies

Test tubes with streptomycin

4.

Velveteen has been
imprinted with bacteria
in same pattern
as colonies on plate.

3.

The colony
that grew on
the replica
plate also
grows here.

The other colonies
cannot grow in
streptomycin.

Replica plate
containing
streptomycin

Incubation of
replica plate

Only this
colony grows.

FIGURE 1.2
The Lederbergs' replica-plating experiment.

Naturalistic Science

Another type of hypothesis testing is one in which direct experimental manipulation is either impossible or is held to be undesirable. For example, if an animal behaviorist wishes to study mating behavior *under natural conditions,* then any experimental manipulation that alters these natural conditions is generally to be avoided. Thus we see ornithologists hiding in blinds to study birds and other species, while other naturalists photograph their subjects using telephoto lenses. There are also sciences like astronomy, whose subject matter includes stars too large or too far away to be experimentally manipulated. The extinct species studied by paleontologists cannot be recreated in the laboratory to suit the needs of some experimenter. We can refer to all these other sciences as **naturalistic sciences** because their method is based primarily upon *naturalistic observation* rather than experiment. Naturalists still attempt to falsify hypotheses, but they do so by patient observation and accurate record keeping. The major difference between the experimental and the naturalistic sciences is that experimentalists set up and control the experiments themselves, while naturalists can only observe those "experiments" that occur in nature.

The naturalistic sciences, moreover, are *historical sciences*. Any scientist seeking to understand why mammals differ from reptiles, or why the U.S. economy differs from the Japanese economy, will soon realize that the histories of the types of animals or of the particular economies form an important part of the explanation. Each of the examples that they deal with is historically unique to some extent.

The most characteristic question in the naturalistic or historical sciences is, "How did X get to be that way?" For example, Cann, Stoneking, and Wilson (1987) examined the DNA inside the mitochondria (the major energy-producing cell parts) of a large number of human populations. Mitochondrial DNA is always inherited from the mother, never from the father. Some populations had mitochondrial DNA whose chemical structure was very similar to that of other populations, allowing groups of related populations to be recognized. Cann and her co-workers hypothesized that populations with similar mitochondrial DNA sequences shared a close common descent through female lineages. This hypothesis explains the patterns of similarity among mitochondrial DNA sequences by a series of progressively "smaller" hypotheses about the past histories of a given set of populations: that the populations of the Americas all share a common descent, that the populations of New Guinea all share a common descent, and so on. Some of these smaller hypotheses are falsified by the data and must be replaced by modified hypotheses. New Guinea, for example, forms two clusters, and we can set up the hypothesis that it was colonized twice, with each line of descent forming a separate cluster. According to a widely accepted interpretation of the data, all of the smaller hypotheses of geographic dispersal fall into a common pattern of descent (see Fig. 5.4), with an area of origin in Africa. That is, the data are consistent with a hypothesis that all human populations are descended from an ancestral African population, or from a single ancestral female, nicknamed "Eve" in the popular press. Like most other explanations in the natural sciences, the "Eve hypothesis" explains present conditions on the basis of their past history, an evolutionary or historical mode of explanation.

To summarize, both experimental sciences and naturalistic sciences use the scientific method of posing hypotheses and then testing those hypotheses. Both methods are used in biology, but neither is confined to biology because both are used in many other scientific disciplines.

THOUGHT QUESTIONS

Which of the following are experimental tests, and which are naturalistic observations?

1. A scientist measures the light that reaches the Earth during a solar eclipse and compares this with similar measurements from other eclipses.

2. A scientist measures the number of times that adult male deer, adult female deer, and fawns forage in a certain feeding area and records whether they do so singly or together.

3. A scientist measures the amount of precipitate formed when 0.1 ml (milliliters) of a particular chemical solution is added to 4 ml each of a number of test solutions.

4. Biologists use an electron microscope to compare the structure of mitochondria in the cells of rats that have been fed with those of rats that have not been fed.

5. Measurements made on the bones of an extinct species are compared with similar measurements made on the bones of a related living species.

6. The activity of white blood cells in a blood sample taken from stressed rats is compared to the activity of white blood cells taken from unstressed rats.

7. A group of animals is fed a certain chemical to see if they will get cancer as a result.

C. Revolutionary Science Differs from "Normal" Science

In a book that was itself considered revolutionary when it was first published, Thomas Kuhn (1968) proposed a new way of looking at the ways in which science accommodates to new discoveries. Kuhn's observations were based on his studies of historical revolutions in science.

According to Kuhn, everything that we have described thus far is part of **normal science**—science that proceeds by the piecemeal discovery and gradual accumulation of new but small findings. Normal science in Kuhn's theory is always channeled by what he calls a **paradigm** (pronounced "para-dime"). A paradigm is much more than a theory; it includes a strong belief in the truth (and the future research potential) of one or more theories and the associated theoretical (unseen) entities. A paradigm thus includes its own specialized vocabulary, which may well involve words that have other meanings in other contexts, but which have specialized meanings within the paradigm. A paradigm also includes a series of value-laden beliefs, including shared opinions as to what problems are important, what problems are unimportant or uninteresting, what techniques and research methods are useful, and so on. The research methodology and sometimes the instrumentation are important parts of the paradigm.

Paradigms, according to Kuhn, are best represented by science textbooks, which are written for the purpose of training new scientists within the paradigm. Students trained by these textbooks are taught not just facts, they are taught attitudes, approaches, values, and a vocabulary that teach them to think in certain preferred ways. The telling of the history of science usually emphasizes the "founders" of the paradigm, who are generally lionized as heroic figures. *Normal science always proceeds cumulatively, in small steps, within the context of an existing paradigm.*

Scientific Revolutions

Once in a great while, says Kuhn, science proceeds in a very different way. A **scientific revolution** occurs: One entire paradigm is discarded and a new one replaces it. Since most of the older scientists were educated in the old paradigm, few of them will support the new paradigm at first. Most of the support for the new paradigm comes from the recruitment of new scientists just beginning their careers, and the founder of the revolution is usually either young or a new entrant into that particular scientific field. Once a scientific revolution occurs, its new paradigm opens up a new field of investigation or rejuvenates an old one. Such an infrequent event is called a *paradigm shift*. A paradigm shift requires that the new paradigm explain everything that the older paradigm explained, and more besides. Usually this means that the vocabulary of the old paradigm must either be adopted by the new paradigm or translated into newer terms.

According to Kuhn, a paradigm shift is not just a triumph of logic or of experimental evidence. It is decided, at least in some measure, by a political-style process involving shifting allegiances and influences. New paradigms succeed when scientists find them to be fruitful or productive of new approaches to research. In modern contexts, papers published, research grants awarded, and professional advancement are often used as measures of research productivity. The shift to the new paradigm becomes inevitable when an entire generation of scientists has been trained in its lessons and the older generation has retired or died off. The triumph of the Darwinian paradigm over its competitors, described in Chapter 4, is a very good example of a paradigm shift.

Paradigms are sometimes so powerful as to allow bothersome observations that "don't fit," or *anomalies*, to be ignored. To scientists working within such a paradigm, anomalies are small problems that they agree to ignore, believing that the integrity and success of the paradigm are more important than trying to accommodate the unexplained anomaly. Scientists who become interested in the anomaly must work outside the paradigm, which means that they may become founders of their own new paradigms.

Within each paradigm, scientists must make many *personal* decisions: What research project should *I* pursue? Under whom should *I* study or work? To whom should *I* go for help? To what agency should *I* apply for funding? Science students quickly find out where the opportunities lie, and those who do not usually do not succeed. Thus, each paradigm recruits a new generation of students to continue research within the paradigm, partly on the strength of the appeal of the research problems, but also partly on the strength of the personality of the leaders in the field. Paradigms are made successful in large measure by the students they attract. Paradigms that no longer attract students die out.

Molecular Genetics: A Paradigm in Biology

As an example of a scientific paradigm in biology, we will describe the field of molecular genetics as it has existed since about 1950. Other examples of scientific paradigms are described in subsequent chapters, including Darwinian evolution in Chapter 4, sociobiology in Chapter 7, and the mind-body connection in Chapter 12.

The *molecular genetics paradigm* (or *molecular biology*) emerged in the decades following the determination of the structure of DNA by Watson and Crick in 1953. The structure of DNA was itself simply a hypothesis whose increasing ability to make falsifiable predictions made it into a theory. The *central dogma* of molecular biology (so named by the molecular biologists themselves) was that DNA was used to make RNA and RNA was used to make protein. (Further details of this process are described in Chapter 3.) Both DNA and RNA were said to contain information, and the making of one molecule from another was said to be a form of information transfer. The central dogma was more than just a theory because it also suggested a new vocabulary and drove a new research program. How did DNA make copies of itself? This was called *replication* long before any of its details became known. How was information from DNA transferred to RNA? This was called *transcription*. How was information from RNA transferred to protein? This was called *translation*. How replication, transcription, and translation occurred were among the major problems to be solved. This terminology was part of an elaborate analogy that drew its inspiration from a comparison with language and included such new vocabulary words as *code* (the language itself), *codons* (items in the code), and *reading frame* (intervals in which the code was processed). Also, words like *transcription* (rewriting within the same language) and *translation* (changing from one language to another) were

deliberately chosen since their literal meaning matched the biological theory. Textbook descriptions were replete with verbs like *read, copy,* and *translate*. There were also a number of laboratory methods, inherited from the field of biochemistry, plus a few extra technical advances, such as the use of high-speed centrifuges. Together, this all formed an orderly paradigm that outlined not only what was known, but also what remained to be discovered, what was thought to be important, and how the details were to be investigated and described. DNA was championed as the most important "master molecule," RNA was almost as important, and protein was important only until its synthesis was completed. Protein that was completely synthesized was no longer deemed interesting, except for a few enzymes that helped in the working of DNA or RNA. The paradigm thus defined the boundaries of the field.

The paradigm of molecular genetics guided research on DNA, RNA, and protein synthesis throughout the 1950s and 1960s; much of the work begun in those decades continues today. For its workers, the paradigm defined a set of shared beliefs (including the central dogma), a vocabulary, a set of research techniques, and, above all, a set of problems to be solved. These problems included the mechanisms by which replication, transcription, and translation took place, as well as how to crack the genetic code. Once this last problem had been solved, the "coding dictionary" (i.e., the list of correspondences between RNA sequences and protein sequences) was given a prominent place in every genetics book and most general biology texts. It was truly a celebration of one of the many great achievements made under this paradigm.

As the molecular genetics paradigm matured, some of its early tenets were modified. Information flow, once thought to be unidirectional, is now believed to be bidirectional. The idea of "master" molecules that "make" or "control" other molecules is slowly being replaced by a vocabulary that speaks in terms of cells "communicating" with other cells (sending and receiving signals) or "influencing" other cells (in both directions now). Likewise, attention has shifted to new questions, such as how the environment of a cell influences that cell to transcribe certain portions of its DNA at certain times. The molecular biology paradigm, like other paradigms before it, has gradually changed over time, although its core beliefs remain unshaken.

D. Science Occurs in a Cultural Context

Is science something that only scientists can do? Can only biologists do biology? On the contrary, many people use scientific methods in their everyday lives. For example, if my car fails to start, I might formulate one hypothesis after another as to the possible cause. To test the hypothesis that the car is out of gas, I would examine the gas gauge. Now, if the gas gauge is very low, the tank might be empty, or it might have a small but adequate amount of gas, or the gauge may be malfunctioning. To distinguish among these hypotheses, I might add a small amount of gasoline to the tank and then try to start the car. If the car now starts, I conclude that it was out of gas. If I want to find out what foods give my child a stomachache, I can keep careful written records of each food she eats and of the occurrence of stomachaches. When comparison of these data has shown that eating a particular food always precedes the onset of a stomachache, I can formulate a hypothesis that the food brought on the stomachache. I can test that hypothesis further by observing what happens the next time she has that food. The child psychologist Jean Piaget has written that children often behave as little scientists, formulating possibilities (hypotheses) in their minds and then testing them. "I can take the toy away from my little brother" can be tested by trying to take it away; the hypothesis would be falsified if brother successfully resisted or if an adult intervened.

BOX 1.2   "TURF WARS" AND OTHER FORMS OF COMPETITION IN SCIENCE

Within the same college or university department, individual scientists often compete with one another for laboratory space, department funds, or the like. They also compete in a more subtle way to attract more and better students to do research *their* way. On a national or worldwide scale, scientists also compete for research grants from various agencies, both government and private.

Sometimes scientists let the turf-war competition itself become the overriding concern. Here are some examples:

COPE VERSUS MARSH

In the last third of the nineteenth century, Edward D. Cope and Othneil C. Marsh competed with one another for the discovery of dinosaur bones and other large fossils in the western United States, which was being rapidly settled at the time. They occasionally stole each other's specimens or hid the specimens that they could not bring back home with them from the field. They spied on each other's field parties, and sometimes dug up their hidden specimens and stole them as well.

WATSON AND CRICK VERSUS PAULING

In his widely read account of the discovery of the double helix structure of DNA, James Watson (1968) admits that he and Francis Crick purposely withheld information from Nobel Prize–winning chemist Linus Pauling, who was also thought to be working on the problem. On one occasion, they even passed misleading information to Pauling's son in the hopes that he would throw his father off the right track.

LYSENKO VERSUS GENE THEORY

Beginning in the 1930s, a Soviet agricultural researcher named Trofim D. Lysenko developed a way of encouraging certain varieties of wheat to germinate earlier. Lysenko incorrectly hypothesized that he had changed the genetic makeup of his wheat. When other researchers failed to duplicate his results, he used his political position to discredit their motives, calling them (in essence) threats to Soviet agriculture. As he became increasingly powerful, Lysenko founded a new journal devoted to his theory that all of Western genetics was a capitalist anti-Soviet plot. Stocks of fruit flies and other research or-

As these examples serve to show, investigation along scientific lines is used in everyday life by children and adults alike. Of course, scientists are generally methodical and rigorous in their record keeping and testing of hypotheses, while nonscientists may or may not be. Scientists also have more formal ways of communicating their results to one another. However, the scientific methods used by ordinary citizens do not differ in any fundamental way from the methods used by scientists.

Science: A Cross-Cultural Discipline

The testing of hypotheses is an ancient discipline. Examples of hypothesis testing can be found in the historical writings of Herodotus (fifth century B.C.). Agricultural practices developed by trial and error or by more systematic observation and experience were put to the test by peoples who depended on farming for their livelihood. Practices that gave poor results were abandoned; those that seemed to give good results were adopted more widely. Non-Western civilizations have frequently developed sophisticated systems of medicine, metallurgy, and other technologies on the basis of a solid scientific foundation. In those cultures where written communication did not exist, other mechanisms ensured transmission of the information across the generations: traditions of apprenticeship, formulations that were chanted or sung (and were thus more easily memorized), and so on.

The Scientific Community

Because it is unusual for a single person to suggest a hypothesis, test the hypothesis, and then criticize the tests, it is important for scientists to communicate with one another. Historians often date the

ganisms were destroyed on Lysenko's recommendation. Beginning in 1948, Lysenko persuaded Stalin to exile several prominent Soviet agricultural geneticists to Siberia, including N. I. Vavilov, who died a prisoner in one of Stalin's labor camps. Lysenko's theories were never accepted outside the Soviet Union.

GALLO VERSUS MONTAIGNIER
Early in the 1980s, the cause of AIDS was not yet known (see Chapter 13). An American team, led by Robert Gallo, found that AIDS patients had antibodies against a new virus that they called HTLV-III. Meanwhile, a French team, led by Luc Montaignier, isolated a virus that they called LAV. (Both viruses are now considered to be the same as HIV.) The American team claimed that they had discovered the AIDS virus because they had data linking the virus to the disease. The French team claimed that they should be honored as the discoverers of the virus because they were the first to isolate it. Claims and counterclaims grew bitter, and the French and U.S. governments got involved because of a patent dispute over rights to a blood test to detect infection with the virus (Chapter 13). The patent dispute was finally settled with a 50-50 split of royalties, but some people from both camps still resent the way their opponents behaved during this feud.

DO TURF WARS MATTER?
A bigger question than "Do turf wars occur?" is, "Do turf wars matter?" Such battles certainly matter to the individuals involved, but will the course of science be changed as a result? To some extent, theories are initially evaluated on the credentials of their source. In cases like that of Lysenko, where people have gotten entirely away from hypothesis testing, they are no longer doing science. The pace of biological science was certainly slowed in Soviet Russia, but the field continued elsewhere. If science is operating as science, it is self-correcting. It should not matter who champions a theory; hypothesis testing and free exchange of information will determine the outcome. To many scientists, a well-fought controversy is a form of cooperation because it moves everyone's ideas forward.

beginning of modern Western science from seventeenth-century England, specifically from the founding of the Royal Society around 1660. Individual scientists certainly existed in many countries before this time: Copernicus, Galileo, Descartes, and Newton were all earlier. Also, much older traditions of scientific investigation existed in India, China, and elsewhere. However, the formation of the Royal Society marked the first time in history that a permanent, *organized community* of scientists had communicated with one another and shared their results. They established the world's first scientific journal, *Philosophical Transactions of the Royal Society*. For the first time, there was a written and permanent record of experiments performed and conclusions reached—a shared record that encouraged scientists to check one another's work in a systematic way.

Many of the ways in which today's scientists behave toward one another may be viewed as efforts to maintain their ability to do such systematic checking, their ability to falsify hypotheses. Every test must be conducted in such a way as to make it possible for the hypothesis to be falsified, if indeed it is false, and the testing of hypotheses should be described as publicly as possible so as to permit the test to be repeated by other scientists. As David Hull points out, "Scientists rarely refute their own pet hypotheses, especially after they have appeared in print, but that is all right. Their fellow scientists will be happy to expose these hypotheses to severe testing" (Hull, 1988: 4).

The process of science is conducted in the public forum as well as in the laboratory; the publishing and dissemination of results and the repetition of observations and experiments by others are thus valued among scientists. Scientists are expected not to work in isolation, but to discuss their results with other interested scientists, allowing them to build upon the results of previous scientists. They can repeat experiments and confirm the results, but they don't need to start from scratch and repeat *all* earlier work in their field. Skeptics who doubt a particular result unless and until they have seen it themselves can best be won over by a tradition that allows them to hear about repetitions of the test (or to repeat it themselves) and to make their own observations. For example, Galileo, the astronomer and early scientist, invited critics who doubted his observations to look for themselves through his telescope.

Collective experience guides individual scientists in deciding where to collect data, what species or habitats to investigate, what animals to use as experimental subjects, what kind (or what brand name) of equipment to use, and what sample size (e.g., the number of specimens to collect or the number of experimental animals to test) will give reliable results in a particular type of investigation. Thus, science is a cumulative process in which it pays for individual scientists to begin with some of the groundwork laid by others, rather than to start always from scratch. As Isaac Newton, a seventeenth-century century physicist and mathematician, once said, "If I have seen further than others who have gone before me, it is because I have stood on the shoulders of giants."

Scientists are generally willing to have their ideas tested by others, not only because such a system gives them the ability to test other people's hypotheses but because they realize that no one person can see all the implications or inconsistencies of their own hypotheses. The same data may suggest different hypotheses to different scientists; thus, observations that are not shared publicly are not contributing to the process of science. Cooperation among scientists from different countries and different language traditions is always considered desirable, and the nonacceptance of certain scientists because of their religion, nationality, race, or gender is generally lamented if not condemned. Certainly, there are scientists who attempt to marginalize their competitors and their theories, or to engage in "turf wars" or other forms of competition (Box 1.2), but most scientists will express disapproval for these practices when they are brought to light. In general, members of the scientific community share a commitment to moving knowledge forward that fosters a spirit of communication and cooperation.

THOUGHT QUESTIONS

1. In what ways has Western science progressed more rapidly and more efficiently since the seventeenth century than it did before that time?

2. Why is learning how to write a scientific paper an important part of learning how to do science?
3. Many companies conduct what they call research and development, yet many of these companies zealously guard their results and do not publish them. Are they doing science?

CHAPTER SUMMARY

Science is the testing of falsifiable hypotheses. Biology is a science because biologists use scientific methods to study living systems. Biologists test hypotheses either by studying natural conditions as they occur or by experimenting with conditions that they help to create. Interpreting experimental results always involves comparison with control situations and sometimes involves the use of statistical methods. Normal science proceeds by testing hypotheses one at a time, but scientific revolutions occur when many widely held ideas are discarded at once and a new paradigm emerges to take their place. Molecular genetics is an example of an important paradigm in biology. Scientific methods are used often in everyday life, but scientists use these methods more often and more systematically. Science has existed for centuries in various countries. In seventeenth-century England, scientists first began to organize as a community into scientific societies and to publish their findings in scientific journals. Certain values are held by members of the scientific community that ensure them the ability to test, falsify, and change each other's hypotheses.

KEY TERMS TO KNOW

biology (p. 1)
control group (p. 7)
data (p. 2)
deduction (p. 3)
experiments (p. 7)

experimental sciences (p. 7)
falsifiable (p. 2)
hypotheses (p. 1)
induction (p. 3)

model (p. 6)
naturalistic sciences (p. 9)
normal science (p. 10)
paradigm (p. 10)

science (p. 3)
scientific revolution (p. 10)
theory (p. 5)
verifiable (p. 2)

CONNECTIONS TO OTHER CHAPTERS

This chapter and the next connect to one another because members of the scientific community share many norms and values, especially when they support the same paradigm. Ethical conflicts may arise when scientific findings are applied to social problems.

This chapter connects to the remainder of the book because the methods of discovery outlined in this chapter were used to explore all of the topics described in subsequent chapters. In addition, the characteristics of life listed in Box 1.1 are referred to throughout this book.

Ethics and Social Decision Making

Animals are used in many forms of biological research. Some of these research experiments cause pain to the animals. Is this right? Is it justifiable? How do we tell? These are among the moral and ethical questions that this chapter addresses. Many other ethical questions arise in the social applications of biology. What legal restrictions should society place on any scientific research? Many, if not all, of the topics raised in the remaining chapters of this book are issues with some moral or ethical dimension. Are some applications of specific biological research morally right and other applications morally wrong? Should certain conclusions drawn from research involving other species be applied to humans? This short chapter will not give you answers to these questions, which many people spend their entire careers attempting to resolve. Rather, our aim is to suggest some of the ways in which individuals and societies make decisions regarding these subjects. What kinds of questions do we need to ask? What kinds of information can help provide answers? What are some of the barriers to ethical decision making by groups of people?

A. Morals Are Guidelines for Everyday Conduct

Moral Codes

Each of us is guided in our own personal behavior by a more or less elaborate set of beliefs concerning what is right or wrong, proper or improper. It is right to come to class at the scheduled time and in general to keep appointments that one has agreed to. It is wrong to steal, to lie, to murder, or to park in the NO PARKING zone. It is proper to wait for the traffic light to turn green and to wait for one's turn in line. All these and more are **moral rules** (or simply "morals"), and most of them belong to *moral codes,* which are sets of such rules. Moral codes are products of societies. Anthropologists who have compared societies from around the world tell us that moral codes differ from one society to the next. They also evolve slowly over time and change as the society changes.

Any personal decision as to whether or not to abide by some part of a moral code may be called a *moral decision*: Should I park in the NO PARKING zone right now? Suppose I am doing so in order to take my son, who has the flu, to the doctor? Moral decisions are often made with the knowledge that society will attempt to enforce the moral code with a set of formal or informal penalties, or *sanctions*. The formal sanctions include fines and jail time. The informal sanctions, which operate much more often, include being criticized or avoided by others, being passed over for promotions or awards, having fewer friends, or not being elected or appointed to positions of esteem, prestige, or power.

Ethics. The logical or philosophical analysis of moral codes and moral systems is a discipline known as **ethics.** In particular, ethics is concerned with the basis for moral judgment, or the ways in which moral judgments are made and justified. Philosophers often distinguish *normative ethics,* the study of how ethical judgments *should* be made, from *descriptive ethics,* the study of how ethical judgments *are* in fact made. Descriptive ethics is studied by observing human behavior using the scientific methods discussed in Chapter 1. Normative ethics is a theoretical discipline rooted in logical analysis, an analysis for which observational data are insufficient. Normative ethics cannot be reduced to a set of data, and no amount of data can either confirm or refute a moral law like "Thou shalt not kill."

In its simplest form, normative ethics can be described as an attempt to express moral codes in terms of a set of basic rules, sometimes called *maxims.* For example, I should come to class on time because my signing up for a course, which includes classes at a certain time, is basically like an appointment I have made. Appointments should generally be kept because they are promises, or contracts: If you come to this particular place at this particular time, I will also come here, and we can carry out business together. An ethic of keeping appointment promises can, in turn, be contained within a larger ethic of keeping other kinds of promises: promises to render services or to pay for services. Lying may be thought of as the opposite of promise keeping, and most arguments in favor of promise keeping are also arguments against lying.

Some rules of conduct are simply convenient inventions of a society, such as the observing of NO PARKING zones, waiting for the green light, or driving (in North America) on the right side of the street. It is convenient that certain places be set aside for bus stops, pedestrian crosswalks, loading docks, handicapped vehicles, and emergency access to fire hydrants and other safety measures. NO PARKING zones are simply a convenient way of setting aside certain places for these and other purposes. We cannot all drive through the intersection at the same time, and traffic lights are a convenient (if arbitrary) contractual way of arranging whose turn is next. The contractual nature of such arrangements is obvious because there are usually publicly controlled processes (like city council meetings) to decide where to put NO PARKING zones. The fact that the choices regulating such conventions are also arbitrary becomes evident when we realize that traffic lights could just as well use a different color code, or that there is no universal requirement that drivers drive on the right-hand side of the street. Clearly, traffic will flow more smoothly and more safely if each traffic lane goes only one way or the other, but the left side of the street works just as well in Britain as the right side does in America. In either case, it is generally preferable (from a safety standpoint at the very least) to observe the conventions established by the country we are in. An American driving in London would do well to observe British traffic customs, and vice versa. In a sense, we promise to observe traffic laws when we apply for a driver's license, so following these laws may be viewed as another form of promise keeping.

Waiting one's turn in a line or queue and waiting for a green traffic light are both ways of introducing order and fairness into a situation that would otherwise be chaotic and conducive to unnecessary disputes. A major difference, however, is that waiting one's turn in line is a practice that is not enforced by law or traffic code—it is enforced informally by tacit and temporary agreement of those who happen to be present.

What other morals can be generalized this way into ethical rules? Stealing can be viewed as an offense against the property rights of others and murder as an offense against their right to go on living. Clearly, a society in which stealing and murder occur is much less safe than one in which these practices do not occur. Not just safety, but also freedom, happiness, and other desirable features will exist in greater supply in a society without murder or theft. Ethical philosophers may debate which condition (safety, freedom, happiness, unhindered commerce, or something else) is the most important, but it is satisfactory for our purposes to notice that ensuring any of them leads to the same general conclusion, that theft and murder are both wrong.

We have thus developed a simple moral code: Keep your promises, don't interfere with the rights of others, and observe the common social conventions. This could easily be expanded into a more general code of benevolence, cooperation, and mutual aid.

Resolving Moral Conflicts

There are occasions when conflicts arise within a moral code. I know I should obey the traffic laws, but what if I am taking an injured person to the hospital and the person's life is in danger? Does the need to save a life justify driving above the speed limit, driving through a red light, or parking in a loading zone? Can I justify disobeying the traffic laws in order to keep an appointment? Does it matter how important I think the appointment is? Resolving conflicts of this kind is one of the major goals of ethics.

In most cases, the resolution of such moral conflicts is made by determining that one rule or goal is more important than another: Saving a life is more important than obeying traffic rules, for example. Thus, there are *exceptions* to most moral rules: Obey traffic rules and other useful conventions, *except* when obeying them violates a more important rule. Notice that this involves ranking certain rules as more important than other rules, allowing us to justify an exception to one rule by invoking a "higher" rule.

Although ethics is generally considered a branch of philosophy, ethical arguments arise in everyday life and also in science. Ethical issues are not just of theoretical interest to philosophers; more and more scientific endeavors are raising ethical issues that are of practical interest to many people in all walks of life. The United States government has sponsored research, meetings, and publications in the field of biological ethics. Many government programs, notably the Human Genome Project (see Chapter 3), have set aside portions of their budgets for the examination of the ethical implications of science. A large number of issues today have far-reaching ethical implications, which is the main reason we have devoted an entire chapter to ethics.

THOUGHT QUESTION

Together with several other students, make a list of five to ten laws or rules that are generally followed on your campus or in your community. For each one, try to discover: (a) why such a rule is considered important to follow (or why it *was* considered important when the rule was adopted), (b) whether there is a more general moral concept of which this rule is just a special case, and (c) whether society is better off with rules of this general kind than without them, and, if so, why?

B. Ethical Systems Are Used to Analyze Moral Rules

An *ethical system* is a set of rules for resolving ethical questions or for judging moral rules. We will first describe two major types of ethical systems, then touch very briefly on some others.

Deontological Systems

A **deontological** system is one in which the rightness or wrongness of an act depends on the act itself and not on its consequences. To a deontologist, the wrongness of murder stems from the nature of the act itself, not from its results or its effects on society. Similarly, a deontologist who believes in keeping promises does so apart from the effects of such a rule on commerce or human affairs.

Deontology based on religion. Traditionally, most deontologists have developed moral codes that are based on one or more religious traditions. The Bible, the Koran, and the sacred texts of other religions have been the source of many moral codes. Deontology grounded in religion has the advantage of commanding agreement among those who belong to the same religious tradition, though differences in the interpretation of sacred texts often arise. A deontological system based on a particular religion will be less likely to command the agreement of people not belonging to that religious tradition.

Deontology based on philosophy.   The German philosopher Immanuel Kant (1724–1804) devised a deontological system not based closely on any religion. According to Kant, all ethical statements are based on a single precept: Act only according to rules that you could want everyone to adopt as general legislation. This rule is called the *categorical imperative,* and it is formulated (as are most ethical rules) as a universal law. Kant's entire ethical system, based on this rule, does have the advantage of being consistent with a variety of religious and other traditions. Under Kant's system, the test of the morality of an act is whether or not the act can be universalized, that is, applied to all people at all times and under all conditions. Thus, killing is (always) a wrongful act because I could not possibly want people always to kill one another—I would be willing my own death and the death of my loved ones. Keeping promises can be universalized, and promise keeping is therefore (always) moral. Kant said that lying could not be moral because you could never want everyone to lie.

Concepts of "rights."   Most deontological systems include a formulation of certain **rights,** based on respect for the dignity and autonomy of all persons. Respect for all human beings is an important part of Kant's system. If you respect the dignity of all human beings, then you cannot ever will the death of any person, nor can you deny them their fundamental rights, nor can you use them as objects for your own personal gratification in any way. If you respect their selfhood, then you cannot morally abridge their freedom. Of course, it remains to be determined exactly what rights we do and don't have.

There has been considerable disagreement among philosophers, and even more variations in historical practice, over the types of beings to which various rights apply. At various times in the past, a variety of people (including women, children, slaves, the lower classes of stratified societies, impoverished people, foreigners, members of other races, mental patients, and persons unable to speak for themselves) were denied the rights that were afforded other members of society. Many people now invoke dignity and autonomy criteria in discussions on whether certain rights should also now be extended to unborn fetuses or to animals.

Criticisms of deontology.   A criticism of rights-based deontology is that there are many circumstances in which one right conflicts with another, resulting in a moral dilemma. Unless there is a clear way of deciding among conflicting rights, moral dilemmas of this sort are inevitable. One could choose among conflicting rights by comparing the consequences of violating one with the consequences of violating another, but then the system would no longer be deontological. An obvious way out is to declare one particular right (like the right to life or the right to freedom of action) supreme over all others. Aside from the problem that different deontologists would choose different rights to take precedence over the others, there is the more serious objection that insistence on a single right leads to the dangers of absolutism. Historically, many atrocities have been perpetrated by the followers of systems that put absolute adherence to a single principle above all others.

Utilitarian Systems

In a **utilitarian** system of ethics, acts are judged right or wrong according to their consequences: Rightful acts are those whose consequences are usually or generally beneficial, while wrongful acts are those whose consequences are usually or generally harmful. To a utilitarian, murder is wrong because the death of the victim is an undesirable outcome under most circumstances. Also, on a larger scale, murder is additionally wrong because a society in which murder occurred would be undesirable, for people would be continually fearful of their lives.

Judging good and bad consequences in a utilitarian system.   A challenge to all utilitarian systems is finding a way to measure the goodness or badness of consequences. Over the years, utilitarian philosophers have come up with different criteria on which to judge consequences: the greatest happiness for the greatest number of individuals, the greatest excess of pleasure over pain, and so on. All utilitarian systems require value judgments to be made between outcomes that are difficult to measure and quantify. Utilitarian ethics stand in contrast to strict deontological ethics, in which outcomes are ignored.

A distinction is sometimes made between act utilitarianism and rule utilitarianism. In *act utilitarianism*, it is individual acts that are judged as right or wrong according to utilitarian principles on a case-by-case basis. In *rule utilitarianism*, it is general rules that are judged on utilitarian principles, and individual acts are judged only as to whether or not they are justified by a general rule. Under act utilitarianism, each act of murder is judged wrong by itself, a process that in theory allows for an occasional exception. The murder of a traitor (or a gangster or a drug dealer) might be judged as a rightful act if it could be shown that greater harm would result from the victim's continued activities than from his or her death. Under rule utilitarianism, individual acts of murder are judged wrong, and this rule is justified by pointing out that society as a whole is better off without murders. Exceptions to a rule can only be justified by appeal to a "higher" rule.

Under either form of utilitarianism, ethical decisions are made by comparing, in what is essentially a cost-benefit analysis, all the good and bad consequences of an act with the consequences of other alternative courses of action (including the failure to act).

Utilitarianism can be summarized as a moral rule or maxim: Always act so as to maximize the amount of good in the universe. The first major utilitarian philosopher was Jeremy Bentham (1748–1832), who said that we should always strive to bring about "the greatest good for the greatest number [of people]." In order to decide which actions produced the greatest good, Bentham suggested a type of cost-benefit analysis that he called "a calculus of pleasures and pains." Other notable utilitarian philosophers include John Stuart Mill (1806–1873) and G. E. Moore (1873–1958).

Criticisms of utilitarianism.   One of the major criticisms of utilitarianism is that its cost-benefit approach treats human beings merely as means to an end, thus reducing their status and dignity, and in some cases violating their rights.

Most deontologists argue that certain individual rights must be protected regardless of whether society as a whole benefits from recognizing and protecting those rights. To do any less, they argue, deprives human beings of their fundamental dignity as individuals and makes them nothing more than a cluster of costs and benefits. Even if a larger benefit can be demonstrated in a given instance, say these critics, it is still unethical to violate an individual's fundamental right, because "the ends do not justify the means."

A related criticism is that freedom of choice is limited under utilitarian systems. For example, if I felt I could contribute to the greater good of humankind by doing charitable work, under a strict utilitarian system I would be *obligated* to quit my job and devote myself to charitable work, since to do otherwise would be to choose a course of action that resulted in less total benefit.

Other criticisms of utilitarianism have been raised from time to time. One criticism, raised by John Rawls (1972), among others, deals with the matter of fairness in the distribution of benefits (goodness, utility, or whatever is measured in calculating costs and benefits). Rawls pointed out that when only total gain to society is considered in a cost-benefit analysis, this will lead to instances in which gain is unequally distributed. For example, all the benefits might accrue to one or a few individuals only, while others suffered from the costs. Rawls prefers instead what he calls the *maximin* solution, in which we are obligated to choose the option that maximizes the outcome (and minimizes the suffering) for the most people, even if that option produces less "total gain."

A final criticism of utilitarianism frequently raised by philosophers deals with the probability that, under most formulations of utilitarian ethics, the killing or torturing of one person would be justified (if not required) if it resulted in saving the lives of many other people. The discussions often center around certain hypothetical situations. In one situation, several people exploring a cave are trapped inside because their leader is stuck in the mouth of the cave; they could escape by blasting him free with a stick of dynamite they happen to have with them, but, if they do nothing, the rising tide will drown them all. Do the calculations of costs and benefits demand that they kill their own leader in order to avoid the deaths of the entire team? In a second situation, a terrorist (whose past actions cause us to believe both his capability and his intent to carry out his stated plan) refuses to reveal where he has planted a nuclear bomb that will soon explode and cause thousands of deaths; the terrorist is known to be very fond of his young daughter, who is in our custody. If we think that we can get the terrorist to reveal the location of the bomb, thus saving countless lives, by torturing his daughter in front of him, should we do so? In both of these situations, there is a choice between some act that would harm or kill an innocent person in order to save the lives of numerous others. Admittedly, the scenarios are unusual and somewhat unlikely situations, but the essential point is that *some situations exist* in which utilitarian ethics might require us to violate even our most strongly held ethical norms.

Of course, adherents of utilitarianism might counter by pointing out that greater harm to society at large would come *in the long run* from the violation of strongly held norms (e.g., that innocent people should not be tortured or put to death), and that the world would be a better place if we defended the rights of innocent people (and various other rights), even if there were some rare situations in which violating these rights would save many lives. History and folklore would, for example, honor the police who refused to torture the terrorist's daughter, or the cave explorers who perished rather than blow up their leader, and the examples of their stories would encourage more lives to be saved in future generations by people who are brought up to honor these important norms.

Other Ethical Systems

Although deontological and utilitarian ethical systems have generally attracted the most followers, other ethical systems also exist.

Egoism.   **Egoism** is an ethical system in which acts are judged solely in terms of their consequences for one individual or group. Under *individual egoism,* acts are judged according to whether they have beneficial or harmful consequences for *me*. Under *group egoism,* acts are

judged according to whether their consequences are beneficial or harmful for the *group* (family, tribe, nation) to which I belong. In a world where only one group exists, group egoism may become indistinguishable from utilitarianism. The differences arise according to whether we seek to maximize goodness just for ourselves, or for ourselves and our family, or our entire tribe or ethnic group, our nation, all humankind, all intelligent species, all animals (or "sentient beings"), all life forms (including trees and bacteria), or the entire cosmos. On this spectrum, individual egoism occupies one pole and utilitarianism most of the middle. The opposite pole comes close to the ethical teachings of certain Eastern religions such as Buddhism, Hinduism, and especially Jainism, which seeks to maximize goodness in the cosmos.

Natural-Law ethics.   Nature-based ethics, or **natural-law ethics,** is based on the general idea that people should imitate nature, or that whatever is natural is always best. There are several ways of looking to nature for ethical guidance, including the following:

1. Whatever occurs in nature is necessarily good.
2. Whatever occurs in nature is always to be preferred over what does not.
3. Whatever occurs in nature is in harmony with the rest of nature.
4. Whatever occurs in nature is the product of natural selection (Chapter 4), and is therefore favored by nature.
5. Whatever maximizes fitness (i.e., results in more offspring) will be favored in the course of evolution, and is therefore to be preferred in human affairs also.

According to this type of ethics, we should look to the animal and plant world for ethical guidance. Acts that occur in nature are judged right; artificial or unnatural acts are judged to be wrong. By this criterion, one could argue that eating meat raw is preferable to eating meat cooked. Even more problematic is the natural occurrence of incompatible forms of behavior: Should parents guard and protect their children, abandon their children, or cannibalize and eat their children? Nature-based ethics provide no guidance in this decision, for natural examples can be found for all these types of behavior.

Scottish philosopher David Hume (1711–1776) made a careful distinction between things that *are* and things that *ought to be*. Things that *are* can be investigated by scientific means, which includes setting up falsifiable hypotheses and examining evidence. On the other hand, things that *ought to be* are not subject to this kind of examination. Under this premise, nature offers an insufficient basis for any ethical system.

Ethical relativism.   Adherents of ethical relativism take the position that each society may have its own system of ethical judgment, different from the rest, all such ethical systems being equally valid. This position is often urged upon those about to visit societies very different from their own, admonishing them not to judge other societies by the (probably different) ethical standards of their own society. The empirically testable theory that different cultures do in fact sometimes make different ethical judgments is called *cultural relativism*, an important finding of descriptive ethics. Ethical relativism, as a form of normative ethics, goes one step further in proclaiming that values *should* vary (ought to vary) from one society to the next according to each society's needs and circumstances. Under ethical relativism, any cultural trait can be judged socially "good" if it operates harmoniously within its cultural setting and serves to promote that culture's goals.

THOUGHT QUESTIONS

1. Try to justify the wrongness of the following acts under each of the ethical systems discussed in this section:

    murder
    rape
    bank robbery
    failure to repay a debt
    racial segregation
    driving over the speed limit

    Which ethical systems make it easy to explain the wrongness of these acts? Under which ethical systems are these explanations difficult?

2. Discuss with other students the cases of the cave explorers and the terrorist's daughter as de-

scribed in this section. What alternative courses of action are possible? How might one justify one course of action over another? What role does *uncertainty* play in any of these scenarios (what if the terrorist's actions were less than fully predictable)? What other courses of action might be open in each case? How would future generations be affected by each possible action and its consequences?

3. Is there an important ethical distinction between causing harm by our actions and allowing harm to occur by our inaction? Does this distinction allow other solutions to the dilemmas posed by the scenarios involving the terrorist's daughter or the explorers trapped in the cave?

4. Is it ethical to infect a few people with a deadly disease in order to study its effects in the hopes of saving many more lives in the future? How do you justify your answer?

5. Is it right or wrong to rob from the rich in order to give to the poor as long as no personal injury takes place? Discuss this situation from both deontological and utilitarian perspectives.

C. Democratic Societies Make Ethical Decisions Collectively

Ethical systems differ from scientific theories in that scientific hypotheses are made to be falsifiable (Chapter 1), while moral judgments and religious concepts are not. This means that a truly devout person's belief in God cannot be shaken by any demonstration of an empirical fact or observation, nor will such a person admit that any such demonstration is even possible. The same goes for strongly held beliefs about the goodness of human equality or the wrongfulness of inflicting death.

In culturally homogeneous societies in which all people (or nearly all) share a common set of values and a common set of religious beliefs, it is often possible (it may even be easy) to reach a consensus as to which acts are wrong and which are right. However, most societies today are *pluralistic* in the sense that their populations include people with differing cultural or religious backgrounds. Reaching collective decisions on ethical issues presents far greater challenges in such pluralistic societies. The majority of pluralistic societies are also democracies, and most of those that are not are striving to become more democratic. We will therefore discuss some of the general problems of reaching agreement on issues with ethical dimensions in modern pluralistic societies with at least some democratic traditions.

Ethical Decision Making in Pluralistic Societies

Not all of the major ethical theories discussed in the previous section of this chapter are suitable as guidelines for collective ethical decisions in pluralistic societies. For example, natural law ethics can often be used to argue opposite points of view simply because natural examples of so many disparate practices can be found. Ethical relativism teaches us that various types of practice can be made to work in the appropriate social context. Since a wide variety of ethical practices *can* be made to work, this type of ethical theory provides very little guidance to a society seeking to choose one practice over another. Egoism may guide the ethical opinions of individuals, but it hardly provides guidance for public policy or for public resolution of conflicting views.

The deontological theories provide guidance to those who can agree on the basic values on which ethical decisions are made. In a public setting, deontological theories can produce broad consensus only in societies in which a single ethical or religious tradition prevails. Remember that deontological theories depend upon certain acts being considered right or wrong independently of their consequences. Only if people of all kinds can agree on what is right and what is wrong can deontological theories alone provide guidance for public debates.

Utilitarian decisions require calculation of the good and bad consequences of each act, so utilitarians come to the public forum with lists of good and bad consequences. Arguments made from the premise of utilitarianism attempt to assess the amount of good or harm that will ensue from various possible courses of action, including the possibility of no action at all. Under a strictly utilitarian system, the details of social decision making are guided by a cost-benefit (sometimes called risk-benefit) analysis. In most actual decision-making

processes, the cost-benefit analysis is used not to make the final decision, but rather to provide the necessary data upon which a decision can be intelligently based.

John Rawls (1972) provides a well-respected analysis of how modern pluralistic societies reach decisions in actual practice and why most such pluralistic societies have evolved some form of democratic traditions. In a pluralistic society, some people will come to the public forum with utilitarian assumptions, others will come with deontological assumptions, and still others will argue from a position of individual or group egoism. Even within these ethical traditions, different utilitarians will come with different evaluations of costs and benefits, different deontologists (and many utilitarians, too) with different lists or rankings of rights, and different egoists with arguments benefiting different groups or segments of society. One way of reconciling these different views is to allow a public debate (so that everyone knows that their views have been heard) and then engage in some kind of voting process. The voting procedure should be structured, says Rawls, in a way that all parties will recognize as fair. **Fairness** is a way of ensuring that certain rights are not violated or trampled upon by others. In most cases, a system that is set up to work in this way ends up achieving the least displeasure with the decision, or the highest possible minimum value among a multiplicity of cost-benefit analyses. Of course, total agreement with any decision is hard to achieve in any large pluralistic society.

Making Social Policy Decisions on Issues Involving Science and Technology

Social policy includes all those laws, rules, and customs that people follow in making individual decisions. The making or changing of a social policy is called a *policy decision*. All or nearly all social policies involve ethical considerations, and so, too, do policy decisions. An increasing number of policy decisions today also involve some aspect of science or technology.

It is often convenient to divide policy decisions involving science into three phases: scientific issues, science policy issues, and policy issues.

Scientific issues. What possible explanations (hypotheses) are there to explain the available data? Can these hypotheses be tested? Do the tests support or falsify each hypothesis? Are alternative explanations available? What additional data are needed to evaluate these hypotheses? These are often characterized as "purely" scientific issues on which scientists of different political or ethical persuasions may be expected to agree if sufficient data are available.

Science policy issues. What would be the consequences of this or that particular legislation or policy change? What would be lost or gained from each proposed plan of action? In the case of uncertain consequences, can we estimate their probability of occurrence? The probability of any outcome, especially one not desired, is called a **risk.** Can we calculate or estimate these risks? In a cost-benefit analysis, what are the costs and what are the benefits? How certain are we of the estimated values? These are still scientific issues in the sense that they are evaluated using data, but disagreements on the data or its significance are expected to occur among experts with different political or ethical viewpoints. If these experts disagree, how can we evaluate their respective claims?

Policy issues. Once we have evaluated the possible consequences of various possible policies, which one should we choose? These are ethical decisions in which values play a prominent role: Is it worth spending billions of dollars and risking the lives of crew members in order to explore a distant continent or an even more distant planet? Is it worth the pain and suffering of a certain number of experimental animals in order to develop a drug that could save a certain number of human lives each year? In general, is some predicted but uncertain benefit worth the calculated risks? If a proposed change can only be made by incurring a certain cost to society, is this social cost worth the intended benefit?

Who makes the decisions?   In most societies, scientific issues are frequently decided by scientists with little input from interested citizens.

Science policy issues are often decided in the court of public opinion by an interplay of scientists and other "experts" under the scrutiny of policy advocates for one side or another. Although evidence is used, it is more like courtroom evidence, obtained by cross-examining witnesses, than like the evidence of the laboratory, obtained by the formulation and testing of alternative hypotheses.

As for the final policy decisions, who makes them—the scientists, the public, the media, or the government? In most democracies, the decisions are made either by the public or by government agencies acting (in theory at least) on the public's behalf and in the public interest. Decision makers are often influenced by scientists, by the media, and by particular interest groups. Decision makers may also be influenced by public pressures that take a variety of forms: marketplace decisions (decisions by individuals over where to spend their money), letters and telegrams, enthusiasm at public gatherings, the results of public opinion surveys, and direct votes on referendum questions. All of these influences certainly play a role in what is essentially a political process.

Misinformation.   *Misinformation* is an important hazard in public policy discussions. For this reason, many of the rules of encounter are designed to expose and eliminate any faulty information. At all stages, for example, open hearings, publication of results, and similar provisions are designed to ensure that the evidence is made openly available to all sides of any dispute; at the very least, this affords an opportunity for data or interpretations to be challenged by the opposing sides.

THOUGHT QUESTIONS

1. Tissues taken from human fetuses, including those derived from both voluntary abortions and spontaneous abortions (stillbirths), are a resource for certain kinds of medical research. For example, fetal brain cells still have the capacity to grow and divide, an ability that most brain cells lose in the first few months following birth. Research into the causation of certain brain diseases (see Chapter 10) requires the use of embryonic brain cells that still have the capacity to divide. Try to use the concepts explained in this chapter to present:

Deontological arguments for fetal tissue research

Deontological arguments against fetal tissue research

Utilitarian arguments for fetal tissue research

Utilitarian arguments against fetal tissue research

2. For the previous thought question, identify some examples of science issues, science policy issues, and policy issues. Why is it important to distinguish these separate types of issues from one another?

D. Animal Experimentation Raises Ethical Questions

Thus far, we have discussed ethical issues in very general terms. We now turn, for illustrative purposes, to a specific and timely issue, that of animal rights and the uses of animals in scientific experiments. Of the many ethical issues surrounding biology today, few are as divisive as this one. We will start by taking notice of the many ways in which animals are used by human societies.

Uses and Abuses of Animals

Human societies have kept animals at least since the origin of agriculture. There are few societies of any kind that do not have some tradition of keeping animals as pets, as workmates, or as food. Most societies that practice agriculture use animals for all three purposes. Love of animals and use of animals can go hand in hand.

By far the largest number of animals used by any society are raised for use as food for humans. Animal products are used for clothing. Animals are also used, even in many industrial societies, for recreational hunting, fishing, and trapping. An estimated 7 percent of the United States population has a hunting or fishing license. Many people keep pets or "companion animals." Pets can offer benefits beyond companionship. For example, studies have shown that heart attack victims who are pet owners live longer and suffer fewer repeat attacks than heart attack victims who do not own pets (Chapter 12). Work animals include animals that are used for riding, for pulling and carrying loads,

for police work, and for helping handicapped people in various ways, such as seeing eye dogs for the blind and also a small but growing number of monkeys and other animals trained to assist patients confined to wheelchairs. Finally, animals are often used in research, although the number of animals used annually in research is only a tiny fraction of the numbers used each day as food and for other purposes.

Justifications for the use of animals. A variety of reasons have been given to justify the use of animals by humans, including (1) saving human lives (e.g., serving as stand-ins for humans in dangerous situations), (2) improving human health (e.g., testing medical and surgical procedures), (3) providing food for people, (4) providing nonmedical information, (5) serving our recreational needs (hunting other than for food, entertainment uses such as circus acts), and (6) serving as status symbols (e.g., wearing furs). A large number of people will accept reasons 1 to 4 as adequate to justify the use of animals, or their more extensive use, while fewer people will accept reasons 5 and 6.

Animals as test subjects. Nearly all new drugs, cosmetics, food additives, and new forms of therapy and surgery are tested first on animals before they are tested on humans. The use of animals in research is considered critical to continued progress in human health (Williams, 1991:63). Over forty Nobel prizes have been awarded in medicine or physiology for research using experimental animals. Organ transplants, open-heart surgery, and various other surgical techniques were first performed and perfected on animals before they were performed on humans. All major vaccines, including those for smallpox, polio, mumps, measles, rubella, and diphtheria, were tested on animals before they were used on human patients. Despite the vociferous opposition of the animal rights movement, the general concept of using animals in research has received widespread support in most industrial societies.

In most cases, animals are used in research as stand-ins for humans. If we were to abolish the use of animals for these purposes, we would be using many more procedures on humans without benefit of prior animal testing. In effect, the first few dozen or few hundred human patients would be serving as human guinea pigs—the experimental subjects on which the new technique is tested. Most people favor the use of animals as substitutes for humans in those cases where the use of animals can lessen the suffering of human beings. This is especially true in those cases (the majority) in which the animal testing is limited to the initial development of a drug but the human benefit continues for many generations or longer.

Regulation of animal use. Very few people would object to the use of animals if human lives were saved as a consequence. Most people would also agree that we should alleviate unnecessary suffering among animals, whether the animals are pets, work animals, or research animals. It is in the best interests of science for scientists to conduct their tests on healthy, well-treated animals, and the U.S. Guide for the Care and Use of Laboratory Animals reflects this concern. The United States Department of Agriculture keeps statistics on the use of experimental animals in research. According to these statistics, 62 percent of the animals used in research experienced no pain, and another 32 percent were given anesthesia, painkillers, or both, to alleviate pain. Only in 6 percent of cases were animals made to suffer pain without benefit of anesthesia. Federal law in the United States requires the use of anesthesia and/or painkillers in animal research wherever possible. Exceptions are allowed only when the experimental design would be compromised by the use of anesthesia and when no other alternative method is available for conducting the test. Each such exception must be approved by the same institutional animal care and use committee that supervises animal research in general. All research using live animals must, by law, be scrutinized and approved by such committees, and the committees are required to minimize both the number of animals used and the amount of pain that those animals experience. Although it is assumed that researchers will design their experiments with these criteria in mind, the review process ensures that the researcher(s) are not the ones making the final decision on whether their experiments conform to ethical guidelines.

The Animal Rights Movement

Like many other movements, the animal rights movement is a heterogeneous mixture of believers,

partial believers, zealots, and sympathizers. Bernard Rollin, a philosopher on the faculty at the College of Veterinary Medicine at Colorado State University, focused on the animal rights issue in a 1992 book:

The main problem, which continues to concern me the most . . . , is polarization and irrationality on the animal ethics issues by both sides. . . . The American Medical Association's recent paper on animal rights labels all animal advocates as "terrorists," and scientific and medical researchers continue to equate animal rights supporters with lab trashers, Luddites, misanthropes, and opponents of science and civilization. Animal rights activists continue to label all scientists as sadists and psychopaths. Thus an unhealthy *pas de deux* is created that blocks rather than accelerates the discovery of rational solutions to animal ethics issues. (1992:10)

Those concerned with animal rights vary from traditional humane societies like the Society for the Prevention of Cruelty to Animals (SPCA) and various national, state, and local humane societies, through groups like People for the Ethical Treatment of Animals (PETA, founded in 1980), to groups like the Animal Liberation Front (ALF, founded in 1972). From the start, the ALF concentrated on such tactics as breaking into animal research laboratories, wrecking their facilities and equipment (often beyond repair), and "liberating" the animals. Because these activities are illegal, leading members of the ALF are sought by law enforcement authorities, and most of them are now in hiding. This is the organization most often labeled as "terrorist."

Some animal rights advocates use a utilitarian ethic to support their position; others write as deontologists. An example of a deontological position would be that animals have inviolable rights, or that it is always wrong to do harm to them, regardless of the circumstances or consequences. A utilitarian animal rights position might insist on a comparison of costs and benefits (or good and bad consequences) but with equal value placed on the lives (or the pain) of humans and animals.

Do Animals Have Rights?

Debates about animal rights will remain fruitless as long as the disputants disagree about basic premises. According to Rollin, the most basic matter in these debates is: Do animals have rights? Separate issues are whether animals are entitled to moral consideration (i.e., do they "count" when evaluating the morality of a decision?), and, if so, do they "count" as much as people? According to Rollin's long and well-reasoned argument, there is no good reason for drawing an ethical distinction between mentally competent adult humans, other human beings (including children, comatose patients, mentally ill or brain-damaged persons), and animals. Animals are therefore, in his opinion, worthy of any moral consideration that would normally be given to babies, comatose patients, and other people unable to speak for themselves.

How broadly can the concept of rights be applied?

Historically, animals have been treated legally as property. Animal owners, in other words, have been treated as having property rights over their animals, but the animals themselves have often been afforded no rights. In case of injury or death to an animal, the animal owner often had the right to sue for damages (in the same way as if the damage had been to a car or a house), but the animal itself had no legal rights.

The question of animal rights is actually part of a broader question: How far do we extend the scope of any rights that we recognize? As we mentioned previously, many societies have denied even the most basic of rights to certain classes of persons on the basis of economics, gender, race, ethnicity, or religious beliefs. The extension of certain basic rights to all humans, including children and convicted criminals, is now considered so fundamental to the ethical sensibilities of most people that we refer to these as *human rights*. International agreements, such as the Geneva Convention on the treatment of prisoners of war and the International Convention on Human Rights (the Helsinki Accord), attest to the importance given to these human rights in world affairs. But should we stop there? Animal rights advocates say that we should extend these same rights to all beings capable of sensing pleasure and pain. A few people go even further, asserting that even trees have such rights as the right to go on living or not to have their air and soil poisoned. To go still further, a few people even assert that habitats themselves, including mountains and forests, have rights not to

be despoiled. The arguments made by people who wish to restrict human rights to humans only center on the premises that our rights derive from our ability to make moral decisions, to distinguish right from wrong, or to enter into contractual agreements.

According to Rollin, animals are worthy of moral consideration because they have interests, "needs that matter" to the animals. Rollin argues that any living thing with interests deserves to be treated as an end in itself. Rollin does not maintain that animals' rights are always overriding. He says that animals' competing interests must be calculated in cost-benefit terms, basing his ethic on what he calls the "utilitarian principle": The benefit to humans must outweigh the costs in pain and suffering to the animals (1992:96). He qualifies this, however, with a "rights principle": Animals should be treated so as to maximize their ability to realize their "nature" or "telos" (their "goal" in life), and the animal's fundamental rights, including the right to live, should be preserved as far as possible, regardless of cost.

Speciesism.

Granting moral standing to animals is not necessarily the same as granting them *equal* moral standing. Some animal rights advocates equate the value of an animal life with the value of a human life. If we follow this attitude to its logical conclusion, then testing a new drug or some other substance on animals is no different ethically from testing it on human volunteers. Animal rights advocate Peter Singer coined the term *speciesism* (intentionally parallel to terms like *racism* and *sexism*) to refer to the attitude that humans are fundamentally different from all other species in ethical standing. Various reasons have been given to justify the distinction between humans and other species: Only humans understand right and wrong; only humans are "rational"; only humans can speak; only humans can enter into legal or moral contracts; and so on. Animal rights advocates have challenged these distinctions by citing exceptions—for example, apes that communicate using sign language and infants and comatose patients who do not use any language. Finally, Singer and his supporters simply argue that none of the distinctions between humans and other species are relevant in moral arguments, and that animals deserve equal moral consideration with humans. A logical extension of this form of reasoning would cause us to consider meat eating as the moral equivalent of cannibalism.

Human moral claims.   Philosopher Carl Cohen proclaims that he believes in speciesism and is proud to assert that animals have no rights. Rights, he argues, can only arise from *claims* or from potential claims: My right to hold you to the terms of a mutual agreement arises from the fact that I can make a claim against you under the agreement. My right to walk the streets without being attacked arises from the possibility of my making an accusation against my attacker. We can extend these rights to infants, comatose patients, and people unable to speak for themselves because they are human, and we can thus assume that they have had in the past, or will have in the future, the ability to make moral claims and moral judgments. However, none of this applies to animals because they lack the capacity, no matter what the hypothetical circumstances, to make moral claims or judgments. In Cohen's view, animals have no moral standing because they have no concept of right and wrong and cannot be held responsible for their actions. We could not grant "rights" to gazelles unless we were prepared to criminally prosecute lions and other predators that chase them down, kill them, and eat them. If animal predators like lions and wolves are not prosecuted, this indicates that their prey have no recognized rights, not even the right to life itself. Humans are predators, too, and almost every society on Earth recognizes that killing animals for food is a permissible and not an immoral act. Most traditional philosophers, from St. Augustine to Kant, recognize the special moral status of humans as distinct from animals.

Prevention of cruelty to animals.   One area in which animals generally *have* been given legal and moral consideration is in freedom from whatever society regards as *cruelty*. Most legislation regarding animal welfare has been directed toward the prevention of cruelty, but criteria as to what constitutes cruelty vary from one society to another. In most cases they are vaguely defined, and even legal criteria regarding cruelty to animals are frequently vague. Rollin uses this fact to argue that anticruelty laws are generally weak or inadequate in safeguarding the rights of animals. Carl Cohen argues that,

even if animals don't have rights, we still have a *responsibility* for those animals in our care, to treat them humanely and to minimize their suffering.

Current Debates

In this section, we will provide background material for ethical debates in four specific areas: fur clothing, LD_{50} testing, drug testing, and the Draize test.

Fur clothing.   The wearing of furs has long been one of the major targets of animal rights activists, and the elimination of fur clothing has consistently been one of their strongest aims. In most cases, fur clothing is made from very small animals like mink or chinchilla, so it takes many dozens of animals to produce a jacket and many more to produce a coat. These animals are either trapped in the wild or are ranch-raised.

What would a cost-benefit analysis of fur clothing reveal? In addition to the death and suffering of the fur-bearing animals themselves, the death and suffering of animals inadvertently caught in the traps must be calculated on the cost side. On the benefit side, there is the utility for humans of having warm clothing. A number of people (fur trappers, shippers, processors, sales people) are economically dependent on the fur industry, but many of these people could use their skills equally well in other jobs.

None of these costs and benefits are easily measured in a way that permits comparison in the same units. Pain and suffering are particularly difficult to measure or quantify. As in many other cases, after listing all the costs and all the benefits, we then leave it up to each of us as individuals to decide whether or not the benefits outweigh the costs. If we approach this question instead from a rights perspective, we must balance the rights of animals to go on living and to be free from pain against the rights of people to wear whatever they please.

Among the alternatives to animal-derived clothing are synthetic fabrics such as polyester, which are made from, and therefore use up, nonrenewable petroleum products. Other forms of warm clothing may not require using a nonrenewable resource. For example, wool can be obtained as a renewable resource from sheep; the shearing causes them no apparent harm, and they simply regrow a new coat of wool. Other fibers for fabrics can be made from plants such as cotton. Leather is generally obtained from animals that are killed for other purposes (such as food). Most of these are much less expensive than fur and raise none of the ethical issues mentioned here.

Toxicology testing.   Thanks in part to the animal rights movement, certain forms of testing on animals are frequently criticized and no longer receive the automatic support of the informed public. One of these is the lethal dose (LD_{50}) test. In a lethal dose test, animals are exposed to varying doses of a substance. From these data, scientists calculate the amount that would kill half of the animal subjects to which it was administered; this amount is called the *50% lethal dose,* or LD_{50}. As several animal rights advocates have pointed out, this number tells us very little by itself, making the use of experimental animals to determine its value uninformative and therefore wasteful. In particular, knowledge of the LD_{50} tells us very little if anything about the effects of chronic low levels of exposure, which is what most human consumers will eventually experience if they come in contact with the test substance. A cost-benefit approach must try to specify what otherwise unobtainable benefit to society will result from the knowledge of LD_{50} for a particular substance. Only if such a benefit can be identified, and only if it can be judged to be more important than the lives of the animals that die and the suffering of those that do not die, can the use of the LD_{50} test be justified from a cost-benefit standpoint. Some people would argue that people's needs outweigh animals' rights to avoid death and suffering. Can other tests, using bacteria or laboratory-grown human cells, provide the same answers as the LD_{50} test? We would then need to extend the above argument and ask if people's needs outweigh the rights of the bacteria used in alternative tests.

Testing of pharmaceuticals.   Of all the types of experiments to which animals are subjected, none are as often justified in the eyes of the public as the testing of medicines intended for human use. In fact, a pharmaceutical company would be considered remiss if it marketed a new drug without first testing it on animals. In many countries, including the United States, animal testing is required by law

before a new drug can be brought to market. If a drug causes adverse effects in even a small fraction of humans who use it, then the failure of the drug company to identify such adverse effects in animal testing could be used against them in a very expensive lawsuit.

New drugs are tested every year, and most of the tests use experimental animals. In fact, a good deal of the expense involved in bringing a new drug to market is the cost of animal testing. In addition to the lives of the animals, the costs of the experimental testing of new drugs include the salaries of the experimenters and animal handlers. On the other side of the cost-benefit equation are the human lives saved or symptoms relieved. If the drug is successful, its benefits may continue far into the future.

Those people who value human life above the lives of animals are only being consistent when they insist that drugs or new procedures be tested on animals before they are used on humans. Some animal rights advocates, such as Ingrid Newkirk of PETA, have adopted the viewpoint that a human life is no more valuable than an animal life, or, in her words, "a rat is a pig is a dog is a boy." A direct logical consequence of this viewpoint is that the pain and suffering of animals in any experiment can no more be justified than an equivalent amount of pain and suffering for a human subject. On this issue, as on many others, the cost-benefit equation can come out differently according to the relative values placed on the lives of humans and the lives of nonhuman animals.

Cosmetic testing.   One test that animal rights advocates have frequently criticized is the Draize test. In a Draize test, cosmetics or other chemical substances are sprayed into the eyes of rabbits in an attempt to assess the irritation caused to the cornea. If the substance sprayed causes corneal damage, blindness frequently results. Dozens of rabbits were used in the testing of each new cosmetic. Does the cost-benefit ratio for the testing of cosmetics differ from the ratio for the testing of new medicines? On the benefit side, profits and employment will accrue in either case. Possible benefits to society from cosmetics include the availability of one more perfume or hair spray, which many people may not consider to be as important as the benefits of new medicines. In addition, Draize tests are often used to test new *batches* of an already tested cosmetic rather than in basic research on new chemicals, leading many people to regard the potential benefits of such tests as extremely low. The costs to humans from the use of an untested medicine are possibly greater than the costs from the use of untested cosmetics. As with the testing of pharmaceuticals, the availability of any alternative testing mechanisms, such as bacteria or laboratory-grown human cells, also enters the cost-benefit analysis. In addition, allergy testing could conceivably be done with human subjects, since the cosmetic is only applied externally, and any allergic response is temporary and does not produce serious illness.

Improving the Treatment of Animals

Attempts to improve the status of animals in research include attempts to prevent animal abuse and neglect and to minimize pain and suffering. Most current legislation deals with the prevention of abuse and neglect by setting minimum standards for housing and care. For example, the U.S. Animal Welfare Act sets standards for the housing of various species (including minimum cage sizes and similar details); the provision of adequate food, water, and sanitation; and such other matters as ventilation, protection from temperature extremes, veterinary care, and the use of anesthetics, painkillers, and tranquilizers whenever it is appropriate.

Animal rights groups have advocated what are known as the three Rs: reduction, refinement, and replacement. *Reduction* would mean using methods that require fewer animals; such measures would also in most cases reduce costs. *Refinement* would mean using methods that get more information from a given amount of experimentation. Among other refinement measures, researchers should always make sure they are not repeating earlier work. *Replacement* would mean using tissue culture and other in vitro methods in preference to whole animals, or avoiding the use of animals entirely whenever this can be done without compromising experimental goals.

Humans as Experimental Subjects

Many animal experiments are undertaken to determine the effects of some new drug or other ther-

apy. Often, we are really interested in knowing what the effects would be in humans, and the animals are merely being used as stand-ins. No matter how the animal experiments turn out, one could always ask whether results obtained from a nonhuman species can safely be extrapolated to humans. In most cases in which adequate data exist to answer this question, physiological reactions have turned out to be comparable. Even when differences between humans and other species are known, they are often known in sufficient detail that the different responses to testing can help us understand the human system better, which still makes the animal tests valuable.

A few instances are known in which humans and certain commonly used experimental animal species respond differently. (Saccharin, for example, causes cancer in rats, but has never been shown to cause harm to humans.)

Advantages to using human subjects. Direct experimentation on humans avoids the question of comparability between species, meaning that the results can be used more directly than results obtained from other species. Also, results obtained by psychologists, epidemiologists, and others who study humans using naturalistic methods can be applied even more directly. For example, one could not ethically force-feed cholesterol to an experimental group of human subjects, but one could observe the diets that different people choose on their own and study how people with high-cholesterol diets differ from people with low-cholesterol diets. The diets in such a study are more directly comparable to the diets experienced by other humans than are the diets of experimental animals fed different amounts of cholesterol in their food.

Among possible experimental subjects, humans have a special status. On the one hand, inflicting pain on human subjects, or exposing them to experimental risks, raises ethical objections (or stronger ethical objections) from more people than does similar treatment of nonhuman animals. On the other hand, human subjects can tell us how they feel, or when and where they experience discomfort or pain. Humans can also be given psychological tests, including tests that measure their perception or their ability to solve problems under varying conditions. Certain types of drug side effects, like headaches or impairment of problem-solving ability, are difficult to assess without using human subjects.

Voluntary informed consent.   A purely ethical consideration is that human subjects can voluntarily consent to serve as experimental subjects, which is something that nonhuman animals cannot do. Humans are considered to be autonomous beings who have the right to consent to putting themselves at risk, whether in a space capsule, a bungee jump, or an experiment. An important consideration, however, is that the person serving as a subject must have given her or his consent voluntarily. This is a legal as well as a moral issue, because anyone who has not consented voluntarily can sue for damages if any harm comes to them. It has become customary (and, in government-sponsored research, mandatory) that voluntary consent be obtained in writing and that the consent form outline both the possible benefits and possible risks of the experimental procedure to the subject. If it is possible that an experiment may bring direct benefit to the human subjects (say, in an experiment in which a disease or its symptoms are being treated), potential subjects may be more willing to undergo certain risks than they would if they saw no possibility of direct benefit to themselves.

Special questions arise in the case of persons who may not have the full capacity to understand all the possible risks and benefits, including mentally deficient persons, unconscious persons, and children. Most people would now consider it unethical to take advantage in any way of a person not in full possession of their mental faculties by subjecting them to an experiment unless there was obvious great promise of direct benefit to them and only minimal risk of harm. In most jurisdictions, parents are considered to have the legal right to make such decisions on behalf of their children. In past decades, prison populations were often used as sources for experimental subjects, but this practice is now usually frowned upon because no one in such a situation can give voluntary consent if they think—rightly or wrongly—that cooperating may result in a sentence reduction.

Formal review procedures.   As a safeguard against possible abuses, research on human subjects is now usually reviewed by institutional committees set up for that purpose. As is the case in reviews of animal experimentation, the review process ensures that someone other than the researcher(s) is involved in evaluating the ethics of the proposed experiment. Federally sponsored research in the United States and in many other countries requires that such committees authorize all experiments in which humans are used as subjects. In addition to ensuring that proper voluntary consent has been obtained, such committees also have the obligation to suggest ways in which risks can be reduced or benefits increased without impairing the validity of the experiment.

Guidelines for experimentation.   Many scientists work within the ethical tradition that exposing humans to experimental risks is more objectionable than exposing animals of other species to those same risks. Guidelines have been developed that specify testing to be carried out on animals first (wherever possible), then on small numbers of carefully chosen and carefully monitored human subjects, and only last of all on large and diverse human populations. In the United States, federally sponsored research and research on new drugs seeking federal approval are required to follow this procedure. For example, if scientists are testing a new drug to treat a disease, they first try the drug on animals that have the disease. Later experiments might test the drug on small and carefully matched human samples: If the few dozen people who receive the drug are adults aged 18 to 50, averaging about 35 years old, and with males and females equally represented, then the control sample should also include an equal number of males and females in the same age range and with about the same average age. Only if the drug proves safe and effective is it tested on large groups of patients who vary more widely in age and other relevant characteristics. At any stage in the testing, if the new drug proves to be far superior to whatever it is being compared to (either an inactive substance or another drug), then all patients are switched to the new drug.

Avoiding gender bias.   Before the 1990s, many animal and clinical trials were done only on male subjects; one reason frequently given for this practice was to avoid hormonal fluctuations (from reproductive cycles) as a variable. Many test results based on studies that included only men were extrapolated to women, with the result that women got drug doses or other treatments that had been effective for men without any evidence to show whether or not these treatments were appropriate for women patients. In several cardiovascular conditions, including heart attacks and strokes, it now appears that men and women respond differently to certain drugs, and that drug doses calculated for men may be inappropriate for many women. When she was head of the National Institutes of Health, Dr. Bernadette Healy criticized a number of studies that had been done on men only that she thought would have been more appropriately done on women alone or on both sexes. One of the studies, for example, was based on the observation that pregnant women almost never have heart attacks. In order to test whether or not estrogen is the cause of this protective effect, the effects of estrogen therapy on heart attack rates were measured in *men*! Largely as the results of Dr. Healy's efforts, National Institutes of Health guidelines now require studies to be done on both sexes when appropriate. Although this is an issue of good experimental procedure, the issue was first raised as an ethical question of unfairness to women. Dr. Healy and others claimed that women were getting potentially substandard medical care if they were treated with drugs that had only been tested on men and whose doses were calibrated using male patients only.

THOUGHT QUESTIONS

1. How would a deontologist go about establishing the rightness or wrongness of the following acts? How would a utilitarian judge these same acts?

   Beating your horse
   Taking a canary into a coal mine so that, if it dies from toxic gases, miners could be warned to evacuate
   Raising broiler chickens or beef cattle for human consumption
   Testing a drug on rats (or cats) before giving it to humans
   Testing a drug on human prison convicts

2. Give some examples of each of the following:
   Reduction in the use of laboratory animals
   Refinement in animal experimentation
   Replacement of animal tests by other forms of testing

3. Discuss the benefits and limitations of each of the following proposed alternatives to the use of nonhuman mammals in research.
   Use of invertebrate animals (insects, lobsters, snails)
   Use of animal parts (eyes, hearts, muscles) maintained in vitro after an animal's death.
   Use of animal embryos rather than adult animals
   Use of cells or tissues maintained in vitro and taken from living animals without killing them
   Use of computer models

4. How widely can experimental results be extrapolated? If a drug is tested on inbred male rats, is it certain that the results are applicable to humans? Is it likely? Is the drug likely to have similar effects on both sexes? What issues of methodology or of ethics are raised by experiments that included only inbred male rats?

CHAPTER SUMMARY

Moral codes guide our conduct. Ethics is the discipline that examines moral codes and attempts to explain or justify them. The two major types of ethical systems are deontological and utilitarian. Deontologists judge the rightness or wrongness of an act by characteristics of the act itself, apart from its consequences. Utilitarians judge the rightness or wrongness of an act on the basis of its consequences. Utilitarian analysis often includes a comparison of the undesirable effects (costs) of an act with its desirable effects (benefits). In pluralistic societies, ethical decisions are often difficult. In facing ethical issues involving scientific questions, it is often useful to distinguish scientific issues, science policy issues, and policy issues that are purely ethical. Animals are used in our society for food, for labor, for companionship, and for laboratory experimentation. In many cases, laboratory animals are used as stand-ins for humans, and their use is often justified on a utilitarian basis (the cost-benefit ratio is lower if animals are tested before humans) or on a deontological basis (humans have rights and animals do not, or human rights supersede animals' rights). Some animal rights activists wish to stop all such experimentation because they consider the rights of animals to be equal to the rights of humans. Milder reforms include reduction in the number of animals used, refinement of experiments to get maximum information from the use of animals,

and replacement of animal testing by other forms of testing whenever possible. Before any experimentation on humans can take place, the proposed experiments must pass an ethics review and the voluntary informed consent of the subjects must be obtained.

KEY TERMS TO KNOW

deontological (p. 18)

egoism (p. 21)

ethics (p. 17)

fairness (p. 23)

moral rules (morals)
  (p. 16)

natural-law ethics
  (p. 21)

rights (p. 19)

risk (p. 24)

utilitarian
  (p. 19)

CONNECTIONS TO OTHER CHAPTERS

Chapter 1   Ethical questions are separate from questions within science. Many applications of science have ethical dimensions.

Chapter 3   Human genetic testing, the human genome project, and new reproductive technologies all raise ethical questions.

Chapter 4   The question of whether nonscience (e.g., creationist) viewpoints should be heard in science classes raises ethical issues.

Chapter 5   Ethical objections have been raised against the ways that many people have treated members of other human populations.

Chapter 6   The ethic of allowing reproductive freedom to individuals conflicts with the need to control population.

Chapter 7   Acts that we consider moral and acts that we consider immoral are both products of evolution.

Chapter 8   Patterns of food consumption raise ethical issues.

Chapter 9   Cancer research often involves animal experimentation.

Chapter 10   Brain research often involves animal experimentation and also the use of fetal tissues.

Chapter 11   Drugs are usually tested on animals before giving them to people.

Chapter 12   Our treatment of other people influences their immune functions and disease resistance.

Chapter 13   Many ethical issues surround the transmission of AIDS, the testing for AIDS, and the prevention of AIDS.

Chapter 15   Ethical objections have been raised to genetically engineered foods.

Chapter 16   Ethical objections have been raised to pollution, to habitat destruction, and to species extinctions that reduce biodiversity.

Human Genes and Genomes

A nervous couple sits in the waiting room, anxiously anticipating the results of a test. Their last child lived her short life in almost constant pain, and died, blind, at age 3, a victim of Tay-Sachs disease. The couple wants another child, but their previous experience was a heart-wrenching nightmare that they don't want to go through again. They are awaiting the results of an amniocentesis, a technique you will read about later in this chapter. A doctor enters the room with good news: The enzyme that her technicians were testing for is present in the amniotic fluid. The mother-to-be is carrying a child who will not get Tay-Sachs disease. The couple can look forward to raising a healthy child.

Scenes like the one just described are happening more often with each passing year. An increasing number of couples are undergoing medical procedures that did not exist when they themselves were born, seeking assurances that would have been unthinkable a mere 20 years ago. Tay-Sachs disease is one of a growing number of conditions that can now be diagnosed before birth. Along with physical characteristics, these conditions (*traits*) of organisms are passed on from one generation to the next. How does this come about? How do offspring come to physically resemble their parents? These are the questions posed by the field of biology called **genetics,** the study of inherited traits. A cornerstone of the first major paradigm in this field is the unifying assumption that biological inheritance is carried by structures called **genes.** The discovery of what genes are and how they work has been the subject of many years of research. Among the earliest findings was the fact that many human traits, including several that cause disease or impair health, follow the same basic patterns of inheritance that apply to most other organisms.

A. Inheritance Follows the Same Laws in Most Organisms

No two human beings are exactly alike. ("Identical" twins seem to be an exception, and even they are not identical at the molecular level.) People who are closely related to each other generally look more like each other than like people to whom they are not closely related or are not related to at all. Some traits, like height at adulthood and hair color, vary along a continuum. Others, like certain diseases, are either present or absent in a given individual. However, even the presence or absence of a disease like diabetes shows variation: Some cases of diabetes are mild while others are severe, and there are several kinds of diabetes as well, some of which may be capable of modification by changes in lifestyle. Thus all human traits are subject to considerable variation among individuals.

Folk wisdom going back to ancient times taught people that a child or an animal sometimes resembles its mother and sometimes its father; it often shows a mixture of traits derived from each side of the family. This suggested a concept that came to be called *blending inheritance,* in which heredity was compared to a mixing of fluids, often identified as "blood." The research we are about to describe caused this theory to be abandoned.

Genes

During the nineteenth century, Gregor Mendel, a Czech scientist living under Austrian rule, worked out the principles of inheritance for simple traits that he described in "either/or" terms. Mendel was a priest who raised pea plants (*Pisum sativum*) in the garden of his monastery. Mendel was curious about how the several types of peas differed, so he decided to breed them and keep careful records. He organized his work so as to answer specific questions, a procedure that we recognize today as good experimental design. Unlike most of his predecessors, Mendel also followed certain careful procedures:

1. He started with plants belonging to pure lines. A *pure line* is one that breeds true for several generations, always producing offspring that closely resemble their parents.

2. He chose which plants would mate, either cross-fertilizing (crossing) the flowers with one another or closing up the flower parts to ensure self-fertilization (mating them with themselves) (Box 3.1). Mendel became skillful at sewing up he flower parts inside one of the enlarged petals to make sure that no unwanted crosses occurred.

3. He studied only one trait at a time, until he understood its pattern of inheritance, then he studied two and three traits at a time. His predecessors, in contrast, often examined several or many traits at once.

4. He counted the offspring of each cross, and was thus able to recognize ratios among them. (Those few of his predecessors who looked at single traits never counted the offspring of each type and thus failed to find ratios.)

5. He continued each experiment through several generations of pea plants, not just one.

Dominant and recessive genes. Mendel's peas differed in a number of discrete traits, like white versus colored flowers, or tall versus short

(dwarfed) plants. First, Mendel raised peas belonging to pure lines for each of these traits; for example, a line in which tall parents produced tall offspring and another line in which short parents produced short offspring. Then he crossed plants differing in one trait at a time. For example, he crossed plants having white flowers with plants having violet-colored flowers, and found that all the first-generation offspring (symbolized as F_1) had violet-colored flowers (Fig. 3.1). When he crossed tall and short plants, all the offspring were tall. Mendel introduced the term **dominant** for

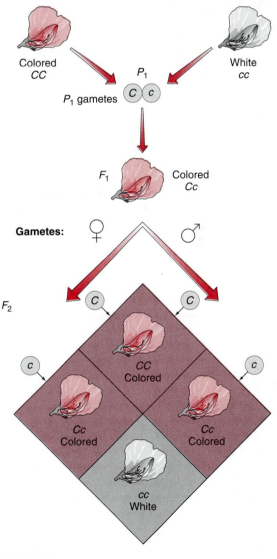

FIGURE 3.1
One of Mendel's crosses for peas differing in a single trait.

BOX 3.1   THE CHOICE OF SPECIES IN AN EXPERIMENT

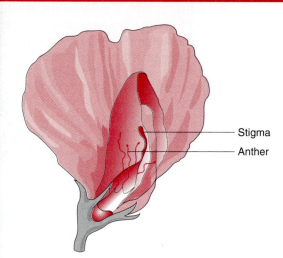

Stigma
Anther

Throughout the history of genetics, the choice of experimental organisms and the questions raised have changed. It could be argued that the choice of questions favored the use of certain species; the reverse—that the choice of species guided the questions that would or could be investigated—could also be argued.

MENDEL'S PEAS

Why were pea plants a good choice of experimental species for Mendel's experiments? Pea plants have many distinctive traits, and other practical advantages: Mendel could easily grow them in the monastery garden, and many vari-

The seven pairs of traits studied by Mendel in peas

	Seeds			Pods		Stems	
	Round	Yellow	Gray coat (red flowers)	Inflated	Green	Axial flowers	Tall
Dominant							
	Wrinkled	Green	White coat (white flowers)	Pinched	Yellow	Terminal flowers	Short
Recessive							

eties were locally available, including those with yellow versus green peas or round versus wrinkled peas. Mendel knew that peas reproduced sexually by means of flowers: Within the pea flower are structures that produce pollen and another structure, called a *stigma*, on which the pollen falls. Mendel knew that pollination could occur by *self-fertilization*, pollen falling onto a stigma in the same plant. He also knew that pollen could instead be supplied by a different plant (*cross-fertilization*), offering a way to investigate the contributions of two parents from different plants.

MORGAN'S FRUIT FLIES

Early in the twentieth century, T. H. Morgan and others introduced the fruit fly (*Drosophila*) as an experimental animal in genetics. Like peas, fruit flies have many phenotypes that can be studied. They also reproduce sexually, though the sexes are separate, so there is no possibility of self-fertilization. Fruit flies produce large numbers of offspring that can easily be raised in a small bottle. Perhaps the greatest advantage of fruit flies over peas is their generation time: Where a new generation of peas takes a year (and requires the patience of a monk in a monastery), a new generation of fruit flies takes as little as two weeks. Also, fruit flies have very few chromosomes (only eight, compared to fourteen in peas or forty-six in humans), so it was easier to study the relations between genes and chromosomes in fruit flies.

MICROORGANISMS

Beginning around 1940, several geneticists turned from fruit flies to bread molds (*Neurospora*) and bacteria (*Escherichia coli*). One great advantage of these microorganisms was that even more of them (trillions) could be grown, and could be grown even faster, with generation times measured in hours instead of weeks. Also, microorganisms are haploid throughout most of their lives, so that it was possible to study the effects of one allele at a time without worrying about dominance or the influence of the other allele in a diploid pair. Most of the phenotypes studied were those with the ability to make a particular chemical or enzyme. These conditions allowed investigation of the mechanisms (transcription and translation) by which genes produce their phenotypes, studies that would have been much more difficult using peas or fruit flies.

WHAT ABOUT HUMANS?

Humans are poor subjects for experiments in genetics: They reproduce very slowly (a new generation takes over 20 years), and family sizes are usually too small to reveal any reliable ratios. Humans, moreover, cannot be mated at the experimenter's will.

Humans do have a few features that are of aid to geneticists. Above all, they keep good records. Records of health and illness allow geneticists to study many disease conditions (including many rare ones), and this is one reason why geneticists have studied so many disease-related traits. Humans also keep good genealogical records, allowing geneticists to trace the inheritance of a trait over several generations in each family. When we study past generations, we can only study those traits that people have recorded. The fact that the cause of death of an individual is usually recorded or remembered accounts for why human genetics has paid so much attention to conditions that can result in early death.

the trait that appeared in the first-generation offspring of his initial cross; the trait that did not show up he called **recessive.** Thus, tall was dominant to short, and colored flowers were dominant to white. In all, Mendel studied seven pairs of either/or traits, and found that in each pair one trait was dominant and one was recessive. The traits did not blend. Violet mated with white produced violet, not pale violet. Tall mated with short produced plants the height of the tall parents, not halfway between.

Phenotype and genotype. Mendel noticed that the tall plants of the F_1 were just as tall as their tall parents, and had the same appearance, also called **phenotype.** Their hereditary makeup, however, was different: The tall parents had come from a pure line, so all their hereditary makeup was tall, but the tall F_1 plants had both tall and short parents. Could this difference in hereditary makeup, or **genotype,** be made visible? Would it show up in future generations? Mendel found out by mating the plants of the F_1 generation with themselves (self-fertilizing them) and raising a second generation, symbolized as F_2. He got both tall and short plants in the F_2. When he counted them, he found that the tall plants were approximately three times as numerous as the short ones. Mendel conducted similar experiments with other traits, such as white versus colored flowers, and in each experiment he raised an F_2 generation. In each case he discovered the same thing: In the F_2 generation, the dominant phenotype outnumbered the recessive phenotype in the approximate ratio of 3:1, meaning that ¾ of the F_2 had the dominant phenotype and ¼ had the recessive one.

Mendel's explanation for inheritance of single traits. Mendel proposed a multipart hypothesis to explain his results:

1. The inheritance of traits is controlled by hereditary "factors." (We now use the term *gene* for these factors.)

2. Each individual has two genes for each trait. If each gene is represented by a letter, then the genetic makeup (genotype) of an individual can be represented as two letters for each trait.

3. Each gene exists in different forms; these variant forms of the same gene are called **alleles.** For example, the gene for flower color in peas has an allele that produces white flowers and another allele that produces colored flowers. Alleles that are dominant produce a visible effect even when other alleles are present; alleles that are recessive are often masked by the dominant alleles. Mendel designated the dominant and recessive alleles controlling the same trait by a single letter of the alphabet, using a capital letter for the dominant allele and a lowercase letter for the recessive allele. For example, the allele T for tall plants was dominant to the allele t for short plants. (Today, many genes are designated by two- and three-letter combinations, and different alleles of the same gene by superscripts.)

4. An individual whose genotype contains two identical alleles, like TT or tt, is said to be **homozygous.** An individual whose genotype combines dissimilar alleles, like Tt, is called **heterozygous** for that trait.

5. Dominant alleles always show up in the phenotype, but recessive alleles are easily masked by their dominant counterparts. When a dominant and a recessive allele for the same trait occur together, only the dominant allele produces a visible phenotype. Recessive alleles produce a phenotypic effect only when the corresponding dominant allele is absent, as in the homozygous recessive condition.

6. The genes behave as particles that remain separate instead of blending. Recessive genes remain hidden during the F_1 generation, unchanged by their passage through individuals who show the dominant phenotype.

7. When the sex cells (egg cells or pollen) of the F_1 are produced, the dominant and recessive alleles separate from one another, or "segregate." During fertilization, the sex cells, which are also called **gametes,** recombine in all possible combinations to form the F_2, which exhibits the 3:1 ratio of dominant to recessive phenotypes (Fig. 3.1). (We now understand that segregation is a consequence of chromosome movements during the cell divisions that give rise to the gametes; see Box 3.2, pp. 41–43.) The phenomenon of segregation is often called the **law of segregation,** or Mendel's first law.

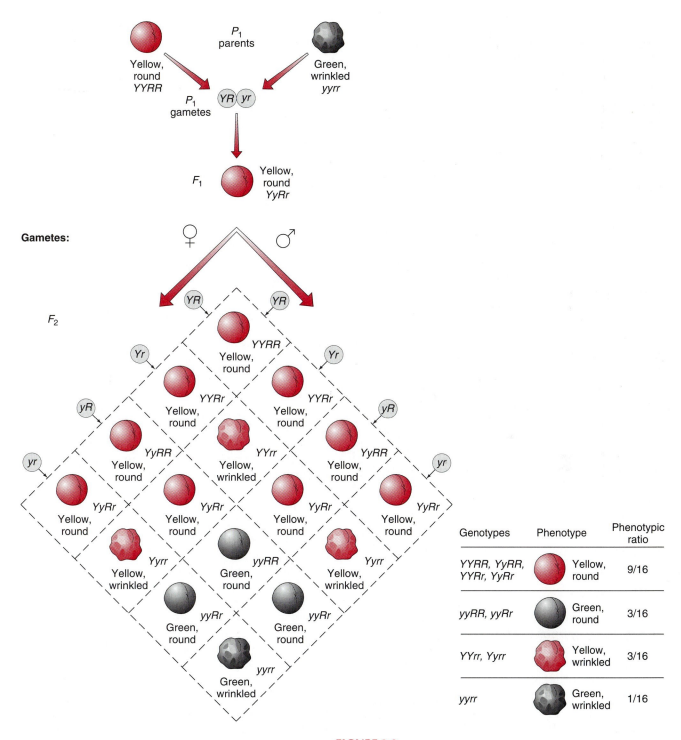

Genotypes	Phenotype	Phenotypic ratio
YYRR, YyRR, *YYRr, YyRr*	Yellow, round	9/16
yyRR, yyRr	Green, round	3/16
YYrr, Yyrr	Yellow, wrinkled	3/16
yyrr	Green, wrinkled	1/16

FIGURE 3.2
One of Mendel's crosses for peas differing in two traits at once.

Independent assortment in the inheritance of multiple traits.   After Mendel had investigated the inheritance of one trait at a time, he proceeded to study the inheritance of two traits at a time. One of his crosses is shown in Fig. 3.2. In this cross, the original parents differ in two traits. One set of parents have yellow, round seeds, and are homozygous for both traits (*YYRR*). The other parents have green, wrinkled seeds (*yyrr*). The first generation offspring (F_1) are all *YyRr*, heterozygous for both traits. All these F_1 plants have round seeds (because round is dominant to wrinkled) and all have yellow seeds (because yellow is dominant to green). No new principles are involved thus far.

Mendel expected half of the gametes to contain the dominant allele *Y* and the remainder the recessive allele *y*. He also expected half of the gametes to contain allele *R* and the other half the allele *r*. But would the choice of *Y* versus *y* influence the choice of *R* versus *r*? To find out, Mendel raised an F_2 generation, and obtained the 9:3:3:1 ratio of phenotypes shown in Fig. 3.2, meaning that $9/16$ of the F_2 individuals showed both dominant traits (they were both round and yellow), $3/16$ were round but not yellow, $3/16$ were yellow but not round, and only $1/16$ showed both recessive traits (wrinkled and green). Mendel then reasoned that this ratio could be explained if he assumed that all four possible types of gametes (*YR, Yr, yR,* and *yr*) were produced in equal proportions. This meant that the inheritance of one trait (round versus wrinkled) had no influence on the inheritance of the other (green versus yellow). This principle is called the **law of independent assortment,** or Mendel's second law.

All the traits that Mendel worked with assorted independently in this way, but as we will soon see, there are exceptions to Mendel's second law.

Mendel's results, published in 1865, were ignored by most scientists. The reasons for the lack of impact of his theories are many, but one contributing factor was that he presented his theories in the language of mathematics, which was not a language in which his fellow scientists were fluent. The same problem blocks interdisciplinary efforts in many fields today (see Chapter 14). In 1900 each of three other scientists conducted experiments similar to Mendel's, reached similar conclusions, and then subsequently discovered Mendel's earlier work.

Chromosomes

The reader may notice that some of Mendel's assumptions raise questions that Mendel himself did not answer:

1. Why do the genes exist in pairs?
2. Why do different traits assort independently?
3. Where are the genes located?
4. What are the genes made of?

Genes are located on chromosomes.   Answers to the first three of these questions were suggested by a young American scientist, Walter Sutton, who read about the rediscovery of Mendel's work in 1900. By this time, it was already well known that all animal and plant cells contained a central portion called the **nucleus** and a surrounding portion called the **cytoplasm.** Division of the cytoplasm was a very simple affair. The nucleus, however, was known to undergo **mitosis** (Box 3.2), a complex rearrangement of the rod-shaped bodies called **chromosomes,** which were seen to exist in pairs. Gametes (sex cells such as eggs and sperm) are **haploid,** with one chromosome from each pair. All other body cells, called somatic cells, have a **diploid** chromosome number in which all chromosomes are paired. Sutton noticed that eggs in most species are many times larger than sperm because of a great difference in the amount of cytoplasm. The nuclei of egg and sperm are approximately equal in size, and these nuclei fuse during fertilization. From these facts, Sutton reasoned as follows:

1. The genes are probably in the nucleus, not the cytoplasm, because the nucleus divides carefully and exactly (Box 3.2), while the cytoplasm divides inexactly. Also, if genes were in the cytoplasm, one would expect the mother's contribution always to be much greater than the father's, contrary to the observation that parental contributions to heredity are usually equal.

2. Of all parts of the cell, only the chromosomes are known to exist in pairs. Chromosomes

BOX 3.2   CELL LIFE CYCLES; MITOSIS AND MEIOSIS

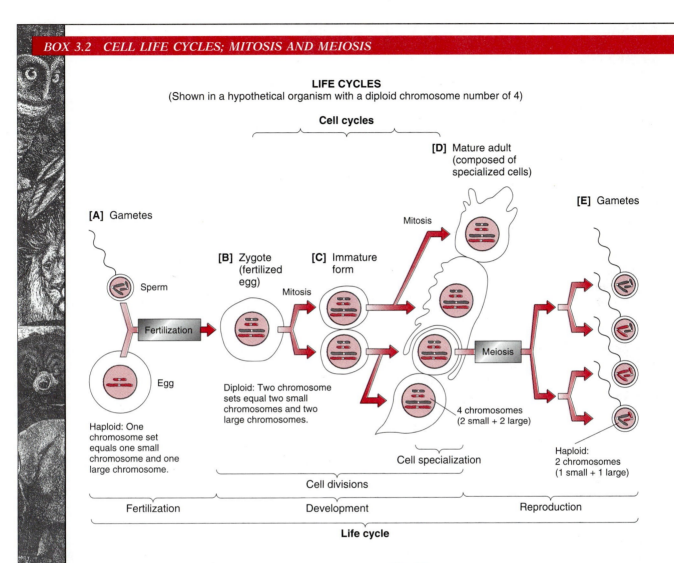

LIFE CYCLES
(Shown in a hypothetical organism with a diploid chromosome number of 4)

LIFE CYCLES

All animal and plant cells contain a nucleus inside of which are chromosomes. The nucleus is surrounded by the cytoplasm. At certain stages of cell division, the chromosomes are visible under a microscope as rodlike structures of different lengths.

[A] Gametes (eggs and pollen in plants, or eggs and sperm in animals) have one of each chromosome, two different chromosomes in this example.

[B] Gametes join during fertilization and the zygote (fertilized egg) then has two of each chromosome, one from each gamete. Since each chromosome exists in pairs, the zygote is said to have the **diploid number** of chromosomes.

[C, D] Cells then divide many times during the lifetime of the organism. In each of these divisions, the chromosomes are first replicated, and then the cell divides; half of its cytoplasm and half of its chromosomes goes to each cell. This process is called **mitosis** and produces two diploid cells, each genetically the same as the parent cell.

[E] A very small fraction of these cells develop into gametes: The chromosomes replicate as usual to four chromosomes, but then the cell divides and divides again, resulting in four gamete cells, each with a single chromosome of each type (the **haploid number** of

(box continues)

BOX 3.2   (continued)

chromosomes). Double division producing haploid gametes is called **meiosis.** Each gamete therefore has half of the genetic material of the parent cell, one chromosome of each type.

MITOSIS AND MEIOSIS
Paired chromosomes are duplicated. The four copies of DNA divide once in mitosis and twice in meiosis.

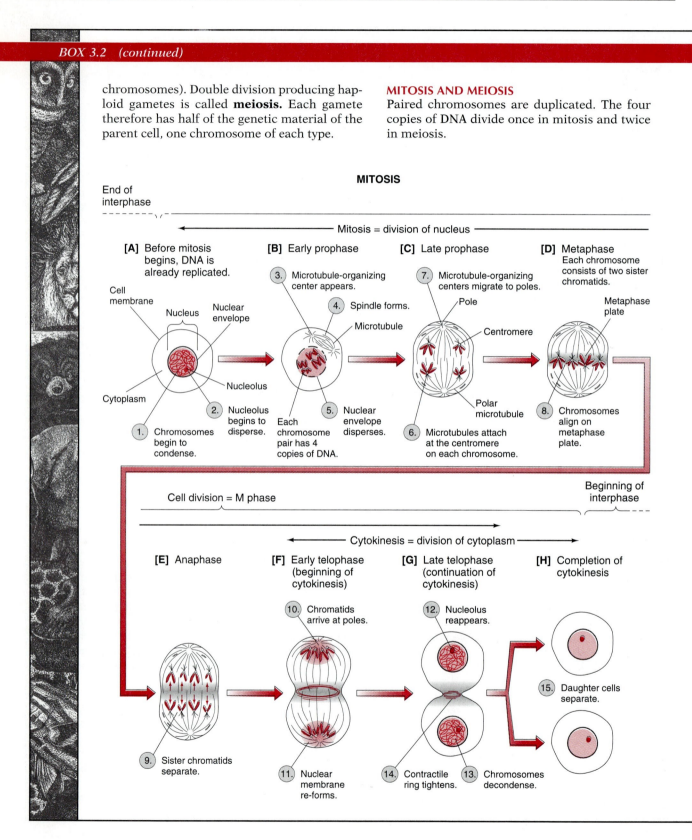

MITOSIS

End of interphase

Mitosis = division of nucleus

[A] Before mitosis begins, DNA is already replicated.

Cell membrane
Nucleus
Nuclear envelope
Nucleolus
Cytoplasm

1. Chromosomes begin to condense.
2. Nucleolus begins to disperse.

[B] Early prophase

3. Microtubule-organizing center appears.
4. Spindle forms.
Microtubule
5. Nuclear envelope disperses.
Each chromosome pair has 4 copies of DNA.

[C] Late prophase

7. Microtubule-organizing centers migrate to poles.
Pole
Centromere
Polar microtubule
6. Microtubules attach at the centromere on each chromosome.

[D] Metaphase
Each chromosome consists of two sister chromatids.

Metaphase plate
8. Chromosomes align on metaphase plate.

Cell division = M phase

Cytokinesis = division of cytoplasm

Beginning of interphase

[E] Anaphase

9. Sister chromatids separate.

[F] Early telophase (beginning of cytokinesis)

10. Chromatids arrive at poles.
11. Nuclear membrane re-forms.

[G] Late telophase (continuation of cytokinesis)

12. Nucleolus reappears.
14. Contractile ring tightens.
13. Chromosomes decondense.

[H] Completion of cytokinesis

15. Daughter cells separate.

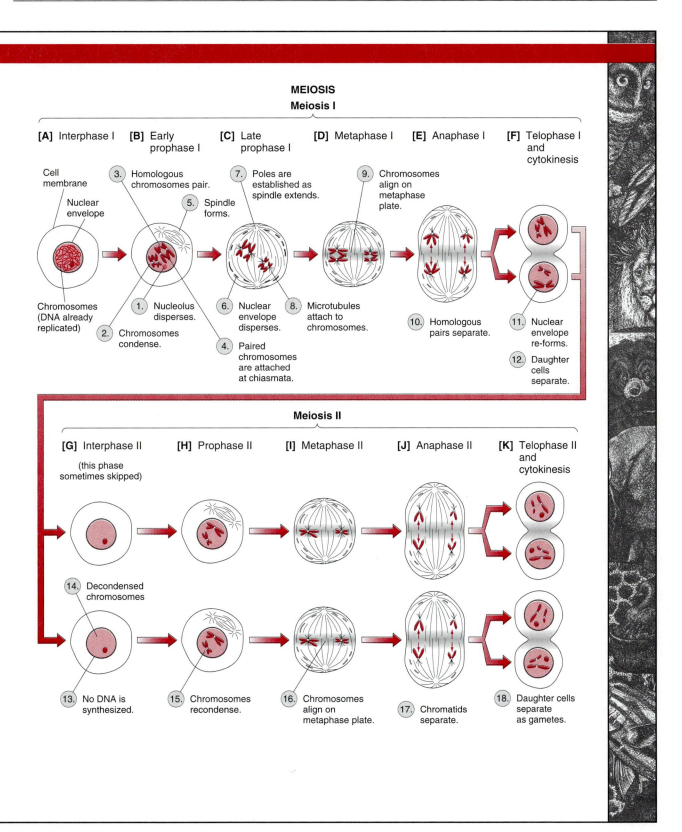

MEIOSIS

Meiosis I

[A] Interphase I   **[B]** Early prophase I   **[C]** Late prophase I   **[D]** Metaphase I   **[E]** Anaphase I   **[F]** Telophase I and cytokinesis

Cell membrane

Nuclear envelope

3. Homologous chromosomes pair.

5. Spindle forms.

7. Poles are established as spindle extends.

9. Chromosomes align on metaphase plate.

Chromosomes (DNA already replicated)

1. Nucleolus disperses.

2. Chromosomes condense.

6. Nuclear envelope disperses.

8. Microtubules attach to chromosomes.

4. Paired chromosomes are attached at chiasmata.

10. Homologous pairs separate.

11. Nuclear envelope re-forms.

12. Daughter cells separate.

Meiosis II

[G] Interphase II   **[H]** Prophase II   **[I]** Metaphase II   **[J]** Anaphase II   **[K]** Telophase II and cytokinesis

(this phase sometimes skipped)

14. Decondensed chromosomes

13. No DNA is synthesized.

15. Chromosomes recondense.

16. Chromosomes align on metaphase plate.

17. Chromatids separate.

18. Daughter cells separate as gametes.

exist singly in gametes. If genes are located on the chromosomes, it explains why chromosomes exist in pairs except in gametes. In fact, *the known behavior of chromosomes exactly parallels the postulated behavior of Mendel's genes.* (This part of Sutton's hypothesis, that genes were located on chromosomes, came to be called the *chromosomal theory of inheritance.*)

3. According to Sutton's hypothesis, Mendel's genes assorted independently because they were located on different chromosomes. However, there are only a limited number of chromosomes (eight in fruit flies, forty-six in humans), while there are hundreds or thousands of genes. Sutton therefore predicted that Mendel's law of independent assortment would only apply to genes located on different chromosomes. Genes located on the same chromosome would be inherited together as a unit. (This phenomenon is now known as **linkage.**)

Gene linkage.   Sutton had predicted the existence of linked genes before other investigators had adequately described the phenomenon. A British geneticist named Bateson soon described a cross involving linked genes in garden peas, and other investigators soon discovered similar examples among fruit flies (*Drosophila*), corn (*Zea mays*), and other species. Figure 3.3 shows a cross involving two linked genes in corn. In this type of cross, *CCSS* (colored, full seeds) × *ccss* (colorless, shrunken seeds) produces mostly colored-full and colorless-shrunken seeds (the two parental combinations of phenotypes) among the F_2, while the cross *CCss* × *ccSS* produces the opposite combinations, with colorless-full and colored-shrunken phenotypes predominating. Notice that a few F_2 plants have new (nonparental) combinations; these recombinant genotypes were thought to arise from a process called **crossing-over,** in which the chromosomes break and then recombine. Some microscopists thought they had observed X-shaped arrangements of the chromosomes that looked like crossing-over, but many scientists were unsure.

Confirmation of the chromosomal hypothesis.
Harriet Creighton and Barbara McClintock first confirmed the chromosomal theory of inheritance in corn, in 1931; later that year Curt Stern observed the same thing in fruit flies. Creighton and

McClintock used plants whose chromosomes had structural abnormalities on either end, enabling them to recognize the chromosomes under the microscope. What they were able to demonstrate was that *genetic recombination* (the rearranging of genes) *was always accompanied by crossing-over* (the rearranging of chromosomes). Barbara McClintock went on to discover genes that move from place to place, the so-called *transposable elements,* or *jumping genes,* a discovery for which she later received the Nobel Prize. As we will see in Chapter 9, genes that move about in this manner may be important in the onset of certain types of cancer.

The frequency of recombination between linked genes is very roughly a measure of the distance between them along the chromosome—recombination between closely linked genes is a rare event, while recombination between genes farther apart is more common. By means of crosses involving linked genes, geneticists were able to determine the linear arrangement of many genes and the approximate distances between these genes on the chromosomes of many species.

An interesting footnote to Mendel's work was provided in 1936 by the British geneticist R. A. Fisher, who noticed that garden peas have seven pairs of chromosomes ($2N = 14$, where N stands for the haploid chromosome number). Mendel had picked seven traits that assorted independently! Since the probability of this outcome occurring by chance is extremely remote, Fisher concluded that Mendel may have studied many more traits and only reported the results for the seven independently assorting traits whose inheritance he could understand (Box 3.1, p. 36).

DNA

We will now address an important question raised earlier: What are the genes made of? Early in the twentieth century, most researchers thought that proteins were the most likely candidates because proteins were known to be complex and varied, while most other molecules were thought not to be.

Transformation in bacteria.   The first indication that the genes might not be made of protein began with a curious experiment in bacteriology by Frederick Griffith in 1928. Griffith worked with two

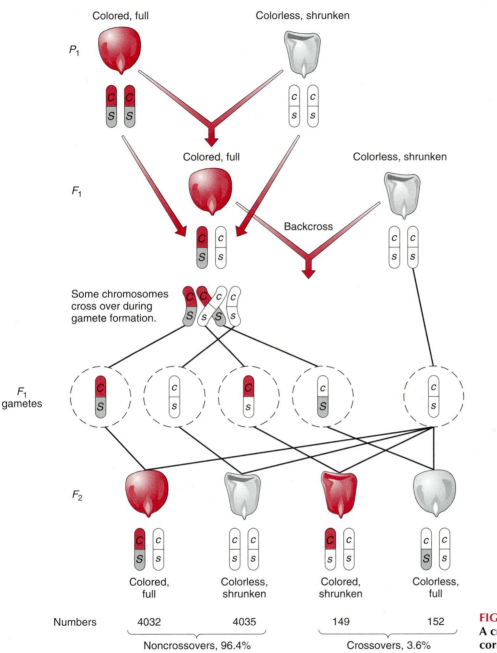

FIGURE 3.3
A cross between linked genes in corn (*Zea mays*).

different strains of bacteria that differed in the structure of their outer coating. Strain IIIS had an outer coat that gave their colonies a smooth appearance when grown on Petri dishes. The smooth-colony bacteria were *virulent* (Chapter 13), meaning that they caused a disease (pneumonia) that killed their hosts. Strain IIR of the same bacterial species lacked the outer coat, which gave

their colonies a "rough" appearance, and they were nonvirulent. (*S* stands for "smooth," *R* for "rough.") When strain IIIS was injected into mice, all the mice died of pneumonia. Strain IIR did not kill mice, nor did bacteria of strain IIIS that had been killed by heat. The surprising result was that a combination of live bacteria of strain IIR and heat-killed bacteria of strain IIIS did kill the mice and that living IIIS bacteria were isolated from all the mice that died this way (Fig. 3.4). Griffith interpreted this experiment as showing that something in the dead IIIS bacteria had somehow transformed the living IIR bacteria into the virulent type, IIIS. Moreover, the bacteria resulting from this **transformation** had been altered genetically, not just phenotypically. A change that was only phenotypic would not be passed on to future generations, but Griffith could demonstrate that descendants of the transformed bacteria continued to be of type IIIS.

DNA and genetic transformation.

The search for the cause of bacterial transformation led Avery, MacLeod, and McCarty (1944) to repeat Griffith's experiment without mice, using bacteria alone. First, they were able to show that a *chemical extract* of heat-killed IIIS bacteria contained what they called a "transforming principle," which transformed strain IIR into IIIS. They then separated the extract into different fractions, each containing different types of chemical molecules. They found that the fraction containing the **nucleic acids** was able to transform strain IIR into IIIS, but that the protein fraction was not. Finally, they distinguished between the two major types of nucleic acid (DNA and RNA) by using enzymes. **Enzymes** are biological molecules (nearly always proteins) capable of speeding up chemical reactions without themselves getting used up in those reactions. Enzymes control many biological processes, and the action of most enzymes is very restricted in that specific enzymes act on specific types of molecules. The enzyme deoxyribonuclease (DNAse) that acts specifically on DNA destroyed the transforming principle, demonstrating that this transforming principle—the genetic material—was, in fact, **deoxyribonucleic acid (DNA).** The enzyme ribonuclease, on the other hand, had no effect; **ribonucleic acid (RNA)** was therefore not a transforming principle.

Despite Avery and his co-workers' startling discovery, it took about a decade for other scientists to pay much attention to DNA. Doubters still remained. The focus of attention had meanwhile shifted from bacteria to the viruses that infected bacteria. In 1952, Alfred Hershey and Martha Chase published the results of a landmark experiment that confirmed the finding that DNA, not protein, was the genetic material, this time in a

virus. The virus that they used was a **bacteriophage,** meaning a virus that infects bacteria and reproduces within them. What Hershey and Chase did was to infect bacteria with viruses and study the viral offspring. It had been found that viruses consisted of a protein capsule, with a nucleic acid interior (Fig. 3.5). In their experiment, Hershey and Chase grew some viruses on a medium containing radioactive phosphorus (^{32}P) and others on a medium containing radioactive sulfur (^{35}S), atoms that the virus used in synthesizing new molecules. Since phosphorus occurs in DNA but not in proteins, the viruses grown with ^{32}P had radioactively labeled nucleic acids (including DNA). In contrast, the viruses grown with ^{35}S had radioactively labeled proteins because proteins contain sulfur and nucleic acids don't. Hershey and Chase exposed bacteria to radioactively labeled viruses long enough to permit viral nucleic acids to be injected from the virus into the bacterial interior. (See Chapter 13 for more on viral life cycles.) At this point, they interrupted the process using a kitchen blender to knock the attached virus capsules off of the bacterial surfaces. These detached virus capsules could easily be separated from the bacteria in a centrifuge. Viral material that had been injected into the bacteria continued the process of viral reproduction, eventually killing the bacteria and breaking them open (lysing them) to release thousands of new virus particles. When ^{35}S-labeled viruses were used, the radioactive proteins remained in the capsules outside and did not enter the bacteria; the viruses released after they had finished reproducing were not radioactive. However, when the ^{32}P-labeled viruses were used, the radioactive DNA entered the bacteria, and the resulting viral offspring were slightly radioactive, showing that they had used some of the radioactive DNA in their reproduction. This experiment upheld the hypothesis that the genetic material of the virus was DNA, while falsifying the hypothesis that the viral genetic material was made of protein.

The chemical composition of DNA.

The structure of DNA was still a mystery, however. Chemical breakdown of DNA into its parts showed that it was made of *phosphate* groups, a sugar (*deoxyribose*), and a series of nitrogen-containing bases called *adenine* (A), *guanine* (G), *cytosine* (C), and *thymine* (T). Biochemists soon realized that the deoxyribose sugar could form a middle link between

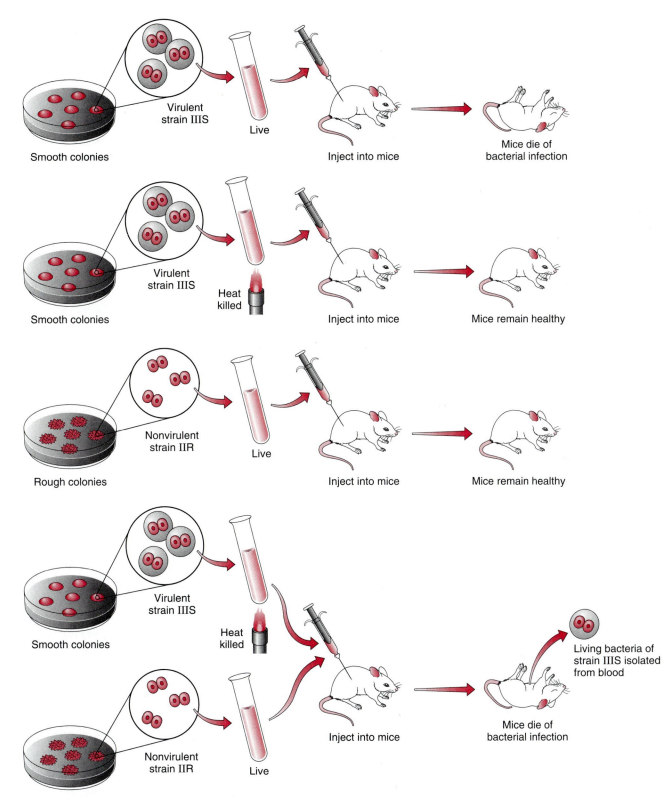

FIGURE 3.4
Griffith's experiment demonstrating transformation in pneumonia bacteria.

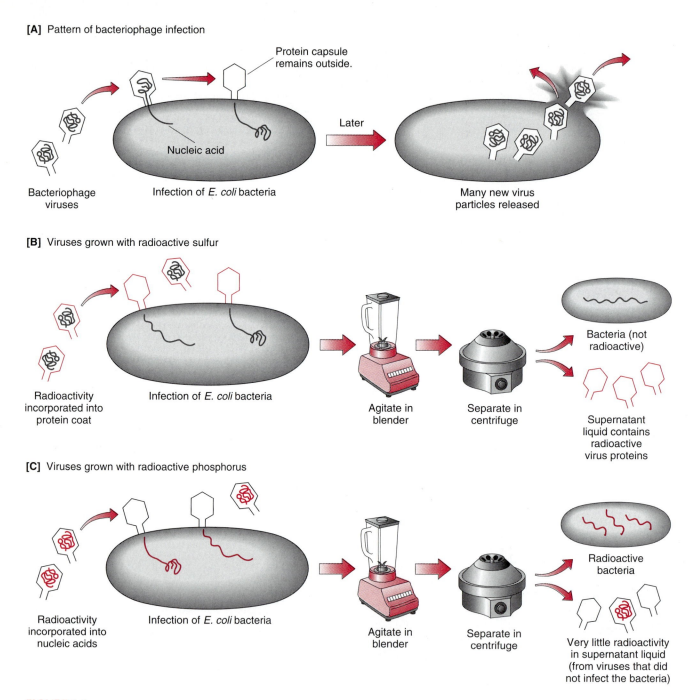

[A] Pattern of bacteriophage infection

Protein capsule remains outside.

Nucleic acid

Bacteriophage viruses

Infection of *E. coli* bacteria

Later

Many new virus particles released

[B] Viruses grown with radioactive sulfur

Radioactivity incorporated into protein coat

Infection of *E. coli* bacteria

Agitate in blender

Separate in centrifuge

Bacteria (not radioactive)

Supernatant liquid contains radioactive virus proteins

[C] Viruses grown with radioactive phosphorus

Radioactivity incorporated into nucleic acids

Infection of *E. coli* bacteria

Agitate in blender

Separate in centrifuge

Radioactive bacteria

Very little radioactivity in supernatant liquid (from viruses that did not infect the bacteria)

FIGURE 3.5
The Hershey-Chase experiment.

phosphate groups and nitrogenous bases, but the rest of the structure was unclear. In particular, it was not at all obvious how DNA's structure gave it the ability to carry genetic information.

Some researchers suggested that the bases in DNA occurred in a regularly repeating pattern: AGTCAGTCAGTC . . . . If this were true, then the amounts of all four nitrogenous bases should always be the same. In order to test this hypothesis, Erwin Chargaff of Columbia University took DNA from various sources, broke it down partially using enzymes, and measured the relative amounts of the four nitrogenous bases. His findings were as follows:

1. The proportions of the four nitrogenous bases were constant for all cell types within an individual. For example, all cells taken from one person contained about 31 percent adenine (A), 19 percent guanine (G), 19 percent cytosine (C), and 31 percent thymine (T), regardless of whether one sampled brain cells, liver cells, kidney cells, or skin cells.

2. The proportions were also remarkably constant within a species, though they varied from one species to another. All humans, for example, showed the same proportions of the four bases. Proportions were different in rats and bread molds, but all rats had the same proportions as one another and so did all bread molds.

3. The most unexpected finding, and the hardest to explain, was that the proportion of adenine was always the same as the proportion of thymine (within the limits of experimental error), and the levels of cytosine and guanine were also equal. These findings (symbolized as A = T and G = C) became known as Chargaff's rules.

The three-dimensional structure of DNA. It remained for James Watson and Francis Crick to propose the structure for DNA that explained all of these findings (Fig. 3.6). They did this in 1953 with the help of some X-ray diffraction information obtained from the laboratory of Maurice H. F. Wilkins, with whom they later shared a Nobel Prize. The data from Wilkins's laboratory were gathered and interpreted by Rosalind Franklin, whose contribution never received the recognition it deserved (Box 3.3).

The X-ray diffraction information pointed to certain dimensions and distances for the repetition of structures within the DNA molecule. From this information, Watson and Crick were able to construct models of the DNA structure that explained all the available data:

1. Each phosphate is attached to a deoxyribose sugar, which in turn is attached to a

nitrogenous base. The three parts together constitute a **nucleotide.**

2. The phosphate group of one nucleotide is also connected to the deoxyribose sugar of the next nucleotide. The alternation of phosphates and sugar units thus forms a backbone that holds the entire strand together, while the nitrogen base points inward.

3. Each strand is a linear sequence of nucleotides (it does not branch) twisted in the shape of a corkscrew (a helix).

4. There are two strands, running in opposite directions but wound around each other, forming a *double helix,* with the bases arranged in the interior like steps in a spiral staircase.

5. The strands are so arranged that an adenine on one strand always matches a thymine on the other strand, and vice versa. Also, cytosine on one strand always matches guanine on the other strand, and vice versa. These pairings of *complementary* (matching) bases explain Chargaff's rules.

6. Because of the base pairings, either strand contains all the information necessary to determine the structure of its complementary DNA strand (Fig. 3.6).

DNA is the material that contains hereditary information on the chromosomes. Each chromosome contains two very long strands of DNA, surrounded (in most organisms) by some protein. A gene is a segment of the DNA, or a subset of bases within the linear base sequence of the whole DNA. Because each person or pea plant (or any diploid individual) has two chromosomes of each type, each individual has two genes for each hereditary trait. The location of the gene on the DNA is called its **locus.** A gene at a given locus may have any number of possible alleles, but each individual can only have two.

DNA replication. Subsequent research confirmed these findings and also revealed the mechanism of **replication** (Fig. 3.7, p. 52), the process by which DNA molecules make copies of themselves. The two strands of the double helix unwind partially, and each strand is then bound by a protein called *DNA polymerase,* which actually begins the replication process. Replication begins at small initial DNA sequences called initiation sites. The new strands of DNA are synthesized one

A large part of the effort to reveal the molecular structure of DNA centered around a technique known as **X-ray diffraction.** In this technique, the molecules being examined must first be crystalized; in the case of DNA, the sodium salt of deoxyribonucleic acid was used instead of the acid itself because it was more easily crystalized. The crystals are then exposed to X-rays and the reflections produced by the atoms in the crystal are examined on photographic film. The films produced by X-ray diffraction are very difficult to interpret and require a sharp mind and a strong mathematical background. Rosalind Franklin (1920–1958), who worked in the laboratory led by Maurice H. F. Wilkins, was an expert in reading and interpreting such X-ray diffraction photographs, a process that required many hours of careful measurement and calculation. (Today, there are powerful computers that process X-ray diffraction patterns, but in the 1950s this work was all done by hand.) Watson and Crick used Rosalind Franklin's data to figure out the structure of DNA. The 1962 Nobel Prize for this discovery was shared by Watson, Crick, and Wilkins; Rosalind Franklin did not share in the prize.

Watson (1968) admits that he broke into the lab where Rosalind Franklin worked and studied her results without her permission; Sayre (1975) provides another account of the same incident. Watson claims that his action was justified because of the highly competitive nature of the race to find the structure of DNA and because Franklin was, in his estimation, proceeding too slowly. Undoubtedly, there were other issues involved. Observers say that Franklin was not well liked by her male colleagues not only because of her gender but also because of her Jewish heritage, socialist politics, and her personality.

Among the questions you might want to ponder are the following:

1. Would you describe Watson's actions as "theft" of Franklin's data, or just "looking" at her data? Is there a difference? In either case, do you think that competition in the race for the structure of DNA justified his actions?

2. Does high-pressure competition do more good for science or more harm? Try to list both good and bad consequences before you decide.

3. Although everyone agreed that Franklin was a brilliant researcher, she was often criticized for not wearing makeup, and both Watson and Sayre mentioned Franklin's apparent lack of concern about her appearance. What kind of treatment would a scientist like Rosalind Franklin likely receive today? Do a scientist's looks, religion, politics, or personality affect how her or his data are regarded?

4. Can the science that scientists produce be separated from the other aspects of their lives? Can we monitor discriminatory behavior without restricting academic freedom?

---

nucleotide at a time, using one of the existing strands as the *template* (the pattern to be copied). If the next unmatched base on the template is adenine (A), then a thymine (T) nucleotide (adenine's complementary base) is added on the growing new strand opposite the adenine. In like manner, G is added to match C, C to match G, and A to match T. Each nucleotide is actually added in the form of a triphosphate such as ATP (adenosine triphosphate). Two of the phosphate groups are split off to provide energy for the replication process, while the remaining phosphate becomes part of the growing DNA molecule by joining to the deoxyribose sugar of the adjacent nucleotide (Fig. 3.7).

One new strand of DNA is synthesized continuously, with the direction of synthesis running toward the point where the parent strands are separating, the so-called *replication fork.* The other strand, however, must be synthesized in the opposite direction, in short fragments that are later joined together.

Transcription and Translation

Nucleic acids like DNA and RNA may be compared to blueprints that contain instructions for how proteins are built. Genes (made of DNA) are expressed as a phenotype by first providing information for

[A] Diagrammatic representation of the larger nitrogenous bases *adenine* (A) and *guanine* (G) and the smaller nitrogenous bases *thymine* (T) and *cytosine* (C).

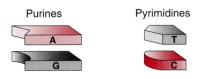

[B] With the addition of deoxyribose sugar (a five-sided molecule) and a phosphate group, each of these nitrogenous bases can form a *nucleotide*.

[C] Nucleotides can be strung together by a phosphate-sugar backbone to form a polynucleotide strand containing many thousands of nucleotides, only a portion of which is shown here.

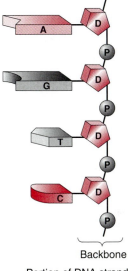

Backbone

Portion of DNA strand

[D] Two nucleotide sequences running in opposite directions can pair with one another if each adenine (A) pairs with a thymine (T) (and vice versa) and if each guanine (G) pairs with a cytosine (C) (and vice versa).

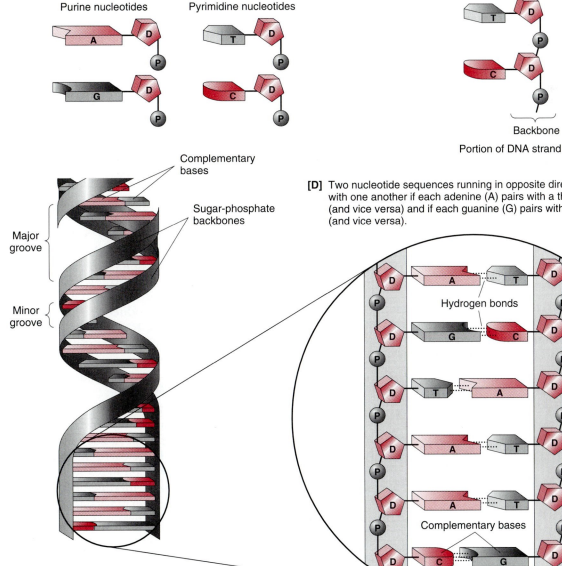

[E] Two strands of DNA twist around one another to form a double helix. A straightened portion of this double helix would resemble a ladder with the paired (complementary) bases forming the rungs.

FIGURE 3.6
The structure of DNA. The letter *D* indicates the sugar deoxyribose, the *D* in DNA.

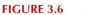

[A] Unwinding
Opposing strands of the DNA double helix separate.

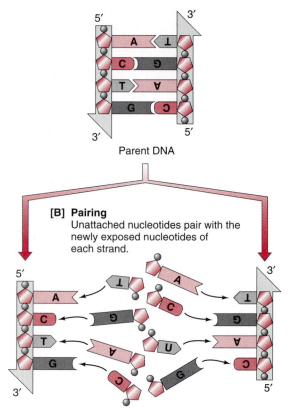

Parent DNA

[B] Pairing
Unattached nucleotides pair with the newly exposed nucleotides of each strand.

[C] Joining
Each new row of bases is linked into a continuous strand by joining adjacent sugars and phosphates. Each double helix contains one new strand and one preexisting (parental) strand.

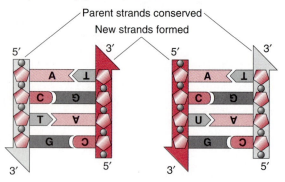

[D] The replication fork

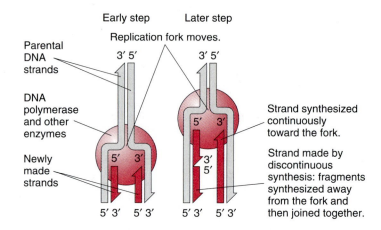

the synthesis of RNA, which in turn provides information for the synthesis of protein. RNA (Fig. 3.8) differs from DNA in that (1) the backbone contains the sugar ribose rather than deoxyribose and (2) the nitrogen-containing bases include uracil instead of thymine (the other three nitrogen bases, adenine, guanine, and cytosine, are the same). All genes undergo **transcription** into RNA, meaning that information is transferred from DNA to RNA, still within the language of nucleic acids with their linear sequence of bases. *The linear sequence of nucleotides in DNA determines a linear sequence of nucleotides in RNA.* Some gene products stop here, resulting in special types of RNA that have functions described below. The product of transcription for most genes is **messenger RNA (mRNA),** which then goes through a second information transfer (*translation*), in which the information is changed into another language, the language of amino acids. *The linear sequence of nucleotides in messenger RNA determines the linear sequence of amino acids in the resulting protein chain.* Knowledge of this process resulted in the central dogma of molecular biology: Information flows from DNA to RNA to protein (Fig. 3.9). As we will see in Chapter 9, this central dogma has been considerably modified by the finding that proteins can affect the expression of DNA, so that information flow is more accurately represented as a network:

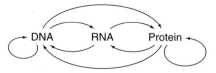

FIGURE 3.7
DNA replication. Note that either strand contains all the information needed to synthesize a complementary strand.

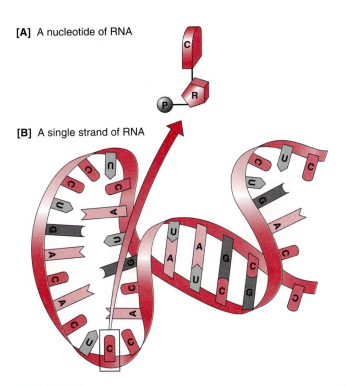

[A] A nucleotide of RNA

[B] A single strand of RNA

FIGURE 3.8
The structure of RNA. The letter *R* indicates the sugar ribose, the *R* in RNA. The molecule as a whole is usually single-stranded, but short portions of some RNA molecules can base-pair with other portions of the same molecule. In RNA, C pairs with G and A pairs with U.

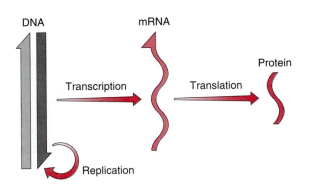

FIGURE 3.9
The central dogma of molecular biology, as it was understood in the 1960s: Information flows from DNA to RNA and then to proteins. Many exceptions are now known. For example, some viruses, like the one that causes AIDS, have a reverse transcription process in which RNA is used to make DNA. Also, many proteins influence the timing and amount of gene transcription.

Transcription to RNA. During transcription (Fig. 3.10), a portion of DNA is used as a template to make a single-stranded mRNA. The portion of the DNA strand that contains the necessary information to make the protein is what constitutes the gene. At any given region of the DNA molecule, the gene is on only one of the two strands. The transcription process begins in the nucleus when a molecule of *RNA polymerase* combines with a portion of the DNA molecule known as a *promoter.* Several proteins are known that can either inhibit or enhance this process, providing a means by which the expression of the gene can be controlled (see Chapter 9). The product of transcription is usually mRNA, which leaves the nucleus and carries the information into the cytoplasm, where the next step occurs. Two other forms of RNA, transfer RNA and ribosomal RNA, are also transcribed from DNA, but are never translated into proteins.

Translation to protein. Much more complex is the process of **translation,** during which the sequence of nitrogenous bases is translated into a sequence of amino acids that make up a protein chain. Translation uses a group of three successive nitrogenous bases on the mRNA as a coding unit, or **codon.** Each codon is paired up with a complementary three-base sequence called an **anticodon,** which is part of a specially twisted form of RNA called **transfer RNA (tRNA)** (Fig. 3.11). Each tRNA molecule carries a specific amino acid molecule that will become the next amino acid added to the growing protein chain. As each three-base codon is read and translated, one amino acid is added at a time to the growing protein until its sequence is complete (Fig. 3.12, p. 56). The mRNA and tRNAs are held in the proper relation by a particle called a *ribosome,* containing both protein and ribosomal RNA.

Mutations

DNA sequences occasionally undergo sudden but permanent heritable changes known as **mutations.** Some of these can result from mistakes during DNA replication, which may be caused in turn by chemicals or by radiation (see Chapter 9). Most mistakes and damage are immediately fixed by various self-correction mechanisms and so do not

[A] A parent molecule of double-stranded DNA

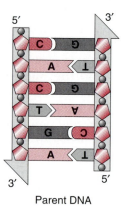

Parent DNA

[B] A portion of the double helix unwinds and the strands separate locally. RNA polymerase binds to a promoter on one strand. Unattached RNA nucleotides can now pair with the exposed strand.

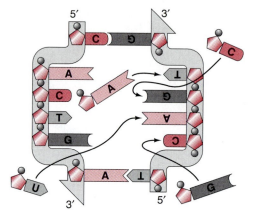

[C] The RNA nucleotides are joined to one another to form a continuous RNA strand.

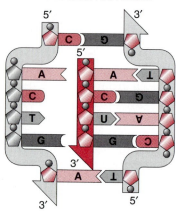

[D] The fully synthesized RNA molecule separates, leaving the DNA parent molecule unchanged.

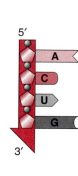

New RNA strand

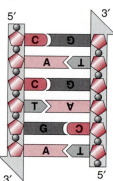

Parent DNA totally conserved

FIGURE 3.10
The transcription of DNA into RNA.

persist. Mistakes that persist in somatic (body) cells can cause problems in the individual carrying the affected genetic material (see Chapter 9), but unless the mistake has affected the cells in reproductive tissues that give rise to haploid gametes, the change will not be passed on to the next generation.

Point mutations. Ultimately, the different forms (alleles) of each gene originate as mutations. The simplest kind of mutation is a single-base mutation, or **point mutation,** such as a substitution of one nitrogenous base (A, G, C, T) for another. Substitutions of this kind may result in the wrong amino acid being inserted into a protein sequence;

phenotypic consequences run the gamut from those that are undetectable to those that are fatal. Many proteins function as enzymes, in which case the substituted amino acid may alter the enzyme shape and therefore change or impair the enzyme's function.

Frame-shift mutations. **Frame-shift mutations** occur when one or two extra nitrogenous bases are inserted into a DNA sequence or when one or more bases are deleted from the sequence. The three bases that form the first codon determine the *reading frame,* or starting point, for each codon. The DNA/RNA code contains no "commas" or other

[A] A molecule of tRNA coding for the amino acid serine

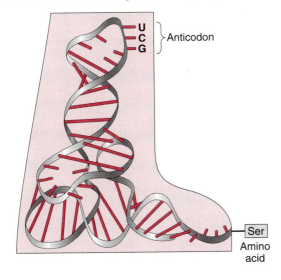

[B] The codon, part of an mRNA molecule, matches with the anticodon of a tRNA molecule, thus determining which amino acid is next added to a growing protein.

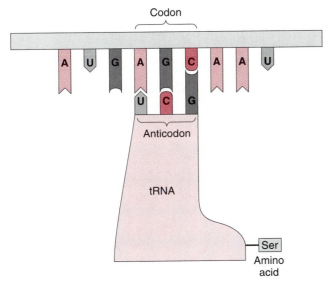

FIGURE 3.11
The structure of transfer RNA (tRNA) and the role of tRNA in translation. The mRNA codon (AGC in this case) matches the tRNA anticodon (UCG), which corresponds to the amino acid carried by the tRNA (serine in this case, abbreviated *ser*).

"punctuation" that signify where a new codon starts—each new codon is the next three bases after the previous codon. If an extra base is inserted or a base is deleted, the reading frame is changed, and all codons that follow the mutation are changed (Fig. 3.13). When the codons are altered, the amino acids added to the growing protein are different and most frame-shift mutations result in nonfunctional proteins.

Chromosomal aberrations. In addition to these small-scale mutations, there are several kinds of large-scale changes involving chromosome fragments, including both DNA strands and the surrounding protein molecules. Chromosome fragments may become duplicated (repeated); they may become attached to a new location, possibly on a different chromosome; or they may be lost entirely. A chromosome fragment may also be turned end-to-end and reinserted at its former location, forming an *inversion*. Of these four types of chromosomal rearrangements, inversions are the most frequent and have the most limited effects, while the other three types may result in nonviable phenotypes when long sections of DNA are involved. Although these chromosomal rearrangements are often classified as mutations, many geneticists prefer to limit use of that term to single-gene mutations and to refer to the larger changes as *chromosomal aberrations*.

Changes in chromosomes can also affect chromosome numbers. Several examples of such changes are explained in the next section.

THOUGHT QUESTIONS

1. Mendel's experiments distinguished between two alternative hypotheses: Either traits blend, so that the offspring have traits intermediate to those of the two parent plants, or traits are inherited as discrete particles, which do not blend. Which hypothesis is favored by Mendel's results? Did he disprove one hypothesis? Did he prove one hypothesis?

2. An experimenter cross-fertilizes tall pea plant flowers with pollen from short pea plant flowers. She harvests the seeds and plants them, and all of the plants that grow in the F_1 generation are as tall as the tall parent plants. The F_1 plants produce flowers, each producing many pollen grains and each having many eggs in its stigma. Each pollen grain and each egg is a haploid gamete. What are the possible genotypes for this trait that can exist in the pollen grains? What are the genotypes that can exist in the eggs? What fraction of the gametes possesses each

[A] Initiation of protein synthesis begins with assembly of ribosome, tRNA, mRNA, and other components.

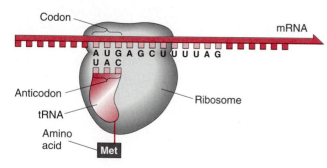

Codon

mRNA

Anticodon

tRNA

Ribosome

Amino acid

Met

[B] Elongation begins when a new tRNA pairs with the next codon on mRNA.

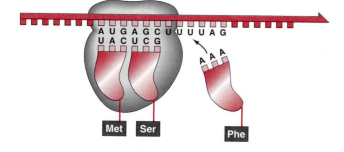

Met   Ser   Phe

[C] Elongation continues when the ribosome moves over, releasing the oldest tRNA after joining its amino acid to the next amino acid, forming a growing protein chain. Steps B and C are repeated over and over.

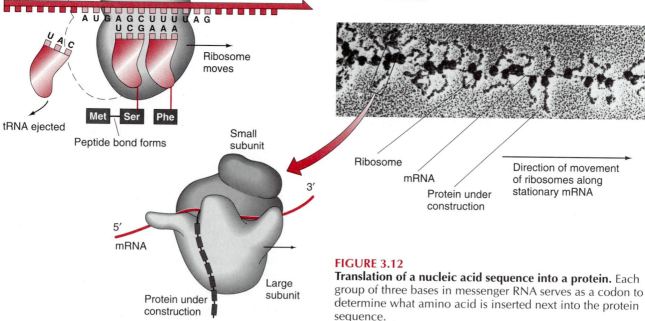

Ribosome moves

tRNA ejected

Peptide bond forms

Met   Ser   Phe

Small subunit

3'

5' mRNA

Large subunit

Protein under construction

[D] Termination occurs when a "stop" codon (UAG in this case) binds to a termination factor and allows the ribosome to move over without binding another tRNA.

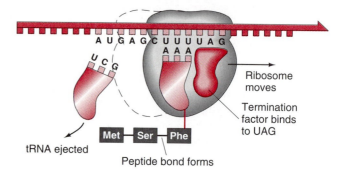

Ribosome moves

Termination factor binds to UAG

tRNA ejected

Met   Ser   Phe

Peptide bond forms

[E] Disassembly of the components concludes the process.

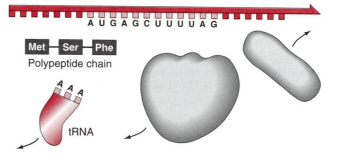

Met   Ser   Phe

Polypeptide chain

tRNA

[F] Many ribosomes synthesize proteins simultaneously along the same mRNA strand. Each ribosome is made of a small and a large subunit, with the mRNA sliding through the groove between them.

Ribosome

mRNA

Protein under construction

Direction of movement of ribosomes along stationary mRNA

FIGURE 3.12

Translation of a nucleic acid sequence into a protein. Each group of three bases in messenger RNA serves as a codon to determine what amino acid is inserted next into the protein sequence.

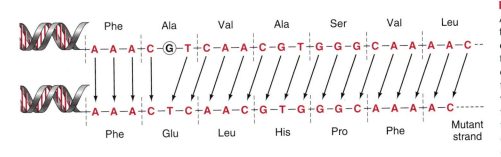

FIGURE 3.13
A frame-shift mutation causes the misreading of subsequent codons. In this example, deletion of a guanine nucleotide (G) causes a change in DNA codons from that point on, leading to wrong mRNA codons, then to wrong amino acids being added to a growing protein. The names of the amino acids added for each codon are shown.

genotype? Will the distribution of types of gametes be the same or different if short pea plants are cross-fertilized with pollen from tall pea plants?

3. To study two pea traits at a time, Mendel first had to have one line of pea plants that bred true for two traits and another line of pea plants that bred true for two different alleles of the same two genes. F_1 pea flowers were then self-fertilized and the F_2 peas produced as before. If you counted sixteen of the F_2 peas, how many would you expect to be the doubly dominant phenotype, round and yellow? If you counted 1600 F_2 peas, how many round and yellow ones would you expect?

4. DNA has been called the "master molecule" because it controls (or determines) RNA sequences, which control protein sequences, which (as enzymes) control all the cell's other activities. Is the term *master molecule* an accurate description of the data? Does the language of *control* (e.g., DNA *controls* the type of RNA produced) say more about the molecules or about the scientists? Does the use of a word like *master* suggest a hierarchical approach in which information flows in one direction only?

B. Genes and Chromosomes Control Many Human Traits, Including Inherited Diseases

Mendelian Traits in Humans

Early in the twentieth century, a number of pioneering geneticists discovered that Mendel's rules, formulated on the basis of experiments with pea plants, could also explain the inheritance of many human traits. For example, *albinism* is a total lack of melanin pigment in the skin, eyes, hair, and the body's internal organs. The skin of albinos is white and their hair is white as well. Since one of the normal functions of melanin is to block ultraviolet light, albino individuals sunburn easily and are very sensitive to bright lights. Albinism occurs in all geographical races of humans and in many other species.

The inheritance of albinism is shown in Fig. 3.14. The recessive allele responsible for this condition can be transmitted without detection through many successive generations of normally pigmented individuals (Fig. 3.14A). However, matings between heterozygous individuals may produce albino children (Fig. 3.14B), and the frequency of albino offspring is increased if matings take place among related individuals such as cousins (Fig. 3.14C). Diagrams showing the pattern of mating and descent, as in Fig. 3.14C, are called **pedigrees.**

Metabolic diseases. Shortly after the rediscovery of Mendel's laws, an English physician, Archibald Garrod, made an important discovery: Mendel's laws applied not only to visible structures like eye colors, but also to certain medical conditions. *Alkaptonuria*, for example, is a rare condition in which a patient's face and ears may be discolored and in which the urine turns black upon exposure to air. Garrod tested the urine of these patients and discovered that the color is caused by an acid. We now know that this substance, homogentisic acid, is formed in the course of breaking down the amino acid tyrosine (Fig. 3.15). In most individuals, the homogentisic acid can be broken down

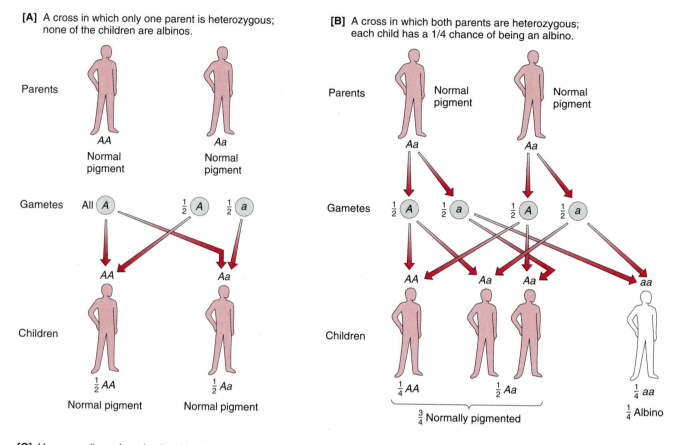

[A] A cross in which only one parent is heterozygous; none of the children are albinos.

[B] A cross in which both parents are heterozygous; each child has a 1/4 chance of being an albino.

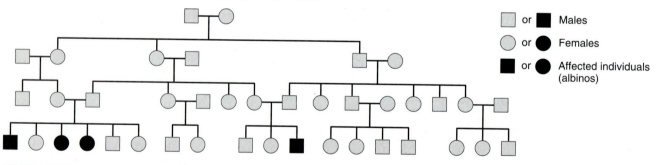

[C] Human pedigree for a family with albinism; the horizontal lines joining a male and a female represent matings from which children are descended.

or ■ Males

or ● Females

or ■ or ● Affected individuals (albinos)

FIGURE 3.14

An example of simple Mendelian inheritance in humans. Albinism, a recessive trait, arises in most cases from matings between heterozygotes.

harmlessly with the help of an enzyme. However, in patients with alkaptonuria, the necessary enzyme is missing or defective. Garrod realized that an error in an important biochemical (or metabolic) process was responsible, and he called this type of condition an **inborn error of metabolism.**

Garrod studied the families of individuals with alkaptonuria and three other such conditions, including albinism, and found a common pattern: Each of these inborn errors of metabolism was inherited as a simple Mendelian trait, and in each case the lack of a functional enzyme was recessive.

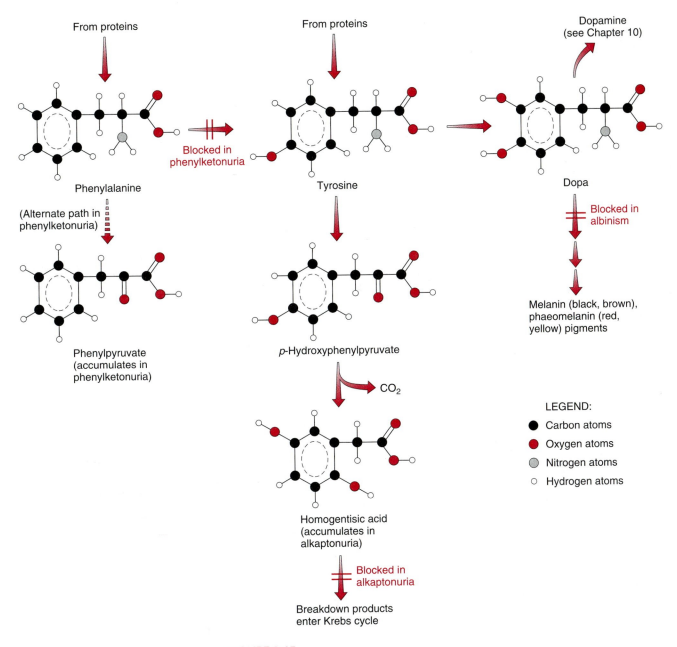

From proteins

Blocked in
phenylketonuria

Phenylalanine

(Alternate path in
phenylketonuria)

Phenylpyruvate
(accumulates in
phenylketonuria)

From proteins

Tyrosine

p-Hydroxyphenylpyruvate

CO_2

Homogentisic acid
(accumulates in
alkaptonuria)

Blocked in
alkaptonuria

Breakdown products
enter Krebs cycle

Dopamine
(see Chapter 10)

Dopa

Blocked in
albinism

Melanin (black, brown),
phaeomelanin (red,
yellow) pigments

LEGEND:

● Carbon atoms
● Oxygen atoms
● Nitrogen atoms
○ Hydrogen atoms

FIGURE 3.15
Biochemical pathways for three inborn errors of metabolism: albinism, alkaptonuria, and phenylketonuria.

Many other inborn errors of metabolism have since been discovered and their biochemical defects identified. Each of these inborn errors is caused by a recessive allele, the product of a mutation that changed a functional gene into a nonfunctional one. Alkaptonuria, albinism, and a condition called phenylketonuria (described below) all arise from errors in a series of closely related metabolic pathways (Fig. 3.15).

Phenylketonuria (PKU) is a genetically controlled defect in amino acid metabolism. The amino acid phenylalanine, occurring in most proteins, is normally converted by an enzyme into another amino acid, tyrosine; the tyrosine is then broken down by the pathway shown in Fig. 3.15. A defect in the enzyme that usually converts phenylalanine to tyrosine causes the phenylalanine to be processed by an alternate pathway. A product of this alternate pathway accumulates in the blood and in all cells, acting as a poison that causes most of the debilitating symptoms of the disease: insufficient development of the insulating layer (myelin) around nerve cells, uncoordinated and hyperactive muscle movements, mental retardation, defective tooth enamel, retarded bone growth, and a life expectancy of thirty years or less.

Fortunately for people carrying this genetic defect, a simple test for its presence exists and it is possible to avoid the symptoms of the disease by greatly limiting those foods that contain phenylalanine—nearly all proteins, including breast milk, and diet foods and soda containing the artificial sweetener aspartame. Small amounts of phenylalanine are essential in protein synthesis (Chapter 8), but without larger amounts, the toxic breakdown product that causes the symptoms never forms.

Diseases carried by dominant genes.   Not every genetic disease is caused by a recessive gene. *Huntington's disease* (or Huntington's chorea) is a neurological disorder that begins between the ages of 40 and 50 with uncontrollable spasms or twitches of the hands or feet. As the disease progresses, the spasms become more pronounced, and the patient gradually loses conscious control of all motor functions and of mental processes. The disease progresses slowly, but is invariably fatal. American songwriter and balladeer Woody Guthrie died of Huntington's disease. Although it is always lethal, Huntington's disease is not eliminated by natural selection because it does not appear until *after* its victims have lived through their prime reproductive years, during which they have often passed the gene to the next generation. Studies of family trees show that Huntington's disease is inherited as a dominant trait. The gene responsible for Huntington's disease was located on chromosome 4 in 1983, using procedures described later in this chapter. A test for the allele responsible for Huntington's disease has since been devised. The gene was fully isolated and identified in 1993. Sub-

sequent studies have shown that this gene differs from its alleles in the presence of many extra repetitions of the three-nucleotide sequence CAG.

The transmission of human traits controlled by one gene follows the rules that Mendel developed for peas. The situation is more complex when we study phenotypic traits like height or skin color, which are controlled by many genes at once and which are also influenced by environmental variables.

Human Sex Chromosomes

In humans and most other animals, one special pair of chromosomes carries the genes for sex determination. Females generally have two similar sex chromosomes, symbolized as *XX*. Males generally have one *X* chromosome and one different sex chromosome, the *Y* chromosome, and thus are symbolized as *XY*. The pattern of chromosomes is visible at certain stages of the cell cycle and is called the **karyotype** (Fig. 3.16).

The *sry* gene.   The determination of *XX* as female and *XY* as male is not absolute. Unusual situations can occur when there is a crossover between an *X* and a *Y* chromosome, followed by an exchange of pieces known as a translocation. Approximately 1 in 20,000 normal males is chromosomally *XX*, but one of the *X* chromosomes contains a small translocated piece of the *Y* chromosome. About the same frequency of normal females are *XY* individuals who are *missing* the same small piece of *Y* chromosome. One such *XY* female had 99.8 percent of the *Y* chromosomal DNA, indicating that the gene for a male-determining factor, or testis-determining factor (TDF), was located in the 0.2 percent portion of DNA she did not have. Examination of this 0.2 percent portion of the DNA led to identification of a gene now called *sry*. Other genes had previously been hypothesized to code for the TDF until male individuals lacking these genes were found. This story is typical of the way in which science proceeds, with today's "facts" being supplanted by additional evidence. The current hypothesis of sex determination by *sry* should be considered just that—a hypothesis, one that explains all of the current evidence but may be supplanted when new information becomes available.

The *sry* gene appears to be responsible for sex determination by producing a protein that binds to another DNA region, perhaps controlling the production of an enzyme that converts the female hormone estrogen into the male hormone testosterone (Fig. 3.16C). The hormone, in turn, triggers the development of the testis in embryos. Embryos that are not signaled to become male will develop as females.

Realization that there are *XX* males and *XY* females forced the International Olympic Committee to reexamine the stipulation that only *XX* individuals could compete in female sporting events. If chromosome appearance is not sufficient to determine which individuals are male, should the presence of the functional allele of the *sry* gene or the male hormone testosterone be used as the test? That turns out not to be infallible because some *XY* individuals who have the functional *sry* allele and produce testosterone are nevertheless phenotypically female because they lack the gene for the receptor for testosterone. Cells cannot respond to a hormone during development or during adult life unless they possess receptor molecules to bind that hormone (Chapter 9). These are not defective individuals; they just do not fit a previously held view of what determines the sex of an individual. The International Olympic Committee now uses the presence or absence of the functional *sry* allele to decide the sex of Olympic athletes, but ambiguous cases of sex determination continue to be a challenge to those who want to develop clear-cut rules for participation of individuals in male or female sporting events.

Sex-linked traits.

Very few genes are, like *sry*, located on the *Y* chromosome. Many more genes are located on the *X* chromosome, and many of these have been assigned to definite locations along that chromosome. Genes on the *X* chromosome, such as the gene for *red-green color blindness*, are said to be **sex-linked** (Fig. 3.17). Females can be either homozygous or heterozygous for sex-linked traits because they have two *X* chromosomes and therefore two genes from every gene pair. Males, on the other hand, have only a single *X* chromosome and therefore only a single allele (dominant or recessive) for each sex-linked trait. As a consequence, sex-linked genes are phenotypically displayed in

[A] The *XX* karyotype (female)

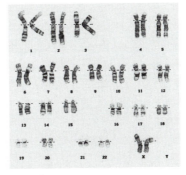

[B] The *XY* karyotype (male)

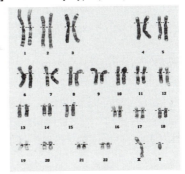

[C] The hypothesized relationship between the *sry* gene, the testis-determining factor, testosterone, and male sexual characteristics

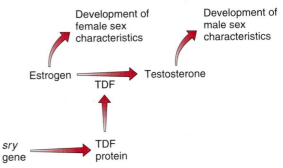

FIGURE 3.16
Chromosomes and hormones in human sex determination.
[A] Normal female karyotype. [B] Normal male karyotype. [C] A possible way in which a gene like *sry* can bring about a hormonal change that determines sex.

males regardless of whether a dominant or recessive allele is present.

Mosaics.

Females generally possess two *X* chromosomes, but in a given cell only one of them has

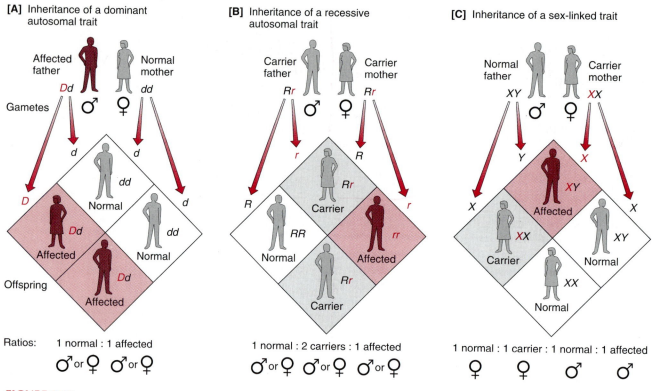

[A] Inheritance of a dominant autosomal trait

[B] Inheritance of a recessive autosomal trait

[C] Inheritance of a sex-linked trait

FIGURE 3.17

Comparison of two forms of autosomal inheritance with a sex-linked recessive trait such as red-green color blindness. [A] In inheritance of a dominant autosomal (not sex linked) trait, all affected offspring have at least one affected parent. [B] In inheritance of a recessive autosomal trait, affected offspring usually have two parents who are heterozygous carriers with normal phenotype. [C] In inheritance of sex-linked traits, affected individuals are usually males who inherited the trait from a mother who was a heterozygous carrier. Father-to-son inheritance never occurs for such traits.

active genes (genes that make a product or express a phenotype). In a random event that takes place very early in embryonic development, the other X chromosome is inactivated and forms a structure known as a *Barr body,* just inside the nuclear envelope. Thus all females express one phenotype (from their mother's X chromosome) in some cells and another phenotype (from their father's X chromosome) in other cells. Difference in phenotype at a cellular level is called *mosaicism.* All females are mosaics. For example, in a female whose father carries an X-linked allele for color blindness, patches of cells within the retina of the eye will not respond to color. Other patches of cells, which express the normal allele and X-chromosome from the mother, will respond normally.

Human Chromosomal Defects

Several abnormalities of the sex chromosomes are known, including those that result in changes in the chromosome number. The XXY chromosomal type results in *Klinefelter's syndrome* (Fig. 3.18); individuals with Klinefelter's syndrome have male phenotypes, but are sterile. Their cells have Barr bodies (inactive X chromosomes) similar to those found in females. Some of the symptoms of Klinefelter's syndrome can be successfully treated with hormones. In contrast, *Turner's syndrome* (Fig. 3.18) results from the XO chromosomal type in which only one X chromosome is present, the O representing its missing partner. Individuals with Turner's syndrome develop as females; however,

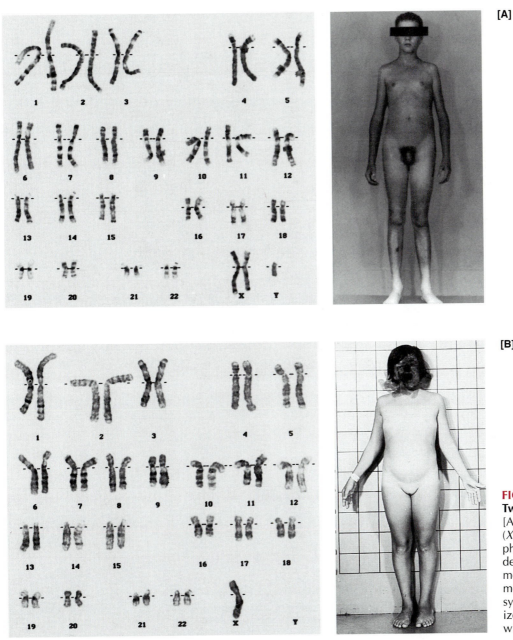

FIGURE 3.18
Two chromosomal defects.
[A] Klinefelter's syndrome (*XXY*). Note the tall, thin phenotype, long legs, underdeveloped testes, and moderate breast enlargement. [B] Turner's syndrome (*XO*), characterized by short stature, widely spaced breasts.

the ovaries do not produce female hormones; puberty does not take place and gametes do not develop, resulting in infertility. The infertility cannot be overcome at present, but the other symptoms of Turner's syndrome are treated hormonally with much success.

Turner's and Klinefelter's syndromes are believed to result from the same cause, a

nondisjunction (abnormal cell division) of the sex chromosomes in which the two sex chromosomes fail to separate, resulting in some egg cells with two of these sex chromosomes and some with none (Fig. 3.19). Nondisjunction during gamete production has in fact been observed, partially confirming the hypothesized series of events shown in Fig. 3.19. Also supporting the hypothesis is the very

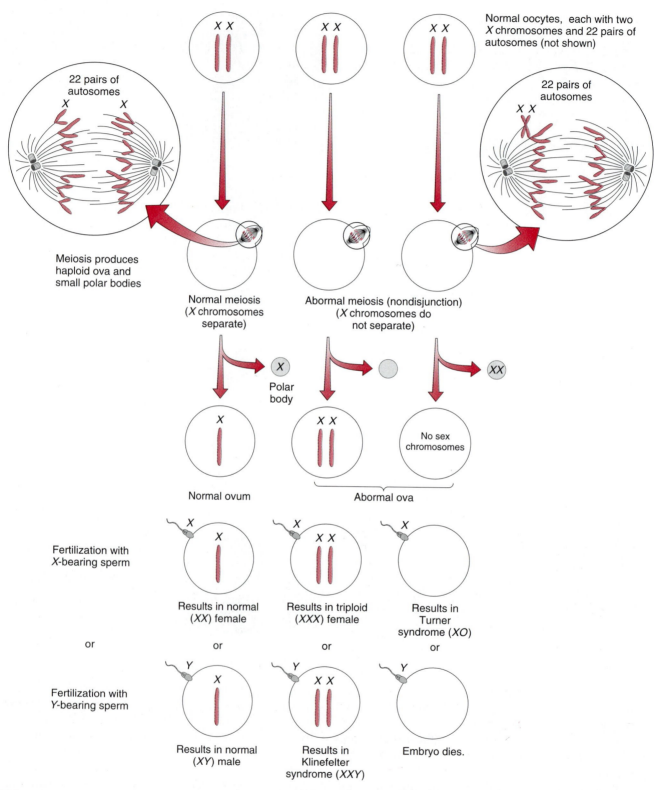

FIGURE 3.19
Nondisjunction during egg production, showing how certain chromosomal abnormalities may arise.

rare occurrence of the *XXX* chromosomal abnormality, which results in females of variable phenotype who are often mentally retarded and often sterile. Nondisjunction can also take place during the formation of sperm cells, resulting in other unusual chromosome arrangements, including *XYY.* Males with an extra *Y* chromosome were once thought to have subnormal or borderline intelligence and a predisposition toward violent and often criminal behavior, but these claims have largely been falsified. The *Y*-only type of embryo, also predicted by the hypothesis, has never been observed, presumably because it is inviable at a very early stage of development.

Chromosomes other than the sex chromosomes may also experience chromosomal changes. The most common of these is *Down's syndrome* (Fig. 3.20), which is marked by facial changes (including an epicanthic fold over the eyes), heart abnormalities, and a variable amount of mental retardation. Down's syndrome usually results from an extra chromosome 21, a condition known as trisomy 21, in which three of these chromosomes are present instead of two. Other chromosome abnormalities are less common. For example, the cri du chat syndrome (caused by deletion of the short arm of chromosome 5) results in a small head, a catlike cry, and mental retardation. Patau's syndrome (trisomy 13) results in severe mental retardation, a small head, extra fingers and toes, and usually death by one year of age.

THOUGHT QUESTIONS

1. In what ways are humans poor subjects for genetics research? In what ways are humans good subjects? Which of your reasons are purely biological, and which have ethical components?
2. Why are certain traits studied in some species and not in others? Why is human genetics studied primarily through traits that produce diseases?
3. Sports officials have repeatedly asked female athletes to submit to testing to confirm their femaleness, but comparable proof is seldom demanded of males. Why do you think this disparity exists? Can you think of other ways besides sex in which athletes might be classified? Could we use age as a criterion? Could we use fat-to-muscle ratios?

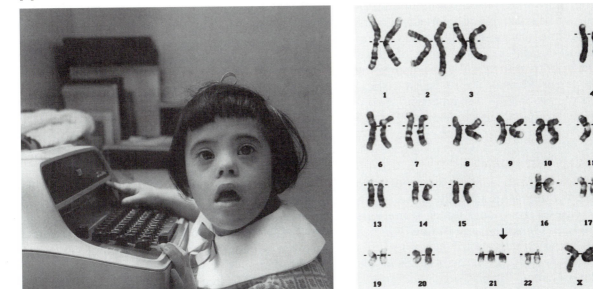

[A]

[B]

FIGURE 3.20
Down's syndrome. [A] A child with Down's syndrome. [B] The karyotype responsible for most cases of this syndrome, with an extra chromosome 21 (trisomy).

C. There Are Many Ways of Dealing with Hereditary Risks and Other Medical Conditions

Identifying Genetic Causes for Traits

Pedigrees. Geneticists have several ways of studying human hereditary traits. One of the most basic methods is to present the available data in pedigrees, as in Fig. 3.14C. Pedigrees are most useful when they span many generations and hundreds of people, or when separate pedigrees are available for hundreds of separate families. Many of the hereditary traits studied by this method are diseases—conditions like Huntington's disease or diabetes that impair health or reduce fitness. (The *fitness* of a particular genotype is measured by its capacity to leave genetic copies of itself in the next generation; see Chapter 4.) Other hereditary traits have only minimal impact on fitness or else no obvious impact at all.

The study of pedigrees can add to our knowledge of human genetics, permitting us to identify which traits are dominant, which are recessive, and which have a more complex genetic basis. If the genetic basis of a trait is complex, or not fully known, then pedigrees can also help in an empirical determination of risks, including medical risks. For example, a child has a greatly increased risk of having insulin-dependent diabetes mellitus (IDDM) if one or both of the child's parents has the disease. The term *risk* has a precise statistical meaning: It is the probability that a particular condition will occur or that a particular condition will be inherited.

Studies of twins and adopted children.

Studies of twins are sometimes useful in suggesting the extent to which a trait is genetically, versus environmentally, caused. In such studies, numerous twin pairs are found in which at least one twin has the condition being investigated. For pairs of this kind, the frequency with which the other twin has the condition is called the *rate of concordance*. Traits under strong genetic control usually have higher concordance rates for *monozygous twins* (identical twins, derived from a single fertilized egg) than for *dizygous twins* (fraternal twins, derived from two separate eggs). In contrast, traits with mainly environmental causes have similar concordance rates for both types of twins.

Adoption studies can also provide important clues on the heritability of particular traits: If adopted children show a higher rate of concordance with their birth parents than with their adoptive parents, then the hypothesis of a genetic cause for the trait under study is made stronger. Some researchers have criticized this type of study because adoption agencies do not place children at random but purposely try to place children with adopting parents whose backgrounds are not dissimilar to the backgrounds of the children's birth parents. This practice introduces a bias that raises the concordance rates between adopted children and the parents who adopted them. Another complication is that many children are adopted by relatives, and such adoptions make it very difficult to sort out which similarities are environmental and which are genetic. These and other criticisms of heritability studies are explained in much greater detail by Lewontin, Rose, and Kamin (1984) and by Kamin (1974). Earlier studies on the heritability of IQ scores were based in large measure on data published over a 40-year period by a British researcher named Cyril Burt, but when Kamin (1974) tried to locate and verify Burt's data, he discovered that most of the data had been fabricated. The extent to which complex traits like intelligence are genetically or environmentally controlled is a scientific issue on which experts continue to disagree.

Use of genetic information. After a genetic basis has been identified for a particular trait, what happens next depends in a large part on the values that individuals and society place on that trait. The rest of this chapter deals with many conditions that at least some people consider undesirable, though not all of them impair health or longevity. In many cases, but not others, we can identify a specific gene that fails to make a functional enzyme of some sort. These are often called *genetic defects*, a category that includes all inborn errors of metabolism such as those described in the previous section. The term *genetic defect* thus means that a specific allele and its product are defective; it does not mean that the person bearing the gene is defective. For this reason, many people prefer terms like *genetic disease* or *genetic condition*.

There are many ways of dealing with hereditary conditions and hereditary risks in humans. We can conveniently describe four broad categories of response: gathering and sharing information through diagnosis and counseling, changing individual genotypes, changing the gene pool at the population level, and changing the balance between genetic and environmental factors. Many of the methods in all four of these categories raise important ethical questions. The ethical questions can be summarized as follows:

1. Who decides who should be tested?
2. Who has access to the results of the test?
3. What are the responsibilities of a person who carries a gene for a hereditary condition? Do children have the right not to be born with a harmful genetic condition if this can be avoided? Do people with certain genotypes have a responsibility not to reproduce?
4. Who determines what traits (if any) are called "defects"?

Think about these ethical issues as you read the following sections.

Diagnosis and Counseling

Advances in medical genetics lead to better ways of detecting genetic diseases and to ways of detecting them earlier. Identification of a specific enzyme responsible for a particular condition usually allows detection of the defect or disease at the earliest possible stage.

Prenatal detection of genetic conditions. Some conditions can be detected before birth, *in utero* (literally, "in the womb"), through the technique of **amniocentesis** (Fig. 3.21), in which a small amount of amniotic fluid is withdrawn from the sack in which the fetus is developing in the mother's uterus (Chapter 6). The fluid itself can be analyzed for the presence or absence of certain enzymes that might indicate a genetic defect in the fetus. Amniotic fluid also contains cells that have been shed from the surface of the fetus. Chromosomes from these cells can be analyzed for changes such as those responsible for Down's, Turner's, or Klinefelter's syndromes. Culturing the cells in the laboratory can reveal additional information. Sev-

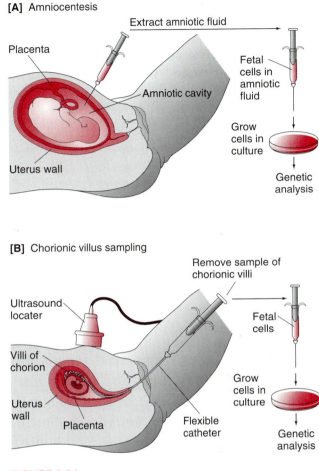

FIGURE 3.21
Amniocentesis and chorionic villus sampling.

eral dozen genetic diseases are now detectable through amniocentesis, including Tay-Sachs disease, cystic fibrosis, phenylketonuria, and type I diabetes. Certain low but nonzero risks are associated with amniocentesis, including a risk of mechanical injury to the growing fetus and a risk that the pregnancy will be prematurely terminated. These risks are one reason that amniocentesis is not routinely performed on every expectant mother.

For conditions that cannot be detected in amniotic fluid, **chorionic villus sampling** is sometimes used. This technique is a type of biopsy (removal of living tissue for examination) using part of the placenta by which the fetus attaches to the wall of the uterus (Chapter 6). Since this part of the placenta is composed of tissue derived from the fetus, not from the mother, the cells sampled by this technique are genetically identical to other fetal cells. Chorionic villus sampling can be performed earlier in pregnancy than amniocentesis can.

A newer type of genetic screening is to test the embryonic DNA by comparing it with a known DNA sequence. This cannot be done on the minute amounts of DNA that exist in most cells unless these amounts are first increased, or *amplified*. This amplification is usually accomplished by using the **polymerase chain reaction (PCR).** In this reaction, the DNA fragments to be tested are mixed with DNA polymerase to promote replication; the new DNA molecules synthesized by this technique are simply copies of the sequences that were originally supplied. By repeating the procedure as needed, one copy of a sequence can be made into 2, then 4, 8, 16, 32, 64, 128, and so on, until a sufficient quantity exists to permit testing (see also Chapter 13). When a sufficient quantity exists, the amplified DNA fragments can be compared with a known DNA sequence to see if they are identical. The polymerase chain reaction is often used to detect genetic conditions using DNA from eight-cell embryos prior to implantation. These embryos are derived from *in vitro* fertilization (literally "in glass"), meaning that the process took place in laboratory glassware rather than inside the body (*in vivo*).

Testing newborns or adults.   Other tests are done on newborns or on adults. Tests that are simple and inexpensive can be used for mass screening. Many hospitals routinely screen all infants at birth for conditions like phenylketonuria. Such screening is considered ethical because it is done on all infants and it is a clear benefit to the infant for the information to be known. Some tests are capable of detecting heterozygotes (e.g., testing for a carrier of an allele causing sickle-cell anemia). These tests can be performed on adults before they become parents or even before they marry. Persons undergoing this type of screening must first give their **informed consent** and sign a form stating that they understand the nature of the test, the possible outcomes (including the nature of the conditions that the test can detect), the possible risks of the procedure, and the possible benefits. Heterozygous individuals screened in this way can

be advised by persons trained in genetic counseling of the prospects they face should they marry one another and bear children. For example, if two individuals marry who are heterozygous for the same genetic trait, then each of their offspring has a 25 percent chance of having the recessive condition. (To understand why, refer back to Fig. 3.14.)

When a genetic defect or disease is detected, the decision about what to do is left up to the individuals themselves, or to their parents if the individual is a child. Possibilities include a decision not to marry or not to bear children (in some cases opting to adopt instead), to abort a fetus or carry it to term, or to intervene with medical treatments that may sometimes be expensive and may not always prolong life or relieve suffering. These decisions are often difficult for the people making them. Genetic counselors should never make a decision on behalf of their clients: Clients will rightfully resent any counselor who has pressured them into a decision. A patient's decision should be based on a clear knowledge of the possible choices and their consequences. For this and other reasons, the obtaining of the patient's informed consent to a genetic test is now considered an important matter of professional ethics and is also a legal requirement in many jurisdictions.

Who should be tested?   Family history of a genetic disorder may prompt people to be tested. Rare alleles are not equally rare in all populations or ethnic groups (see Chapter 5), so diagnostic tests for particular genetic traits are sometimes recommended specifically to those ethnic groups known to have a greater prevalence of the trait.

- Many African-Americans now seek testing to see if they carry a sickle-cell allele (see Chapter 5); screening for such an allele can take place at any time of life, including before marriage.
- People of Mediterranean or southeast Asian descent may seek testing to see if they carry the allele for the inherited blood disorder thalassemia (see Chapter 5).
- Ashkenazi Jews (those of eastern European descent) commonly seek testing for Tay-Sachs disease, a fatal disorder of brain chemistry caused by a recessive allele.
- People of western European (especially Irish) descent may seek testing for cystic fibrosis, the

most common genetic disorder among whites in the United States.

Testing of this sort should only be done on a voluntary, informed consent basis and, in general, only when it is of potential benefit to those being tested or to their children. Community leaders of various ethnic groups, including many clergy, have in many cases helped organize genetic testing programs and have encouraged as many people as possible to participate.

The results of genetic testing can be used in various ways. Couples who know that they both carry an allele for one of the disorders previously listed may decide to adopt children rather than bear their own. In some cases, knowledge of a genetic condition permits medical intervention at the earliest possible stages, when chances of successful treatment may be better. For conditions that cannot be treated, some couples may choose to abort the fetus bearing the condition. However, people committed to a pro-life position believe that the potential benefits to those being tested or to any future children can never justify what they view as murder of a fetus.

Genetic testing has already led to some highly inventive mixtures of tradition and modern technology. The Hasidic Jews of Brooklyn, New York, who are mostly descended from the Ashkenazi Jews of Eastern Europe, have a relatively high population frequency of the gene for Tay-Sachs disease. Marriages are traditionally arranged within the Hasidic community, and marriages outside the community are rare; this type of inbreeding generally increases the rate at which recessive genes will come together and produce a recessive phenotype. Because they are ethically opposed to all abortions, the Hasidim will not permit genetic testing in utero. The availability of a test that detects Tay-Sachs heterozygotes has, however, allowed the Hasidic community to set up a computerized registry under their strict control. Testing of all individuals within the community is encouraged, and the results are entered into the registry under a code number that guarantees confidentiality. The registry permits the traditional matchmakers to check potential couples before proposing a match; if both partners are carriers for Tay-Sachs disease, the matchmaker is warned of this fact and the match is never made. Before this

registry was set up in 1984, the Kingsbrook Jewish Medical Center in Brooklyn, which services the Hasidic community, had an average of thirteen Tay-Sachs children under treatment at any one time; after just five years, the number of Tay-Sachs children under treatment in the hospital dropped to two or three.

When there is a known risk for a genetic disease within a particular family, additional, more expensive testing methods may be used, in contrast to the less expensive tests used to screen large populations.

The ethics of genetic testing. Genetic testing is not altogether an unmixed blessing. If a genetic defect can either be cured or phenotypically suppressed (possibilities discussed later in this chapter), or if heterozygote detection permits couples to take precautions against pregnancy, then a genetic test can be justified on the grounds that it relieves future suffering. However, most genetic defects cannot be cured in this way. What is the point of testing an individual for a condition like Huntington's disease that can neither be controlled or cured? One reason is that it permits people who carry a genetic condition to decide whether or not to bear children. Will a person who tests positive for such a genetic disease be denied insurance or employment on the basis of the test results? Will a woman choose to abort a fetus if a genetic disease is detected in utero?

Box 3.4 examines some of the ethical questions that arise in connection with various forms of prenatal testing. Another ethical issue involves the use of prenatal screening not for the purpose of detecting a disease-associated allele but to find out whether the fetus is a boy or a girl, something easily determined from examining the chromosomes. Will couples use this technology to try to select the sex of their offspring? This already happens in India, where abortion is legal and determination of the sex of the fetus by ultrasound is widely available to those who can afford (or can borrow) a fee of a few hundred dollars. A 1988 study of 8000 abortions in India's clinics showed that 7997 were female and only 3 were male (Ridley, 1993). In the United States, clinics offering prenatal genetic testing have found that over one-fourth of the couples who come to them are motivated by the possibility of choosing their baby's sex.

BOX 3.4   *ETHICAL ISSUES IN MEDICAL DECISION MAKING REGARDING GENETIC TESTING*

Should society influence the private decisions of individuals? To what extent do (or should) financial considerations limit the choices available to individuals?

Suppose a child is born with a birth defect or other congenital condition. Is it ever ethical to withhold treatment? (Some of these conditions are caused by defective genes, but others are caused by injuries, infectious diseases, poor maternal nutrition, or other causes, but the ethical issues raised are much the same regardless of the cause.)

What if the same disease is diagnosed on a fetus in utero—is it ethical to abort the fetus? The decision to abort a fetus or to withhold treatment from a child with a genetic disease raises important ethical questions. Here are some questions to consider:

1. Tay-Sachs disease is a genetically controlled disease whose victims are in constant pain and never survive beyond approximately 4 years of age. Does it make sense to spend thousands of dollars on the medical care of a child who has no chance of living beyond age 4, or even of enjoying those few years free from pain? Would it make a difference if a few people with the disease were capable of surviving? What if we were dealing with a disease that people could survive, but only with some disability?

2. The involvement of third-party insurance policies raises yet other issues. Should insurance policies pay medical expenses for genetic diseases that could have been avoided by screening? Some insurance policies will pay for medical treatment, but not for the testing that might have avoided the need for the treatment. Do you think insurance policies should cover genetic testing?

3. Should genetic screening be covered for certain ethnic groups but not others, just because the risks differ? For example, if a condition like thalassemia is more prevalent in certain ethnic groups (such as Italians, Greeks, and certain Southeast Asians), should insurance cover testing for this condition in a person of Italian descent, but not in a person of English or Danish descent?

4. If a genetic disease is detected during pregnancy, should insurance policies cover termination of the pregnancy if desired by the parents? If parents elect not to terminate a pregnancy, and a child is born with a genetic disease, should insurance policies cover any specialized medical care that might be necessary? Should insurance companies be allowed to deny coverage or raise premiums if a genetic disease is discovered?

5. What role might science play in answering questions of medical ethics that are related to genetics? For example, can science help in assessing the benefits and risks of genetic screening? To what extent does the detection of an allele predict a harmful phenotype? What constitutes sufficient evidence that a condition is genetically determined? What are the limits of the ability of science to contribute answers to these questions of medical ethics?

As genetic testing becomes more common, it is inevitable that test results will occasionally be misused. In one case, school officials were told that a child needed to be kept on a special diet because he had phenylketonuria (PKU). Although he was functioning normally, the child was placed in a class for the learning disabled. The school officials apparently knew that PKU could cause mental retardation, but were unaware that this outcome could be averted by a special diet. As this case shows, the misuse of genetic information may result from ignorance.

Discrimination by employers or insurers represents another possible misuse of genetic information. Employers may refuse to hire or promote, or insurance companies may refuse to insure, persons who have or who are suspected of having a number of possible genetic conditions. In some cases, benefits have been denied to heterozygous carriers of recessive conditions or to other persons whose phenotype was unaffected. This practice is still uncommon, but it is growing and is likely to continue to grow as more and more genetic information becomes available through medical testing. One of the best safeguards against this kind of discrimination is a strict adherence to rules governing confidentiality of medical records and other personal information held by health care providers and medical testing laboratories.

Altering Individual Genotypes

Some rare genetic traits impair health. In the future it may be possible to correct certain genetic defects by direct alteration of the individual genotype. A more realistic possibility is a form of gene splicing known as **recombinant DNA therapy,** in which the defective gene is replaced with a functional counterpart in some cells that are inserted into a patient. The genetic engineering techniques involved in recombination therapy are described later in this chapter.

Altering the Gene Pool of Populations

Instead of treating people one at a time, some people have proposed altering the genetic make-up of populations (the entire gene pool) over time to change the frequencies of certain genotypes. One difference between this approach and the approaches already described has to do with who is perceived to reap the benefits. Diagnosis, counseling, and the altering of individual genotypes are justified in terms of the pain and suffering that may be spared to individuals. On the other hand, all attempts to alter the gene pool carry with them notions of harm and benefit to society rather than to the individual.

Positive eugenics. Altering the gene pool through selection is called **eugenics,** from the Greek words meaning "good birth." This idea is not at all new—Plato's *Republic* (book 5) suggests that the best and healthiest individuals of both sexes be selected for breeding purposes, much as we do with horses and cattle. Plato's type of eugenics is called *positive eugenics,* meaning an attempt to alter the gene pool by selectively increasing the genetic contributions of certain chosen individuals or genotypes. Positive eugenics was also proposed in the twentieth century by the Nobel Prize–winning geneticist H. J. Muller, who advocated setting up sperm banks to which selected male donors would contribute. Muller thought that women

would eagerly seek artificial insemination with these sperm in the hopes of producing genetically superior children. Several entrepreneurs have established sperm banks (though usually not egg banks) of people whom they consider to be genetically superior. The system is not regulated, however, by any government agency.

Ethical and other questions raised by positive eugenics usually center around the lack of an agreed standard for human excellence. The traits most often discussed by those who favor eugenics are intelligence and athletic ability. However, these traits are genetically complex and are highly influenced by education, training, and other environmental variables. Studies attempting to demonstrate a genetic influence on these and other traits were in many cases poorly done, leading Lewontin, Rose, and Kamin (1984) to doubt the existence of any reliable evidence concerning the genetic control of human intelligence and other complex attributes.

The complexity of the human genotype raises other issues: What if Einstein had been heterozygous for some genetic disease? If a society wanted to breed people of superior intelligence, they would also unwittingly be breeding whatever other traits the intelligent people happened to possess, possibly including a genetic defect in the process. Suppose a person inherited the manic-depressive disorder of Robert Schumann or Vincent van Gogh, instead of their creative talents? What liability or what responsibility would a sperm bank face if a descendent was born with a genetic defect? What constitutes "superiority" in an individual, and who should have the power to make such choices?

Negative eugenics. Most discussions of eugenics have centered on *negative eugenics,* the prevention of breeding among people thought to be genetically defective or inferior. Founded by Francis Galton (1822–1911), a cousin of Charles Darwin, the modern eugenics movement has generally tended to emphasize negative measures. Galton and his supporters were very much interested in measuring intelligence, and they developed some of the early versions of what we now call IQ tests. Through the use of these and other tests, supporters of eugenics have long sought scientific respectability for their attempts to label certain people as genetically defective or inferior.

The Nazis instituted a program of negative eugenics in Germany, beginning with the forced sterilization of mental "defectives," deaf people, homosexuals, and others. The eugenics program soon grew into a program for the mass killing of all those millions who did not belong to Hitler's "master race." By 1945 the Nazis had killed millions in the name of racial purity and Aryan superiority. The Nazis also practiced positive eugenics by encouraging German women with certain traits to have more children.

In the United States, the eugenics movement started as a series of attempts to identify, segregate, and sterilize mental "defectives." The movement soon found allies among racists and especially among those who sought to curb the new waves of immigration during the period from about 1890 to 1920. During the 1890s, one Kansas doctor sterilized forty-four boys and fourteen girls at the Kansas State Home for the Feeble-Minded, while Connecticut passed a law prohibiting marriage or sexual relations between any two people "either of whom is epileptic, or imbecile, or feeble-minded." A 1907 Indiana law required the sterilization of "confessed criminals, idiots, imbeciles, and rapists in state institutions when recommended by a board of experts." Fifteen other states passed similar laws, as often for punitive as for eugenic reasons. From 1909 to 1929, 6255 people were sterilized under such laws in California alone (Haller, 1963).

The writings of the eugenicists became increasingly racist and anti-immigrationist in tone during this period, as amply documented by Gould (1981). One eugenicist, for example, wrote in 1910 that "the same arguments which induce us to segregate criminals and feebleminded and thus prevent breeding apply to excluding from our borders individuals whose multiplying here is likely to lower the average [intelligence] of our people." In the 1960s, H. J. Muller wrote several articles warning against the practice of protecting and extending the lives of the "genetically unfit," those whom natural selection would tend to eliminate from the population. According to Muller, our medical intervention would only perpetuate genetic defects in our gene pool. Muller spoke pessimistically of a population divided into two groups, one so enfeebled from genetic defects that their very lives had to be sustained by extraordinary means, and the other group consisting of phenotypically normal people who had to devote their entire lives to the care and sustenance of the first group.

Biological objections to eugenics. Biological arguments against eugenics are based on the realization that eugenic measures could only be expected to produce small changes at great cost. Most known genetic diseases are both rare and recessive, and selection against rare, recessive traits can only proceed very slowly no matter what the circumstances. As the trait gets increasingly rare, selection against it becomes increasingly ineffective. For example, the allele for albinism has a frequency of about 1 in 2000 in many human populations. If a eugenic dictator ordered all albinos to be killed or sterilized, theoretical calculations show that it would require about 2000 generations (about 50,000 years) of constant vigilance just to reduce the frequency of this trait to half of its present value. The reason the process works so slowly is that most individuals carrying the allele for a rare, recessive trait are heterozygous and their phenotype does not reveal the presence of the allele. Any type of gene-gene or gene-environment interaction—such as a developmental process that allows the gene to express its phenotype only in certain environments—would further diminish the effectiveness of any eugenic selection program where selection is based on phenotype. On the other hand, modern techniques that allow the detection of alleles in heterozygous form would greatly increase the effectiveness (and hence the dangers) of negative eugenic measures.

For characteristics like height or IQ, controlled by many genes and influenced strongly by environmental factors, estimates are that eugenic selection would be so slow as to be barely perceptible. One geneticist calculated that it would take about 400 years of constant, unrelenting, and totally efficient selection to raise IQs by about four points; the same improvement could be achieved through education in as little as four years, and with far less cost (Goldsby, 1971). This topic will be addressed in more detail in Chapter 5.

Finally, eugenic measures can at best address only a small percentage of undesirable conditions, as most physical disabilities and medical conditions result from accidents, from infectious illnesses, or from exposure to toxic substances in the environment, not from inherited genetic makeup, and eugenic measures are powerless to alter these causes. Even some genetic conditions may result from new mutations, rather than from the inheritance of defective genes, and eugenic measures have no capacity to eliminate newly mutated genes or to depress the frequency of any gene below the mutation rate.

There is no biological basis for the claims of any eugenics movement that their methods could in any way improve humankind other than at great cost. The risks of negative eugenics are especially great, and include the possibility of genocide, the attempted extermination of a race or ethnic group. Given the alternative measures described later for dealing with genetic diseases, the cost-benefit ratio for eugenics is especially unfavorable.

Cloning.   Another possible technology for achieving eugenic aims involves clones. A **clone** is defined as a cell or individual and all its asexually produced offspring (i.e., those produced without sexual recombination). All members of a clone are genetically identical, except when a mutation occurs. No mammal or other complex animal reproduces naturally by cloning, though several insects and plants do in some circumstances and a few do so regularly. Some biologists have proposed that it might be possible to remove the nucleus from a fertilized egg (zygote) and insert the nucleus from a fully developed individual. The altered cell would then be implanted in a suitable womb and brought to term. The new individual formed in this way would be a genetically identical clone of the individual whose nucleus was used, though its cytoplasm would be inherited from the original egg, as cytoplasm always is. In theory, cloning could make multiple copies of a desired genotype without need of recombination, but the technology to do so does not currently exist.

Another type of cloning is the division of a single egg or early embryo into two or more separate embryos, which is essentially what happens when identical twins occur naturally. In 1993, two scientists from George Washington University succeeded for the first time in cloning a human embryo under laboratory conditions (in vitro) by

stripping away a jellylike protective coating separating the individual cells and then furnishing a new artificial coating for each cell. The resulting embryos were never implanted in a uterus or brought to term, but the mere possibility has set off wild speculations and a series of ethical debates. If cloning ever became possible on a large scale, it would raise many of the same ethical problems as positive eugenics. Many people, frightened by these prospects, staged a demonstration near the laboratory where the cloning experiments were carried out. Various ethicists have pointed out an array of possible misuses or dubious uses of the technology, such as raising a child for the sole purpose of having the child serve as transplant donor to an identical twin born years earlier. Public opinion polls also show strong opposition to unrestricted cloning: 75 percent of people questioned disapproved of human cloning, and 58 percent thought that cloning was morally wrong (*Time*, November 8, 1993:65–70).

Cloning procedures have already been used to produce genetically identical cattle and other farm animals by separating the cells of early embryos, as described above. The technique is expensive and has a low success rate. It has been available for over a decade in domestic cattle and has become an accepted part of breeding practice for these animals, though it has never proved popular; one company founded to take advantage of the cloning technology has gone out of business. Cloning techniques are most likely to be used for valuable animals of known pedigree, such as race horses.

Changing the Balance Between Genetic and Environmental Factors

While many traits are inherited, most are also influenced by the environment. Phenotypes resulting from the expression of genes, from environmental influences (like accidents and illnesses), or from the interactions between genes and their environment can be modified in several different ways.

Euphenics.   **Euphenics** (literally, "good appearance") includes all those techniques that either modify genetic expression or alter the phenotype to produce a modified phenotype known as a **phenocopy.** Most forms of medical intervention are included, such as plastic surgery to repair defective body parts. Other examples include the installation of pacemakers in people with defective hearts (Chapter 14), the giving of insulin to diabetics, or dietary control of phenylketonuria (described earlier). Though the genes remain unchanged, their phenotypic expression is modified or compensated for in such a way that it no longer causes harm.

As an alternative to eugenics, many leading geneticists have argued that there is nothing wrong with altering the phenotype or the environment so that formerly disabling genotypes are no longer quite so harmful or debilitating. A leading advocate of this viewpoint was Theodosius Dobzhansky (1893–1975), who favored measures to permit people with hereditary "defects" to overcome their handicaps and become phenotypic copies (phenocopies) of normal, healthy human beings. Once phenotypes could be controlled culturally, said Dobzhansky, the presence of formerly defective genotypes would cease to be the subject of any great concern. Euphenic intervention is already common practice for a number of genetically controlled conditions. As our ability to modify phenotypes increases (e.g., with advances in corrective surgery), this type of practice is likely to become more common.

Euthenics. Another type of intervention is called **euthenics.** In this form of intervention, both genotype and phenotype remain unchanged, but the environment is modified or manipulated so that the phenotype is no longer as disabling as before. (In euphenics, by contrast, the phenotype is always modified.) Examples of euthenic measures include canes, crutches, wheelchairs, and wheelchair ramps for those who cannot walk unaided, guide dogs or Braille for the sight-impaired, eyeglasses for the nearsighted, and so on. Some of the important applications of bioengineering discussed in Chapter 14 lie in the field of euthenics. Conditions that are improved or assisted by euthenics may either be genetic or not.

Many of the people who use euthenic devices feel that they would be better served by simple improvements in the devices (e.g., better wheelchairs) than by an exclusive emphasis on medical research. Medical research is expensive, its results are uncertain, and its benefits may take many years to become widely available. Mechanical aids are often less expensive and more quickly made available once they are developed.

Eupsychics. Many people with uncommon traits or conditions (whether genetic or not) feel that they are best served by being accepted as they are and do not necessarily want to be "cured" (see Box 3.5). Social and behavioral measures, or **eupsychics,** may lessen the impact of or compensate for disabling conditions. Included are the special education of handicapped individuals, mainstreaming (education of the handicapped in a regular public school setting), and the education and social conditioning of nonhandicapped members of society so they will better understand and accommodate to the needs of all citizens. Society may be better served by an emphasis on the abilities, rather than the disabilities, of each individual. All individuals should be encouraged to develop their talents and abilities to the fullest. Whenever a young person is discouraged from trying to develop a certain skill, ability, or talent, both the individual and the society are the losers in the long run.

THOUGHT QUESTIONS

1. Are all "birth defects" genetic defects? Do the same ethical questions regarding diagnosis and counseling apply to both genetic and nongenetic traits?
2. In some hospitals, screening for phenylketonuria is performed on *all* infants. Does this violate the principle of informed consent? Is this practice ethical or not? How is it commonly justified? Do you feel that the justification is adequate?
3. If cloning were ever perfected, how identical to the original would cloned individuals actually be? How strong an influence would maternal cytoplasm have? How strong a role would upbringing play?
4. Some groups opposed to abortions have also begun to object to certain kinds of genetic testing. What good, they ask, can come from knowing that a fetus suffers from a particular genetic or chromosomal defect if the parents are opposed to abortion of the fetus on religious or similar grounds? For such situations, discuss the costs,

BOX 3.5 WHO DECIDES WHAT IS CONSIDERED A "DEFECT"?

Very often, groups of people that society wants to "help" are given little or no voice in how society will treat them or "help" them. The deaf are a case in point. Some forms of deafness are inherited, yet many deaf people consider it abhorrent if genetic counseling is used to avoid having deaf children.

The following excerpt is from a speech by I. King Jordan delivered in 1990 to an international symposium on the Genetics of Hearing Impairment. A deaf man both culturally and medically, Jordan is president of Gallaudet University in Washington, D.C., which was established in the 1800s to educate deaf people.

For about 18 years, I have taught a course on the psychology of deafness. One of the first things we discuss in the class is the difference between viewing deafness as a pathology that should be cured or prevented and viewing it as a human condition to be understood. I call these two perspectives the medical and cultural points of view. Individuals from these two groups agree on audiological definitions, but disagree on the emphasis that should be given to social and rehabilitative services. I adhere to the social or cultural point of view.

What I mean by this is that I, personally, and many of the people I know well, have accepted the fact that deafness is one aspect of my individuality. I do not spend any time or energy thinking about curing my deafness or restoring my hearing, but I do spend substantial time and energy trying to improve the quality of life for all people who are deaf.

For some reason, people who hear have a very difficult time understanding this concept. If you will permit me to digress for a moment, I will give you an example. I was interviewed by Ms. Meredith Vieira for the television show *60 Minutes*. During the interview, she asked me this question: "If there was a pill that you could take and you would wake up with normal hearing, would you take it?" I told her that her question upset me. I told her that it was something I spent virtually no time at all thinking about, and I asked her if she would ask me the same question about a "white" pill if I were a black man. Then I asked if, as a woman, she would take a "man" pill. Our conversation continued long after the videotaping was done, and we have had several subsequent conversations. But she never understood. She still does not. She still thinks only from her own frame of reference and imagines that not hearing would be a terrible thing. Deafness is not simply the opposite of hearing. It is much more than that, and those of us who live and work and play and lead full lives as deaf people try very hard to communicate this fact.

As you can see, this is an emotional issue for me. It is a much more emotional issue for many other deaf people. Is that relevant here? Yes, I believe it is, because the genetic study of deafness and genetic counseling have a great deal of significance for the deaf community generally. As Dr. Christiansen pointed out, many deaf people, particularly those who consider themselves members of the deaf community, do not consider themselves to be defective, rather, they consider themselves to be different, normal but different. In particular, this difference has a cultural or sociological basis and is expressed most saliently in the use of sign language. If deaf people are not defective or dysfunctional then, at least in their own eyes, it follows that they would be suspicious of attempts to eradicate deafness. . . .

Genetic counseling and screening with respect to potential deafness must differ, therefore, in a fundamental way from screening for "birth defects."

(I. King Jordan, "Ethical Issues in the Genetic Study of Deafness," *Annals N.Y. Acad. Sci.*, 630:236–237.)

benefits, and ethical status of genetic testing. Does it matter what kind of testing is performed? Does it matter what genetic or chromosomal defect is involved?

5. In what sense is "positive" eugenics positive? In what sense is it negative? Should the same amount of control be applied to human populations as are applied to domesticated animals?

6. If public funds will be spent for the care and treatment of a person with a genetic condition, does that alter the ethical balance between the rights of the individual and the rights of society?

7. How much access to genetic information *should* an insurance company or an employer have? How much access *do* they have now?

D. Newer Technologies Offer Additional Choices

New Reproductive Technologies

Humans have been manipulating the reproduction of domesticated animals and plants for at least 4000 years. Our modern strains of hybrid corn or highly productive dairy cattle are the result of many generations of genetic manipulation, mostly in the form of selection. Some other forms of genetic manipulation are discussed in Chapter 15.

Several newer reproductive technologies are now available for human use as well as for domestic animals and plants. These techniques include artificial insemination, in vitro fertilization, surrogate pregnancy, and cloning (described earlier). These are all euthenic measures that get around a variety of reproductive problems, from genetic sterility to conditions that would make pregnancy inadvisable, conditions that may be genetic in origin, but are more often nongenetic.

Artificial insemination.   **Artificial insemination** means introducing sperm into a female's reproductive tract other than through sexual intercourse. The sperm could be derived from a woman's husband or from another man. The procedure is fairly simple and is routinely performed on cattle and certain other domesticated species (as are in vitro fertilization and surrogate pregnancy). However,

the legal rights and responsibilities of a sperm donor other than the recipient's husband are unclear in a number of jurisdictions.

In vitro fertilization.   An in vitro process is one that takes place in laboratory glassware rather than inside the body (in vivo). In the case of **in vitro fertilization,** eggs are harvested from a woman patient and fertilized in a glass or plastic dish, using sperm contributed either by her husband or by another man. Fertilized eggs are then allowed to develop to approximately the 64-cell stage, after which one or more of these embryos are implanted in the woman's uterus and allowed to develop to term. Using sperm donated by someone other than the woman's husband raises the same kinds of legal issues (including custody and financial responsibility) as in cases of artificial insemination. Individuals who possess genotypes that they do not wish to pass on to their offspring may seek either in vitro fertilization or artificial insemination using donated sperm.

Some researchers have been experimenting with techniques that test eight-cell embryos for genetic diseases. Because the testing process is usually destructive, only one of the eight cells is separated from the rest for testing, leaving the other seven cells available for implantation if desired; the organism resulting from a seven-cell embryo is just as normal as if all eight cells had been used. Embryos obtained by in vitro fertilization can thus be tested *before* they are implanted into a uterus for gestation. This technique, sometimes called *BABI* (blastomere analysis before implantation), has already been used to test human embryos in vitro for cystic fibrosis, allowing selection of only those embryos that are free of the disease. Selected embryos can then be implanted, and the parents raising these embryos can be free of the fear that their child will be born with cystic fibrosis, a disease that is usually fatal. Once this technique becomes readily available for a wider variety of human conditions, it will become possible to avoid certain genetic diseases without performing any abortions or to choose certain other characteristics, such as the child's sex.

Surrogate pregnancy.   **Surrogate pregnancy** is simply the use of another woman's womb to carry a baby to term on behalf of a woman who cannot undergo the pregnancy herself, usually for medical reasons. In most cases, the baby is conceived by in vitro fertilization, using egg and sperm cells donated by a married couple. The resulting embryo, which is the genetic offspring of the donor couple, is then implanted into another woman who agrees to act as a surrogate mother, usually for a fee. In addition, medical expenses are generally paid by the donor couple. The legal status and rights of the surrogate mother are subject to many ethical and legal questions (see Field, 1988; Shannon, 1988; and Gostin, 1990). Surrogacy contracts have been outlawed or held invalid in a number of jurisdictions that view the birth mother (i.e., the surrogate) as the legal parent, who is therefore "selling" her baby if she receives any payment. Among the ethical issues raised are the exploitation of poor women by wealthy couples. Financial need is often a factor (one of many) in a woman's decision to become a surrogate. Other issues include the amount of compensation that can ethically or legally be given to the surrogate and the ways in which this situation is distinguished from baby selling. A final issue concerns the available alternative of adoption, which is generally less expensive and raises fewer legal and ethical objections.

Genetic Engineering

If an individual with a genetic disease is identified, it may become possible in the near future to "cure" the genetic disease by altering the genotype. Any such direct alteration of individual genotypes is called **genetic engineering.**

Recombinant DNA technology. The first step in inserting a gene into a genome through recombinant DNA technology is to use what is known as a **restriction enzyme** to snip out a desired segment of DNA. (Several enzymes are known that can break apart a DNA molecule, but an enzyme that acts indiscriminately is of little use in genetic engineering. Restriction enzymes act specifically.) Each restriction enzyme cuts the DNA at specific places, defined by their DNA sequences. The most useful restriction enzymes are those that cut the two DNA strands at slightly different locations, producing short sequences of unpaired DNA known as **sticky ends** (Fig. 3.22). For example, the

commonly used restriction enzyme *Eco*RI always attacks the sequence GAATTC, breaking it between G and AATTC. Since the complementary strand (opposite GAATTC) also reads GAATTC, this sequence is considered a *palindrome,* a sequence that reads the same in either direction. The *Eco*RI enzyme recognizes and selectively attacks the GAATTC sequences on each strand, breaking the two-stranded sequence into fragments that have overlapping single-stranded sequences (Fig. 3.22). The matching single-stranded end sequences are called "sticky" because they are able to stick together spontaneously to form complementary double strands. In this case, it is the unpaired AATT sequences that form each of the single-stranded sticky ends that can pair with one another, stick together, and then join permanently. (An enzyme that cut GAATTC between GAA and TTC would form "blunt," rather than sticky, ends.) The unpaired complementary sequences at these sticky ends are predictable for each enzyme, which is why *fragments* cut out with a particular restriction enzyme can always be replaced with another fragment cut with the same enzyme. This makes it possible to use restriction enzymes to cut a DNA sequence and insert a functional gene with matching sticky ends.

Cutting an entire chromosome with a restriction enzyme produces many fragments, only one of which contains the gene being sought. In order to isolate only those fragments containing the gene of interest, a DNA probe is used; such a probe is a complementary DNA (cDNA) strand that is radioactive. If a radioactive cDNA probe is added to some DNA that has been digested into fragments with a restriction enzyme, only those fragments that can pair with the cDNA will pick up the radioactivity. This allows geneticists to isolate the radioactive sequences, then separate the desired genes from the cDNA probes that paired with them. A functional gene isolated in this way can then be inserted into a chromosome that lacks a functional version of that gene.

To date, most genetic engineering of human genes has involved the introduction of these genes into bacteria. The reasons for this are largely practical: Many human gene products are useful in medicine but are more readily produced in large amounts inside genetically engineered bacteria than inside people. The hormone *somatostatin,* for

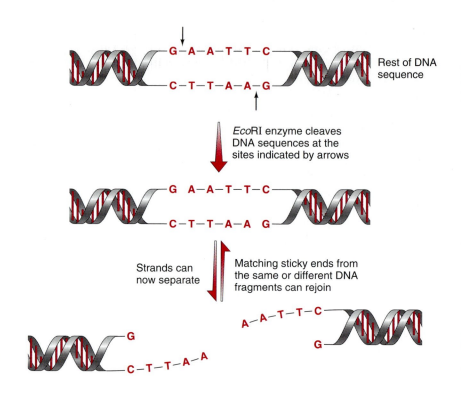

FIGURE 3.22
How restriction enzymes produce DNA fragments with sticky ends.

example, also called *growth hormone,* is highly valued for the treatment of certain types of dwarfism. The hormone is, however, difficult to obtain from human sources (the traditional way is to extract it from the pituitary glands of dozens of cadavers) and is therefore very expensive. *Insulin,* the hormone needed by diabetics, is another example of a human gene product. Both of these hormones could be obtained from sheep or pigs or other animals, but the animal hormones are not as active in humans as the human hormones, and some patients are allergic to hormones obtained from other species. Genetic engineering provides a cost-effective way of manufacturing large amounts of these human hormones in bacteria.

Genetically engineered insulin.   Human insulin was the first commercially produced genetically engineered product. The first step in making genetically engineered insulin is to grow human cells in tissue culture (Fig. 3.23). An extract of the cell nuclei of several cells is then exposed to a restriction enzyme, and the same restriction enzyme is used on nonchromosomal DNA molecules, called *plasmids,* from the recipient bacteria.

Bacteria have a single chromosome and many also have a number of **plasmids,** detachable DNA fragments that can lead an independent existence for a long while and then become incorporated into the bacterial chromosome. Plasmids are used in genetic engineering because, being fragments, they have fewer sites at which a given restriction enzyme can cut. Cutting a DNA sequence from the plasmid with the same restriction enzyme that was used on the human DNA creates sticky ends that match the DNA fragment taken from the human cell. This allows incorporation of the human DNA fragment identified by a cDNA probe into the bacterial plasmid. After the plasmid has been taken up by the bacteria, it will insert itself into the main bacterial chromosome. In most cases, the plasmid either contains or is given another DNA sequence that can be used to select those bacteria that have incorporated the plasmid. For example, the plasmid might contain the gene for an enzyme that gives the bacteria resistance to a common antibiotic; the antibiotic can then be used to select the bacteria that have incorporated this gene, while killing the majority that are still susceptible. The procedures sound straightforward, but each step

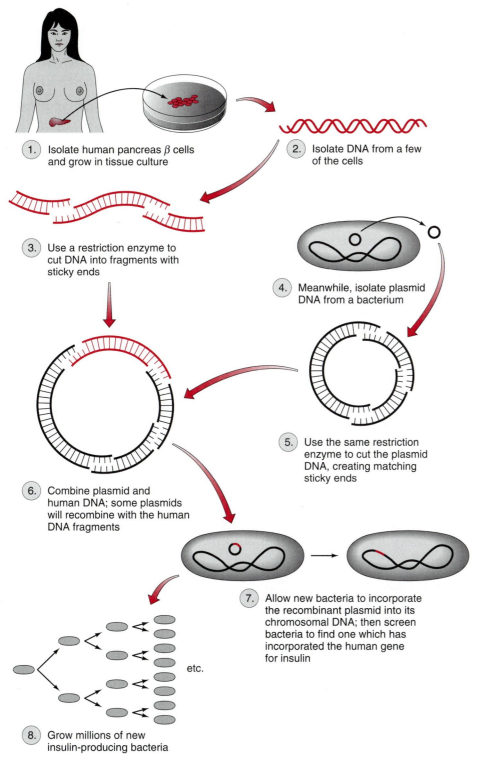

1. Isolate human pancreas β cells and grow in tissue culture

2. Isolate DNA from a few of the cells

3. Use a restriction enzyme to cut DNA into fragments with sticky ends

4. Meanwhile, isolate plasmid DNA from a bacterium

5. Use the same restriction enzyme to cut the plasmid DNA, creating matching sticky ends

6. Combine plasmid and human DNA; some plasmids will recombine with the human DNA fragments

7. Allow new bacteria to incorporate the recombinant plasmid into its chromosomal DNA; then screen bacteria to find one which has incorporated the human gene for insulin

etc.

8. Grow millions of new insulin-producing bacteria

FIGURE 3.23
How genetically engineered insulin is produced.

of the process is technically difficult and only a small proportion of the attempts succeed.

The bacteria can now be cloned by allowing them to multiply asexually in large numbers. The human gene is transcribed and translated by the bacterial cell to produce human insulin. Since DNA has the same chemical structure in all species, a gene from one species can be used in protein synthesis by the cells of another species. The human insulin extracted from these bacteria, called *recombinant human insulin,* can now be given to diabetic patients.

Gene therapy.   Instead of growing human insulin in bacteria, gene splicing could theoretically be used to introduce the insulin gene into human cells that do not possess a functional copy. (That would still not cure diabetes unless these cells were also capable of appropriately increasing or decreasing their output of insulin according to conditions.) This type of genetic engineering is called **gene therapy**—the introduction of genetically engineered human cells into a human body for the purpose of curing a disease or a genetic defect. Human gene therapy has been used successfully to treat *severe combined immune deficiency syndrome* (*SCIDS*), a severe and usually fatal disease in which a child is born without a functional immune system. Unable to fight infections, these children will die from the slightest minor childhood disease unless they are raised in total isolation—the "boy [or girl] in a bubble" treatment. The enzyme that controls one form of SCIDS has been identified—it is called *adenosine deaminase* (*ADA*) and is located on chromosome 20. A rare homozygous recessive condition results in a deficiency of this enzyme, which in turn causes the disease. Gene therapy to cure this condition was first successful in 1990. It consists of the following procedural steps shown in Fig. 3.24:

1. Normal human cells are isolated. The cells most often used are T lymphocytes, a type of blood cell that is easy to obtain from blood and easy to grow in tissue culture.

2. The isolated cells are grown in tissue culture.

3. The DNA from these cells is isolated.

4. A restriction enzyme is used to cut out a DNA fragment containing the functional gene for

ADA and two sticky ends. A probe with a complementary DNA sequence is then used to isolate and identify fragments bearing the gene.

5. The same restriction enzyme is used to create matching sticky ends in viral DNA isolated from a virus known as LASN. This virus was chosen because it can be used as a vector to transfer the gene into the desired host cells.

6. The viral DNA is then mixed with the human DNA fragments and allowed to combine with them.

7. The virus is allowed to reassemble itself. It is now ready for further use.

8. Blood is drawn from the patient to be treated and T lymphocytes are isolated from this blood. These lymphocytes are ADA-deficient because they do not possess a functional ADA allele.

9. The virus is now used as a *vector* to transfer the functional gene into the lymphocytes.

10. The lymphocytes are tested to see which ones are able to produce a functional ADA enzyme, showing that they have successfully incorporated the functional ADA allele.

11. The genetically engineered lymphocytes are injected back into the patient, where they are expected to outgrow the genetically defective cells because the ADA-deficient cells do not divide as fast as cells with the ADA enzyme.

The gene therapy described above provides a functional gene that is transcribed and translated by the body cells, producing the missing enzyme in lymphocytes. Since lymphocytes are not the only cells that need the ADA enzyme, the patient must also receive injections of the ADA enzyme coupled to a molecule that permits it to enter cells. (This last step might not be necessary for the treatment of other enzyme defects.) The enzyme controls the symptoms of the disease, but it is not a cure because the underlying disease is still present. Gene therapy for ADA was first successfully used on a 4-year-old girl in 1990 (Anderson, 1992). A second patient, a 9-year-old girl, began receiving treatments in 1991. Both patients are being closely monitored, and their immune systems are now working properly (Ferrari, 1992; Grompe, 1991). However, since the genetically engineered cells are mature lymphocytes, which have only a limited lifetime, repeated injections of genetically engineered cells are needed. To get around this problem, and

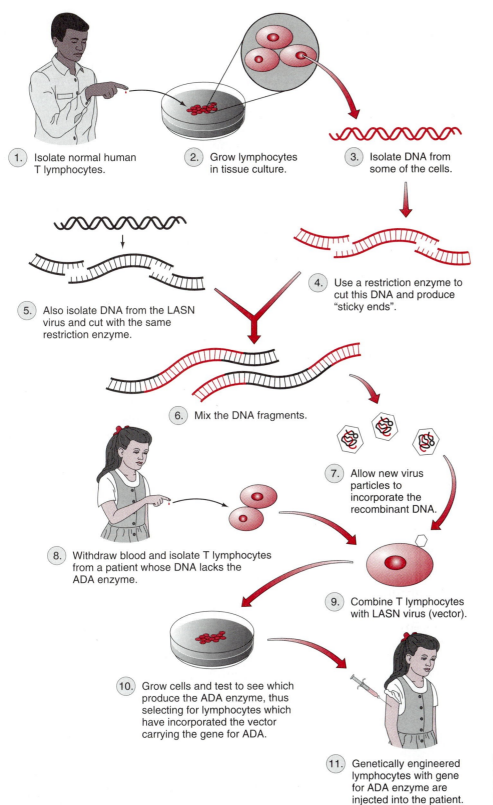

1. Isolate normal human T lymphocytes.

2. Grow lymphocytes in tissue culture.

3. Isolate DNA from some of the cells.

4. Use a restriction enzyme to cut this DNA and produce "sticky ends".

5. Also isolate DNA from the LASN virus and cut with the same restriction enzyme.

6. Mix the DNA fragments.

7. Allow new virus particles to incorporate the recombinant DNA.

8. Withdraw blood and isolate T lymphocytes from a patient whose DNA lacks the ADA enzyme.

9. Combine T lymphocytes with LASN virus (vector).

10. Grow cells and test to see which produce the ADA enzyme, thus selecting for lymphocytes which have incorporated the vector carrying the gene for ADA.

11. Genetically engineered lymphocytes with gene for ADA enzyme are injected into the patient.

FIGURE 3.24
Gene therapy to replace the human gene for adenosine deaminase (ADA).

hopefully to bring about a more lasting cure, some Italian researchers have tried using both genetically engineered lymphocytes (as described above) and genetically engineered bone marrow stem cells. Stem cells divide to form all the developed types of blood cells, and they maintain this ability throughout life. Therefore, after repaired lymphocytes die off, stem cells with repaired DNA could divide to provide new, ADA functional lymphocytes, possibly for the lifetime of the individual. This type of therapy was begun on a 5-year-old boy in 1992 (Anderson, 1992) and since then several other children have received this treatment.

Changing genotypes in gametes.   The above examples all involve somatic cells, the diploid body cells that do not give rise to gametes. Whatever happens to these somatic cells, they will eventually die, but the germ cells that are passed on to succeeding generations will still contain the defective genes that produced the troublesome genotype in the first place. Still in the future is the possibility of gene therapy to the germ cells, those that give rise to gametes. If gene therapy is ever successfully used to repair genetic defects in germ cells, the result will be a permanent change in genotype that will be transmitted to all future generations.

Questions of safety and ethics.   There are legitimate safety concerns with human gene therapy. For example, any virus used as a vector must be capable of entering human cells. Might such a virus cause a disease of its own? To preclude this possibility, most of the viruses used to date in most human gene therapy have been from strains with genetic defects that render them incapable of reproducing and spreading to other cells.

Gene therapy also raises certain ethical concerns. The price of genetically engineered proteins, initially high when they were first introduced, has fallen as they have become more widely available. New recombinant DNA procedures continue to be very expensive to develop. This raises ethical issues of fairness: Will the benefits of genetic engineering be available only to those who can afford them? Should government programs provide genetic therapies through Medicare and Medicaid? Should insurance cover their use? How can society's health

care resources best be distributed? If medical resources are limited, should an expensive procedure used on one person take up needed resources that could cover inexpensive treatments of other diseases for many people? These particular questions are not unique to recombinant DNA; they apply to any expensive form of medical treatment.

Recombinant DNA therapy may someday become commonplace in human cells as well as bacteria. In theory, recombination therapy could be practiced either on somatic cells or on gametes. If it is performed on somatic cells, the effects of the recombination therapy would last as much as a lifetime, but no longer. For example, insertion of the functional allele for insulin into the pancreatic cells of patients with diabetes may cure them of the disease, but they will still pass on the defective alleles to their children. A general consensus has been reached concerning the ethical value of using gene therapy on somatic cells for the purpose of treating a serious disease.

If successful recombination therapy is performed on germ cells, then the genetic defect will be cured in all future generations derived from those germ cells. In addition to all the ethical questions raised earlier, gene therapy on germ cells raises many additional ethical questions, some of which are listed in thought question 5 below.

Most medical ethicists today advise caution and waiting in the case of germ-cell gene therapy on humans until we have more experience with gene therapy on somatic cells or in other species.

THOUGHT QUESTIONS

1. In what ways do in vitro fertilization, artificial insemination, and surrogate parenting raise new ethical questions? What rights and what duties do the various participants have in each case? How can we best avoid the ethical and legal problems that may sometimes arise?
2. The use of growth hormone for the treatment of shortness (not dwarfism) in otherwise healthy children is controversial, but its testing for this purpose was approved in 1993 by the Food and Drug Administration. When does a phenotypic condition unwanted by its bearer become a disease? Who decides? Is gene therapy to increase someone's height simply another form of cos-

metic surgery, similar to breast implants or face-lifts?

3. If a person dissatisfied with their phenotype suffers from lack of self-esteem on that account, does the lack of self-esteem justify a procedure to correct the phenotype? (This same argument is raised to justify traditional forms of cosmetic surgery.) In the case of a child, do the parents have the right to anticipate what the future effects on self-esteem will be with and without corrective procedures? In the case of a phenotype like height that develops over a period of years, at what age is it appropriate (if ever) to evaluate the phenotype and decide upon corrective measures?

4. A procedure like gene therapy is expensive. Who should pay for it? Is gene therapy a limited resource? Does giving gene therapy to one patient thereby deprive another of medical care?

5. Do unborn children have a "right" to inherit an unmanipulated set of genes? Do they have a right to inherit "corrected" genes if such a possibility exists? What kind of informed consent can we expect on behalf of unborn generations? Can an individual make decisions that affect all their progeny? Do we need safeguards to protect future generations against the selfish interests of the present generation?

E. The Human Genome Project Will Greatly Advance Our Knowledge

The complete hereditary material of an entire organism is known as its **genome.** There are an estimated 50,000 to 100,000 human genes (Watson, 1990). Some types of human variation have been known for centuries, before their genetic basis was known. Many more human genes have been identified throughout the twentieth century, but chromosomal locations of these genes were not always known. The *Human Genome Project* was first proposed in 1986 as a program to make a genetic map, or catalog, of a prototypical *Homo sapiens*, including the chromosomal location of all human genes and the complete DNA sequence. Af-

ter some initial competition between the National Institutes of Health (NIH) and the Department of Energy, the Human Genome Project was funded by Congress to begin work in the fall of 1989, and James Watson was appointed as the first director. Watson stated his belief that the Human Genome Project will tell us what it means to be human.

Restriction Fragment Length Polymorphisms (RFLPs)

Mapping a gene in a slow-breeding species like humans poses several problems. Researchers cannot perform crosses at will, nor can they produce the thousands of offspring needed to determine linkage in the traditional manner. They can study inheritance in families, but this provides reliable information on gene location only for genes that are closely linked to something else that can be easily detected—a **marker.** Before 1980 very few markers were known that could identify with certainty the chromosomal location of any human gene. The use of restriction enzymes changed this situation drastically, beginning with the development of a new technique by Botstein et al. (1980). The technique uses restriction enzymes to break a DNA sequence in several places, forming a series of **restriction fragments.** Some parts of the DNA vary from one healthy individual to another and these variations produce fragments of altered length. Variations (also called polymorphisms) in the lengths of these fragments are known as **restriction fragment length polymorphisms**, or **RFLPs** ("riflips"; see Fig. 3.25). RFLPs are numerous and are scattered throughout the genome. RFLPs are used as genetic markers. Once their chromosomal locations and genetic map positions have been established, geneticists can determine the positions of any genes located near the RFLPs by finding a pattern of linkage (Fig. 3.3) between the two. In this way, an increasing number of gene locations and gene sequences are being discovered at an accelerating rate.

Identifying a specific gene as the cause of a trait. In the past, scientists seeking the cause of a particular inherited trait began their search by looking for a defective protein. The search for the location of the gene coding for that protein was generally a long, arduous process of discarding one possibility after another. Now it is possible to begin the search by looking first for RFLPs that are inherited along with a certain trait. Since RFLPs occur at known chromosomal locations, the process now begins with the DNA directly. The first gene to be identified in this manner was the gene for Duchenne's muscular dystrophy, a sex-linked genetic disorder that causes muscles to become weak and nonfunctional. In most cases, the inability of the muscles of the diaphragm to keep the patient breathing leads to death during the teenage years or in the early twenties. The product of the gene responsible for this disease is a protein called *dystrophin*. With a molecular weight of over 400,000, dystrophin is one of the largest proteins known. A test can be devised to screen people and see if they are carriers for the disease. Researchers hope someday to treat muscular dystrophy by using genetic engineering to provide this protein to persons who cannot make their own dystrophin molecules.

Using RFLPs to map genes. The Human Genome Project was begun by biologists who realized that the RFLP technique allowed the rapid identification and mapping of gene locations at a rate far beyond what had previously been possible. Thus far, over a thousand genes have been mapped to their locations on particular chromosomes. Diseases whose genes were located and identified during the 1980s and 1990s include Huntington's disease (chromosome 4), Duchenne muscular dystrophy (chromosome 23X), cystic fibrosis (chromosome 7), Alzheimer's disease (chromosome 21), one form of colon cancer (chromosome 2), and two forms of manic depression (chromosomes 11 and 23X). Most scientists working in this area see the medical and other benefits that could flow from knowing the location and sequence of all the genes. However, the genes, meaning the DNA sequences that are transcribed and then translated into protein sequences, constitute only about 3 percent of the entire genome. The rest of the DNA consists of "spacer" sequences that are never transcribed or other kinds of sequences that are transcribed but never translated. The function of these nongene sequences is currently unknown, and the wisdom of spending an estimated $15 bil-

lion on their sequencing is a question on which opinion, even among scientists, differs widely.

Ethical and Legal Issues

Legal and ethical issues associated with the project include questions of ownership and patent rights. Who owns the human genome or the sequence of any particular gene? If a researcher localizes a gene to a particular chromosome, can that researcher patent the information? Can a gene sequence be copyrighted in the manner of a book? Can the genes themselves be patented? Certain biotechnology companies stand to profit greatly from the marketing of gene sequences, tests for gene sequences, or cures for various genetic diseases, but the sharing of information on gene sequences appears at first glance to threaten their competitive position. One corporation, the Genome Corporation, intends to determine as many gene sequences as possible and then copyright them and sell the information at a profit. Other scientists feel that the human genome is (or should be) public information, and that scientists should share this information cooperatively. A few scientists have tried to seek a middle ground by allowing certain types of patent rights. They argue that in the absence of patent rights that will protect their profits, biotechnology corporations are unlikely to cooperate in the Human Genome Project (or similar projects) and will instead withhold any information that they may have, requiring other researchers to repeat the research already done.

The medical uses of genetic information are many, especially in the cases that involve genetically controlled diseases. Most obvious is the repeated experience that identification of the gene and its product usually leads to a better understanding of the mechanism that produces the disease, and this understanding may lead (within a few decades) to a treatment or a cure, or both. The possibility of using recombinant DNA techniques to repair human cells raises new ethical dilemmas and gives added importance to the search for a total map of the human genome.

THOUGHT QUESTIONS

1. To what extent do you agree with Watson's statement that sequencing the human genome will

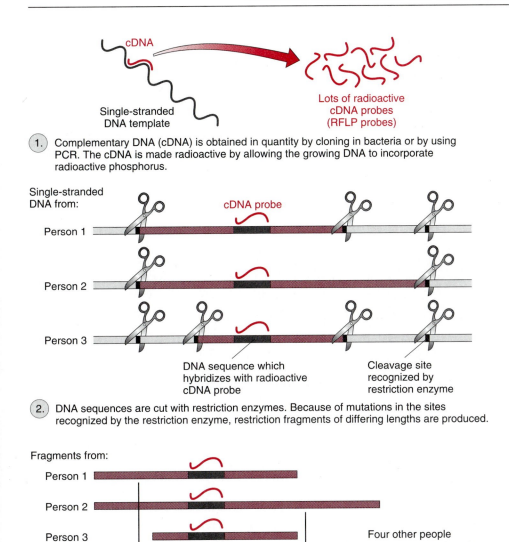

1. Complementary DNA (cDNA) is obtained in quantity by cloning in bacteria or by using PCR. The cDNA is made radioactive by allowing the growing DNA to incorporate radioactive phosphorus.

2. DNA sequences are cut with restriction enzymes. Because of mutations in the sites recognized by the restriction enzyme, restriction fragments of differing lengths are produced.

3. Radioactively labeled fragments are allowed to migrate through an electric field (electrophoresis). The shorter fragments travel faster down the electrophoresis gel than the longer fragments because the gel retards the movement of the longer fragments more than it retards the movement of the shorter fragments. The fragments therefore separate (or sort) according to their size, and each band in the resulting electrophoresis pattern represents a restriction fragment of different length. The bands are detected by means of the radioactive probes.

FIGURE 3.25
Restriction fragment length polymorphisms, or RFLPs.

tell us what it means to be human? Suppose you knew the exact gene sequence of part or all of your genome—what would you really know about yourself?

2. Will the DNA sequence of the human genome tell us what traits are controlled by each part of the sequence? Will it tell us which sequences represent genes and which sequences represent spacers?

3. If you have a certain rare genetic condition, and scientists use cell samples from your body to determine the gene's DNA sequence, what rights (if any) does this give you to the information? Do the scientists have the right to publish your gene sequence, or any part of it? Is it an invasion of your privacy? Can the scientists sell the information? If they do, are you entitled to a share of the profits?

CHAPTER SUMMARY

Hereditary information is carried in the form of DNA segments known as genes, which are arranged on chromosomes. Most human cells have $2N = 46$ chromosomes arranged in 23 pairs, except for the sex cells (gametes) that have only one chromosome of each pair. The sex chromosomes are *XX* in most females and *XY* in most males. Genes function by making proteins, many of which function as enzymes. Many genetic diseases result from alleles that make nonfunctional proteins. Genetic diseases can be diagnosed by amniocentesis, by chorionic villus sampling, and by other techniques, including those that use the polymerase chain reaction to amplify DNA sequences into many copies. For some genetic diseases, an enzyme product can be supplied artificially, or some substitute or mechanical compensation can be engineered. Genetic engineering consists of inserting functional genes into cells. The recipient cells may be bacterial cells that may then acquire the ability to make certain human proteins, or they may be human cells that are injected into a patient in the hopes of curing a disease. Altering the gene pool at the population level is called eugenics. Restriction enzymes have been used to break apart DNA into fragments; variations in the lengths of these fragments are called restriction fragment length polymorphisms, or RFLPs. RFLPs have helped in the localization of many genes. Human genes are now being identified at an increasingly rapid pace, and a project has been started to map the entire human genome.

KEY TERMS TO KNOW

alleles (p. 38)
amniocentesis (p. 67)
anticodon (p. 53)
artificial insemination (p. 76)
bacteriophage (p. 46)
chorionic villus sampling (p. 67)
chromosomes (p. 40)
clone (p. 73)
codon (p. 53)
crossing-over (p. 44)
cytoplasm (p. 40)
deoxyribonucleic acid (DNA) (p. 44, 46)
diploid (p. 40)
diploid number (p. 41)

dominant (p. 35)
enzymes (p. 46)
eugenics (p. 71)
euphenics (p. 73)
eupsychics (p. 74)
euthenics (p. 74)
frame-shift mutations (p. 54)
gametes (p. 38)
gene (p. 34–35)
gene therapy (p. 80)
genetic engineering (p. 77)
genetics (p. 34)
genome (p. 83)
genotype (p. 38)
haploid (p. 40)

haploid number (p. 41)
heterozygous (p. 38)
homozygous (p. 38)
inborn error of metabolism (p. 58)
independent assortment, law of (p. 40)
informed consent (p. 68)
in vitro fertilization (p. 76)
karyotype (p. 60)
linkage (p. 44)
locus (p. 49)
marker (p. 83)
meiosis (p. 42–43)
messenger RNA (mRNA) (p. 52)

mitosis (p. 40–42)
mutations (p. 53)
nucleic acids (p. 46)
nucleotide (p. 49)
nucleus (p. 40)
pedigrees (p. 57, 66)
phenocopy (p. 73)
phenotype (p. 38)
phenylketonuria (PKU) (p. 59)
plasmids (p. 78)
point mutation (p. 54)
polymerase chain reaction (PCR) (p. 68)
recessive (p. 38)
recombinant DNA therapy (p. 71)

replication (p. 49)

restriction enzyme (p. 77)

restriction fragments (p. 83)

RFLPs (p. 83)

ribonucleic acid (RNA) (p. 46)

segregation, law of (p. 38)

sex-linked (p. 61)

sticky ends (p. 77)

surrogate pregnancy (p. 76)

transcription (p. 52)

transfer RNA (tRNA) (p. 53)

transformation (p. 46)

translation (p. 53)

x-ray diffraction (p. 50)

CONNECTIONS TO OTHER CHAPTERS

Chapter 1   Mendel gave genetics its first paradigm; the structure of DNA is the basis for the current paradigm.

Chapter 2   Manipulating human heredity has raised several ethical concerns.

Chapter 3   Mutations and other genetic changes supply the raw material for evolutionary change.

Chapter 4   Evolution takes place whenever gene frequencies in populations change.

Chapter 5   Human populations differ in the frequencies of many alleles.

Chapter 6   Many reproductive technologies offer new options in both human genetics and population control.

Chapter 7   Mating patterns and sexual strategies can alter gene frequencies.

Chapter 9   You can inherit predispositions for many cancers. Retinoblastoma is a cancer caused by a genetic defect.

Chapter 10   Huntington's disease is one of several brain disorders for which a gene has been identified.

Chapter 14   Many bioengineering applications are examples of euthenics.

Chapter 15   Genetic engineering can be used to improve the traits of domesticated plant species.

Chapter 16   Conserving genetic diversity is an important aspect of protecting biodiversity.

Evolution

Ask any biologist to name the most important unifying concepts in all of biology, and the theory of evolution is likely to be high on the list. The world-renowned geneticist Theodosius Dobzhansky once declared, "Nothing in biology makes sense, except in the light of evolution." The United States population at large, however, is remarkably unaware of the importance of evolution as a unifying concept: Public opinion surveys reveal that 25 percent to 40 percent of Americans either do not believe in evolution or consider the matter unresolved. (The percentage varies depending on how the question is worded.) In this chapter we will examine both the theory of evolution and the opposition to it.

As we explained in Chapter 1, scientists use the word *theory* to designate a coherent cluster of hypotheses that have in most cases withstood many years of testing. In this sense, evolution is a thoroughly tested theory that has withstood over a century of rigorous testing. Scientific evidence for evolution is as abundant as and considerably more varied than the evidence for nearly any other scientific idea. To refer to evolution as "just a theory" is thus a grave misunderstanding of scientific theories in general and evolutionary theory in particular. When physicists speak of the *atomic theory* or the *theory of relativity*, or when medical professionals speak of the *germ theory of disease*, they are speaking of the greatest unifying principles of their respective professions, principles that have withstood repeated testing for somewhat

fewer years than has the theory of evolution. Educated people no longer doubt the existence of atoms or of germs, and nobody refers to any of these concepts as "just a theory." In the way that the atomic theory is a unifying principle for much of physics and chemistry, the theory of evolution is a unifying principle for all biological sciences.

Biology attempts to answer two great questions. In physiology, biochemistry, cell biology, molecular biology, and genetics, the overriding question is usually some version of "How does it work?" However, biologists also often ask, "Why that way instead of some other way?" or "How did it come to be that way and not the other way?" As Dobzhansky's remark above properly suggests, questions of this type require explanations related to the past history—including the evolutionary history—of the species involved.

A. The Darwinian Paradigm Reorganized Biological Thought

The most influential biology book of all time was published in 1859. *On the Origin of Species by Means of Natural Selection*, written by Charles Darwin (1809–1882), contained at least two major hypotheses and numerous smaller ones, along with an array of evidence that Darwin had already used to test these hypotheses. The first major hypothesis, *branching descent,* was that present species are descended from species that lived earlier in time and that the lines of descent form a branched pattern resembling a tree (a favorite analogy). Similarities among groups of related species were explained under this hypothesis as resulting from common inheritance. The second major hypothesis was that much of the change has occurred with the help of *natural selection,* the consistently unequal contribution of different genetic forms (genotypes) to each succeeding generation. Both of these hypotheses are falsifiable, and have been tested hundreds if not thousands of times, without being falsified, since Darwin first proposed them in 1859. Darwin also suggested how his hypotheses made sense of several previously noticed but unexplained regularities in anatomy, classification, and geographic distribution (as we will explain), and he suggested what further regularities might be found. As both a unifying theory and a stimulus to further research, Darwin's *Origin of Species* fits the concept of a scientific paradigm expounded by Thomas Kuhn and explained in Chapter 1 of this book. Modern evolutionary thought is still largely based on the Darwinian paradigm, expanded to include the findings of genetics.

Pre-Darwinian Thought

Darwin's evolutionary theory was not the first. An earlier theory had been proposed by the French zoologist Jean-Baptiste Lamarck in 1809. Lamarck's theory was a modification of an earlier nonevolutionary concept, that of a **scale of being** (also variously called *chain of being, échelle des êtres,* or *scala naturae*).

The scale of being.   The scale of being was a linear hierarchy in which each species had its assigned place. Originally conceived as a never-changing order, the scale of being was as much a theological and social concept as it was biological. God had placed man above the animals and below the angels, plants below animals, and so forth. In the conventional sexism and class bias of the time, the scale of being was used to argue that God had placed men above women and the ruling classes over the lower classes. As Europeans discovered the nonwhite races of the world, each was assigned its place in the scale. The scale of being was thus an important social concept, one that was used to justify every sort of social inequality as part of the "natural" order of things, and everyone was expected to stay in their own proper place.

As a theological concept, the scale of being was conceived as a continuum, celebrated in verse by Alexander Pope:

Vast chain of being, which from God began,
Natures aetherial, human, angel, man,
Beast, bird, fish, insect! what no eye can see,
No glass [microscope] can reach! from Infinite to thee,
From thee to Nothing!—On superior pow'rs
Were we to press, inferior might on ours:
Or in the full creation leave a void,
Where, one step broken, the great scale's destroy'd:
From Nature's chain, whatever link you strike,
Tenth or ten thousandth, breaks the chain alike.

Since the continuum was created by God, any challenge to the unbroken chain was a challenge to the established religion. In biology, the scale of being became a unifying theory describing nature and predicting that each newly discovered species would somehow fit into its assigned place in the continuum, filling one of the "missing links" in the chain. In fact, the scale of being was the single most important unifying concept in biology from the time of Aristotle into the early part of the nineteenth century. It was a great stimulus to exploration and research by those seeking to fill in the missing links. The terms *higher* and *lower,* still occasionally heard in biological discussions, are really references to this preevolutionary concept.

Lamarckism.   Lamarck converted the scale of being from a static scale into a moving escalator. As each species ascended the scale, Lamarck insisted that the continuum was maintained because all the other species ascended in tandem. Spontaneous generation, said Lamarck, was continually taking place to fill in the lowest level. Extinction, said Lamarck, was impossible because the continuity of the scale had to be maintained. To the simultaneous ascent of the scale of being by all of life's forms, Lamarck gave the name *la marche de la nature* ("the parade of nature"). Lamarck's major hypothesis, like several other hypotheses proposed later, was thus a hypothesis of *unilineal evolution* in a single line of ascent that would later be called "progress." Such unilineal theories of evolution are no longer accepted by biologists.

Lamarck became an expert on invertebrate animals (those without backbones) and tried mightily to place them in a single ascending scale, according to the chain of being, each group connected to the next through intermediates. He recognized, however, that the transitions in his scale were by no means smooth. In order to explain these deviations from his grand scheme, Lamarck supplemented his grand hypothesis of a unilineal evolution with a second evolutionary hypothesis: local adaptation to the environment, a force which several of his contemporaries also recognized. An **adaptation** is any feature that enables a species to survive under circumstances in which it could not survive (or could not survive as well) without the adaptation. To explain adaptation, Lamarck further hypothesized a mechanism of "use and disuse," by which organs that were often used would strengthen and enlarge, while those that remained unused would atrophy and might eventually be lost altogether. Physiological changes of this kind were already known in Lamarck's time, and Lamarck adopted an incorrect folk belief that these **acquired characteristics** would be inherited.

Some 80 years later, an Austrian zoologist named August Weismann demonstrated experimentally that acquired characteristics were not inherited. Weismann raised twenty-two generations of mice, measuring all their tails before he cut them off. Despite the fact that the mice were tailless when they reproduced, Weismann found no decrease in tail length over twenty-two generations. He concluded that Lamarck's evolutionary mechanism would never work because acquired characteristics were not inherited. Weismann's refutation of the inheritance of acquired characteristics was devastating to Lamarckism. Except for the importance of behavior in evolution (see Chapter 7), Lamarck's ideas have all been falsified and are now rejected by most biologists. It is perhaps ironic that Lamarck is best remembered nowadays not for his major hypothesis of a unilineal ascent of the chain of being, or even for his secondary hypothesis of local adaptation through use and disuse, but for the inheritance of acquired charac-

teristics, which was only an assumption that he used to explain his secondary hypothesis.

Other environmentalist theories. During Lamarck's lifetime, his secondary hypothesis of local adaptation through use and disuse was joined by several rival explanations of environmental adaptation. Most of these environmentalist theories were French in origin, and most of them agreed with Lamarck in attributing to species almost limitless capacity for adapting to their environments by a variety of mechanisms. Isidore Geoffroy, for example, hypothesized that the environment could directly induce physiological changes in living organisms. Geoffroy's explanation, like Lamarck's, relied on the assumption (falsified by Weismann's experiment) that acquired characteristics would have to be inherited. Georges Buffon proposed an alternative hypothesis, that organisms simply migrated to wherever the environment suited their needs. All of these theories of environmental determinism viewed the adaptations of living organisms as completely plastic, capable of being perfectly matched or molded to their environments, without any restrictions. As we will see, Darwinian explanations proved superior to all such theories of environmental determinism when it became clear that adaptation was often imperfect and could only take place within certain limits.

Paley's natural theology. While Lamarck and the other environmental determinists struggled to explain adaptation to the environment by a variety of mechanisms, British naturalists were examining adaptations from a different perspective. A movement led by the Reverend William Paley sought to prove the existence of God by examining the natural world for evidence of perfection. By careful examination and description, British scientists were discovering case after case of various organisms with anatomical structures so well constructed, so harmoniously combined with one another, and so well suited in every detail to the functions that they served that one could only marvel at the degree of perfection achieved. Such harmony, design, and detail, they argued, could only have come from God. Paley himself went so far as to offer well-planned adaptation as proof of God's existence: "The marks of *design* are too strong to be gotten over. Design must have a designer. That designer must have been a person. That person is God." In a nation in which many clergymen were also amateur scientists, it became quite the thing to dissect organisms down to the smallest detail, all the better to marvel at the wondrously detailed perfection of God's design. A large series of intricate and sometimes amazing adaptations were thus described, which Darwin would later use as examples to argue for an evolutionary explanation based on natural selection.

The Development of Darwin's Ideas

South America. From 1831 to 1836, Charles Darwin was a passenger aboard H.M.S. *Beagle,* a ship that traveled to many parts of South America and then around the globe. His observations in South America convinced him that the animals and plants of that continent were vastly different from those inhabiting comparable latitudes or comparable environments in Africa or Australia. For example, all South American rodents were relatives of the guinea pig and chinchilla, a group found on no other continent. South America also had its llamas, anteaters, monkeys, parrots, and numerous other groups of animals, each with many species inhabiting different environments throughout the continent, but different from comparable species elsewhere. This was definitely not what Darwin had expected! Environmental theories similar to Lamarck's had led Darwin to expect that the forests of South America and of Africa would have many of the same inhabitants, as would the deserts of the two continents, the mountain regions, the plains, and so forth. What he found instead was that most of the inhabitants of South America had close relatives living elsewhere on the continent under strikingly different climatic conditions, but they showed no relationship to the inhabitants of Africa or Australia, even when the climates they inhabited were similar (Fig. 4.1). The Andes, for instance, had rodents related to those of the South American rainforests and the Pampas of Argentina. Moreover, Darwin observed, "If we look to the islands off the American shore, however much they may differ in geological structure, the inhabitants, though they may all be peculiar species, are essentially American." When Darwin discovered South American fossils, they belonged

to many of the same groups still inhabiting that continent. "We see in these facts some deep organic bond, prevailing throughout space and time, over the same areas of land and water, and independent of their physical conditions. The naturalist must feel little curiosity, who is not led to inquire what this bond is. This bond, on my theory, is simply inheritance, that cause which alone, as far as we positively know, produces organisms quite like, or . . . nearly like each other."

The Galapagos Islands. Darwin's visit to the Galapagos Islands proved especially enlightening. Here, on a series of small volcanic islands in the Pacific, a very restricted assortment of animals greeted him. No native mammals or amphibians were present; there were, instead, several species of large tortoises and a species of crab-eating lizards. Most striking were the land birds, now often known as "Darwin's finches," which constituted a cluster of more than a dozen closely related species that differed from one island to the next (Fig. 4.2) but which all belonged to a single subfamily with South American relationships. The tortoises also differed from island to island, despite the clear similarities of climate throughout the archipelago. Each species cluster, Darwin hypothesized, had arisen as a series of modifications from a single species that had originally colonized the islands. The islands, Darwin noted, were very similar to the equally volcanic and equally tropical Cape Verde Islands in the Atlantic, which Darwin had also visited, but the inhabitants were altogether different. Darwin concluded that the Galapagos had received its colonizing species (including the finches) from South America, while the Cape Verde Islands had received theirs from Africa, and that each group of colonists had given rise to a cluster of related species. Darwin was the first evolutionary theorist to emphasize that clusters of related species indicated a branching pattern of descent, a pattern that Darwin called "descent with modification."

Patterns of distribution. Darwin continued to find examples in which different continents were inhabited by their own large clusters of species, unrelated to those of other land masses. The mammals of Australia, for example, were mostly marsupials, pouched mammals whose babies are

FIGURE 4.1
An assortment of South American mammals.
All of the species shown have relatives throughout the continent in different habitats, but few have any close relatives on other continents, even where climates are similar.

born in a fetal stage and complete their gestation in the mother's pouch, unlike the typical placental mammals of other continents, which retain their fetuses inside the mother's womb throughout gestation. Several land areas had large, flightless birds, but they differed strikingly from one continent or island to the next: rheas in South America, kiwis and extinct moas in New Zealand, emus and cassowaries in Australia, extinct elephant birds on Madagascar, and ostriches in Africa. (Darwin observed most of the living species in his travels, and

he was shown the bones of the extinct species.) The climates inhabited by these birds were all similar, but each land mass had produced its own distinct type. Theories of environmental adaptation (such as Lamarck's) could not explain these differences, nor could theories of divine creation explain why God had seen fit to create half a dozen distinct types of flightless birds where one might have sufficed. (Creationist theories at the time generally assumed that God would populate each environment with the best possible species for that

FIGURE 4.2
Some of Darwin's finches from the Galapagos Islands.

environment, and thus similar environments would have similar species.)

Natural Selection

Darwin's search for a mechanism. When Darwin returned to England, he began reading all he could on the ways in which species could be modi-

fied. How, he wanted to know, could a single colonizing species produce a whole cluster of related species on a group of islands? During the preceding hundred years, Darwin observed, British animal breeders had produced many new varieties of dogs, sheep, and pigeons, and had greatly improved wool yields in sheep and milk yields in cattle by careful breeding practices. By methodically selecting the best individuals in each generation (that is, the ones with the most desired traits) and breeding these preferentially, British animal breeders had brought about great changes in a number of domestic species through a process that Darwin called **artificial selection.** This historically documented process simply took advantage of the natural variation that was present in each species, yet it produced breeds that were strikingly different from their wild progenitors. Darwin remarked that some of the domestic varieties of pigeons or dogs differed from one another as much as natural species or even genera, despite the fact they had been produced within historical time from a known group of common ancestors. Could a similar process be at work in nature?

At about this same time, Darwin read Malthus's *Essay on Population* (see Chapter 6). Malthus emphasized that, in the natural world, each species reproduces much more than is necessary to maintain its numbers. The result of this overpopulation is the death and destruction of many individuals in each generation, and the survival of relatively few. When Darwin compared this process to the actions of the animal breeders, he concluded that nature was slowly bringing about change in each and every species. Since every species varied, those individuals that died in each generation would, in general, differ from those that survived to maturity and mated to produce the next generation. In this "struggle for existence," Darwin hypothesized,

. . . individuals having any advantage, however slight, over others, would have the best chance of surviving and of procreating their kind. . . . On the other hand, we may feel sure that any variation in the least degree injurious would be rigidly destroyed. This preservation of favourable variations and the rejection of injurious variations, I call Natural Selection. . . .

Natural selection . . . is a process incessantly ready for action, and is as immeasurably superior to man's feeble efforts, as the works of Nature are to those of Art.

Darwin's extended comparison between natural selection and artificial selection emphasized how much the animal breeders had already done, and how nature had the capacity to do so much more:

. . . As man can produce and certainly has produced a great result by his methodical and unconscious means of selection, what may not nature effect? . . . Man selects only for his own good; Nature only for that of the being which she tends. . . . How fleeting are the wishes and efforts of man! how short his time! and consequently how poor will his products be, compared with those accumulated by nature during whole geological periods. Can we wonder, then, that nature's productions should be far "truer" in character than man's productions; that they should be infinitely better adapted to the most complex conditions of life, and should plainly bear the stamp of far higher workmanship?

The "far higher workmanship" that Darwin was referring to was a pointed reference to Paley's ideas, but Darwin attributed this higher workmanship to the process of natural selection.

Natural selection defined.   All modern descriptions of natural selection are stated in terms of the concepts of genetics outlined in Chapter 3. New genotypes originate by mutation and recombination, both of which act prior to any selection. Darwin, of course, knew nothing of mutations or of modern genetics, but he did realize that heritable variation had to come first and that "any variation which is not heritable is unimportant to us."

Natural selection may be defined as consistent differences in what Darwin called "success in leaving progeny," meaning the contributions of different genotypes to future generations. The relative number of viable individuals that each genotype contributes to the next generation is called its **fitness.** *Natural selection will favor any trait that increases fitness*, while weeding out any trait that lowers fitness. Darwin's theory of natural selection is the basis for all modern explanations of adaptation.

Mimicry.   In the years since Darwin first proposed the hypothesis of natural selection as an evolutionary mechanism, many tests of the hypothesis have been conducted. One of the earliest tests involved the phenomenon of **mimicry,** in which one species of organism deceptively resembles another. The example of mimicry most familiar to North American readers involves the monarch (Fig. 4.3C), a distasteful species of butterfly that feeds on milkweed plants, and the viceroy (Fig. 4.3B), a palatable species that is unrelated to the monarch but is similar in superficial appearance. (See Color Fig. A, following Chapter 16.) In this type of mimicry

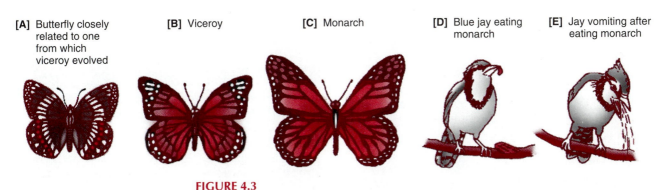

[A] Butterfly closely related to one from which viceroy evolved   **[B]** Viceroy   **[C]** Monarch   **[D]** Blue jay eating monarch   **[E]** Jay vomiting after eating monarch

FIGURE 4.3
Warning coloration and Batesian mimicry. [A] *Limenitis arthemis,* a nonmimetic relative of [B] *Limenitis archippus,* the viceroy. The viceroy resembles the unrelated monarch butterfly ([C], *Danaus plexippus*), which is avoided by predators following just a single unpleasant experience ([D], [E]). (See also Color Figure A.)

(called *Batesian mimicry* after Henry W. Bates, who discovered many instances in tropical South America), a distasteful or dangerous species, called the *model*, gives a very unpleasant and memorable experience to any predator that attempts to eat it. Predators always avoid the model following such an unpleasant experience (Figs. 4.3D and E), and a palatable species, the *mimic*, secures an advantage if it resembles the model enough to fool predators into avoiding it as well. In the case of the two species of butterfly, the monarch is considered the model; the viceroy, its mimic, has evolved over time to resemble the monarch.

Selection by predators explains Batesian mimicry rather easily, because any slight resemblance that might cause a predator to avoid the mimic as well as its model would be favored by selection and passed on to future generations, while individuals not protected in this way would be eaten in greater numbers. Any advantage that increased the number of predators fooled would be favored by selection, so we should expect closer and closer resemblance to evolve with the passage of time.

Batesian mimicry often varies geographically (Color Fig. B at end of book). In South America, Bates discovered several cases in which one mimic species mimicked different model species in different geographic areas. Whenever Bates discovered a species that deceptively mimicked another, the deceptive resemblance was always to a species that occurred in the same area, never to a faraway species. Most strikingly, some cases of mimicry involved only one sex, which varied geographically from place to place according to the models present in each locality, while the other sex retained the same appearance everywhere. Environmentalist theories such as Lamarck's had no way to account for the evolution of mimicry, and the patterns of geographic variation could not be explained by either environmentalist theories or by Paley's natural theology. Natural selection, however, could explain the variation as resulting from predator selection.

Another common type of mimicry is *Müllerian mimicry,* in which several distasteful or harmful species resemble one another. Predators will learn to avoid distasteful species, but a certain number of prey individuals are killed for each predator individual that learns its lesson. Without mimicry, each prey species must sustain this loss separately. Müllerian mimicry allows all the species that resemble each other to benefit by being avoided; fewer individuals of each species must die in order for predators to learn the lesson that all should be avoided.

Industrial melanism.   The twentieth century has given further demonstrations of the power of natural selection. In the British Isles, a species known as the peppered moth (*Biston betularia*) had long been recognized by an overall light gray coloration with a salt-and-pepper pattern of irregular spots (Fig. 4.4). Beginning in the 1890s, black varieties of the same moth species were first discovered, and these increased in numbers until they came to outnumber the original forms in some localities.

[A]

[B]

[C]

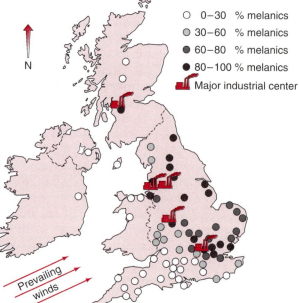

FIGURE 4.4
Industrial melanism among peppered moths in Great Britain. [A] The melanic (black) variety and the original "peppered" variety (below the right wing tip of the melanic moth) on a light, lichen-covered tree trunk. [B] The same two varieties on a dark, soot-covered tree trunk. [C] Geographic variation in the frequency of melanic moths in the 1950s, which reached as high as 100% in polluted localities downwind from major industrial centers.

E. B. Ford and H. B. D. Kettlewell studied the geographic variation in these moths. Downwind from the major industrial areas, the woods had become polluted with a black soot that killed the lichens growing on the tree trunks. The moths living on the darkened tree trunks in these regions were mostly black. However, where the woods were unaffected by pollution, the tree trunks were still covered with lichens and the moths kept their original color pattern. Ford and Kettlewell suggested as a hypothesis that the moths that resembled the color of their background would be camouflaged and thus harder for predators to recognize. To test this hypothesis, they pinned both light and dark moths on dark tree trunks in polluted woods, and both again on lichen-covered tree trunks in unpolluted woods. They observed that the local birds ate a higher proportion of the dark moths in the unpolluted woods (favoring the survival of the original salt-and-pepper pattern), while birds in the polluted woods ate more of the light-colored moths, sparing the dark ones. These experiments, and the geographical patterns of variation (Fig. 4.4C), were easily explained in terms of natural selection by predators. In addition, since the experiments were first conducted, laws to control smokestack emissions and other forms of pollution have been passed and enforced, and many of the woods affected by pollution have returned to their former state. In these woods, the lichens have returned to the tree trunks, and the moths have returned to their original color pattern.

Agents of selection.   In the preceding examples of natural selection, the selecting agents have been predators. There are many other kinds of selecting agents operating in nature. Deaths caused by diseases may be less obvious than those caused by predators, though they may be a good deal more frequent. In many cases, a disease or invasion by a parasitic species may not be fatal by itself but may cause death indirectly by weakening the host organism so greatly as to render it susceptible to death by another cause. Another large class of selecting agents includes climate- and weather-related extremes, including storms, droughts, floods, periods of extreme cold, and so forth. Other selecting agents are falls and other accidents, starvation, and rare events such as landslides and earthquakes.

Another type of selection is called **sexual selection,** which may be defined as selection on the basis of success in attracting a mate and reproducing. For example, animals of many species attract their mates with mating calls (like bird songs), visual displays (as in peacocks, Fig. 7.2 on p. 195), or olfactory displays (as in silkworm moths or many other invertebrates). Individuals that do not perform these tasks sufficiently well to please their potential mates may live long lives but leave few or no offspring.

Descent with Modification

Geographical occurrence of species.   Darwin's concept of descent with modification has been used to make sense out of a variety of observations not easily explained by other means. Darwin used this explanation to account for the species clusters found on the Galapagos Islands and elsewhere. In many instances, Darwin recognized that the species clusters found on islands had their closest relatives on the nearest continent, not on geologically similar or climatically similar but distant islands. Geographic proximity, in other words, was often more important than climate or other environmental variables in determining which species occurred in a particular place.

Earlier scientists, wrote Darwin, "will have to admit, that a sufficient number of the best adapted plants and animals have not been created on oceanic islands; for man has stocked them [with introduced species] . . . far more fully and perfectly than has nature," even though an omnipotent God would not be limited by great distances or by geographical barriers. Darwin made a particular point about the geographic distribution of frogs, salamanders, and other amphibians, which never occur on truly oceanic islands (those beyond the margins of the continental shelf, usually over 150 kilometers, or 100 miles, away from the nearest larger landmass):

The general absence of frogs, toads, and newts on so many oceanic islands cannot be accounted for by their physical conditions; indeed it seems that islands are particularly well fitted for these animals; for frogs have been introduced into Madeida, the Azores, and Mauritius, and have multiplied so as to become a nuisance. As these animals and their spawn are known to be immediately killed by sea-water, on my view we can see that there would be great difficulty in their transportal across the sea, and therefore why they do not exist on any oceanic island. But why, on the theory of creation, they should not have been created there, it would be very difficult to explain.

Evolutionary classification.   At least a century before Darwin, biological classifications had already taken their familiar, modern form (described later in this chapter) of smaller groups within successively larger groups. Darwin explained this pattern as the natural result of descent with modification, a process that results in a series of treelike branchings in which species correspond to the finest twigs, groups of species to the branches from which these twigs arise, larger groups to larger branches, and so forth. Darwin predicted that classifications would increasingly become genealogies (that is, maps of descent) as more and more details about the evolution of each group of organisms became known. As an example, Darwin wrote a two-volume treatise on barnacles, in which he put all fossil and living barnacles into a classification based on genealogical principles.

Homologies.   The construction of family trees is based in large measure on the study of shared structures or of common gene sequences. Structures or gene sequences that are shared within a group of organisms are called **homologies.** Under Darwin's paradigm, homologies are important as evidence that the organisms in question share a common ancestry. In another sense, a homology is a falsifiable hypothesis about which species are related to one another by descent. By itself, one homology reveals very little, but a large number of such homologies must fit together into a consistent pattern. If one resemblance is inconsistent with the rest (by uniting species that are dissimilar in other respects), then the hypothesis of common descent for the species sharing the resemblance is falsified, and the resemblance can no longer be considered a homology. As more and more homologies are fit into a consistent framework, confidence grows in the family tree of relationships based on these homologies. Fossils provide information about species from the geologic past that allows scientists to fit those species into the family tree

somewhere. In this way, every fossil discovery adds a detail that may support some hypotheses while it falsifies others.

Darwin noted the principle of homology among the forelimbs of mammals: "What can be more curious than that the hand of a man, formed for grasping, that of a mole for digging, the leg of the horse, the paddle of the porpoise, and the wing of the bat, should all be constructed on the same pattern, and should include the same bones, in the same relative positions?" (See Fig. 4.5.) Preevolutionary German naturalists had described such similarities in terms of a common *Bauplan,* or structural plan, but the Darwinian paradigm reinterpreted such homologies as the result of common descent. Likewise, Darwin noted that repeated structures in the same organism can also exhibit a form of homology if they were evolved as modifications of a single basic pattern. Goethe, the German poet and playwright, also noted this pattern: He interpreted petals, sepals, stamens, ovaries, and other flower parts as modified leaves. Darwin wondered why similar bone structures appeared in the wings and legs of a bat, used as they are for such totally different purposes. Why should crustacean species with many mouthparts have correspondingly fewer legs, and why should crustaceans with more legs have fewer mouthparts? Darwin's answer was that all these structures arose as homologies by modification of the same basic type of repeated part. Crustaceans, for example, have evolved their mouthparts, legs, and certain other structures from a set of common leglike appendages (similar to those found among extinct trilobites) that were modified by natural selection according to how they were used. Thus, animals that evolved more mouthparts, because more appendages were used for mouth-related activities, had fewer appendages to be used as legs. An omnipotent God, however, would be subject to no such limitation, leaving Darwin to declare, "How inexplicable are these facts on the ordinary view of creation!"

Darwin's evolutionary paradigm brought with it evolutionary explanations of previously described phenomena, such as adaptations and homologies. Remember that an *adaptation* is any feature that enables a species to survive where it might not survive otherwise. Adaptations were favored by natural selection, said Darwin, and they

[A] Human **[B]** Cheetah **[C]** Whale **[D]** Bat

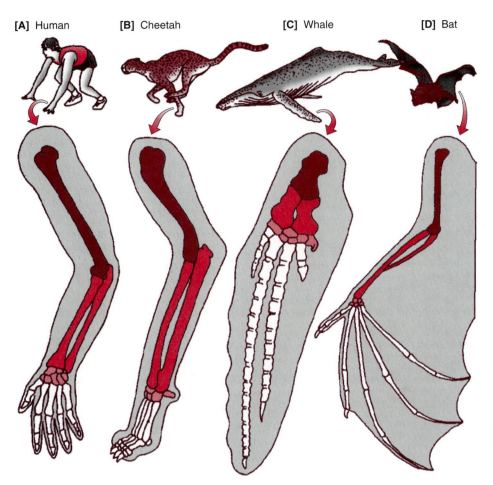

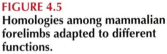

FIGURE 4.5
Homologies among mammalian forelimbs adapted to different functions.

persisted because they permitted survival and reproduction while the nonadaptive forms died out. Darwin's explanation stimulated research into finding out how each adaptation contributed to fitness, either in terms of survival or reproduction, or both. Homologies, which were similarities among differently adapted but related organisms, were presumed to be inherited from common ancestors. Textbooks of the period not only presented evidence to support the paradigm, they also showed students (by way of example) how the evolutionary paradigm allowed this or that shared similarity to be interpreted as a homology. The organisms sharing a homology were then declared to be a *natural group,* and these natural groups, in turn, were depicted on family trees as branches, and in classifications as orders or classes, each with its own scientific name. The textbooks served as training manuals for the next generation of young scientists, who would examine *other* similarities among organisms and show how most could be interpreted as homologies. Each homology was then used to help construct new family trees and new classifications and was hailed as further evidence to support the underlying evolutionary paradigm.

Vestigial structures.   Organs whose function has been lost in one branch of a family tree may be present nevertheless as functionless vestiges. Such **vestigial** organs exhibit homology in the similarity of their position, their attachment to nearby structures, and so forth. A good human example is the *coccyx,* a set of two or three vestigial tail bones at the base of the spinal column; another human example is the *caecum,* or blind-ended beginning portion of the large intestine, including the tiny *appendix* (vermiform appendix). The Darwinian paradigm explains these structures as the remnants of formerly useful organs. Neither Lamarck nor the creationists had any explanation for the presence of functionless vestiges, and certainly not for the fact that many such structures exhibit resemblance by homology to the larger, functional organs of related species.

Convergence. In cases in which two sets of supposed homologies show inconsistent patterns, evolutionists reexamine all the similarities more closely to see if a reinterpretation is possible for one set of resemblances. Similarities that result from common ancestry (that is, true homologies) should also be similar at a smaller level of detail, and they should be similar in embryological derivation as well. Resemblances that are falsified by more careful scrutiny are often reinterpreted as cases of **convergence,** or adaptations evolved independently in unrelated lineages. Distinguishing homology from convergent resemblance (also called **analogy**) is therefore one of the recurrent aims of evolutionary classification. The wings of bats, birds, and insects are similar only by analogy: They are constructed in different ways and of different materials, and their common shapes (which they also share with airplane wings) reflect adaptation to the aerodynamic requirements of flying. Bat wings (and bird wings to a lesser extent) are homologous to human arms, whale flippers, and the front legs of horses and elephants—all have similar bones, muscles, and other parts in similar positions *despite* their very different shapes and uses.

Evolution and embryology. In many cases, homologies are easier to recognize when earlier embryological stages are examined; this is especially true for highly modified organs. The Darwinian paradigm thus stimulated a good deal of research in embryology. In addition, the late nineteenth-century theory of **recapitulation,** championed by Darwin's greatest German supporter, Ernst Haeckel, held that embryos would usually look like ancestral evolutionary stages. Haeckel believed that the sequence of embryonic through adult stages in development were a brief summary (a "recapitulation") of the sequence of

ancestors. Though this hypothesis was eventually rejected, it did stimulate a great deal of research in embryology.

Evolution of complex structures. Paley and his supporters had paid much attention to complex organs such as the human eye. The eye, they pointed out, was composed of many parts, each exquisitely fashioned to match the characteristics of the other parts. What use would the lens be without the retina, or the retina without a transparent cornea? An eye, they argued, would be of no use until all its parts were present, thus it could never have evolved in a series of small steps, but must have been created, all at once, by God. To counter this sort of argument, Darwin pointed out that eyes among the invertebrates can be arranged into a series of gradations, ranging in complexity from "an optic nerve merely coated with pigment" to the most elaborate structures of squids, approaching those of vertebrates in form and complexity. A large range of variation in the complexity of visual structures could be found within a single group, the Arthropoda, the group that includes barnacles, shrimp, crabs, spiders, insects, and their many relatives. All degrees of complexity were fully functional adaptations, advantageous to their possessors. It would therefore be quite reasonable, argued Darwin, to imagine each stage to have evolved from one of the structurally simpler conditions found among related animals. Eyes, in other words, could have evolved as a progression through a series of small gradations.

The explanatory power of branching descent. Tests of the hypothesis of branching descent have taken several other forms. One type of test identifies a group of organisms that share some particular character, such as an anatomical peculiarity. The hypothesis of branching descent now gives rise to a more specific hypothesis, that this particular group of organisms share a common descent from a common ancestor. An example of this type of hypothesis testing is shown in Box 4.1.

In the same way that a hypothesis of common descent has been tested for the group of mollusks shown in Box 4.1, similar hypotheses have been tested by the hundreds for other families, orders, and classes of animals and plants. In general, branching patterns of descent can be hypothesized for each group of organisms, and at least one such family tree can always be reconciled with the observed data. This increases our confidence in the larger hypothesis, that all species of organisms have evolved from earlier species in patterns of branching descent.

Supporters of the evolutionary paradigm have always claimed that evolution is the scientific hypothesis with the greatest explanatory power, making sense out of many disparate observations. The many facts of comparative anatomy, comparative physiology, embryology, biogeography, and animal classification all make much more sense if we hypothesize that modern species have evolved from ancestors that lived in the remote past. This sweeping statement is really a collection of thousands of smaller hypotheses: The occurrence of so many species of ground finches on the Galapagos Islands (or honeycreepers in Hawaii, or canaries on the Canary Islands) can best be explained by hypothesizing that an ancestor arrived from the nearest continent at some time in the past and diversified there, filling a variety of available niches that would already have been filled by other organisms in other localities. The occurrence of teeth in the Jurassic bird *Archaeopteryx* (Fig. 4.6, p. 104) and the ability of hens' normally toothless jaws to produce teeth if a few critical genes are supplied to chick embryos from mice can best be explained by hypothesizing that birds evolved from toothed ancestors similar to certain reptiles. The patterns of relationships among families, orders, and classes of living organisms can best be explained by hypothesizing that every such group corresponds to the descendents of a common ancestor, giving the evolutionary explanations strong explanatory power. Since Charles Darwin proposed his theory of evolution in 1859, thousands of tests have been made of his twin hypotheses of branching descent and natural selection. Since these thousands of tests have failed to falsify either hypothesis, both now qualify as scientific theories that enjoy widespread support. The Darwinian paradigm continues to this day as a major guide to scientific research.

Fossils and the Fossil Record

The history of life on Earth is measured on a time scale encompassing millions, if not billions, of years. This *geological time scale* (Fig. 4.7, p. 105) was worked out, mostly in the nineteenth century by studying **fossils,** the remains and other evidence of past life forms.

Stratigraphy. The geological time scale was established first through *stratigraphy*, the study of layered rocks. One of the first observations in stratigraphy was the *principle of superposition*: When sequences have not been drastically disturbed, the oldest layers are on the bottom and successively newer layers are on top of them. Using this principle, it is possible to place several rock formations into a local sequence, with the time sequence running from bottom to top.

Local sequences can be matched with one another in several ways, but the most reliable of these is the study of their fossil contents. *Correlation by fossils* is a technique for judging that two rock formations are from the same time period (contemporaneous) because they contain many of the same fossil species. The rocks do not need to be similar in composition or rock type—one can be a limestone and the other can be a shale—but if their fossil assemblages are similar, they are judged to be contemporaneous. (Notice that a single species of fossils is never sufficient; a *collection* or *assemblage* of fossil species is needed.) Using this technique of correlation, nineteenth-century geologists and *paleontologists* (scientists who study fossils) were able to assemble the world's various local sequences into a "standard" worldwide sequence, as shown in Fig. 4.8, p. 106. Notice that each judgment of contemporaneity is really a falsifiable hypothesis: If a local bed sequence in one place is A, B, C (from the bottom up) and P, Q, R, S in another place, then judging beds A and Q to be contemporaneous would allow further matching of bed B with either R or S, but not with P. Assembling local sequences into a worldwide sequence must be done consistently, that is, such that inconsistencies (such as jumbled-up sequences) do not occur.

Relative and absolute dating. Using the worldwide sequence of geological formations to specify a time period for a rock bed or a fossil assemblage is an example of *relative dating*. For example, if we find a particular assemblage of fossil shells, we

BOX 4.1   THE EVOLUTION OF CEPHALOPOD MOLLUSKS

The class of mollusks known as the *Cephalopoda* includes the squids and their relatives. All cephalopods can be recognized by the presence of a well-developed head and a *mantle cavity* beneath it. The mantle cavity occurs in all mollusks and is recognizable by the fact that it contains the gills, the anus, and certain other anatomical structures. Only in the Cephalopoda, however, is the mantle cavity located beneath the head and prolonged into a nozzlelike opening known as the *hyponome*. The specific hypothesis now is that all cephalopods share a common descent. If this is true, we should be able to find some additional similarities among cephalopods that they do not necessarily share with other mollusks. This is indeed the case, for all cephalopods have a beak at the front of their mouth and a muscular part, called the *foot* in other mollusks, subdivided into a series of *tentacles*.

Moreover, all cephalopods have an *ink gland* that secretes a very dark, inky fluid. When a squid or octopus feels threatened by a predator, it quickly squirts this fluid into its mantle cavity and then squirts the contents of its mantle cavity forward through its nozzlelike hyponome. The squid or octopus is rapidly propelled backward in a direction not expected by its predator, while the predator's attention is held by the puff of black, inky fluid. By the time the inky fluid dissipates, the squid or octopus has vanished. This elaborate and unusual escape mechanism occurs among all members of the Cephalopoda, including squids, cuttlefish, octopuses, and the chambered nautilus. The hypothesis of a common descent for all the Cephalopoda is thus consistent with the known data, meaning that the hypothesis has been tested and not falsified.

Further tests of this hypothesis include the finding of fossil cephalopods belonging to a variety of groups; some extinct cephalopods are

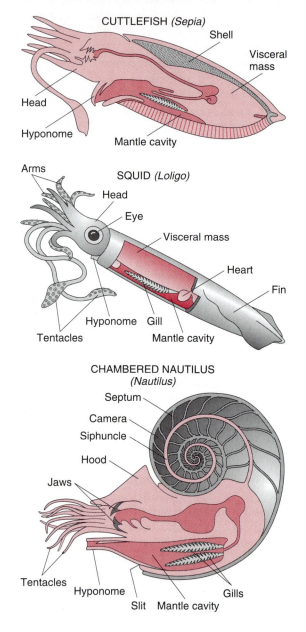

might be able to identify them as representing the same time period as the shells preserved in the White Cliffs of Dover on the English coast, a time period called *Cretaceous* (from the Latin word for chalk, referring to these white cliffs). If we then identify another assemblage as belonging to the Devonian period (after Devonshire province), we can determine that the Devonian assemblage is older because Devonian rocks lie beneath other rocks that in turn underlie Cretaceous rocks. The complete sequence, as worked out by the end of the nineteenth century, is shown in Fig. 4.7.

shown in Fig. 16.4. Known fossil species can be arranged into a family tree that includes the modern descendents, again showing that the hypothesis of common descent has been tested and not falsified. Differences among the living cephalopods, like the presence of a shell in *Nautilus* and the absence of a shell in octopuses, can be explained by pointing to squids, which have a small internal shell, and to various fossil lineages in which the shell underwent reduction.

Family tree of the Cephalopoda, showing branching descent over time

Knowing that one species is Devonian and another is Cretaceous tells you which is older, but not how much older. The determination of ages in years requires *absolute dating,* meaning dates that are measured numerically. In most cases, fossils are placed into a sequence by relative dating techniques, and numerical values are then assigned to this sequence by absolute dating methods. The most important method of absolute dating involves measuring the rate of radioactive decay. Most of the rocks that can be dated by measuring radioactive decay are *igneous rocks,* meaning rocks that have solidified from the molten state, a process which would destroy any fossils that may have been present. Most fossils are contained in *sedimentary rocks*, meaning rocks that are formed by the accumulation of small particles of sediment. Only infrequently do these two kinds of rocks occur together in such a way that the igneous rock (which can be dated radioactively) can be used to assign an absolute date to part of the sedimentary sequence. Most of the numerical dates shown in Fig. 4.7 are now based on radioactive dating.

Phylogeny.   The age of a fossil, by itself, tells us very little about its place in any family tree. It is only when we study a group of organisms represented by many fossils that the relative ages of these fossils begin to assume some meaning. This is because the family tree or genealogy of the group,

also called its *phylogeny*, is an elaborate hypothesis that biologists will use to explain the classification of the species in the group as well as the anatomy, physiology, and other characteristics of each species. In any family tree, the known fossils must fit into a consistent framework. For example, the cephalopod mollusks illustrated in Box 4.1 have a good fossil record. Among living species, the chambered nautilus is very different from the rest because it is fully housed within a coiled shell and has four gills, while the squids and octopus have only two gills and a very small, reduced shell or else none at all. One would therefore imagine a family tree in which octopus and squids had a common ancestor that *Nautilus* did not share. The fossil Cephalopoda do, in fact, conform to these expectations. The group of cephalopods with the oldest fossil record are the nautiloids, of which *Nautilus* is the only living remnant. A second group of cephalopods, called the ammonoids, flourished in

[A]   **[B]**

FIGURE 4.6
[A] *Archaeopteryx,* the first bird, compared with [B] a modern pigeon. Modern birds have lost the teeth, enlarged the braincase, and strengthened other parts (wing, rib, breastbone, pelvis, tail) highlighted here.

Eon	Era	Period	Epoch	Millions of years ago	Geological events	Plants and microbial	Millions of years ago	Animal life
Phanerozoic	Cenozoic	Quaternary	Recent	0.015	North and South America join.	Gymnosperms, angiosperms widespread. Temperate grasslands, forests expand.	0.135	Modern *Homo sapiens* arises.
			Pleistocene	2			2	Genus *Homo* arises.
		Tertiary	Pliocene	7	Cooling after middle of period; series of glaciations. South America, Antarctica separate. Continents separate; mountain building in Asia, Europe, Western North America.		4	Australopithecines present.
			Miocene	26				
			Oligocene	38				
			Eocene	54				
			Paleocene	65			65	Mammals diversify. Primates arise.
	Mesozoic	Cretaceous		136	Laurasia and Gondwana present; last expansion of shallow seas (as over Europe). Rocky Mountains arise and climate cools worldwide at end of period. Africa, South America split.	Angiosperms arise and diversify.		Major extinction event. Most large reptiles, ancient birds extinct.
		Jurassic		195	Extensive inland seas; climate with small seasonal or latitudinal variation.			Teleost fish diversify. Dinosaurs dominant. Modern crustaceans common.
		Triassic		225	Land high; few shallow seas; deserts widespread. Pangaea drifts apart.	Gymnosperms, ferns dominant.		Dinosaur ancestors common. First mammals, birds.
	Paleozoic	Permian		280	Land higher than previously; cold climate that slowly warms.	Conifers appeared.		Mammal-like reptiles common. Major extinction of invertebrates, amphibia.
		Carboniferous — Pennsylvanian		321	Warm, humid; glaciation in Southern Hemisphere. Lowland forests, coal swamps, mountain building.	Forests widespread. Coal deposits form.		First reptiles. Amphibia diversify. Major extinction event.
		Carboniferous — Mississippian		345				
		Devonian		395	Continents rising; Appalachian Mountains forming; cooler climates. Freshwater basins.	First forests. Vascular plants and seeds present.		First insects, sharks, amphibians. Fish diversify.
		Silurian		435	Land uplifting slowly; still extensive shallow seas.	Green, red, brown algae common.		First land arthropods. Jawed fish arise.
		Ordovician		500	Warm, shallow seas reach maximum size. Warming continues.	First vascular land plants probably appeared.		Second major extinction event. Jawless fish diversify; large invertebrates present; mollusks diversify. First tracks left by land animals.
		Cambrian		570	Warm climate; huge equatorial shallow seas.	Algae dominant.		First major extinction event. Trilobites common; onychophorans, first jawless fish at end of period. Evolution of many phyla.
Proterozoic		Precambrian		1,500	Atmosphere oxidizing, not reducing. Major land masses; shallow seas.	Algae abundant. Multicellular organisms: algae, fungi. Cyanobacteria diversified. Eucaryotes present: green algae, protists.		Wormlike animals, cnidarians present.
Archean				3,000		Photosynthetic cells liberate oxygen.		
				3,500	Earth's crust hardens, rocks form. Seas, atmosphere develop. Solar system, Earth form.	Origin of life. Procaryotic heterotrophs. Chemical evolution.		
		Origin of Earth						

Millions of years ago

FIGURE 4.7 The geological time scale.

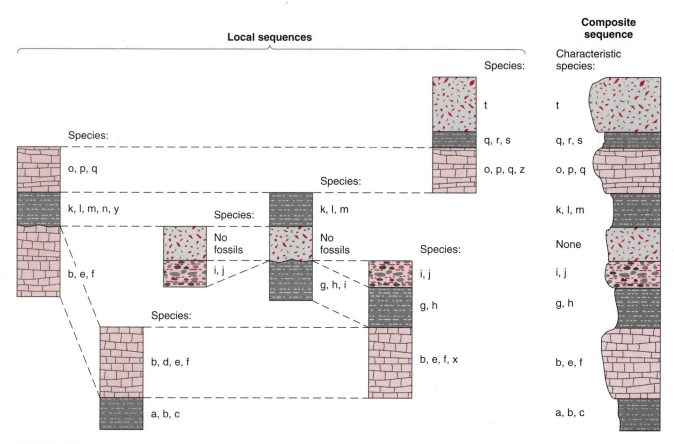

FIGURE 4.8
How correlation by fossils (a relative dating method) is used to establish a single world-wide stratigraphic sequence.

Mesozoic times, during the age of dinosaurs. A small third group, derived from the ammonoids, had an internal shell that was reduced in size. When the ammonoids became extinct, this third group persisted and is represented today by the squids and octopus. Thus, the fossil record of the cephalopod mollusks, including both the anatomy and age relationships of fossil forms, confirms the relationships hypothesized on the basis of the anatomy of the living forms.

The fossil record has repeatedly confirmed hypotheses of descent for particular living species. For example, Thomas Henry Huxley, one of Darwin's early supporters, studied the anatomy of birds and declared them to be "glorified reptiles." The interpretation of birds as reptile descendants was strengthened by the discovery of *Archae-*

opteryx, a fossil with many birdlike and also many reptilian features. Among the reptilian features of *Archaeopteryx* were a long tail, simple ribs, a simple breastbone, and a skull with a small brain and tooth-bearing jaws (Fig. 4.6). Despite these reptilian features, *Archaeopteryx* had well-developed feathers and was probably capable of sustained flight. The discovery of transitional forms like *Archaeopteryx* strengthens our confidence in the hypothesis that birds evolved from reptiles. Other transitional forms are known between older and more modern bony fishes, between fishes and amphibians, and between reptiles and mammals. Instead of being exactly intermediate in each trait, transitional forms like *Archaeopteryx* usually exhibit a mosaic of very advanced and persistently primitive characteristics.

Punctuated equilibria.   Where Darwin had emphasized the gradual nature of transitions from one species to another, Niles Eldredge and Stephen J. Gould have instead claimed that evolution is marked by an alternation between long periods of stability and short bursts of rapid change. Supporters of this concept of *punctuated equilibria* have shown that many species spend long periods of their evolution undergoing little or no change. One study of East African snails shows that most of their evolution occurred by the geographic spread of some species and the replacement of others. On the other hand, paleontologist Philip Gingerich has described several examples of mammalian species that he believes to have evolved more gradually. In most cases, gradualist and punctuationalist explanations fit the known fossil record equally well. Only a long period of continuous deposition in which many fossils are preserved will allow us to falsify one of these two hypotheses, and the resolution of this controversy lies in the future.

<hr>

THOUGHT QUESTIONS

1. For a family tree like the one shown in Box 4.1, what kinds of fossil evidence (be specific) would falsify the descent pattern shown? What kinds of evidence would cause paleontologists to modify the family tree but continue to believe in a process of descent with modification? What kinds of evidence would falsify the hypothesis of descent with modification?

2. One of the Galapagos finches studied by Darwin (species number 11 on Fig. 4.2) has woodpeckerlike habits and certain woodpeckerlike features: It uses stiff tail feathers to prop itself up on vertical tree trunks in the manner of true woodpeckers, then it uses a chisel-like bill to drill holes for insects. However, it lacks the long, barbed tongue that true woodpeckers use to spear insects, so it uses cactus thorns instead. How would Lamarck have accounted for this set of adaptations? How would Paley? How would Darwin? Which of these explanations accounts for the absence of the barbed tongue in the woodpecker finch? How would each hypothesis account for the absence of true woodpeckers on the Galapagos?

B. Creationists Challenge Evolutionary Thought

Pre-Darwinian Creationism

Opposition to evolution has come from various quarters. In the eighteenth and nineteenth centuries, there were many scientists who proposed alternative hypotheses or who suggested ways in which evolutionary hypotheses could be falsified. For example, Reverend William Paley and his supporters proposed that biological adaptations were the work of a benevolent God. Paley himself pointed to the structure of the heart in human fetuses as containing features that adaptation to the local environment could not account for. In adult mammals, including humans, the blood on the left side of the heart is kept separate from the blood on the right side of the heart. In fetal mammals, the blood runs across the heart from the right side to the left, bypassing the lungs, which are collapsed and nonfunctional prior to birth. As the blood enters the left atrium of the heart, it passes beneath a flap that is sticky on one side. When the baby is born, its lungs fill, and blood flows through them. The blood returning to the heart from the lungs now builds up sufficient pressure in the left atrium that the flap closes. Since it is sticky on the side that closes, it seals shut. No amount of adaptation to environment, said Paley, could endow a fetus with a valve that was sticky on one side, designed to seal shut at birth. Only a power with foresight could have realized that the fetus would need a heart whose pattern of blood flow would change at birth, and the sticky valve was part of this plan. Paley attributed the foresight to God, and he insisted that no other hypothesis could explain such an adaptation to future conditions. What is most interesting to a modern reader is that Paley and his many supporters understood the nature of science and used the methods of science to argue their case. Paley in particular sought scientific proof of God's existence and benevolence by arguing that *no other hypothesis could explain the evidence as well.* This example shows that good science is certainly compatible with a belief in God or a rejection of evolution. In fact, the best scientists of the period from 1700–1859 were, with few exceptions, devout men who rejected the evolutionary paradigm on scientific grounds.

Darwin was quite familiar with Reverend Paley's arguments, and he offered evolutionary explanations for many of the intricate and marvelous adaptations that Paley's supporters had described. In each case, Darwin argued that the hypothesis of natural selection could account for the adaptation as well or better than the hypothesis of God's design could. There were also some adaptations that were *less* than perfect, or that seemed to be "making do" with the materials at hand. The gills in barnacles are modified from a brooding pouch that once held the eggs. The milk glands of mammals are modified sweat glands. The thyroid gland is modified from what was once a feeding structure (called an *endostyle*) that produced a sticky mucus that trapped food particles along the gills and aided in their movement into the digestive tract. The giant panda, evolved from an ancestor that had lost the true thumb, developed a new thumblike structure made from a little-used wrist bone. (This last example was not known in Darwin's time, but fits well into Darwin's argument.) These many adaptations seemed more easily explained by natural selection than by God's design because God could presumably have "done better." Natural selection is limited to the use of the materials at hand, and then only if there is variation; an omnipotent God could have made barnacle gills from entirely new material without taking away the brood pouches, and could have given pandas a true thumb instead of a wrist bone. Darwin and his supporters used examples like these to show that the evolutionary explanation fit the available evidence better than Paley's explanation of divine planning. For example, natural selection perpetuates only those hearts whose flaps seal properly at birth.

The course of evolution is constrained by its own past history. Pandas evolved from animals that had lost the true thumb. When giant pandas began eating bamboo and needed to strip leaves from bamboo stalks, they used the thumb side of their wrists, and the wrist bones that they used were modified by natural selection into a substitute for the thumb. The original thumb had long been lost; much of embryology would have to be drastically reorganized in order to restore the loss. Such a reorganization, while not impossible, is

quite unlikely because it would mean undoing many past changes that were each adaptive when they evolved and replacing them with some better alternative. Once a structure is lost in the course of evolution, it is usually lost forever in that particular form because natural selection cannot make something out of nothing. It is for this reason that Belgian paleontologist Louis Dollo declared that evolution is often irreversible, a concept sometimes called *Dollo's law*. This is also the basis for what Stephen J. Gould has called *contingency* in evolution. When several alternative events are possible, the one that is chosen by natural selection may constrain or limit future evolution to only certain possibilities. For example, if something that could possibly have evolved into X evolves instead into B, then none of the possible descendents of X could ever evolve, even though they may have been far superior to the descendents of B.

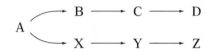

In this sense, evolution may be said to follow a path of least resistance or short-range opportunity instead of being guided by any long-range plan.

Early Twentieth-Century Creationism

In the early twentieth century, most opposition to evolution came from certain Protestants in the United States (but not in Europe), who declared that evolution conflicted with the account of creation given in the Bible. These people founded a number of societies, including the Society for Christian Fundamentals (the origin of the term *fundamentalist*). Since they insisted on the biblical account of creation, they were also creationists. The fundamentalists persuaded several state legislatures to pass laws restricting or forbidding the teaching of evolution in schools. Some of these state laws remained on the books until the 1960s.

In 1925, a famous court case was defended in Tennessee by the fledgling American Civil Liberties Union (ACLU). A teacher, John H. Scopes, was arrested for reading in his high school class a passage from a textbook that dealt with evolution. The trial attracted worldwide attention. Scopes lost and was assessed a $100 fine. Upon appeal, the case was thrown out because of the way the fine had been assessed; the merits of the case were never really debated. The Scopes trial did, however, have a chilling effect on the textbook publishing industry: Books that mentioned evolution were revised to take the subject out, and most high school biology texts published between 1925 and 1960 made only the barest reference, or else none at all, to Charles Darwin and his theories.

Creationism Today

The launch of the Soviet Earth-orbiting satellite Sputnik in 1958 set off a wave of self-examination in American education. Groups of college and university scientists began examining high school curricula with renewed vigor, and several new high school science texts were written. Most of the new biology texts emphasized evolution, or at least gave it prominent mention.

Alarmed in part by the new textbooks, a new generation of creationists began a series of attacks on the teaching of evolution. These new creationists tried to portray themselves as scientists, calling their new approach "creation science," even though they never conducted experiments or tested hypotheses. Instead of making their studies falsifiable, the new creationists claimed that they held the absolute truth:

Biblical revelation is absolutely authoritative. . . . There is not the slightest possibility that the *facts* of science can contradict the Bible and, therefore, there is no need to fear that a truly scientific comparison . . . can ever yield a verdict in favor of evolution. (Morris, 1974:15–16)

The processes of creation . . . are no longer in operation today, and are therefore not accessible for scientific measurement and study. (Morris, 1974:104)

In contrast, Charles Darwin knew that his theories were—rightly—subject to the principle of falsifiability:

If it could be demonstrated that any complex organ existed, which could not possibly have been formed by numerous successive, slight modifications, my theory would absolutely break down. (Darwin, 1859:189)

Some creationist writings also contain faulty explanations of many scientific concepts, including the second law of thermodynamics. According to this law, a closed system (one in which energy neither leaves nor enters the system) can only change in the direction of less order and greater randomness. Thus, a building may crumble into a pile of stones, but a pile of stones cannot be made into a building without the expenditure of energy. Creationists have claimed that this law precludes the possibility of anything complex ever evolving from something simpler. The second law of thermodynamics *does* apply to all biological processes. If the Earth were a thermodynamically closed system, life itself would soon cease. However, the Earth is not a thermodynamically closed system because energy is constantly being received from the sun, and this energy allows life to persist and evolution to occur. The second law of thermodynamics does not rule out the building up of complexity; rather, it states that making something complex out of something simple requires an input of energy.

In the 1960s, because many of the laws forbidding the teaching of evolution had been declared unconstitutional, the new creationists, led by Henry Morris, Duane Gish, and John Slusher, decided on a new approach. Evolution could be taught in the schools, they argued, but only if "creation science" was taught along with it and given equal time. A few state legislatures passed laws inspired by this new group of creationists. An Arkansas law known as the *Balanced Treatment Act* (Public Law 590) was finally declared unconstitutional in 1981, and a similar Louisiana law was declared unconstitutional a few years later. Interestingly, in these trials, the scientific issues *were* raised in court, and prominent scientists were called upon to testify. The court was asked, in particular, to rule on what was scientific and what was not. The court finally ruled that evolution was a scientific theory and could legitimately be taught, while "creation science" was not a science at all, since it involved no testing of hypotheses and since its truths were considered to be absolute rather than provisional. Instead, "creation science" was found to be a religion, or to include so many religious concepts (creation by God, Noah's flood, original sin, redemption, and so forth) that it could not be taught in a public school without violating the U.S. Constitution's historical separation of church and state.

Even though these state laws have been declared unconstitutional, the creationists have vowed to continue pressuring each local school board and each state legislature to support their approach. Their major arguments continue to revolve around the concept of "equal time," which was originally a measure to ensure fairness in political campaigns.

1. In what ways did William Paley use scientific evidence? Did he use falsifiable hypotheses? Do today's creationists use falsifiable hypotheses to support their claims?
2. How much time should be devoted in science classes to alternative explanations or theories that have been falsified? Should time be given to explanations that are not falsifiable hypotheses? Should all explanations be given equal time? How much (if any) of a science curriculum would you devote to divine creation as an alternative to evolution? To astrology as an alternative to astronomy? To the theory that disease is caused by demons or evil spirits?
3. Does the teaching of unpopular or rejected theories encourage students to think critically? Does it encourage attitudes of fairness? Does it increase students' understanding of what science is and how science works?

C. Species Are Central to the Modern Evolutionary Paradigm

Darwin's paradigm continues to guide biological research to this day. Research within the paradigm continues to be successful, and the methods and concepts of the paradigm enjoy the continued support of scientists. The paradigm has, however, been expanded by twentieth-century findings, especially those from the field of genetics (see Chapter 3).

The early 1940s saw the birth of a new evolutionary paradigm, known at the time as the *modern synthesis*. Most of the Darwinian paradigm was incorporated into this newer paradigm, but the findings of genetics were also incorporated into the paradigm as a source of variation. The new cornerstone of the modern evolutionary paradigm was a

theory of **speciation,** the process by which a species branches into two species.

Populations and Species

Populations.   Biological **populations** are defined as all those individuals within a species who can and do breed or mate with one another, contributing their genes to the next generation. Looking backward in time, any two individuals in a population are assumed to share at least some of their genes because of common descent in the past, usually within the past 100 generations. Looking into the future, any two opposite-sex individuals in a population are potential mates. The population is thus a cluster of interbreeding individuals. Membership in a population is determined by descent or by the capacity to interbreed, not by physical characteristics.

Biological populations within a species will receive hereditary information (genes) from one another whenever conditions permit. This passage of genetic information between individuals or its exchange between populations is called **interbreeding.** The existence of biological barriers to this exchange is called **reproductive isolation.** Interbreeding between populations takes place whenever mating occurs between members of different populations; reproductive isolation inhibits such matings to varying degrees.

Species.   The term *species* has a precise meaning to biologists. **Species** are defined as *reproductively isolated groups of interbreeding natural populations.* Several aspects of this definition are noteworthy:

1. Populations belonging to the same species will interbreed whenever conditions allow them to. The opportunity may not exist if the populations are geographically separated.

2. Populations belonging to different species are reproductively isolated from one another and will thus not interbreed. Any biological mechanism that hinders the interbreeding of these populations is called a **reproductive isolating mechanism** (Box 4.2). Two species of plants, for example, may flower in different seasons, thus preventing pollen exchange and therefore mating.

3. Physical characteristics (morphology) are not part of the definition of species. Such characters are useful only as evidence that hereditary information is or is not being exchanged.

4. Species are composed of natural populations, not of isolated individuals. An individual organism belongs to a population first, and this population belongs to a species. One consequence is that the mating behavior of individuals in captivity does not mark the definition of species. The behavior of individuals can only serve as *evidence* for the ability of *natural populations* to interbreed under natural conditions. Alternatively, the discovery that certain individuals will not mate may serve as evidence for the existence of a reproductive isolating mechanism. Such a discovery may become the first research step in investigating that reproductive isolating mechanism. The *last* step of such an investigation is generally a field study undertaken to confirm that certain laboratory findings can be duplicated in the wild.

How Do New Species Originate?

In order to explain how a new biological species has come into existence, we need to explain how it has become reproductively isolated from closely related species. The origin of a species is thus the origin of one or more reproductive isolating mechanisms.

The geographic theory of speciation. In the vast majority of cases, new species have come into existence through a process, called **speciation,** that includes a period of **geographic isolation** by some sort of extrinsic barrier. Such a barrier may take the form of a mountain range, or it may simply be an uninhabited area separating the geographic ranges of two related groups of populations. The essence of the theory is that new reproductive isolating mechanisms originate during times when such extrinsic barriers separate populations geographically. Geographic isolation is not by itself considered to be a reproductive isolating mechanism; rather, it sets up the conditions under which the separated populations may evolve along different lines.

This process depends in part on the length of time that passes—more time allows more chance for a reproductive isolating mechanism or mechanisms to evolve. Another very important factor is the extent of the difference between what natural selection favors on either side of the geographic barrier. If the populations on opposite sides of the barrier are subjected to different types of selective forces for a long enough period of time, then one or more reproductive isolating mechanisms may evolve between the two groups of populations, therefore separating them into new species (Fig. 4.9, p. 114). If the populations later come in geographic contact again, the reproductive isolating mechanisms that have evolved during their separation will keep them genetically separate as two species. For example, birds with larger bills may become more frequent in one population because larger bills may be more efficient as feeding devices on certain foods and thus increase fitness. Thus, bird populations isolated on opposite sides of a mountain chain or a large body of water may evolve different bill sizes if feeding patterns differ by locality. If the birds choose their mates on the basis of bill size, they might become reproductively isolated as a result. The same types of change may occur in the timing of the breeding season or the nature of the mating calls. Frog or cricket populations isolated on opposite sides of a mountain chain or a large body of water may develop different mating calls. Since the animals would respond only to the mating calls of their own population, the two populations would be reproductively isolated.

Populations on opposite sides of a barrier may also come to differ as a result of *genetic drift*, the statistically greater probability of abrupt, accidental genetic changes in small populations (see Chapter 5). If larger bills became more frequent purely by accident, because several small-billed birds happened to die for reasons that had nothing to do with bill size, then the change in bill size would be due to genetic drift. Accidental change of this sort has only minimal effects in large populations. Regardless of the ways in which differences between populations originate, the differences may inhibit interbreeding, and the populations may become reproductively isolated.

Incomplete speciation. One important prediction of the geographic theory of speciation is that cases of incomplete or imperfect speciation should occasionally occur. If populations are separated for a very long time (or if selective forces on opposite sides of a barrier differ greatly), then speciation is

BOX 4.2 TYPES OF REPRODUCTIVE ISOLATING MECHANISMS

Any biological mechanism that hinders the interbreeding of populations belonging to different species is called a reproductive isolating mechanism. Mechanisms that prevent the exchange of gametes are more efficient and are therefore favored by natural selection over mechanisms that act later in the lifetime of the individual.

MECHANISMS THAT OPERATE PRIOR TO MATING (*PREMATING MECHANISMS*)
Ecological Isolation
Potential mates never encounter each other, either because they live in different habitats, because they are active at different times of day or in different seasons, or because they are not physiologically capable of reproduction at the same time. As the graph shows, wood frogs are fully isolated from tree frogs and bullfrogs by seasonal differences in breeding; they are partially isolated from pickerel frogs because the breeding seasons overlap slightly.

Behavioral Isolation
Potential mates meet but do not mate because their courtship rituals differ. This causes individuals of at least one of the species to reject potential mates of the other species. For example, fireflies (family Lampyridae) recognize their mates on the basis of the various flashing patterns and flight patterns. As shown, nine different species can coexist in the same area without interbreeding.

Mechanical Isolation
In insects and other species with hardened and inflexible sexual parts (genitalia), a lock-and-key fit is necessary in order for mating to take place. Closely related species of insects often have differing genitalia, preventing mating between species from taking place.

MECHANISMS THAT OPERATE AFTER MATING HAS TAKEN PLACE (*POSTMATING MECHANISMS*)
Gametic Mortality
In animals, the sperm may die within the female reproductive tract before fertilization takes place. In plants, the pollen may fail to germinate on the female parts of another species.

Zygotic Mortality
The fertilized egg may die following fertilization.

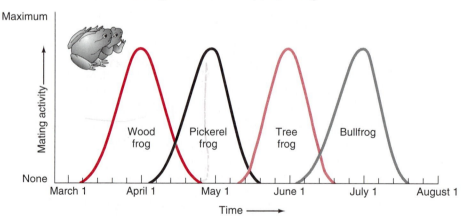

Mating seasons for four species of frogs

Embryonic and Larval Mortality

Developmental patterns may differ so much that normal cell divisions and developmental rearrangements cannot take place, and the embryo or larva dies at an early developmental stage.

Hybrid Inviability

First-generation (F_1) hybrid individuals are produced but do not live to maturity.

Hybrid Sterility

First-generation hybrids live but are sterile. The mule, a sterile hybrid between a horse and a donkey, is an example.

F_2 Breakdown

The F_1 hybrids can mate, but no viable F_2 hybrids are produced.

Reproductive isolating mechanisms may evolve initially as the result of genetic drift, or as the result of either sexual selection or natural selection that differs from place to place. Once such a mechanism exists, even if it is an imperfect one, any attempted mating with the wrong species will result in fewer offspring and in reduced fitness. When multiple isolating mechanisms evolve, one mechanism can back another up in case one of them fails to keep the species separated. Natural selection also favors isolating mechanisms that operate earlier, sparing both species the energetic costs and genetic costs of wasting time, energy, and gametes in unfruitful attempts at mating. In particular, selection favors premating isolating mechanisms, even when postmating mechanisms are already present.

Light patterns of nine species of fireflies

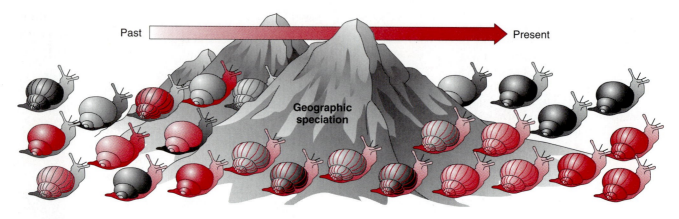

Past   Present

Geographic
speciation

FIGURE 4.9

Geographic speciation: the evolution of reproductive isolation. Genetically variable populations that spread geographically can develop locally different populations that are capable of interbreeding with one another initially. When the populations are separated by an extrinsic barrier such as a mountain range or a deep canyon, they may develop further differences that prevent interbreeding even after contact is resumed.

likely to occur. If the separation, however, is very brief, then speciation will not occur. These two situations lie at opposite ends of a continuum. Somewhere along this continuum lies the situation in which species have been separated by a geographic barrier long enough for reproductive isolation to begin evolving, but not yet long enough for the reproductive isolation to be complete. The expected result is an imperfect or partial degree of reproductive isolation between two species, something that would lessen the chances of interbreeding between them but not prohibit it entirely. Such situations have indeed been found, for example, among the South American fruit flies known as *Drosophila paulistorum*. Crosses among the different incipient species (semispecies) of *D. paulistorum* produce fertile hybrid females but sterile hybrid males, which is a partial degree of reproductive isolation. The geneticists studying these flies referred to them as "a cluster of species *in statu nascendi*" ("in the process of being born").

Another form of incomplete speciation is provided by cases of *circular overlap,* in which a series of populations are arranged in a circular pattern (Fig. 4.10). Even though all populations in the circle are able to exchange genes through intermediate populations, reproductive isolation may evolve between populations at the beginning and end of the range, and these populations may come to share part of a geographic range without interbreeding. Cases of circular overlap have been documented among a variety of amphibians, birds, insects, mammals, and higher plants. As often happens in science, a hypothesis receives strong support (and thus becomes a theory) if it predicts something never previously observed or expected, and the prediction is then confirmed. The observed cases of incomplete speciation confirm the predictions of the theory of geographic speciation.

Other theories of speciation.   Although the geographical theory of speciation has become the majority viewpoint, some biologists have proposed additional models of speciation to account for certain unusual cases. Some theories claim that speciation can take place much faster than the geographic theory predicts. One such mechanism that is known to work in certain types of flowering plants is speciation by polyploidy. **Polyploidy** means that the chromosomes exist in *more than two complete sets*, with chromosome numbers such as $4N$ (*tetraploid*), $6N$ (*hexaploid*), $8N$ (*octaploid*), and so on. Polyploidy can arise rapidly in certain plant species by a doubling or tripling of the

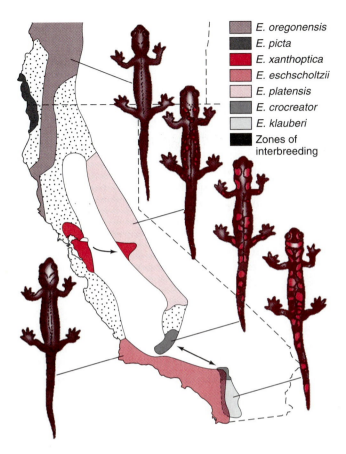

▨	E. oregonensis
◼	E. picta
◼	E. xanthoptica
▨	E. eschscholtzii
▨	E. platensis
▨	E. crocreator
▨	E. klauberi
◼	Zones of interbreeding

FIGURE 4.10

Circular overlap in salamanders of the genus *Ensatina*. The northern subspecies are connected to one another by zones of interbreeding, but *E. croceator* and *E. klauberi* each overlap the range of *E. eschscholtzii* in southern California without interbreeding with it, and are therefore considered distinct species.

chromosome numbers. Most polyploids still belong to the same species as their parents; in fact, many common food plants (including oats, corn, wheat, and strawberries) and several ornamental plants (e.g., day lilies) are polyploids. In a few instances, however, the differences in chromosome number prevent the polyploids from interbreeding with their parents while they can breed with each other, and this reproductive isolation immediately sets them apart as a new species. Although this process is seldom observed in progress, many naturally occurring plant species differ from one

another by a doubled or tripled chromosome number and are therefore believed to have arisen through polyploidy.

Higher Taxa

All human societies, but especially those living close to nature, have names for collective groups of similar species: birds, snakes, insects, pines, orchids, and so forth. Biologists formalize this process and communicate these ideas in the form of a **classification** (see Appendix), which is an arrangement of larger groups that are subdivided into smaller groups. Any one of these collective groups, such as the insects or the rose family, is called a **taxon,** and **taxonomy** is the study of how these taxa are recognized and how classifications are made. In a biological classification, species are grouped into successively larger (more inclusive) groups, or *higher taxa*: Related species are grouped into *genera* (singular, *genus*), related genera into *families*, related families into *orders*, related orders into *classes*, related classes into *phyla* (called *divisions* in plants), and related phyla into *kingdoms* such as the animal or plant kingdoms. All these are arranged as groups within groups, with the smaller groups sharing many more characters and the larger groups sharing only the few most general characters. For example, human beings constitute the species *Homo sapiens*. This species is grouped together with *Homo erectus* and certain other fossil species into the genus *Homo*. (A genus always has a one-word name that is capitalized; a species has a two-word name in which the first word is the name of the genus.) The genus *Homo* is grouped together with the extinct genus *Australopithecus* into the family Hominidae. This family is included in the order Primates, which also includes apes, monkeys, and lemurs. The primates are grouped together with rodents, carnivores, bats, whales, and over two dozen additional orders into the class *Mammalia*, including all warm-blooded animals with hair or fur that feed milk to their young. Mammals are one of several classes in the phylum Chordata, a group that includes all vertebrates (animals with backbones) and a few aquatic relatives such as the sea squirts and amphioxus. The Chordata and several dozen other phyla are together placed in the animal kingdom (*Animalia*). Animals are one of the several kingdoms currently recognized, as we will describe later.

Several kinds of organisms reproduce asexually, producing offspring without combining their genes with those of any other individual. Can asexually reproducing organisms belong to species? Can the definition of species be modified to apply to asexual as well as sexually reproducing organisms?

D. Life Originated on Earth by Natural Processes

In addition to dealing with the changes that occur within species and during the evolution from one species into others, modern evolutionary theory also includes an account of the origins of living organisms on Earth.

Oparin and Miller

The development of Oparin's theory. Modern thoughts on the origin of life on Earth began with the biological theories of the early twentieth-century Russian biochemist Aleksandr Oparin. Most scientists at the time accepted Louis Pasteur's conclusions that life had always come from preexisting life. But where had the first living organisms come from? Oparin, a Marxist, rejected the possibility of a divine or other miraculous creation. Had life always existed? Astronomers of Oparin's time and earlier believed that the Earth originated as a molten mass that was much too hot to support any life, so Oparin reasoned that life had not always existed on Earth. Oparin also knew that the conditions in interplanetary space, such as extreme cold (around $-270°C$), utter dryness, and constant bombardment by high levels of ultraviolet radiation, were incompatible with all forms of life. These facts convinced Oparin that life could not have come through space from anywhere else. Life must have originated on planet Earth.

Looking for some way that life might have originated in the first place, Oparin carefully reexamined all earlier experiments, especially those of

Louis Pasteur. Pasteur had established that living organisms always came from preexisting organisms, rather than originating by spontaneous generation from nonliving materials. According to Pasteur's carefully worded conclusion, "There is now no circumstance known in which it can be affirmed that microscopic beings came into the world without germs, without parents similar to themselves." From this, Oparin reasoned that, if present conditions do not permit organisms to originate from nonliving matter, then life must have originated under conditions very different from those that prevail today.

Oparin began searching for clues as to what those "primitive Earth" conditions might have been. The planetary atmospheres of Jupiter and Saturn had just been discovered to consist mostly of hydrogen (H_2), with smaller amounts of ammonia, methane, and other gases, but no oxygen. The sun was also known to contain mostly hydrogen. Hydrogen was thus the most abundant material in the solar system. An abundance of hydrogen and a total absence of free oxygen (O_2) characterize what chemists call *reducing conditions*. With such a great abundance of hydrogen, any other element would have existed in combination with a maximum amount of hydrogen: Oxygen would be most stable as H_2O (water vapor), nitrogen as ammonia (NH_3), and carbon as methane (CH_4). Oparin therefore postulated an early atmosphere consisting of the gases *hydrogen, ammonia, methane,* and *water vapor.* On the assumption of such an atmosphere, Oparin worked out on paper some of the chemical reactions that he thought might have produced simple biological molecules such as sugars and organic acids. Oparin theorized that these molecules would have built up in the primitive oceans after the Earth had cooled sufficiently to permit water to become liquid. By the slow accumulation of these molecules, the world's oceans came to have a consistency approximating what the British geneticist J. B. S. Haldane called a "hot, dilute soup." Oparin published this entire theory in book form in 1935, and the book was soon translated into English under the title *The Origin of Life.*

Miller's experiments testing Oparin's theory. Oparin's theory was ignored (or overlooked) by most scientists for several decades. In 1952, a biochemist named Stanley Miller decided to test

Oparin's model experimentally. He built several types of apparatus, such as the one shown in Fig. 4.11. The apparatus was filled with hydrogen, ammonia, methane, and water vapor, the four gases that had been postulated by Oparin. These gases were allowed to react in a 5-liter (5 L) chamber in which electric sparks could be generated to simulate atmospheric lightning. Reaction products were cooled in the condenser, where the water became liquid and other compounds that had formed would dissolve in it. This simulated "rain" was al-

lowed to collect in the lower part of the apparatus, simulating the ponds and oceans of the primitive Earth. Finally, a heat supply vaporized some of the liquid and returned it to the sparking chamber. Miller circulated his reaction mixture for several days and withdrew samples to analyze the results. Among the compounds that had been formed, he found amino acids and some simple peptides. Similar experiments were repeated by Miller and other investigators, with the following results:

1. The reaction products were produced by the experimental reactions, and not by any biological contaminants. Miller showed this by using a control experiment: He repeated all experimental conditions except for the electric sparks. Without the energy source provided by these electric sparks, no reaction products were detected.

2. Although the reactions required an energy source, the particular kind of energy was not important. An ultraviolet light source could successfully substitute for the electric sparks; so could natural sunlight, even with no other heat source.

3. As long as a carbon compound is present initially, it doesn't have to be methane; carbon dioxide (CO_2), for example, will work just as well. Likewise, nitrogen gas (N_2) or various nitrogen oxides could substitute for ammonia.

4. In at least some of the reactions, simple sugars, amino acids, glycerol, purines, and pyrimidines occurred among the products. The building blocks for all major biological molecules were thus included.

Experiments like Miller's have succeeded in producing so many biologically important products that the following conclusion is now inescapable: All of the molecules important to life *could have been produced* in a lifeless environment on the primitive Earth, starting with nothing more than a few basic gases mixed together in a reducing atmosphere.

Oparin also hypothesized that more complex molecules could also have formed from the constituents of the early atmosphere. The first fatlike molecules (lipids) formed tiny droplets that may have contained proteins. Some of the proteins (or perhaps RNA molecules) may have catalyzed (speeded up) reactions that led to the growth of the

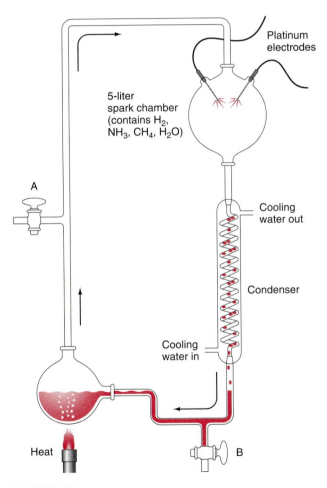

FIGURE 4.11
Stanley Miller's experiment, in which amino acids and simple peptides were produced. Heating the flask at the lower left boils the water and keeps the mixture circulating in the direction shown by the arrows. Reactions take place in the spark chamber and reaction products are condensed and recirculated. Valve A is used to sterilize the apparatus and to introduce the starting materials; valve B is used to withdraw samples of the reaction products.

Labels in figure: Platinum electrodes; 5-liter spark chamber (contains H_2, NH_3, CH_4, H_2O); A; Cooling water out; Condenser; Cooling water in; Heat; B droplets that contained them, and these particular droplets grew, divided, and multiplied. Proteins that lacked this "autocatalytic" ability didn't survive long; Oparin referred to this process as *protoselection*. The droplet systems that grew did so by incorporating material taken from the "hot, dilute soup" that surrounded them. These growing systems increased in size and complexity and eventually became the first organisms.

Evidence of Early Life on Earth

Miller-style experiments show us a model of what *could* have happened. Is there any evidence that such events actually *did* happen? There is some evidence, but it is incomplete and indirect. For example, meteorites falling to Earth from elsewhere in the solar system sometimes contain many compounds indicative of Miller-style synthesis in other parts of the universe, including compounds that organisms now on Earth produce only rarely or not at all. Ancient rocks on Earth contain further evidence. If we search the Precambrian portion of the geological record, we find various kinds of "chemical fossils," compounds that give us clues to the conditions that prevailed at the time when these compounds were formed. Our present atmosphere differs greatly from the primordial atmosphere postulated by Oparin because free oxygen (O_2) is abundant, producing conditions that chemists refer to as *oxidizing conditions*. The oldest rocks on Earth seem to have been deposited under reducing rather than oxidizing conditions.

Evolution of our atmosphere.   How did the present oxygen-rich atmosphere of the Earth originate? Most scientists who have investigated this question have concluded that life itself was primarily responsible. Most important is the finding that chemicals formed from the breakdown of chlorophyll molecules are not present in the oldest rocks. Chlorophyll is a key molecule in photosynthesis, a reaction by which plants and blue-green bacteria (Cyanobacteria) produce O_2 (Chapter 15). The breakdown products of chlorophyll first appeared in the geological record at about the time that the reducing atmosphere began to change slowly, over a period of about a billion years, to the oxidizing conditions that now prevail (Schopf, 1974).

The first forms of life had to live under reducing conditions in which no free oxygen was present, conditions also called *anaerobic*. A variety of bacteria are capable of living under anaerobic conditions. Scientists now hypothesize that the first bacteria would have had to derive all their energy from the high-energy molecules in their surroundings. Nowadays, such molecules are in most cases produced by other organisms, but the first organisms would have had to rely on the molecules that had formed without life (abiotically), under conditions like the ones simulated in experiments like Miller's. For many thousands or maybe millions of years, the supply of these molecules may have been adequate for life to expand and perhaps to flourish. (We can't tell for sure, because such organisms leave very few fossil traces of their existence.) Eventually, however, the organisms expanded to the point that the limited amount of energy-rich chemicals in the environment were just not enough. This was possibly the first global environmental crisis in the history of life on Earth.

We can imagine several possible responses to this crisis. Some organisms may have discovered a way to attack and devour other organisms, getting their nutrients from their prey. Organisms of this type would have continued to eat one another, but the total quantity of living organisms (the total *biomass*) that the planet could support would have remained limited. The problem may even have gotten worse because some of the waste products of metabolism included gases like CO_2, which would simply have escaped into the atmosphere, taking carbon out of reach of most organisms.

Evolution of photosynthesis.   A more permanent solution occurred much later, when some organisms produced the first chlorophyll-like molecules, about 4 billion years ago. Such a molecule would enable life forms to use solar energy in the form of light, which would avoid the problem of deriving energy from the limited number of other organisms. All forms of photosynthesis require the use of some chemical to supply hydrogen atoms and to act as an electron acceptor (see Chapter 15). Most bacterial forms of photosynthesis use hydrogen sulfide (H_2S) for this purpose, or iron compounds, or organic molecules derived from nucleic acids, but none of these compounds was abundant.

The greatest change to the Earth and its atmosphere resulted from the evolution of a new and more efficient kind of photosynthesis, using a new source of hydrogen: water, the most abundant hydrogen source on Earth, a source readily available in most environments. The splitting of water in photosynthesis generated a new atmospheric gas: oxygen (O_2). The first organisms to evolve this kind of photosynthesis were blue-green bacteria (*Cyanobacteria*). During the next 2 billion years or so, these blue-green bacteria became the dominant form of life on Earth, reducing the abundance of atmospheric CO_2 to a small fraction of its former level and slowly generating more and more O_2. Calculations of the photosynthetic capabilities of these blue-green bacteria show that they were quite capable of generating all the oxygen gas (O_2) in the Earth's atmosphere within half a billion years or so. Not only did all of the life forms on Earth arise by evolution, but our very atmosphere is a product of biological evolution.

Procaryotic and Eucaryotic Cells

Procaryotic cells.   The first organisms had simple cells with no internal compartments. A *cell membrane* (also called *plasma membrane*) formed the outer boundary of the cell and kept its contents inside, but there were seldom any internal membranes capable of forming separate compartments. Simple cells of this type are called **procaryotic** cells. All bacteria and cyanobacteria have procaryotic cells. Procaryotic cells lack most of the complex internal structures possessed by more advanced (eucaryotic) cells. For example, procaryotic cells have only a single chromosome containing nucleic acid only and no protein. This chromosome, usually circular in shape, is located in a simple nuclear region that is never surrounded by a nuclear envelope or set apart from the rest of the cell in any other way.

Eucaryotic cells.   Plants, animals, fungi, and certain other organisms called *protists* are composed of more complex cells, called **eucaryotic** cells (Box 4.3), which contain nuclear envelopes, multiple chromosomes, and a variety of membrane-enclosed cellular structures called **organelles.** Most notably, eucaryotic cells contain many internal membranes that form compartments, for example, a *nucleus* that is always surrounded by a nuclear envelope. Separate internal compartments also exist in such organelles as the mitochondria, golgi, plastids, and endoplasmic reticulum. Eucaryotic cells have a cytoskeleton composed of structural and/or contractile fibers made of protein. The possession of eucaryotic cells sets animals, plants, fungi, and protists apart from procaryotic organisms.

Endosymbiosis.   How did these types of cells evolve? According to a theory first championed by Lynn Margulis in the 1970s, eucaryotic cells arose from procaryotic cells by a process called **endosymbiosis** (literally, "living together inside"). According to this theory, large procaryotic cells incapable of performing certain energy-producing chemical reactions (those of the Krebs cycle, described in Chapter 8) engulfed smaller procaryotic cells able to carry out these reactions. The larger (host) cells could obtain energy by digesting the smaller cells, but they could obtain even more energy if they allowed the smaller cells to go on living inside of them and used the products of the energy-producing reactions. In this situation, host cells that allowed the smaller cells to persist were favored by natural selection over host cells that digested the smaller cells. Over time, the relationship between the host cells and the smaller cells grew more intimate, and the smaller cells became cellular organelles now known as *mitochondria*.

Certain other organelles, such as the *plastids* that occur inside plant cells (see Chapter 15), are believed to have evolved by a similar process, in which symbiotic cyanobacteria were incorporated into the original plant cells. The plant cells then achieved greater growth potential by harboring these symbionts rather than by digesting them; plant cells with symbiotic plastids did better and reproduced in greater numbers than plant cells that had no plastids, and those with plastids persisted, while those without plastids died out. In support of this theory is the fact that both plastids and mitochondria have their own types of membranes and their own DNA, separate from the DNA of the host cell. The membranes and DNA of both mitochondria and plastids have certain similarities to the membranes and DNA of procaryotic organisms instead of to the membranes and DNA of their

BOX 4.3   PROCARYOTIC AND EUCARYOTIC ORGANISMS

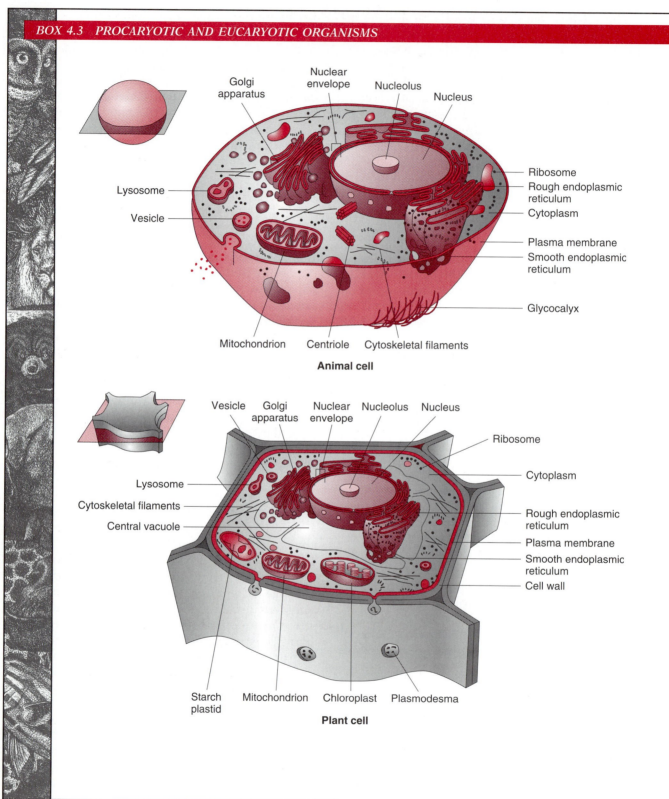

Animal cell

Golgi apparatus
Nuclear envelope
Nucleolus
Nucleus
Lysosome
Vesicle
Ribosome
Rough endoplasmic reticulum
Cytoplasm
Plasma membrane
Smooth endoplasmic reticulum
Glycocalyx
Mitochondrion
Centriole
Cytoskeletal filaments

Plant cell

Vesicle
Golgi apparatus
Nuclear envelope
Nucleolus
Nucleus
Ribosome
Lysosome
Cytoplasm
Cytoskeletal filaments
Central vacuole
Rough endoplasmic reticulum
Plasma membrane
Smooth endoplasmic reticulum
Cell wall
Starch plastid
Mitochondrion
Chloroplast
Plasmodesma

A great gulf separates the major types of organisms based on the structure of their cells. Bacteria and related blue-green photosynthetic organisms (Cyanobacteria) have a simple type of cell called *procaryotic.* Plants, animals, fungi, true algae, and certain one-celled organisms like amoebas have a more complex type of cell called *eucaryotic.* All procaryotic organisms are single-celled, while eucaryotic organisms can be either single celled or multicellular; thus the distinction between the two groups is based not on single-celled versus multicellular organization but rather on the structure of the cells themselves.

Eucaryotic cells have true nuclei (the name *eucaryotic* means "true nucleus") and various other internal parts, called *organelles,* which are delimited by intracellular membranes. These intracellular membranes separate the various functions of the cell into different compartments. Several of these types of organelles are shown in the accompanying illustrations of a typical animal cell and a typical plant cell. Keep in mind that there are a vast number of variations on these cell types, both among species and among the various specialized cells within multicelled organisms; there is probably no actual cell that exactly matches these diagrams.

In the cells of eucaryotic organisms, DNA is associated with protein and is arranged in multiple chromosomes. The DNA is enclosed in a membrane-delimited nucleus, except during cell division. Eucaryotic cells have a *cytoskeleton,* an internal network of protein filaments that guide the position and movement of the organelles and, in some cases, movement of the cell. Those eucaryotic organisms that are photosynthetic have their photosynthetic pigments contained in membrane-enclosed organelles called *plastids.*

The term *procaryotic* means "first nucleus," and it reflects the theory that some organisms of this type existed before the evolution of the true (eucaryotic) nucleus. Procaryotic cells lack true nuclei, nuclear envelopes, mitochondria, or any of the other internal membrane-enclosed structures illustrated at left. Procaryotic cells thus have a simpler type of structure, not compartmentalized internally, as shown in the accompanying diagram of a bacterial cell.

Each procaryotic organism has a single chromosome, containing DNA but no protein. Procaryotic chromosomes are often circular, like a closed necklace. Procaryotes also have detached DNA fragments, called *plasmids,* that can lead an independent existence for a long while and then reincorporate into the procaryotic chromosome. Procaryotic cells lack cytoskeletal proteins. Procaryotic organisms that are photosynthetic never have plastids; their pigment molecules are simply dispersed throughout the cytoplasm.

The table on p. 122 compares structure and function in procaryotic and eucaryotic cells. (Note that the terms *procaryotic* and *eucaryotic* (and related nouns) are also sometimes spelled with a *k,* that is, *prokaryotic* and *eukaryotic.*)

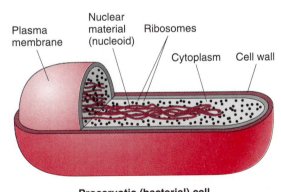

Procarvotic (bacterial) cell

Plasma membrane · Nuclear material (nucleoid) · Ribosomes · Cytoplasm · Cell wall

BOX 4.3 *(continued)*

Structure	Function	Present in Procaryotic Cells	Present in Eucaryotic Cells	
			Plant Cells	Animal Cells
Plasma membrane	Protection; communication; regulates passage of materials	✓	✓	✓
DNA	Contains genetic information	✓	✓	✓
Nuclear envelope	Surrounds genetic material		✓	✓
Several linear chromosomes	Contain genes that govern cell structure and activity		✓	✓
Cytoplasm	Gel-like interior of cell	✓	✓	✓
Cytoskeleton	Aids in cell movement and in maintaining cell shape		✓	✓
Endoplasmic reticulum	Transport and processing of many proteins		✓	✓
Golgi apparatus	Adds sugar groups to proteins and packages them into vesicles		✓	✓
Ribosomes	Protein synthesis (translation) along mRNA	✓	✓	✓
Lysosomes	Contain enzymes; aid in cell digestion; play a role in programmed cell death		✓	✓
Mitochondria	Provide cellular energy		✓	✓
Chloroplasts	Capture sunlight; produce energy for cell		✓	
Central vacuole	Maintains cell shape; stores materials and water		✓	
Flagella (whiplike appendages)	Cell movement; only found in a few types of cells in each group	✓	✓	✓
Cilia (hairlike appendages)	Cell movement; present only in certain types of cells			✓
Cell wall	Protects cell; maintains cell shape	✓	✓	
Glycocalyx	Surrounds and protects cell			✓
Intercellular links Pili (hairlike appendages)	Mating, adhesion	✓		
Plasmodesmata	Cell-to-cell communication		✓	

eucaryotic hosts. The presence of plastids is used in this book (and many others) as the defining attribute that determines the boundaries of the plant kingdom.

Kingdoms of Organisms

Early classifications of organisms recognized only two Kingdoms: Plants and Animals. Plants were distinguished as being nonmotile organisms with rigid cell walls, capable of using sunlight as a source of energy. Animals, in contrast, were recognized for their ability to move, their lack of cell walls, and their inability to derive energy directly from sunlight. This two-kingdom classification remained standard for many years, despite the discovery of animals that do not move and many bacteria and other organisms that do not fit well into either the plant or animal kingdom.

Advances in our knowledge made possible by electron microscopy, particularly the discovery of the profound structural differences between procaryotic and eucaryotic cells, led to major classification changes. A five-kingdom classification system that is widely followed today was first proposed in 1963.

Organisms did not change, but our arrangement of them changed because classifications are *socially constructed*, that is, devised by humans and agreed upon as a matter of social convention. To say that a classification scheme is socially constructed does not mean that the process of establishing a classification scheme is arbitrary or that all schemes are equally valid. Classifications are now usually understood as hypotheses about how organisms are related by patterns of descent. As more and more knowledge is accumulated about organisms, that knowledge is used to test the hypotheses and to replace rejected hypotheses with newer ones. The division of living things into five kingdoms must be viewed as a widely accepted theory, not a set of facts. Indeed, a number of biologists have proposed adding a sixth kingdom, the Archaebacteria (see Box 4.4 and Appendix).

1. Which step in the origin of life do you think was the most important? Which step marks the boundary between the nonliving and living worlds? How much do we know about each step in the process of life's origin? What means of investigation do you think will bring us additional knowledge?

2. Find four or more books on botany or general biology. List the divisions of the plant kingdom that each book recognizes. What similarities do you find? What differences do you find? How do you account for the differences?

3. How would a biologist decide which of several possible classifications is best? What would she or he look for? What kind of evidence is relevant? (These are among the most basic questions of all in taxonomy.)

E. Humans Are Products of Evolution

Charles Darwin barely mentioned human evolution in his *Origin of Species*, but public debates soon centered about questions concerning our own ancestry. Although very few fossils of humans or apes were known at the time, Darwin responded with a book entitled *The Descent of Man, and Selection in Relation to Sex*. In this book, Darwin attempted to explain certain human traits, including behavior, as resulting from individuals' choices of mates, a phenomenon that Darwin called *sexual selection*. **Sexual selection** may be defined as selection acting directly on reproductive processes, including selection by choice of the opposite sex. According to Darwin, the traits that evolved among men were those favored by women, and the traits that evolved among women were those favored by men. The issues that he raised are still being debated (see Chapter 7).

Australopithecus

In 1925 a fossilized child's skull was discovered in a cave near Taung, South Africa, and was named *Australopithecus africanus*. Although it had both apelike and human features, most experts treated it as just another ape. The discovery of additional fossils in the 1940s and 1950s enabled the anatomist W. E. LeGros Clark to show that these primates

BOX 4.4   THE KINGDOMS OF ORGANISMS

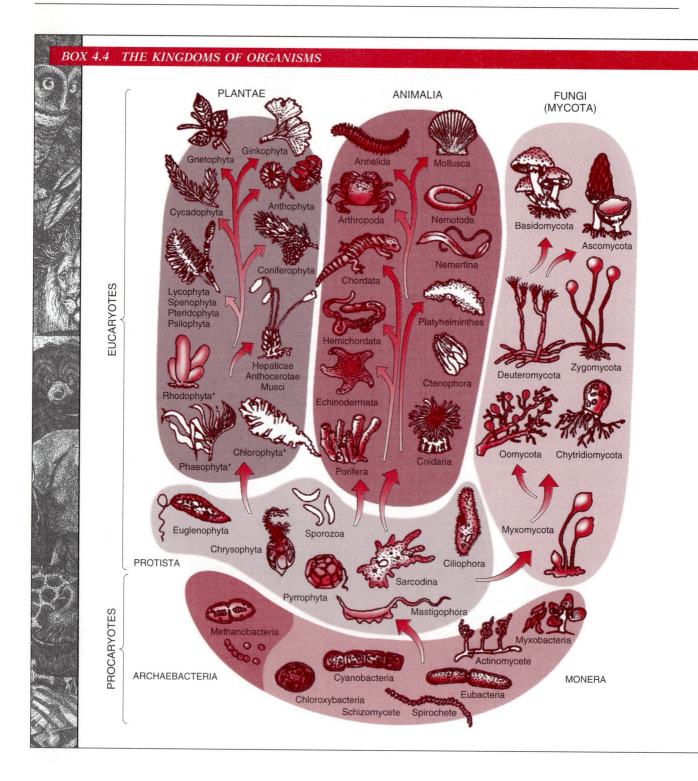

PLANTAE

ANIMALIA

FUNGI
(MYCOTA)

EUCARYOTES

Gnetophyta

Ginkophyta

Annelida

Mollusca

Basidomycota

Ascomycota

Cycadophyta

Anthophyta

Arthropoda

Nemotoda

Coniferophyta

Chordata

Nemertina

Nemertina

Deuteromycota

Zygomycota

Lycophyta
Spenophyta
Pteridophyta
Psilophyta

Hemichordata

Platyhelminthes

Oomycota

Chytridiomycota

Hepaticae
Anthocerotae
Musci

Rhodophyta*

Echinodermata

Ctenophora

Myxomycota

Phaeophyta*

Chlorophyta*

Porifera

Cnidaria

PROTISTA

Euglenophyta

Chrysophyta

Sporozoa

Ciliophora

Pyrrophyta

Sarcodina

PROCARYOTES

Mastigophora

Myxobacteria

Methanobacteria

Actinomycete

ARCHAEBACTERIA

Cyanobacteria

Eubacteria

MONERA

Chloroxybacteria
Schizomycete   Spirochete

For a more detailed classification of the king-doms of organisms, see the Appendix.

Archaebacteria

A small group of methane-producing bacteria-like organisms, some of which live at extremely hot temperatures. Nucleic acid sequences of these organisms show them to be only distantly related to the more typical bacteria with which they share the procaryotic type of cell structure explained in Box 4.3.

Monera

The majority of procaryotic organisms (see Box 4.3), including the typical bacteria and also the Cyanobacteria or blue-green organisms. Mon-era do not have a well-defined nucleus, a nuclear envelope, or any type of organelle (such as mitochondria or endoplasmic reticulum) that requires internal membranes.

Protista

Eucaryotic unicells without plastids or cell walls. Various locomotor adaptations may be present (cilia in one group, whiplike flagella in a second group, protoplasmic extensions called pseudopods in the largest group), but one group lacks motility and resembles the fungi in repro-ducing by spores. Some authorities list the algae here rather than among the plants. This and the remaining three kingdoms all have eu-caryotic cells of the type described in Box 4.3.

Mycota (Fungi)

Nonphotosynthetic eucaryotic organisms with cell walls and absorptive nutrition, reproducing by means of spores. Includes slime molds, yeasts, mushrooms, and various other forms.

Plantae (Plants)

Eucaryotic organisms with plastids, including various algae, mosses, liverworts, ferns and fern allies, conifers, and a vast array of flowering plants, from buttercups to orchids and from grasses to trees. Most plants have nonmotile life stages and cells surrounded by cell walls, whose presence strengthens plant tissues. Nearly all possess chlorophyll *a* and are capable of carry-ing out photosynthesis using sunlight.

Animalia (Animals)

Eucaryotic organisms without plastids, usually possessing a life stage with at least some loco-motor capabilities, and developing by means of an embryonic stage in which a hollow ball (blas-tula) is formed. Animals have no chlorophyll, do not carry out photosynthesis, and have no cell walls.

walked upright and were therefore more like humans than like apes. Scientists have since unearthed remains of several other species of *Australopithecus* (Fig. 4.12). The oldest species is the recently discovered *A. anamensis*, which lived about 4 million years ago in Kenya (Leakey et al, 1995). *A. anamensis* is thought to be the ancestor of all later species of *Australopithecus* and a close relative of a newly discovered species of small hominids now known as *Ardipithecus ramidus*. A somewhat later species, *Australopithecus afarensis*, is represented by the well-known skeleton known as Lucy, a female about 1.3 meters in height, a bit over 4 feet. Enough of the dimensions of Lucy's brain are known to permit us to say that these dimensions are consistent with the hypotheses that *A. afarensis* was the common ancestor of all later *Australopithecus* species and of the genus *Homo* as well. *A. robustus* and *A. boisei*, which lived from 2.3 to 1.7 million years ago, were considerably larger in size.

It is now certain that *Australopithecus* walked upright, that their brains were about as big as those of modern chimpanzees, and that at least some of them used tools. Evidence for upright walking comes from the anatomical structure of the foot, the pelvis, and the lower part of the spinal column, and also from the discovery of a set of fossil footprints at Laetoli, Kenya. The earliest *Australopithecus* came well before the earliest known *Homo*, but later *Australopithecus* persisted side by side with *Homo*, at least in East Africa.

Homo

Modern humans (*Homo sapiens*) and two extinct species are placed in the genus *Homo* (Fig. 4.13). The oldest of these species was *Homo habilis*, an East African species that coexisted with *Australopithecus boisei* and perhaps with other *Australopithecus* species as well. *Homo habilis* had a brain that was small in absolute terms (about 400 cm^3 (cubic centimeters), compared with 1200 to 1500 cm^3 for most modern humans), but the proportions of the brain to body size were more comparable to those of *Homo* than to those of *Australopithecus*. *Homo habilis* was found to be

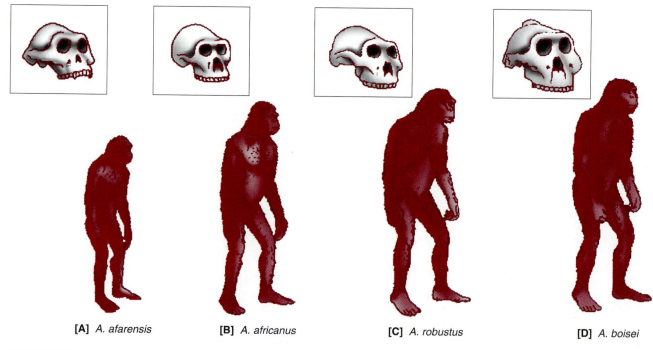

[A] *A. afarensis*   [B] *A. africanus*   [C] *A. robustus*   [D] *A. boisei*

FIGURE 4.12
Representative species of the genus *Australopithecus*.

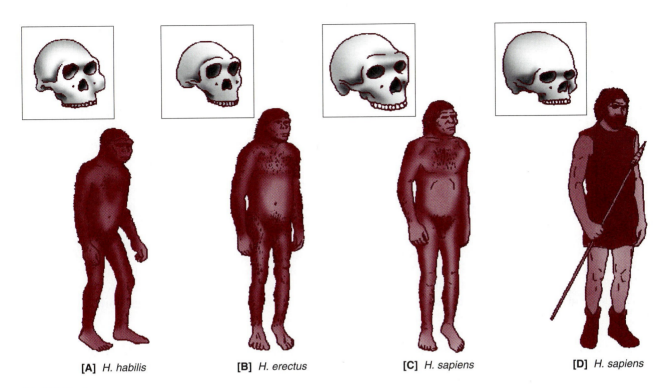

[A] *H. habilis*   **[B]** *H. erectus*   **[C]** *H. sapiens*   **[D]** *H. sapiens*

FIGURE 4.13
Representative species of the genus *Homo*.

contemporaneous with certain types of tools, including simple stone tools. It is generally presumed that *H. habilis* was the maker of these tools.

A later species, *Homo erectus*, is now known from China, Java, Europe, and several parts of Africa (Fig. 4.14). Tools associated with *H. erectus* belong to at least two distinct traditions: a hand-ax tradition in Europe and Africa, and a chopper tradition in Asia. A cave site at Choukoudian, China (near Beijing), has heat-fractured stones indicative of the use of fire. There is also evidence of round or oval tents supported by poles and held down along the margins by a circle of stones.

Homo sapiens, the modern species, was descended from *H. erectus*. With the advent of *H. sapiens*, we have evidence of even more sophisticated tools, in many cases mounted upon wooden shafts. *Homo sapiens* of the last ice age in Europe (the so-called *Neanderthals*) hunted deer, horses, and other species as large as rhinoceroses. Healed surgical wounds show that these skilled hunters

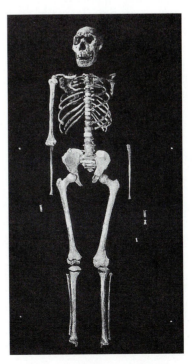

FIGURE 4.14
Skeleton of a 12-year-old *Homo erectus* boy who lived 1.6 million years ago in east Africa.

took care of sick companions, set broken bones, and even practiced simple brain surgery. They performed burial of the dead and decorated the graves with flowers of preferred colors, mostly white or cream-colored. The decoration of graves is thought by several anthropologists to indicate a belief in some form of an afterlife.

The Cro-Magnons and other more modern (Upper Paleolithic) *H. sapiens* who replaced the Neanderthals had an even greater variety of tools, including fishhooks and harpoons. They hunted wooly mammoths and large herd animals. They also left records of their activities in the form of cave paintings, showing their interest in hunting and their understanding of both animal anatomy and physiology. By prominently drawing the heart and singling it out as a target, these hunters showed that they understood how vital this organ was. Their drawings of pregnant deer and of mating rituals show that they knew enough reproductive biology to understand the relationships between mating, birth, and subsequent herd sizes.

The Neolithic stage of history begins with the discovery of agriculture. With the planting and harvesting of crops, humans began to settle down into villages. Most of the crops planted were grasses such as wheat, rye, oats, rice, and corn. All of these require grinding, mixing with water, and baking or cooking in order to make them usable as foods. Agricultural technology therefore required grinding stones (mortars), baskets and other vessels for carrying water, cooking vessels, and a variety of tools for digging, for harvesting, and for planting.

F. Evolution Continues Today

Continuing Evolution within Species

Evolution is a process that takes place within species as well as between species. The process continues in the present as it has in the past. Within the twentieth century, the peppered moths of England changed from predominantly light-colored to almost all dark (in some locations at least) and back again. In one species of Galapagos ground finches, *Geospiza fortis* (number 2 in Fig. 4.2), bill size has changed. Small-beaked birds eating soft seeds survive and proliferate best in years when rainfall is adequate, but birds with larger beaks are at an advantage in drought years because they can open large, tough old seeds. The average bill size of birds within the population thus increases in drought years and decreases in wet years. A similar phenomenon occurs among fruit flies in the southwestern United States, where one chromosomal variant is favored (and increases in frequency) during the summer months and a different variant in the winter. Not only does evolution continue, it also responds adaptively to fluctuating environmental conditions. Different alleles are selected by different environmental conditions at different times because their phenotypes are more adaptive in those conditions.

Cultural Change and Biological Evolution

Among human beings, nobody questions that cultural changes have far outstripped biological ones as the most rapid and far-reaching changes taking place today. Cultural innovation spreads rapidly, in part because there are no species barriers to prevent transmission from one society to the next. (Language barriers and geographic barriers can always be crossed, especially in the age of jet travel and television.) Cultural change is also more rapid than biological evolution because new inventions and other culturally acquired characteristics *are* inherited, although not genetically. Each generation inherits the stored knowledge of past generations (in libraries and museums, for example), along with tools (from tractors to telephones to satellites) and the technology needed to design and build new and better tools in the future.

The rapid pace and awesome power of cultural change leave many people wondering whether biological evolution of *H. sapiens* has become a thing of the past. If we need to travel faster, the argument goes, our species tames horses or builds automobiles instead of evolving longer legs by any mechanism. Evolution by natural selection is much slower than cultural innovation. In this view, the future development of our species resides more in our technology than in our bodies. Also, many selection forces that shaped human evolution in the past, including famines, epidemics, and predators, have been greatly diminished in modern times (see Chapter 6).

Continued Biological Evolution in Humans

Although it now interacts with cultural evolution and technological revolution, biological evolution continues to act today. Natural selection (and therefore continued evolution) occurs in every situation in which some people have an increased (or a decreased) chance of death, survival, or reproduction as a consequence of their genotypes. A few examples follow.

Many genetic traits, including cystic fibrosis, Tay-Sachs disease, muscular dystrophy, and others continue to cause numerous deaths prior to reproductive age, despite the best that medical technology has to offer. Even diseases that are generally survivable are often marked by a reduction in the number of offspring and thus in diminished fitness. For example, chondrodystrophy is a rare disease, controlled by a dominant gene, in which the cartilage tissue turns bony at an early age, resulting in a form of dwarfism. Most chondrodystrophic dwarfs enjoy fairly normal health as adults, but have only about one-fifth as many children as their nondwarf siblings. Lowered reproductive rates are also found among diabetics. As these examples show, natural selection as a consequence of differences in fitness continues to affect the human species.

Genetically determined blood groups such as A, B, and O have an influence on various diseases and therefore on survival. A smallpox epidemic in rural India in the 1960s killed more people belonging to blood groups A and AB, and the proportions of blood groups B and O in northern India rose as a result. Certain cancers, ulcers, and endocrine disorders also occur at different rates among the different ABO blood types. We do not know the causal mechanisms involved in these effects, but they show that natural selection may still control the frequencies of ABO blood groups in human populations (see Chapter 5).

Interaction between Cultural and Biological Evolution

Many human traits are influenced by a complex interaction of cultural and biological influences. Height, weight, body proportions, facial features, and hair are subject to both genetic and environmental influences. To the extent that people choose their mates by these criteria, evolution will continue to occur by means of sexual selection. The interaction is complex, however, because standards of beauty are in large measure culturally determined and because the people judged by any society to be the most attractive are often not the ones who bear the most children. Studies at Harvard show that the average height of students has increased over the past century, and an abundance of other evidence (e.g., clothing from past centuries) shows that the increase in height was widespread throughout the United States, Europe, and Japan. Improved nutrition may have caused some of the increase in height, so it is difficult to determine how much of the change resulted from ongoing evolution.

Traits that may once have been disadvantageous have become much less so. Poor eyesight is no longer an important barrier to survival and reproduction in societies that supply eyeglasses (Chapter 3). The ability to avoid animal predators is of little use in industrial societies. The ability to go without food for long periods may have become a disadvantage in some well-fed societies because it predisposes some people to diabetes under these conditions (see Chapter 5). For the several genetic traits that protect against malaria (see Chapter 5), the draining of swamps and the elimination of mosquitoes will shift the balance of selection among genotypes. At the same time, new qualities have become important that were once much less so: the ability to read or to graduate from high school, the ability to drive carefully and avoid accidents, and the capacity to avoid alcohol and various other drugs. These traits are subject to very strong environmental influences, but certain genetic backgrounds predispose people to take risks or to drink excessively, and these are selected against. In these and other ways, selection and evolution continue to operate and to interact with cultural evolution in human populations.

THOUGHT QUESTIONS

1. In Europe, Upper Paleolithic culture replaced the culture of the earlier Neanderthal populations rather suddenly. Do you think that the replacement of one set of tools and traditions by another took place mostly by conquest, by intermarriage, or by some combination of the two? What evidence would you look for to test one hypothesis against the others?

2. Which would have a greater influence in terms of natural selection: 100 deaths of people in their seventies or 100 deaths of people in their twenties? Why? List five or more causes of death among humans aged 30 or younger. What traits or abilities might raise or lower someone's odds of dying from each of these causes? How do you think natural selection might work on the traits you have listed?

CHAPTER SUMMARY

Evolution is one of the most important paradigms in biology. It is the central or unifying concept that provides a context into which all other aspects of biology fit. Darwin was confronted with two important earlier theories: Lamarckian and other environmentalist theories, which emphasized rapid adaptation to local conditions, and Paley's natural theology, which emphasized purposeful design by God. Darwin's paradigm encompassed two new and important theories. Of these, the theory of branching descent (descent with modification) states that each group of related species is descended, with modifications, from a common ancestor. One major mechanism of evolutionary change is described by Darwin's theory of natural selection: Inherited variations occur in all species, and different genotypes leave different numbers of surviving offspring, resulting in the eventual disappearance of those genotypes that leave relatively fewer offspring.

Descent with modification explains the known facts of geographic distribution and less-than-perfect adaptation better than the earlier theories. Darwin argued for natural selection largely by comparison with artificial selection. His supporters have shown that natural selection occurs in the natural world and that it accounts for phenomena like mimicry much better than competing theories do. Twentieth-century versions of creationism owe much to Paley's earlier views and to religious ideas that stand outside of the realm of science.

The modern evolutionary paradigm extends Darwinian concepts. New variation arises by genetic mutation. Species formation occurs through reproductive isolation arising primarily during times of geographic isolation.

Life originated on Earth under reducing conditions that included such gases as hydrogen, methane, ammonia, and water vapor. The first organisms used chemicals as sources of energy. The subsequent evolution of photosynthetic organisms able to use sunlight as an energy source brought about changes that included the production of oxygen and the development of our present atmosphere. Early cells contained no internal membrane-limited compartments, but the evolution of eucaryotic cells produced a variety of intracellular organelles containing internal compartments. We now think that eucaryotic cells originated by endosymbiosis.

The fossil record of human evolution suggests a common ancestry with apes. Upright walking evolved much earlier than brain enlargement or the extensive use of tools. Known human fossils are placed in an earlier genus, *Australopithecus*, and a later genus, *Homo*, which includes all living people. Evolution continues today in many species. In humans, biological evolution interacts with cultural evolution.

KEY TERMS TO KNOW

acquired characteristics (p. 90)

adaptation (p. 90)

analogy (p. 100)

artificial selection (p. 94)

classification (p. 115)

convergence (p. 100)

endosymbiosis (p. 119)

eucaryotic (p. 119)

fitness (p. 94)

fossils (p. 101)

geographical isolation (p. 111)

homologies (p. 98)

interbreeding (p. 110)

mimicry (p. 94)

natural selection
(pp. 93–94)

organelles (p. 119)

polyploidy (p. 114)

populations (p. 110)

procaryotic (p. 119)

recapitulation (p. 100)

reproductive isolating
mechanism (p. 110)

reproductive isolation
(p. 110)

scale of being (p. 89)

sexual selection (p. 123)

speciation (p. 111)

species (p. 110)

taxon (p. 115)

taxonomy (p. 115)

vestigial (p. 99)

CONNECTIONS TO OTHER CHAPTERS

Chapter 1 Darwinian evolution and modern evolutionary theory are both good examples of successful paradigms.

Chapter 2 Presenting creationist ideas in school classrooms raises several social policy issues.

Chapter 3 Gene mutations provide the raw material for evolution.

Chapter 5 Differences have evolved among human populations.

Chapter 6 Successful species may increase so rapidly in numbers that they outstrip the available resources.

Chapter 7 Social behavior and reproductive strategies are, in part, products of evolution.

Chapter 7 Mate choice is an important form of sexual selection.

Chapter 9 Cancer developed as a consequence of the evolution of multicellular organization.

Chapter 10 Differences in brain anatomy provide good evidence of evolution.

Chapter 13 Viruses and other microorganisms may evolve disease-causing strains.

Chapter 14 Organisms exhibit good engineering design as a result of evolution.

Chapter 15 Plant characteristics resulting from evolution include the presence of alkaloids and vascular tissues.

Chapter 16 Speciation increases biodiversity, while extinction diminishes biodiversity.

Variation among Human Populations

T he human species is highly variable. Human variation includes many physiological features, body proportions, skin color, and gene frequencies. Many of these features influence susceptibility to disease and other selective forces. Continued selection over time has produced adaptations of local populations to the environments in which they occur. Much of human variation is geographic. There are differences between geographic population groups—for example, between northern European peoples, those from eastern Africa, those indigenous to Japan, and those indigenous to, say, the mountains of Peru. Between these populations, however, lie many other populations that fill in all degrees of variation between the populations we have named, and there is also a lot of variation *within* each of these groups.

Central to the study of human variation is the concept of a biological *population*, as defined in Chapter 4 and as explained again below. Both physical features and genes vary within populations, but a good deal of variation also occurs geographically between human populations as the result of evolutionary processes. How do populations come to differ from one another? How do genes spread through populations? How do environmental factors like infectious diseases influence the spread? Why do light complexions occur more often in northern Europe than in tropical Africa, instead of the other way around? These are some of the questions that are explored in this chapter.

A. *Human Populations Vary at Many Levels*

Variation within Populations

All genetic traits in humans and other species are subject to considerable variation. Some of this variation consists of different alleles at each gene locus; other variation results from the interaction of genotypes with the environment. The simplest type of variation governs traits like those genetic diseases discussed in Chapter 3, in which an enzyme may either be functional or nonfunctional. The inheritance of these traits follows the patterns described in Chapter 3, which you may want to review at this time. In particular, be sure that you understand the meaning of *dominant* and *recessive* alleles and of *homozygous* and *heterozygous* genotypes.

Discontinuous variation. Albinism, Tay-Sachs disease, and Huntington's disease (all described in Chapter 3) are a few of the many traits known to have a simple genetic basis. For each of these conditions, only a few possible genotypes are known, and we can therefore count the number of individuals in a population that have a particular condition or genotype. Any number expressed as a fraction of the whole population is called a *frequency*. The frequency of a particular allele in the population is called its **gene frequency.** Gene frequencies are most easily studied for traits whose inheritance is known and is simple. It is important to realize that while individuals have genotypes, only populations can have gene frequencies.

The study of variation among genes in populations is called *population genetics*. The measuring of gene frequencies requires that the different genotypes, and the alleles responsible for them, can be readily distinguished from one another. It is for this reason that population geneticists often concentrate on those genes whose phenotypic effects are easy to tell apart from one another. Most of those genes control *discontinuously variable traits* that are either present or not present, such as a particular blood group or the presence or absence of certain medical conditions.

Continuous variation. While some genetic traits are either present or absent, many more traits vary over a range of values (like different heights), with all intermediate values being possible. Some of these *continuously variable* traits, like height, can be measured in an individual and expressed as a numerical value; others, like hair curliness or skin color, are seldom expressed numerically, although they theoretically could be. A description of continuous variation in a population requires the use of statistical concepts such as average (mean) values.

Continuous variation can result from the cumulative effects of multiple genes, each of which by itself contributes a small effect. Dozens of known genes, perhaps hundreds, influence height in one direction or another. If we make the simplifying assumption that these effects are independent of one another and that they add up, we can predict that a population of individuals will show a range of variation in height similar to the bell-shaped curve of Fig. 5.1. When we measure heights in any large population, we do in fact get a curve that closely matches this predicted curve. Many other continuous traits vary in much the same way as height. For most of these traits, a strong environmental component also exists; for example, height is strongly influenced by childhood nutrition as well as by genetic background. Environmental components of traits can consist of thousands of small influences (like a few extra glasses of milk or a few meals missed each week). In a population, the varying numbers of such small environmental and genetic influences in different individuals contributes to the formation of the bell-shaped curve for each population.

For a particular group of people, we can calculate an *average* height, weight, or head breadth, but these averages are just statistical abstractions—there are perfectly normal individuals that differ from the average, perhaps even greatly. Average values conceal a large amount of variation for each

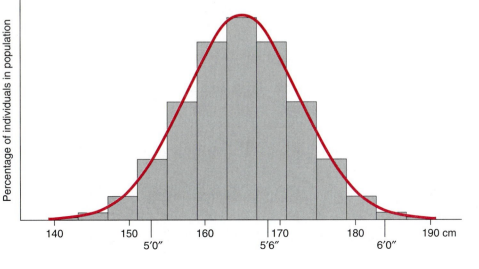

FIGURE 5.1
Continuous variation: distribution of individuals of various heights in a population whose average height is 165 cm (about 5 ft 5 in.).

trait; any statement about the height, hair, or skin color of any group of people is at best an average value surrounded by lots of variation.

For most continuous traits like height, the group average tells us little about any individual. Your own individual traits result from the influences of your mother's genes, your father's genes, and the environmental factors to which you are exposed. What you inherit from your parents is a predisposition for a range of possible future variation. For example, when you are born, your exact height as an adult cannot be predicted, but if your mother and father are both significantly taller than average, you will, if you receive adequate nutrition, probably be taller than average, also. While height can actually be measured, concepts like "tall" are relative: A height that is average in England would probably be considered tall in India or the Philippines.

Variation between Populations

Populations and species. One of the central tenets of modern biology is that evolution can only occur if populations are variable. However, biologists did not always think in terms of evolving and variable populations. For over 2000 years, biologists believed that species were constant, unvarying entities. Plato and Aristotle had declared that each species was designed according to an ideal form that they called an *eidos,* often translated as "type" or "archetype." Biologists following this view developed the **typological species concept.** For example, the eighteenth-century biologist Carl Linnaeus thought that each species had been created by God according to the model of a heavenly form, or type. Since the heavenly forms were permanent and unchanging, the species were thought to be permanent and unchanging as well. Each species was characterized by describing its *essential* features, the ones thought to typify the heavenly ideal. Because these descriptions emphasized physical features (*morphology*), the typological species concept also became a *morphological species concept*: Each species was described as having certain fixed and invariant physical characteristics. The whole "type" of that species was believed to be a cluster of "essential" characteristics inherited as a single unit.

Biologists now recognize that species are constantly evolving and that the variation that can lead to speciation begins with populations (Chapter 4). In order to describe variation among human populations, we must first have some clear way of recognizing population groups. We could group people by some physical trait, such as distinguishing people who are tall, short, or average in height. If we chose some other physical trait, like eye color or hair curliness, we would find that *each physical characteristic results in a different grouping of the same people.* In addition, we find that groupings based exclusively and strictly on any single trait will group together (especially on a worldwide basis) people who are quite dissimilar in many other respects. For these reasons primarily, biologists prefer not to base the definition of population groups on physical characteristics.

Instead of using physical characteristics to define groups of organisms, biologists separate groups based on whether or not members of the group mate with one another under natural conditions. Naturally occurring groups of organisms that are able to mate with one another in nature are said to belong to the same *species* (Chapter 4). A biological **population** is part of a species; it consists of those members of a species from which mates will actually be chosen. Thus, a biological population is more than just a collection of individuals. Biologists use the term *population* to refer to all members of a species who live in a given area and therefore can and do interbreed with one another (Chapter 4). Membership in a population is determined by mating behavior, not by physical characteristics. Population membership depends very strongly upon geographical location.

Populations that interbreed with one another under natural conditions belong to the same species (Chapter 4). All humans have the capacity to mate with one another and produce fertile offspring; for this reason, all humans are placed in a single species, *Homo sapiens.* Not all humans belong to the same population, however. Genetic variation within any population is usually less than in the species as a whole. In past centuries, geographic isolation kept many human populations more distinct than is the case now with worldwide transportation and migration. Population boundaries are not the same as national boundaries.

Several different populations often live in the same geographic area; sometimes, these populations are distinguishable by cultural factors or by their derivation from geographically separate earlier populations.

Geographic variation among populations. Human populations in different places differ from one another. The average Canadian is taller than the average Southeast Asian, and the average African has darker skin than the average European. In terms of natural selection, however, the characteristics that matter the most are those with the greatest impact on health and disease (or life and death). For example, cystic fibrosis and skin cancer are more frequent among people of European descent, while people of African descent have a higher risk of sickle-cell anemia and are more susceptible to frostbite if exposed to very cold temperatures. Most single-gene traits that are examined closely show some small average difference in gene frequency among human populations. For continuous traits, the *difference between population averages is much less than the variation within either population*: The average height in the United States is taller than in China, but many Americans are shorter than the Chinese average and many Chinese are taller than the American average.

Although it is easy to find human populations that differ from one another in both physical features (morphology) and genetic traits, it is usually very difficult to find sharp boundary lines dividing these populations from one another. If you were to walk from Asia to Europe and then to Africa, you would see populations differing only slightly, in most cases imperceptibly, from their neighbors, and you would meet with representatives of the three largest population groups on Earth *without finding any abrupt boundaries between them.*

Concepts of Race

Humans have developed various ways of describing both themselves and the other human populations with which they have had contact. *Biologists* (who study all forms of life) and *anthropologists* (social scientists who study human populations and human cultures) have assisted in these descriptions by studying and measuring certain physical traits and gene frequencies. There are many ways in which human variation can be described, and there are many uses to which these descriptions have been put. One of the most problematic has been the attempt to separate people into different **races.** As we will soon see, there are various different meanings to this term, all of them different from the term *population*, which deals with smaller and more cohesive local units. No physical features are used in defining populations, while some race concepts have been based on physical features. We will describe four different approaches to the concept of race in the order in which they originated. The older concepts have not entirely died out; they have in many cases persisted side by side with the concepts that came later.

Race as a social construct. In the Bantu languages of Africa, the word for "people" is *Bantu*. Likewise, the Inuit word for "people" is *Inuit*. Every group of people has a name for itself and its members, and the name often means *people* or *human*. Names that one people applies to other groups of people may simply be descriptive, but value judgments may be implied as well. In some instances the value judgment implicit in the choice of name has been used to justify widespread abuses against the negatively labeled population. Large-scale cereal agriculture developed independently in Egypt, Mexico, Peru, Mesopotamia, the Indus Valley, and the Huang Ho and Yangtse valleys of China. A commonly developed solution to the resulting land and labor shortages was to conquer neighboring people (the "other") and confiscate their land. Slavery and several other systems were developed to secure the labor of conquered peoples. Slavery, oppression, and conquest all call upon the victorious people to practice certain atrocities on others that they would never tolerate within their own group. In order to justify these atrocities to themselves, and to protect their own members from practicing similar atrocities on one another, just about every conquering group has found it expedient to distinguish itself from the "other," and furthermore to depict the outsiders as somehow inferior, subhuman, or deserving of their fate.

The imposition of social inequalities between "us" and "them" is now recognized as racism. **Racism** has many connotations, but all of them are based on the idea that some groups of people are better than others, and that it is somehow justified or proper for the more powerful group to subdue and oppress the less powerful. Racism has often arisen from politically or economically motivated conquest and oppression of other peoples. In most cases, the motivation to conquer and oppress others came first; the racist ideology came later and was usually an attempt to justify atrocities that would otherwise be considered reprehensible behavior within any society. The "races" identified by such conquering groups are socially constructed to serve the interests of the oppressors only. The distinctions and values of the oppressors are forcibly imposed on the oppressed, who are often taught to believe in their own inferiority.

Many of the groups that were socially defined as races in the past are really language groups, cultural groups, or national groups that are hardly distinguishable on any biological basis from the group that traditionally oppressed them. Customs, languages, and religious beliefs are among the nonbiological criteria that have been used in distinguishing socially constructed races. Since at least the eighteenth century, scientists have attempted to abandon these criteria and rely only on biological differences.

Separation based on race serves better the political and economic causes that have engendered it if the distinctions recognized are declared to be biologically based and therefore "natural" and irremediable, as opposed to characteristics that can easily be changed by education or religious conversion (and also in contrast to the acceptance of the legitimacy and equality of humanity in all of its variations). Scientists belonging to racist societies have therefore sometimes attempted to "prove" the **hereditarian** assertion that the traits characteristic of an "inferior" race have an inherited basis that cannot easily be changed. Behind these assertions is the view that a group identity (an "essence" or Platonic *eidos*) can be inherited, a view for which there is no basis in genetics. Shanklin (1994) documents several instances in which scientists conducted "scientific" studies to help "prove" the values and prejudices of their own social group. In their genocidal campaigns of the 1940s, the Nazis exterminated many millions of Jews, gypsies, Slavs, and other groups, but not until they had declared each of them to be an inferior "race."

Racism and hereditarianism are not synonymous, but they often go together as attitudes shared by many of the same people. The supporters of eugenics (see Chapter 3) had many followers, including Nazis in Germany and anti-immigrationists in the United States. These followers sought ways to prove the inferiority, especially the biologically unchangeable inferiority, of other people. Lest one think that science has long since banished such attitudes among educated people, it is only necessary to point to the great storm of controversy over the subject of race and IQ in the 1970s. Arthur Jensen (1968, 1972) attempted to convince his readers that the mental abilities of African-Americans were below those of other races and that these differences were fixed by heredity and unchangeable by educational means. His claims and the fallacies on which they were based were shown to be unsupportable by Kamin (1974), Lewontin, Rose, and Kamin (1976), and Gould (1981), among others. Racist and hereditarian attitudes die hard; they surface anew in each generation, especially during times of political uncertainty and economic hardship (Shanklin, 1994). The recent book by Herrnstein and Murray (1994) once again brings up hereditarian arguments (see Box 5.1).

One of the strange ironies of a racist past is that many attempts at remediation, such as affirmative action, continue to require, at least for a time, the identification and naming of the same groups that were used previously for racially divisive purposes. Attempts to ensure fair and nondiscriminatory treatment for members of different socially recognized racial groups (in housing, employment, schooling, and so forth) require that we first identify and study the groups that we wish to compare. In this way, societies trying to overcome a history of racism find themselves using the very racial classifications of their racist past in order to redress the injustices of past generations.

The other race concepts we will discuss differ from this earliest concept, and resemble one another, in their avoidance of language, customs, and other cultural traits in the delineation of races.

The morphological or typological concept of race. Biologists who study plant and animal species often describe the geographical variation within a species by subdividing many species into smaller and more compact subgroups, each of which is less variable than the species as a whole. These subgroups are generally called subspecies, but within our own species they are called races. In order to bring the treatment of human variation more in conformity to the treatment of other species, scientists began to restrict their attention to characteristics that could be studied biologically and to exclude personality traits, languages, religions, and customs more influenced by culture than by biology.

Before the days of oceangoing vessels, most of the world's people had only a limited awareness of human variation on a worldwide scale. Each population, of course, knew about other populations nearby, but in most cases adjacent populations differ only slightly from one another. When trade extended over great distances, it usually did so in stages, so that none of the traders ever had to go more than a few hundred miles from home. Trade routes were also in most cases traditional, meaning that traders and migrants had generally come and gone over the same routes for centuries. This contributed to a gene flow, or mixing of genes, that lessened the degree of difference between populations that would be noticed along the trade routes.

When explorers began to sail directly to other continents, they found people in other lands who differed from themselves in many physical features, and many scientists subsequently became curious about the origin of these physical differences. Discussions of racial origins from about 1750 to 1940 tended to dwell on the origin of physical differences; a *morphological definition* of each race, based on physical features (morphology), was an outgrowth of this thinking. At least initially, the major scientific founders of this tradition were people who had no interest in oppressing the newly discovered peoples, so an excuse for racial oppression was less of a motive than was scientific curiosity. The emphasis was not on distinguishing "us" from "them," but on distinguishing many different racial groups.

By the 1700s, biologists were actively describing and categorizing the variation in all living things. The eighteenth-century naturalist Carl Linnaeus divided the biological world into kingdoms, classes, orders, genera, and species (Chapter 4). He also divided humans into four subspecies: white

Europeans, yellow Asians, black Africans, and red (native) Americans. The use of physical features such as skin color and hair form to define subspecies was common among biologists using a morphological basis for race. Other scientists in this same tradition recognized more races or fewer, but each race was always described on the basis of skin color, hair color, curly or straight hair, and other morphological characteristics, such as folds of skin (epicanthic folds) over the eyes.

Under the morphological concept of race, each race was defined by listing its common physical features *as if* they were invariant. For example, when describing a feature such as color, only one color was given, as if this color were invariant throughout the group and throughout time. This approach, which classifies races on the basis of "typical" or "ideal" characteristics, and ignores variation, is called *typology. Morphological definitions of race were always typological.* Africans, for example, were declared to have black skins and curly hair, overlooking the fact that both skin color and hair form vary considerably from place to place within Africa and even within many African populations. All of the morphological characteristics were assumed to be inherited as a whole; a person was assumed to inherit a Platonic *eidos* (a "type") for whiteness or redness, not just a white or red skin.

Years after morphological races had been defined, closer scrutiny revealed both variation within the morphological races and intergradation between them across their common boundaries. A few Europeans tried to save the morphological definitions by proposing that each race had originally been "pure" and invariant, and that present-day variation within any population was the result of mixture with other races. One zoologist, Johann Blumenbach (1752–1840), divided up humans into American, Ethiopian, Caucasian, Mongolian, and Malayan races. He thought that each of these races was originally homogeneous (that is, pure), and he named each after the place that he identified as its ancestral homeland. For example, white-skinned people are called Caucasian because Blumenbach thought that this race originated in the Caucasus mountains, east of the Black Sea. There is no scientific support nowadays for the concept of originally pure races or for the concept of ancestral centers of origin; human populations have never

BOX 5.1   IS INTELLIGENCE HERITABLE?

To address a question such as this, we must first define intelligence. Intelligence is not easily defined, but it includes the ability to reason and the ability to learn new ideas and new forms of behavior, the measurement of which is far from simple. The biological bases for these abilities are likely to be multifaceted (Chapter 10), and genetic factors are likely to be the result of the interaction of many, many genes. Most discussions on the inheritance of human intelligence deal only with a single measure of this very complex trait, the IQ score, obtained from a test. IQ is not the same thing as intelligence and is at best an imperfect measure of mental abilities.

Also, to address this question, we must define the word *heritable.* Heritability is defined in statistical terms as the proportion of the *variation* in some trait associated with genetic as opposed to environmental variation. Statistical association, or correlation, does not imply causation, and it certainly cannot be used to justify the claim that "there is a gene for" the trait in question. One way to determine heritability of a trait in a domesticated species is to compare the variability of that trait in the population at large with the variability of the trait among highly inbred, genetically uniform individuals. Another

way to determine heritability is to compare the variability that a trait exhibits at large with the variability of that trait among individuals raised in a standardized, experimentally controlled environment. Neither of these methods can be applied to humans, and the measures that are used to study humans are all indirect, complicated, and subject to criticism on technical grounds. For these reasons, *there is no agreement* on the heritability of any important human ability, including "intelligence." Moreover, since variation is a characteristic of populations and not of individuals, a term such as "60% heritable" would simply be a ratio of variation in one population to variation in another and would not tell us anything about the genes or phenotype of any individual.

Numerous studies on IQ scores have shown the following:

- It is difficult to devise IQ tests that are free from cultural bias and from bias based on the language of the test, the gender and race of the test subjects, and the circumstances under which the test is administered.
- IQ scores seem to have both genetic and nongenetic components. Children's IQ scores

been homogeneous. In some cases, however, Europeans who feared for the "purity" of their own group sought to pass laws limiting contacts, especially sexual contacts, between the races that they recognized. Most of these laws were brutal but ineffective in stopping what were viewed as interracial matings. There is no scientific basis for the belief that such matings are in any way harmful. On the contrary, variation within any species confers a long-term evolutionary advantage because it provides the raw material that natural selection can use to adjust to changing environmental conditions.

The population genetics approach.   Biologists who study geographic variation are well aware that all boundaries between groups of populations are arbitrary and all transitions gradual. In order to

describe the geographic variation in *Homo sapiens* or any other species, we could draw one map showing geographic variation in average body height, another showing variation in skin color or hair form, and so on. For a dozen characteristics, a dozen different maps would be needed, for the patterns of variation would in general not coincide. Each of these mapped patterns of variation in one character is called a **cline.** Clines are an accurate (but lengthy) way of depicting the variation in each trait, one trait at a time. (See, for example, the maps in Fig. 5.2, which show the clinal distribution, or worldwide frequencies, of three blood group alleles.) Local populations must be identified and sampled *before* clinal maps can be drawn; for example, blood groups must be studied in many local populations before drawing maps such as those in Fig. 5.2. Notice that the use of clines correlate strongly with those of their parents. The IQ scores of adopted children usually agree more closely with their adoptive parents than with their birth parents, although studies on adopted children have been criticized for a variety of reasons.

- IQ scores can be greatly improved by environmental enrichment. They can also be adversely affected by poor nutrition, poor prenatal conditions, and a number of other environmental circumstances.

- Populations historically subject to discrimination, such as African-Americans in the United States, Maoris in New Zealand, and Buraku-Min in Japan, have average IQ scores about 15 points below that of the surrounding majority populations. However, these lower average scores do not always persist among people who migrate elsewhere: Descendants of Buraku-Min living in the United States have, on average, IQ scores on a par with those of other people of Japanese descent.

- In the United States, IQ scores of whites and also of blacks (African-Americans) vary from state to state, in some cases more than the average 15-point difference between blacks and whites. Among African-Americans born in the South but now living in the North, IQ scores vary in proportion to the number of years spent in northern school systems.

- Transracial adoption studies show that African-American children adopted at birth and raised by white families had IQ scores close to (in fact, slightly higher than) the white average.

- Careful studies of matched samples in Philadelphia schools failed to show significant average differences in IQ scores between black and white schoolchildren if differences in background were controlled. "Matched samples" mean that children in the study were compared only with other children of comparable age, gender, family income level, parents' occupational status, and similar variables.

Taken together, these data indicate, at most, a small degree of heritability for IQ. They provide little support for the hereditarian claim that IQ is fixed and immutable, or that observed differences in scores cannot be diminished. They provide no support whatever for predicting any individual's IQ score on the basis of their inclusion in any group.

does not lead to the establishment of names or labels that can be applied to particular populations or groups of populations.

Since the clinal variation concept was introduced in 1939, it has become customary to describe human variation by drawing maps of one cline after another. Clines can be described for morphological traits like height or skin color, but an increasing number of descriptions are of clines for gene frequencies, including frequencies of the genes determining blood groups. The techniques of molecular genetics (such as the RFLP technique described in Chapter 3) are now being used to study clines at the molecular level (e.g., Papiha et al., 1991). The population genetics approach encourages the scientific study and description of populations, including studies concerning the origins and former migrations of populations.

Following the Holocaust (1933–1945), the fledgling United Nations felt the need to refute many Nazi claims about race. The result was the 1948 *Statement on Race and Racism*, written by a committee that included several prominent anthropologists and geneticists. The statement, which has been revised several times since 1948, correctly pointed out that nations, language groups, and religions have nothing to do with race, and that no group of people can claim any sort of superiority over another. The statement went further, however, and proclaimed a new definition of race that replaced older, morphological explanations based on the inheritance of Platonic "ideal types" with a new definition based on population genetics.

Under the population genetics definition, *a race is a geographic subdivision of a species distin-*

[A] The disribution of allele *A*.

Frequencies of *A*

	< 0.05
	0.05-0.10
	0.10-0.15
	0.15-0.20
	0.20-0.25
	0.25-0.30
	0.30-0.35
	0.35-0.40
	0.40-0.50
	0.50-0.55

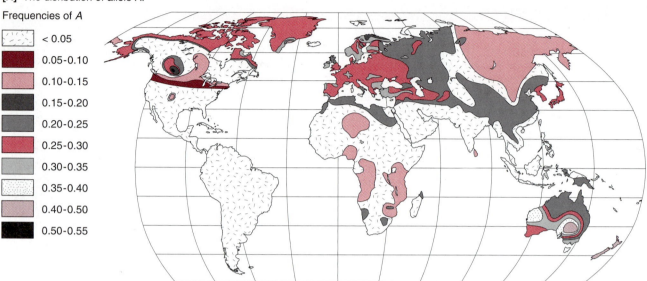

[B] The distribution of allele *B*.

Frequencies of *B*

	< 0.05
	0.05-0.10
	0.10-0.15
	0.15-0.20
	0.20-0.25
	0.25-0.30

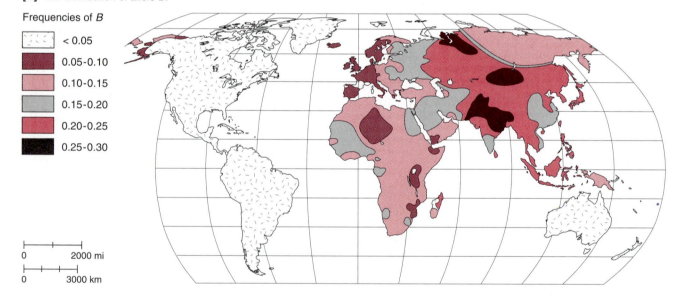

0	2000 mi
0	3000 km

FIGURE 5.2

The distribution of ABO blood groups in indigenous populations of the world, that is, populations that have lived for hundreds of years in approximately the same region, to which they have presumably become adapted to some extent. This includes Native Americans in the Western Hemisphere, Bantu and Xhoisan peoples in southern Africa, and Aborigines in Australia, but not the European colonists who came to these places after 1500.

[C] The distribution of allele *o*.

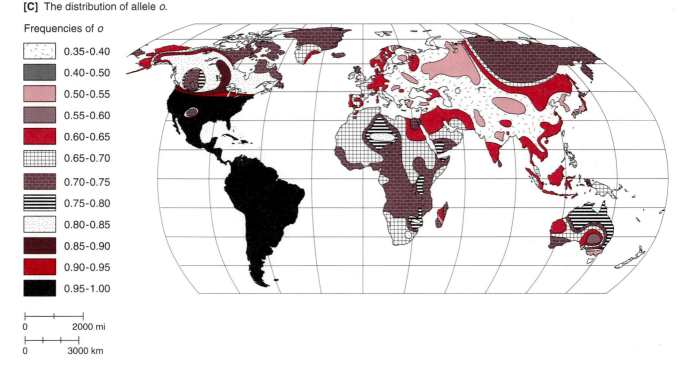

Frequencies of *o*

	0.35-0.40
	0.40-0.50
	0.50-0.55
	0.55-0.60
	0.60-0.65
	0.65-0.70
	0.70-0.75
	0.75-0.80
	0.80-0.85
	0.85-0.90
	0.90-0.95
	0.95-1.00

0   2000 mi

0   3000 km

guished from others by the allele frequencies of a number of genes. A race could also be defined as *a genetically distinct group of populations possessing less genetic variability than the species as a whole.* Either definition means that blood group frequencies are now considered more important than skin color in determining race, and that races are groups of similar populations whose boundaries are usually ill-defined. It also means that one cannot assign an individual to a race without first knowing what interbreeding population that individual belongs to. "Race" is no longer a characteristic feature of any individual, since gene frequencies characterize populations only, not individuals. Gene frequencies are consequences of population membership; they cannot be used to assign someone to a particular population or group. For this reason, racially discriminatory laws cannot and do not use population genetics; such laws rely invariably on the older morphological definitions or the still older social definitions of race.

The "no races" approach.

Some scientists went even further in rejecting the heritage of the racist past: Led by the British anthropologist M. F. Ashley Montagu (who had contributed to the United Nations definition), they declared that they would not recognize races at all. Among their arguments, one of the most compelling is that race concepts have always been misused by racists of the past and that the only way to rid the world of racism was to reject the entire concept of race. History is replete with examples of slavery, apartheid, discrimination, genocide, and warfare between racial groups. It is therefore easy to argue that the naming of races has in past generations done far more harm than good.

One stimulus to the "no races" approach arises from the great increase in international travel and migration that has occurred especially since World War II. Migrations have always brought the peoples of different continents into more frequent contact with each other. Dissimilar populations have in some cases been brought into close contact with one another, and several new populations have formed as a result.

To a certain extent, human populations have always mated with one another whenever there has been geographic contact between them. This is one reason why human population groups do not differ more than they do and why neighboring populations are so often similar. Since the advent of the jet age, frequent migrations have allowed more extensive contact and more opportunities for mating between people of different genetic backgrounds than ever existed before. Such matings have always occurred and always will; they even occur in societies that have tried to outlaw them. This type of mating will slowly but inevitably diminish the differences in the mix of genes (the so-called *gene pools*) of individual populations, making it progressively more difficult to identify *any* significant differences between populations.

THOUGHT QUESTIONS

1. Twentieth-century approaches to the description of human variation have in large measure been revolts against the earlier approaches. Against which of the earlier approaches was the "no races" approach primarily directed? Against which earlier approach was the population genetics approach directed?

2. African-Americans more often have high blood pressure and more often die from their first heart attacks than do white Americans. How would you decide whether this is the result of a difference in genes, in diets, in the availability of medical care, or in the lasting effects of discrimination in U.S. society? If people in rural Africa seldom have heart attacks or high blood pressure, what possible hypotheses are falsified?

3. To produce research results of the kind referred to in thought question 2, one must have a way of assigning an individual to a population group. How does one determine a person's membership in a biological population? Is it sufficient to know that they live in a particular place? Will asking people to name the racial or ethnic group in which they claim membership (self-identification) produce biologically meaningful results?

4. Group the students in your class on different physical criteria, such as height, eye color, hair color, and so forth. Are there clearly separated groups of "tall" and "short" people, or "brown-haired" and "red-haired" people? Do the same individuals end up together in a group regardless of what criterion is used to the delineate the groups?

B. Genetics Can Help Us Understand Human Variation

Human Blood Groups and Geography

The example of clinal variation shown in Fig. 5.2 dealt with variation in blood groups. We know a lot about the genetic basis of blood groups, and a person's blood groups are easily determined, making blood groups good candidates for study by population geneticists.

ABO blood groups.   In the days before reliable blood banks, blood transfusions were much riskier than they are today. Soldiers wounded in battle were generally treated in the field. If a transfusion was needed, it was done directly from the donor to a patient lying on an adjacent stretcher. Some transfusions were successful, but others resulted in the death of the patient. During the Crimean war (1854–1856), a British army surgeon kept careful records of which transfusions succeeded and which did not. From his notes, he was able to identify several types of soldiers, including two types that he called A and B. Transfusions from type A to type A were nearly always successful, as were transfusions from type B to type B, but transfusions from A to B or B to A were always fatal. Also discovered at this time was a third blood type, O, which was initially called *universal donor* because people with this blood type could give transfusions to anyone. These results were put to immediate practical use in treating battlefield injuries.

A German doctor, Karl Landsteiner, discovered the reason for these distinctions. Persons with blood type A make a carbohydrate of type A, which appears on the surfaces of their blood cells. Persons with blood type B make a carbohydrate of type B; persons with type AB make both type A and B carbohydrates; and people with blood type O make neither of these carbohydrates. The A and B carbohydrates are also called *antigens* because they are capable of being recognized by the immune system. The immune system of each individual also makes *antibodies* against the blood group antigens that their own body does not make (Box 5.2). In a person receiving a transfusion with incorrectly matched blood, these antibodies will bind to the

BOX 5.2   THE ABO BLOOD GROUPS

The antigens of the ABO blood group system are carbohydrates on the surface of the red blood cells. People of blood type A have antigen A on the surface of their blood cells, and people of blood type B have antigen B. People of blood type AB have both antigens; people of blood type O have neither.

People who do not produce antigens of type A will produce antibodies against A, and people who do not produce antigens of type B will produce antibodies against B.

If a person who receives a blood transfusion has antibodies against the antigens on the surface of the donor's blood cells, then a reaction occurs in the recipient that causes the donated blood cells to clump together, a process called *agglutination*. Agglutinated blood cells clog the blood vessels and the recipient often dies within a short time.

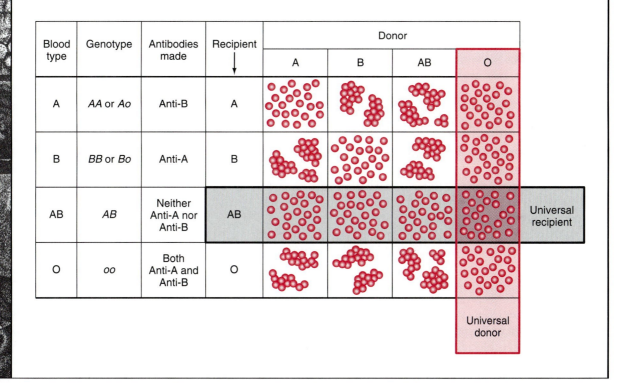

Blood type	Genotype	Antibodies made	Recipient ↓	Donor			
				A	B	AB	O
A	*AA* or *Ao*	Anti-B	A				
B	*BB* or *Bo*	Anti-A	B				
AB	*AB*	Neither Anti-A nor Anti-B	AB				Universal recipient
O	*oo*	Both Anti-A and Anti-B	O				Universal donor

type A or B antigens, causing the blood cells to clump together within the blood vessels, often with fatal results. For explaining these immune reactions, Landsteiner received the Nobel Prize in 1930.

The A and B antigens allow all people to be classified into the four blood groups A, B, AB, and O. We now also know that these blood groups are controlled by a gene that includes three possible forms (called *alleles*) instead of two: Allele *A* is dominant and it contains information for the cells

to produce a protein enzyme that adds the carbohydrate antigen A (its phenotype) to the cell surface; allele *B* is dominant and it contains information for producing antigen B (its phenotype); allele *o* is recessive, and it functions as a "placeholder" on the DNA while producing neither functional antigen. The *AA* and *Ao* genotypes both produce antigen A and are therefore assigned to blood group A. Likewise, both *BB* and *Bo* genotypes produce antigen B and result in the B blood type. Genotype *oo* produces neither A nor B antigens, which results in the O blood type (universal donor). Finally, genotype *AB* allows both alleles *A* and *B* to produce their respective antigens, resulting in the AB blood type. When they occur together, the *A* and *B* alleles are said to be *codominant* because the heterozygote shows the effects of *both* phenotypes.

For the purpose of matching blood donors and recipients, any person who shares your blood type is a good donor. It is therefore possible to collect blood in advance from many donors, sort the blood by blood type, and store it under refrigeration for use in an emergency. It is ironic that the doctor who developed this concept, an African-American named Dr. Charles Drew (1904–1950), was denied its full benefits because many hospitals kept separate blood banks for whites and nonwhite patients, a practice that has no biological foundation. Because the chemical composition of the allele products does not vary, type A antigen from an African-American is identical to type A antigen from a Native American or from anyone else. A person with blood type A is therefore a good donor to any other person with blood type A.

The frequency of the ABO blood groups varies greatly from one human population to another. The variations, however, do not necessarily coincide with other traits or with the groups that had been recognized on the basis of morphology. Type B, for example, reaches its highest frequency on mainland Asia, but is nearly absent among Native American populations or among Australian aborigines. The frequency of type A increases from east to west across Asia and Europe. Among Native American populations, blood group A occurs mostly in Canada, while almost none of this blood group occurs among Central or South American native populations. The allele for blood group O has a frequency of 50 percent or more in most human populations, but its frequency approaches 100 percent among Native American populations south of the United States. African populations generally have all three of the ABO blood group alleles at moderate levels.

Other human blood groups. Another totally independent system of blood groups, called the Rh system, actually has three genes located very close together on the same chromosome: The first is either *C* or *c*, the second is either *D* or *d*, and the third is either *E* or *e*. In all, there are eight possibilities, of which *cde* is sometimes called *Rh negative* and the others *Rh positive*. Genotype *CDe* is the most frequent genotype in most populations, except in Africa south of the Sahara, where *cDe* predominates. The Rh-negative genotype, *cde*, is the second most common Rh genotype in Europe and Africa, but is rare elsewhere. The *cde* genotype causes problems when a mother who is Rh negative (homozygous for *cde*) is pregnant with a baby who is Rh positive. In this case, the mother makes antibodies against antigens in the baby's blood, especially in response to the tearing of blood vessels during the process of birth. Once these antibodies are made, the mother's immune system will attack any subsequent pregnancy with an Rh-positive fetus (Fig. 5.3). This problem can now be prevented by giving the Rh-negative mother gamma globulin (e.g., RhoGAM) at the time of the birth of any Rh-positive child; the globulin inhibits the formation of antibodies against Rh.

Separate from the ABO and Rh blood group systems are an MN system (with M most frequent among Native Americans and N among Australian Aborigines), a Duffy blood group system (with alleles *Fy*, *Fy^a*, and *Fy^b*), and many others.

Geographic variation in blood group frequencies. Table 5.1 shows how the major geographic subgroups of *Homo sapiens* differ in the frequencies of various blood groups and other genetic traits. It is important to remember that a gene *frequency* is something that an individual cannot have. Gene frequencies characterize populations only, not individuals—each of us has only a single genotype for each gene pair, or a single blood type for each blood group system.

Geographic variations in blood group frequencies take place on a smaller scale also. This is especially true among rural people who remain in their native villages or districts all their lives. The geneticist Cavalli-Sforza has documented variations in the ABO, MN, and Rh blood group frequencies from one locality to another across rural Italy. Similar results have been observed among rural populations in the valleys of Wales, among African-Americans from city to city across the United States, and among the castes and tribes of a single province in India. These studies empha-

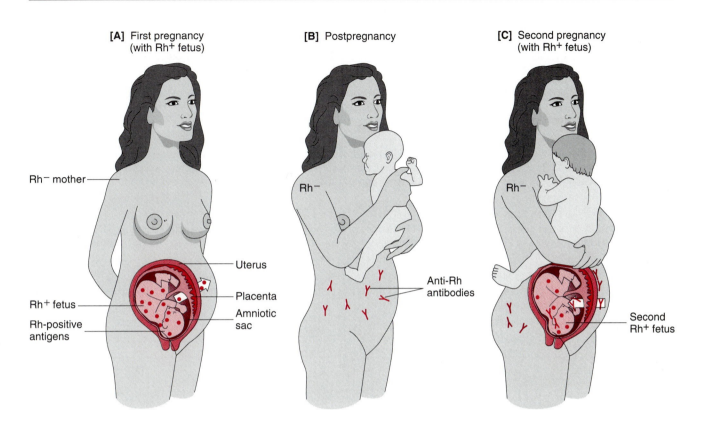

FIGURE 5.3

Rh incompatibility arising in an Rh-negative mother pregnant with an Rh-positive child.
When an Rh-negative mother has her first Rh-positive pregnancy [A], some of the baby's
antigens may leak across the placental barrier, and many more will enter the mother's
circulation during the detachment of the placenta following birth. The mother's immune
system will then make antibodies against Rh [B], which can endanger any subsequent Rh-
positive fetus [C] unless protective measures are taken.

size the hazards of assigning all people in a single
country to a single population, especially when
cultural barriers discourage random mating. However, populations that have become more mobile
experience less of this microgeographic variation.
As stated earlier, these are variations in gene frequencies that cannot be used to establish clear-cut
boundaries between populations.

Isolated Populations and Genetic Drift

In large, randomly mating populations in which
selection and migration are not operating, the frequencies of the genotypes in the population tend to
remain the same. This principle, which operates in
all biological species, is called the **Hardy-Weinberg principle** (or law), and the predicted
equilibrium is called the *Hardy-Weinberg equilibrium* (Box 5.3, pp. 148–149).

One of the criteria for a Hardy-Weinberg equilibrium is that the population be large. In small
populations, gene frequencies tend to vary erratically, in unpredictable directions, from the
expectations of the Hardy-Weinberg equilibrium.
This phenomenon is called **genetic drift,** defined
as changes in gene frequencies in small to
medium-size populations due to chance alone. (We
can now define a *large* population as one of sufficient size that genetic drift does not occur, and
hence each generation receives a representative
sample of genes from the previous generation.)

TABLE 5.1   GENE FREQUENCIES THAT VARY AMONG THE MAJOR GEOGRAPHICAL POPULATION GROUPS

African Populations

Frequencies of ABO and MN blood group alleles close to world averages; Rh blood group system with alleles *cDe* most frequent and *cde* second; allele *Fy* most common in Duffy system; allele P_1 more common than P_2 in P system; more Hb^S and Hb^C than in most other populations.

European or Caucasian Populations

Frequencies of MN blood group alleles close to world averages; blood group A somewhat more frequent than in African or Asian populations; Rh blood group system with alleles *CDe* most frequent and *cde* second; Duffy system with allele Fy^a most frequent and Fy^b second; alleles P_1 and P_2 both common; more G6PD deficiency and thalassemia than in most other populations.

Asian Populations

High frequencies of blood group B (and correspondingly less of A); MN frequencies close to world averages; Rh blood groups with *CDe* most frequent and *cde* rare or absent; Fy^a especially common; allele P_2 more common than P_1; some populations with high rates of thalassemia or ovalocytosis.

Native American or Amerindian Populations

Very high frequencies of O and M and virtually no B; high frequencies of Di^a in the Diego system; Rh blood groups with *CDe* most frequent and *cde* rare or absent.

Australian and Pacific Island Populations

High frequencies of blood group N; Rh blood groups with *CDe* most frequent and *cde* rare or absent.

The original model of genetic drift dealt with populations that remained small all the time, but other types of genetic drift were found to apply in particular situations. For example, a large population that became temporarily small and then large again would experience a *bottleneck effect*: The random changes that occurred when the population was small—the bottleneck—would be reflected in the gene frequencies of subsequent generations. Another type of genetic drift is known as the **founder effect.** If a small number of individuals become the founders of a new population, then the gene frequencies in the new population—whatever its subsequent size—will reflect the composition of this small group of founders.

Several cases of genetic drift have been studied in isolated human populations. One well-studied example concerns the German Baptist Brethren, or Dunkers, a religious sect that originated in Germany during the Protestant Reformation. Forced to flee their native Germany, a few dozen Dunkers came to Pennsylvania in the 1600s and started a colony that now numbers several thousands, mostly in rural Pennsylvania and neighboring Ohio. Because their strict religious code forbids marriage outside the group, they have remained a genetically distinct population.

Gene frequencies among the Dunkers have been influenced by genetic drift, particularly by the founder effect. If the Dunkers were a representative sample of seventeenth-century German populations, we would expect similar gene frequencies to those of present German populations derived from the same source. If, on the other hand, natural selection had changed the Dunker populations as the result of adaptation to their new location, then we would expect their gene frequencies to come closer to those of neighboring populations of rural Pennsylvania. Neither of these predictions is correct. Gene frequencies among the Dunkers differ from populations of *both* western Germany and rural Pennsylvania in a number of traits that have been studied. Blood group B, for example, hardly occurs at all among the Dunkers, although its gene frequency is around 6 to 8 percent in most European populations, including those of both Germany and Pennsylvania. Other genetically determined traits show similar patterns, including the nearly total absence of the Fy^a allele (from the Duffy blood group system) among

Dunkers. The explanation that best agrees with the data is that the original founder population, known to have been made up of only a few dozen individuals, happened not to include anyone carrying the genes for blood group B or Fy^a. Additional alleles may have been lost by genetic drift while the population remained small. The result was a population that derived its gene frequencies from the assortment of genes that happened to be present among the founders. We can test this assumption by looking for the rare Dunkers who do possess an allele such as Fy^a. In every case that has been investigated, the occurrence of such an allele among the Dunkers can be traced to a documented religious convert who joined the group within the last few generations.

Because they are genetically isolated, except for occasional religious conversions, the Dunkers have kept a unique combination of unusual gene frequencies. In the absence of blood group B, they resemble Native American populations; in the absence of Fy^a, they resemble African populations. In most traits, however, their derivation from a European source population is evident. These findings show that populations resemblances based on a single blood group or gene system may often be misleading, and that *distinctions among human populations, if used at all, should be based on a multiplicity of gene frequencies.*

The bottleneck effect has been used as a hypothesis to explain the near-total absence of blood group B among Native Americans and of *cde* (in the Rh blood groups) among Pacific islanders. When these groups first migrated from Asia, the random changes in gene frequency that occurred when the groups were small gave rise to distinct, isolated populations whose gene frequencies differed from those of the ancestral populations. Genetic drift of this kind would apply primarily to groups of people, like the Polynesians or Native Americans, whose founder populations were initially small. The effects of genetic drift are minimal among the larger and more widespread population groups of Africa, Europe, or mainland Asia.

Reconstructing the History of Human Populations

Gene frequencies in modern populations can be used as clues to their origins. For example, Cann, Stoneking, and Wilson (1987) worked out a family tree of human populations on the basis of mitochondrial DNA. Mitochondria are organelles in the cytoplasm of eucaryotic cells (Chapter 4) that produce much of the cell's energy and that also contain small strands of DNA, independent of the DNA in the nucleus. Mitochondrial DNA is transmitted only maternally, from mother to both male and female offspring. Sperm from the father contain almost no cytoplasm and do not transmit mitochondrial DNA. Because mitochondrial DNA is smaller than nuclear DNA, it is ideal for tracing evolutionary patterns. One feature of this family tree is that several geographic areas (such as New Guinea) appeared in two or more parts of the tree, suggesting that these geographic areas were colonized more than once. Cavalli-Sforza et al. (1988) used 120 alleles to study the genetic similarities among 42 populations representing all the world's major population groups and many small ones as well. The findings of these two studies support the hypothesis of a primordial divergence in prehistoric times between African and non-African populations, with the non-African populations later splitting into Northeurasian and Southeast Asian subgroups (Fig. 5.4, p. 150). Australian Aborigines and Pacific islanders are descended from the Southeast Asian subgroup, while Caucasians (Europeans, West Asians) and Native Americans are both descended from the North Eurasian group, which also includes Arctic peoples. The groups suggested by this study are geographically coherent and confirm certain well-documented patterns of migration. Existing linguistic evidence also matches these groupings, except for a few cases of cultural borrowing that can be documented historically. Cavalli-Sforza's group estimates, largely on the basis of archaeological evidence, that the split between African and non-African populations took place 92,000 or more years ago.

The study of gene frequencies has also been used to determine the origins of particular groups of people. For example, Saha and Tay (1992) have shown that Koreans are derived from a group that includes the Mongolians and Japanese but not Chinese. Also, separate studies by Rickards et al. (1992) and by Walter et al. (1991) have provided evidence for a Middle Eastern contribution (perhaps via Phoenician sailors) to the populations of both Sicily and Sardinia.

BOX 5.3   THE HARDY-WEINBERG EQUILIBRIUM

The Hardy-Weinberg law may be stated as follows:

> In a large, random-mating population characterized by no immigration, no emigration, no unbalanced mutation, and no differential survival or reproduction (that is, no selection), *the frequencies of the several genotypes will tend to remain the same.*

The law, in other words, specifies the conditions under which there will be a simple equilibrium of unchanging gene frequencies.

Let us consider the case of a gene locus that contains two alleles, A and a. If the frequency of allele A is called p and the frequency of allele a is called q (where $p + q = 1$, because the two must add up to 100% of the population's gene pool), then the equilibrium frequencies of all three diploid genotypes is given by the Hardy-Weinberg formula:

Genotypes	AA		Aa		aa		
Frequencies	p^2	+	$2pq$	+	q^2	=	1

This formula predicts that the frequency of genotype AA will be p^2, the frequency of genotype Aa will be $2pq$, and the frequency of genotype aa will be q^2.

To show that these equilibrium frequencies will remain stable over successive generations and will not tend to change in either direction, consider the production of gametes in a population already at equilibrium. The proportion of gametes that carry gene A will include all the gametes (p^2 of them) produced by the dominant homozygotes AA (p^2 in the equation), plus half of the gametes (half of $2pq$, which equals pq) produced by the heterozygotes Aa ($2pq$ in the equation). We can then use simple algebra, separating out the common factor and then applying the equation $p + q = 1$ to calculate the frequency of each type of gamete:

Genotypes	AA		Aa		aa		
Frequencies	p^2	+	$2pq$	+	q^2	=	1

$$p^2 + pq = p(p + q)$$
$$= p(1)$$
$$= p$$

Frequency of A gametes

$$pq + q^2 = (p + q)q$$
$$= (1)q$$
$$= q$$

Frequency of a gametes

How did these various population differences come about? The next two sections attempt to provide some of the answers to this question.

THOUGHT QUESTIONS

1. *Random mating* in a sexual species means that any two opposite-sex individuals have the same chance of mating as any other two. If there are a million individuals of the opposite sex, then each should have an identical chance (one in a million) of being chosen as a mate. Do you think human populations mate at random? Why or why not?

2. Is there ever a real population (of any species) in which the conditions specified by the Hardy-Weinberg equilibrium exist? How close do particular populations come?

3. If language has nothing to do with race, why do you suppose researchers attempting to reconstruct the past history of human populations use linguistic evidence?

C. Many Human Traits Help Protect against Malaria and Other Diseases

Diseases are among the selective forces that can result in differences between populations. In this

So the frequency of gametes corresponds to the frequency of alleles. Combining the gametes at random produces the following results:

Female gametes

		A	a	← Gametes
		p	q	← Frequencies
A		AA	Aa	← Genotypes
p		p^2	pq	← Frequencies
a		Aa	aa	
q		pq	q^2	

Male gametes

Taking the resulting genotypes from the chart above (and adding the two heterozygous combinations together), we obtain:

$$\underset{p^2}{AA} + \underset{2pq}{Aa} + \underset{q^2}{aa} = 1$$

which shows that the frequencies have not changed. It can also be shown that a population that does not start out at equilibrium will establish an equilibrium in a single generation of random mating.

Notice all the assumptions of the model: The population must be closed to both emigration and immigration, and there must be no unbalanced mutation and no selection. The population must be large enough to permit accurate statistical predictions, and the population members must mate at random. In reality, most natural populations are subject to mutation, selection, and nonrandom mating (including inbreeding), and those that are not geographically isolated (on islands, for example) usually experience emigration and immigration as well. The Hardy-Weinberg model, in other words, describes an idealized situation that is seldom realized in practice. The Hardy-Weinberg equilibrium is important to population genetics as an ideal situation to which real situations can be compared; if a population is *not* in Hardy-Weinberg equilibrium, one can ask *why* and then seek to measure the extent of the deviation from equilibrium. The same procedure is followed in other sciences as well. For example, "freely falling bodies without air resistance" are an ideal situation in physics, and air resistance can be measured as a deviation from this ideal.

section, we consider some genetic traits that confer partial resistance to malaria. In malaria-ridden areas, natural selection acts to increase the frequency of genes that confer partial resistance to malaria while decreasing the frequency of genes that leave people susceptible to malaria. Many other selective forces have also operated over the course of human history, but resistance to malaria provides a series of well-studied examples.

New traits are produced by mutation (Chapter 3) and are then subjected to natural selection, a process in which many traits die out. The traits that survive natural selection are adaptive traits, or **adaptations** (Chapter 4), that is, traits that increase a population's ability to persist successfully in a particular environment. A good deal of human variation consists of adaptations that have resulted from natural selection operating over time, disease being a significant agent of that selection process.

Malaria

On a worldwide basis, malaria causes over 110 million cases of illness each year and causes close to 2 million deaths, more than almost any other single disease except for malnutrition. Malaria also has the greatest impact of any disease on the average human life span, since most of its victims are young and more years of life are lost for each death that occurs. Malaria is more prevalent in tropical and subtropical regions than in temperate climates. The threat of malaria has largely been

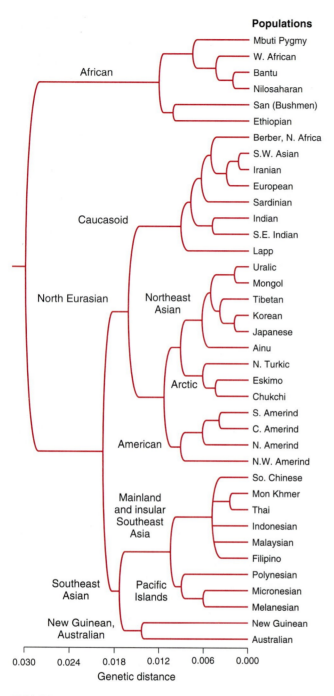

Populations

- Mbuti Pygmy
- W. African
- Bantu
- Nilosaharan
- San (Bushmen)
- Ethiopian
- Berber, N. Africa
- S.W. Asian
- Iranian
- European
- Sardinian
- Indian
- S.E. Indian
- Lapp
- Uralic
- Mongol
- Tibetan
- Korean
- Japanese
- Ainu
- N. Turkic
- Eskimo
- Chukchi
- S. Amerind
- C. Amerind
- N. Amerind
- N.W. Amerind
- So. Chinese
- Mon Khmer
- Thai
- Indonesian
- Malaysian
- Filipino
- Polynesian
- Micronesian
- Melanesian
- New Guinean
- Australian

African

Caucasoid

North Eurasian

Northeast Asian

Arctic

American

Mainland and insular Southeast Asia

Southeast Asian

Pacific Islands

New Guinean, Australian

0.030   0.024   0.018   0.012   0.006   0.000

Genetic distance

FIGURE 5.4
A family tree of human populations constructed on the basis of mitochondrial DNA sequences. "Genetic distance" refers to the fraction of mitochondrial DNA sequence not shared between any two populations.

eliminated in the industrially developed countries through mosquito eradication programs and the draining of swamps, but in the nineteenth century and even into the first half of the twentieth, malaria claimed many thousands of victims in Florida, Louisiana, Mississippi, and Virginia.

Historical and anthropological evidence confirms that malaria was rare (and therefore not a significant selective force) before the invention of agriculture. Even today, the disease is rare in undisturbed forests or among hunting and gathering societies. The clearing of forests for agricultural use opens up more swampy areas, and the building of irrigation canals or drainage ditches creates additional pools of stagnant water. The mosquitoes that carry malaria breed best in stagnant water open to direct sunlight. Agriculture therefore did much to change the agents of death (and thus the selective pressures) that act on human populations, and many of these changes were in unintended directions.

Life cycle of *Plasmodium.* Malaria is caused by one-celled protozoan parasites belonging to the phylum Sporozoa, genus *Plasmodium*. Of the four species of *Plasmodium* that cause malaria, *Plasmodium falciparum* is the most virulent. All species of *Plasmodium* have a complex life cycle (Fig. 5.5), spending different parts of their cycle in two different host species, humans and mosquitoes. The infectious male and female sexual stages (gametocytes) are intracellular parasites that inhabit human red blood cells. When a female mosquito of the genus *Anopheles* takes her blood meal, she ingests large numbers of these parasites into her gut. Gametes combine here to form *zygotes* (fertilized eggs). The zygotes develop through several stages, culminating in the infective forms (sporozoites) that migrate into the mosquito's salivary glands.

The mosquito's thin mouthparts function like a tiny soda straw or hypodermic needle. Shortly before consuming its blood meal, the mosquito injects saliva containing anticoagulants into the victim; the anticoagulants prevent the human blood from clotting inside its mouthparts. When the mosquito injects its saliva into a new human host, the parasites enter the human bloodstream and are taken up by the liver. Each parasite then develops into thousands more, which may remain

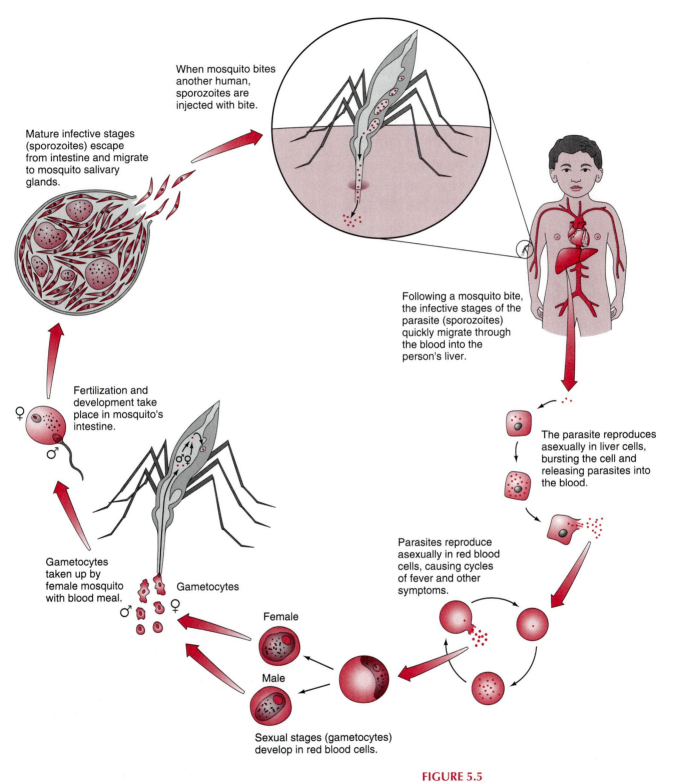

When mosquito bites another human, sporozoites are injected with bite.

Mature infective stages (sporozoites) escape from intestine and migrate to mosquito salivary glands.

Following a mosquito bite, the infective stages of the parasite (sporozoites) quickly migrate through the blood into the person's liver.

Fertilization and development take place in mosquito's intestine.

The parasite reproduces asexually in liver cells, bursting the cell and releasing parasites into the blood.

Gametocytes taken up by female mosquito with blood meal.

Gametocytes

Parasites reproduce asexually in red blood cells, causing cycles of fever and other symptoms.

Female

Male

Sexual stages (gametocytes) develop in red blood cells.

FIGURE 5.5
Life cycle of the malaria parasite *Plasmodium*.

in the liver for years, causing periodic relapses in infected patients. Some parasites escape into the bloodstream and invade the red blood cells. The parasites reproduce asexually within the red blood cells, producing the disease symptoms. The parasites digest the cell's oxygen-carrying molecule, hemoglobin, and one stage also ruptures the red blood cells. Any impairment of the ability of the blood to carry oxygen to the body's tissues is called an **anemia;** all anemias leave their victims run-down and weakened. In the case of malaria, the anemia is caused by destruction of both the hemoglobin and the red blood cells. Cell rupture also brings on fevers, headache, muscular pains, and liver and kidney damage. Within a given host, the asexual cycle of *Plasmodium* continues again and again until the patient either recovers or dies. In the red cells, the parasites can also develop into the sexually reproducing gametocytes, which may be picked up by another mosquito in its next blood meal, spreading the disease.

Malarial drugs and drug resistance.   There is no cure for malaria, but quinine and chloroquine are two drugs commonly given. Both these drugs suppress the asexual cycles of the parasite and thus relieve the symptoms, but relapses may occur months or years later because parasites are still present. Many strains of the parasite have become chloroquine-resistant, however, as a result of natural selection. Spontaneous mutations for resistance to a particular drug are always occurring among microorganisms. Those microorganisms (and their offspring) that are able to survive in the presence of the drug are selected for, and they rapidly outnumber nonresistant organisms that are killed off by the drug.

Sickle-Cell Anemia and Resistance to Malaria

A very serious type of anemia was first discovered in 1910 by a Chicago physician named Charles Herrick. This strange and usually fatal disease also produced abnormally shaped red blood cells that sometimes resembled sickles. For this reason, Herrick called the disease *sickle-cell anemia.*

A simple blood test was soon devised to test for the condition: A thick slide containing a bowl-shaped depression was used, and a drop of the patient's blood was placed inside the depression. A ring of vaseline was placed around the margins of the depression and a cover glass was then applied, forming an airtight seal with the vaseline. As soon as the cells used up the available oxygen in the depression, a condition of low oxygen tension prevailed and, under these conditions, the red blood cells of a person with the disease would assume their characteristic sickle shape, while normal red blood cells retained their circular, biconcave shape (Fig. 5.6). This blood test also allowed the recognition of heterozygous carriers, half of whose blood cells would sickle, while the other half remained round.

Normal and abnormal hemoglobins.   Sickle-cell anemia is caused by an abnormality in the **hemoglobin** molecules that carry oxygen within the red blood cells. The hemoglobin molecule consists of four protein chains (two each of two different proteins) surrounding a ringlike *heme* portion. The heme structure is responsible for hemoglobin's red color. Suspended in the middle of this ring is an iron atom that can switch from the Fe^{2+} to the Fe^{3+} oxidation state by losing an electron. The Fe^{2+} form of the iron atom can bind one oxygen molecule (O_2), giving the hemoglobin its ability to transport oxygen.

A change in a single amino acid, number 6 in one of the protein chains, is responsible for sickle-

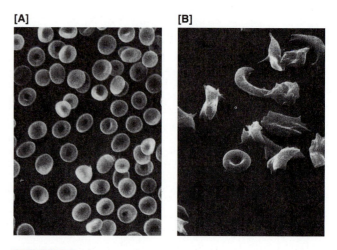

[A]   **[B]**

FIGURE 5.6
Normal red blood cells [A] and red blood cells from a patient with sickle-cell anemia [B].

cell anemia. Normal adult hemoglobin (*hemoglobin A*) has glutamic acid in this position in the chain, while sickle-cell hemoglobin (*hemoglobin S*) has valine instead. This minute change makes the hemoglobin S molecules stickier; these molecules adhere to one another and also to the inside of the red cell membrane, deforming the cells into the characteristic sickled shapes. The altered shape strains the ringlike part of the molecule so that hemoglobin S does not carry oxygen as well as hemoglobin A. The difference in the proteins is hereditary and is caused by an altered codon in the hemoglobin gene on the DNA.

The genetics of sickle-cell hemoglobin.

Sickle-cell anemia is inherited as a simple Mendelian trait. People who die from sickle-cell anemia are always homozygous and their parents are almost always heterozygous, as are a certain number of siblings and other relatives. The gene for hemoglobin is designated *Hb* and the different variants of the gene (alleles) are designated by superscripts, Hb^A being the gene code for normal hemoglobin and Hb^S coding for sickle-cell hemoglobin.

In U.S. and Caribbean populations, the vast majority of people carrying the gene for sickle-cell anemia are blacks of African ancestry. Tests of African populations also show high frequencies of the sickling allele, up to 25 percent in certain populations. In homozygous individuals ($Hb^S Hb^S$), all the red blood cells are deformed. Heterozygous individuals ($Hb^A Hb^S$) have both types of hemoglobin and about half of their red blood cells are sickled, and the rest are normal in shape. Since both alleles produce a phenotypic result among heterozygotes, they are codominant, as we described earlier in connection with the AB blood type.

Symptoms of sickle-cell anemia.

Most of the debilitating symptoms of the disease are consequences of the deformed, sickle-shaped cells. The smallest blood vessels, capillaries, have a diameter only slightly larger than the diameter of blood cells (Chapter 14). Because of their sickle shape and changed diameter, sickled cells cause flow resistance in the capillaries and thus impair microcirculation. In most of the body's organs, impaired microcirculation brings about *hypoxia* (reduced oxygen levels), which results immediately in a severely painful condition known as *sickle-cell crisis*.

These crises begin in infancy. Damaged cells collect in the capillaries of the joints and result in painful swelling. The sickled cells are also more rigid than normal and so are more readily destroyed than the normal-shaped round ones, resulting in a reduced number of red blood cells. The anemia that results from the reduced blood cell count and reduced O_2-carrying capacity brings about weakness, poor physical development, impaired mental function, enlargement of the heart, and impaired immune function and consequent susceptibility to infection. The bone marrow responds to the loss of blood cells by proliferating and enlarging, stepping up its production of new blood cells; an abnormal, "tower" shape of the skull is one result of these growth patterns. The sickled cells accumulate in the spleen, which thus enlarges. Repeated or chronic hypoxia also results in tissue damage to many organs, eventually resulting in death (Fig. 5.7). In African populations, the death of homozygous $Hb^S Hb^S$ individuals often occurs before adulthood, but in the United States and the Caribbean, survival to reproductive age is now increasingly common. The reduction in red blood cell number and the sickle-cell crises also occur among heterozygotes, but is not as severe.

Treatment for sickle-cell anemia.

Treatment for the disease is difficult and expensive, requiring frequent blood transfusions to augment the number of normal blood cells, administration of additional oxygen, and the use of potent pain-killing drugs during a sickling crisis. Patients are advised to avoid all conditions in which tissues have extra oxygen demands, such as cold temperatures, high altitudes, and heavy manual labor.

Some researchers have tried developing drugs that inhibit the grouping together of the hemoglobin S molecules to prevent the sickling. Most of these medicines have serious side effects that prevent their widespread use. Cyanate, which once looked promising, is too toxic. Another anti-sickling drug is being developed in Nigeria from the neem tree, twigs from which are widely used in West Africa as chewing sticks, which have antibacterial effects and which therefore substitute for both toothbrushes and toothpaste. In 1995 a new anti-sickling drug, hydroxyurea, was approved in the United States.

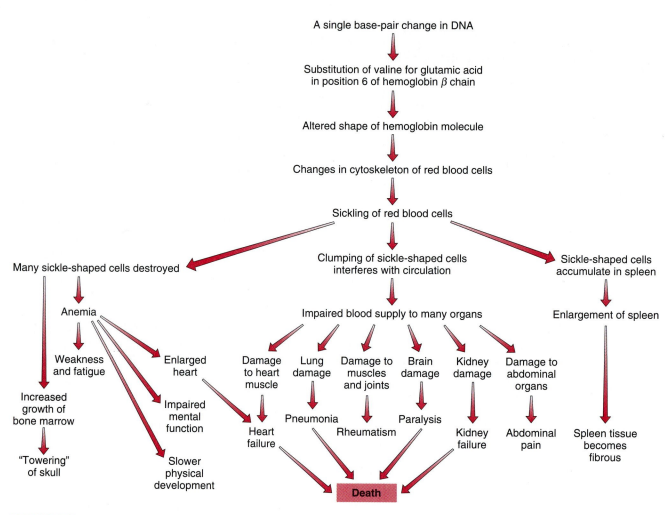

FIGURE 5.7
Development of the consequences of hemoglobin S. A small change in a gene can have
great and many phenotypic consequences.

Population genetics of sickle-cell anemia.

When geneticists realized that sickle-cell anemia in the United States and Jamaica was largely confined to black people, they began to investigate the black populations of Africa. Using the blood test described earlier in this chapter, researchers investigated the frequency of the allele for hemoglobin S in many African and Eurasian populations. Over large parts of tropical Africa, researchers found remarkably high rates of the Hb^S allele, approximating 25 percent or more. At first, this appeared puzzling, because sickle-cell anemia was nearly always fatal before reproductive age. An allele whose effects are fatal in homozygous form should long ago have been eliminated by natural selection because people having sickle-cell children would have fewer children surviving to reproductive age.

Maps were made of the frequency of the sickle-cell gene. From these maps and from other evidence, it was noticed that the areas in which the sickle-cell gene was frequent were also areas with a high incidence of malaria, particularly the variety caused by *P. falciparum* (Fig. 5.8A, B).

Subsequent research confirmed the basic fact that the Hb^S allele, even in heterozygous form, con-

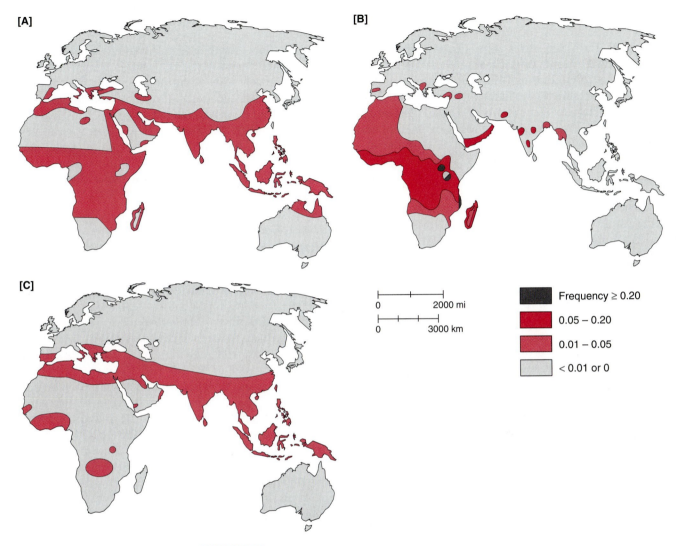

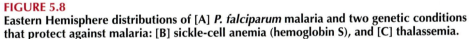

FIGURE 5.8
Eastern Hemisphere distributions of [A] *P. falciparum* malaria and two genetic conditions that protect against malaria: [B] sickle-cell anemia (hemoglobin S), and [C] thalassemia.

fers important resistance to the most virulent form of malaria. Tests in which volunteers were exposed to *Anopheles* mosquitoes showed that the mosquitoes were far less likely to bite heterozygous Hb^A/Hb^S individuals than homozygous Hb^A/Hb^A individuals. Tests with the *P. falciparum* parasites showed that they thrived on the red blood cells of Hb^A/Hb^A individuals, who nearly always came down with a serious case of malaria upon infection. However, when Hb^A/Hb^S heterozygotes or Hb^S/Hb^S individuals with sickle-cell anemia were

infected with *P. falciparum,* their symptoms were mild and they quickly recovered because the parasite cannot complete its asexual cycle in their sickled blood cells. The protection that the Hb^S allele affords against malaria is sufficient to explain its persistence in those populations in which the incidence of malaria is high.

Hemoglobin S thus decreases the fitness of homozygotes by causing sickle-cell disease but increases the fitness of heterozygotes in areas in which malaria occurs. In this way, malaria acts as an instrument of natural selection and has a dramatic influence on the gene frequencies of populations. Fatal diseases are among the most striking instruments of natural selection.

In addition to hemoglobin A and hemoglobin S, several other genetic variants of hemoglobin have also been discovered. Some of these also occur principally in areas in which malaria is present and are thought to confer some resistance to malaria.

Thalassemia

In many regions bordering the Mediterranean Sea (including Spain, Italy, Greece, North Africa, Turkey, Lebanon, Israel, and Cyprus), many people have suffered from another debilitating type of anemia known as *thalassemia* (literally, "sea blood" in Greek). The disease also occurs further east, especially in Southeast Asian countries like Laos and Thailand (Fig. 5.8).

Thalassemia is marked by a reduced amount of one or more of the protein chains in the hemoglobin molecule. The disease exists in a more serious, often fatal, homozygous form called *thalassemia major* and a less severe heterozygous form called *thalassemia minor*. Red blood cells containing nonfunctional hemoglobin are destroyed in the spleen, producing an anemia.

The symptoms of thalassemia vary, but all forms result in some reduction of oxygen flow in the blood; blood cell volume and hemoglobin levels are also usually reduced, and the overproduction of red blood cells robs the body of much-needed protein and results in stunted growth and smaller stature.

Populations at risk can now be screened for the genotypes that cause thalassemia, with genetic counseling provided to those found to carry the trait (e.g., Rosatelli, 1992). Heterozygotes can be cautioned against the risks of marrying one another, or, when they do marry, the genotypes of their children can be tested at an early age. Screening programs and newer methods of treatment have greatly reduced the problems caused by this disease in Italy, Greece, and elsewhere in the Mediterranean (Weatherall, 1993).

The geographical distribution of thalassemia follows closely the distribution of malaria in countries in which sickle-cell anemia is infrequent or absent. For this reason, it has long been suspected that thalassemia confers a protective resistance to malaria, similar to that conferred by sickle-cell anemia. The evidence is indirect: If heterozygous individuals (those with thalassemia minor) did not have *some* selective advantage such as malaria resistance, then the deaths caused by thalassemia major would have caused the genes for this trait to die out long ago. Because there are different alleles for different forms of thalassemia, several researchers (e.g., Chifu et al., 1992) have suggested multiple independent origins for these alleles.

Other Genetic Traits that Protect against Malaria

Blood sugar (glucose) is normally broken down within each cell in a series of reactions that begins with the formation of *glucose 6-phosphate*. Most of the glucose 6-phosphate is broken down into pyruvate (Chapter 8) in a series of energy-producing reactions, but some is also used to make ribose (the sugar used in RNA) and to make reducing agents such as NADPH and glutathione. The removal of two hydrogens from the glucose 6-phosphate molecule requires the enzyme *glucose 6-phosphate dehydrogenase* (G6PD). There are many people who have reduced amounts of this enzyme, a condition known as *G6PD deficiency*, or *favism*. G6PD deficiency results from a mutation in the gene that encodes the G6PD enzyme. It affects some 10 million people, and is thus the most common disorder offering protection against malaria.

Under many or most conditions, people with G6PD deficiency remain perfectly healthy, but they occasionally suffer from a *hemolytic anemia* in which the red blood cells rupture, spilling their hemoglobin into the blood plasma (where it is physiologically useless, but easy to detect by simple laboratory tests). Hemolytic anemia, which is potentially fatal, can occur in susceptible people as a reaction to certain drugs (aspirin, quinine, quinidine, chloroquine, chloramphenicol, sulfanilamide, and others), in response to certain illnesses, or after eating fava beans (*Vicia faba*), a common legume of the Eastern Mediterranean and Middle East. The anemia may also exist chronically in a nonfatal form in people with G6PD deficiency.

G6PD deficiency has been shown to offer protection against *P. falciparum* malaria. Most important, heterozygous carriers of the deficiency are also malaria-resistant, but the exact mechanism of the resistance has yet to be figured out.

G6PD deficiency occurs mostly in Mediterranean populations from Greece to Turkey, and from Tunisia to the Middle East, and among Sephardic Jews. It also occurs south of this area into Africa and eastward across Iran and Pakistan to Southeast Asia and southern China. The Greek mathematician Pythagoras may have suffered from this disorder, for his aversion to beans became legendary, and the avoidance of beans was an important belief of his religious cult. Rivals once captured Pythagoras by chasing him toward a bean field, which they knew he would not cross.

Some members of Southeast Asian populations have a condition called *ovalocytosis*, in which 50 to 90 percent of the red blood cells are oval or rod-shaped rather than circular. Experiments have shown that malaria parasites do not survive as well in the oval red blood cells as they do in normal, circular red cells, suggesting that ovalocytosis offers protective resistance to malaria.

Population Genetics of Malaria Resistance

The genetic nonuniformity of a population is known as **polymorphism,** a condition in which two or more alleles of a gene are present at frequencies higher than new mutations could possibly explain. If the polymorphism persists for many generations, it is likely to be a **balanced polymorphism.** The conditions of a balanced polymorphism are that all homozygous genotypes suffer from some selective disadvantage or reduction in fitness, while the heterozygotes have the maximum fitness. The alleles of polymorphic genes often have harmful effects when homozygous, but they persist in populations because they also confer some important benefit (like malaria resistance) when heterozygous. For example, in a country in which malaria is present, $Hb^A Hb^A$ homozygotes have lower fitness because they are susceptible to malaria, while $Hb^S Hb^S$ homozygotes will generally die from sickle-cell anemia. The $Hb^A Hb^S$ heterozygotes have maximal fitness because they are malaria-resistant and because they have enough normal red blood cells that they do

not suffer from severe sickle-cell anemia. Under conditions like these, natural selection will bring about and perpetuate a situation in which both alleles persist.

The selection by malaria for genetic traits that offer resistance to it is at least as old as the open, swampy conditions (ideal for the breeding of mosquitoes) brought about by agriculture in warm climates. Evidence for this exists in the form of human bones found at a Neolithic archaeological site along the coast of Israel (Hershkovitz et al., 1991). Cultural remains found at this site show that it was an early farming community, one of the first in the area. Pollen analysis shows the presence of many plants characteristic of swampy areas. Some of the bones show characteristic increases in porosity (due to the increased production of red blood cells in the bone marrow) indicative of thalassemia.

Other Diseases as Agents of Selection

Hereditary diseases that confer some advantage in the heterozygous state are not confined to those that protect against malaria. In European populations of past centuries, *tuberculosis*, an infection caused by a type of bacteria called *Mycobacterium tuberculosis*, was an important force of selection, especially in crowded cities from the Middle Ages to the early twentieth century. Meindl (1987) has proposed that people heterozygous for alleles that cause *cystic fibrosis*, an inherited lung disorder, were protected against tuberculosis; they therefore survived tuberculosis epidemics in greater numbers than did people without a cystic fibrosis allele. As the heterozygotes increased in number, some of them married one another, and, on average, one out of four of their children became afflicted with cystic fibrosis. Cystic fibrosis is the most common hereditary disorder in many Western European countries (especially Ireland) and also in the United States, especially the eastern half. The normal gene for this trait codes for a protein channel that controls ion diffusion in lung cells. There are many known mutations of this gene that can lead to channel proteins that are changed in different ways and produce forms of cystic fibrosis whose symptoms range from mild to fatal.

What about the worldwide variation in blood groups and other genetic traits? There is evidence that at least some of this variation may also result from the natural selection brought about by various medical conditions. In a smallpox epidemic in Bihar province, India, researchers found that those who died were more often of blood group A, while survivors were more often of blood group B. In similar fashion, cholera selects against blood group O and favors blood group B. (Note that these studies demonstrated a difference in fitness, but did not explain the mechanism.) Other studies have shown statistical correlations of various blood types to other diseases: Blood group O has an increased risk of duodenal ulcers and ovarian cancers, while blood group A has a slightly increased risk of stomach cancer. Associations of particular blood groups with cancers of the duodenum and the colon have also been postulated. Such statistical associations do not necessarily indicate a cause-and-effect relationship between the associated factors.

THOUGHT QUESTIONS

1. How is an *average* lifespan measured? Why is the average lifespan of a population more affected by the deaths of children (e.g., from malaria) than by the deaths of elderly people?
2. All heterozygous carriers of the allele for G6PD deficiency are female. What does this tell you about the location of the G6PD gene? (You may need to review Chapter 3 in order to answer this question.)

D. Natural Selection Causes Variations in Physiology

There are other agents of natural selection in addition to diseases. Among them are climate factors such as temperature or sunlight, as well as climatic variation that makes food more scarce at some times of year or from one year to another.

Human Variation in Physiology and Physique

Like the genetically based traits that confer protection against disease, other variation between populations has arisen in response to these other selective factors. Examples include the genetically regulated aspects of physiology and of body shape and size.

Performance under different climatic extremes. During part of the Korean War (1950–1953), American soldiers were exposed to the fierce, frigid conditions of the Manchurian winter. Many soldiers were treated for frostbite. Most of the Euro-American (Caucasian) soldiers responded well to the medical treatment that was given, but a disproportionate number of African-American soldiers did not and many of them lost fingers and toes as a result. Disturbed by these findings, the U.S. Army ordered tests on resistance to environmental extremes among soldiers of different racial backgrounds.

In one series of tests, army recruits were required to perform strenuous tasks (like chopping wood) under a variety of climatic conditions. In a hot, humid climate, the African-American soldiers were able to continue working the longest and performed the best as a group, Asian-American and Native American soldiers performed nearly as well as the African-Americans, and Euro-American recruits lost excessive fluids through sweating and became easily fatigued and dehydrated. Under dry, desert conditions, the Asian-American and Native American soldiers did best, the African-Americans were second best, and again the Euro-American soldiers became dehydrated. Under extremes of cold, it was the Euro-American soldiers who did best, followed close behind by the Native American and Asian-American soldiers; the African-Americans shivered the most and some became too cold to continue. These tests demonstrated definite differences between groups in bodily resistance to physiological stress under a variety of environmental extremes. The significance of these differences was enhanced by the fact that, in other respects, the recruits represented a fairly homogeneous population: 18- to 25-year-old males who had all been screened by the army as being physically fit and free from disease and who had passed the same army physical and mental exams.

Other physiologists outside the Army conducted tests in which adult male volunteers immersed their arms in ice water almost to the shoulders. African-Americans in general shivered the most and suffered the most rapid loss of body heat, as measured by a decline in body temperature. Euro-Americans and Asian-Americans lasted longer without shivering, but they, too, eventually suffered loss of body heat. Only the Inuit (Eskimo) volunteers were able to keep their arms immersed indefinitely without any discomfort and without shivering. Subsequent studies that replicated these results made the additional finding that diet is also a factor: Inuits who ate traditional high-protein, high-fat Inuit diets did far better than Inuits who had become acculturated to American dietary habits. It would be a mistake, however, to extrapolate findings from studies such as these beyond the groups used for the tests (adult males) without further investigation. Many traits vary with age or sex, or both.

The greater susceptibility of African-Americans to frostbite may be related to a generally low capacity for blood vessel dilation, as recently demonstrated by Lang et al. (1995). The low dilation capacity may explain why blood pressure (Chapter 14), which rises in all people under stress, remains high longer among African-Americans, a condition that may, in turn, contribute to their higher rate of high blood pressure (hypertension).

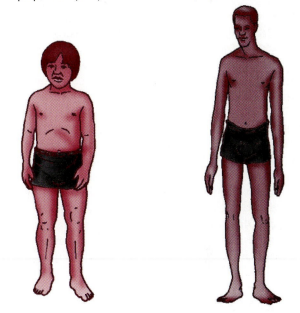

[A] Arctic body proportions (Inuit)

[B] Hot climate body proportions (Sudanese)

FIGURE 5.9
Bergmann's and Allen's rules illustrated by comparisons between arctic and tropical body forms.

Bergmann's rule. Genetically based differences in physiology that correlate with climate are the basis for a number of *ecogeographic rules*. Adaptations can also work indirectly, through variables such as body physique. Biologists have long noticed certain general patterns of geographic variation among mammals and birds. In one such pattern, called **Bergmann's rule,** body sizes tend to be larger in cold parts of the range and smaller in warm parts. This can be explained by the relationship of body size to mechanisms of heat generation and heat loss. For example, an animal twice as long in all directions as another animal has eight times the volume of muscle tissue generating heat (2 × 2 × 2) as the smaller animal but only four times the surface area over which heat is lost (2 × 2). Thus, overall, the larger animal is twice as efficient at conserving heat under cold conditions. A survey of human variation confirms that the largest average body masses are found among people living in cold places (like Siberia), while most tropical peoples within all racial groups are of small body mass, even when their limbs are long (Fig. 5.9).

Allen's rule. Another broad, general phenotypic pattern in most geographically variable species of mammals and birds is **Allen's rule:** Protruding parts like arms, legs, ears, and tails are longer and thinner in the warm parts of the range and shorter and thicker in cold regions. This rule is usually explained as an adaptation that conserves heat in cold places by reducing surface area and dissipates heat more effectively in warm places by increasing surface area. Human populations generally follow this rule: Inuits have shorter, thicker limbs, while most tropical Africans have longer, thinner limbs. There are exceptions, however. A number of forest-dwelling populations along the Equator are much smaller than Allen's rule would predict, although they are usually thin-legged. And the tallest (Tutsi) and shortest (Mbuti) people on Earth live near one another in Zaire, showing that climate is not the only factor governing limb length or overall height.

Diabetes and thrifty genes. Diabetes, a potentially life-threatening illness in many populations, may be an indirect result of one or more of the so-called thrifty genes that protected certain people from starvation in past centuries. Ancestral Polynesians, for example, had to endure uncertain journeys over vast stretches of Pacific Ocean waters. Uncertain food supplies during such voyages selected for people who could withstand longer and longer periods of starvation and still remain active. The postulated thrifty gene or genes may have caused excess food, when it was available, to be converted into body fat that could be used for energy in times of famine. The result was a population that was stocky in build and resistant to starvation in periods when food supplies were low but that was also more susceptible to diabetes under modern conditions, when physical exhaustion is rare and the availability of food is no longer uncertain. Diabetics fed "ordinary" diets have excess sugar in their blood, much of which is converted to fat and stored. While diabetes is itself an unhealthy condition, the storage of fat may have been, under conditions like those described for the early Polynesians, an adaptive trait. Perhaps diabetes is an unfortunate modern consequence of having one or more genes originally selected for their ability to convert sugar to body fat.

A similar history of selection for thrifty genes (not necessarily the same ones) might also explain the late twentieth-century upsurge of diabetes among certain Native American populations, notably the Navajo and Pima of the southwestern United States. The risks that selected for thrifty genes in the past were more significant in barren environments than in places in which the food supply was more assured. However, the commercial introduction of sugar-rich foods, and a change from an active to a sedentary lifestyle (the risks of diabetes are higher for sedentary peoples) changed the environment so that genes that were once advantageous in some cases turned into a liability, putting the carriers at greater risk for diabetes. The Navajo and Pima have discovered that a return to frequent long-distance foot racing (a traditional activity they had nearly abandoned) has kept their populations healthier and significantly lowered the incidence of diabetes among the runners. Although the gene frequencies of the "thrifty genes" within this population have not changed, the partial return to an earlier lifestyle changed the environmental stresses and decreased the incidence of disease.

Natural Selection, Skin Color, and Disease Resistance

The skin is the largest organ of the body and a major surface across which the body makes contact with the forces of natural selection in its environment. Human populations vary widely in skin color. Could these differences in skin color be adaptive?

Geographic variation in skin color.   Skin color is one of the most visible human characteristics, and the one to which Americans have always paid the most attention when identifying race. Why would it be adaptive for people to be light-skinned in Europe but dark in Africa, Sri Lanka, or New Guinea? Notice that there are some very dark-skinned people outside of Africa, and they generally have few other physical or genetic characteristics in common with Africans other than their dark colors. The natives of Sri Lanka, for example, have very straight hair and blood group frequencies totally different from those of Africa. One clue to this puzzle is that all very dark-skinned peoples originated in tropical latitudes.

Among human populations, we find that, before A.D. 1500, Europe was inhabited mostly by light-skinned peoples, Africa and tropical southern Asia by dark-skinned peoples, and the drier, desert regions of Asia and the Americas by people with reddish or yellowish complexions. What is even more remarkable is that we find geographic variation along the same pattern *within* most continents, and in fact greater variation within the larger population groups than between such groups. For example, among the group of populations spreading continuously from Europe across Western Asia to India, we find the lightest skin colors (also eye and hair colors) in Scandinavia and Scotland, progressively darker average colors (and darker hair) closer to the Mediterranean Sea, further darkening as we move through the Middle East and across Iran to Pakistan and India, and an extreme of darkness at the southern tip of India and on the island of Sri Lanka. A similar gradient (a cline) for skin color can be found among Asians, from northern Japan south through China into the Philippines and Indonesia.

Geographic variation in exposure to sunlight.
Tropical regions receive on a year-round basis more direct sunlight than do temperate regions. In fact, the amount of sunlight received at ground level decreases with increases in latitude and follows belts of latitude more closely than it follows temperature. This is especially true for light in the ultraviolet region of the spectrum.

Of the world's densely inhabited regions, Europe receives the least amount of sunlight. First, Europe is thickly populated at higher latitudes than is any other continent: London and fourteen other European capitals are located north of latitude 50°, while North America above this latitude contains few large cities and a great deal of sparsely inhabited land. Second, Europe also has a frequent cloud cover that screens out even more of the sun's rays. As a result of both high latitude and cloud cover, people in Europe receive much less exposure than most other people to ultraviolet light.

Gloger's rule. While Bergmann's and Allen's rules, described earlier, deal only with temperature, a third ecogeographic rule, **Gloger's rule,** takes into account sunlight and humidity as well. Under Gloger's rule, most geographically variable species of birds and mammals have pale-colored or white populations in cold, moist regions, dark-colored or black populations in warm, moist regions, and reddish and yellowish colors in arid regions. We do not know all the reasons for this variation. Camouflage has been suggested as a cause, but vitamin D synthesis also plays an important role, as the next section describes.

Sunlight, vitamin D, and bone formation. Vitamin D is needed for the proper formation of bone (Chapter 8). Vitamin D deficiency produces rickets in children, a disease of bone formation that may result in crippling bone deformities if untreated. Sunlight is necessary for vitamin D synthesis. Many foods are rich in vitamin D, such as egg yolks and whole milk, but most vitamin D found in foods is in a biologically inactive form. The final step of vitamin D biosynthesis takes place just beneath the skin, with the aid of the ultraviolet rays of natural sunlight. This is why vitamin D is sometimes called the "sunshine vitamin." In order for a population to get adequate amounts of vitamin D, it must have both adequate intake of the vitamin in the diet and adequate exposure to sunlight. European populations have the lightest skin colors (and they get lighter the farther north you go) as an adaptation that allows maximum sunlight penetration into the skin. Europeans also have many cultural adaptations related to vitamin D intake, like the eating of cheeses and other fat-rich milk products containing vitamin D. Northern Europeans place great value on outdoor activity at all times of the year, including such occasional extremes as nude dashes into the snow after the traditional sauna.

In northern Europe, people with dark skins could be at a very high risk for vitamin D deficiency because melanin pigment blocks out a large proportion of the sun's ultraviolet rays. Very few dark-skinned people were indigenous to northern Europe or migrated there from other countries until after World War II, when synthetically prepared, biologically active vitamin D first became widely available. Since prepared vitamin D is already in its active form, sunlight is no longer needed for its synthesis, so the risks for dark-skinned people in northern latitudes are greatly diminished, provided they get enough of the synthetic vitamin D.

Sunlight and skin cancer. At latitudes closer to the Equator, another problem exists: The same wavelengths of ultraviolet that are needed in the final step of vitamin D synthesis are also cancer-causing. Skin cancer (malignant melanoma, Chapter 9) is generally a disease of white-skinned people when they are overexposed to the sun's direct rays. Populations of all racial groups living closer to the Equator have been selected over the centuries to have darker skins. Those individuals who had lighter skins in the past more often got skin cancer and died, in many cases before their reproductive years had ended.

Inuit get vitamin D from fish. For the reasons given in the preceding sections, populations living in the high latitudes are generally light-skinned and populations that are adapted to living in tropical latitudes are generally dark-skinned. There is one very interesting exception: the Inuit populations of Arctic regions, sometimes known as Eskimos. (These people have always called themselves Inuit; the name "Eskimo" was a pejorative term attached to them by their enemies.) The Inuit are not very light-skinned, nor do they expose themselves much to the sun. Most Inuit people live in places so cold that the exposure of bare skin poses a greater danger than any benefit of ultraviolet rays could overcome, and most Inuit are fully protected by clothing that offers hardly any exposure to the sun. So how do they get enough vitamin D? The Inuit have discovered their own way of staying healthy. One of the world's richest sources of vitamin D is in fish livers, especially in cold-water fish. (Cod liver oil is a very rich source of both A and D vitamins.) Moreover, the vitamin D in fish oils is fully synthesized and needs no sunlight to activate it. So, instead of having pale skins and traditions of exposing their skins to the sun, the Inuit have traditions of catching cold-water fish and eating them whole, liver and all. These traditions have allowed them to stay healthy in a climate that is too cold and too sunless for most other populations.

THOUGHT QUESTIONS

1. If people differ in their resistance to extreme cold or heat, does this mean that the difference is genetic? What would you need to know in order to answer this question? How could an experiment be arranged to test this?

2. Blood type O is statistically associated with duodenal ulcers, one of many such correlations between a blood type and a disease. Does a correlation demonstrate a cause? Does a correlation imply a mechanism of some kind? Does a correlation suggest new hypotheses? How can scientists learn more about whether there is a causal connection between the blood type and the disease?

CHAPTER SUMMARY

Human populations vary geographically. Differences among populations have historically been described in terms of socially constructed or morphological races. Population genetics allows us to describe groups of populations that differ from one another by certain characteristic gene frequencies. Gene frequencies vary gradually and continuously among populations. Continuous variation of this sort is best described in terms of gradual character gradients, also called clines. Variation within populations usually exceeds variation between them.

Populations that were at one time small may have gene frequencies that have been shaped in part by genetic drift. Aside from such instances, most geographic variation among human populations has an adaptive basis. This adaptive basis includes protection of heterozygous individuals against malaria, a widespread parasitic infection that causes more deaths (especially among the young) than nearly any other single disease. Genetic diseases believed to confer resistance to malaria among heterozygotes include sickle-cell anemia, thalassemia, G6PD deficiency, and ovalocytosis. Such diseases will result in a balanced polymorphism whenever the heterozygous genotype enjoys maximum fitness. Temperature selects for body size and shape. Ultraviolet light selects for different skin color at different latitudes. Pale skin is favored in high latitudes as an adaptation to absorb more ultraviolet light and prevent vitamin D deficiency (rickets). Dark skins are favored near the Equator as a protection against skin cancer from too much ultraviolet exposure.

KEY TERMS TO KNOW

adaptation (p. 149)

Allen's rule (p. 159)

anemia (p. 152)

balanced polymorphism (p. 157)

Bergmann's rule (p. 159)

cline (p. 138)

founder effect (p. 146)

gene frequency (p. 133)

genetic drift (p. 145)

Gloger's rule (p. 161)

Hardy-Weinberg principle (p. 145)

hemoglobin (p. 152)

hereditarian (p. 136)

polymorphism (p. 157)

population (p. 134)

races (p. 135)

racism (p. 135)

typological species concept (p. 134)

C O N N E C T I O N S   T O   O T H E R   C H A P T E R S

Chapter 1   Every study of human variation is conducted in a cultural context.

Chapter 2   Studies of human variation have ethical implications, including those arising from inappropriate use of the results.

Chapter 3   Many human variations have a genetic basis; such variations arise ultimately from mutations.

Chapter 4   Human variations reflect evolutionary processes, including mutation, natural selection, and genetic drift, all of which continue to work in modern populations.

Chapter 6   Nearly all human populations are growing, and some are growing much faster than others. Population growth and migrations will change various gene frequencies.

Chapter 8   Different populations sometimes have different ways of meeting their nutritional requirements.

Chapter 9   Some types of cancer are more frequent in some human populations and less frequent in others.

Chapter 16   Human variation is an example of biodiversity at the population level.

The Population Explosion

I magine a world where people must share a room with 4 to 12 others. A room of one's own is a rare luxury. In fact, people who have any housing at all consider themselves fortunate, because so many people have none. Drinking water is in short supply each summer, and overworked sewer systems are breaking down all the time; many millions have no sewer system at all. Jobs are scarce, and well-paying jobs are almost unheard of. Beggars crowd every street, and each garbage can is searched through over and over again by starving people looking for something to sustain them.

Some parts of the world already experience these conditions. Some experts predict that a future like this may be in store for all of us unless something is done soon, and on a massive scale, to control population growth.

The Earth is currently experiencing the most rapid population increase in all of human history. From 2.5 billion people in 1950, the total world population more than doubled to about 5.6 billion in 1994. At current rates of increase, the global population will double again in about 38 years. Each year, the world's population increases by some 94 million people.

In this chapter, we consider the factors that control the size and the rate of growth of populations, including the biological controls on populations that operate independently of any conscious planning. We will focus on human populations, but much of what we discuss applies equally well to populations of other organisms.

As with many of the other issues in this book, population growth cannot be looked at as a purely biological issue. There are political, religious, and ethical dimensions to population growth and its control. Individual decision making in family planning is sometimes at odds with government decisions aimed at population control and also sometimes with various religious teachings. There are also many other reasons that people might resent strangers urging them to modify their personal behavior in one way or another. All of these factors, moreover, vary from place to place, and the lessons learned in one country or population cannot necessarily be applied uncritically to other populations elsewhere. As we have seen earlier (Chapter 2), the boundaries between the individual good and the social good are one of the subjects of ethics. Biology can inform ethical debate by assessing, for different scenarios, the biological risks to the individual and to society. However, any attempts to implement change based on biological data will necessarily take place in a context of many, often competing, social values.

A. People Have Been Concerned about Population Growth for about 200 Years

The world's population has long been on the increase, but the *rate* of increase was slow before modern times. During the seventeenth and eighteenth centuries, several European countries began to experience a great upswing in their populations.

Malthus's Views on Population

It was during the early part of the Industrial Revolution that philosophers and economists began to pay attention to the phenomenon of population growth. David Hume and Benjamin Franklin each wrote about populations, generally with the attitude that a population increase was a blessing for civilization. The first person to emphasize the negative consequences of population growth was Thomas Robert Malthus. In his *Essay on the Principle of Population* (1798), Malthus explained the following dilemma:

1. Population tends to increase *geometrically* if its growth is unchecked. (A geometric sequence, like 1, 2, 4, 8, 16, . . . , is one in which each number is *multiplied* by a constant, in this case 2, to obtain the next number.)

2. The available food supply, in Malthus's view, increased only *arithmetically*. (An arithmetic sequence, like 3, 4, 5, 6, 7, . . . , is one in which a constant, in this case 1, is *added* to each number to obtain the next number.)

3. Since the population increases faster than the food supply, an increasing population compounds human misery and poverty, especially among those segments of the population with limited resources.

Malthus's assumption of an arithmetical increase in the food supply has been questioned. Although the limited data available to Malthus were consistent with an arithmetical increase, there is no theoretical justification for saying that the food supply can *only* increase in this manner. The important point, however, is that the growth in food supplies is too slow to keep up with the rise in population, and this point has become painfully evident in many countries.

Positive checks and preventive checks. Malthus divided the factors controlling population increase into two broad categories that he called preventive checks and positive checks. *Preventive checks* were those that could prevent births from occurring. These were usually voluntary measures, operating on an individual level. They included delayed marriage (also called "moral restraint"), reduced family size, and several forms of "vice." The *positive checks* on population were those that would operate automatically after births had taken place, whenever the preventive checks were not sufficient. Malthus identified as positive checks overcrowding, poverty, epidemic diseases, rising crime rates, and warfare, and, if these checks were insufficient to control population, starvation, and famine. One logical outgrowth of this type of thought was the nineteenth-century theory attributing warfare to the economic needs confronting the population of each nation, the so-called economic theory of war.

Malthus noticed that, even in European countries with stable populations, the rate of population increase temporarily rose in the years following a famine or plague until the population was restored to its predisaster level. This phenomenon also takes place after many wars and economic hard times (witness the "baby boom" in the years following World War II). Such events show that a population's potential for increase is much greater than is usually realized. The *actual rate* of population growth is kept lower than the *potential rate* of growth by positive and preventive checks.

Gathering data to test theories of population growth. Malthus's theory of population growth cannot be tested without census data. A *census* is, at minimum, a head count of all the people living in a specified area, usually within recognized political boundaries; modern censuses also record information about age, sex, marital status, and often income and employment status as well. From biblical times until Malthus's time, censuses were rare, sporadic events. The United States implemented the first nationwide census of any modern nation in 1790 for the purpose of achieving proportional representation of the different states in Congress.

Testing Malthus's theory also requires information called *vital statistics*, meaning the registration and counting of births, deaths, marriages, and divorces as they occurred. Modern record-keeping now also includes such additional details as age at marriage and cause of death.

Malthus included some early census figures and vital statistics in the later revisions of his essay. Although crude by modern standards, these statistics nevertheless did demonstrate that populations were increasing. In a few cases, Malthus was able to distinguish the relative effects of positive and preventive checks by comparing changes during epidemics and wars with changes in more normal years.

Recent Changes in Population Pressures

Throughout the nineteenth and early twentieth centuries, technological progress in agriculture, especially in mechanized farming and in the use of chemical fertilizers, increased crop yields among the wealthy nations. European population crises were in many cases also relieved by large-scale emigration to other continents. Malthus and his gloomy predictions of starvation and misery were largely forgotten.

Following World War II, the colonial era came to a close in most parts of the world. Former colonies became independent nations, especially in Africa and Asia. People attempting to deal with poverty, disease, and food shortages in these and other nations faced the harsh reality that burgeoning populations were exacerbating all these problems. Public health improvements (in public

sanitation, in mosquito control, in vaccination against infectious diseases, and in the delivery of medical care generally) were diminishing the death rates, especially among the young. The result was in many cases a staggering population growth.

In several African countries, famine and mass starvation followed closely on the heels of internal warfare: southern Nigeria in the 1960s, Ethiopia (and parts of Sudan) in the 1980s, Somalia in the 1990s. Though these crises were caused by civil strife, they also reminded the world that population was outstripping the available food resources, and that other critical resources (housing, drinking water, sanitation, and medical care) were also in limited supply. The positive checks that Malthus had foreseen were operating. The developing world was coming face to face with a population crisis.

The population crisis in developed countries takes a different form: Wealth can be diverted into providing additional housing, roads, sewers, and needed services. However, this development may not be sustainable in the long run because it depends on the use of nonrenewable resources such as fossil fuels, products imported from less developed countries, or both. Even stable population sizes cannot be supported indefinitely by the diversion of nonrenewable resources. By diverting resources, wealthy nations can postpone, but not avoid, the effects of global population growth.

THOUGHT QUESTIONS

1. The text mentions four African countries that have suffered famines subsequent to civil unrest since the 1960s. Do you think that these countries would have been able to feed their people adequately had it not been for the internal conflicts? What evidence would you seek in order to decide? What other factors may have been involved in addition to population size and political turmoil?

2. Is an increase in population the only factor that puts resources in limited supply? Does population growth affect the availability of housing or medical care in the same way that it affects the availability of drinking water, sanitation controls, and food?

B. Many Factors Influence Population Growth

A **population** is defined as a set of potentially interbreeding individuals in a certain geographical location at a certain time (see also Chapter 4). Our ability to understand population growth depends on our ability to make predictions. Since populations are large aggregates, we need mathematical models to make these predictions and to study the factors that might influence and possibly curb population growth. The numerical study of populations is a field known as **demographics.** The study of the biological factors that affect the size of populations is called *population ecology.*

Exponential (Geometric) Growth

The observation that populations grow geometrically was made by several observers independently of one another long before anyone understood why this was so. The explanation of the geometric increase was an early twentieth-century discovery to which several mathematicians and bacteriologists contributed, among them Raymond Pearl and Alfred Lotka.

The **birth rate,** B, of a population for a given period of time is found by dividing the number of births during that time period by the population size, N:

$$B(\text{each year}) = \frac{\text{Number of births per year}}{N}$$

Birth rate, like any other rate, is always a fraction; the word *rate* always means that one number is divided by another. To illustrate with some actual numbers, if there are 10,000 people in a population ($N = 10,000$) and 1000 babies are born that year, then the birth rate is

$$B = \frac{1000}{10,000} = \frac{1}{10} = 0.1 \text{ per year}$$

Death rate is found in a similar manner. Given that a certain fraction, D, of the population dies within the time interval in question, we have:

$$D = \frac{\text{Number of deaths per year}}{N}$$

The term D is thus the death rate. If 100 people die that year from our sample population of 10,000, what is the death rate? Notice that N, the population size, appears in both equations because the number of births and number of deaths are both proportional to the size of the population, and the birth rate and death rate are both expressed as fractions of N.

The equations can be rearranged to give:

$$\text{Number of births per year} = BN$$

and

$$\text{Number of deaths per year} = DN$$

At the end of the year, the size of N will have increased by the number of births and decreased by the number of deaths. In equation form:

$$\text{Change in } N \text{ per year} = (BN) - (DN)$$

We can rewrite this equation using the notation standard in mathematics for rates of change: If the change in N, called dN, is divided by the change in time, dT, we have

$$\frac{dN}{dT} = BN - DN$$

or

$$\frac{dN}{dT} = (B - D)N$$

The quantity $(B - D)$ is thus the difference between the birth rate and the death rate. This quantity is given a new name, r, which is also called the **intrinsic rate of natural increase** (or Malthusian parameter). The equation then simplifies further to

$$\frac{dN}{dT} = rN$$

The type of growth described by this equation is an example of *geometric growth* because each new value of N results from multiplying the previous value by $1 + r$. (Another familiar example of geometric increase is the growth of money by compound interest.)

Calculus methods can now be used to solve this last equation and obtain

$$N = N_0 e^{rT}$$

where N is the population at time T, N_0 is the initial population size (at time $T = 0$), e is the base of natural logarithms (approximately 2.71828), and r is the intrinsic rate of natural increase. Geometric growth of this kind is also called **exponential growth** because time appears in the exponent. Graphing such an equation gives a growth curve such as the one shown in Fig. 6.1. The same equation can be used to describe populations of bacteria, fish, or humans, though different time units may be used in each case.

Any population will grow if its birth rate, B, exceeds the death rate, D, for in this case r will be positive. If $B = D$, then $r = 0$, and the population is stable, neither increasing nor decreasing. A population whose death rate exceeds its birth rate will be a declining population, with a negative value of r.

Immigration and emigration. The preceding discussion assumes a "closed" population unaffected by migration. This is a safe assumption for the world as a whole, but for a single country or region we must also add terms for both immigration and emigration. The United States, for example, currently has a birth rate approximately in balance with the death rate, but overall population growth continues because more people move to the United States each year from other countries than move away from the United States. To include migration rates in the calculation of r, we must write

$$r = B - D + i - m$$

where i is the rate of immigration (the number of immigrants in a year divided by the population size) and m is the similarly defined rate of emigration. In most nations, population growth results more from the excess of births over deaths than it does from the excess of immigration over emigration.

Doubling time. For growing populations, we can calculate the **doubling time,** the length of time it

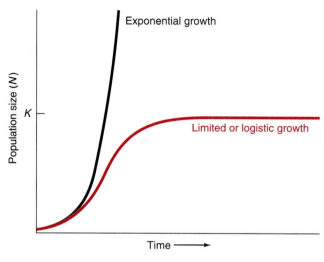

FIGURE 6.1
Exponential growth compared to logistic growth. A steeper (more vertical) slope indicates a more rapid rate of growth.

takes for the population to double. Mathematically, doubling time is expressed as

$$\text{Doubling time} = \frac{0.69315}{r}$$

where the number 0.69315 is the natural logarithm of 2. Malthus calculated that the population of the United States was doubling every 25 years. Using the above equation, we have

$$25 \text{ years} = \frac{0.69315}{r}$$

$$r = \frac{0.69315}{25} = 0.028$$

Expressed as a percentage (0.028 multiplied by 100), this is a 2.8 percent annual increase in the population size N. Money growing at this rate of compound interest will also double in 25 years, because the same equations apply.

In 1993, the United Nations calculated that the growth rate (r) for the world's population was 1.8 percent per year, which will cause a doubling every

$$\frac{0.69315}{0.018} = 38.5 \text{ years}$$

The value of r also varies from place to place (Fig. 6.2). The fastest-growing nations have growth

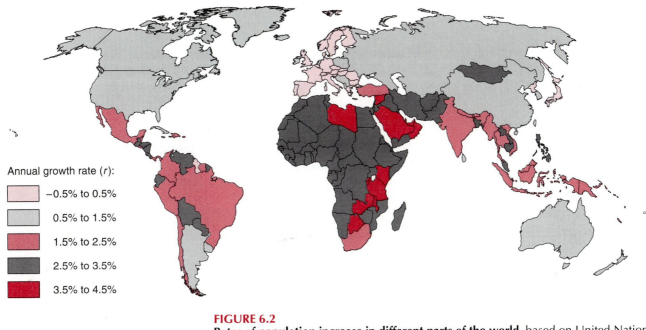

FIGURE 6.2
Rates of population increase in different parts of the world, based on United Nations data (1991a, tables 30 and 31).

Annual growth rate (r):

- −0.5% to 0.5%
- 0.5% to 1.5%
- 1.5% to 2.5%
- 2.5% to 3.5%
- 3.5% to 4.5%

rates near 4 percent, which will cause them to double their population every 17.3 years; many more nations are growing at 3 percent per year and will thus double their population in 23.1 years.

Logistic Growth

Because food and other resources increase more slowly than population, any population growing exponentially will outstrip its food supply and other resources, including the available space in its habitat. Clearly, no population can continue growing exponentially. What usually occurs, at least in experimental populations, is called **logistic growth** (Figs. 6.1, 6.3). It can be modeled mathematically by the equation

$$\frac{dN}{dT} = rN\left(\frac{K-N}{K}\right)$$

In this equation, K is a new quantity called the **carrying capacity** of the environment, which refers to the size of the population when it reaches the level at which the birth rates and death rates are balanced. As population size approaches this carrying capacity, the population growth (symbolized by dN/dT) slows down, and when $N = K$ the population growth is zero (Figs. 6.1 and 6.3). If the population size should ever overshoot K, then there follows a population crash in which many deaths occur and the population size declines.

K is related to the amount of space in an environment and to the other resources available, including the amount of energy which is in a form that can be used by the organisms living there. For animal or plant species living in an unchanging environment, K is generally constant. For our own species, K will vary with changes in technology, especially technology that gets more usable energy from the same environment, for example, by increases in the efficiency of energy use or of food production. A given amount of land may support a particular population size of hunters and gatherers at a low carrying capacity (a low value of K). The development of agriculture will generally result in an increased carrying capacity, and agricultural mechanization (including the use of tractors and chemical fertilizers) will in most cases increase it still further. An increase in carrying capacity is the basis for the theory of a demographic transition, explained below.

[A] Different rates of increase (*r*)

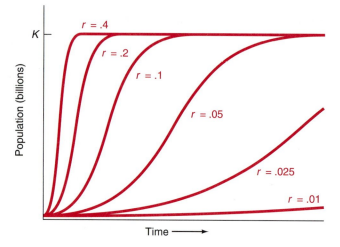

[B] Different initial population sizes (N_0)

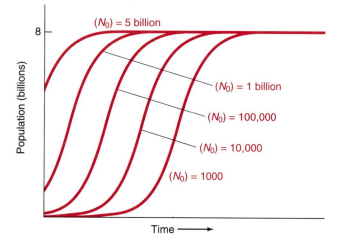

[C] Different values of *K*

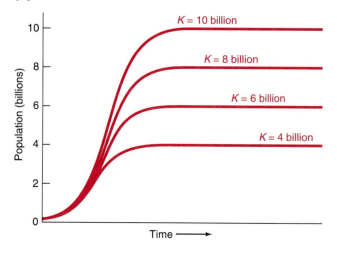

Age Structure of Populations

The models we have examined so far treat all members of a population the same. However, we know that the probability of death and the probability of reproduction both vary with age. Thus, the true rate of population growth may vary, depending upon the ages of individuals within the population, also called the *age structure* of that population. The age structure can best be characterized by constructing a population pyramid or **age pyramid** (Fig. 6.4A). Each horizontal layer on such a diagram represents the fraction of the population in a particular age group, with the youngest age groups on the bottom. Altogether, the age pyramid shows the *distribution* of individuals among the various age groups. Most pyramids are divided by a vertical midline, with male age distribution shown on one side and female age distribution on the other. Among human populations, a pyramid with sloping sides and a wide base (many children) characterizes an expanding population. A pyramid maintaining more or less the same width throughout (except for the oldest few age classes) indicates a stable population. A *stable age distribution* is reached when the pyramid keeps the same shape as each age group grows older.

Predictions of future values of *r* can sometimes be made on the basis of age structure. Clearly, a population of 10,000 individuals with 4000 females of reproductive age has much more potential for increase than one with only 400 females of reproductive age. Calculations of the potential for future increase are often carried out by multiplying the number of females in each age group by the number of children that each of those females is likely to bear, then adding up these products for all age groups.

The importance of demography as a research tool goes beyond the collection of vital statistics. Those statistics are analyzed using mathematical models to predict future events under different hypothetical conditions. Population ecologists use demography to test hypotheses regarding the factors that influence population size and growth. What are the factors that affect birth rates and

FIGURE 6.3
Logistic growth curves under a variety of conditions.

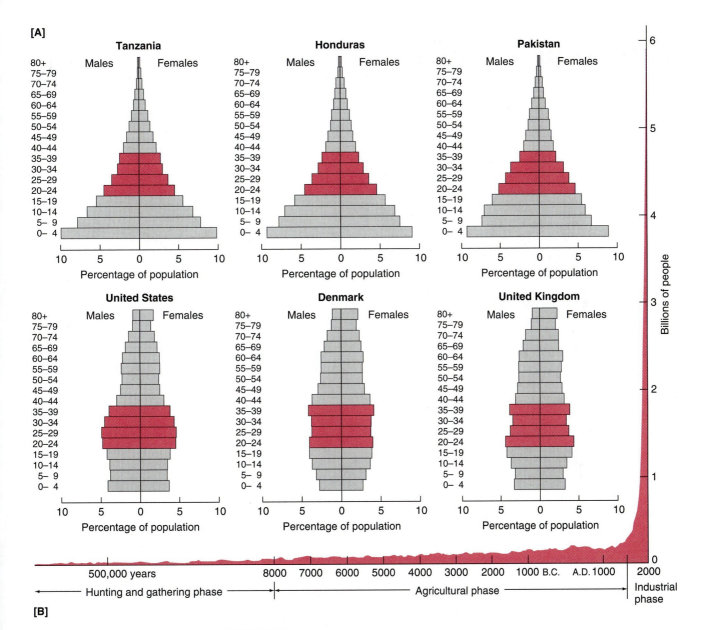

FIGURE 6.4
Population graphs. [A] Age pyramids for three rapidly growing populations (top row), a slowly growing population (United States), and two very stable populations (Denmark and United Kingdom), from United Nations (1993) data. Age groups from 20 to 39 are shaded to emphasize that these are the years when most reproduction occurs. [B] Graph showing the growth of the world's human population.

death rates? What are the results of changes in age structure? What effects does an increase in the population of one species have on its interaction with other species and their relation to the food supply?

Demographic Transition

Population increases in the past have coincided with major advances in technology. The development of agriculture made it possible for human populations to increase well above the size permitted by hunting and gathering (Fig. 6.4B). Song and Yu (1988) estimate that the world's population stood at some 50 million in 7000 B.C. and increased to around 250 million at the time of Christ. The earliest date for which there are reliable estimates is 1650, at which time the world's population stood at 500 million. By 1850, when actual census figures were available for most of the industrialized nations, the world's population stood at an estimated 1 billion and has increased rapidly since (Fig. 6.4B).

Most of our understanding of the demographic trends that accompany a population increase come from studying the increases brought about by the industrial revolution or by the spread of industrially based technology to the nonindustrial world. Our current model of this process is that it occurs in an orderly succession of stages known as a **demographic transition** (Fig. 6.5).

Stages of a demographic transition.

The first stage of a demographic transition is viewed as a stable population in which a high death rate is balanced by a high birth rate. Traditional societies that have had high death rates (especially from infant mortality and childhood infections) have over the centuries developed customs that encourage high birth rates. In other words, high mortality rates encourage high fertility rates. The second stage of the process is brought about by technological changes that result in falling mortality rates. However, it always takes at least a few generations for the cultural values and customs to change so as to permit a matching decline in birth rates. In the meantime, the memory of high mortality rates in the recent past continues to encourage high fertility rates. In some cases, the birth rate may even rise as the result of better nutrition and the im-

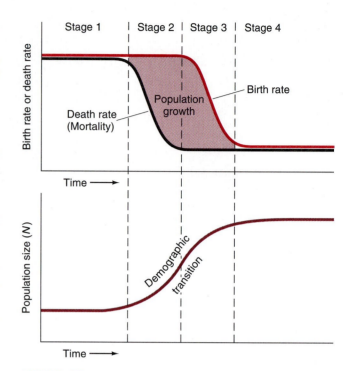

FIGURE 6.5

Idealized stages of a demographic transition. Some authorities recognize only three stages by combining the middle two.

proved physiological condition and reproductive health of prospective mothers. The combination of high birth rates and lower death rates causes the demographic transition proper, a period in which population levels climb to a new high. The third stage of the transition is marked by a decline in birth rates as the population adjusts to new conditions and as the incentives for high fertility are removed. As the birth rate declines to match the new lower mortality rate, the population once again stabilizes, but at a much larger population size (larger K) than before. The demographic transition is complete.

Demographers estimate that overall mortality rates held steady or declined slowly in preindustrial Europe, with occasional but temporary upsurges during wars and epidemics such as the bubonic plague (the black death) that decimated European populations in the 1300s. England's demographic transition began in the early 1700s and took some 250 to 300 years to complete. Other industrialized countries in North America, Europe, and Japan took closer to 200 years to complete their demographic transitions. The remaining countries of the world began their demographic transitions only during the twentieth century, and did so much more suddenly, often going from high traditional mortality rates to low modern rates in a single generation. Country-by-country analysis of the data on demographic transitions is given by Chesnais (1992).

Is the United States in the final stage of demographic transition? It now appears that the United States has completed the transition because birth rates and death rates are now approximately equal, a condition known as *zero population growth*. The U.S. population continues to increase, however, because of the excess of immigration over emigration. In addition, there can be short-term increases, for example, the relatively recent trend for baby boomers (noted earlier) to have additional children a decade or more after their first children were born and for people to remarry and start second families.

Demographic momentum. Even after the birth rate falls to the level of the mortality rate, the population may continue to increase for another generation or two because of a **demographic momentum.** The momentum is caused by an age structure in which a large fraction of individuals are not yet of reproductive age, while only a small fraction are past the age of reproduction. Since the younger age groups are more likely to reproduce in the next 20 years and less likely to die, a temporary increase in the birth rate (and a temporary decrease in the death rate) can easily be predicted. The population will continue to grow until the age distribution readjusts itself more evenly, that is, until a stable age distribution is reached. With a stable age distribution, the birth and death rates will no longer change, unless some external factor disrupts this stable situation.

The Effects of Population Growth

Demographic transition brings about a dramatic increase in human population. In most parts of the world, the excess population tends to migrate to the cities, producing an overcrowding that strains the resources of those urban areas. In Europe and Japan, this process occurred gradually over a period of several hundred years (roughly, 1600 to 1900), giving cities a chance to adjust to their changing conditions. Many cities have accommodated high population densities (*density* refers to the number of people per square mile) without widespread misery. Since World War II, urbanization in the third world has taken place much more rapidly than it did in Europe. Rapid, unplanned growth has strained most urban services to the point where many of the newly arrived migrants have inadequate housing. Crowded slum areas or shantytowns often lack safe drinking water and may also suffer from chronic water shortage; sanitation and waste disposal are also frequent problems. It is often not crowding per se that results in these problems, but crowding without sufficient facilities to support the population. Crime often increases and may become difficult to control, although many other factors besides population contribute to crime rates. Unemployment and economic hardship, when they occur, may compound these problems. The hardships of urban crowding usually fall disproportionately on the poor.

Pollution (Chapter 16) tends to increase approximately in proportion to population, most obviously because of increases in household garbage and waste water. Densely populated areas are dependent on food, water, and fuel coming in from a much wider radius; consequently, their environmental impact is felt far beyond their political borders. As population increases, more forests are cleared for agricultural use and more trees are cut down to build houses. The destruction of habitat, particularly of forests, is one result (Chapter 16). The loss of arable land (through topsoil erosion, desertification, and other processes) is accelerated by population growth, as is the depletion of nonrenewable resources such as minerals or fossil fuels.

Effects of consumption patterns. The impact on the environment is not, however, solely a function of the *number* of people. The amount of the world's resources that each person consumes is not equal around the globe. On average, the amount of resources consumed by a person in the United States is 54 times as great as that consumed by a person in a developing country. The impact of this consumption is magnified still further by the fact that much of this consumption is of nonrenewable resources. In addition, resources that might be renewable are often consumed or discarded in ways that make them nonrenewable. The enormous size of municipal solid waste disposal sites in industrialized countries is testament to these consumption patterns. These landfills are the largest structures ever built by humans, and the materials within them are unavailable for reuse or for biodegradation.

Both energy and materials flow through all ecosystems from producer organisms to consumer organisms to decomposer organisms (Chapter 15). Carbon compounds and other materials are recycled by *decomposer organisms* like fungi and bacteria, which return these materials to biological systems, thus making certain resources renewable. The term *biodegradable* as applied to materials means that decomposer organisms will be able to recycle those materials. Nonbiodegradable trash cannot be recycled by decomposer organisms (although it may be by humans) and becomes inaccessible.

Energy, on the other hand, is not recycled: It passes through the ecosystem in one direction only and is lost as heat during each energy-transfer process. Vast amounts of energy are stored in the chemical bonds of organic materials in plants, animals, and fuels. When these fuels are burned, or when the animals and plants are eaten, or when organic materials are broken down by decomposer organisms, much of this energy is lost as heat. Radiant energy from the sun replenishes some of the loss; plant photosynthesis (Chapter 15) represents the principal means by which this radiant energy is captured. If the rate of energy use does not exceed the rate at which energy is captured from the sun, then the energy use is considered *sustainable*. In many industrialized countries, however, present patterns of energy consumption are already unsustainable because they remove far more energy from global ecosystems than they produce.

Discussions of the world's population crisis frequently become linked to discussions of the environmental crisis. Many people, especially in the third world, believe that the population crisis is only a small part of a greater environmental crisis. This environmental crisis, they say, is made worse by the industrial world's overconsumption more than by the third world's population increase. Frances Moore Lappé is one of several American writers holding such views. Some analysts even question whether the industrial world's concerns over population are misdirected (and possibly racist). Third world countries, they say, could well support far larger populations than they do now if it were not for the export of so many of their resources to support the patterns of overconsumption that have become so typical of the industrial world. If the industrial countries, they say, were to give up their lavish patterns of consumption, then the third world could well support larger human populations (at a higher carrying capacity) than it does under current circumstances.

Others take what might be called a neo-Malthusian position. Paul Ehrlich, for example, views most other problems as consequences of overpopulation. If the world's population were smaller, he argues, most environmental problems would diminish or even disappear. Since some countries (mostly in Europe) have already limited their population growth, the greatest efforts should be directed at those nations (mostly in the third world, see Fig. 6.2) that have the highest population growth rates.

Overpopulation and overconsumption need not be viewed as opposing viewpoints. Population growth and profligate consumption are both widely recognized as problems, and each makes the other worse. Some people see one of these as the bigger problem; some people see the other. Efforts directed at addressing either problem can only help ameliorate both.

Limits on carrying capacity.   Many scientists tell us that we will soon reach or even exceed the carrying capacity of the planet. In fact, this is one point on which people concerned with overpopulation and those concerned with overconsumption agree, though they postulate different causes for this condition. One of the few dissenters, economist Julian Simon, observes that the technological revolutions of past centuries have repeatedly brought about demographic shifts, each of which has increased the carrying capacity. He predicts that future technological revolutions will continue to enlarge the planet's carrying capacity indefinitely. Nearly all other scientists and writers who have contemplated the subject of population believe instead that the planet's carrying capacity has a limit.

Can the global carrying capacity be increased further? The answer is not known with certainty, but it depends in part on whether we assume the Earth's natural resources to be renewable and unlimited (Julian Simon's view) or limited and nonrenewable (the majority viewpoint). Those who accept the limits imposed by nonrenewable resources will be driven to the conclusion that carrying capacity cannot be increased very much. In fact, if we maintain our present patterns of consumption, we may not even be able to sustain the present population levels forever. Even well-planned efforts to address social, economic, or environmental problems may prove to be inadequate as resources are stretched to the breaking point in the face of increasing population pressure. "Whatever your cause," says one slogan, "it's a lost cause unless we can control population."

1. What changes (biological, social, or economic) could increase the carrying capacity of the entire world or of one nation? Is it more important to modify r or to modify K?

2. Suppose that you count the number of people in a given town every year for 5 years and you discover that N has stayed about the same during that time period, say, at 10,000. You also know that no one has moved into the town or away from the town in that time. What can you deduce about B and D? Confirm this for yourself by solving the equations given earlier in this chapter. Can you tell from this information how many people were born in the town in any of the 5 years?

3. Study the age pyramids in Fig. 6.4A. Does the age distribution of the females or the males have a greater impact on future population size? Why?

4. What values of r are typical during the several stages of a demographic transition? See Fig. 6.5, and review the way in which r is defined.

5. Which is likely to produce a greater increase in the *number of people*, a small population with a high r or a large population with a small r? Try some calculations with any values of N and r that you would like to examine. Use the simplest model that seems appropriate, then ask what changes the more complex models would bring.

6. From the data presented in this chapter, can we estimate the value of K (the carrying capacity) for the human population on planet Earth? Can we make a minimum or maximum estimate? Have we already reached the carrying capacity of the planet?

C. Controlling Population Growth Raises Many Issues

Most mechanisms that operate to control population are one of two types: those which control the birth rate, B (**birth control**), and those which control the death rate, D (**death control**). Our modern population problems stem in large measure from the great advances that have been made in death control over the last century or two. We will not discuss efforts to control immigration or emigration; these efforts may strongly affect population in a particular country for a time, but the effect on a global level is small and in most cases temporary.

We should observe another distinction at this point: **Population control** is usually understood to operate on the level of populations, while birth control mechanisms generally operate by preventing births *one at a time*. A birth control method does not control population unless it is widely adopted. Death control measures can be at either the individual or the population levels: Medical advances in disease control generally operate on an individual level to decrease the death rate, while genocide, war, and epidemics increase the death rate for whole populations.

Human populations, like all other populations, are biological entities, requiring energy flow to survive. Populations are therefore subject to the laws of physics (energy is neither created nor destroyed), and populations cannot exceed the availability of energy. When they approach K, populations will be controlled by biological factors, such as starvation and disease. So the question "Should populations be controlled?" is academic because populations will be controlled by the forces of biology and physics, regardless of our answer to such a question. The relevant question is, "Should we exercise *preventive* efforts at population control?"

Death Control

War, starvation, violence, and disease all cause death. Even when they do not cause death directly, they add to the risk of death from other causes. The same may also be said of poverty, crime, overcrowding, unsanitary living conditions, and crop failures. In Malthusian terms, they are positive checks to population size, meaning factors that increase the death rate, D. To the extent that any of these factors are controlled, the consequent rate of death is controlled also. In modern times, the emphasis is on alleviation of positive checks and thus a lowering of D.

Decreasing death rates. The present worldwide population increases are in large measure the result of our success in controlling death. Epidemic diseases, infant mortality, famine, and other causes of widespread mortality result in far fewer deaths than they did in past centuries. In the industrialized countries, these reductions in mortality occurred gradually, over the course of 200 years or more, and the birth rate had a chance to become adjusted downward. Many nations of the third world have experienced a sharp reduction in death rates over the last 30 to 50 years, while their birth rates have in most cases remained high. In past centuries, high birth rates compensated for the traditionally high death rates. Parents often justified their desire to bear many children by citing the need for more helping hands and the hope that at least one or two would survive to take care of their parents in old age.

Life expectancy. In the United States and in most other industrialized nations, the control of many infectious diseases since the late 1800s and improvements in sanitation have resulted in decreased infant mortality. A few twentieth-century changes, like increases in smoking, auto accidents, and handguns, have increased death rates, but these have generally been offset by much greater declines in death rates due to control of famines and infectious diseases. Therapies for chronic diseases and better nutrition have increased the maximum age to which people live, also called *longevity*.

A few modern social scientists have found an evolutionary explanation for increased longevity:

Adults who support their elderly parents usually receive much more in return, such as help with child-raising and agricultural chores. By helping their adult children, elderly parents who live long can contribute both directly and indirectly to the raising of their own grandchildren. Long-living people therefore leave more healthy grandchildren than people who do not live as long, a difference that may also be expressed by saying that long-living people are more fit, or that natural selection (Chapter 4) favors greater longevity.

Life expectancy is the age to which a person born in a particular place and at a particular time may expect to live. It is calculated on a statistical basis, taking into account the probability at any age of a person proceeding on to the next age group. Therefore, both decreased mortality and increased longevity contribute to an increased life expectancy, with the largest increases resulting from reductions in childhood mortality. Life expectancy in the United States has risen from somewhere near 50 years in colonial times to over 75 years today, and the age group over 80 is the most rapidly growing segment of the U.S. population. This "graying" of America and of many other countries has resulted in a decrease in the segment of the population belonging to the most fertile age group and an increase in the older segment of the population, which, in many cases, is dependent on the younger generation for care. Numerous sociological changes occur when a population is composed of families with aging parents and fewer young children—more emphasis on medical care and less emphasis on schools, for example.

Birth Control

Birth control methods allow the spacing and timing of the birth of children and so are also called *family planning methods*. The majority of these methods seek to control births by preventing pregnancy. Many prevent pregnancy by interfering with the reproductive anatomy or physiology of either the female or her male partner. Therefore, in order to understand how these methods work, we need to understand the biology of reproduction. Human reproductive anatomy is covered in Box 6.1 (pp. 178–179), while the reproductive physiology of the female hormonal cycles is covered in Box 6.2 (p. 180).

Contraceptive measures. The various birth control methods form a spectrum of possibilities that are listed in Table 6.1 (p. 181). Of these methods, those that act to prevent pregnancy prior to conception, that is, prior to the joining of an egg and a sperm, can be called **contraceptive** measures. For each method, a series of questions may be asked:

1. How does it work?
2. How effective is it in birth control?
3. What costs or risks are involved?
4. What kinds of objections have been raised against it?
5. Does it have any benefits apart from birth control, for example, in the prevention of sexually transmitted disease?

Sterilization methods. *Sterilization,* the elimination of reproductive capacity, usually involves surgery and is usually permanent, although some methods are potentially reversible. One method of permanent male sterilization is the removal of the testes (castration). The testes (Box 6.1) produce the male gametes or sperm (see Chapter 3), so the removal of the testes permanently prevents reproduction by that individual. The testes also produce the hormone *testosterone,* so that removal of the testes also has many other consequences, depending on the age of the individual. One way to achieve male sterilization but allow hormone secretion to continue is by *vasectomy,* the surgical cutting and tying off of the sperm duct (the *vas deferens,* Box 6.1). Males with vasectomies continue to produce both testicular hormones and sperm for some time, but the sperm cannot reach the penis for release.

Among female sterilization methods, *tubal ligation* (tying off of the uterine tubes, Box 6.1) is the only one done primarily as a birth control measure. Tubal ligation is analogous to male vasectomy; eggs and hormones are still produced, but the eggs are blocked from traveling to the uterus. Surgical removal of the uterus (hysterectomy) is performed for medical reasons other than birth control, but the removal of the uterus results in permanent sterility because the uterus is where the developing embryo grows. Surgical removal of the ovaries, the organs that produce the female gametes (eggs), also results in sterility, but this is usually avoided for the same reason as male cas-

tration is avoided: The ovaries produce many hormones and their removal has widespread effects on the individual.

All these sterilization methods involve surgery, which makes them expensive to implement on a very large scale, especially in poor or medically underserved areas. As with all forms of surgery, there are risks, such as those of infection or from the use of anesthesia. However, both costs and risks are experienced on a one-time basis only and do not recur. All sterilization methods are completely effective as birth control methods without any further action on the part of the patient. One of the greatest objections to all these methods are that they are permanent. Many people, even people who want birth control, avoid these methods because they do not want to become permanently sterile. In the United States, about 60 percent of men who have undergone vasectomy later regret their decision. Vasectomy and tubal ligation can in some cases be reversed, but success depends on microsurgical techniques, and reported success rates vary greatly.

Abstinence methods. *Abstinence* methods have the distinct advantage of being available to all people free of charge, but their success depends upon the determination of the people using them. Total voluntary abstinence (celibacy) has long been practiced as part of a regimen of religious devotion, but only by small numbers of people. Delayed marriage (with no other sexual activity) greatly reduces the birth rate, especially inasmuch as the years before age 30 are the most fertile period for a majority of women.

The so-called *rhythm method* is a form of partial abstinence, based on the fact that a woman is fertile only for a few days following ovulation (Box 6.2); sexual intercourse performed at other times will generally not result in conception. *If practiced correctly,* this method is highly effective, but its effectiveness depends on several conditions, including the regularity of a woman's menstrual cycle (this varies individually), the ability of the couple to keep a calendar and count the days without making a mistake, and the willingness of the couple to refrain from sex (or to practice another method of birth control) during the woman's fertile period, when a mature egg has been released from the ovary. The effectiveness of the rhythm method can

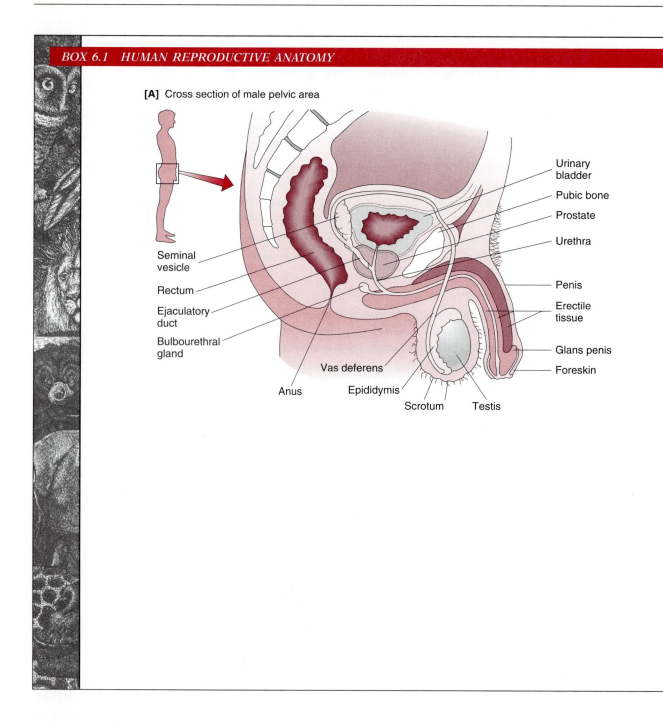

[A] Cross section of male pelvic area

Urinary bladder

Pubic bone

Prostate

Urethra

Penis

Erectile tissue

Glans penis

Foreskin

Seminal vesicle

Rectum

Ejaculatory duct

Bulbourethral gland

Vas deferens

Anus

Epididymis

Scrotum

Testis

be increased by monitoring the woman's vaginal temperature, since a rise in temperature determines the time of ovulation more precisely.

Coitus interruptus is the withdrawal of the male before ejaculation occurs. Some couples are able to use this method effectively, but the majority find it unsatisfying or difficult to follow. Since some sperm can be released prior to ejaculation, coitus interruptus is not reliable. On a population-wide scale, it is generally not as effective as other methods.

Barrier methods.   *Barrier methods* are those that impose a barrier to the passage of sperm. Most

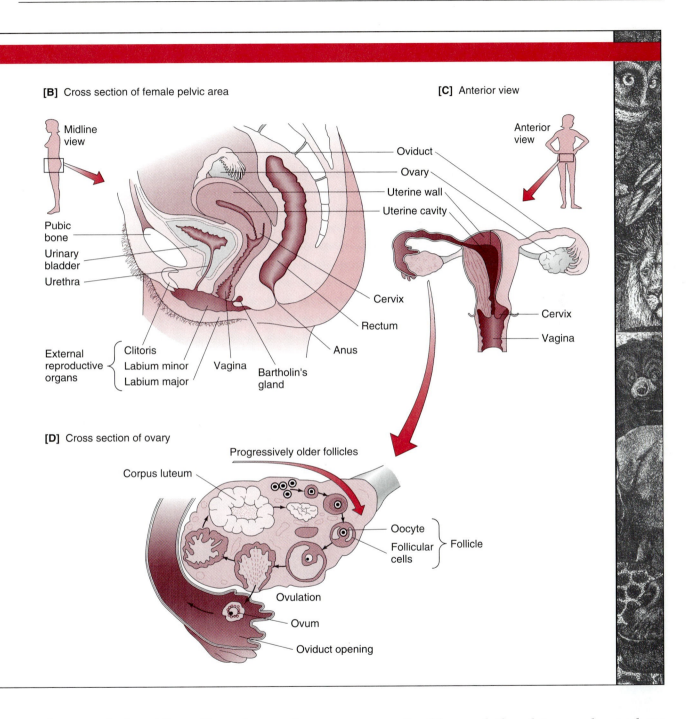

[B] Cross section of female pelvic area

Midline view

Pubic bone
Urinary bladder
Urethra

External reproductive organs
{ Clitoris
Labium minor
Labium major

Vagina

Bartholin's gland

Anus

Rectum

Cervix

[C] Anterior view

Anterior view

Oviduct
Ovary
Uterine wall
Uterine cavity

Cervix

Vagina

[D] Cross section of ovary

Progressively older follicles

Corpus luteum

Oocyte
Follicular cells
} Follicle

Ovulation

Ovum

Oviduct opening

condoms are designed for males and cover the penis, but a condom that is worn by women has also been developed. The male condom is the oldest of the barrier methods. First developed in England, traditional condoms were constructed of animal membranes (usually sheep intestine) and were therefore considered a luxury item. The develop-ment of rubber and then latex made condoms more widely available and also more reliable, and led to the development of the other barrier methods. These other barrier methods include such vaginal inserts as cervical caps, vaginal di-aphragms, and sponges, all worn by the woman. Vaginal diaphragms must initially be individually

BOX 6.2   HUMAN FEMALE REPRODUCTIVE CYCLES

Hormones are small molecules that are used for chemical communication throughout the body (see Chapters 10 and 12). The rise and fall of several hormones control the growth of egg-containing follicles in the ovary and the cycle of events in the uterine lining.

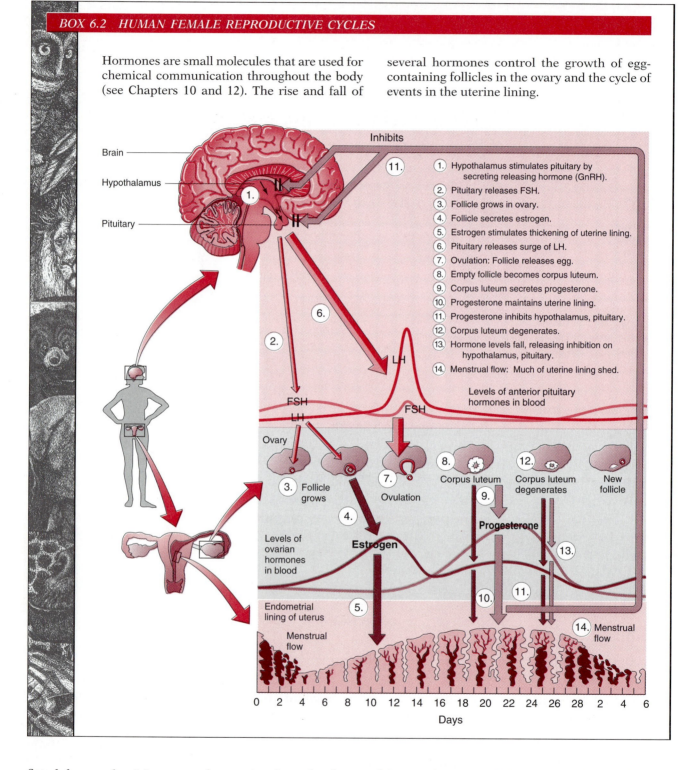

1. Hypothalamus stimulates pituitary by secreting releasing hormone (GnRH).
2. Pituitary releases FSH.
3. Follicle grows in ovary.
4. Follicle secretes estrogen.
5. Estrogen stimulates thickening of uterine lining.
6. Pituitary releases surge of LH.
7. Ovulation: Follicle releases egg.
8. Empty follicle becomes corpus luteum.
9. Corpus luteum secretes progesterone.
10. Progesterone maintains uterine lining.
11. Progesterone inhibits hypothalamus, pituitary.
12. Corpus luteum degenerates.
13. Hormone levels fall, releasing inhibition on hypothalamus, pituitary.
14. Menstrual flow: Much of uterine lining shed.

fitted by a physician or other trained medical worker, and must be inserted correctly into the vagina prior to intercourse and left in place for several hours thereafter. When properly placed, vaginal diaphragms block the movement of sperm from the vagina to the uterus, thereby preventing

TABLE 6.1   BIRTH CONTROL METHODS

METHOD	% EFFECTIVENESS*	RELATIVE COST
I. Contraceptive		
A. Sterilization		
1. Irreversible sterilization (castration, hysterectomy, or ovariectomy)	100	High
2. Semipermanent sterilization		
Tubal ligation	99.6	High
Vasectomy	99.6	Medium
B. Abstinence		
3. Long-term sexual abstinence (celibacy, delayed marriage)	100	None
4. Timed abstinence ("rhythm" methods)	Varies (76–98)	None
5. Withdrawal during intercourse (coitus interruptus)	Varies (77–84)	None
C. Barrier		
6. Condom	90–98	Low
7. Vaginal diaphragm with spermicide	81–98	Low
8. Cervical cap with spermicide	87–98	Low
9. Sponge with spermicide	80–98	Low
D. Spermicidal		
10. Spermicidal creams, foams, and jellies alone	82–97	Low
E. Hormonal		
11. Estrogen alone	98–99.5	Medium
12. Estrogen plus progesterone	98–99.5	Medium
13. Progestin only (injectable or "minipill")	97.5–99	Medium
II. Postfertilization		
14. Postcoital pills	Not known	Medium
15. Intrauterine devices (IUDs)	95–98.5	Medium
16. Abortion	100	High

*When two figures are given, the first generally represents populations at large in the industrialized world, while the second represents experienced or carefully instructed users only. A 99% rate of effectiveness means that only 1% of couples using that method will become pregnant.

their joining with the egg. (One recent study in Brazil reported a higher-than-usual success rate if the diaphragm was left in place nearly all the time, but this result remains to be confirmed in other populations.) Cervical caps are made to fit over the narrow portion (cervix) of the uterus, where they also block sperm.

Spermicidal agents, chemicals that can kill sperm, in creams, foams, jellies, or suppositories, are often used together with a barrier method; the combination of barrier plus spermicide is much more effective than either method used alone. One of the newest methods is a sponge impregnated with spermicidal fluid. Barrier methods used with spermicides have extremely low failure rates when used by people familiar with their proper use; most pregnancies occurring with barrier methods are the result of improper use. Barrier methods are widely used in many countries. Condoms have the added advantage of protecting against AIDS and other sexually transmitted diseases.

Hormonal administration methods. Several birth control methods depend on alterations of the female reproductive cycle. During the normal female cycle, which lasts about 28 days, an egg matures in an ovary, is released into the uterine (fallopian) tube, and awaits the presence of sperm. If no sperm arrive or if fertilization of the egg does not occur, the egg and the uterine lining are sloughed off in the form of menstrual bleeding, and the cycle repeats. This cycle is controlled by two ovarian hormones, estrogen and progesterone, and by three hormones secreted by the pituitary gland at the base of the brain (see Chapter 12). At the start of each uterine cycle, the pituitary secretes *follicle-stimulating hormone (FSH)*, which stimulates the growth of an egg-containing region in the ovary called the Graafian or ovarian follicle. In response, the ovary produces the hormone **estrogen,** which reaches a peak concentration during the second week of the cycle (Box 6.2). The estrogen stimulates the release of a second pituitary hormone (luteinizing hormone), which induces release of the egg (ovulation), after which the tissue that surrounded the egg is left behind to form a scar tissue called the *corpus luteum.* The third pituitary hormone stimulates the corpus luteum to secrete the hormone **progesterone,** which maintains the uterine lining in a thickened and receptive condition, ready for the implantation of an embryo should the egg be fertilized. If no implantation takes place, the corpus luteum degenerates and the supply of progesterone drops sharply, causing the uterine lining to break down and the menstrual flow to begin. The absence of progesterone also releases the pituitary to begin secreting follicle-stimulating hormone once again, and a new cycle resumes. The concentration of each hormone rises and falls, stimulating the rise of the next. In several cases, the presence of one hormone has an inhibiting effect on the secretion of the previous hormone. This *feedback mechanism* prevents the overproduction of any hormone and the continued production of a hormone once it has done its job.

Hormonal birth control methods interfere with the reproductive cycles, often by taking advantage of a feedback mechanism. For example, estrogen and progesterone both inhibit the secretion of FSH, so that supplying these hormones (or a similar compound) will prevent ovarian follicles from reaching maturity and releasing their eggs. The hormones can be given as *birth control pills,* as injections, as implants (such as Norplant) just under the skin, or as patches on the skin. Regardless of the method of delivery, all hormonal methods work by preventing the egg from maturing and being released. Because hormones have many effects throughout the body, hormones used in birth control have many side effects, which include the possibility of medical problems such as blood clots. For this reason, medical supervision is recommended. Hormonal methods are available in the United States and most industrialized countries only by prescription.

Early birth control pills contained estrogen alone, but progesterone was later added (producing the combination birth control pills) in order to reduce the levels of estrogen and its side effects. The continuous levels of female hormones prevent the usual cycling from taking place. The expense and the requirement of obtaining a prescription limits the use of birth control pills in many populations. (Some developing countries make birth control pills available without prescription in order to encourage their more widespread use, but cost is still a problem.) Birth control pills have, however, become a commonly used contraceptive method among the middle and upper classes in many countries. Newer types of birth control pills were developed in the 1980s. These include the minipill, which uses progestin (a progesterone-like compound) only.

One of the newest methods of birth control is a male contraceptive pill that uses a drug called *gossypol,* derived from cottonseed. This drug was first developed around 1970 in China and is said to be about 99 percent effective. Although it does not affect the hormone testosterone, gossypol does somehow interfere with sperm production. Tests of gossypol have reported some toxic side effects, so an effort is now being made to develop a synthetic substitute.

Delayed weaning.   An older hormonal method of birth control is the practice of delayed weaning, a common practice in many African countries. Children in Africa are almost always breast-fed, and many are not weaned until they are from 4 to 6 years old. While a woman is breast-feeding, she is producing hormones that stimulate milk production. These same hormones also inhibit the rise and fall of the hormones produced by the ovary, thus interrupting the menstrual cycle and preventing egg maturation. African women who use this method of birth control do not wean their youngest child from the breast until they feel they are ready to have another child. This method of birth spacing is a widespread and seemingly effective practice in many parts of Africa, although studies have shown that it is unreliable among well-nourished women in the industrialized world. For example, prolonged breast-feeding has a contraceptive effect among the !Kung San (Bushmen) of South Africa and Namibia. These women are poorly nourished and walk from 4 to 6 miles a day, conditions that seldom occur among North American women. Among Hutterite women (belonging to a Protestant sect of mostly farming communities in the north central United States and Canada), breast-feeding has been shown to have a delaying effect on the interval between pregnancies, but the effect is less than among African women. Since the ability of breast-feeding to delay the return of the menstrual cycle is greatest among women who are physically active but who have a low caloric intake, improvements in maternal nutrition may actually increase birth rates. Birth spacing by prolonged breast-feeding is also most effective when babies suck vigorously and often. Any contribution to infant nutrition other than breast milk (e.g., by bottled milk or cereal) reduces the effect. In most third world countries, the use of bottled milk results in a reduced effectiveness of birth spacing by delayed weaning. The closer spacing of births leads to an increase in fertility at the population level, possibly offsetting the effects of birth control programs. Breast milk has many benefits for infants, including a reduction in diseases because of the antibodies carried in the mother's milk.

Postfertilization methods. Another category of birth control methods work after fertilization of the egg but before it is implanted in the uterus. The morning-after pill is an example; it hormonally prevents implantation. One of the newest methods is a drug called RU-486, developed in France and now available also in Great Britain, Sweden, and China. In the United States, experimental testing of RU-486 on women volunteers was begun in 1994, and the drug is available only as part of such experiments. RU-486 is a hormone-like drug that induces uterine contractions and expulsion of the uterine contents. RU-486 has its greatest potential use as a morning-after pill to prevent implantation from occurring during the first five days following intercourse. Prostaglandins (another type of hormone) are usually given with RU-486 to avoid some of the possible side effects.

An *intrauterine device* (IUD) is a small piece of plastic and/or wire that can be any one of a variety of shapes (e.g, a loop or a coil) and that is inserted by a physician into the uterus of the woman, where it remains until removed by the physician. IUDs prevent pregnancy by preventing implantation, although the exact mechanism by which this occurs is not known. (Desert Bedouins have long practiced a similar method of birth control on their camels, by inserting stones into the uteri of female camels to prevent pregnancy and removing the stone when breeding was again wanted.) A major advantage to this method is that, once inserted, the IUD works on its own with no need of further action on the part of the woman. The use of IUDs trails far behind the use of birth control pills and even sterilizations, and the method should not be used by women who have never been pregnant. However, women who use IUDs have a higher rate of satisfaction than with any other form of birth control, including pills. There are some possible side effects with IUDs, such as the possibility of uterine bleeding, but the method is effective and the failure rate is low. Earlier IUDs were not as safe and one early model, the Dalkon Shield, caused infections and severe bleeding problems in many women, resulting in hysterectomies, large lawsuits, and the corporate bankruptcy of its manufacturer. Many people have shown renewed interest in IUDs in the last decade because of the development of newer, safer types.

Abortion is the termination of a pregnancy, including the cleaning out of the uterine lining and the expulsion of the embryo or fetus. The traditional method of dilation and currettage (enlarging the cervix, then scraping out the uterine interior with a spoonlike instrument) has now been supplemented by newer techniques such as vacuum aspiration (using a machine that uses suction to clean out the uterine contents). There are medical risks to the woman, including excessive bleeding, the chance of infection, and uterine injury, which may result in sterility. The sum total of risks to the life and health of the woman is less than the sum total of risks associated with completing the pregnancy and giving birth. The medical risks are especially low if the abortion is done early, during the first trimester (three months of a nine-month pregnancy). Second trimester abortions using saline injections can also be done safely in most cases. At whatever time an abortion is performed, however, the risks are much higher when it is being done by an untrained person. The drug RU-486, previously mentioned as a postfertilization method, can also be used after implantation, to induce abortion. Thus one name for RU-486 is the "abortion pill." Successful abortion has also been achieved in 96 percent of a group of 178 women given two drugs (methotrexate, followed several days later by misoprostol) that are already legally available (by prescription) in the United States (Hausknecht, 1995).

Of course, birth control methods can be combined, and a later-acting method can be used if an earlier-acting method fails in a particular case. Most population planners have advocated that abortion be made a widely available option as a backup when other, less costly methods have failed.

Infanticide.   *Infanticide*, though not technically a means of birth control, has long been practiced as a means of population control in many parts of the world, especially in times of famine. There are records of infanticide from Medieval Europe, and the practice was still widespread in China, India, and other parts of Asia well into the twentieth century. In most cases, the infant is not directly killed, but is rather allowed to die through lack of care. In societies in which infanticide is practiced, it is done more often on females. In China, infanticide is now officially outlawed. However, the government's strict population control policy allows only one child per couple, and social scientists suspect that female infanticide is still widely practiced by couples who want a boy but instead have a girl.

Cultural and Ethical Opposition to Birth Control

No single method of population control is best for all societies. Abortions and sterilizations, for example, require medically trained personnel. They are more expensive and more labor-intensive than other methods, and are therefore unlikely to become the methods most widely used, even among populations that have no objections to them. All methods need to be adapted to the customs of the people using them, and education in the use of certain methods will meet with resistance of various kinds. For instance, women in many Muslim societies are generally forbidden to discuss reproductive matters with anybody outside their families, including health care workers.

The attitudes of the Catholic church toward birth control have varied over the centuries; official opposition to most forms of birth control is historically recent; see Petersen (1975) and especially Noonan (1965) for reviews of this complex subject. The writings of Saint Augustine (A.D. 354–430) have long influenced Church attitudes on sex and reproduction. A considerable debate about birth control took place within the Catholic hierarchy in the 1960s, resulting in two Papal encyclicals, *Populorum progressio* (1967) and *Humanae vitae* (1968). The first of these acknowledges the population problem and the need for family planning in underdeveloped areas; the second denounces abortion, sterilization, and all forms of birth control except for the rhythm method. Surveys in many countries show that a large majority of Catholics use various forms of birth control despite the Church's official position. Attempts to spread birth control information have often been opposed by the Church, especially in Latin America, but Church teachings have not stopped Italy, a country over 98 percent Catholic, from achieving a stable (nongrowing) population, with one of the lowest birth rates in Europe.

Other religious groups have generally been more tolerant of contraceptive methods. Abortion, however, is opposed by Catholics, Protestant fundamentalists, Orthodox Jews, and Muslims. In its simplest terms, the principal argument voiced by these groups is that a fetus is a living human being and that killing it is an act of murder. People who wish to keep abortion legally available have argued several major points, including a woman's right to choose, a child's right to be wanted by his or her parents, and the need to control the world's population. This highly charged issue is further examined in Box 6.3.

Population Control Movements

Organizations to promote birth control and control population growth were first formed in the nineteenth century. In England, several of these organizations called themselves "Malthusian," even though Malthus, a curate in the Anglican church, was opposed to nearly all of the birth control methods available in his day. Knowledge about reproduction and birth control was not widely available; publications on the subject were banned as immoral in the United States and elsewhere. Two British reformers, Charles Bradlaugh and Annie Besant, and an American, Margaret Sanger, worked through the late 1800s and early 1900s to make information and contraceptives more widely available. Besant strongly influenced Mahatma Gandhi and later migrated to India. Sanger always viewed birth control from the perspective of giving individual women more control over their own lives. The availability of birth control in the United States owes much to her tireless campaigns. She also traveled widely, spreading the message of birth control to India, China, and Japan.

The influence of Gandhi, Besant, and Sanger led India to become the first modern nation to

BOX 6.3   SOME ASPECTS OF THE ABORTION DEBATE

Laws on abortion vary greatly from place to place and sometimes from one time period to another. Abortions were illegal in many parts of the United States until 1973, when the Supreme Court ruling in *Roe* v. *Wade* recognized a woman's right to an abortion under certain conditions. During the first trimester of pregnancy, this right can be exercised by the woman in consultation with her physician without any interference from state laws. During the second trimester of pregnancy, state governments can impose waiting periods or certain other conditions, and during the third trimester they can limit abortions more strictly or outlaw them entirely. As a result, abortion practices vary from state to state. Practices also vary in other parts of the world: Ireland and most Moslem countries outlaw abortions, while most other European and Asian countries permit them.

Many of the questions usually raised in the course of the abortion debate revolve around matters of definition: When does life begin? Is a fetus a human being? The biological definitions of life mention properties like motility, metabolism (internal chemical reactions), homeostasis (the ability to maintain certain conditions), irritability (the ability to respond to stimuli), and the presence of genetic material that is inherited (Chapter 1). By these criteria, a fetus has the characteristics of life as long as it remains in the womb, but it does not have an independent life.

As to whether or not a fetus is human, we must first decide what we mean by *human*. If we mean anything that possesses human genetic material in the form of DNA, then a fetus is definitely human, but so are white blood cells, haploid sperm and egg cells, and the many organs discarded each year as medical waste. If we distinguish the fetus from these others as being potentially able to form a human life capable of independent existence, then we need to examine what we mean by *potential* and when an *independent existence* begins.

Can biology help us define "humanness"? Can this subject be studied through the formulation of falsifiable hypotheses? One recent book by Morowitz and Trefil (1992) examines several possible definitions and argues that all of them point to an "acquisition of humanness" at around the 25th week of fetal life, the time at which most of the connections between nerve cells in the cerebral cortex are made. Electrical waves in the brain (detected by EEGs; Chapter 10) begin at about this time, providing evidence that the fetus can respond to certain stimuli and can be described as having experiences. Only after the 25th week are the fetal lungs sufficiently well-endowed with their own vascular blood supply to allow the tissues to receive enough oxygen if birth occurs prematurely. Prior to the 25th week, the lungs tend to collapse when empty, and their internal mucous linings tend to stick together, preventing the lung from refilling. Under current technology, the 25th week approximates to the minimum age of independent viability. The survival of premature babies is 80% or higher beginning with the 26th week of gestation, but 50% or less in the 25th week or earlier because of the underdeveloped state of the lungs and brain. Morowitz and Trefil conclude that, prior to the 25th week, the fetus does not have the potential for an independent, human existence, and that abortions prior to the 25th week are thus not terminating a human life. According to the Centers for Disease Control, only 1% of abortions are performed past week 21, and only a small fraction of these beyond the 24th week.

Much of the frustration of the abortion debate is that different people start with such different premises, use such different terms, and rely on such different arguments. People arguing from a deontological perspective or from a utilitarian perspective (Chapter 2) may not be capable of swaying one another's opinions because their arguments are based on such very different premises. Will biological research such as is summarized by Morowitz and Trefil add clarity to this debate? Will all biologists agree with Morowitz and Trefil?

institute a government-funded campaign to control its population. Beginning in 1951, India implemented population control measures that featured easy access to both contraception and abortion and an information campaign to encourage their widespread voluntary use. During the 1960s, nine other nations, including China, Egypt, and Pakistan, also implemented population control programs.

The most ambitious population control program in history was adopted in China in 1962, at a time when their population was around 700 million and was growing at just above a 2 percent annual rate. The campaign for "only one child for one family" was waged with special vigor, and parents who bore more children than this were fined and sometimes imprisoned. The goal of this campaign was not just to limit population growth, but to reduce the population to its 1962 level of 700 million as quickly as possible. Because of a large demographic momentum (that is, an age structure with many children), China's officials realized that they would have to cut the birth rate to somewhat *below* the mortality rate for a time in order to achieve a stable population. China's 1990 population was around 1.1 billion, but the annual growth rate has been cut to 1.45 percent.

The Education of Women

One of the most effective methods of reducing the number of children borne by each woman (termed the *fecundity* of populations) is to educate women. Studies in many parts of the world have shown that the population birth rate falls with each rise in the education level of women, even in the absence of any program aimed specifically at birth control (Fig. 6.6). Around the world, the countries with the lowest rates of female literacy also have the highest population growth rates, while those with higher female literacy have lower growth rates. Countries with female literacy rates below 15 percent include Mali, Yemen, and Afghanistan, with population growth rates from 3.0 percent to 4.5 percent. Third world countries with female literacy rates above 80 percent, including Jamaica, Sri Lanka, Thailand, and Colombia, all have population growth rates below 2.0 percent.

Third world women with a seventh-grade education or higher tend to marry later than other

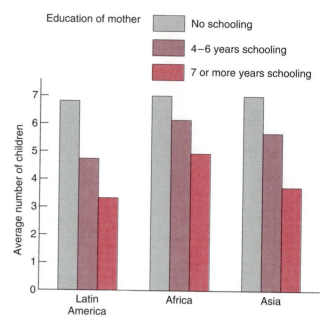

FIGURE 6.6
The education of women reduces the average number of children per family. Based on data from the World Bank.

women (four years later, on average, and these are among the most fertile years). They also use voluntary means of birth control more often, have fewer (and healthier) children, and suffer far less often from either maternal or infant mortality in childbirth. The empowering of women (by giving them more education and more control over their reproductive lives) also raises the educational level of their children and results in more rapid economic development (at lower cost) than many other programs aimed more specifically at development.

In the United States, educational efforts are also an important part of most efforts to reduce teen pregnancy rates. The rate of teenage pregnancy is lowest among women with the most years of schooling and highest among those with only a grade-school education or less.

Although most population control programs are carried out by national governments on their own populations, several programs are international in scope, including those run by the World Health Organization (a branch of the United Nations) and by the U.S. Agency for International Development (USAID). The United Nations has sponsored many conferences on population.

Some past conferences emphasized the environmental impacts of population growth, but the 1994 conference in Cairo, Egypt, shifted the attention to the education of women. Women's rights advocates from a variety of countries stressed the need to improve both the education and legal status of women. Many people cautioned that overzealous government-sponsored programs aimed at population control could restrict the reproductive freedom of individual women as much as the earlier lack of birth control information. Although these people generally see the need to reduce the rate of population growth, they are deeply suspicious of programs that coerce individual women or restrict their freedom, and they are especially suspicious of programs urged upon third world nations by male-dominated institutions in the industrial world. Instead, they favor programs to educate women and improve their legal status and reproductive choices. They point out that lower birth rates have resulted in all cases in which the education, legal rights, and reproductive choices of women have improved.

THOUGHT QUESTIONS

1. Which would you think would contribute more to an increase in life expectancy: decreased infant mortality or increased longevity?
2. Find out if your college makes birth control information and birth control itself available to the student population. Are certain methods favored over others? Why?
3. Why do most contraceptive measures involve medical action on women rather than men? Why are most population control campaigns aimed at women? In societies in which women have little or no control over their lives, are they likely to be able to carry out family planning? Will family planning give them greater control?
4. Do you think it is proper to view birth control information as a freedom-of-speech issue? Do you think birth control methods should be taught in the public schools? Why or why not?
5. If cloning (see Chapter 3) becomes a reliable procedure, then any cell from an early embryo would have the potential to form a new individual. Under these conditions, would a failure to clone a cell be a failure to realize a potential life, and thus comparable to performing an abortion? How are these situations morally comparable? How are they different?
6. What, in your opinion, is the best way for a government to control its population growth without restricting the reproductive freedom of its women? What social and ethical problems need to be considered?
7. Can biology play any useful role in the abortion debate? Would any biological data be persuasive to a person who opposed abortion based on deontological principles? Would data be persuasive to someone using utilitarian ethics?

CHAPTER SUMMARY

Each major advance in technology has resulted in a dramatic increase in human populations. Demographic transition and population increase usually begin with declining mortality and end when the birth rate declines to match the death rate. Demographic momentum results from an age structure with many younger and fewer older people. The effects of population growth are numerous: Pressure on the utilization of resources is increased, and any inefficiency results in starvation and death. Large numbers of people flock to cities and strain urban resources. Crowding may result in more crime, warfare, poverty, and disease. Reduction in mortality will result in population increase unless the birth rate also is controlled. Many methods of birth control are available. They differ in their biological mechanism, their costs, their medical risks, and their acceptance by different groups of people. Methods that work for some people may be unsuitable for others. Many studies have found that improving the education and legal status of women lowers the birth rate in a cost-effective manner and brings other benefits besides.

KEY TERMS TO KNOW

abortion (p. 183)

age pyramid (p. 170)

birth control (p. 175)

birth rate (p. 167)

carrying capacity (*K*) (p. 168)

contraceptive (p.177)

death control (p. 175)

death rate (p. 167)

demographic momentum (p. 173)

demographic transition (p. 172)

demographics (p. 167)

doubling time (p. 168)

estrogen (p. 181)

exponential growth (p. 168)

intrinsic rate of natural increase (*r*) (p. 167)

logistic growth (p. 168)

population (p. 167)

population control (p. 175)

progesterone (p. 181)

CONNECTIONS TO OTHER CHAPTERS

Chapter 2 Abortion and other forms of birth control raise important ethical issues.

Chapter 4 Population size responds to evolutionary forces, such as natural selection and competition from other species.

Chapter 7 Different organisms have different reproductive strategies that are either *r*-selected or *K*-selected.

Chapter 7 Reproductive strategies control fertility.

Chapter 8 Undernutrition and malnutrition increase mortality.

Chapter 12 Overcrowded conditions cause stress in many species.

Chapter 13 AIDS has increased the death rate in many countries.

Chapter 14 Water and sewage treatment needs increase with population growth.

Chapter 15 Plants are the main biological energy producers on which the life of humans and other consumer organisms depends.

Chapter 16 Human population growth threatens other species and their habitats.

CHAPTER 7

Sociobiology and Reproductive Strategies

Any behavior is called **social behavior** if it influences the behavior of other individuals of the same species. Examples of social behavior in animals include cooperative feeding, cooperative defense, aggression within the species, courtship, mating, and various forms of parental care. People also practice many forms of social behavior: nurturing their young, helping their neighbors, defending their possessions, and providing both material help and emotional support to their loved ones and to others. Some types of social behavior (including "antisocial" behavior) result in problems for society; examples include violence, crime, racist acts, sexist acts, and child abuse and neglect. The population crisis discussed in Chapter 6 is a direct result of reproductive behavior. All of the above are social behaviors because they affect the future behavior of other individuals. **Sociobiology** is the comparative study of social behavior and social groupings among different species.

Can behaviors that cause problems be changed easily? Can beneficial behaviors (however defined or recognized) be substituted for destructive behaviors? Is behavior in general, and human behavior in particular, rigid and unchangeable, or plastic and easily molded? Are we governed more strongly by our genetic background (nature) or by our upbringing and environment (nurture)? If human behavior is strongly determined by genes, then cultural influences, including education and training, will have

only limited power to bring about changes in human behavior. Social reformers, on the other hand, pin their hopes on the opposite belief, that human behavior can be modified almost at will, subject to few if any restrictions. Debates about alcoholism or homosexuality are often unproductive because some people assume that these traits are behaviors that could easily and voluntarily be changed, while others assume that these traits are permanent and deeply rooted in biological differences that may or may not be genetic. Differences in behavior between the sexes are likewise seen by some researchers as genetically constrained and by others as culturally controlled and easily changeable.

We have spoken of *human* behaviors, which provoke the most heated discussions. However, sociobiology is a broad field of study, and humans are but a single species. The majority of people doing research in sociobiology have nonhuman animals as the focus of their research. Altruism, for example, is a major research question in the sociobiology of all species. Among other broad-spectrum issues within sociobiology are the advantages of sociality itself, the kind of social organization found in a species and how it evolved, and the social relations between the sexes of each species. The category *relations between the sexes* includes the concept of a reproductive strategy; we will show that parental care, infanticide, adultery, and altruism can all be viewed as important components of reproductive strategies. The evolution

189

of these strategies is an important field of investigation for sociobiology.

A. Sociobiology Deals with Social Behavior

Sociobiology means different things to different people. To scientists working in sociobiology, it is a field of study that deals with social behavior and its evolution. The explanations favored by sociobiologists are usually couched in evolutionary terms. However, a good deal of behavior is modified by learning, while only those components of behavior that are inherited can be modified by natural selection. Therefore, one of the important research goals of sociobiology has become investigation of the relative importance of learned and inherited influences on particular behaviors.

Sociobiology has a number of critics who challenge the viewpoint that any behavior can be inherited and who contend that learning (including cultural learning in humans) is the most important influence on behavior. We will examine both viewpoints.

Learned and Inherited Behavior

Many behavioral patterns may be strongly influenced by experience in dealing with the environment, i.e., by **learning.** Learned behavior may increase fitness, but only the genetic components (or predispositions) underlying the behavior can be influenced by natural selection (Chapter 4). Natural selection can operate on the capacity for learning particular kinds of things, like how to find one's way through the maze of one's surroundings. The character favored by selection in such cases is not the behavior itself, but rather the capacity to learn the behavior. One of the important generalizations in sociobiology is that behavioral characteristics are subject to natural selection only to the extent that these characters are under genetic influence.

Nearly every behavior that has been carefully investigated has at least *some* genetic component. This is even true of the ability to run through mazes, one of the most often-studied types of learned behavior. Rats were tested for their ability to learn certain mazes; then their littermates, who were never tested themselves, were selectively bred over several generations (this is called *kin selection*). Breeding littermates of fast maze-learners resulted in a strain of fast learners, while a strain of slow learners was bred from the littermates of low-scoring individuals. The use of kin selection in this experiment was crucial in eliminating learning experience or other possible influences besides inheritance in determining the differences between the two selected strains. Notice that the behavior was not fully determined by inheritance, but the *difference* between the two strains resulted from the buildup of heritable traits. As we pointed out in Box 5.1, *heritable* is a descriptor of populations, not of individuals, and therefore makes no predictions as to the phenotype of any individual.

The observation that a particular behavior has evolved, or that it can be changed by selection, can be taken as strong evidence that at least *some* genetically controlled component exists. *This does not mean that the behavior of an individual is fixed and unchangeable,* because there are extremely few behaviors in any species (and none at all in humans) that are not subject to modification through learning.

It is important to emphasize that making the issue one of learned *versus* innate behavior creates a false dichotomy. Every learned behavior is based in part on some innate capacity to learn, including natural abilities to learn certain kinds of behaviors and not others, to respond to some stimuli and not others, to learn up to a certain level of complexity, and so on. Likewise, just about every innate behavior pattern can be modified to some extent by learning. These observations give rise to the falsifiable hypothesis that *nearly every behavior pattern is at least partly learned and at least partly innate.* No behavior is 100 percent learned, and few are 100 percent innate in any species (Fig. 7.1). The methods used to distinguish learned and innate components of behavior are described later in this chapter.

The Paradigm of Sociobiology

Sociobiology, the study of social behavior among different species, uses a scientific paradigm of the kind described in Chapter 1: one or more theories, plus a set of value-laden assumptions, a vocabu-

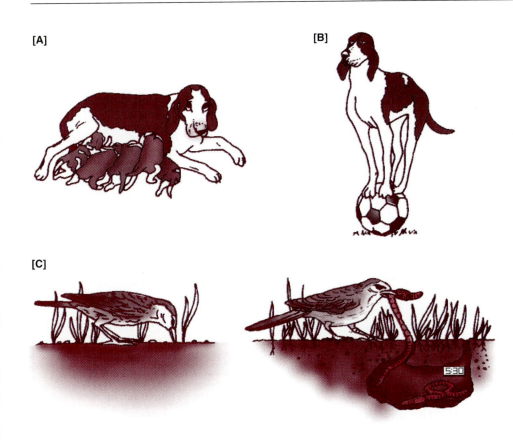

FIGURE 7.1
Learned versus instinctive behavior patterns. [A] Nursing and suckling behaviors have high survival value and have strong instinctive components in most mammals (but this doesn't prevent bottle-feeding of many human infants). [B] An example of a learned behavior in a dog. [C] Most forms of behavior show both learned and innate components. Robins are genetically programmed to peck at certain stimuli, exposing them to a learning experience from which they learn how to hunt more effectively and how to distinguish food objects from other objects.

lary, and a methodological approach (Box 7.1). The formulation of sociobiology as a paradigm dates from the publication of the book *Sociobiology: the New Synthesis,* by E. O. Wilson (1975). Nothing in this paradigm was without antecedents—every idea had been expressed before, including the use of the term *sociobiology*. Many of the ideas could be traced to Charles Darwin's writings. What was new in 1975 was the way in which these ideas were put together to form the paradigm.

If people outside the paradigm saw sociobiology as no more than the study of social behavior and its evolution, few objections would have been raised to it. However, sociobiology was frequently criticized by sociologists, philosophers, feminists, and many others for what these critics saw as a hereditarian bias (see below). Critics such as Michael Ruse and Marshall Sahlins take exception to the sociobiology approach in general and to its application to humans in particular. The many critics of sociobiology have raised the following points:

1. Much of behavior is learned, and nearly all behavior can be modified by learning. This is particularly so in mammals.

2. Critics contend that by presupposing that innate components of behavior *must* exist and must be sought for (see Box 7.1), sociobiologists have already decided the issue in favor of innate control in every case. By making innate behavior the focus of their research, sociobiologists have devalued the importance of learning.

3. The sociobiology paradigm makes no distinctions between the study of social behavior of insects and the social behavior of fish or birds or people. This offends many critics, and that is one of the major points that makes sociobiology controversial. People have both language and culture, and our behavior is strongly influenced by both. Because language and culture are considered distinctly human traits, even many scientists who are otherwise sympathetic to sociobiology consider any extrapolation to humans of findings based on other animals strongly suspect.

BOX 7.1   THE SOCIOBIOLOGY PARADIGM

Research activity in science is often organized around paradigms (see Chapter 1). Listed here are some of the major points of the sociobiology paradigm:

1. Behavior is interesting to observe and to study. (This is a value judgment; people who do not share it will never be attracted to the paradigm.)

2. Much of the interesting behavior influences the behavior of other individuals and is called social behavior. (This is a definition with an implied value judgment that people within the paradigm are expected to share.)

3. Social behavior has evolved and continues to evolve. (This is a central theory which, if rejected, would bring down the entire paradigm.)

4. The evolution of social behavior takes place by natural selection, along the lines outlined by Darwin: Variations occur, and the variations that increase fitness persist more often than those that do not. (This is again a theory; it includes theoretical concepts like *fitness* and *variation*.)

5. Behavior is often modified by individual experience (learning). However, this learning takes place within limits set by the biology of the organism: The eyes limit what can be seen (likewise with other sense organs); the muscles and skeleton limit the possible responses; the structure of the brain limits the learning capacity; and so on. There are also many preexisting predispositions to respond to certain types of stimuli, to react in certain ways, and so on. These predispositions may have been learned at an earlier time, but *at least some of them* precede any learning and may be called *innate*. (This is a central tenet of the paradigm, forming the basis for its further research.)

6. In the evolution of behavior, learned modifications are not directly inherited. Learned behaviors can contribute to fitness but cannot be inherited. Only the innate predispositions and their biological underpinnings can be inherited, and only these inherited components can evolve. Natural selection can only work on the inherited aspects of behavior. (These ideas follow in part from the ways in which the terms *learned* and *innate* are defined, and in part from the findings of evolutionary theory.)

7. It is therefore important to distinguish the learned and innate components of behavior and to focus attention on the latter. This is a value judgment about the aims of research within the paradigm. It does not mean that learned behaviors are unimportant; it just means that sociobiologists would rather identify what is learned so that they can ignore it and spend the rest of their time studying the innate components. *It is this preference for studying the innate components of behavior that makes the sociobiology paradigm so controversial; most critics of sociobiology have the opposite preference.*

8. We can use (modifications of) Darwinian methods of investigation to study those components of behavior that evolve. One method is to measure variations in fitness by observing many individuals and studying the number of viable offspring successfully reared by each. Another method is to study the results of past evolution by comparing social behaviors among different populations or different species. (These are the basic research methods.)

9. Before behaviors can be compared, there must first be an often lengthy period of observation and description. However, the presence of observers might modify the behavior that we wish to study. Since we are interested in behavior under "natural conditions," it follows that we should conduct most observations at a distance and interfere as little as possible. (These are more research methods.)

The objections just listed arise within science. Other criticisms originate outside the sciences, from those who assume that anything "innate" is unmodifiable, a view that sociobiologists do not generally share. Thus some critics of sociobiology equate sociobiology with genetic determinism, the assertion that our individual characteristics are determined before birth and cannot be changed. Genetic determinism is feared for at least two reasons:

1. Throughout history, people in power have often sought to control other people (other social classes, other races, and women) by teaching that existing inequalities were "natural" because they were based on innate and unchangeable differences.

2. Many people fear that the mere claim that some behavior is innate will discourage people from trying to improve the behavior through education or similar means. The claim that a particular behavior might be innate is particularly disappointing to social reformers whose hopes for the future depend on the ability of people to modify their behavior.

Among biologists, including sociobiologists, those who believe in genetic determinism are decidedly in the minority. Most biologists, especially those who study animal behavior, are impressed with the plasticity of behavior, meaning its ability to change in response to environmental circumstances, including the behavior of other individuals. To be sure, there are genetic constraints on what can and cannot be learned, but, within these limits, behavior is remarkably plastic in most animal species, especially those that are similar to ourselves. It is misleading to say that a particular behavior is "determined," either by genetics or by environment—almost every behavior is influenced by both of these factors throughout the lifetime of the individual.

We will now examine research methods used by sociobiologists and the results they have produced. Think about whether these results support the sociobiology paradigm (Box 7.1). Think also about whether these results negate the points raised by the critics of sociobiology.

Research Methods in Sociobiology

Distinguishing learned from innate behaviors in a particular species. How does one distinguish the learned and innate components of behavior? Sociobiologists use several methods to address these questions:

1. Rearing animals in isolation. A classic type of experiment is to raise an animal in isolation, in a soundproof room with bare walls and minimal opportunities for learning, including no opportu-

nity to learn the behavior from others. Behavior that the animal exhibits under these conditions is assumed to be largely innate. Experiments of this sort cannot ethically be performed on humans, and experiments that could be done on humans would usually take decades to yield their results.

2. Rearing animals under different conditions. If the behavior is performed the same way by animals or humans reared under strikingly different circumstances, then the behavior is largely innate. If, on the other hand, the behavior varies according to the circumstances of rearing, then the variation can be attributed to environmental influences. (This does not rule out inherited influences, which may also be present.) *Cross-cultural studies* are used to compare the behaviors of people raised in different societies or under different customs; innate behaviors are expected to be constant across various cultures, while learned behavior patterns are expected to vary.

3. Studying behavior in different genetic strains. If different strains or breeds differ behaviorally in a consistent and characteristic way, then a strong inherited component exists. (This does not rule out learned components, which might also be present.)

4. Conducting adoption studies. If two populations differ in a particular behavior, it may be useful to study individuals from one group who are adopted early in life and raised by the other group. Under these conditions, behavior consistently resembling the population of birth demonstrates an inherited influence, while behavior resembling the population of rearing demonstrates a learned influence. Mixed or inconsistent results may indicate that both influences are present.

5. Conducting twin studies. If a trait is under strong genetic control, then identical twins should usually both exhibit the trait whenever either one does, while fraternal twins will more often exhibit differences. Twin studies in humans are frequently criticized because the effects of learning cannot easily be separated from those of inheritance unless the twins are reared separately in families randomly chosen, conditions that are rarely even approximated.

Field studies. No behavior can be analyzed by any method until it has been adequately described. Sociobiology therefore includes a great many observational field studies of nonhuman animals.

Most field studies conducted before the 1970s consisted of unplanned observation in which the observer recorded all behavior that seemed at the time to be important. Studies of this kind are useful in sampling rare or unusual events, or in preliminary surveys whose intent is largely to identify interesting problems for more careful subsequent study. However, as Jeanne Altmann has emphasized, field studies of this kind contain a serious weakness: The data would not be usable in testing most hypotheses or in making comparisons from one group or one occasion to another, let alone from species to species. A better strategy, said Altmann, would be to focus on a single individual subject and to record the starting time and ending time of each activity, data that could later be used to compare the amount of time different individuals spent on different activities. Data gathered in this way, she emphasized, could be used to compare results from different groups or different species, but data collected in the more traditional way could not. Data are more useful if gathered with attention to the requirements of hypothesis testing.

Instincts

Complex behavior patterns that are under strong genetic control are called **instincts.** The classical test for whether or not a particular behavior is an instinct is the appearance or nonappearance of the behavior at the appropriate time of life in an animal reared in isolation since birth or hatching. For example, if a songbird reared in a soundproof room sings the song of its species and sex upon reaching maturity, the song is considered to be instinctive. By this test, many behaviors that have been studied in fish, birds, and many invertebrates (including insects) have been shown to be largely instinctive. In general, behaviors related to courtship and mating have strong instinctive components in most species. Other behaviors that are frequently instinctive include automatic "escape" behavior, nest-building behavior, orb-weaving in spiders, and the gestures used to threaten other individuals of the same species. When instinctive behavior leaves a lasting product, such as a nest or a spider's web, these products are often so distinctive that they can be used as aids in species identification.

Mammals generally rely more on learned behavior than on instinct. This is especially true of primates; many behaviors, such as mating behaviors and aggressive threats that are instinctive in other species, have strong learned components in monkeys and apes and may vary greatly among human societies.

Advantages of instincts.   Short-lived animals rely heavily on instincts. Mayflies, for example, have an adult life span of less than 24 hours. During this brief period, they do not feed, but have just enough time to find a mate, copulate, lay their eggs, and die. There is no time for learning to take place, nor is there any time for mistakes. The mayflies that accomplish their mission successfully are those that can perform their behavior correctly on the first try; they will probably never get a second chance. Selection over millions of years has therefore produced a series of adult behaviors that are instinctive and automatic, allowing no room for diversity or innovation. This is typical of instincts generally: They occur in situations where uniformity and automatic response are adaptive and where innovation and diversity might be inadaptive. (*Adaptation* is explained in Chapter 4.) A greater complexity of behavior is possible with a simpler brain if the behavior is instinctive; learned behavior of equal complexity requires a more elaborate nervous system and also a long learning period during which many mistakes will be made.

Mating behavior.   Mating behavior includes both courtship (attracting a mate and becoming accepted as a mate) and the actual release or transfer of gametes. Mating behavior has a strong instinctive component in nearly all species, except in higher primates. Scientists can demonstrate the instinctive component of most forms of mating behavior by raising individuals in isolation until they are sexually mature, then testing them to see if they perform the behavior typical of their species.

As we pointed out above, instinctive behaviors tend to occur in situations where natural selection favors uniformity rather than diversity. Such unvarying behavior is called *stereotyped* behavior and is ideal for mate location and recognition. The behavior that evolved in each species matches the type of signal that each is able to sense, so that visual mating signals are used by species with good vision, chemical signals in species with good chemical reception, and sounds in species with good sound discrimination. Male birds of many species show conspicuous colors or conspicuously colored parts that are advertised or displayed during mating (Fig. 7.2). Many species of birds, frogs, and insects use sounds as mating signals, and the opposite sex responds only to mating calls of the proper pitch, duration, and pattern of repetition. Members of each sex know exactly what to listen for in the other sex, and nonconformists will fail to mate, a form of sexual selection in which individual callers that deviate from expectations will leave no offspring. Closely related species often differ in their mating calls and courtship patterns, and the noncalling sex is very sensitive to the differences. Such differences can therefore serve as reproductive isolating mechanisms that prevent interbreeding between species (Chapter 4).

Mating rituals evolve as a consequence of sexual selection in those species where the discriminating sex (the one doing the choosing) consistently prefers the most conspicuous displays. Peacocks, lyrebirds, and birds of paradise are renowned for their beautiful and ornate male plumage (Fig. 7.2). Male birds of some species with less conspicuous plumage may concentrate instead on building an elaborate nest. The South Pacific bowerbird builds its nest within a large framework (or bower) that also serves as a nuptial chamber. A few species even build an "avenue" lined with colorful stones leading to the bower's entrance. Generally, the species with fancy and ornate plumage do not build elaborate bowers, and the species that build impressive bowers do not have elaborate plumage.

Territorial behavior. In many species, one or both sexes may show **territorial behavior** by defending a *territory*, either throughout the year or during the mating season only. The defense of a territory against intruders of the same species is common among many animal species. In some species, only males are territorial. Territorial behavior spaces individuals apart and encourages the losers to strike out in search of new territory, thus extending the range of the species wherever possible. Each territory must have sufficient food resources for a mating couple and their offspring, places for hiding and refuge, and at least one suitable nest site. Males without any territory are usually unable to attract mates and thus leave no offspring in that particular season.

Territorial species may use gestures to threaten rivals, or they may mark their territory with their own scent. The intimidation of rivals by gestures or by the presence of odors serves to space individuals apart without causing injury or loss of life. Such ritualized forms of territorial defense are much more common than any form of fighting in which injuries are likely.

Nesting behavior. Choice of a nesting site may be an important part of territorial behavior. In some bird species, the male builds the nest and then offers it to the female as part of the mating ritual. In other species, male and female may cooperate in building the nest together as part of

FIGURE 7.2
An example of a conspicuous mating display in a peacock. Females of this species prefer the males with the most conspicuous displays.

the mating ritual. Females may incubate the eggs alone, but males may provide other forms of assistance by bringing food or by defending the area against predators. In other species, the males and females take turns guarding the nest and sitting on the eggs. Feeding the hatchlings may, likewise, be either a solitary or a shared task. In the Florida scrub jay, the offspring of the previous year are not old enough to mate that year. Instead, they help feed the nestlings who are their own brothers and sisters (see Fig. 7.6, p. 200). In so doing, they contribute to the survival (and thus the fitness) of these near relatives, who share a portion of their own genotype. This is a form of kin selection, which is explained later in this chapter.

THOUGHT QUESTIONS

1. Does "antisocial" behavior (such as assaulting others and causing them injury) fit the definition of social behavior? Do you think the definition of social behavior given in this chapter should be modified? In what way?
2. Is sociobiology a subject area with room for many viewpoints, or is it a single viewpoint that is hereditarian and sexist? Can sociobiology be studied without hereditarian assumptions?
3. Can the methods used for gathering or analyzing data in sociobiology be the same for different species? To what extent do size (small versus large animals) and habitat (aboveground, underground, underwater, in trees, and so forth) require differences in field methods? What special problems in methodology arise when humans are being studied? Can the methods used for other species be applied to *Homo sapiens*?

B. Social Organization Is Adaptive

Very few animal species consist of solitary individuals that lead all their lives by themselves. Even in species where solitary life is frequent, individuals must come together for sexual reproduction. Most species, however, are far more social than this, and species that contain social groups greatly outnumber those that consist primarily of solitary individuals. Even bacteria have been shown to influence one another in the timing of their cell cycles and metabolic events. Social groups vary greatly in both size and cohesiveness. Simple pairs and family groups have only a few individuals. Larger social groupings include antelope herds, baboon troops, and fish schools, all of which may include up to a few hundred members. Still larger are the colonies of social insects, which may include many thousands or in some cases millions of individuals. Some social groupings are loosely organized groups in which individuals simply stay together but seldom interact, while others are organized into social hierarchies within which social interactions can be quite complex, as they are among humans or social insects.

The Biological Advantages of Social Groups

Some advantages of social grouping are related to the obtaining of food. A large group of individuals searching for food together has a higher probability of finding food than a single individual. If food tends to be discovered in quantities much greater than the needs of a single individual, then social groups will be favored by selection.

Other advantages of social grouping relate to defense against predation. Social groups can often defend themselves more effectively than individuals can. Musk oxen, for example, respond to threats by forming a circle with all individuals facing outward. Even in species that do not practice group defense, members of the group may warn one another by giving *alarm signals*, or simply by fleeing as soon as the predator is spotted. Thus, belonging to a group gives all group members the advantage of greater (and earlier) alertness against predator attacks. For this reason, large but loosely organized herds or flocks are common among *ungulates* (hoofed mammals such as antelopes) and also in many bird species (Fig. 7.3). Other advantages to group membership arise from the sharing of risks: A predator attacking the entire herd may capture one of its members at most, while the rest escape, so that each individual in a herd of 500 has

FIGURE 7.3
Social groups are formed in various species. [A] Pelicans. [B] Minnows, showing schooling behavior in the presence of predators. [C] Wildebeest and zebras around an African water hole. [D] Gannets on the coast of Quebec.

been exposed to only 1/500 of the risk of capture faced by a solitary individual. Actually, the risk may be even smaller because predators can easily capture solitary individuals one by one, but herds are less vulnerable to attack because predators cannot surround or easily attack herd members. Most casualties among herd animals occur when individuals stray from the herd.

Simple Forms of Social Organization

Social organization refers to the ways in which social groupings are structured. The fact that social organization sometimes varies among closely related species suggests that to some extent it was modified by evolutionary processes. The few studies on the inheritance of social status (dominance)

within organized social groups all point to a complex interplay between learned and inherited behavioral components.

Groups without dominant individuals. Perhaps the simplest form of social organization is shown by brittle stars (Fig. 7.4), marine organisms distantly related to starfish. Upon encountering one another, brittle stars tend to stay together in clumps, even though there is no evidence of any more complex interaction. Adding more brittle stars to the population results in larger clumps.

The *schooling behavior* of fish is another very simple form of social organization. There are some minor hydrodynamic advantages to schooling—swimming is made a bit easier by certain changes in water pressure caused by the swimming

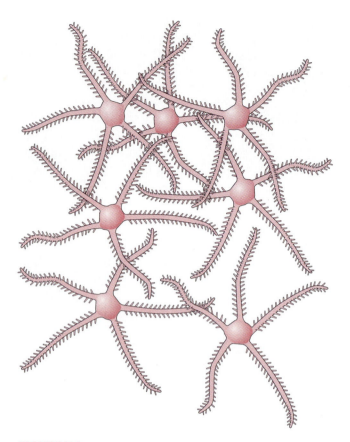

FIGURE 7.5
Domestic fowl form a linear dominance hierarchy, or pecking order.

FIGURE 7.4
Brittle stars form simple aggregates.

activities of other fish—but these effects are rather small. The major advantage to schooling behavior seems to be that the fish hide behind one another in such a way that most of them escape the attention of predators. Most fish school closer together when they notice a predator nearby (Fig. 7.3B). When attacked, schools of fish or flocks of birds tend to scatter in every direction, a reaction that confuses many predators and that may give all individuals a chance to escape.

Groups with dominant individuals. One form of social organization is called a **linear dominance hierarchy,** or *pecking order*. Pecking orders are found among domestic fowl and certain other captive animals (Fig. 7.5). The top-ranked individual, usually a strong male, can successfully bully or threaten all the other individuals in the group, literally pecking at them in the case of birds. The second-ranked individual can intimidate all others

except for the top-ranked individual. The third-ranked individual can intimidate all except for the first two, and so on. Some feminist critics of sociobiology suggest that such male-dominated forms of social organization exist more in the minds of male sociobiologists than in the animals they study. In at least some species, linear hierarchies may reflect the artificial conditions of captivity and confinement.

Altruism: An Evolutionary Puzzle

Efforts to solve social problems such as pollution often call for individuals or corporations to sacrifice their own interests for the common good, a practice called **altruism.** Altruistic behavior exists in other species as well. As an example, consider the "broken wing" display of certain female birds such as nighthawks. When guarding her nest from predators, a female nighthawk may sometimes lead the predator away from the nest location, distracting the predator's attention by pretending to limp or to have a broken wing. Once she has drawn the predator's attention sufficiently far from the nest, she takes off and flies away, leaving the predator confused. Although she has protected her young, she has exposed herself to an increased danger from the predator. Thus altruism is more precisely defined as behavior that *decreases* the fitness of the performer while it increases the fitness of another individual. In this example, the female bird has apparently decreased her own fitness by putting her life at risk for the sake of her offspring. Remember (Chapter 4) that fitness is defined as the relative number of fertile offspring produced by an individual. Only changes that increase fitness are favored by natural selection.

Units of selection.
Altruistic behavior is a problem in evolutionary theory because natural selection might be expected to work against it. How could altruism evolve if it decreases fitness? The definition above highlights the problem: If my behavior benefits both you and me, then its evolution is easy to explain because it increases my own fitness. More difficult to explain is behavior defined as altruistic—behavior that decreases my own fitness while it improves yours. (It is not necessary to assume that altruists are consciously aware of any goals.)

Various hypotheses have been developed to explain how altruistic behavior could be favored by natural selection. These hypotheses are also part of a larger, long-standing controversy among evolutionary biologists over the "units of selection," meaning the level(s) at which natural selection acts (genes versus individuals versus groups versus species). We will examine several of these hypotheses both as explanations of altruism and also as examples of selection acting at levels other than the individual organism.

Possible benefit to the species.
One early hypothesis for the evolution of altruism is that it benefits the species as a whole. However, careful examination of this hypothesis shows it to be unsatisfactory. If a species had both altruists and selfish individuals ("cheaters"), and if some part of the supposedly altruistic behavior was controlled genetically, then selection would work against the altruists and in favor of the cheaters. Altruism may benefit all recipients of another individual's altruistic behavior, but the advantage is greater to selfish individuals than to other altruists. Under these conditions, natural selection should favor selfishness and eliminate altruism from the population.

Group selection.
One explanation for altruistic behavior, favored by V. C. Wynne-Edwards (1962), assumes that each species is subdivided into populations or other groups. Selection among these groups, called **group selection,** may favor one group over another. In particular, a group containing altruists will be favored *as a group* over other groups composed of selfish individuals only.

As we explained earlier, the defense of territory prevents excessive population density by spacing individuals apart and limiting population size. Wynne-Edwards argues that the losers of territorial disputes are altruists who forgo mating for the benefit of the group as a whole. According to Wynne-Edwards, the mating of unterritoried individuals would lead to overpopulation, increased mortality, and a smaller resulting population size. Selection between groups would thus favor altruism. Other biologists who have examined this claim algebraically and through the use of computer models have shown that a loser who cheats (mates anyway) would greatly increase its fitness over one that does not mate, and that cheating behavior would thus be favored over altruism in territorial species.

Kin selection.
Many biologists dissented from the group-selection hypothesis because they thought a simpler and therefore more satisfactory explanation would be based on individual selection instead of group selection. An explanation of altruism that is acceptable to many of these biologists is based on the concept of **inclusive fitness**—that is, the total fitness of all copies of a particular genotype, including those that exist among relatives. Relatives are listed according to their degree of relationship, symbolized by R. For sexually reproducing organisms with the common types of mating systems, an individual shares half of its genotype ($R = \frac{1}{2}$) with its parents, its children, and, on average, with its brothers or sisters (who share two parents). Also, an individual shares one-fourth of its genotype ($R = \frac{1}{4}$) with grandchildren, half-siblings (who share one parent only), uncles, aunts, nieces, and nephews. The total fitness of your genotype, your inclusive fitness, is now the sum total of your individual fitness *plus* one-half the fitness of your parents, children, and full siblings ($R = \frac{1}{2}$), *plus* one-fourth the fitness of those relatives who share one-fourth of your genotype, *plus* one-eighth of the fitness of your cousins ($R = \frac{1}{8}$), and so on. This concept allows us to define **kin selection** as the increased occurrence of a genotype in the next generation on the basis of its inclusive fitness.

The conditions under which kin selection favors the evolution of altruism were specified by W. D. Hamilton (Axelrod and Hamilton, 1981). Assume that altruistic behavior results in a certain reduction in fitness or "cost" (*c*) to the altruist, and a corresponding gain in fitness or "benefit" (*b*) to another individual who shares a fraction (*R*) of the altruist's genotype. Hamilton reasoned that natural selection would favor altruism whenever *Rb*, the gain in inclusive fitness to the altruist's genotype, exceeds the cost *c*. If I perform an altruistic act that diminishes my individual fitness by a certain cost, but which raises by child's fitness or my sister's fitness (with whom I share half my genotype) by more than twice that cost, then the net effect on my inclusive fitness is positive. The probability that my genotype will be represented in future generations is increased because the benefit to my relatives (or to the fraction of my genotype that they share) exceeds the cost, so the net result is an increase in my inclusive fitness.

The above explanation, however, gives rise to an interesting prediction: Kin selection will only favor altruism if close relatives are more likely to benefit from altruistic acts than more distant relatives or nonrelatives. Studies of many species have confirmed this prediction: The beneficiaries of altruism are often close relatives of the altruist (Fig. 7.6), and the frequency of altruistic acts varies in almost direct proportion to the degree of the relationship (Trivers, 1985). Note that it is not necessary that the altruist be able to distinguish relatives from nonrelatives; it is only necessary that close relatives are more likely to benefit from altruistic acts.

Kin recognition.   Although kin selection does not *require* kin recognition, can animals and plants assess the degree to which other organisms are related to themselves? Among social animals, individual recognition (based on growing up together) can be used. Some plants can chemically assess the "match" between proteins or other molecules derived from (male) pollen and (female) stigma. Other species, including mice, use odor cues. The odor of each animal is genetically influenced, and the diversity of genotypes results in a diversity of odors. Mice can detect by odor which individuals are the most closely related to themselves. Mice seem to use odor-based kin recognition when they

[A]

[B]

FIGURE 7.6
Two examples of altruism favored by kin selection. [A] In scrub jays, older siblings (left) will assist in the care and feeding of their younger siblings. [B] A female ground squirrel (*Spermophilus beldingi*) stands guard against predators. If a coyote or hawk is spotted, the guard female emits an alarm call, which attracts the predator and thus endangers the caller, but the alarm also warns the caller's next of kin and thus raises her inclusive fitness.

establish communal nests. Females who share a nest nurse each others' young. A mother's inclusive fitness is maximized if she nurses only individuals that are closely related to her. Females who share communal nests are usually related genetically.

Reciprocal altruism and game theory.   Another explanation for altruism, one that is completely different from hypotheses based on group or kin selection, is based on a concept called **reciprocal altruism.** If you have behaved kindly toward me in the past, I will be inclined to perform altruistic acts that benefit you. It is not at all necessary that the altruist expects anything in return; it is only important that altruists do generally benefit in the long run from reciprocal acts of kindness. Researchers seeking to provide a possible basis for the evolution of reciprocal altruism have turned to mathematical models that are part of what is known as **game theory.** As with other mathematical models, game models have the form of an elaborate hypothesis—that the participants, viewed as players in a game, will behave in the manner that the model predicts. In the mathematical theory of games, the underlying hypothesis is that each player will make choices that will maximize the player's chances of winning. The guidelines by which the choices are made are called a *strategy,* and the strategy that maximizes the chances of winning (or getting the highest score) is considered a winning strategy. For example, in one of the most well known of these models, called prisoner's dilemma, two captured prisoners find out that, if they both defect to the police (confessing and also implicating one another), they will both receive jail sentences. If neither confesses, the prisoners hope that they might be released for lack of evidence. However, they are told, if only one of them defects, then the defector will receive a medal, while the other will be shot. The dilemma is that since neither knows what the other will do, the only way to avoid the worst outcome is to defect.

Evolutionarily stable strategies.   In the biological applications of game theory, different individuals (or their genotypes) are compared to players, and the biological concept of fitness is used to keep score: Any strategy that increases my fitness will result in my leaving more offspring, and the strategy will thus spread more widely throughout the population.

Axelrod and Hamilton (1981) used mathematical game theory to determine what would happen in a population in which many individuals were "playing" prisoner's dilemma repeatedly against one another. They tested a variety of gaming strategies, allowing the more successful ones occasionally to "reproduce" others like themselves and the least successful ones to die. Over time, certain strategies became more common, while others diminished and died out. In such a situation, any

strategy that can successfully spread (increase in frequency) when it is initially rare is known as an **evolutionarily stable strategy (ESS).** Such strategies are stable because they resist the spread of alternative strategies.

Tit for Tat.   By testing a variety of strategies for repeatedly playing prisoner's dilemma, Axelrod and Hamilton were able to show that a strategy called *Tit for Tat* was evolutionarily stable and that it was the most successful strategy among those tested. The Tit for Tat strategy is to cooperate with any other individual who cooperated with you on your last encounter and with newcomers with no previous track record of behavior, but to act nasty toward anyone who was nasty to you the last time. If other members of the population practice the same Tit for Tat strategy, then individuals using this strategy will become reciprocal altruists toward one another, and they will all benefit from the altruism. The benefits of reciprocal altruism will ensure that its frequency will increase in the population, even from an initially low value; in other words, Tit for Tat is an evolutionarily stable strategy in this model.

"If successful, repeat; if unsuccessful, switch." In the early 1990s, several researchers discovered that another strategy, which some have nicknamed "Pavlov," can increase fitness even more than Tit for Tat. A simple description of the Pavlov strategy is: "If you were successful last time, repeat whatever you did; if not, switch your behavior." As a strategy, Pavlov resembles Tit for Tat in encouraging reciprocal altruism, but it differs from Tit for Tat in more frequently taking advantage of other individuals, called "suckers," who are willingly victimized or exploited. The superiority of the Pavlov strategy over Tit for Tat thus depends on the presence of "suckers" in the population. Observations of natural populations have confirmed that Pavlov is in fact a widely used strategy.

Reciprocal altruism begets deception: If I can take advantage of you, but make you think that I have acted kindly, then I receive *both* the direct advantage of my action and also the future benefit of your acting to return the imagined kindness. Among mammals especially, reciprocal altruism selects both for deception and for the ability to detect the deception or true motives of others. Often, it is the higher-ranked (more dominant) individuals that make a career out of deceiving others, while lower-ranked (more subordinate) individuals, including many females, become skilled at discerning the true motives of others so that they are not deceived. Both deception and the ability to detect deceptive behavior among other individuals are favored by natural selection in the models made by game theorists.

The Evolution of Eusociality

The highest degree of social cooperation is developed among the truly social, or eusocial, insects. **Eusocial** species are recognized by the possession of three characteristics: strictly defined subgroups (castes), cooperative care of the eggs and young larvae (cooperative brood care), and an overlap between generations. Eusociality evolved over a dozen times among the members of two insect orders: the Isoptera (termites) and the Hymenoptera (bees, wasps, and ants). A few bird species and one mammal (the naked mole rat, a burrowing type of rodent) approach eusociality in the existence of "helper" individuals that assist in parental care, but these helpers do not form a distinct caste.

Eusociality in termites. Termites (Isoptera) are a group of insects related to the cockroaches. Termite colonies are founded by a single reproductive pair, the king and queen. The queen grows many times larger than the other individuals, her offspring, who continually feed her and raise her additional offspring. An important termite characteristic is their chewing and digesting of wood. Termites can digest wood only with the help of symbiotic microorganisms (mostly flagellated protists) that live in their guts. Individual termites transmit these protists throughout the colony by passing samples of regurgitated food and anal secretions to other members of the colony. This habit, favored by selection, also provides a food source for those members of the colony, such as the queen, that do not feed themselves.

Along with food and microorganisms, termites can also pass their chemical secretions to other colony members. Chemicals that communicate social information among members of a species are called **pheromones.** Some pheromones are similar to hormones in their effects, except that the pheromone secreted by one individual produces its effects in other individuals. One such pheromone, secreted by the queen, inhibits most other individuals in the colony from becoming reproductively mature. The passing of symbiotic microorganisms from one termite individual to the next was a precondition that probably led to the evolution of termite eusociality by providing termite queens with an opportunity to secrete pheromones that would suppress the reproduction of other individuals.

At seasonally timed intervals, winged reproductive individuals of both sexes are produced; these winged individuals emerge from the colony all at once and embark on nuptial flights during which mating takes place. Newly mated pairs become the founders of new colonies. Meanwhile, the original colony persists for the lifetime of the queen, a period of some 10 to 12 years.

Eusociality in the Hymenoptera. The insect order Hymenoptera (bees, wasps, and ants) has a much larger number of social species than the Isoptera, and an estimated 12,000 of these social species are ants. Wilson (1975) estimates that eusociality has evolved among the Hymenoptera as many as a dozen times. Why has eusociality evolved so many times in this one insect order and so seldom among other animals? The clue seems to be found in a unique form of sex determination and in its effects on inclusive fitness. The social Hymenoptera have a unique form of sex determination called *haplodiploidy*, in which fertilized eggs produce diploid females and unfertilized eggs produce haploid males. Each reproductive female mates only once, for life, with a single male, who contributes the same haploid set of chromosomes to all his offspring. The daughters are therefore very closely related to one another, for they share 100 percent of their father's set of chromosomes, plus an average of 50 percent of the genes that they inherit from their common mother. Since the average of 50 percent and 100 percent is 75 percent, a female shares an average of three-fourths of her genotype with her sisters, much more than the half of her genotype that she shares with her daughter or her mother (Fig. 7.7). By neglecting her own daughters (to whom she is only half related) and raising her mother's daughters instead (to whom she is three-fourths related), she is increasing her genetic fitness. For this reason, haplodiploidy

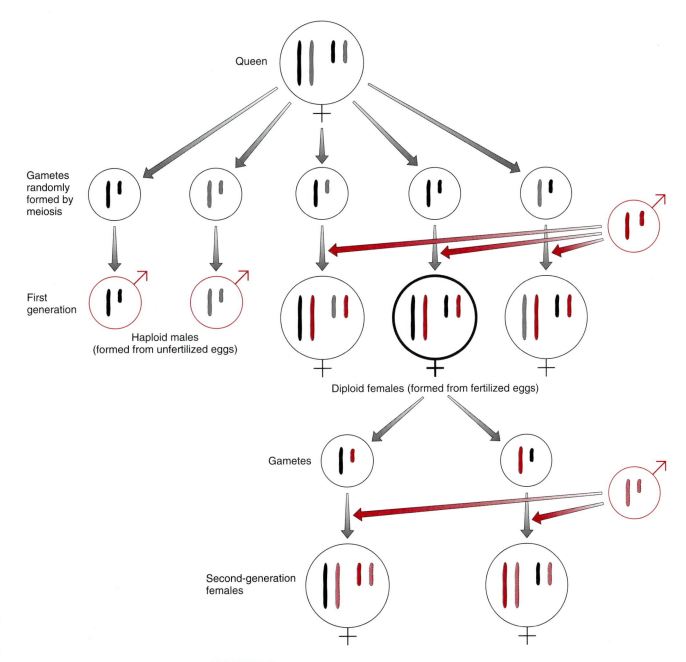

FIGURE 7.7
Haplodiploidy in a species with a haploid chromosome number of 2. Notice that the female with the heavy black border on average shares three-fourths of her chromosomes ($R = \frac{3}{4}$) with her sisters but only one-half of her chromosomes with her own daughters ($R = \frac{1}{2}$). Any two sisters may share more or less than three-fourths of their chromosomes; the fraction $\frac{3}{4}$ is an average across all individuals.

favors the evolution of sociality among the Hymenoptera because most females can gain greater inclusive fitness by becoming sterile workers and by helping their mother (the queen) to raise her offspring (their sisters) than by raising offspring of their own. Ancestral Hymenoptera were solitary (and many solitary species still exist), but sociality has evolved repeatedly and independently among this group of insects (Fig. 7.8).

The queen bee or wasp usually secretes pheromones that inhibit the sexual development of other females in the colony. There are other mechanisms that distinguish which larvae will develop into queens and which into sterile workers: For example, future queens are fed a more nutritious "royal jelly" that also contains pheromones that stimulate her reproductive development. Also, whenever new queens emerge, one of them (usually the one emerging first) stings the others to death and thus emerges as the undisputed queen.

The eusocial insects represent the height of complexity that instinctive behavior can reach, for most of their social behavior is under instinctive control. Antisocial behavior (meaning behavior that decreases the fitness of others) does not exist in these societies because antisocial individuals are quickly eliminated. By contrast, social behavior among humans consists largely of learned behavior patterns, and antisocial behavior is among the many possibilities that behavioral plasticity permits.

[A]

THOUGHT QUESTIONS

1. Are the behavior patterns of individuals within a species more alike than the behavior patterns of individuals from different species?
2. Compare the sociobiological definition of altruism given in this section with the common usage of this term outside biology. Are they the same or different? Does one include the other? Would differences in meaning result in different estimations of the importance of altruism?
3. Think of five or more examples of altruism among humans. Do you think that sociobiology provides an adequate explanation for these examples? If not, what other explanations are possible?

[B]

FIGURE 7.8
Eusocial insects. [A] Honeybees swarming. [B] Ants moving their larvae.

4. It was stated above that individuals share, on average, one half of their genotype with their siblings. Refer to Box 3.2 and deduce how this would come to be.

5. Why should humans be interested in the social behavior of insects?

C. Reproductive Strategies Can Alter Fitness

Natural selection favors those genotypes that are able to leave more copies of themselves in subsequent generations. The manner in which these copies are produced may be called a **reproductive strategy.** Reproductive strategies include such behaviors as the laying of many eggs or few, the presence or absence of parental care, the presence or absence of sexual recombination, and, if there is a mating system, whether it is predominantly monogamous, polygamous, or promiscuous. Sexual behavior is an important part of reproductive strategy in many social species.

Asexual versus Sexual Reproduction

Asexual reproduction. **Asexual reproduction** may be defined as reproduction without any genetic recombination. This type of reproduction has certain advantages over sexual reproduction. Within a group of organisms that includes both sexually and asexually reproducing species, those reproducing asexually generally do so faster and at lower energy costs. Asexual reproduction allows reproduction at an earlier age and a smaller body size, and it also avoids the costs (detailed below) associated with sexual reproduction. For an individual that discovers a large but finite and/or perishable supply of food or some other resource, asexual reproduction is an advantage because more offspring, and more generations of offspring, can be produced in a minimum of time, without any need of finding or courting a mate. Moreover, each of the numerous offspring are genetically identical to the original parent or founder, assuring that favorable combinations of genes will be perpetuated exactly. All the genetically identical asexual offspring of a single individual are referred to as a *clone.*

Sexual reproduction. **Sexual reproduction** is more costly than asexual reproduction because of the time and energy expended in searching, finding, and courting a mate, and in actually transferring sperm or pollen. Energy is also used in synthesizing structures that attract mates, and in the mating act itself. Mate attraction also makes sexually reproducing individuals more visible to predators, exposing them to increased risks. A major genetic cost results from giving up half of one's genes (during meiosis) in favor of someone else's. In view of these costs, why is sexual reproduction so widespread in both the animal and plant kingdoms?

The great adaptive advantage of sexual reproduction is genetic variety among the offspring. In the most common type of sexual reproduction, males produce sperm cells that contain half of their genetic material, and females produce eggs that also contain half of their genetic material. The sex cells (*gametes*) produced by a single individual vary among themselves, and their combination with the gametes of the opposite sex are nearly random. (Mate choice will cause the combinations of gametes to be nonrandom, but they will still vary much more than will asexually produced offspring.) The result is that *sexually produced offspring vary greatly in nearly all genetically controlled traits.* This may be a disadvantage if tomorrow's (or next year's) conditions are identical to today's, and unchanging conditions do in fact favor asexual reproduction. However, if tomorrow's (or next year's) environmental conditions are uncertain, then the best hedge against this type of uncertainty is to produce many *different kinds* of offspring. Sexual reproduction achieves this type of variation very efficiently. What we have just said pertains not only to the common forms of sexual reproduction, but to other forms as well (such as the haplodiploidy described earlier): However much they differ in detail, all forms of sexual reproduction are characterized by greater variation among offspring, compared to any form of asexual reproduction.

Environmental control of sexual or asexual reproduction.   The hypothesis that sexual reproduction derives its adaptive advantage from the greater variation among the resultant offspring receives support from the study of certain insect species (such as aphids, also called plant lice) that are capable of either sexual or asexual generations. During the summer, when maturing crops offer dependable food supplies for several months in a row, these insects produce several asexual generations in quick succession. For successful individuals, asexual generations are favored by their ability to produce very rapidly large numbers of individuals with exactly the same genotype as the stem mother, the female founder of the clone. At the end of the season, however, these insects reproduce sexually, and the sexually produced eggs overwinter. When they emerge the following spring, diverse genotypes of offspring find their way to the new stands of plants under new weather conditions, neither of which could have been predicted during the previous fall. Many genotypes will perish, but a few will survive and prosper by reproducing asexually during the new season. The important point is that the genotype that proves most fit each spring is not necessarily the same one that produced the most offspring asexually in the previous year. *Sexual reproduction is favored whenever future conditions are uncertain,* and experiments confirm that individuals laying eggs in the fall will have more surviving offspring the next year if they reproduce sexually than if they reproduce asexually.

Notice that nothing in the above argument specifies how the choice between sexual and asexual reproduction is made—whether it is a conscious individual decision, a learned response to stimuli, or simply the result of physiological changes. All that is necessary is that somehow individuals switch from asexual to sexual reproduction according to the season, and that the ability to do so (or to learn to do so) has at least some genetic component capable of being favored by natural selection.

r-*Selection and* K-*Selection*

Species that have high reproductive rates are said to be *r*-**selected.** This term refers to r, the intrinsic rate of population increase, which was discussed in Chapter 6. High values of r lead to very rapid population growth under favorable conditions. Most r-selected species, whether sexually or asexually reproducing, live in unstable environments in which very favorable and very unfavorable conditions alternate with one another erratically and unpredictably. Their populations suffer frequent and unpredictable devastating losses, from which the population must rebound with rapid population growth. Population sizes in these species undergo rapid change, with explosive increase under favorable conditions compensating for the rapid devastation that occurs under unfavorable conditions. Since conditions are not always uniform geographically, species living under these conditions usually have a high capacity for *dispersal* (spread of the population) to new habitats and locations. Most new locations may be unsuitable, but the occasional favorable one results in a rapid explosion in numbers. Reproduction in such r-selected species is prodigious and may be either sexual or asexual. Since the emphasis is on rapid reproduction, selection favors reproduction at a small body size and a young age. Eggs, seeds, or other reproductive stages are produced in great numbers, are released at a small size, and are widely scattered, with no provisions made for their care. Most will die or be eaten, but r-selected species tend to compensate by producing offspring in even greater numbers.

The opposite of r-selection is called **K-selection.** K-selected species live in populations of stable size at or near the carrying capacity, which is symbolized by K (see Chapter 6). Selection in these species favors reproduction (always sexual) at a large body size, and a small number of offspring for which parental care or some other form of "parental investment" (see below) is provided. Offspring may themselves be large, and each of them represents a greater proportion of its parents' reproductive output. K-selection favors efficient use of resources, especially energy. The advantage generally goes to whomever can most efficiently convert their food into new adults of the next generation.

Differences between the Sexes

Isogamy.   In sexually reproducing species, the two sexes are not necessarily different. Some species, like the green alga *Chlamydomonas*, have male and female haploid cells (gametes) that look identical, a condition called **isogamy** ("equal gametes"). Gametes in such isogamous species are usually so similar to one another that they are simply called "+" and "−" mating types rather than "male" or "female."

Anisogamy.   A pair of gametes may be at an advantage if at least one of them is capable of finding the other over greater distances, thus allowing more mating or mating from a wider choice of potential mates. In some cases, there may also be an advantage for the resultant fertilized egg (*zygote*) if it contains stored food or protective layers, either of which increases bulk. In most cases, the advantage of motility and the advantages of large size can best be balanced if one of the gametes is large and the other one is small and motile, a condition called **anisogamy** ("unequal gametes"). The larger, nonmotile gamete is generally called an **egg,** and the smaller, motile gamete is known as a **sperm.**

Males and females.   Different-size gametes could be produced by identical organisms, but this is not usually the case. Instead, reproductive anatomy and behavior differ between the sexes in most species of animals and plants.

Most of the familiar differences between males and females are explained within evolutionary theory as the consequences of anisogamy. Selection among sperm producers, or **males,** favors the release of numerous gametes, each of which is of minimal size and thus of maximum motility. The minimal size means that each individual sperm represents a trivial investment (in energy and materials) for the male that produces it, for he can easily produce thousands (or millions) more, and can compensate for a poor choice of mates by mating more often. Competition among males usually favors whichever one can produce the most gametes that combine successfully with the most eggs.

Selection among egg-producers, or **females,** generally favors a larger investment of parental resources, such as stored food, in each egg. Among numerous eggs, the ones with the most stored food or the strongest protective layers generally have the best chance of survival. This necessarily limits the number of eggs that a female can produce and

places a premium on egg *quality* rather than number. A further consequence is that females, having fewer eggs, can produce more surviving offspring if they invest more care and protection in each one. This is especially true in mammalian females, where female investment includes a long period of gestation and intrauterine feeding, plus an added period of lactation and breast-feeding.

Parental investment in offspring has further consequences. Because male parental investment is low (in terms of both energy and material costs), the price that a male pays for mating with a given female is very small. If their offspring are low in fitness (i.e., have a small chance of survival), the male can simply mate again with other females. Low parental investment produces nondiscriminating males. Although this sounds a bit like *r*-selection, the terms r- and K-*selection* are applied only to whole populations or species, not to sexes or individuals. The predictions of *r*-selection (such as smaller adult body size and reproduction at an earlier age) are not observed in males of most species.

On the other hand, a female's parental investment is high, and each of her offspring is therefore more costly to her. The price that she pays for mating with a low-fitness male is a great reduction in her own fitness. She cannot simply make up for a poor choice by mating again because her capacity for repeated mating is generally limited by the large investment she must make in each of her offspring. Females thus have more at stake in each mating, and stand to gain more by choosing a mate who will father offspring who are more fit, or to lose more if the father of her offspring is less fit. In social species where males can vary in social status, a female will generally maximize her fitness by mating with a high-status male who will provide her (and her offspring) with a greater degree of protection. Females, in other words, tend to become more discriminating in their mate choices, both as to social status and genetic fitness. Females of many species have a remarkable capacity to discern variation in male fitness (and social status, if it also varies), and to mate preferentially with high-fitness and high-status males.

The differences between the sexes usually follow the patterns described above, but evolution has also produced some interesting exceptions to these generalizations. Among Mormon crickets, males mate only once by offering a large clump of sperm to a female of their choice. Since a male's parental investment is high, he has reason to be careful and discriminating in his choice of mates. Male Mormon crickets will lift the females during the mating ritual, estimating the number of eggs they contain by weighing them! Heavier, more egg-laden females are more likely to be chosen as mates and to receive all the sperm.

Mating Systems

In species in which care of the young requires the cooperation of both parents, parental investment tends to be high for both sexes. These conditions favor **monogamy,** or mating between one male and one female. If the rearing of their common off-spring takes a long time, formation of a permanent pair-bond (that is, mating for life) is favored. Another common situation is one in which females alone rear their young, but males provide protection to both female and offspring. This situation generally favors the development of **polygyny,** a mating system in which one male mates with several females (Fig. 7.9). In addition to monogamy and polygyny, other types of mating systems include **polyandry** (an uncommon type in which one female mates with multiple males) and **promiscuity** (in which members of both sexes mate with multiple partners and generally avoid forming permanent partnerships). The term *polygamy* is sometimes used to include both polygyny and polyandry.

Social organization in some polygynous species.   Male fur seals come ashore in the breeding season and establish territories that they defend against other males (Fig. 7.9B). The strongest male defends the best territory, an area where females can rear their pups within easy reach of the sea. Females nest in these territories and mate with the resident male. Males that lose territorial contests go off in search of other suitable territories. If they find none, they will not mate during that season.

Red deer, bighorn sheep, and certain other species of hoofed mammals (ungulates) form polygynous mating units in a different way. Adult males establish a dominance hierarchy, either through ritualized threat displays or through actual fighting. The dominant male gathers together

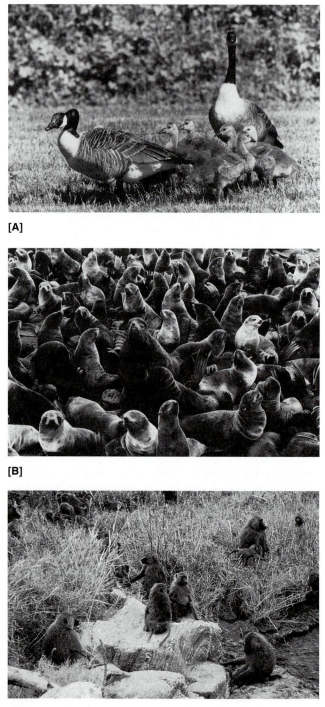

[A]

[B]

[C]

FIGURE 7.9

Examples of different mating systems. [A] A monogamous family group of Canada geese. [B] Polygyny: a dominant male fur seal (center) and his harem. [C] Baboons, a promiscuously mating species.

as many females as he can, forming a *harem*. Male social status in harem-forming species often correlates with fighting ability and with the size of horns, antlers, or other conspicuous features, so females can see at a glance which male is dominant. Females will assure better protection against predators for themselves and their offspring by following and mating with the dominant male. Nondominant (subordinate) males may trail along the margins of the harem and wait for an opportunity to mate quickly with as many females as they can while the dominant male is distracted. Though they do have occasional opportunities to mate, the subordinate males have far fewer opportunities than the dominant leader.

1. What are the biological definitions of *male* and *female*? How do these compare with cultural definitions of the same words? Do *male* and *female* (or *masculine* and *feminine*) mean different things in different cultures, or at different times in history?
2. In humans and other species, males tend to have greater muscle mass than females. Under what conditions would you expect anatomical differences (in muscle mass, antlers, or size) to evolve? Is there a reason why such differences would be favored by natural selection?
3. Does anisogamy explain differences in sexual behavior among mammals? Are human males "destined" to be promiscuous?

D. Primate Sociobiology Presents Added Complexities

The *Primates* are an order of mammals that includes monkeys, apes, lemurs, tarsiers, and humans. Primates are all extremely social animals. Primates are so interested in interacting with other members of their species that they will go to great lengths to maintain the ability to interact, or even merely to look at the interaction of others. We can experimentally set up a window, a partition that completely obstructs a primate's view through the window, and a lever that raises the partition for a predetermined length of time, affording a temporary view through the window. Most primates will spend hours repeatedly pressing the lever and looking through the window in such situations, especially if the window affords a view of other individuals interacting. Likewise, people spend hours looking through windows or at television screens.

The Development of Primate Social Behavior

Early development of behavior. Social skills in both human and nonhuman primates depend strongly on learning that takes place early in life. Early childhood experiences are of paramount importance in the shaping of future behavior and personality. Experiments showing the effects of social deprivation on young rhesus monkeys (*Macaca mulatta*) were conducted by Harry Harlow of the University of Wisconsin. Harlow's original motivation was to test one of Siegmund Freud's hypotheses, that a baby's attachment to its mother is based initially on its need for nutrition. Harlow raised infant rhesus monkeys with various forms of care but with no live mothers. Instead, dummy "mothers" with colorful wooden "heads" held baby bottles mounted in wire frames. Although the infant monkeys drank the milk, their behavior grew progressively more abnormal with time. They frequently cowered in the corner and were easily frightened. They formed no emotional attachments and seemed to ignore their "mothers" except when they were hungry. Freud's hypothesis was falsified because the young monkeys failed to treat the wire model as a mother. Something more than milk was needed for infants to form a bond with their mothers.

Harlow noticed that young monkeys liked the feel of terry-cloth towels. He tried wrapping the wire mothers in a few layers of terry cloth to make them soft and clingy. The terry cloth retained the infant's own body heat during periods of clinging. The infant monkeys enjoyed clinging to these cloth-covered dummies. Harlow raised several infant monkeys with two dummy "mothers," with and without terry cloth, one of them holding a baby bottle. Young monkeys spent countless hours clinging to the terry-cloth mother, regardless of which dummy held the bottle (Fig. 7.10). When

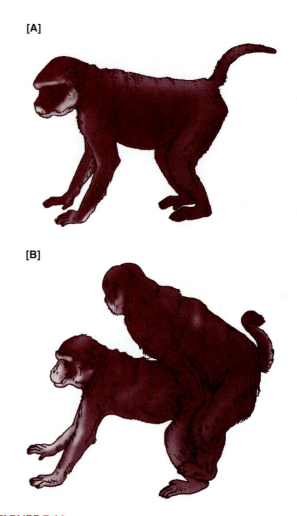

FIGURE 7.10
An infant rhesus monkey raised by two dummy "mothers," one made of bare wire and the other covered with soft terry cloth. Note that the infant maintains contact with the terry-cloth mother even while nursing from the wire dummy.

FIGURE 7.11
Normal mating behavior in rhesus monkeys. [A] An adult female presenting. [B] Mounting and copulation.

exposed to a novel or frightening stimulus, the infant monkeys always ran to their terry-cloth mothers to cling for reassurance. Wire dummies, on the other hand, were never reassuring, and the frightened monkeys would avoid them rather than cling to them.

Development of adult behavior.   Rhesus monkeys raised with terry-cloth mothers appeared to function normally until they became sexually mature, but behavioral deficits eventually did appear. A normally raised male rhesus monkey will mount an estrous female if she presents to him (Fig. 7.11), but the motherless males never mounted any estrous females to which they were exposed. Motherless females did come into their estrous periods (their genitals swelled up and became bright

pink), but they never presented (exposed their genital area) to any test males and they consistently rejected all sexual advances. A few such females were artificially inseminated under anesthesia and became pregnant. When their babies were born, they showed no signs of maternal behavior, such as picking up their infants and holding them to the breast. Instead, they either ignored or rejected (pushed away) their infants, in some cases so forcibly that the infants had to be removed for their own safety. Sexual and maternal behavior had never been learned in these monkeys. Adult social behavior has very strong learned components in rhesus monkeys and in other higher primate species as well. Harlow later discovered that motherless monkeys could develop normal adult social behavior if they had opportunities (1) to cling to a terry-cloth dummy whenever they wanted to and (2) to play socially with age mates for as little as half an hour per day during their upbringing. Harlow concluded that instincts were not sufficient to produce proper sexual behavior or maternal behavior in monkeys, but that a youthful period of social learning was also required.

Rough-and-tumble play. The nature of social learning during play is only indirectly related to sexual activity or to maternal care. Most play in primates is of a type known as **rough-and-tumble,** in which there is frequent and repeated body contact, including pushing, pulling, and climbing—just watch young children in a schoolyard to see examples. Primate play also includes a good deal of chasing and dodging behavior, usually followed by more rough-and-tumble play. Although this play is neither sexual nor maternal, it seems to teach many lessons, such as how to handle and perhaps restrain other individuals without hurting them. In the context of play, the players learn how strong or weak other individuals are and how much rough play each will tolerate. These lessons are later refined into dominance and submission relationships with other individuals and into sexual behaviors such as those in which male monkeys mount females. Mounting behavior arises during rough-and-tumble play, without regard to the sex of either individual; only after sexual maturity does it take on an explicitly sexual meaning. The defense and protection of smaller individuals, sometimes picking them up and delicately cradling them, is also learned in play.

There are parallels in human behavior. Children learn many lessons in play, including cooperation, taking turns, role-playing, counting, and keeping score; setting and following rules; and settling arguments and disputes. They also learn a good deal about each other's personalities: who plays fair, who cheats, who is a bully, who cries if they don't get their way, and so forth. Children often imitate adult roles in play, practicing many of the skills that they see adults using and that they may themselves use later in life. Abused children and those deprived of opportunities for exploratory and rough-and-tumble play with other children will often fail to develop the proper adult social behaviors, including both parental care and behavior (much more than just sexual) in committed couple relationships such as marriage.

Social Organization among Primates

Most primates are extremely social, but the size and complexity of social groupings vary greatly.

Savanna baboons. A complex type of social organization is found among the savanna baboons of Africa, *Papio cynocephalus* and related species. In captivity, savanna baboons often fight and establish a linear dominance hierarchy similar to the pecking order seen in domesticated chickens. In the wild, however, they hardly ever fight; they express dominance largely through gestures such as staring at an opponent or slapping the ground. We can study dominance by watching pairwise encounters (two individuals interacting at a time) and noting how many times one individual succeeds in threatening another or making the other give way. Dominance status generally follows size and fighting strength, although it is rarely contested and outright fighting is rare.

The situation becomes more complex when we examine encounters of more than two individuals. One group of males, called the *central males,* will support each other, in effect ganging up on any threat to one of their number or to an infant or juvenile member of the group. Their mere arrival breaks up fights, and their superior fighting abilities protect the entire group from external threats such as predators.

The females help hold the group together in other ways. Baboons are forever grooming one another—picking burrs and parasites from each other's fur. Any baboon may groom any other, but females generally do the most (Fig. 7.12). As a gesture of friendliness, grooming is generally reciprocated, with groomer and groomee taking turns. Infants and juveniles are groomed often by their mothers. Females who are not yet mothers themselves often practice at grooming behavior and infant care. This mother-in-training behavior is generally called *aunt behavior* or *aunting*; it can also be compared to baby-sitting.

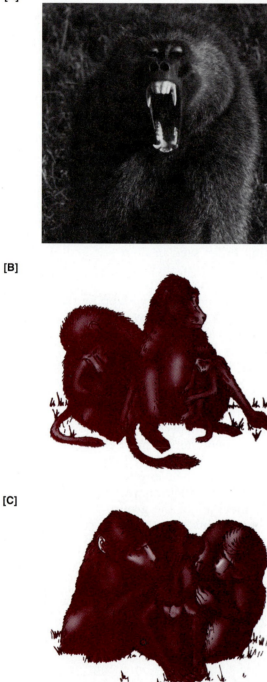

FIGURE 7.12
Examples of primate social behavior. [A] Threat display of a male baboon, showing his large canine teeth. [B] Grooming behavior among two female baboons and a juvenile. [C] Grooming behavior in rhesus monkeys.

Females go through reproductive cycles when they are not pregnant or nursing. These reproductive cycles are marked, as in most female mammals other than humans, by a conspicuous **estrous period** that coincides with the time of ovulation. The female's sexual status is advertised to males by swelling and reddening of her genital area and by presenting, a behavior in which the female presents her genital area to the inspection of interested males, one after another.

The characteristics of the central males are perpetuated by a form of selection in which they gain access to estrous females at the time when sexual swellings are maximal and ovulation is most likely. Other males "take what they can get," meaning that their access to estrous females occurs mostly at times when ovulation is less likely. As a consequence, high-ranking males are likely to leave more offspring than low-ranking ones, and any genes found more often among these high-ranking males will be favored by selection. If we study dominance interactions among female baboons, we find that high-ranking females generally have high-ranking offspring. Juvenile baboons at play will frequently look back to their mothers for backup, and a higher-ranking mother generally provides more reassurance.

Reproductive Strategies among Primates

Primatologists of the 1960s and earlier decades usually emphasized social dominance relationships among males. Beginning with the early work of Jeanne Altmann, Phyllis Jay, and Jane Goodall, relationships among female primates began to receive equal or greater attention. Primatologists of the subsequent generation conducted many important new studies that focused on the social behavior of female primates as well as males.

Man the hunter and woman the gatherer.   During the 1960s, many primatologists hypothesized that hunting, a predominantly male activity, had played a key role in human evolution. Hunting had encouraged both bipedalism and tool use, two of the features that set humans apart from apes. This idea was expanded into a theory expressed by the title of a symposium volume by Lee and DeVore (1968), *Man the Hunter*. Beginning in the 1970s, several female primatologists began asking whether simultaneous or earlier ecological and behavioral changes had also occurred among women. One such primatologist was Adrienne Zihlmann, an expert in both primate anatomy and paleontology who had done her doctoral work at Berkeley on the muscular rearrangements that accompanied the evolution of upright posture, a key event in human evolution. Zihlmann emphasized the possibility that the critical change leading to the evolution of humans was a shift, about 5 million years ago, toward the digging of more underground plant parts (roots and tubers), not toward hunting. She knew that **sexual dimorphism,** meaning anatomical differences between the sexes, results whenever the two sexes have markedly different behaviors, as in societies that encourage hunting in males only. Gathering, however, is an activity carried out by both men and women and requires few anatomical changes, so it would not result in dimorphism. Zihlmann was able to use her knowledge of both anatomy and paleontology to demonstrate that, contrary to some earlier speculations, early humans were considerably less sexually dimorphic than the hunting and gathering societies of the last million years. "Woman the gatherer" may have antedated "man the hunter" by several million years. Zihlmann's research led to a realization of the inadequacy of any explanation of human evolution that does not deal with both sexes.

Female reproductive strategies.   Another primatologist who has changed our views of primate sexual biology is Sarah Blaffer Hrdy. Hrdy's work was influenced both by feminism and by sociobiology. In her work on langur monkeys in India, Hrdy (1977) discovered important ways in which female monkeys, though subordinate in power and strength to males, nevertheless managed to influence male reproductive choices and male social behavior to the female's own advantage. Male primates differ from one another in the number of offspring that they leave, but female primates frequently modify what males must do to achieve reproductive success. Female primates can often maximize their own reproductive success by the

ways in which they influence male social behavior. Hrdy identified at least five ways in which female primates can maximize their reproductive fitness:

1. By choosing their mates
2. By influencing males to support and protect them
3. By competing with other females for resources
4. By cooperating with other females (usually close relatives)
5. By increased efficiency in daily activities such as locomotion and obtaining food

Male-female interactions.   Hrdy (1986) pointed out that studies of animal behavior and parental investment prior to about 1970 were in most cases authored by male scientists and tended to emphasize male behavior and dominance relations among males. Males were often described as making choices, while females were often depicted as either passive or "coy." Hrdy observed that females make important choices of their own, and solicit male attention for a variety of reasons not always related to the production of offspring. For example, females have often been observed to mate when they were already pregnant or otherwise unable to produce new offspring. Males can generally increase their reproductive fitness by mating with as many females as they can, indiscriminately. The optimal behavior for a female, however, depends on her own fitness and social status, as well as that of her possible mates. If a female is of high fitness herself and is mated to a high-status male, then she has nothing to gain from mating with a lower-ranking male. Since higher-status male offspring have more opportunities to mate, females will maximize their fitness (leave more grandchildren) if they raise high-ranking male offspring. On the other hand, a female of low status, or one mated to a low-status male, could potentially increase her fitness by mating additionally with a high-status male. If he sires one of her offspring, then she has produced a higher-status offspring and raised her own fitness as a result. Moreover, even if their mating produces no offspring at all, the high-status male that has mated with her will maximize *his* fitness by protecting any female that he has mated, as well as her offspring, who might be his. Females can therefore gain important advantages from liaisons with high-ranking males.

Hrdy also discovered that male langurs were sometimes infanticidal and that female willingness to mate with powerful males was sometimes a strategy to discourage their infanticide. Infanticide may occur among certain primate species whenever a new dominant male takes over a group. The new male can increase his fitness if he kills infants that are not his, especially if their mothers are lactating. Lactation inhibits the female reproductive cycle in most mammalian species; infanticide causes lactation to end, and the male gains access sooner to estrous females. Once he has mated and produced offspring, however, the male will maximize his fitness if he defends all his mates and their offspring.

Female primates, according to Hrdy, are much more sophisticated than previous researchers had ever imagined. While the adult males use rather obvious means to maximize their inclusive fitness, Hrdy discovered that the means used by females were considerably more subtle and usually involved influencing the behavior of the males.

Female-female interactions.   Both Hrdy and Jane Goodall have observed several instances in which competition between female primates produced outright hostility, even murder and infanticide. Arguing from a sociobiological perspective, Hrdy explained that competition among unrelated females should be expected when their genetic self-interests are in conflict. A universal sisterhood, in which all females cooperate as a unit, would therefore never evolve. In evolutionary terms, such a universal sisterhood would not be a stable strategy, because an individual female would always be able to "cheat" by refusing to cooperate, and by doing so she would raise her fitness and be favored by natural selection. Because evolution would never be expected to produce cooperative sisterhoods among unrelated females, Hrdy suggests that women who share her desire for such cooperative sisterhoods should strive to create them socially. Many of Hrdy's views are further described in her book *The Woman Who Never Evolved* (1981), which successfully combines a feminist outlook with a sociobiological account of primate evolution. Humans are not prisoners of biological destiny and are able to create social groupings and social behaviors that have not evolved.

Behavioral consequences of reproductive strategies.   One of the many consequences of primate reproductive strategies is a sex-differentiated behavior in paying attention to social rankings. Males in socially ranked species (including humans) must pay attention to their own rank and status—they must remember who has ever threatened them or been intimidated by them. Females, however, must know much more, because each female must not only know her own status, but also that of every male in the group. In order to know if one potential mate ranks higher than another, she must pay attention to *all* the social interactions among the males. In social species, females will therefore generally take more interest than males in knowing about the social interactions and social status of all other members of the group. Those who are better at paying attention to male-male interactions and correctly judging each male's social status and genetic fitness are at a selective advantage because they are better able to maximize their fitness by their behavior toward these males.

Some Examples of Human Behavior

Many people are interested in assessing the extent to which human behavior is learned or instinctive, and several historical controversies have centered around this question. We will cover only two of the myriad human behaviors, alcoholism and sexual orientation. We will look at the research methods that sociobiologists and others have used to investigate the question of heritability in these behaviors.

Alcoholism.   Alcoholism is a complex form of behavior that seems to have both learned and inherited components. To complicate matters, there is great heterogeneity in the disorder itself: There are varying degrees of alcoholism, and the disorder manifests itself differently in men than it does in women, and it may also have different characteristics in different social classes. Recent studies show that alcoholism exists in two or more separate forms. Type I alcoholism, also called late-onset or milieu-limited, typically arises after age 25 and is common in both sexes. It is characterized by psychological or emotional dependence (or loss of control), by guilt, and by fear of further dependence. This type of alcoholism frequently responds well to treatment. By contrast, type II alcoholism, also called male-limited, early-onset, or antisocial alcoholism, typically arises during the teenage years and is much more common in males than in females. It manifests itself in novelty-seeking or risk-taking behavior and in frequent fighting and other antisocial behavior, as measured using police records. This type of alcoholism responds poorly to conventional forms of treatment. Adoption studies in Denmark, Sweden, and the United States show that the biological children of alcoholics have a high rate of type II alcoholism even when raised in nonalcoholic families, suggesting that a predisposition for type II alcoholism may be inherited. The largest study, of 1775 adoptees in Sweden, found that the rate of alcoholism among the biological sons of type II alcoholic fathers was nine times the rate among other adoptees, including those with nonalcoholic biological parents adopted into type II alcoholic households. Type I alcoholism, however, shows a much smaller hereditary influence and may instead be subject to strong environmental influences. Some experts suggest that type I alcoholism is still heterogeneous and should be subdivided further.

Sexual orientation. A similar question exists with regard to homosexuality and other variations of sexual orientation: Some people regard such variations as innate, while others view them as learned behavior patterns. The available evidence, which is not very extensive, is summarized and reviewed by LeVay (1993). Some small differences have been observed in the brain structure of heterosexual men as compared to homosexual men, but many of the homosexual men in the study had died from AIDS, so it is uncertain whether these differences resulted from AIDS or predated the onset of that disease. If a difference in brain structure could be demonstrated between homosexual and heterosexual men, other questions would remain to be answered: Did the structural difference precede the sexual orientation, or might the structural change have been the result (rather than the cause) of some aspect of a behavioral difference? Scientists are only just beginning to examine such questions dealing with homosexuality among men; studies examining lesbian women are much rarer.

Twin studies have been conducted on homosexual males who have twin brothers. The rate of concordance (see Chapter 3) is higher for identical twins than for fraternal twins, meaning that, if one twin is homosexual, the other has a much higher probability of being homosexual if he is an identical (monozygous) twin than if he is just a fraternal (dizygous) twin. Such a result is suggestive of at least some genetic influence, but the very real methodological problems of such twin studies makes it very difficult to rule out other possible influences. The biggest shortcoming of twin studies is that the environments in which the twins are raised are never chosen at random and are usually very similar, even in cases of adoption.

THOUGHT QUESTIONS

1. To what extent can sociobiological findings on animals be extrapolated to humans? Are animal studies relevant at all to the study of human behaviors such as alcoholism or sexual orientation?

2. On the basis of your own knowledge of human behavior, how important do you think fathers are in early childhood development? What important *social* skills do children learn from interacting with their mothers? With their fathers? What do children learn from watching their parents interact with one another? What happens in families in which no father is present? What happens when no mother is present?

3. Think of the many ways in which humans learn (and subsequently practice) the social skill of evaluating the social status and motives of others. How much do we learn (or what skills do we exercise and practice) from play, from small-group discussions, from gossip, from novels, or from television? Do males and females participate in these activities in the same way? Why or why not?

CHAPTER SUMMARY

Sociobiology is the biological study of social behavior and social organization among all types of organisms. Sociobiologists tend to emphasize the inherited components of behavior that can be modifed by natural selection. All behavioral characteristics that have been closely studied are influenced by both genetic and environmental influences to varying degrees. In general, behaviors related to mating and courtship are more often instinctive, while locomotor behaviors are more often learned. Among higher primates, the importance of learned behavior is greatly increased, especially among humans. Organisms live in social groups because it affords such advantages as group defense, help in finding and exploiting food resources, and greater reproductive opportunities. Altruistic behavior is favored by natural selection if it contributes to inclusive fitness, or if the altruist can later benefit from a reciprocal act of altruism.

Among reproductive strategies, asexual reproduction is favored in situations where a quickly produced series of uniform offspring are advantageous, but sexual reproduction is favored whenever future conditions are uncertain and a greater variety of offspring are a greater advantage. Organisms living in unstable environments are often r-selected, meaning that they reproduce rapidly and prolifically but provide little or no parental care. By contrast, K-selected species living in stable environments provide more parental care and are selected for greater efficiency in exploiting environmental resources. Different levels of parental investment favor different reproductive strategies in the two sexes. Female primates often pursue complex strategies in which they maximize their own fitness by modifying the behavior of adult males.

Twin studies, adoption studies, cross-cultural studies, and studies of other species can all provide important clues to the understanding of human behavior patterns. Many human behaviors vary across cultures; others are strongly influenced by early childhood experiences. Studies of alcoholism, sexual orientation, and other traits show that we still have a great deal to learn about human social behavior.

KEY TERMS TO KNOW

altruism (p. 198)
anisogamy (p. 207)
asexual reproduction (p. 205)
egg (p. 207)
estrous period (p. 212)
eusocial (p. 202)
evolutionarily stable strategy (ESS) (p. 201)
females (p. 207)
game theory (p. 201)
group selection (p. 199)

inclusive fitness (p. 199)
instincts (p. 194)
isogamy (p. 207)
K-selection (p. 206)
kin selection (p. 199)
learning (p. 190)
linear dominance hierarchy (p. 198)
males (p. 207)
monogamy (p. 208)
parental investment (p. 207)

pheromones (p. 202)
polyandry (p. 208)
polygyny (p. 208)
promiscuity (p. 208)
r-selection (p. 206)
reciprocal altruism (p. 200)
reproductive strategy (p. 205)
rough-and-tumble play (p. 211)

sexual dimorphism (p. 213)
sexual reproduction (p. 205)
social behavior (p. 189)
social organization (p. 197)
sociobiology (p. 189)
sperm (p. 207)
territorial behavior (p. 195)

CONNECTIONS TO OTHER CHAPTERS

Chapter 1   Sociobiology is a good example of a paradigm.

Chapter 3   Social behavior can differ among different genotypes.

Chapter 3   Social behavior can affect the fitness of each genotype.

Chapter 4   Social behavior evolves, and the evolution of behavior is of prime interest to sociobiologists.

Chapter 4   Social behavior can greatly affect fitness and thus alter gene frequencies.

Chapter 6   Mating is one of the most important kinds of social behavior. Population size is a result of social behavior en masse.

Chapter 8   Access to good nutrition is one important motivating force in social behavior.

Chapter 10   Social behavior in most species results from brain activity.

Chapter 11   Drugs can alter social behavior.

Chapter 12   Social support can promote healing; stress can interfere with healing.

Chapter 13   AIDS is spread by certain social behaviors.

Nutrition Affects Health

When she weighed 140 pounds, Melanie thought of herself as fat and ugly. Her menstrual periods stopped when she weighed 100 pounds. Now that she weighs 90 pounds, all her friends tell her she is too skinny, but she is sure they are wrong because she still thinks of herself as chubby. She wants to lose even more weight. Melanie has an eating disorder known as anorexia nervosa. Her body is not getting the nutrition it needs. She could die if the situation remains unchanged.

Melanie's father went in last week for a routine checkup. Although he was feeling fine, the doctor told him he had high blood pressure and needed to control his fat intake. Unless he lowers the fat content of his diet, he faces an increased risk of getting a heart attack. He must now learn to eat a low-fat, high-fiber diet, which will lower his chances for getting heart disease, the number one cause of death in most industrialized countries.

All of us need food, but our dietary requirements may vary according to our body size, age, sex, level of activity, and previous state of health. In addition, there are variations caused by hereditary differences in body constitution, metabolic rates, and other factors. In this chapter, we will examine the body's use of food, human dietary requirements, and some of the variations in dietary requirements among different human populations. Different diets are correlated with the incidence of chronic diseases. Malnutrition, resulting from either insufficient food supplies or from conditions such as anorexia, can lead to deficiency diseases and to an increase in infectious disease.

A. Digestion Processes Food into Chemical Substances that the Body Can Absorb

Most of what we call food can be classified chemically into three types of major constituents and several minor constituents. The major constituents, which are called *macronutrients*, include carbohydrates, proteins, and lipids; the minor constituents, which are called *micronutrients*, include vitamins and minerals.

Chemical and Mechanical Processes in Digestion

In order to be useful to our bodies, food must first be converted into substances that the body can absorb. Digestion is the process that breaks down food into absorbable products. **Chemical digestion** breaks foods down chemically using **enzymes,** which are *substances that promote or speed up a chemical reaction without themselves being used up in the reaction.* Since this general process of speeding up reactions is known in chemistry as *catalysis*, enzymes may be described as organic catalysts. Nearly all enzymes are proteins. Chemical digestion is usually effective only over the surfaces of food fragments. **Mechanical digestion** exposes new surface areas to chemical digestion by breaking fragments into smaller fragments and by removing partially digested surface material.

The Digestive System

The mouth.   Mechanical digestion begins in the mouth when the food is chewed. Chemical digestion of starches (a type of carbohydrate) begins in the mouth with the enzyme *salivary amylase*. This enzyme, present in saliva, breaks down starches into smaller units (sugars). There is usually insufficient time in the mouth for the digestion of starch to be completed, however, so a similar enzyme is needed in the intestine to complete the process.

The stomach.   Once food is swallowed, it passes quickly through the *esophagus* and into the *stomach* (Fig. 8.1). The stomach accomplishes an additional amount of mechanical digestion by rhythmic contractions that knead the food back and forth, mixing it thoroughly, rubbing food particles against one another, and exposing new surface areas. The main activity in the stomach is protein digestion, accomplished with the aid of an enzyme (*pepsin*) that breaks up large protein molecules into smaller fragments called **peptides.** Like many other protein-digesting enzymes, pepsin is secreted in an inactive form, which protects the glands that secrete the enzyme from digesting themselves. The inactive form is converted into active pepsin by digestive enzymes. Pepsin works best in an acidic solution, which is provided in the stomach by the production of *hydrochloric acid* (HCl). The stomach also secretes a layer of *mucus* that protects the stomach lining (which is partly protein) from these digestive processes.

The small intestine: processing of fat.   The lower end of the stomach empties into the *duodenum*, which is the first portion of the *small intestine*. The term *small* refers to the diameter, which is about 3 centimeters; the small intestine is actually very long (20 feet, or 6 meters). Here, the food receives the secretions of the *liver*, called *bile*, and a series of digestive enzymes secreted by the

pancreas. The bile pigments give the feces their characteristic brownish color.

The bile performs an important detergent action by breaking up fat globules into smaller droplets and keeping these small droplets separate. Fats are insoluble in water due to their characteristic chemical structure (see Fig. 8.5, p. 232) and the chemical structure of water. Like other detergent molecules, bile has one portion that is **polar** (stable in contact with water) and another portion that is **nonpolar** (unstable in contact with water but stable in contact with fat). In the watery environment of the intestine, fats tend to form large globules that coalesce to form even larger globules whenever they collide. The nonpolar portions of the bile molecules dissolve in the fat droplets, leaving the polar portions of these bile molecules exposed on the surface, in contact with the watery intestinal fluids (Fig. 8.2). The polar coating helps the fat droplets mix with the water and also prevents the small droplets formed by mechanical action from coming back together to form large globules. This overall process is called **emulsification.** Exactly the same process occurs when you wash your hands with soap: Detergent molecules orient around the oily dirt and allow the detergent-dirt complex to be floated away in the water. Emulsification increases the total surface area of the droplets and thus increases the efficiency of digestion because only the molecules at the surface of a droplet are accessible for digestion and absorption.

Bile is secreted by the liver in a steady dribble, but is used in large amounts when lipids (fats or oils) are present in the intestine. This release of bile is achieved by the *gall bladder*, a pouch that accumulates the liver secretions until they are needed, then releases them all at once under the stimulus of a digestive hormone. A **hormone** is a chemical messenger that is secreted into the bloodstream and that causes a specific type of physiological change in one or more target organs. This digestive hormone is secreted by the intestinal lining whenever fats are present in the intestine; it acts on the gall bladder—its target organ—stimulating the release of bile into the intestine.

The small intestine: digestive enzymes.   In the next section of the small intestine, the *jejunum*, chemical digestion is completed, using enzymes

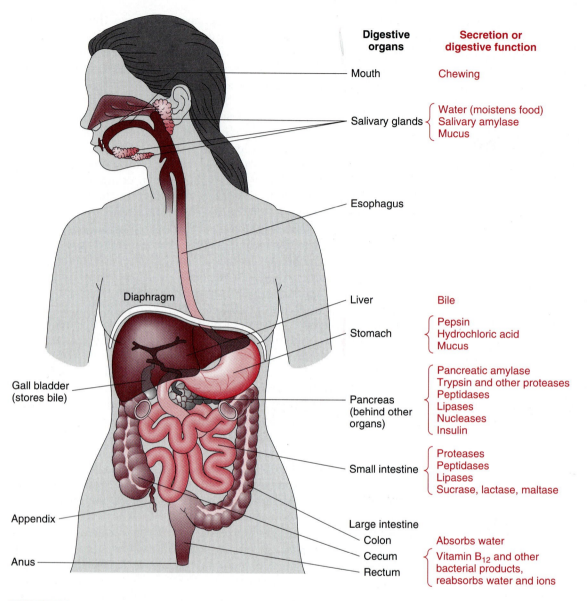

Digestive organs	Secretion or digestive function
Mouth	Chewing
Salivary glands	Water (moistens food) Salivary amylase Mucus
Esophagus	
Liver	Bile
Stomach	Pepsin Hydrochloric acid Mucus
Pancreas (behind other organs)	Pancreatic amylase Trypsin and other proteases Peptidases Lipases Nucleases Insulin
Small intestine	Proteases Peptidases Lipases Sucrase, lactase, maltase
Large intestine Colon Cecum Rectum	Absorbs water Vitamin B$_{12}$ and other bacterial products, reabsorbs water and ions

FIGURE 8.1
The human digestive system.

secreted by the pancreas and enzymes secreted by the intestine's own lining. Enzymes are often named by adding the suffix *-ase* to the end of the type of molecule broken down by the enzyme: Proteases break down proteins, lipases break down lipids, and so forth. Among the intestinal enzymes are the following:

1. Protein-digesting enzymes (*proteases*) like trypsin and chymotrypsin, secreted by the pancreas. Like the pepsin in the stomach, these enzymes break down proteins into peptides. Each of these proteases breaks the chemical bonds between the amino acids of the proteins. Each protease is specific and breaks the bonds only be-

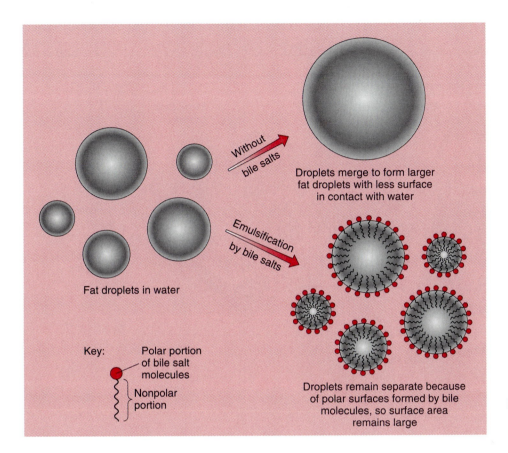

Fat droplets in water

Key:

Polar portion of bile salt molecules

Nonpolar portion

Without bile salts

Droplets merge to form larger fat droplets with less surface in contact with water

Emulsification by bile salts

Droplets remain separate because of polar surfaces formed by bile molecules, so surface area remains large

FIGURE 8.2
The action of bile salts in the emulsification of fats.

tween certain specific amino acids, resulting in shorter peptides.

2. Peptidases, enzymes that complete the final stages of protein digestion by breaking down the peptides into individual amino acids. Both the pancreas and intestinal lining secrete peptidases.

3. Pancreatic amylase, secreted by the pancreas. This enzyme continues the job of breaking down starch into sugars.

4. Fat-digesting enzymes (*lipases*), secreted by both the pancreas and the intestinal lining. These enzymes break down fats and oils into *glycerol* and *fatty acids,* molecules small enough to be absorbed.

5. Sugar-digesting enzymes like *sucrase* and *lactase,* which break down larger sugars (sucrose or lactose) into simple sugars like glucose and fructose.

The small intestine: nutrient absorption. The absorptive part of the intestine is known as the *ileum.* It is lined on the inside with thousands of tiny fingerlike tufts called *villi,* which greatly increase the surface area through which the products of digestion are absorbed. Absorption takes place through the cell membrane (Box 8.1) of the cells lining the intestine. Absorbable products include simple sugars, glycerol, fatty acids, and amino acids. Also absorbed by the villi are water and mineral salts, including dissolved ions (charged atomic particles) of sodium, calcium, and chloride, which did not require digestion to make them absorbable.

The polar chemical structure of most of these products means that they cannot enter the cell directly. Each of these chemicals is absorbed by the cells by one of four mechanisms (Fig. 8.3, p. 224), mechanisms that also occur in other tissues to bring chemicals into cells throughout the body:

1. Some small molecules enter the cells by **diffusion,** a process that requires no added energy and is therefore sometimes called *passive diffusion*

BOX 8.1   THE CELL MEMBRANE

[A] General structure of the plasma membrane

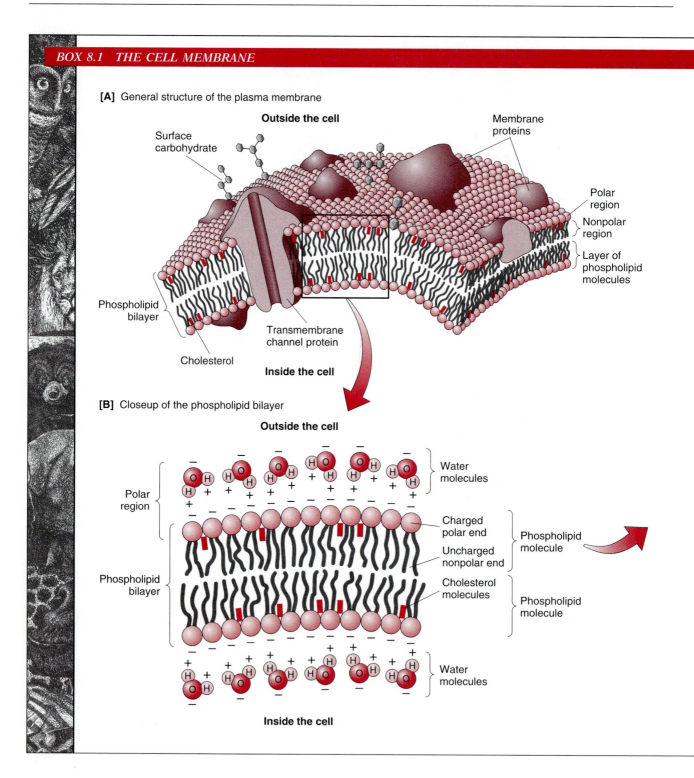

[A] General structure of the plasma membrane

Outside the cell

Surface carbohydrate

Membrane proteins

Polar region

Nonpolar region

Layer of phospholipid molecules

Phospholipid bilayer

Transmembrane channel protein

Cholesterol

Inside the cell

[B] Closeup of the phospholipid bilayer

Outside the cell

Water molecules

Polar region

Charged polar end

Phospholipid molecule

Uncharged nonpolar end

Phospholipid bilayer

Cholesterol molecules

Phospholipid molecule

Water molecules

Inside the cell

All cells are enclosed in membranes that are composed of a double layer (bilayer) of phospholipid molecules. These molecules have a polar region at one end and a long nonpolar tail. The molecules orient in water with their nonpolar tails hidden inside the bilayer, away from the water, and with their polar ends in contact with water on both sides of the bilayer. Membrane proteins also have nonpolar portions embedded in the phospholipid bilayer. The phospholipids and the membrane proteins are free to move around in the bilayer but cannot easily leave the bilayer because their nonpolar portions are repelled by the surrounding water.

[C] Structure of a phospholipid molecule

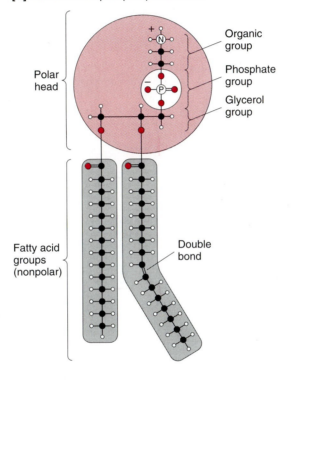

Polar head

Organic group

Phosphate group

Glycerol group

Fatty acid groups (nonpolar)

Double bond

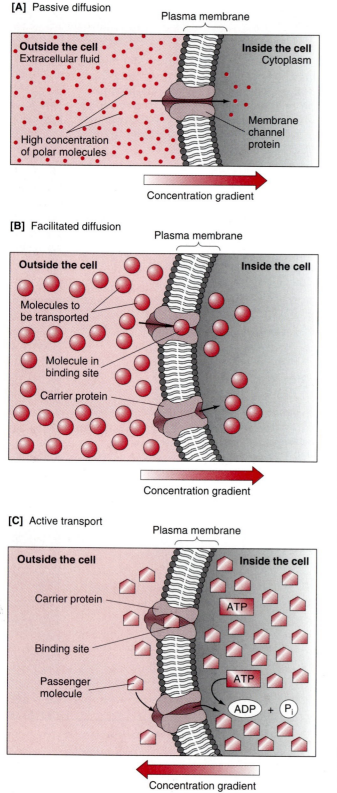

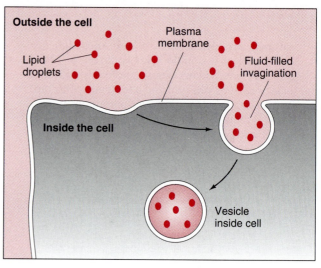

Uptake without receptors

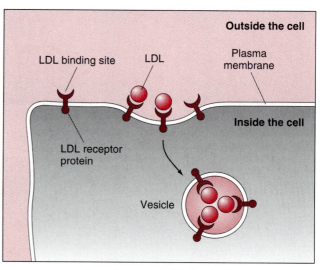

Receptor-mediated endocytosis

FIGURE 8.3

Membrane transport processes in the intestinal lining.
[A] Passive diffusion. [B] Facilitated diffusion.
[C] Active transport. [D] Endocytosis.

(Fig. 8.3A). Diffusion only works if a **concentration gradient** exists, that is, if the substance undergoing diffusion is going from a place where it is more concentrated to a place where it is less concentrated. Small, uncharged molecules such as water (H_2O) or oxygen (O_2) can passively diffuse directly through the cell membrane, which consists mostly of two lipid layers (a *bilayer*, see Box 8.1). Other molecules diffuse through *channels*, protein-lined "holes" in the membrane.

2. Small, polar molecules are internalized with the help of protein molecules (called *carriers*) that extend through the cell membrane. In some cases this transport goes along with a concentration gradient and is called *facilitated diffusion* (Fig. 8.3B).

3. Other molecules are absorbed *against* the concentration gradient, that is, from an area of lower concentration to an area of higher concentration of that type of molecule. This process requires an input of energy (usually from the breakdown of ATP) and is called **active transport** (Fig. 8.3C). Active transport is accomplished by membrane proteins called *transporters*.

4. Large particles are in many cases internalized by a process called **endocytosis** (Fig. 8.3D), in which the plasma membrane is pulled in toward the interior of the cell. In both forms of endocytosis shown in the figure, energy is used to take materials into the cell. The plasma membrane forms a pit, then the margins of the pit draw closed and the pit pinches off to form a vesicle inside the cell. This bulk processing transports many molecules at once, either suspended in liquid or attached to membrane proteins called *receptors*.

Molecules absorbed through the villi enter the bloodstream and flow to the liver, where most of the glucose and fructose are converted into a storage molecule called *glycogen* (see Fig. 8.4, p. 228). Glycogen can be broken down slowly as it is needed, a mechanism that ensures a dependable but moderately low concentration of glucose in the blood.

The storage of glycogen and the efficient use of glucose both require the hormone *insulin*, secreted by special clumps of cells, the *islets of Langerhans*, within the pancreas. Persons in whom the islets have degenerated cannot produce enough insulin; their condition is known as insulin-dependent diabetes mellitus (IDDM, or type I diabetes). The symptoms of diabetes can often be controlled by supplying insulin and/or by controlling weight and diet.

The large intestine: our mutualistic relationship with our intestinal bacteria.

The material that has not been absorbed by the ileum passes into the *large intestine*, most of which is also called the *colon*. In comparison to the small intestine, the large intestine has a larger diameter (2.5 inches, or 6 to 7 centimeters), but is much shorter (4 feet, or 1.2 meters). This part of the intestine is inhabited by bacteria, and certain nutrients produced by the bacteria are absorbed here. Mammals cannot make the enzymes that digest *cellulose*, the major constituent of plant cell walls (see Fig. 8.4, p. 228). Bacteria that live in the intestine, and especially in a small dead-end portion called the *caecum*, must do it for them. The caecum is especially important (and also much larger) in species of plant-eating mammals like horses and rabbits, which consume large amounts of cellulose. We humans cannot make the enzymes that degrade cellulose (no mammals can), and we also do not have the right species of intestinal bacteria to digest cellulose for us, so we cannot digest cellulose at all. We do, however, get some necessary nutrients from our intestinal bacteria. Humans cannot make vitamin K and biotin, needed for synthesis of blood clotting factors and for fatty acid synthesis, but our gut bacteria synthesize them.

Gut bacteria can be considered *symbiotic* with vertebrate organisms. *Symbiosis* means simply that two organisms live together; *mutualism* is the form of symbiosis in which the two species are beneficial to each other. Here, the gut bacteria are deriving nutrients from the food taken in by the human host. In exchange, they synthesize essential vitamins and break many complex molecules into simpler components that are more easily absorbed from the intestine. This symbiosis may be upset by factors such as antibiotics, which kill the bacteria.

The large intestine: water absorption.

The remainder of the large intestine consists of a straight portion called the *rectum*, which leads to a final opening called the *anus*. The major change that takes place throughout the colon and rectum is the absorption of water, mostly by diffusion, making the feces into a material with a firmer consistency.

Much of this material is undigested food, but more than half is intestinal bacteria, which are rapidly replaced by bacterial cell division in the intestine. Cancers of this part of the digestive tract are among the leading causes of death in the United States, but such cancers are far less frequent in populations in which higher quantities of cellulose are consumed. (Possible reasons for this association will be discussed later.)

THOUGHT QUESTIONS

1. Why is it important for food to spend the proper length of time in the stomach? What would happen if food left the stomach too soon?
2. The contents of the digestive tract are pushed along rather slowly by rhythmic muscular action called peristalsis. How does this relate to the body's need for a long intestinal tract?
3. What would be the consequences of a mutation that prevented a person's body from making a membrane protein involved in active transport of sugar from the intestine?
4. Why do food substances need to be digested into smaller molecules before they can be absorbed from the intestine?

B. All Humans Have Certain Dietary Requirements

Caloric Needs

Foremost among our dietary needs is a need for energy, as measured in **kilocalories** (kcal). A kilocalorie is equivalent to the amount of energy required to raise the temperature of a kilogram of water by 1°C. The "calories" that dieters count are actually kilocalories. Your body's need for caloric energy depends on many factors, including your body weight, your level of activity, and whether you are male or female (Table 8.1).

Caloric intake is the most important measure of dietary sufficiency. In most industrialized countries, the majority of the people are adequately

---

TABLE 8.1   CALCULATING YOUR BODY'S CALORIC NEEDS

A. Basal Metabolic Rate

First, find the number of kilocalories required in order to maintain **basal metabolism,** that is, to keep you alive but lying down inactively:

    Average adult woman 21.6 kcal per kg body weight per day

    Average adult man   24.0 kcal per kg body weight per day

    This works out to about 1225 kilocalories daily for a 125-pound woman or 1850 kilocalories for a 170-pound man.
      (1 pound = 0.454 kg)

B. Level of Activity

Multiply the figure obtained above by a factor depending on your normal level of activity:

    1.35 for sedentary activity (e.g., telephone sales, TV viewing)

    1.45 for light activity (e.g., college studies, office work with occasional errands, light housekeeping)

    1.55 for moderate activity (e.g., nursing, vigorous housekeeping, waiting on tables, light carpentry)

    1.65 for heavy activity (e.g., pick-and-shovel work, bricklaying, full-time competitive athletics)

C. Other Factors

Figures obtained above need to be increased by as much as 10% for any of the following conditions:

    Growth (children 15 years old and younger)

    Pregnancy

    Recovery from a major illness or injury

D. Individual Differences

The figures calculated above are only guidelines or averages. Your individual need may either be greater or smaller. If you maintain a steady caloric intake on a day-to-day basis and you gain weight, your caloric intake is greater than your caloric needs. Conversely, if you lose weight, your intake is less than your caloric needs.

nourished or overnourished. This is certainly true in the United States, where one-third of the adult population is estimated to be obese. **Obesity** is defined as a body weight more than 120 percent of the ideal for the particular subject's sex and height.

On a worldwide basis, inadequate caloric intake is the most widespread nutritional problem. The resultant starvation kills millions of people each year, most of them children. Starvation and malnutrition are most noticeable in the nonindustrialized, or third world, countries, but occur as well in certain areas of poverty, both rural and urban, within many industrial nations. Many other nutritional problems, such as vitamin deficiencies, exist among undernourished people; most of these other problems are hard to treat unless the inadequate caloric intake is remedied first.

The varieties of foodstuffs available to meet human dietary requirements can be classified into three major chemical groupings and several smaller ones. The major groupings include carbohydrates, lipids, and proteins; in addition, people require fiber and many micronutrients such as vitamins and minerals.

Carbohydrates

Most people derive the majority of their calories from **carbohydrates,** which include starches and sugars (Fig. 8.4). Cereal grains like wheat, rice, oats, and corn are the most nutritious source of carbohydrates because they also contain important vitamins, protein, and fiber. Breads, pastas, and other foods made from cereal grains retain all their nutritional value as long as the whole grain is used. Fruits and fruit products (including juices) generally contain sugars such as fructose or sucrose, together with important vitamins, minerals, and fibers. However, refined sucrose (table sugar) lacks these other nutrients and can also contribute to tooth decay (Box 8.2). Most vegetables contain carbohydrates but are even more important as sources of vitamins and minerals.

Carbohydrates are molecules formed principally of three types of atoms: carbon, hydrogen, and oxygen (Fig. 8.4). A single carbohydrate unit is called a *monosaccharide* or simple sugar. Simple sugars differ from each other by their number of carbon atoms and the placement of their chemical bonds. More complex carbohydrates are built by hooking these monosaccharides together in groups of two (*disaccharides*) or of many (*polysaccharides*).

Carbohydrates are stable in water. When chemical bonds are made between atoms to form a molecule, some of the electrons (carriers of negative charges) from the individual atoms become shared by the atoms forming the bond. The electrons can be shared equally, in which case the negative charges are evenly distributed and the bond is nonpolar. In a polar bond, the electrons are shared unequally, spending more time around one atom in the bond than around the other atom, making one atom slightly negative in comparison to the other. When certain atoms (oxygen, nitrogen, phosphorous) make bonds, they tend to

---

BOX 8.2   *HOW DOES SUGAR CONTRIBUTE TO TOOTH DECAY?*

The sugar that we add to coffee or cereal is known chemically as sucrose. There are many other sugars: fructose (fruit sugar), lactose (milk sugar), and dextrose (a synonym for glucose). There are many bacteria that live in our mouths and use these dietary sugars for their metabolic energy. One type of mouth bacteria make a gluelike substance with which they attach to the tooth surface, and to make this substance they require sucrose. Once the bacteria are glued to the tooth they can use other sugars as energy sources. When bacteria extract energy from sugars, acids are produced. It is these acids that dissolve the tooth enamel, resulting in cavities. The bacteria can even make acid from sorbitol, the sugar in "sugarless" gum, after they have adhered with the help of sucrose. Without sucrose, the bacteria cannot make the glue, the acids are not trapped so closely against the enamel surface, and tooth decay is lessened.

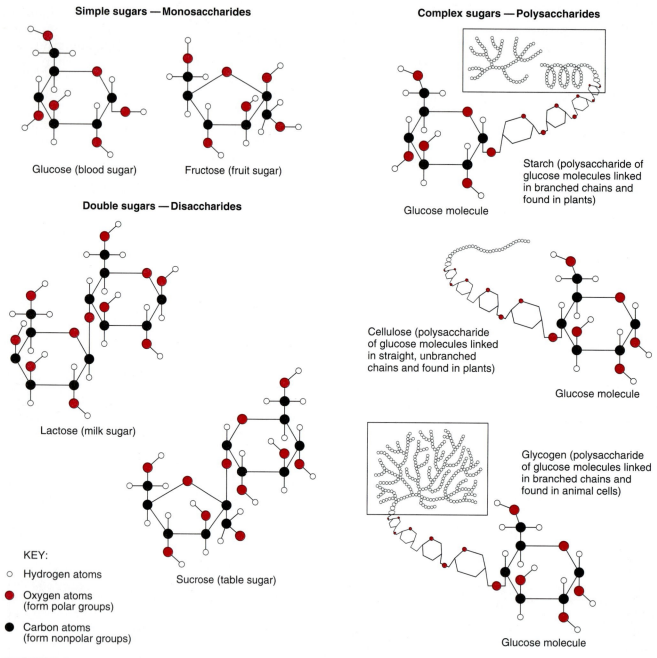

Simple sugars — Monosaccharides

Glucose (blood sugar)

Fructose (fruit sugar)

Double sugars — Disaccharides

Lactose (milk sugar)

Sucrose (table sugar)

KEY:

○ Hydrogen atoms

● Oxygen atoms
(form polar groups)

● Carbon atoms
(form nonpolar groups)

Complex sugars — Polysaccharides

Starch (polysaccharide of glucose molecules linked in branched chains and found in plants)

Glucose molecule

Cellulose (polysaccharide of glucose molecules linked in straight, unbranched chains and found in plants)

Glucose molecule

Glycogen (polysaccharide of glucose molecules linked in branched chains and found in animal cells)

Glucose molecule

FIGURE 8.4

Chemical structure of selected carbohydrates. The abundance of oxygen atoms (especially in OH groups) makes most of these substances polar and therefore stable in water.

attract the electrons toward themselves, making the bond polar. Molecules like carbohydrates that have many oxygen atoms thus have many polar bonds, and we call them polar molecules. Water (H_2O) is also polar, with electrons unequally shared between hydrogen and oxygen atoms, so other polar molecules, including carbohydrates, remain stably dissolved in water.

Membrane transporters for sugars. In the digestive tract, all carbohydrates are digested into simple sugars like glucose and fructose. They are soluble in the watery digestive fluids but are unable to cross the nonpolar interior part of cell membranes and thus cannot enter cells without help. The cells that line the small intestine have *transporter* proteins, which use energy to help move a particular small molecule through the membrane. A different transporter protein exists for each type of simple sugar such as glucose, fructose, or galactose. Starch can only be used after it is broken down into these simple sugars or other small molecules for which the organism has transporter proteins. Each species can only derive energy from and grow on those sugars for which it has specific membrane transporters. Higher animals have many transporters. Bacteria often have only a few, and these can be identified by trying to grow the bacteria in various chemical media containing only one sugar at a time (and nothing that can easily be converted into sugar).

Carbohydrates as a source of energy. After the body absorbs simple sugars, their further processing is essentially the same regardless of which carbohydrate was initially present or which simple sugar was absorbed (Box 8.3). The human body's daily need for carbohydrates is measured in terms of total caloric intake, as indicated in Table 8.1. In terms of caloric content, all carbohydrates, both sugars and starches, are the same, providing 4 kilocalories per gram (kcal/g). From an energy standpoint, it makes little difference if the carbohydrates are eaten in the form of sugar or starch, or whether the sugar is fructose or sucrose. There is, however, a difference in the rate of absorption: Starches generally take a few hours to be digested into absorbable sugars, while dietary sugars are capable of being absorbed within minutes. A meal containing both sugars and starches will therefore maintain the body's energy level (or blood glucose) more evenly over a longer period of time.

In most populations, an increase in the consumption of carbohydrate-rich foods (especially whole grains) is desirable. Many third world diets supply inadequate calories; carbohydrates provide the most efficient and most economical means of improving these diets. Fewer kilocalories of labor, or fewer dollars, are needed to produce a kilocalorie of carbohydrate food, as compared to a kilocalorie of most fat-rich or protein-rich foods. In the United States, replacement of dietary fats by carbohydrates would have certain indirect health benefits, including a reduction in risks for heart attacks and certain forms of cancer.

Lipids

Lipids are organic molecules that do not dissolve in water because they are comprised primarily of hydrogens and carbons organized into nonpolar *hydrocarbon* chains. Dietary lipids are mostly *triglycerides,* molecules in which *glycerol* (a three-carbon molecule) is linked to three long chains of carbons and hydrogens called *fatty acids* (Fig. 8.5, p. 232). Triglycerides that are solid at room temperature are commonly called *fats*; those that are liquid at room temperature are commonly called *oils.* As sources of caloric energy, fats and oils contain almost 9 kcal/g, which is over twice as much as carbohydrates (Box 8.3). A small amount of lipid is a dietary necessity, in part because the fat-soluble vitamins (especially A and D) cannot be absorbed without it. Lipids are also a source of fatty acids, which are the nonpolar portion of the phospholipid molecules that form the cell membrane. Two particular fatty acids (linoleic and arachidonic acids) are required from dietary sources since they cannot be made by the body, but they are only required in very small amounts (about 3 grams per person per day). Most nonstarving people have adequate intake of lipids.

In the United States particularly, a significant fraction of the population overconsumes lipids. Under these conditions, the body tends to store the excess (as well as some excess carbohydrate) as fatty deposits within numerous *adipose* (fat-storing) cells.

Dietary lipids and atherosclerosis. Excess dietary fat can result in fat deposits that build up in the arteries, causing **atherosclerosis** (Fig. 14.10, p. 414). The fat deposits obstruct the blood vessels, making the passages narrower; eventually these deposits may calcify and make the vessels more rigid. Atherosclerosis contributes to high blood

BOX 8.3 HOW FOOD BECOMES ENERGY THAT CELLS CAN USE

The digestive system breaks large macromolecules into simple subunits. Some of this process continues inside cells. Carbohydrate foods are broken down into glucose and other simple sugars, lipids are converted into fatty acids and glycerol, and proteins are broken down into their separate amino acids.

[A] Flow diagram showing the major products of digestion and how they are broken down in energy-yielding reactions.

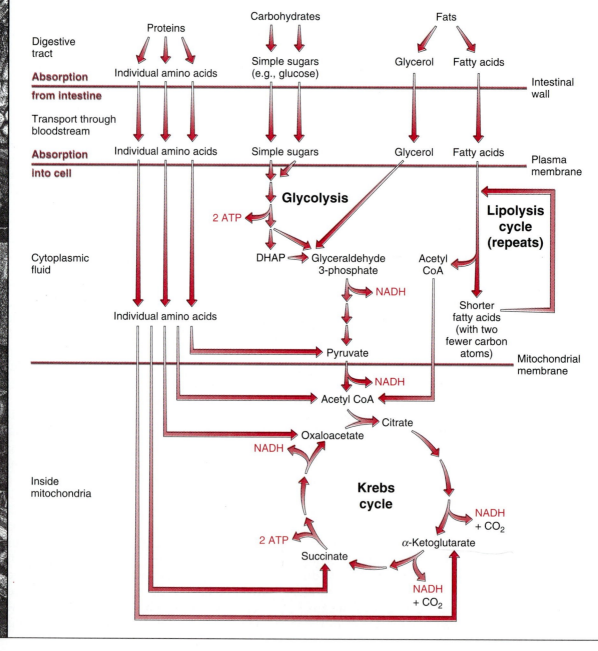

These simple subunits are then broken down into even smaller molecules, as follows:

- The long carbon chains of fatty acids are broken down in stages. In each stage, two carbons at a time are broken off of the long carbon chains to make an *acetyl group*. These acetyl groups are each put onto a carrier molecule called coenzyme A, so that the complex is called acetyl coenzyme A (acetyl CoA).
- Sugars such as glucose are converted by a process called *glycolysis* into a three carbon molecule called *pyruvate*. Pyruvate is then converted into acetyl CoA (diagram A).
- The various amino acids are degraded into either pyruvate or acetyl CoA or one of the molecules in the *Krebs cycle*.

During these reactions, limited amounts of the energy-rich molecule **ATP** (adenosine triphosphate) are synthesized. ATP is the principal form in which cells store chemical energy for later use. During glycolysis, some of the energy that was contained in the chemical bonds of glucose is converted to energy stored in the phosphate chemical bonds of ATP inside cells.

Pyruvate and acetyl CoA are then transported from the cytoplasm into organelles called **mitochondria,** where further energy is extracted from them in a cycle of reactions called the **Krebs cycle,** or citric acid cycle. In each cycle, two ATP molecules are synthesized. Also, several reactions in the cycle extract energy by removing both hydrogen atoms and electrons and passing them to a molecule called NAD. The enzymes for the Krebs cycle are proteins in the interior of the mitochondria, including some that are attached to the inner mitochondrial membrane (diagram B).

The electrons removed by NAD during the Krebs cycle (1) are donated to the *electron transport chain,* a series of electron-carrying proteins in the mitochondrial membrane (2). As electrons pass from one electron carrier to the next carrier in the chain, protons (H^+) are pumped from one side of the mitochondrial membrane to the other (3), creating an unequal distribution of protons on opposite sides of the membrane, called a *proton gradient*. This proton gradient is a way of storing energy that is then used to make additional ATP (4) when the protons leak back across the membrane. The electrons meanwhile pass from carrier to carrier and they, along with protons, finally combine with oxygen to form water (5). This need for oxygen as an electron acceptor is why we need to breathe in oxygen from the atmosphere.

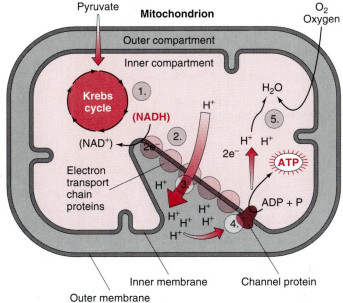

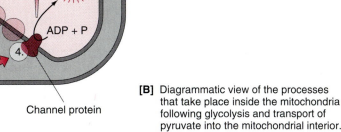

[B] Diagrammatic view of the processes that take place inside the mitochondria following glycolysis and transport of pyruvate into the mitochondrial interior.

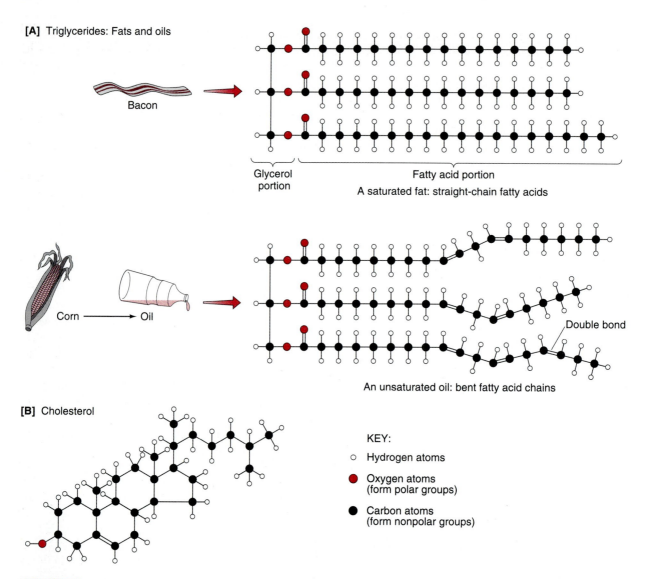

[A] Triglycerides: Fats and oils

Bacon

Glycerol portion

Fatty acid portion

A saturated fat: straight-chain fatty acids

Corn → Oil

Double bond

An unsaturated oil: bent fatty acid chains

[B] Cholesterol

KEY:

○ Hydrogen atoms

🔴 Oxygen atoms (form polar groups)

● Carbon atoms (form nonpolar groups)

FIGURE 8.5

Chemical structure of two types of lipids. [A] Triglycerides and their decomposition into glycerol and fatty acids. [B] Cholesterol. Notice that the general lack of OH groups (or other "polar" groups in which electric charges are unequally shared) makes most parts of these molecules nonpolar, tending to separate away from water.

pressure (hypertension), although hypertension can also occur independently of atherosclerosis.

Several researchers have found that different types of fat in the diet have different effects on health. **Saturated fats** are fats that have only single bonds in their chemical structure (Fig. 8.5); most saturated fats are derived from animal sources (or from a few tropical plants like palm or coconut), and most are solid at room temperature.

There is some evidence that saturated fats contribute more readily to atherosclerosis than unsaturated fats. **Unsaturated fats,** often derived from plant sources, have double bonds as well as single bonds in their chemical structure, causing the molecules to bend (Fig. 8.5). Those containing only one double bond are sometimes called *mo-nounsaturated*, while those with multiple double bonds are *polyunsaturated*. Both types are usually liquid at room temperature because the bends made by the double bonds prevent the molecules from packing too tightly together and solidifying. When unsaturated fatty acids are incorporated into the phospholipid cell membrane, the bends also prevent their tight packing in the membrane. This makes the membrane more fluid and better able to function.

Because saturated fats have been linked to greater risk of atherosclerosis, many experts recommend that saturated fats be replaced with unsaturated fats in most diets. Advertising has convinced many people that unsaturated fats—especially the polyunsaturated kind—are desirable, but this is true only when those unsaturated fats are used as replacements for saturated fats. For people who do not want to keep track of saturated versus unsaturated fats, most experts recommend that the quantities of all dietary fats be reduced in order to lower the risk of heart disease and atherosclerosis.

In addition to fatty acids, dietary lipids include **cholesterol,** a fat-soluble molecule that is the precursor of several important hormones (Fig. 8.5). Cholesterol is also an important constituent of animal cell membranes. Along with unsaturated fatty acids, cholesterol helps keep cell membranes fluid, thereby keeping the cell and the organism functioning properly. Animal cells can synthesize most of the cholesterol they need, so little or none is required from food. Plant cell membranes do not contain cholesterol, so plant products are always cholesterol free, although some (like coconut oil) contain saturated fatty acids that are easily converted into cholesterol by the body. All dietary fatty acids are broken down into several molecules of acetoacetate, one of the major starting materials of cholesterol synthesis; cholesterol synthesis is thus increased by nearly all fatty foods, even if they are advertised as "cholesterol free."

Because the body makes about 75 to 80 percent of its own cholesterol and makes it from dietary fats, most of your serum cholesterol comes from dietary fats (especially saturated fats), not from dietary cholesterol. Excess cholesterol, like excess amounts of other lipids, can build up on blood vessel walls and increase your risk of atherosclerosis. Most foods that contain cholesterol are also high in saturated fats, so avoiding either one also helps you to avoid the other. Eggs are exceptional in having a lot of cholesterol with few fats.

LDLs and HDLs: lipid transport particles. In contrast to carbohydrate molecules, lipid molecules contain few oxygen and nitrogen atoms and have mostly nonpolar bonds in which electrons are shared equally around carbon and hydrogen atoms (Fig. 8.5). Because the bonds in lipids are nonpolar, lipids are not water soluble. Since the liquid part of blood is mainly water, lipids must be transported through the blood from one part of the body to another by transport particles such as **low-density lipoproteins (LDLs)** and **high-density lipoproteins (HDLs).** These transport particles are proteins that can bind lipids in such a way that they can move through body fluids.

People eating identical diets may not have the same cholesterol level. This difference seems to have a genetic component and relates in part to the individual's ability to make LDLs and HDLs. No one has very much free (unbound) cholesterol because cholesterol is very nonpolar, so what is called *serum cholesterol* is actually the total of all the cholesterol found in all the types of lipid transport proteins, including both HDLs and LDLs. Phospholipids and triglycerides are also transported by these particles.

HDLs transport cholesterol and other lipids out of tissues to the liver, where they are used in the synthesis of bile acids, while LDLs transport cholesterol and lipids into tissues and cells. The *HDL/LDL ratio* is the ratio of outbound to inbound lipids. A high HDL/LDL ratio thus indicates that the proportion of "good cholesterol" (lipids on their way out) is higher than the proportion of "bad cholesterol" (lipids on their way into cells). A very low ratio, meaning a preponderance of inbound lipids, correlates strongly with an increased risk of atherosclerosis in the arteries of the heart (coronary arteries), a condition that can precipitate a heart attack.

Some people are genetically prone to high cholesterol levels because their cells lack LDL receptors on their surfaces. When cells need cholesterol, they get the cholesterol from the LDLs in the bloodstream by binding these particles to LDL receptors and internalizing the receptors and LDLs by endocytosis (Fig. 8.3). If the LDL receptors are missing or nonfunctional, the cells manufacture their own cholesterol even when LDL levels are already high, because the ability to take up LDL from the blood has been lost. Thus, diet is not the only factor leading to atherosclerosis; problems with lipid transport and lipid uptake are also factors.

Proteins

The body uses **proteins** for tissue growth and repair, including healing of wounds, replacement of skin and mucous membranes, and manufacture of antibodies. Proteins are important components of all cell membranes, and function as transporters, receptor molecules, and channels (Fig. 8.3). Many proteins of the cellular interior provide structure, motility, and contractility to muscles and other cells. Other proteins outside the cell, such as *collagen* and *elastin*, give connective tissues their strength and thus help support the entire body, while *keratin* is essential for healthy skin and is the main constituent of hair and fingernails. A much larger assortment of proteins function as the enzymes described at the beginning of this chapter. Some enzymes (such as those used in digestion) function extracellularly, while many others function inside cells.

Dietary proteins provide amino acids. Proteins are built from chains of smaller chemicals called amino acids (Chapter 3). The digestive system breaks down the proteins in food into individual amino acids (Fig. 8.6B). After they are absorbed by the body, these amino acids can then be used to build the body's own proteins. The function of each type of protein depends to a large extent on the *shape* of the protein after its linear sequence of amino acids has folded upon itself. The way the protein folds, and whether it is stable in the watery cytoplasm of the cell or in the nonpolar cell membrane, is determined by the arrangement of the polar and nonpolar side groups of its amino acids (Fig. 8.6B). Since proteins are synthesized by adding one amino acid at a time to the end of the chain (Chapter 3), if one type of amino acid is missing from the cell, synthesis of any protein needing that amino acid stops. An amino acid that is present in small quantities and is used up before other amino acids are is called a **limiting amino acid.**

Proteins are necessary in the diet. The daily requirement is 0.8 grams per kilogram of body weight (about 45 grams for a 125-pound woman or 64 grams for a 175-pound man). Each species has its own capacities to make certain of the amino acids and therefore has its own dietary requirements for the ones it cannot make. The assortment of individual amino acids in the diet is thus at least as important as the total amount of dietary protein. Eight of the twenty standard amino acids cannot be synthesized by the human body and are therefore considered essential in the human diet; a ninth amino acid is essential in human infants. The human body can make the remaining amino acids from these essential ones.

Complete and incomplete proteins. Most animal proteins are **complete proteins** in that they contain all the amino acids essential in the human diet. Soy protein is also complete, but most plant proteins lack at least one essential amino acid needed by humans. When an incomplete protein of this kind is eaten, the body uses all the amino acids until one of them, the limiting amino acid, becomes depleted. After the limiting amino acid is used up, the body uses the remaining amino acids to produce energy instead of making proteins, since dietary protein cannot be stored for later use the way carbohydrates and lipids can.

To get around the problem of incomplete proteins in the human diet, plant proteins can be eaten in combinations in which some of the proteins supply an essential amino acid missing in the others. The Iroquois and many other Native Americans commonly obtained complete protein by combining beans, squash, and corn in their diets. Most bean proteins, for example, are deficient in the amino acids valine, cystine, and methionine, while corn proteins are deficient in the amino acids lysine and tryptophan. Alone, neither one of these proteins is nutritionally complete for humans, but in combination (as in succotash, a mixture of beans and corn cooked together), the two plant sources provide a nutritionally complete assortment of amino acids because beans and corn each have the amino acids that the other lacks. Some examples of nutritionally complete protein combinations are shown in Box 8.4.

Vegetarian diets. Strict vegetarians, also called *vegans,* who do not eat food from any animal source, including milk or eggs, must be more care-

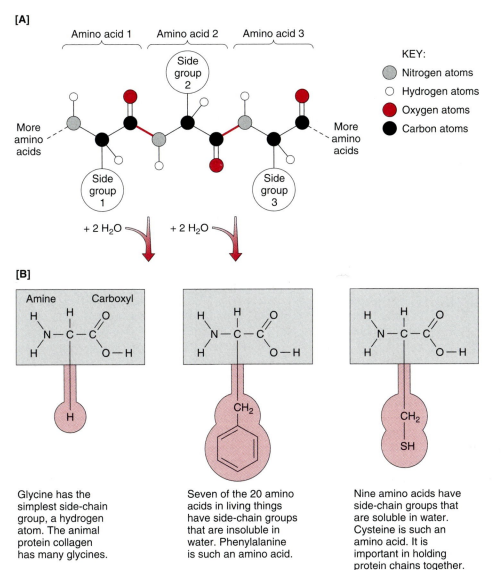

[A]

Amino acid 1   Amino acid 2   Amino acid 3

Side group 2

More amino acids

Side group 1

Side group 3

More amino acids

KEY:
- Nitrogen atoms
- Hydrogen atoms
- Oxygen atoms
- Carbon atoms

+ 2 H₂O

+ 2 H₂O

[B]

Amine   Carboxyl

H   H   O
| | ∥
N — C — C
| | \
H | O — H
     H

Glycine has the simplest side-chain group, a hydrogen atom. The animal protein collagen has many glycines.

H   H   O
| | ∥
N — C — C
| | \
H | O — H
     CH₂

Seven of the 20 amino acids in living things have side-chain groups that are insoluble in water. Phenylalanine is such an amino acid.

H   H   O
| | ∥
N — C — C
| | \
H | O — H
     CH₂
      |
     SH

Nine amino acids have side-chain groups that are soluble in water. Cysteine is such an amino acid. It is important in holding protein chains together.

FIGURE 8.6
Chemical structure of a peptide. [A] A tripeptide, a sequence of three amino acids held together by two peptide bonds (shown in color). The breaking of these peptide bonds releases the individual amino acids, such as those shown in [B]. A *protein* consists of one or more long polypeptide chains, each containing many more amino acids in a row.

ful to include protein sources in their diets. They must also combine several protein sources in order to supply their bodies with nutritionally complete protein. For example, legumes (beans, peas, and peanuts) can be combined with whole grains (like rice, corn, or wheat). Nuts and seeds contain protein and may be used to supplement amino acids missing from plant proteins from other sources. Some vegetables also contain individual amino acids that can serve the same function.

Since animal cells store energy as fat, proteins obtained from animals are accompanied by fat.

Plant cells, in contrast, store energy in the form of complex carbohydrates such as starch, and plant proteins are thus accompanied by very little fat.

Fiber

Material that the body cannot digest and absorb is technically known as **fiber.** There are several kinds of fiber, including *soluble fiber* (pectin, gums, mucilages) and *insoluble fiber* (mostly cellulose). Several studies have shown that an increase in dietary fiber reduces the incidence of several

---

BOX 8.4   TWO VEGETARIAN RECIPES PROVIDING NUTRITIONALLY COMPLETE PROTEINS

The first recipe obtains nutritionally complete protein by combining the lysine-poor protein of rice with the lysine-rich protein of milk. The milk makes it a *lactovegetarian* recipe. The second recipe combines corn (rich in valine but poor in lysine) with beans (poor in valine but rich in lysine). It is a strictly *vegan* recipe that avoids all animal products. Both recipes call for seasoned vegetable broth (stock); bouillon may be substituted if desired.

CARROT AND ONION SOUP
Serves 4
~

3 tablespoons margarine	pinch tarragon
4 to 5 medium carrots, grated	½ cup rice
1 medium onion, minced	4 cups seasoned stock
1 teaspoon salt	1 to 1¼ cups hot milk
	croutons (optional)

Heat margarine in a heavy pot or pressure cooker and gently sauté carrots, onion, salt, and tarragon for about 5 minutes. Add rice and stir into mixture. Add stock and cook until rice is very well done, about 45 minutes (only 25 with a pressure cooker). You may want to sieve the soup or puree it in a blender; I puree only half, so that some chewiness is left. Return it to the pot and add milk to your preferred consistency; do not let it boil. Add a pat of margarine and serve with croutons.

The author of this recipe also writes:

This soup is simple yet has a unique quality that makes it a favorite with my family and with guests. I like it especially because I almost always have the ingredients on hand. Any homemade bread would make this soup into a special supper. I cut the carrots and onions into big chunks and reduce them to tiny slivers in the blender.

CHILI CON ELOTE
Serves 6
~

3 tablespoons oil	1 cup fresh corn
1 onion, chopped	4 cups cooked kidney or pinto beans
1 clove garlic	
1 green pepper, diced	½ teaspoon chili powder
2 cups vegetable stock	¼ teaspoon cumin powder
1 cup chopped tomato *or* 2 tablespoons tomato paste	1½ teaspoons salt
	1 teaspoon oregano

Elote means "corn"—a colorful and tasty addition to a traditional favorite.

Sauté onion and garlic clove in oil until onion is soft. Discard garlic clove. Add green pepper. Sauté another 2 or 3 minutes. Add stock, tomatoes, and corn. Mash 2 cups of the kidney beans and add to pot along with whole beans and seasonings. Simmer 30 minutes. If too watery, remove cover and cook another 10 minutes.

Source: Carrot and Onion Soup from Lappé (1982), *Diet for a Small Planet*, 10th anniversary ed., p. 301, New York: Ballantine Books; Chile con Elote from Robertson, Flinders, and Godfrey (1976), *Laurel's Kitchen, A Handbook for Vegetarian Cookery and Nutrition*, p. 262. Reprinted courtesy Ten-Speed Press.

---

cancers, especially those of the colon and rectum. Scientists are not sure about the exact mechanism of the link between dietary fiber and cancer. One intriguing possibility is that the rate of movement of food through the colon is crucial, and that fiber maintains the optimal rate of food movement. Faster rates cleanse the colon of potentially toxic chemicals, while slower rates allow these chemicals to remain in one place long enough to undergo fermentation by bacteria into cancer-causing substances (*carcinogens*; Chapter 9). Another possibility is that harmful carcinogens are frequently present inside the intestine for various reasons, but mucus protects the intestinal lining against them; the insoluble fiber rubbing against the intestinal lining stimulates the lining to secrete more of this protective mucus, while a diet low in fiber results in less mucus and therefore less protection against cancer. Most of the above findings deal with insoluble fiber such as cellulose, and most of them are based on population studies in which high levels of fiber intake correlate with low levels of cancer.

Separate studies on soluble fiber have also been conducted in recent years. Some of these studies suggest that soluble fiber such as oat bran, for example, can reduce the level of serum cholesterol and the risk of heart disease. The mechanism is not known with certainty, but one hypothesis is that certain soluble fibers bind strongly to the bile secreted into the intestinal tract. Without the soluble fiber, the bile would be reabsorbed by the intestinal lining and be reused, but the soluble fiber prevents this reabsorbtion and ensures that the bile is eliminated with the stools. Without recycled bile, new bile must be synthesized, and it is synthesized from cholesterol, which the body withdraws from the blood.

Most fruits, vegetables, legumes (beans, peas, etc.), and whole grains are high in both soluble and insoluble fiber. Diets that are high in fiber are statistically associated with lower rates of coronary heart disease and stroke.

Vitamins

In addition to the macronutrients mentioned above, there are nutritional factors that are not used as energy sources or as structural building blocks but which still must be present for normal cellular functions, and yet cannot be made by the organism. These factors are collectively called *micronutrients*, and include vitamins and minerals.

In general, **vitamins** are complex nutrients needed only in very small quantities. Most vitamins are **coenzymes**—the nonprotein portions of enzymes needed for the enzyme to function as a catalyst. Since enzymes (and their coenzymes) are used and reused in the chemical reactions that they regulate, they are needed only in very small quantities. The other characteristic that all vitamins have in common is that the body cannot

make them—each vitamin (or its precursor molecules) must be included in the diet as a nutritional requirement. The recommended consumption levels of vitamins for individuals trying to maintain good health are called *recommended dietary allowances (RDAs)*.

Vitamins may be obtained either from pills or from food. Although vitamins obtained from pills are fully functional in the body, there are a number of reasons for preferring vitamins derived from food:

1. They are much less expensive this way.

2. Foods rich in vitamins are also rich in other important substances, including minerals, fiber, and protein. We do not know the complete nutritional requirements of any organism more complex than bacteria, and undoubtedly our food contains many unknown but needed nutrients. These other nutrients, known and unknown, are not obtained from vitamin pills.

3. Some vitamins are more easily absorbed by the body in the combinations with other ingredients that exist in food than they are in the combinations that exist in vitamin pills.

4. Purified vitamins can be toxic if taken in excessive amounts, a danger that arises much less often with vitamins contained in foods.

Vitamin overdoses and vitamin deficiencies. Vitamins are generally grouped into two large classes: water-soluble vitamins and fat-soluble vitamins. *Water-soluble vitamins,* including vitamin C and the B group of vitamins (Table 8.2), are not accumulated by the body. They are stored in very limited quantities; therefore they must be provided regularly. When you eat more than you need, the excess is simply excreted in the urine. Because they are not stored, these vitamins cannot easily build up to toxic overdoses, especially if you get them from foods. It is, however, possible to overdose on water-soluble B vitamins taken in pill form or as concentrated liquids, particularly vitamin B_6. Vitamin B_6 (pyridoxine) is a coenzyme for many of the enzymes of amino acid synthesis; it therefore helps build proteins and is sometimes used by bodybuilders. Daily doses of 500 milligrams or more can be dangerously toxic to the nervous system and liver. *Fat-soluble vitamins,* including A, D, E, and K, are accumulated in the body's fat tissues and can build up over time. Toxic overdoses of these vitamins can therefore occur, especially for vitamins A and D.

Vitamin deficiencies can also occur. People who have disorders of fat absorption (LDL defects, for example) will often be deficient in fat-soluble vitamins since these are transported and absorbed along with dietary fats. Diseases result from deficiencies of each of the vitamins; in fact, research on the cause of these diseases led to the discovery of vitamins.

Vitamin B₁.   **Vitamin B₁** (thiamine) was the first vitamin to be chemically characterized. While stationed on the island of Java in the 1890s, the Dutch physician Christiaan Eijkman noticed that a neurological disease called *polyneuritis* in chickens had many similar symptoms to a human disease called *beri-beri*. Both diseases caused muscle weakness and leg paralysis, resulting in an inability to stand up; both diseases were fatal if they persisted. Eijkman noticed that the chickens got polyneuritis only when they were fed on polished white rice, but the disease cleared up when rice polishings were added to their feed or when whole brown rice was used. From the rice polishings, thiamine was later isolated and was found effective in both treating and preventing beri-beri in humans. Because it was a vital (necessary) substance and also an *amine* (a chemical containing an NH_2 group), it was called a *vital amine*. This term was later shortened to *vitamine* and then *vitamin*, and the name was applied to the entire class of substances needed only in small quantities and capable of curing deficiency diseases such as beri-beri. Beri-beri occurs primarily among people whose dietary carbohydrates come from a single, highly refined source such as white rice or white (unenriched) flour. Many countries now have laws requiring the addition of thiamine (and other B vitamins) to refined flour. For this reason, beri-beri is now rare in the industrialized world, though it does occur among severe alcoholics whose dietary intake is inadequate.

As an example of how a vitamin functions as a coenzyme, consider the role of thiamine in the breakdown of the molecule pyruvate, one of the major products of carbohydrate metabolism and a major entry point into the energy-producing Krebs

cycle (Box 8.3). An enzyme that contains thiamine as one of its constituent parts combines with the pyruvate molecule, releases CO_2, then emerges from a later reaction in its original form. Since the thiamine is not used up, it can participate in the reaction again and again. For this reason, only minute amounts of thiamine are needed to facilitate the breakdown of large quantities of pyruvate formed in carbohydrate metabolism.

Other B vitamins.   Other water-soluble vitamins are described in Table 8.2. Many are coenzymes or form portions of other molecules. Vitamins B₂ and B₃ form parts of the molecules (FMN and NAD, respectively) that carry electrons from the Krebs cycle to the electron transport chain (Box 8.3). Many were found as the result of research on *vitamin deficiency diseases* like beri-beri. Vitamin B₆ deficiency (microcytic anemia) is seen frequently in people whose diets consist mostly of rice. A deficiency of niacin causes pellagra, a disease of the skin and nervous system. Because corn is particularly deficient in niacin, populations in which corn is the only protein source are often subject to pellagra.

Vitamin C.   A deficiency of vitamin C causes *scurvy*, a disease once common among sailors at sea and among prisoners. A British naval surgeon discovered in the 1600s that limes and other fresh fruits would both prevent and cure scurvy. British ships then began carrying limes and became so well known for this practice that British sailors came to be called limeys. An inflammation of the mucous membranes, as in a cold, increases the body's need for vitamin C. Vitamin C may therefore decrease the severity of the symptoms of such an infection, but it cannot on this account be considered a "cure" or a prevention for the common cold, as has sometimes been claimed.

People who take large doses of vitamin C can suffer the symptoms of scurvy when they stop taking the vitamin. In addition, megadoses can produce hemolytic anemia (red blood cell deficiency caused by the rupture of red blood cells) in susceptible people with the G6PD metabolic deficiency (Chapter 5) found in African-American, Asian, and Sephardic Jewish populations. In addition, individuals who are genetically predisposed to gout find that high doses of vitamin C can

TABLE 8.2   VITAMINS AND MINERALS IN HUMAN HEALTH

	IMPORTANCE FOR GOOD HEALTH	GOOD FOOD SOURCE
Water-Soluble Vitamins		
Vitamin B_1 (thiamine)	Helps break down pyruvate; maintains healthy nerves, muscles, and blood vessels; prevents beri-beri	Meats, whole grains, legumes
Vitamin B_2 (riboflavin)	Important in wound healing and in metabolism of carbohydrates; prevents dryness of skin, nose, mouth, and tongue	Yeast, liver, kidneys
Vitamin B_3 (niacin)	Maintains healthy nerves and skin; prevents pellagra	Legumes, fish, whole grains
Vitamin B_6 (pyridoxine)	Coenzyme used in amino acid synthesis; prevents microcytic anemia	Whole grains (except rice), yeast, liver, mackerel, avocado, banana, meats, vegetables, eggs
Vitamin B_{12} (cyanocobalamine)	Required for DNA synthesis and cell division; prevents pernicious anemia (incomplete red blood cell development)	Meat, liver, eggs, dairy products, yeast
Folic acid	Used in synthesis of hemoglobin, DNA, and RNA; prevents megaloblastic anemia and spina bifida	Asparagus, liver, kidney, fresh green vegetables, yeast
Pantothenic acid	Needed to make coenzyme A for carbohydrate and lipid metabolism	Liver, eggs, legumes, dairy products, whole grains
Biotin	Used in fatty acid synthesis and other reactions using CO_2	Eggs, liver, tomatoes, yeast
Vitamin C (ascorbic acid)	Antioxidant; used in synthesis of collagen (in connective tissues) and epinephrine (in nerve cells); promotes wound healing; protects mucous membranes; prevents scurvy	Fresh fruits (especially citrus and strawberries), fresh vegetables, liver, raw meat
Fat-Soluble Vitamins		
Vitamin A (retinol)	Antioxidant; precursor of visual pigments; prevents night blindness and xerophthalmia	Yellow and dark green vegetables, some fruits, fish oils, creamy dairy products
Vitamin D (calciferol)	Promotes calcium absorption and bone formation; prevents rickets and osteomalacia	Eggs, liver, fish, cheese, butter
Vitamin E (tocopherol)	Antioxidant; protects cell membranes against organic peroxides; maintains health of reproductive system	Whole grains, nuts, legumes, vegetable oils
Vitamin K	Essential for blood clotting; prevents hemorrhage	Green leafy vegetables
Minerals		
Electrolytes (Na^+, K^+, Cl^-)	Maintain balance of fluids in body; maintain cell membrane potentials	Raisins, prunes; K^+ also in dates
Calcium	Part of crystal structure of bones and teeth; maintains muscle and nerve membranes	Dairy products, peas, canned fish with bones (sardines, salmon), vegetables
Phosphorus	Part of crystal structure of bones and teeth	Dairy products, corn, broccoli, peas, potatoes, prunes
Magnesium	Maintains muscle and nerve membranes	Meat, milk, fish, green vegetables
Iron	Part of hemoglobin; also used in energy-producing reactions	Meats, egg yolks, whole grains, beans, vegetables
Iodine	Maintains thyroid gland; prevents goiter	Fish and other seafood products
Fluorine	Strengthens crystal structure of tooth enamel	Drinking water
Zinc	Promotes bone growth and wound healing	Seafood, meat, dairy products, whole grains, eggs
Copper	Cofactor for enzymes used to build proteins, including collagen, elastin, and hair	Nuts, raisins, shellfish, liver
Selenium	Statistically associated with lower death rates from heart disease and stroke	Vegetables, meats, grains, seafood

sometimes bring on the condition by raising blood levels of uric acid. Vitamin C megadoses can produce deficiencies of another vitamin, B$_{12}$, in people who are iron deficient. Even in healthy people, vitamin C can irritate the bowel sufficiently to result in diarrhea.

Antioxidant vitamins.   Vitamin A (retinol) is essential in the synthesis of the light-sensitive chemicals (retinal) used in vision. Vitamin A is also an **antioxidant,** meaning that it protects body tissues from chemicals that would rob those tissues of electrons. The removal of electrons from other molecules is a process that chemists call **oxidation.** Chemicals that bring about oxidation by taking up electrons are called oxidizing agents; among the most highly reactive oxidizing agents are a group of chemicals, called *free radicals,* that have one or more unpaired electrons. Electrons are much more stable when they occur in pairs; therefore, agents with unpaired electrons have a strong tendency to remove electrons from other molecules. Free radicals may thus damage many cellular molecules and are hypothesized to play a role in initiating some cancers (Chapter 9). Vitamin A and other antioxidants protect the body by destroying free radicals.

Vitamin A can be obtained from animal sources such as dairy products or fish. Many vegetables also contain a vitamin A precursor, the orange-yellowish pigment *beta carotene,* which is split after ingestion to produce two molecules of vitamin A. High consumption rates of foods rich in beta carotene are statistically associated with lower rates of lung cancer, but the causal link between the two is unclear. Laboratory studies on beta carotene, mostly in rodents, have shown that it suppresses or retards the growth of chemically induced cancers of the skin, breast, bladder, esophagus, pancreas, and colon.

Vitamin E (tocopherol) is an antioxidant vitamin that is especially important in breaking down a group of strong oxidizing agents called peroxides. Vitamin E also helps prevent spontaneous abortions and stillbirths among pregnant rats, and for this reason it has acquired a reputation as an antisterility vitamin. However, health claims related to the effects of this vitamin on sexual function remain unproven. Vitamin E overdose results in low blood sugar and headache, fatigue,

blurred vision, muscle weakness, and intestinal upset. Vitamin E occurs in several forms, of which α-tocopherol is the most potent. Vitamin E is destroyed by freezing and also by cooking food.

Other fat-soluble vitamins.   Vitamin D (calciferol) is discussed at greater length in Chapter 5. It is essential to the body's use of calcium in bone formation. Vitamin K is essential to blood clotting because it serves as a cofactor in reactions that produce blood clotting factors from their inactive precursors. Most people get adequate amounts of vitamin K from the bacteria that live in their intestines. However, newborn infants, whose intestines have not yet been colonized by bacteria, and persons whose intestinal bacteria have been killed off by antibiotics, need more dietary vitamin K until gut bacteria have become established or reestablished.

Minerals

Minerals are the inorganic ions and atoms necessary for proper physiological functioning.

Electrolytes.   *Sodium* (Na$^+$), *potassium* (K$^+$), and *chloride* (Cl$^-$) *ions* are the principal **electrolytes** (charged particles) of the body. Differences in the concentration of ions on opposite sides of a cell membrane are a type of concentration gradient called a *membrane potential.* Ion concentration gradients are one way in which cells store energy in a usable form. All of the body's electrically excitable cells (nerve and muscle cells) are able to respond to changes in the amounts of sodium outside and potassium inside the cells (Chapter 10). Maintenance of electrolytes within a very narrow concentration range is thus very important. If the number of ions, particularly of sodium ions, gets too high, the body compensates by retaining water that would otherwise be excreted in the urine (Chapter 11). Since blood pressure is related to the volume of fluid in the circulatory system (Chapter 14), too high a concentration of sodium in the body tissues results in **high blood pressure (hypertension).** This is an otherwise symptomless condition that increases the risks for vascular (blood vessel) diseases such as stroke and coronary artery disease. Overuse of salt (sodium chloride) makes this condition worse, but is seldom the original cause. Many people in the United States consume too much sodium and not enough potassium. Potassium must be present in the proper amounts; either too much or too little can lead to heart failure and death.

Iron.   Iron is needed for the formation of blood hemoglobin and as a cofactor for many enzymes. Iron deficiency causes an anemia that is more common in older people with poor dietary habits and in menstruating women. Iron deficiency anemia is in fact the single most common nutritional deficiency in most industrialized countries, including the United States. Menstruating women need almost twice as much iron as men, and pregnant women need even more for the proper synthesis of hemoglobin in the baby's blood. Some people may have iron deficiencies resulting from low levels of the proteins involved in the transport of iron in and out of cells. If vitamin C supplies are inadequate, then cellular absorption of iron will be inadequate, too.

Calcium.   Calcium is needed as an intracellular messenger for many processes, including muscle contraction. In addition, the crystal structure of bones and teeth is composed of calcium, along with phosphate. Many older women suffer from low bone density and bone brittleness (*osteoporosis*). Although low calcium intake is involved in osteoporosis, the problem is not so simple as to be solved by increasing dietary intake of calcium later in life. Hormones like estrogen are also involved, and so is vitamin D, which promotes calcium absorption, but the exact processes are poorly understood. Supplementary doses of both calcium and vitamin D are recommended for post-menopausal women, although most bone loss within the first five years after menopause is caused by estrogen withdrawal, not by any nutritional deficiency. There is evidence that higher levels of exercise among women aged 18 to 25 can increase their bone density and forestall the development of osteoporosis later in life.

Fluoride.   Fluoride (the ion F^-) is important in the growth of strong teeth during the childhood years. Insufficient fluoride results in a greater incidence of tooth decay. Drinking water is the most important dietary source of fluoride. Many cities now add fluoride (in carefully measured amounts) to the drinking-water supply as a preventive measure against tooth decay. Fluoride is also available as drops for breast-fed infants and others who do not have access to fluoridated water. Fluoride is toxic in very high doses. If high doses are accidentally ingested, milk can neutralize the fluoride.

Trace minerals.   Most of the remaining mineral nutrients are sometimes called *trace minerals* because they are needed by the body only in very small quantities. Deficiencies of these trace minerals occurred more often in the past, when local vegetables were grown in soils deficient in one or another trace mineral, and when domestic animals grazing on plants growing in the same soil were the main supply of animal food. In the industrial world, such nutritional deficiencies are much less likely to occur because our food supply comes from numerous sources grown in a variety of different soils and climates.

In addition to the trace minerals listed in Table 8.2, *chromium* and *manganese* are needed in carbohydrate metabolism; *cobalt* is an important part of the vitamin B_{12} molecule; *molybdenum* and *nickel* are required in the metabolism of nucleic acids; and *silicon, tin,* and *vanadium* are needed in trace amounts for proper growth, including bone and connective tissue development. Diets adequate in other nutrients usually supply sufficient amounts of all these trace minerals.

THOUGHT QUESTIONS

1. If two people drink soft drinks every day, but one chews sugarless gum and the other chews regular gum, would you expect they would have a difference in the number of cavities at their next dental checkup? Would it matter if the soft drinks were sugar free or not? (See Box 8.2 for help with this question.)
2. Why are sugars absorbed from the intestine faster than starches are?
3. Think about the definition of obesity: weight greater than 120 percent of the ideal weight for a person's sex and height. What is meant by *ideal*? Who sets these ideals? Are there cultural differences in what is considered ideal? How do cultural definitions of *ideal* relate to biological definitions?

4. How is it possible for different species (like rats and people) to have different vitamin requirements? What does this mean on a biochemical level?

5. How is it possible for some studies to show that calcium supplements forestall osteoporosis while other studies do not?

C. Human Diets Vary among Populations

The world's populations have found many different ways of meeting their nutritional needs. Different diets arose in different parts of the world because different kinds of plants grew best in each climate and in each type of soil. However, no culture makes use of all available foodstuffs; every culture has its own preferences and prohibitions that limit their uses of the available foods in their environment. Cultural variations are important because they limit the extent to which we can expect people to modify their diets or improve their food habits.

Links between Diet and Chronic Diseases

During the first half of the twentieth century, infectious diseases and vitamin deficiencies were the major public health concerns. Many infectious diseases have declined in importance in most industrial nations, the result of clean water, good sewage treatment, and regular garbage removal as much as of any medical advances. Nutritional deficiencies have declined as well, the results of better eating habits, more varied diets, and vitamin-fortified foods. Since the 1960s, public health officials and nutritionists have increasingly turned their attention to conditions like heart disease, stroke, and cancer, and also to the chronic health problems like obesity and high blood pressure that can influence the risks for these fatal diseases. Greater interest now centers on whether certain foods can *promote good health* and reduce disease risks.

In order to assess the good health of populations (which is much more than just the absence of a few infectious diseases or vitamin deficiencies), we can look at the rates at which various chronic conditions like obesity and high blood pressure occur among different populations and different types of people within those populations. This type of statistical correlation of factors studied over large populations is called **epidemiological** evidence; such evidence offers no final proof of any hypothesis, but it often suggests where we should look and what types of hypotheses are worth pursuing. For example, the importance of fluoride in preventing tooth decay was deduced from epidemiological evidence when lower rates of tooth decay were found in populations living in areas where groundwater fluoride levels were naturally high.

Epidemiological studies have also pointed to a connection between certain types of dietary fat and cardiovascular diseases. The United States and certain other meat-producing countries like Australia and New Zealand have a high per capita consumption rate of meat products and also high occurrence rates for coronary heart disease and strokes. Mediterranean countries consume much of their fat in the form of olive oil, an unsaturated fat, and their heart disease rates are correspondingly lower. Heart attacks are very rare among the Inuit (Eskimos), whose diet contains large amounts of cold-water fish, a good source of a type of fatty acid called *omega-3 fatty acid*, which has been shown to guard against the production of oxidizing chemicals that damage cell membranes. The Japanese also tend to have low consumption rates of saturated fats and lower rates of cardiovascular disease and stroke.

Epidemiology can also provide clues as to whether a particular association is more closely related to diet or to genetically inherited traits: Japanese people in Japan have much lower rates of heart disease or stroke than do Japanese living in Hawaii or California, who in turn have rates similar to those of their non-Japanese neighbors. These findings (and similar ones involving other immigrant groups) all point to diet, not heredity, as the major difference responsible for the different disease rates between populations. Yet diet is clearly not the only relevant factor: African-Americans, whose diets are generally similar to those of other Americans, have higher rates of hypertension (high blood pressure). Although genetic influences are difficult to rule out in this case, one hypothesis is that the difference results from the statistically greater stress levels that African-Americans continue to suffer in a society with a long history of racism.

Natural Selection and Lactose Intolerance

Some aspects of diet are, however, genetically determined. Dairy products are used only in certain cultures, principally those of Europe and those founded by European colonists. Asian, Pacific, and Native American peoples typically avoid milk beyond early childhood, preferring to drink tea or fruit juices rather than milk with most of their meals. One important reason for this is that many individuals in these populations have the genetic trait of *lactose intolerance,* the inability to digest lactose, the sugar in milk (Fig. 8.4). Lactose helps in the absorption of the calcium needed for good bones and teeth. Outside of Europe, ultraviolet light is usually sufficient for the synthesis of large quantities of vitamin D (see Chapter 5), permitting the absorption of calcium without the help of lactose. Cloud-covered northern Europe, however, was always characterized by less ultraviolet light and therefore less vitamin D, conditions that selected for people who could tolerate lactose in their guts to help in calcium absorption. So, while most of the rest of the world lacks the enzyme (*lactase*) needed to digest lactose, natural selection acted on European populations to favor those individuals who possessed this enzyme. This does not mean that every person who is of northern European descent can tolerate lactose, but the lactase gene is much more frequent in Caucasian populations descended from northern Europeans.

Well-meaning Americans, unaware of this phenomenon, once caused an international incident when they donated large amounts of foreign aid to Brazil in the form of surplus food, including powdered milk. Most of Brazil's population is lactose-intolerant because they are descended from a mixture of Native American, west African, and southern European (mostly Portuguese) populations, all of which have high population frequencies of lactose intolerance. When the Brazilians consumed the milk, they could not digest the lactose. Unused lactose is fermented by gut bacteria, producing painful amounts of carbon dioxide gas. The resulting symptoms (painful cramps, diarrhea, and sometimes vomiting) quickly gave rise to suspicions that the food was somehow poisoned. Researchers finally pinpointed the cause, but not before large amounts of ill will had been created. There may still be thousands of people in Brazil who believe that the Americans tried to poison them.

Regional Variations in Nutrition

The energy requirements of food production. A given amount of arable land can support a larger human population if that land is used for raising crops for human consumption, including sources of plant proteins, than if the same land is used to raise food for animals that humans can eat. It takes between 5 and 16 pounds of grain protein to produce 1 pound of meat protein. This results from a basic ecological principle: Each time that chemical bond energy is used by organisms at one step of a food chain, some of that energy is lost, because energy conversions are never 100 percent efficient. At every successive step in the food chain, more energy is lost due to inefficiency; therefore, there must always be a lesser quantity of "consumer" organisms like beef cattle than of "producer" organisms like photosynthetic plants. In well-fed countries with plenty of land, such as Australia or the United States, large tracts can be used for grazing or for the raising of crops primarily for animal consumption, such as corn eaten by pigs. However, poor countries of high population density can ill afford to feed crops to animals. As a result, most of the world's poor eat little meat and get most of their protein from vegetable sources. All whole-grain cereals contain some amount of protein. If this protein is eaten with beans or other legumes, a high-quality protein source is created that is much less expensive than meat and contains far less saturated fats. Populations located near lakes and oceans can also supplement their diets with fish protein.

Regional variations in micronutrient consumption. Because of regional variations in the mineral content of soils, the mineral content of foods that grow in those soils also varies. Therefore, deficiencies of vitamins and minerals often vary geographically. Zinc deficiency, for example, is common in the Middle East. Iodine deficiency is found primarily in certain inland locations, such as the high Andes, the Himalayas, and parts of central Africa.

Most vitamin and mineral needs can be met economically, even in poor countries, from grain and vegetable sources. Grains contain most B vitamins and also vitamin E and several important minerals, including zinc. Fresh vegetables contain additional vitamins (including A and C) and several important minerals including calcium and iron. Fresh fruits provide additional vitamin C. Legumes provide calcium and iron in addition to proteins. In general, all mineral nutrients can be supplied from plant sources, except vitamin B_{12}, the one important vitamin that cannot be supplied from plant sources alone.

Diversified agriculture.   In several cases, the health of human populations can be greatly improved by diversifying the crop species planted, especially as carbohydrate sources. The major reason for this from a nutritional point of view is that vitamins lacking in one carbohydrate source are often supplied in another. An additional benefit is that diversified planting will usually result in higher quantities and quality of plants using lesser amounts of fertilizer and pesticides. Planting of several crops in one area, especially if the crops are planted in rotation, makes more efficient use of the soil minerals, so that the plants grow better, reducing the need for fertilizers. Plants grown in this way are also of improved quality because they are better dietary sources of minerals. Mixed planting practices also make the plants less susceptible to devastation by insect pests. Plant pests thrive best and cause the most damage on large, pure stands of a single crop (*monocultures*), partly because they can more easily spread from one plant to another and reproduce with devastating speed. If crop losses to insects are reduced, then more food is available for people (Chapter 15).

There are impediments to the diversification of carbohydrate sources: Not all cereal crops will grow in all climates, nor will the populations necessarily accept the introduced plants and add them to their diets. Some success has been achieved with the cultivation of corn (maize), a crop of New World origin, in both Africa and China. The potato

is another crop that has been successfully grown on many continents.

THOUGHT QUESTIONS

1. What are some factors that prevent each crop species from being grown all around the world? How easily could corn (or wheat, or soybeans) be grown in places where it is not currently grown? What impediments may exist to its introduction to new places?

2. In Mexico before European contact, the diet consisted mostly of corn, beans, chili peppers, and squash; these same foods are still the major elements of most Mexican diets, especially in rural areas. Can you think of any biological reasons why this diet has proved so stable?

3. Why would a trait like lactose intolerance be favored by natural selection in some places but not in others?

4. Which will support a larger human population, a vegetarian diet or a diet high in meat? Why?

D. Malnutrition Contributes to Poor Health

Eating Disorders

Anorexia nervosa.   Among the middle and upper classes of the industrialized nations, some people suffer from a condition called **anorexia nervosa.** The ratio of anorexic women to anorexic men is about 9:1 (Gordon, 1990). Anorexic individuals suffer from a mistaken perception of their body size—they imagine themselves to be heavier than they really are, and desire to be thinner as a result, a feature that clearly distinguishes anorexia from all other forms of undernourishment. Anorexics also respond poorly to body cues of hunger and satiety. The misperception of hunger, satiety, and body size are early symptoms that precede the most noticeable feature of the disease, what Dr. Hilde Bruch has called a "relentless pursuit of thinness," a self-imposed undernourishment that borders on starvation. At the same time, there is usually an absorbing or obsessive interest in food, which may include talking or reading about food, preparing food, collecting recipes, or serving food to others, while all the time avoiding eating.

One of the surest signs of the disease in anorexic women is that they usually stop menstruating because of a lack of the cholesterol needed for synthesis of the hormones needed to regulate the menstrual cycle (Chapter 6). Other symptoms include changes in brain activity. The brain must be constantly supplied with glucose to provide cellular energy for nerve cell function, and when it is not, many mental functions may be impaired. These impairments may manifest themselves in anorexic persons as deception (hiding things, keeping secrets) and a distrust of others, but only late in the process, after starvation has already set in. Untreated anorexia is usually fatal.

Anorexia is most common among Caucasian women between the ages of 15 and 30 with an average or above-average level of education. The disease also occurs in Japan and in parts of Southeast Asia, but only among the uppermost social strata. (In Singapore, for example, it occurs principally among the economically powerful Chinese, but rarely among the majority Malay population, which includes most of the poor.) Anorexia is all but unknown among people living in poverty or in undernourished populations anywhere. It never seems to occur where food is scarce, or in times of famine. Even in countries where it occurs, it seems to vanish during economic hard times, such as the Depression of the 1930s.

Anorexia was once extremely rare. Its marked increase in the United States since World War II has been attributed by several experts to a general standard of beauty that has increasingly glorified thinness, as measured by such criteria as waist and hip measurements of Miss America contestants, *Playboy* centerfolds, and models and ballet dancers more generally. In fact, women in professions like modeling and ballet dancing are particularly likely to develop anorexia. Also, female athletes in sports like rowing (where competition is organized by weight classes) are at high risk for developing a *female athlete triad*, consisting of eating disorders like anorexia, combined with loss of menstruation and a loss of bone mass that may cause osteoporosis later in life.

Bulimia.   Many anorexics also suffer from a related condition called **bulimia.** Bulimia is characterized by occasional binge eating of everything in sight, usually including large quantities of high-calorie "forbidden" foods, in total disregard of any concept of a balanced diet. Immediately following a binge, bulimics typically force themselves to vomit or else purge themselves with an overdose of a laxative. Both the binge and the purge are usually done in secret; bulimics become extremely skillful at hiding their condition from others. Persistent bulimia can lead to ulcers and other digestive tract problems and also to chemical erosion of the teeth from the frequent contact of the teeth with the acidic secretions of the stomach. Bulimia occurs in both sexes, but more often in women. Bulimia is especially common among educated women who have easy access to unrestricted amounts of food, a situation common on many college campuses.

Protein Deficiencies

Inadequate caloric intake can result in protein deficiencies that take several forms, including both low total protein and low levels of particular amino acids. If protein intake is inadequate for either reason, a protein deficiency develops called *kwashiorkor*. Kwashiorkor occurs when carbohydrate intake is adequate but protein intake is not.

In all organisms, cells and proteins are constantly being broken down, and in a healthy, adequately nourished body, they are constantly being replenished. When protein intake is insufficient, replenishment does not occur. There is considerable loss of muscle tissue, and the death of many cells releases numerous dissolved ions into the surrounding tissue. These ions retain water and contribute to tissue swelling (*edema*) that makes the loss of muscle tissue harder to see. Children suffering from kwashiorkor also have swollen abdomens, a fact often noticed in photographs from protein-deficient areas. Their large bellies conceal the fact that these children are actually starving to death.

If protein intake is inadequate and carbohydrate intake is also inadequate, the combined deficiency produces a condition called *marasmus*, in which the body slowly digests its own tissues and wastes away. When carbohydrate is not available as an energy source, amino acids are used to produce energy. Since amino acids are not stored except in the form of the body's structural and functional proteins, use of amino acids for metabolic energy degrades these proteins. Once the body's muscle mass falls below a certain minimum, marasmus is always fatal.

Ecological Factors Contributing to Poor Diets

Malnutrition is recognized as a worldwide disease with regional differences in its cause but with similar outcomes everywhere. We will use Africa as an example, although many other examples exist. Many populations in Africa experience either chronic or periodic protein deficiency; Africa has therefore been described as a protein-starved continent. Fresh vegetables are widely available, so diets are high in fiber and most vitamins. Vegetables (including legumes) can fill nearly all of a population's nutritional needs as long as supplies are adequate and as long as some form of nutritionally complete protein is obtainable from animal sources (including fish) or from a combination of plant proteins.

Inadequate grain supplies are due to a combination of ecological and social factors. Tse-tse flies have made large tracts of African land uninhabitable to most domestic animals, because they bite and spread blood-borne diseases. This not only limits the availability of meat proteins, it also limits the supply of draft animals that can pull plows and till the soil. Most farmers cannot afford to buy tractors or to keep them supplied with expensive fuels. Additionally, many places are either too dry or too wet to support agriculture. The world's largest desert occupies most of Africa's northern half; other deserts exist in Somalia and Namibia. Most of these deserts, including the vast Sahara, are growing larger each year as animals such as sheep and goats overgraze and destroy the plants on the desert fringes, a process known as *desertification* (Chapter 16). In certain other parts of Africa, high rainfall leeches important minerals from the soil, and the resulting *lateritic* soils are deficient in the minerals necessary for adequate plant growth. Rice, wheat, and most other grains grow rather poorly in many African soils. Some success has been achieved with millet and with corn (maize) imported from the New World, but raising a sufficient quantity and variety of grains to provide complete protein is difficult.

In most places in Africa, populations can maintain adequate nutrition in years when there are ample harvests and efficient distribution. Unfortunately, these two conditions are not always met. Protein deficiencies are made worse by political and military upheavals that drive people from their farms or that prevent planting, harvesting, or distribution of crops. Protein starvation (marasmus and kwashiorkor) is all too common, particularly among children.

Effects of Poverty and War on Health

Climatic, economic, and political factors all contribute to the unequal production and unequal distribution of food across the planet. Poverty exists in all nations of the world. Some of the aspects of malnutrition associated with chronic poverty were discussed earlier in this chapter. Many populations are marginally nourished and therefore more vulnerable to a year of drought or bad harvest, but malnutrition exists in every part of the globe, especially among the poor. Even in regions in which nourishment is usually adequate, people can become undernourished because of disruptions of life associated with war.

Some efforts have been made to discover the short-term and long-term consequences of undernourishment. Many of these studies have been done on animals because ethical considerations make it difficult to design experiments in which to study the effects of starvation on humans. (The ethics of conducting such studies on animals are discussed in Chapter 2.) Animal experiments show that there can be biological changes in an organism that continue after the period of starvation is over. The studies that have been done on humans have agreed with the animal studies, for the most part.

Wartime starvation. One way to study the effects of starvation on humans is called a **retrospective study.** In a group of people who were subjected to starvation by war, data can be collected at a later date to correlate with the data on undernourishment. Such a study on the short-

and long-term effects of starvation was conducted using the birth records of infants born in the Netherlands in 1944 and 1945. During that winter, people in some of the cities of the Netherlands experienced starvation. The food situation in other cities was less critical, but all of the other effects of war were equal across the country. It was found that below a threshhold level of food rations, fertility in the population decreased and the decrease was greatest among people in the lower classes. Fetuses carried by women who experienced starvation in the first trimester had a higher rate of abnormal development of the central nervous system. Maternal undernourishment in that period also carried forward to premature births, very low birth weight, and an increase in the rate of death immediately after birth. Maternal undernourishment in the third trimester produced the greatest increase in the death rate in the three months after birth. Brain cells were depleted in infants who died. Among survivors, undernourishment in infancy was not correlated with long-term effects on mental development when males were tested at age 18. Other similarly designed retrospective studies have found similar results: Poor maternal nutrition can lead to low-birth-weight babies and an increase in the likelihood of death of newborn babies.

Long-term effects of childhood undernutrition.
There are several indications that undernutrition during childhood decreases brain development and capacity. Infants who died of kwashiorkor or marasmus had less DNA, protein, and lipids in their brain cells compared to infants who died at similar ages of causes not related to nutrition. Evidence suggests that both the severity and length of the period of malnutrition affect intellectual development. Iron deficiency, particularly in the early years of life, can also impair mental function.

Malnutrition can also result not from a lack of nutritionally adequate foods but from children's failure to eat properly as a result of a depressed mental state. Malnutrition of this type is called *failure to thrive*. Infants must learn that their needs will be met, and this learned capacity is termed *basic trust*. When circumstances are such that a child is not regularly fed and nurtured, the child will fail to develop trust. Loss of a caregiver or a diminution or loss of nurturing care can create depression even in infants, who will lose trust and become malnourished. This situation can go hand in hand with parents being overwhelmed by the stress of poverty; medical intervention, particularly of a short-term nature, will not rescue the child unless efforts are made to reestablish a trusting relationship.

When people have been starved, sudden intake of food is actually damaging. Food intake must be increased gradually. Much has been learned about the types of foods that can be tolerated, and in what forms and in what quantities. This lengthy process is called *nutritional replacement*. Nutritional replacement has often been carried out by putting people, and especially children, in hospitals during the renourishment period. In addition to the delayed social development that resulted from starvation, social development was found to be further delayed by lengthy hospitalization, and the delayed social development continued after the child had recovered nutritionally.

It is very difficult to assess the effects of nutrition on later cognitive development because many other social factors and other biological factors, such as the effects of infections, are involved. It seems certain that adults will recover all of their physical and mental capacities if they are rescued from starvation, but this is not yet known for children. What does seem certain is that adequate nutritional replacement alone, in the absence of emotional and psychological support, will not lead to full recovery. Recovery of an individual from famine requires much more than food.

Micronutrient Malnutrition

Malnutrition can still exist when total caloric intake is sufficient or even when it is excessive. Micronutrient malnutrition may be overlooked among people who lose interest in eating, including elderly people, people with chronic diseases including cancer, and people who have mental illnesses such as depression. Limited incomes, limited mobility, and limited availability of fresh foods all make nutritional problems worse. Inadequate nutrition is also a problem in chronic alcoholics, where a large percentage of total caloric intake may be in the form of alcohol.

Nutrition, Health, and the Immune System

The study of nutrition has changed over the years from a focus on prevention of deficiency diseases to a focus on the promotion of health. Nutrition may work on health indirectly through its effects on the immune system. Too much food (overnutrition) may suppress the immune system. Obese adolescents have lowered activity levels of the type of white blood cell that rids the body of bacteria and viruses, and this lowered activity level can lead to disease. Undernutrition and malnutrition are also immunosuppressive. Chronic protein undernourishment, even of a moderate nature, impairs the ability of the immune system to fight off infectious diseases. The immune system becomes less responsive with age, so the effects of protein deficiency or micronutrient malnutrition on the immune function of elderly people may be magnified. Deficiencies of many of the micronutrients, including magnesium, selenium, zinc, copper, and vitamins A and C, impair various aspects of the immune system. Many elderly people become deficient in one or more of these micronutrients, risking additional impairment of their already compromised immune systems. Adherence to good nutrition prolongs the functioning of immunity. Many of the infectious diseases of the elderly can thus be minimized in frequency and severity if nutrition is adequate.

Recent work has identified dietary factors that stimulate the immune system's function. Beta-carotene and vitamin A, which appear to be important in preventing cell damage by acting as antioxidants, also seem to stimulate an increase in the number of immune cells and to stimulate antibody production. Iron stimulates both the production and activity of white blood cells. Eating foods rich in these micronutrients can help to promote health.

THOUGHT QUESTION

1. Why do you suppose anorexia occurs only among well-fed populations and never among the poor or in times of famine? What does this imply about the possible causes of the condition?
2. Will efforts aimed at improving nutrition among people with various forms of malnourishment be readily accepted by the people they are meant to help? What factors determine acceptance?
3. If sufficient food can be produced, will every person have good nutrition? Why or why not?
4. When social factors produce high numbers of refugees, we often hear of widespread hunger among the displaced persons. We also often hear about outbreaks of infectious disease, such as the cholera epidemic among Rwandan refugees in the 1994 civil war. How does inadequate nutrition contribute to these outbreaks?
5. Why might starvation have greater long-term effects in children than in adults after they have been returned to adequate nutrition?

CHAPTER SUMMARY

Digestion is a process in which materials consumed as food are broken down as far as possible into soluble substances that the body can absorb. Carbohydrates are in most cases broken down into simple sugars. Fats and oils are split into fatty acids and glycerol. Proteins are broken up into individual amino acids. Material that the body cannot absorb constitutes fiber. There are certain levels of fiber and of micronutrients (vitamins and minerals) needed for good health. Good nutrition reduces such chronic conditions as high blood pressure and lowers the risks for cardiovascular disease and certain cancers.

Malnutrition can take many forms. Inadequate caloric intake can result from crop failures, from poverty, or from eating disorders. Micronutrient deficiencies can lead to various deficiency diseases, birth defects, or neurological damage. Malnutrition also impairs the proper functioning of the immune system and thus leads to higher rates of infectious diseases. War usually worsens nutrition among civilians. In most cases of widespread malnutrition, it is the children who suffer the most.

KEY TERMS TO KNOW

active transport (p. 225)
anorexia nervosa (p. 244)
antioxidant (p. 240)
atherosclerosis (p. 229)
ATP (p. 231)
basal metabolism
  (p. 226)
bulimia (p. 245)
carbohydrates (p. 227)
chemical digestion
  (p. 218)
cholesterol (p. 233)
coenzymes (p. 237)

complete proteins
  (p. 234)
concentration gradient
  (p. 225)
diffusion (p. 221)
electrolytes (p. 240)
emulsification (p. 219)
endocytosis (p. 225)
enzymes (p. 218)
epidemiological (p. 242)
fiber (p. 235)
high blood pressure
  (hypertension) (p. 240)

high-density lipoproteins
  (HDLs) (p. 233)
hormone (p. 219)
kilocalories (p. 226)
Krebs cycle (p. 231)
limiting amino acid
  (p. 234)
lipids (p. 229)
low-density lipoproteins
  (LDLs) (p. 233)
mechanical digestion
  (p. 219)
minerals (p. 240)

mitochondria (p. 231)
nonpolar (p. 219)
obesity (p. 227)
oxidation (p. 240)
peptides (p. 219)
polar (p. 219)
proteins (p. 234)
retrospective study
  (p. 246)
saturated fats (p. 232)
unsaturated fats (p. 232)
vitamins (pp. 237–238)

CONNECTIONS TO OTHER CHAPTERS

Chapter 3 Some genetic differences exist in the body's ability to digest certain substances (as in phenylketonuria) or to use nutrients after their absorption (as in diabetes).

Chapter 5 Nutritional requirements may differ among human populations for various inherited and environmental reasons.

Chapter 6 Unchecked population growth puts more people at risk for malnutrition.

Chapter 9 Certain cancers have frequencies that vary according to diet: High-fat diets promote certain cancers, while high-fiber diets lower many cancer risks.

Chapter 12 Poor nutritional status impairs the immune system.

Chapter 15 Crop improvements may help alleviate starvation in many populations.

Chapter 16 Biodiversity is threatened by the need to clear more land for farming to feed the world's population. Developing and conserving better and more varied crop plants will feed more people and support greater biodiversity at the same time.

CHAPTER 9

Cancer and Cancer Therapy

C ancer is now the second leading cause of death in most industrialized countries, second only to heart disease. In the Netherlands and some other countries, cancer ranks first. Cancer is also one of the most dreaded illnesses, and people who learn that they have cancer often suffer additionally from fear of the disease.

There are many forms of cancer, so that it is really more accurate to speak of cancers in the plural. Cancers can occur in all multicellular organisms, plants as well as animals. **Cancers** result from cell division that is out of control; the cancerous cells no longer respond to the signals that normally limit the frequency and timing of cell division. These signals keep normal cells functioning in an *integrated* manner, a necessity in all complex organisms where cells are organized into specialized tissues. In many ways we can compare the activity of normal cells in tissues to the behavior of animals in social groups (Chapter 7): The behavior of one cell (or organism) influences the behavior of others. Cancer cells do not behave in this integrated, social way. Since cancers result from uncontrolled, "antisocial" cell division, we must first describe the growth of normal cells and how their growth and behavior is regulated before we can truly understand cancer. We will also consider how we can most effectively reduce our risks for various cancers.

A. Multicellular Organisms Are Organized Groups of Cells and Tissues

Most organisms larger than a certain microscopic size are subdivided into compartments called **cells.** The first organisms consisted of only single cells, carrying out all life functions in that single compartment, with little or no spatial separation among their several functions. Bacteria continue to live very successfully as single-celled organisms. So why should multicellular life forms have evolved at all? Multicellularity evolved because it offers living things several advantages over unicellularity; organisms representing many intermediate stages in this evolution are still living today. Cancer is one of the hazards of multicellularity, for it occurs only in multicellular organisms and results from mutations which affect the mechanisms that integrate cellular processes in those organisms.

Compartmentalization

Compartmentalization into cells gives multicellular organisms a way to become much bigger than they would be able to be as single cells. There are physical restrictions imposed on living things by the ratio of their surface area to their volume. As a cell enlarges, its volume grows faster than its surface area (Chapter 14). The requirements for energy and the production of wastes both increase in proportion to the volume of a cell, while the ability to absorb nutrients and to release wastes vary with the surface area. Compartmentalization of an interior volume into cells divides the organismal volume into smaller, discrete volumes. Compartmentalization thus increases the effective surface area and maintains an efficient ratio of surface area to volume.

During the course of evolution, several kinds of unicellular organisms began to grow as collections or communities. Some of these aggregates were actually collections of separate cells that each functioned more or less independently as if each were a separate living organism. Some present-day species can change back and forth between unicellular and aggregate forms. The *cellular slime molds,* such as *Dictyostelium discoideum* (Fig. 9.1), are independent mobile unicells as long as conditions are favorable for feeding. When conditions are no longer favorable, the individual cells begin to creep toward one another, forming a large multicellular aggregate or colony.

Specialization

As they grew in communities, not every single-celled individual needed to perform every function. As a result, it was possible for some cells to

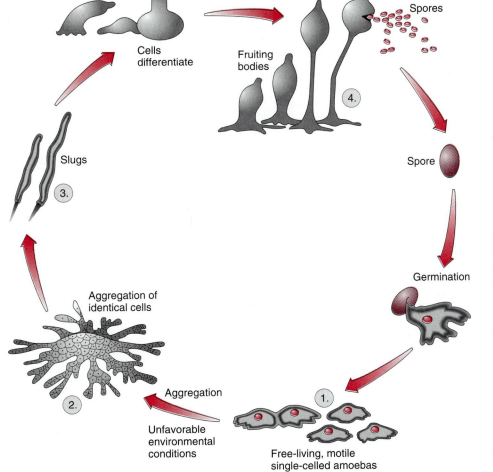

Cells differentiate

Fruiting bodies

Spores

4.

Slugs

3.

Spore

Aggregation of identical cells

Germination

2.

Aggregation

1.

Unfavorable environmental conditions

Free-living, motile single-celled amoebas

FIGURE 9.1

The life cycle of the cellular slime mold *Dictyostelium discoideum.* Cellular slime molds are generally single-celled organisms that can move about and divide (1). When environmental conditions become unfavorable, the separate cells can aggregate (2), forming a slug (3) in which all of the cells are still functionally identical. Some of the cells of the slug then become different from other cells, forming a fruiting body and spores (4). Each spore then gives rise to a new motile cell.

abandon certain functions to their neighbors, a change that resulted in specialization. In sponges, cells are specialized but are not organized into tissues (Fig. 9.2). In different species of the animal phylum Cnidaria, also called *coelenterates*, cells have formed **tissues.** A tissue consists of similar cells and their products located together (*structurally integrated*) and functioning together (*functionally integrated*). The inner and outer cell layers of the Cnidaria are separate tissues with different functions. Although the cells in the two layers are different, each cell is still changeable, so that the Cnidaria are able to regenerate an entire organism from a small piece.

The cellular flexibility of Cnidaria (and other organisms capable of regeneration from parts) stands in marked contrast to the situation in more complex multicellular organisms. In these complex organisms, the fate of a cell becomes more and more restricted as it divides, through a process called **differentiation.** Initially, a cell is capable of performing a variety of functions, but as the cell differentiates (literally, "becomes different"), it progressively loses some of its abilities and becomes specialized at doing only a few things very well (see Fig. 9.8, p. 258). Some cells, such as human muscle or nerve cells, lose so many important abilities during differentiation that they may no longer be capable of further cell division. Other cells, such as those of human bone marrow, retain throughout the individual's lifetime a good deal of the flexibility characteristic of cells in developing organisms. How cells "know" what type of tissue to become, either during embryonic development or later in the adult organism, has long been one of the major questions in biology. Much of what we currently know about normal cell division and differentiation has been aided by the comparison of normal cells and cancer cells.

Cooperation and Homeostasis

Organization of cells into tissues allows multi-celled organisms to specialize their functions while maintaining an efficient ratio of surface area to volume. In order for specialization to be beneficial, the behavior of one type of cell must be integrated with the behavior of other cells. Tissues are further integrated into organs and organ systems, in which two or more types of tissues coordinate to carry

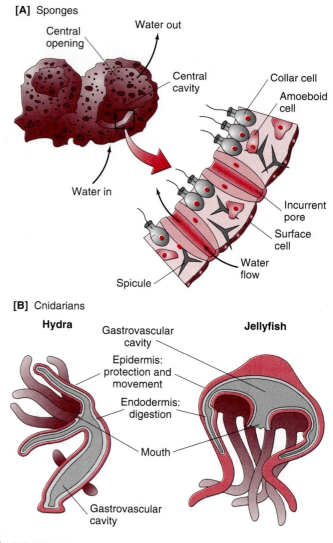

FIGURE 9.2
Specialized cells with and without specialized tissues.
[A] *Sponges* are comprised of specialized cells that are not organized as tissues; each cell acts independently to digest small food particles. [B] *Cnidarians* are composed of cells organized into *tissue* layers organized to carry out specialized functions.

out more complex functions, such as reproduction (Chapter 6), digestion (Chapter 8), circulation (Chapter 14), or communication (Chapters 10 and 12). A multicellular organism can thus be considered a complex ecosystem; a human organism, for example, is an ecosystem composed of some 10 trillion individual cells. In the course of evolution, specialization based on cooperation worked well and was favored by natural selection. Now, all multicellular organisms, both plant and animal, exhibit this basic plan.

The proper functioning of the whole organism depends on the continued integration and cooperation of all of the cells. When this integration is functioning properly, that is, when the ecosystem of cells is stable, we may consider that organism to be in a state of health. According to French physiologist Claude Bernard (1813–1878), cells are responsible for maintaining a "milieu intérieur" within each cell and within the body as a whole. Good health is defined as the maintenance of more or less constant conditions within this milieu intérieur, a process that Bernard named **homeostasis.** This does not mean the absence of change within the organism. Just the opposite is true: Molecules and cells are constantly changing, but they change around a balance point, and homeostasis is the ability to return to that balance point. Disruption of homeostasis produces illness (see also Chapter 12). As we will see, cancer is a disruption of the cellular ecosystem in which some cells no longer respond to the social signals of other cells and begin to divide as if they were not part of the ecosystem.

THOUGHT QUESTIONS

1. How does cell volume influence the concentration of materials?
2. What would an organism be like if all of its cells were the same?
3. Cold-blooded animals do not maintain a constant internal temperature but allow their internal temperature to change with the external temperature. Are these animals in homeostasis? Why or why not? (Many of these animals can regulate their temperature behaviorally by moving to different locations.)

B. Cell Division Is Closely Regulated in Normal Cells

The proper functioning of multicellular organisms depends on the regulation and integration of the processes of all of their cells, particularly the process of cell division.

Cell Division

Normal cells grow only a small fraction of the time. They continually make new proteins and other cellular chemicals to replace ones that have been used or damaged, but they do not necessarily increase in size. When cells do begin to grow larger, they soon reach the size at which surface-volume relationships make them inefficient. Instead of becoming increasingly inefficient, the cells divide. When we talk about how fast cells grow, we really mean how frequently they divide, not how fast they enlarge.

Eucaryotic cell division is a complex process. Different cellular constituents are synthesized at different times. The entire process, which recurs whenever a cell divides, is called the **cell cycle** (Fig. 9.3). The cell cycle includes a synthesis phase, or *S phase*, marked by DNA synthesis and replication of both DNA strands (Chapter 3). The phases before and after DNA synthesis are known as the

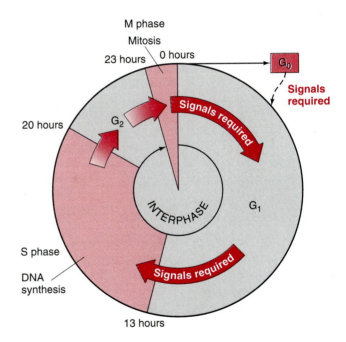

FIGURE 9.3

The cell cycle. Cells spend most of their time in the G_0 stage. They must receive signals to enter the cell cycle and they always enter the cycle at the G_1 phase. More signals are required for cells to progress to the S phase in which they synthesize new DNA. If they enter the S phase, they are committed to complete the cycle and divide, undergoing mitosis in the M phase. Hours shown are the approximate length of time for each phase in a cell with a 24-hour cell cycle, typical of many eucaryotic cells.

first and second *gap phases, G_1* and *G_2*, with the division of the cell (mitosis) occurring in the *M phase*. The decision point is the move from the G_1 to the S phase; if a cell enters the S phase, it becomes committed and must go on to mitosis. There are proteins in the cell nucleus, called *cyclins*, whose rising and falling concentrations coordinate the cycle. The concentration of cyclin proteins in each cell nucleus is affected by conditions in the neighboring cells.

Each time a cell divides, two cells result, never three or more. Since the number of cells doubles, the length of time between successive cell divisions is called the **doubling time.** A cell spends most of its time in a *resting stage*, or *G_0*, between cell divisions, and it does not reenter the cell cycle to divide again unless it is signaled to do so. The duration of the cell cycle (G_1 through M) is fairly constant within a species, but G_0, and thus the doubling time, vary greatly. For single-celled organisms, the exact length of time in the resting stage, and thus the length of the doubling time, depends on the availability of nutrients.

The doubling time of cells within a multicellular organism varies from one tissue to another and also varies with the developmental stage of the organism. When an individual animal or plant is developing, the rate of increase in the number of cells can be very rapid. In most species, cell division slows down for most types of cells once adulthood is reached. For animal nerve cells and some other cell types, cell division stops altogether.

Regulation of Cell Division

Cell division is a tightly controlled process in all types of organisms, both single-celled and multicellular. As an example of how well this process is controlled, consider the human liver. A normal liver grows until it reaches a certain size, and then it stops growing. If a piece of liver is removed, the liver cells divide to replace what's missing until the liver reaches its normal size again; then they stop. Body tissues have some way of determining how big they should be, and they stop growing when they reach that point. We will soon see that the major difference between normal tissues and cancerous tissues is that cancers grow without any such limits.

In multicellular organisms, the size of most cells is restricted not only by surface area but also by the cell being confined to a space of a certain size within a tissue. In adult organisms, cells do not divide unless a previous cell has died or been damaged, opening a space for a new cell. Many types of cells, including mature blood cells, are no longer capable of dividing at all. When old or damaged blood cells are removed, they are replaced by cells exiting from the bone marrow, the hollow interior spaces in bones. The cells that leave the marrow have had their growth suspended by contact with their neighbors in the bone marrow. Once these cells leave, other cells still in the bone marrow divide and produce new cells to take their places.

Contact inhibition and anchorage dependence. The suppression of cell division in normal cells as a result of contact with neighboring cells is called **contact inhibition.** Contact inhibition can most easily be demonstrated by growing cells in artificial tissue cultures (Fig. 9.4). Normal cells from many types of tissue have an additional requirement: **anchorage dependence.** Normal cells will only divide when they are attached to a surface. In multicelled organisms this surface is formed by complex organic molecules outside the cells called the *extracellular matrix*. Changes in the matrix or in the cells' ability to adhere to the matrix prevent normal cells from dividing.

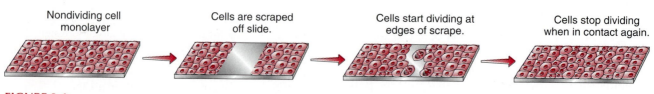

| Nondividing cell monolayer | Cells are scraped off slide. | Cells start dividing at edges of scrape. | Cells stop dividing when in contact again. |

FIGURE 9.4
Contact inhibition stops cell division once cells are in contact with each other.

Growth factors.   We have said that cell division is a tightly controlled process, but *how* is this process controlled? In general, cells receive signals of various kinds from their external environment and do not divide unless they receive a signal or signals that send them out of the resting phase and into the G_1 phase of the cell cycle. These signals are usually small molecules, called *cytokines,* secreted by other cells. Signaling by means of cytokines is one important way in which cells communicate with one another. *Cytokine* is a general term for the molecules that regulate many important body functions, including reproduction and defense against diseases. The cytokines that regulate entry into the cell cycle are called **growth factors.**

External receptors and internal second messengers.   External signals are received by means of *specific receptors,* molecules that extend through the membrane of the cell (Fig. 9.5). The term *specific* means that a given receptor can only bind to one particular type of molecule. Binding of a growth factor to its receptor on the exterior of the membrane changes the portion of the receptor molecule that is on the interior side of the membrane. The interior part of the receptor is in some cases an enzyme. In other cases, a change in the shape of the receptor's interior portion activates another molecule. In either case, *information* has been transmitted across the cellular membrane without the signal molecule passing through the membrane. The changes are then passed along through the cytoplasm to the nucleus of the cell by a network of molecules called **second messengers** (Fig. 9.5). In response to second messengers, the concentrations of cyclins change. When the concentration of cyclin D is high, the cell enters the S phase, committing it to divide.

The growth of cells within tissues is also regulated by anchoring proteins that keep the cells attached to the extracellular matrix; even in the presence of growth factors, these cells will not divide unless they are attached to the matrix. The response of a cell to external signals thus depends on the presence and normal functioning of the signal molecules, the presence of receptors for these signals, the presence of second messengers and of nuclear proteins, and the information from anchoring proteins.

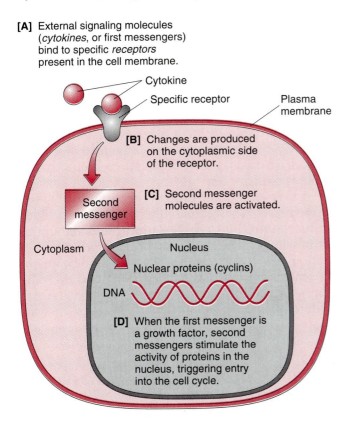

[A] External signaling molecules (*cytokines,* or first messengers) bind to specific *receptors* present in the cell membrane.

Cytokine

Specific receptor

Plasma membrane

[B] Changes are produced on the cytoplasmic side of the receptor.

[C] Second messenger molecules are activated.

Second messenger

Cytoplasm

Nucleus

Nuclear proteins (cyclins)

DNA

[D] When the first messenger is a growth factor, second messengers stimulate the activity of proteins in the nucleus, triggering entry into the cell cycle.

FIGURE 9.5
How information is carried from outside the cell to the cytoplasm and nucleus inside the cell.

Regulation of Gene Expression

Cell division, like other cellular processes, depends on the right proteins being present at the right time and in the right amounts. Proteins are encoded by genes, but not all genes are used by a cell at any particular time. The process of using the DNA of the genes to make RNA and then protein is called **gene expression,** a complex process with two major steps called transcription (using DNA to make RNA) and translation (using RNA to make protein), as we have seen in Chapter 3. Gene expression is regulated at many steps to control how much, if any, of a protein will be produced (Fig. 9.6).

Regulation of transcription.   The process of transcription begins when a molecule called RNA polymerase binds to a special DNA sequence known as a *promoter* sequence (Fig. 9.7A). A promoter sequence serves to indicate the place on the

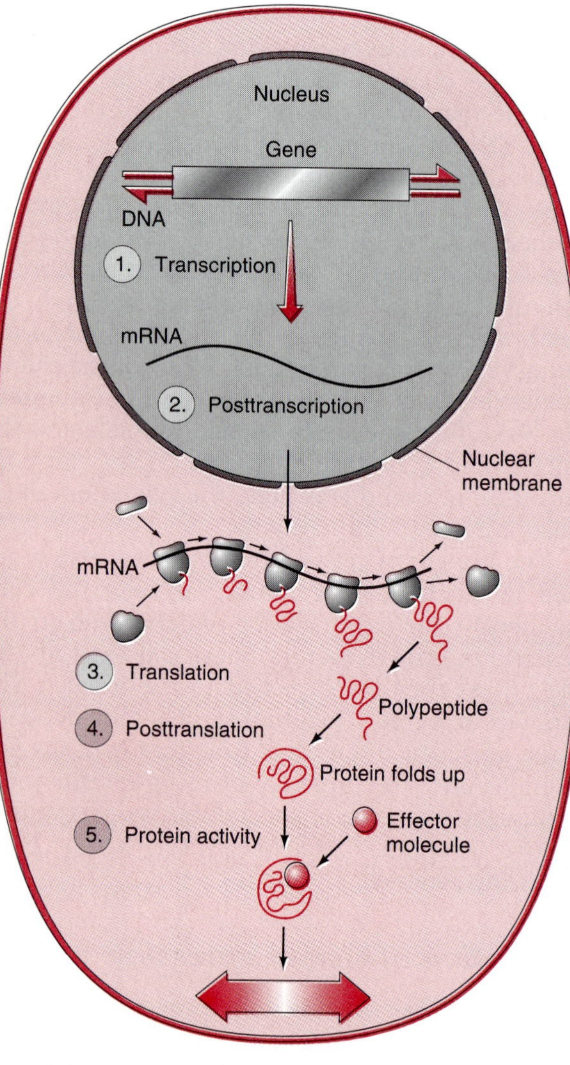

FIGURE 9.6
Five levels of regulation of gene expression. Whether or not a DNA sequence will be expressed as a functional cellular protein is regulated at all levels of the process. (1) A gene may or may not be transcribed into mRNA. (2) Some mRNAs need to be chemically modified before they leave the nucleus; if they are not, they will not be translated. (3) The cell can change the rate of translation from mRNA to protein. (4) The amino acid chain produced from mRNA is usually not a functional protein; this primary translation product may need to be modified by removing some amino acids or by adding chemical groups such as carbohydrates, processes catalyzed by other enzymes. (5) The activity of the protein itself can be altered by the binding of effector molecules.

DNA where transcription of a gene should begin. Each gene region has its own promoter. Promoter sequences can be called strong or weak, according to how strongly they bind RNA polymerase. Proteins that are needed only in small amounts are represented by genes that are present only as single copies and are controlled by weak promoters. Proteins needed in very large amounts are often represented by multiple copies of the same gene, and they may also be controlled by a strong promoter. The stronger the promoter, the faster the rate of transcription, and thus the greater the amount of that messenger RNA (mRNA) produced.

Control of cell division by regulation of gene expression. The transcription of some cell division genes is turned "on" by the binding of RNA polymerase to their promoters after receipt of second messengers resulting from cytokine signals. On the DNA near the promoter region are regulatory gene sequences that can either enhance or repress transcription (Fig. 9.7B,C). If enhancers bind, more mRNA molecules are transcribed from that gene. If repressors bind, RNA polymerase is blocked and transcription is halted. For example, tumor repressors (protein products of tumor repressor genes) slow or prevent cell division by inhibiting the transcription of the genes for some proteins needed in the nucleus for the cell to enter the G_1 or S phases of the cell cycle. If such a repressor is inhibited (Fig. 9.7D), transcription is once again allowed and the cell divides, possibly dividing repeatedly until a tumor results.

Expression of the proteins needed for cell division can also be regulated at several steps after transcription (Fig. 9.6). The timing and amount of expression of other genes involved in cell differentiation (see below) can also be regulated by any of the mechanisms shown in Figs. 9.6 and 9.7. The genes that control and coordinate cell division and differentiation are not all known. Many of the ones that are known were identified by research on cancer cells or on cells in developing organisms.

Cellular Differentiation and Tissue Formation

A fertilized egg (*zygote*) is a single cell whose cellular descendents are capable of forming all the different cell types within the body. The long list of

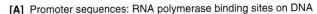

[A] Promoter sequences: RNA polymerase binding sites on DNA

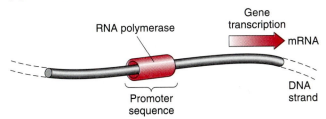

[B] Enhancers: Increase binding of RNA polymerase to promoter

[C] Repressors: Block binding of RNA polymerase

[D] Repressor inhibition: Allows binding of RNA polymerase

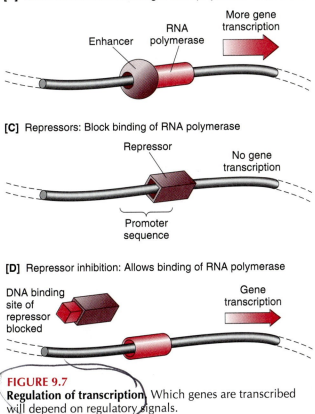

FIGURE 9.7
Regulation of transcription. Which genes are transcribed will depend on regulatory signals.

possibilities includes skin cells, muscle cells, glandular cells, bone cells, liver, and so forth (Fig. 9.8). Such a list constitutes the **potentiality** of a cell or group of cells. Within a developing multicellular organism, cells that are dividing also differentiate. **Differentiation** takes place in steps; at each successive cell division and differentiation, the number of possible future identities for that cell is narrowed until the potentiality narrows to include only a single cell type. Once a cell has differentiated as a muscle cell, for example, all of its progeny cells are committed to being muscle cells.

Just as cell division is tightly regulated by control of gene expression, so is differentiation. Much of what we know about the processes of cell differentiation has come from embryology, the study of the development of an organism from a zygote. Studies of normal differentiation have taught us much about the abnormal conditions that exist in cancer.

Potentiality and determination.   The zygote has maximum potentiality because it gives rise to all cell types and may thus be called *totipotent*. The potentiality of cells has been investigated by transplanting cells from the embryos of experimental animals. Up until the eight-cell stage, each of the cells in a mammalian embryo is totipotent and could develop into a complete organism. If a group of cells is removed from the ectodermal layer of the embryo at a stage called the *gastrula* (Fig. 9.8A) and transplanted elsewhere on the same embryo, it can form a wide variety of tissue types, but only those types that are ectodermal. Such cells are said to be *pluripotent:* Their potentiality is still quite broad, but not as broad as that of the zygote. However, the potentiality of cells transplanted at a later time is narrowed to include only certain kinds of epidermal tissues (Fig. 9.8B). Finally, at a still later stage, the fate of these cells is completely **determined;** for example, they will form only eye lens tissue regardless of where they are transplanted (Fig. 9.8B).

The totipotentiality of cell genomes.   We have seen that a cell becomes differentiated when its potentiality becomes restricted and that the cell somehow seems to "know" what these restrictions are. But where in the cell is this information about restricted potentialities located? One possible hypothesis is that the information is carried in the cytoplasm; another is that the information resides in the nucleus. In order to test these hypotheses, a British researcher named J. B. Gurdon exposed some frog eggs to ultraviolet radiation. Once fertilized, frog eggs are able to grow into a whole tadpole and so are totipotent. Because ultraviolet radiation is absorbed by DNA, a sufficient dose of ultraviolet can be used to destroy the egg nucleus without damaging its cytoplasm. Gurdon then carefully inserted into the egg the nucleus of a

[A] Cell division produces three cell layers.

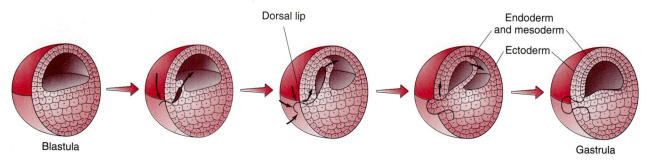

[B] Each cell layer is the source of specific types of differentiated cells.

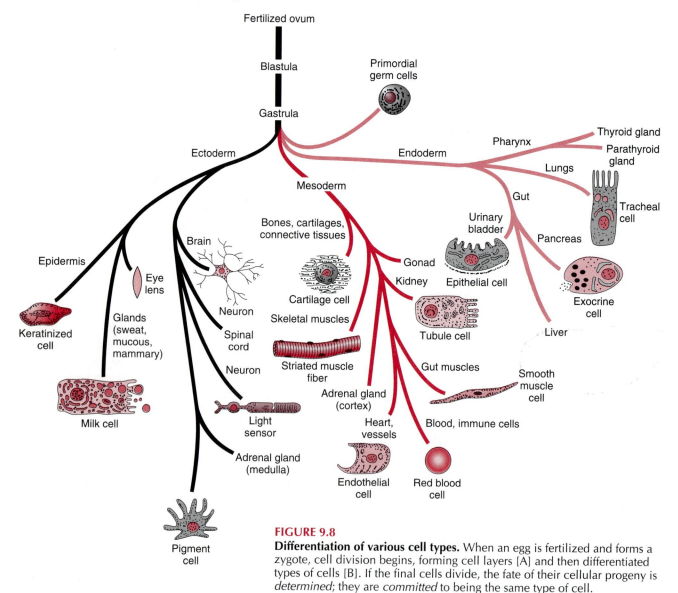

FIGURE 9.8

Differentiation of various cell types. When an egg is fertilized and forms a zygote, cell division begins, forming cell layers [A] and then differentiated types of cells [B]. If the final cells divide, the fate of their cellular progeny is *determined*; they are *committed* to being the same type of cell.

differentiated cell type, such as a skin cell (Fig. 9.9). The resulting cell thus had cytoplasm from a totipotent cell but a nucleus from a differentiated cell. Gurdon was able to show that this cell produced an entire tadpole. The various types of cells of the body thus do not differ because they contain different genes, but instead because they express different genes from their total genome.

Control of tissue differentiation by regulated gene expression.

Tissue differentiation, like cell division, is coordinated by the regulation of gene expression. To a great extent, the process of differentiation is a process of controlling what proteins (including enzymes) are made by a particular cell. Some proteins, like the enzymes needed in cell metabolism, are produced by cells from all parts of the body. Other proteins are produced in most tissues, but in amounts that vary by age and by cell type. Still other proteins are only produced in certain tissues. For example, insulin is only produced in the beta cells of your pancreas, although the gene for insulin is in every cell in your body. Although the DNA in the genes of all cells continues to carry the instructions for building all of the different cell types that make up that organism, each cell is somehow restricted to expressing only certain of its genes. Gene expression also varies during the lifetime of the individual; some proteins are needed only by the developing organism and are not made in later stages of development.

Differentiation is heritable, not by the offspring of the organism but by the cells descended from a differentiated cell. Lung cells in tissue culture remain lung cells even after several cell divisions. While the nucleus contains the genetic information, it is the cytoplasmic and nuclear proteins that regulate the expression of the information, determining the future identity of the cell. These proteins permanently change the expression of genes in the subsequent offspring of differentiated cells.

Maternal effect genes.

One example of the regulation of gene expression by factors in the cytoplasm is a group of regulatory proteins that are in the egg cytoplasm even before its fertilization to produce a zygote. Much of what we know about these proteins and the genes that code for them was studied first in fruit flies of the genus

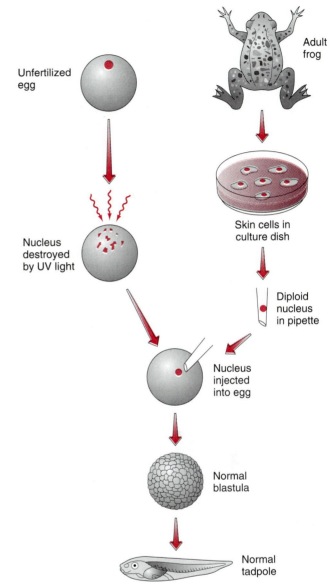

FIGURE 9.9

Differentiated cells contain all the DNA needed to direct the development of a complete organism. The experiment that demonstrated this was as follows: The nucleus of a frog egg was destroyed by ultraviolet irradiation and was then replaced by the nucleus from the fully differentiated skin cell of another frog. The egg with its transplanted nucleus was allowed to grow and it developed into a normal tadpole.

Drosophila. The earliest acting of the developmental control genes that function to establish the basic pattern of the body are the *egg-polarity genes.* As the egg develops in the female fly, meiosis produces a haploid number of chromosomes (Chapter 3). These chromosomes come from the female fly and are called maternal chromosomes. A few genes are transcribed from the maternal chromosomes in the unfertilized egg. The protein products of these genes are present in the cytoplasm of the egg, in unequal concentrations in different locations, before the egg is fertilized (Fig. 9.10). Even before the egg is fertilized to become a zygote, the products of the maternal genes have already defined the "head" and "tail" ends of the embryo. Genes expressed in this way are called **maternal effect genes.**

Tissue induction by organizers.   At each cell division, a developing cell receives cytokine signals that determine whether or not it will differentiate and what type of cell it will form. The region of an embryo which produces cytokines that cause cells to differentiate is called the **organizer.** The effect is local, meaning that most of the influences on cell differentiation come from neighboring cells. The concept of an organizer was first developed by the

work of Hans Spemann and Hilde Mangold, work for which Spemann later won the Nobel Prize. In their early experiments, Spemann and Mangold transplanted various parts of frog gastrula into different positions on other frog gastrulas, as shown in Fig. 9.11. Through such experiments, they were able to show that a part of the embryo called the *notochord* acts as an organizer that *induces* (causes) the overlying tissue to roll up and form a neural tube, the earliest part of the nervous system to be formed.

Organizers do not actually contribute cells to the tissues that they stimulate. Spemann demonstrated this by transplanting the notochord tissue of a chick into the embryo of a duck (Fig. 9.12). Because these species have different chromosome numbers, Spemann was able to show that the resultant neural tube was made of duck cells, even though the transplanted tissue was made of chick cells. The transplanted chick cells, in other words, formed no part of the neural tube that they induced, thus falsifying the hypothesis that the neural tube cells were derived from notochord

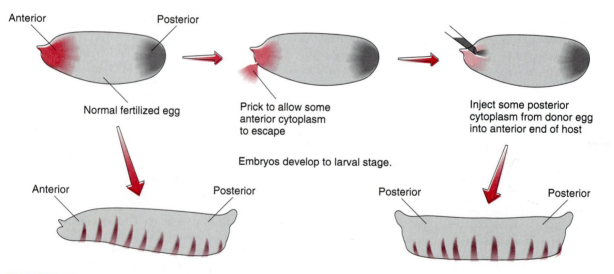

FIGURE 9.10
What determines the tail end of a developing embryo? The tail (posterior) end of a *Drosophila* embryo is determined by the protein product of the egg-polarity gene, a protein that is distributed unevenly in the cytoplasm. Replacement of some anterior cytoplasm by posterior cytoplasm from another egg will induce the beginning of a second tail end. Although the DNA donated by the sperm also contains the egg polarity gene, the sperm do not contain any of its protein product. Therefore it is the preformed proteins already present in the cytoplasm as products of the maternal DNA that determine the tail end of the embryo.

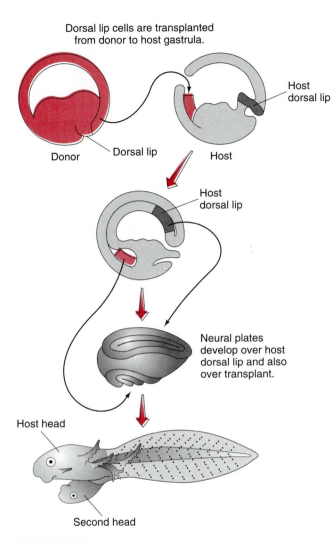

Dorsal lip cells are transplanted from donor to host gastrula.

Donor

Dorsal lip

Host

Host dorsal lip

Host dorsal lip

Neural plates develop over host dorsal lip and also over transplant.

Host head

Second head

FIGURE 9.11

Spemann and Mangold's experiments on tissue formation. Using cells from the gastrula stage of frog embryos, Spemann and Mangold showed that neural plates form from the ectoderm at the location of both the host and the transplanted dorsal lip cells. Both neural plates go on to form neural tubes, then brains and heads.

cells. The duck cells had simply responded to some signal from the organizer region of the chick. In later experiments, a chemical extract of the notochord was combined with egg white; this material was able to induce a neural tube in precisely the way that the notochord had, suggesting that the signal from the organizer was a chemical substance rather than a group of cells.

Hormonal control of differentiation.   Other proteins regulate transcription by functioning as hormone receptors rather than as direct DNA binding proteins. Unlike many receptors, which are cell surface proteins, hormone receptors are located in the cytoplasm. The nonpolar chemical structure (Chapter 8) of some hormones allows them to pass through the cell membrane and bind to their receptors in the cytoplasm. The hormone-receptor complex then binds to specific segments of DNA, thus regulating transcription. The hormone testosterone, for example, normally triggers the differentiation of male reproductive tissues. Some of this differentiation occurs during puberty, but many of the effects occur during fetal development. In humans and several other mammalian species, mutations are known in the gene for the testosterone receptor. Individuals carrying such a mutation do not have functional testosterone receptors and cannot respond to testosterone and thus develop as females, although their chromosomes characterize them as males (see Chapter 3).

Gene regulation in adult animals.   Gene regulation is thus important for development. Is it important in adult animals? The answer is yes, in part because very few types of cells are permanent (Table 9.1). Cell division and differentiation must take place throughout the lifetime of the organism. These processes are normally tightly coordinated in adult organisms so that cells lost from specific tissues are replaced by the correct number and type of cells.

Cancer is the result of the loss of these controls on cell division and differentiation. Cancer cells are often less differentiated than the normal cells around them. As we have seen, there are many levels of regulation, so production of cancer is a multistepped process. Cancer can, however, arise from only a single cell that has stopped responding to normal controls.

Cell migration.   In the development of normal tissues, cells need not arise in their end location; tissue formation relies on many in-migrating cells that travel through the organism until they find their "proper" location, where they adhere and join the tissue. They are generally partially

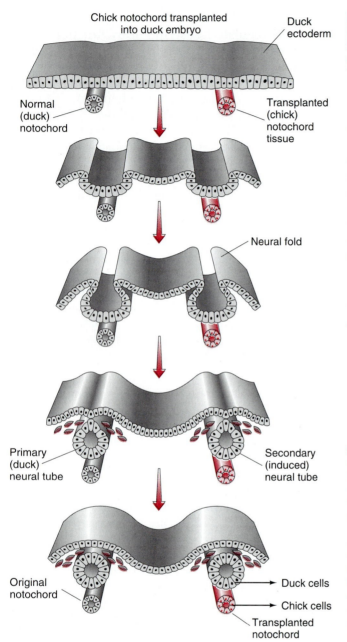

Chick notochord transplanted into duck embryo

Duck ectoderm

Normal (duck) notochord

Transplanted (chick) notochord tissue

Neural fold

Primary (duck) neural tube

Secondary (induced) neural tube

Original notochord

Duck cells

Chick cells

Transplanted notochord

TABLE 9.1 AVERAGE LIFE SPANS OF HUMAN DIFFERENTIATED CELL TYPES	
CELL TYPE	**LIFE SPAN (DAYS)**
Intestinal lining	1.3
Stomach lining	2.9
Tongue surface	3.5
Cervix	5.7
Stomach mucus	6.4
Cornea	7
Epidermis—abdomen	7
Epidermis—cheek	10
Lung alveolus	21
Lung bronchus	167
Kidney	170
Bladder lining	333
Liver	450
Adrenal cortex	750
Brain nerve	27,375+ (75+ years)

FIGURE 9.12

Chick-duck transplants to distinguish between cell donation and cell induction. In a duck embryo, a neural tube develops from the ectoderm above a group of cells called the notochord. Do the notochord cells become the neural tube or do they induce neural tube formation in other cells? This question was answered by transplanting chick notochord into duck embryos. Ducks and chicks are closely enough related that a transplant of embryo cells from one to the embryo of the other will grow. The chromosome number in the two species is different, however, so that after two neural tubes had developed, the cells could be examined to distinguish which species they had originated from. It was discovered that they came from the recipient species and were induced by the donor cells.

differentiated at the time of their migration and become fully differentiated when exposed to the growth factors in their new microenvironment. Such molecular "addresses" can be in the form of membrane receptors that bind specifically to molecules expressed only on certain types of tissues. Abnormalities in cellular adhesion and cell migra-

tion are pertinent to the spread of cancer to other tissues.

Limits to Cell Division

Normal cells that have already differentiated into various cell types seem to have a limit to the number of additional times they can divide. In tissue culture, when cells have formed a monolayer, they stop dividing. At this point they can be removed from their substrate, diluted to a lower density (fewer cells per milliliter), and put into fresh flasks with fresh nutrients. The cells will adhere to the new flask and, since there are no longer enough cells to cover the surface, they begin to divide and will continue to divide until they again have formed a complete monolayer (Fig 9.4). This process cannot be repeated indefinitely; after a certain number of divisions, the cells will die rather than divide, even when optimal conditions exist. The concept that there are limits to the number of times a cell can divide is called *cellular mortality*.

The exact number of divisions possible in culture dishes depends on the species and the cell type. This limit was first discovered by Leonard Hayflick in 1965, and it is sometimes called the Hayflick limit. In general, the maximum **doubling number,** or Hayflick limit, is proportional to the *life span*—the maximum possible length of life—for that species. Life span is different from life expectancy, which is the length of time an individual in a given environment is expected to live. In humans, for example, the life span is around 110 years, while the life expectancy for white males in the United States is 72 years. Normal human fibroblasts will grow for approximately 50 population doublings and then divide no more, after which they age and die. The maximum number of doublings for embryonic mouse cells in culture is 20 (compared to a mouse's 3.5-year life span), while for tortoises it is 90 to 125 doublings (compared to a 150-year life span). If cells are frozen and stored for a prolonged period, then thawed and returned to tissue culture, the maximum doubling number is not shortened or lengthened. There seems to be a biological clock of some sort that keeps track of the number of cell divisions.

All noncancerous cells taken from living organisms have a limit to the maximum number of doublings that they will undergo in tissue culture; it is thought that similar limits exist in the living body. The number of times that cells can divide seems constant for each species, but the rate of cell division may vary. Cells that divide at a faster rate, including basal-layer skin cells and blood-forming cells in the bone marrow, die sooner and must be replaced sooner than more slowly dividing cells (Table 9.1).

THOUGHT QUESTIONS

1. Why is it an adaptive advantage to an organism to have certain proteins like insulin produced by just one type of cell rather than produced by all cells throughout the organism?
2. How does a growth factor influence events inside a cell? A hormone? What are the similarities and differences in their mechanisms?
3. Does a cell's DNA determine what type of cell it will become? What other factors, if any, are involved?
4. If the doubling time of a cell type in tissue culture is 24 hours, how many cells will there be from a starting population of 10,000 cells in 1 week (see also Chapter 6)?
5. If a cell occupies an area 10 micrometers (μm) by 10 μm (100 μm^2) and has a doubling time of 24 hours, how long will it take 10,000 cells to reach contact inhibition in a tissue culture dish that is 5 centimeters (cm) wide and 8 cm long? (1 μm = 10^{-6} m; 1 cm = 10^{-2} m.)

C. Cancer Results When Cell Growth Is Uncontrolled

Now that we have seen how cell division and differentiation are coordinated in normal cells, we can examine these processes in cancer cells. The medical term for cancer is *neoplasm*, meaning "new growth." Almost all tissues contain some newly dividing cells, so cancer is more than new growth; it is uncontrolled growth. Mechanisms for growth control probably evolved along with the evolution of multicellular organisms, because growth control is necessary to integrate cellular activity in organisms that are cooperative multicellular aggregates. Cancer is the escape of some cells from these control signals. All multicelled plants and animals can develop cancer. We will be primarily discussing human cancers, although much of what follows also applies to cancers in other species.

Cancers can arise in any tissue whose cells are dividing. Human cancers are named according to the type of cell from which the cancer is derived. A cancer that arises in epithelial tissue (sheetlike tissue or glandular tissue) is called a *carcinoma*; a cancer that arises in connective tissue is called a *sarcoma*. There are also subtypes of tumors; a *mesothelioma*, for example, is a sarcoma of the lining of the abdominal cavity. A cancer that arises among white blood cells (leukocytes) is called a *leukemia* if the cells are circulating throughout the body via the bloodstream, but it is called a *lymphoma* if it is a solid tumor in lymphoid (leukocyte-containing) tissue.

Cancer at the Cellular Level

In cancer cells, control of cell division can be lost in at least two ways: The cells may be continuously signaled to divide without entering the resting stage of the cell cycle (positive control), or a signal that would have caused normal cells to stop dividing (negative control) may have been removed. Cancer cells do not necessarily grow *faster* than normal cells. Breast cancer cells, for example, do not divide as rapidly as normal skin or blood cells, but their growth is uncontrolled. They do not exhibit the normal kind of cellular mortality (or Hayflick limit) and so are called **immortal.** In many cases, cancer cells are less differentiated than the tissues from which they arose. The process that a cell undergoes in changing from a normal cell to an unregulated, less-differentiated, immortal cell is called **transformation.** The transformed state is traceable to changes in the DNA, and therefore is passed on to all of the progeny cells. Cancer can result from the transformation of just a single normal cell.

After a cell has been transformed, it exhibits many characteristics that differ from normal cells (Table 9.2). The membrane transport systems of transformed cells carry nutrient molecules into the cell at a higher rate, and a transformed cell is therefore able to live on lower levels of nutrients in culture than can normal cells. In the body this gives them a competitive advantage over normal cells. Transformed cells also no longer show contact inhibition. In tissue culture, transformed cells do not stop as monolayers, but continue to grow, forming piles of cells. This also occurs inside or-

TABLE 9.2   CHARACTERISTICS OF NORMAL CELLS AND TRANSFORMED (CANCER) CELLS

CELLULAR BEHAVIOR	NORMAL CELLS	TRANSFORMED CELLS
Hayflick limit	Finite	Immortal
Differentiation	Present	Inhibited
Transport of nutrients across cell membrane	Slower	Faster
Nutrient requirement	Higher	Lower
Contact inhibition	Present	None
Anchorage dependence	Present	None
Adhesiveness	High	Low
Secretion of protein-degrading enzymes	Low	High
Genetic material	Stable	Unstable

ganisms, and these growing piles of cells are called **tumors** (Fig. 9.13).

Transformed cells can also grow without the need to be attached; in fact, this is the characteristic that best predicts whether a particular type of cell will form a tumor if put into an animal. Changes back and forth from attached growth to unattached growth are believed to be instrumental in the spread of some tumors to new locations, a process known as *metastasis* (see below).

The process of transformation always involves cell division; therefore, cells that are terminally differentiated, such as nerve cells in the brain and muscle cells of the heart, will never again divide and cannot become transformed. On the other hand, many types of cancers arise from the transformation of the **stem cells,** whose normal function is to divide and replace cells that are lost through routine physiological processes. These stem cells are located in areas where cells are continually being lost: the skin, gut lining, uterine cervix, and bone marrow. Since stem cells divide often, they respond more easily to cell division signals than do highly differentiated cells. When stem cells divide normally, one daughter cell remains undifferentiated as a stem cell and the other differentiates and is therefore less likely to continue dividing. After cell division in transformed stem cells, however, both daughter cells are undifferentiated stem cells with a greater potential for continued proliferation. The less differentiated (or more pluripotent) the stem

 **[A]** Normal growth

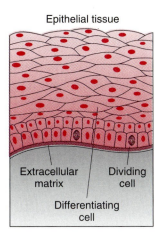

 **[B]** Abnormal growth

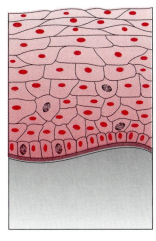

[C] Tumor

[D] Malignant tumor

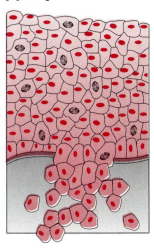

FIGURE 9.13

The growth of an epithelial tumor. [A] *Normal growth:* Dividing cells exist only in the basal layer, anchored to the extracellular matrix. Dead differentiated cells are replaced from cells in the basal layer. [B] *Abnormal growth:* Dividing transformed cells are more numerous and are not confined to the basal layer. [C] *Tumor:* Abnormal growth continues until a mass of cells, called a *tumor,* develops; cells are less differentiated. [D] *Malignant tumor:* Tumor cells break through the extracellular matrix; some cells may lose their attachment to the tumor and spread, a process called *metastasis.*

cell that has been transformed, the more aggressive the cancer. One concept of experimental therapy for stem cell cancers is to give drugs that promote cellular differentiation and thus slow down the progression of the cancer.

Cancer at the Organismal Level

As mentioned earlier, an organism can be considered an ecological system of many billions of cells. The interacting growth and differentiation signals serve as the *feedback loops* that keep the cellular ecosystem stable. Cancer is a disease in which a single cell, as a result of DNA mutations, escapes from normal controls to cell division, giving it a competitive advantage over its neighbors. After cells are transformed, progression to a tumor depends on many ecological factors. Mutated progeny cells may be killed or they may not outgrow the normal cells and thus never progress to a tumor. On the other hand, the transformed cell and its progeny may continue to divide, taking up space and nutrients required by their neighbors, passing on the mutations to each new progeny cell.

Normal cells begin to die off, not because they are killed outright by cancer cells, but because they are deprived of space and food. The declining number of normal cells results in reduction of their normal function, and the organism begins to show signs of illness as a result. The particular symptoms will depend on the type of cancer and the type of normal cells that are lost.

Transformation of cells within organs results in solid tumors within those organs. For a tumor to be visible on X-ray, the original transformed cell must divide repeatedly until there are 10^8 cells in the tumor. For a tumor to be large enough to be felt (about 1 cm in diameter), approximately 10^9 cells are needed. By the time tumors are this size they have begun to influence their environment in other ways. Growth factor genes in the transformed cells have begun to be expressed, so that the tumor cells begin to secrete *angiogenic growth factors* that induce nearby blood vessels to develop new branches that grow into the tumor. A tumor may thus contain normal cells as well as transformed cells.

A tumor is said to be *benign* if it is contained in one location and has not broken through the basement membrane to which normal cells are attached in tissues. Benign tumors, as their name suggests, often cause no health problems for the individual. Benign tumors can become large enough to interrupt the functioning of normal tissues, but their removal by surgery is generally successful since they have not intermingled with normal tissue. If the tumor cells are able to invade normal tissues, rather than just push them out of the way, then they are said to be **malignant** (Fig. 9.13D). For cells to be invasive, they must produce enzymes such as *collagenase,* an enzyme that dissolves the collagen connective tissue separating groups of cells from one another. The term *cancer* is generally reserved for malignant tumors.

Since malignant tumors are able to produce enzymes that allow them to invade other tissue, they often **metastasize.** In this process, one or more of the transformed cells lose their attachment to the other cells of the tumor, break through the basement membrane, and spread via the blood or lymphatic circulation to other areas of the body (Fig. 9.13D). In the new location, they regain attachment and continue to divide, forming a new tumor. The new tumors will be of the same type as the original tumor and thus when viewed with a microscope will appear to be different than the cells around them. Cancers that have begun to metastasize are far more serious and more resistant to treatment than those that have not, because no amount of surgery can eliminate all the cancerous cells that have spread.

Cancer at the Population Level

The incidence and mortality rates for various cancers in the United States are given in Fig. 9.14A. The study of diseases at the population level constitutes the science of *epidemiology*. The basic epidemiological data for various forms of cancer have been compiled for the United States since 1950.

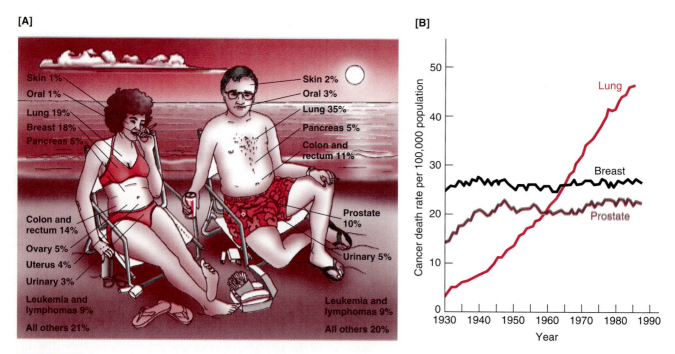

FIGURE 9.14

Deaths from cancers at various body sites (United States). [A] The pattern of cancer deaths by site differ in women and men. Also shown are several behaviors that increase cancer risks: smoking, alcohol consumption, high fat consumption, and exposure to the sun. [B] Deaths from lung cancer have increased steadily since 1930, while deaths from breast and prostate cancer have held relatively steady.

Epidemiology uses descriptive statistics to find patterns in the incidence of diseases. Those patterns indicate possible risk factors. The fact that some disease is prevalent among a certain group may mean that members of that group share some factor that is not immediately apparent. Epidemiology often cannot distinguish between alternative mechanisms; it *can* suggest hypotheses that can be further tested in other ways.

In general, and with a number of exceptions, the causes of adult cancers appear to be mainly environmental, not genetic. Evidence to support this conclusion takes the form of epidemiological data for the United States and several European countries, showing in each case a marked increase in cancer rates throughout the twentieth century. Most of this increase in cancer pertains to a single type, cancer of the lung (Fig. 9.14B). The increase coincided with advancing industrialization and other changes in the environment, but very little change in the gene pool. In Germany, for example, cancer caused only 3.3 percent of all deaths in 1900, but 20.9 percent of all deaths in 1967, over a sixfold increase. From 1950 to 1979, lung cancer rates more than doubled for U.S. males and nearly tripled for U.S. females, with higher rates of increase among nonwhites than among whites. Cancers of the pancreas and large intestine increased more slowly over the same period, while stomach and rectal cancers declined. The rapid change and irregular pattern of change both fit much better the hypothesis of environmental causes than the alternative hypothesis of a genetic cause or causes. We will present the relevant evidence in the next section.

THOUGHT QUESTIONS

1. How do stem cells differ from transformed cells? How do stem cells differ from muscle cells or blood cells?

2. What other kinds of feedback loops exist in the body in addition to those that control cell division?

3. In what ways are benign and malignant tumors the same and in what ways are they different?

4. Does an increase in the percentage of deaths due to cancer necessarily mean that cancer rates have increased? What else could explain such findings? How would you go about determining which of the possible explanations best fits the data?

D. Cancers Have Complex Causes and Multiple Risk Factors

In this section we will examine the evidence from epidemiological studies and animal studies that have suggested many possible causes of cancer. We will see in a later section how these seemingly disparate causes may be working by very similar pathways at the cellular and molecular level.

Childhood cancers and adult cancers may have different sets of causes. Some blood cell cancers (leukemias) are more frequent among children. Approximately 85 percent of childhood cancers are acute lymphocytic leukemias that arise from stem cells in the bone marrow. Although these are very aggressive, they have a good cure rate because children still have many normal cells to take over after therapy (see below).

About 85 percent of adult cancers are epithelial cancers (carcinomas), including cancers of the lungs, breast, colon, rectum, pancreas, skin, prostate, and uterus. The incidence of all these cancers (and many others) increases with age, so that cancers become more and more significant as causes of death with advancing age. Environmental or lifestyle factors are believed to be important in the incidence of most of these adult cancers. The following have been suspected, based on epidemiological evidence, either of causing at least one type of cancer or of increasing the rate at which at least some cancers occur: genes, viruses, ionizing radiation, ultraviolet light, diet, stress, mental state, weak immune systems, sexual behavior, hormones, alcohol, tobacco, and a long list of chemical substances found in the home, in the environment, or in the workplace.

Inherited Cancers

A few cancers are inherited genetically, most often as dominant traits that show up in only a small percentage of the people who have the gene. These cancers are all very rare and they typically appear at an early age. For example, *retinoblastoma* is a rare cancer of the eye, caused by a dominant gene. The gene is about 80 to 90 percent *penetrant*, meaning that it develops into cancer in about 80 to 90 percent of the people who carry the gene. (Most cancer-related genes have much lower rates of penetrance.) Another rare cancer is *xeroderma pigmentosum*, a skin cancer that results from a defect in the mechanism of DNA repair. This disease produces an extreme sensitivity to even normal exposure levels of ultraviolet light, resulting in cancer in almost all persons possessing the gene. There is also a form of colon cancer that is inherited. Very recently, the genetic defect associated with this cancer was located and identified as a defective DNA checking mechanism. When DNA is replicated, mistakes can be made in the bases put into the growing strands. Specific proteins called *checking proteins* check for mistakes, like a spell-checking program on a computer. Enzymes can then repair the mistake. When either the checking proteins or the repair proteins are defective, the chances of the mistake (mutation) being passed along to progeny cells is increased. If the mistake occurred in a growth control gene, the cell may become transformed (see below). Other cancers can be associated with this same defective checking protein; the defect is not confined to colon cells.

There are also genetic predispositions to some types of cancerous transformation. For example, early-onset (premenopausal) breast cancer is correlated with family background. The rate at which this cancer occurs is much higher among women who had at least one female relative with the disease. (The more common type of breast cancer, a late-onset disease of postmenopausal women, is discussed below; its incidence is more closely related to diet than to inheritance.)

Far more cancers appear to be environmentally caused than genetically caused, even after taking into account genetic predispositions for some cancers. The tabulation of data on a county-by-county basis permits epidemiologists to focus on cancers that may differ from place to place. The variations in cancer rates are geographically spotty; they also change with time, even in relatively stable populations. These patterns are much more suggestive of environmental than of genetic causes. Most states in the United States have coun-

ties or other small regions whose industrial activity differs markedly from other counties in the same state, a pattern that matches the spotty patterns of cancer rates. In these same states, populations either differ very little from county to county in genetic or ethnic makeup, or else there is a pattern of ethnic distribution that does not match any pattern of cancer incidence, which argues against genetic control.

Cancers Associated with Infectious Agents

Several cancers are known to be associated with viruses. In 1911, Peyton Rous showed that a tumor of connective tissues (a sarcoma) in chickens was caused by a virus that was later named *Rous sarcoma virus*. This virus can infect chickens, and infected chickens develop sarcomas. Another cancer-causing virus, feline leukemia virus, causes a blood cancer in cats. By the late 1960s and 1970s, cancer-causing viruses had been identified in several species, but not in humans. Then, in 1980, Robert Gallo found a human T cell leukemia virus, HTLV, which is associated with one form of human leukemia. (Many other human leukemias do not show evidence of HTLV or other viral infection.) Viruses appear to be associated with cancer in at least two different ways. Some viruses carry genes that, when inserted into the host DNA, cause the host cell to become transformed into a cancerous cell. These genes are called **oncogenes,** from the Greek word *onkos* meaning "mass" or "tumor" (see below). Other cancers, such as Kaposi's sarcoma, are associated with a decrease in the activity of the immune system, which may result from a viral infection such as AIDS (see Chapter 13).

The high incidence of some cancers is correlated in certain countries with the incidence of infectious disease. The high incidence of Burkitt's lymphoma in equatorial Africa is correlated with infection by Epstein Barr virus (EBV), followed by infection with malaria. EBV is just as common in the United States as it is in Africa, but the incidence of malaria is very low. This pattern suggested a mechanism for Burkitt's lymphoma that will be explored in a later section. EBV was found to carry an oncogene, but viral infection with EBV by itself does not produce cell transformation, as Rous sarcoma virus does.

Another example of a correlation between infection and cancer appears to be the high incidence of liver cancer in third world countries. A very high proportion of the people who develop liver cancer have previously had hepatitis B, a viral infection of the liver. In most people, the acute symptoms of hepatitis are resolved, but some people remain infected chronic carriers of the virus. Over 250 million people worldwide are carriers of hepatitis B virus; carriers have a 100-fold higher risk of developing liver cancer, although this may be as long as 40 years after the hepatitis.

Cancers of the reproductive organs, especially cancer of the uterine cervix, are statistically related to both male and female sexual behavior (Levy, 1985). Incidence rates for cervical cancer are higher among women who were younger at the time of their first intercourse or who have had multiple sexual partners. The rate of this cancer is particularly low among religiously cloistered women (e.g., nuns) or among women belonging to certain religious groups, including Mormon, Amish, and Jewish women. Epidemiological studies among Mormon women find a correlation between church attendance (a convenient measure of adherence to religious teachings) and lower cancer rates of the reproductive organs. However, Levy (1985) points out that the rates of these female cancers are high in those countries in which women tend to have few sexual partners and to marry as virgins, but where a tradition of *machismo* often encourages men to seek multiple sexual partners, including prostitutes. She uses this type of epidemiological evidence to argue that male promiscuity, not female promiscuity, is the important risk factor for cancer of the cervix. It has since been found that some types of the sexually transmitted human papilloma virus are involved. While papilloma viruses more often cause genital warts, they also cause cancer of the cervix, which is thus a sexually transmitted form of cancer.

In the case of each of these viruses, the epidemiological evidence has since been extended by animal and tissue culture experimentation. While the incidence of virally induced tumors is low in the United States, they account for 20 percent of all cancers worldwide.

Physical and Chemical Carcinogens

A large and growing number of external agents are known to cause cancer; such agents are called **carcinogens.** Some carcinogens are physical agents, particularly certain types of energy sources. Ultraviolet (UV) light, for example from sunlight, can cause skin cancers in susceptible people, including malignant melanoma, a cancer of the pigment cells (melanocytes). Light-skinned people are more susceptible to these cancers, which affect some 22,000 Americans annually. Epidemiological evidence shows that even exposures to natural levels of UV radiation will increase the incidence rates for skin cancers, which are much higher in the southern half of the United States than in the northern half. Ionizing radiation, such as that produced by radioactive substances, was clearly shown to be carcinogenic by studies on the Japanese survivors of the 1945 bombings of Hiroshima and Nagasaki. Marie Curie (1867–1934), a two-time Nobel Prize-winner and pioneer in the study of radioactive elements, died of a leukemia induced by her frequent handling of these elements. X-ray machines also produce ionizing radiation, so the level of exposure is carefully controlled to minimize the hazard.

Other carcinogens are chemicals. Tobacco smoke, including secondhand smoke, is statistically the largest single risk factor for cancer in the industrialized world. People who begin to smoke when they are teenagers or in college have more than a 10-fold increase in the risk for developing lung cancer compared to people who never smoked (Fig. 9.15A). Although the risk of cancer is decreased after someone stops smoking, the risk never returns to levels comparable to those for people who never smoked (Fig. 9.15B). Nonsmoking women whose husbands smoke have higher cancer rates than nonsmoking women with nonsmoking husbands. Tobacco smoke is a risk factor for heart disease and emphysema as well as cancer.

A large number of industrial chemicals have been shown to be carcinogenic, including vinyl chloride (used in the making of many plastics), formaldehyde, asbestos, nickel, arsenic, benzene, chromium, cadmium, polychlorinated biphenyls (PCBs), and many others. The aromatic amines used in dye manufacture and in the rubber

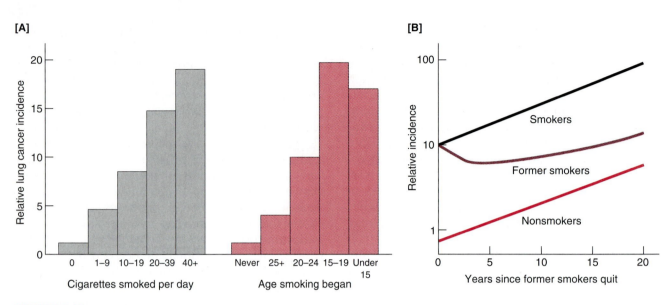

FIGURE 9.15

The effects of smoking on the incidence of cancer. [A] The incidence of lung cancer is greater the more cigarettes a person smokes and the younger they were when they started smoking. [B] Cessation of smoking decreases a person's chances for lung cancer, but never to a level as low as the chances for a person who has never smoked.

industry can cause bladder cancer; benzene can cause bone marrow cancer. Vinyl chloride causes an otherwise unusual liver cancer among plastics workers. Cadmium workers have an increased risk of prostate cancer, while asbestos workers have higher cancer risks at a number of sites, including the lungs, peritoneal lining (mesothelioma), stomach, colon, and esophagus. Smoked, cured, or pickled foods contain nitrites that can be converted into carcinogenic substances in the body.

Evidence that the above chemicals are carcinogens comes from studies in which animals are exposed to the chemicals. Evidence also comes from epidemiological studies of occupationally exposed persons, such as industrial workers who handle the material. Still other carcinogens are discovered by recognizing epidemiological clusters of persons with unusually high incidence rates for particular cancers living in the area surrounding an industrial plant or waste disposal site.

Many of these carcinogens cause mutations in DNA. Does this mean that every mutation will result in cancer? Fortunately, no. Because of the checking proteins and repair enzymes already mentioned, the vast majority of mutations are re-

paired. Many others are in regions of the DNA that have no consequences, or no consequences for the transformation of cells.

Some carcinogens exhibit a **synergistic effect,** meaning that the increased risk due to two factors is much more than the additive combination of their effects taken separately. For example, smokers experience an 11-fold increase in the risk of dying from lung cancer as compared to nonsmokers, and nonsmoking asbestos workers experience a 5-fold increase in this risk. Asbestos workers who smoke face a 53-fold increase in the mortality risk, an effect that is multiplicative and not additive.

Dietary Factors in Cancer

Evidence that dietary factors contribute to the development of cancer is best established for cancers of the digestive tract, including the colon and rectum. The evidence comes from laboratory studies of animals exposed to experimentally controlled diets, from clinical studies on human patients, and from epidemiological studies using statistics involving large populations.

Dietary fiber and fats. Diets high in fiber and low in fats are associated with a lower incidence of cancers of the intestinal tract (including the colon and rectum) and also those of the pancreas and breast. In countries where fiber consumption is high and fat consumption is very low, such as in most of equatorial Africa, the incidences of colon and rectal cancer are only a fraction of what they are in the industrialized world. Australia, New Zealand, and the United States, all of which are beef-raising and beef-eating nations, have high rates of colon and rectal cancers. In fact, diet is more strongly correlated with cancer incidence than is industrial pollution. Studies comparing the cancer rates in Iceland and New Zealand, where diets are similar to those in the United States but where there is far less industrialization, have shown that the cancer incidence is the same in these countries as it is in the United States. On the other hand, cancer rates overall, and for many specific types of cancer, are much lower in Japan,

which, like the United States, is an industrialized nation, but one in which dietary fat intake is much lower than that in the United States (Fig. 9.16). The incidence of cancers among Seventh Day Adventists who are vegetarian and who do not smoke or drink is much lower than the incidence in their neighbors, despite both groups' living under the same conditions and being exposed to the same environmental pollutants. Several studies have also shown that fresh vegetables, particularly those rich in vitamins A, C, E, and beta carotene (a vitamin A precursor), reduce the incidence of many cancers.

Diet also appears to be an important factor in late-onset (postmenopausal) breast cancer. Again, the evidence is mostly epidemiological. Unlike the premenopausal form of breast cancer, which we described earlier as a hereditary form of cancer, postmenopausal breast cancer appears to be linked to the amount of dietary fat and to obesity. The rates of postmenopausal breast cancer are very low in most parts of Asia, about one-sixth of the

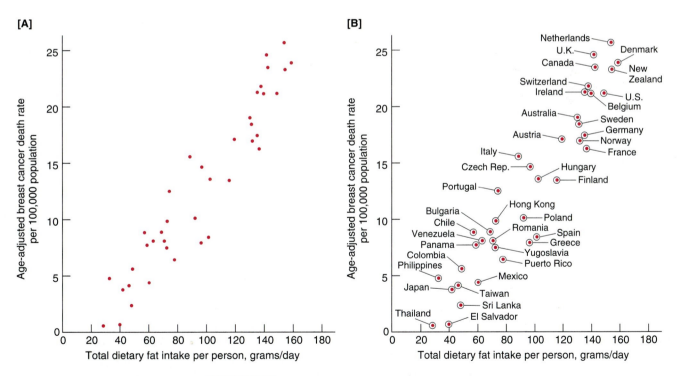

FIGURE 9.16
The relationship between levels of fat intake and mortality from female breast cancer in various countries. Age-adjusted death rate per 100,000 people. [A] Data from the 49 countries identified in [B].

comparable rates in the United States. After Asian women migrate to the United States, their breast cancer rates rise in about two generations to rates that are nearly as high as in other Americans. During the transitional generations, breast cancer rates correlate with the adoption of Western cultural habits, including dietary habits; those women who maintain their traditional cultures have lower breast cancer rates than women who adopt Western cultural practices more quickly.

A clinical study in Canada that followed 56,837 women with different dietary practices over a 5-year period found a higher rate of breast cancer among those with high-fat diets. This study began with an initially healthy population and examined the women's patterns of food consumption prior to the development of any disease. Such studies, called **prospective studies,** are usually considered the most reliable epidemiological method in identifying the causes of disease. Laboratory studies on mice also show that high-fat diets promote mammary cancer, while high-fiber diets have a protective effect.

Salty and pickled foods.   Cancer of the stomach follows a different epidemiological pattern correlated with a different set of dietary factors. This cancer is most frequent in Japan and in certain Latin American countries, where it seems to be correlated with the eating of very salty foods and pickled vegetables. Japanese immigrants to Hawaii and California have a lower incidence, closer to U.S. rates, while their rates for other cancers (those of the colon, rectum, and breast in particular) increase to American levels in about a generation or two. Among Japanese-Americans in Hawaii, the rate of stomach cancers correlates closely with the retention of other aspects of Japanese culture: Japanese-Americans who maintain more of their traditional culture have higher rates of stomach cancer than Japanese-Americans who adopt more Western cultural practices. This evidence suggests that diet plays a larger role than genetic background in the incidence of stomach cancer.

Alcohol and other factors.   Alcohol has been identified as a risk factor for cancer by a number of researchers, but the increased cancer risk is largely confined to people who also smoke. The risk from both smoking and drinking is greater than the sum of the risks from either activity by itself. (As mentioned above, this is called a synergistic effect.) Several food additives, including some artificial colors and at least one artificial sweetener, have been shown to cause cancers in animals.

Internal Resistance to Cancer

A good deal of evidence shows that people vary in their resistance to cancer. People with the same exposure to all known risks still do not get cancer at the same rate. People also vary in their recovery rates once they get cancer. Individual variation in hormones, stress, mental outlook, and immune function may be involved.

Hormones have been implicated in some types of cancer, especially uterine cancer, which occurs more often in women who have been exposed to certain estrogens, including the synthetic hormone diethylstilbestrol (DES). Hormones also influence breast cancer rates, although the process is unclear. The risk of some breast cancers can be reduced by ovariectomy (removal of the ovaries) or by the estrogen-inhibiting drug tamoxifen.

Statistically, cancers of the breast, cervix, and endometrium are less frequent in women who have given birth than among those who never have, although the reasons for this finding are not entirely clear. Women whose first pregnancy occurred after age 30 do not seem to benefit as much from the reduction in cancer risks as do women who first became pregnant before age 25. Additional pregnancies beyond the first do not seem to increase the protective effect. Completion of the pregnancy does seem to be important: Women whose pregnancies have been interrupted do not benefit from any decreased risk. The effect of breast feeding is still under study; some early researchers claimed that mothers who breast-fed their babies had a lowered risk of breast cancer later in life, but other studies in which mothers who breast-fed their babies were compared to women who have given birth but did not breast-feed their babies (rather than being compared to women who had not breast-fed but might or might not have given birth) did not show this protective effect.

Certain cancers are statistically more common among people under chronic stress. Stress is a difficult variable to measure, and stress levels are usually reported simply as "high" or "normal." Many of these studies, however, were poorly controlled. For example, night workers and daytime workers in the same industry may differ in many other ways besides stress levels. Psychologists who have focused their attention on the means by which people deal with their stress have found lower cancer rates among people with better coping skills.

The immune system removes most transformed cells before they are able to establish cancers. Thus, when the immune system is suppressed (as in transplant recipients) or deficient (as in AIDS), certain otherwise rare cancers are more likely to occur, such as Kaposi's sarcoma. Stress can also impair immune functions, thereby increasing certain cancer risks.

Social and Economic Factors

Social and economic factors have also been shown to correlate with incidence rates and especially with survival rates for various cancers. In the United States, some studies show that whites and blacks (African-Americans) have comparable incidence rates for certain cancers, but blacks often have higher mortality (death) rates because they receive far less often the kinds of routine checkups that would detect the common cancers in their earliest and most treatable stages.

More recent studies have found higher incidence rates as well as death rates for urban black populations, compared to urban white populations, for most types of cancer. Much of the excess cancer rate among blacks was among those with low income levels and low educational attainment. In other words, a large amount of the difference in cancer incidence rates could be explained by differences in income and in life factors related to income. When blacks and whites of comparable socioeconomic status were compared, many of these differences disappeared or were reversed. Colon cancer showed no difference in incidence by race, and rectal cancer was more common among whites than among blacks. However, blacks still had higher rates for cancers of the stomach, prostate, and uterine cervix. Female breast cancer showed a higher rate in white women than in black women, and among white women it showed a higher incidence in the higher income brackets

than in the lower ones. As we have seen, there are many factors involved in determining cancer incidence rates, including differences in genetic background, in lifestyle, or in exposure to environmental carcinogens.

Other studies have shown that some groups of patients are less likely to follow health advisories or their doctor's advice, a pattern that may relate more to economic status (poverty) or to education than to other factors. One recent study found that the size of breast tumors at the time of discovery and diagnosis was related to reading scores and educational attainment: Tumors in college-educated women were more often discovered while still very small and treatable, while tumors in women of low reading skill were more likely not to be discovered until they had grown large. Larger and more careful studies are needed to unravel the variables at work. More studies are also needed on the health-care-seeking behavior of various population groups and on ways to increase adaptive behaviors.

THOUGHT QUESTIONS

1. Not everyone who smokes gets cancer. Does this mean that smoking is not a risk factor for cancer?
2. Are cancer risk factors identified by studying the family and environmental histories of individuals or of populations of individuals?
3. What is the difference between a *risk factor* and a *cause*? Can we say what *caused* cancer in a given individual?
4. Why do different individuals respond differently to cancer risk factors?
5. What factors might contribute to the correlation between women's reading level and the size of breast tumors at the time of their discovery and diagnosis?

E. Mechanisms of Carcinogenesis Are Incompletely Known

We have seen that an important early step in carcinogenesis is the transformation of a cell into a less-differentiated "immortal" cell that no longer responds to the signals that normally regulate cell division and cell differentiation. The signals involved in cell transformation are not completely identified, but this much seems certain: Cancerous growth signals are aberrant forms of the normal growth signals, and the aberration is located in the cell's DNA.

Oncogenes and Proto-oncogenes

Certain genes, called oncogenes, were discovered in cells that had become transformed. Later it was discovered that oncogenes are often altered alleles of genes regulating normal cell growth. The normal genes are called proto-oncogenes.

Oncogenes. Discovering the function of growth-regulating genes in normal and transformed cells has depended in a large part on the techniques of molecular biology: moving test pieces of DNA into and out of different test genomes and examining the resultant changes in cell function. Very often in these experiments, viruses were used to carry DNA into or out of the host genome, taking advantage of the fact that insertion into host DNA is part of the normal life cycle of viruses (Chapter 13). In the process of carrying the test DNA into the host, viral DNA is also inserted. It was discovered that some viral genes brought about transformation in host cells; these viral genes were given the name **oncogenes.**

Epidemiologic evidence, however, shows that in most cases cancer incidence does not correlate with viral infection. In 1978, Weinberg showed that cancer can be transmitted to normal cells by taking the DNA from the transformed cells and inserting it into the normal cell DNA without the help of viral carriers. This process could be repeated over and over, each time diluting any virus or chemical that might have been present initially. By such dilution experiments, Weinberg showed that it was the DNA itself that was responsible for transformation, not the intact virus. When viruses are involved in transformation, they simply act to move pieces of DNA around.

Proto-oncogenes. Later, when techniques had been developed to determine the sequences of nucleotide bases in DNA, researchers started comparing the DNA of viruses with the DNA of their hosts. One group discovered that the DNA of untransformed, uninfected chicken cells contained a DNA sequence very similar, but not identical, to that of a known oncogene, the Rous sarcoma, or src oncogene. The oncogene was actually a slightly altered form of a normal growth promoter gene in the chicken. The direction of DNA transfer (from host to virus) was just the opposite of what had been originally assumed: Viruses had not donated oncogenes to the host; instead, host DNA had been removed and "captured" by the virus, becoming slightly altered in the process. This presents a cautionary tale: The location in which we initially notice something is not always the place where it originates or the place where it carries out its major function.

Normal host genes that are closely similar to oncogenes are called **proto-oncogenes,** reflecting the belief that these proto-oncogenes were the original forms. Oncogenes are now designated by the prefix v- to designate their viral origin (v-src, for example), while the proto-oncogenes are given the prefix c- (c-src). The three letters designating the gene itself are derived from the name of the oncogenic (cancer-causing) virus in which they were first discovered. J. Michael Bishop and Harold Varmus received the 1989 Nobel Prize for their discovery of proto-oncogenes.

Activation of cell division by oncogenes in humans. Several oncogenes and their normal counterpart human proto-oncogenes are now known. Oncogenes differ from proto-oncogenes in any of three basic ways: the timing and quantity of their expression, the structure of their protein products, and the degree to which the function of the protein products is regulated by cellular signals. The expression of a proto-oncogene responds to cellular controls (Fig. 9.6), while the expression of an oncogene does not. The protein product of an oncogene may differ by as little as a single amino acid from the protein product of a proto-oncogene, but this small change in structure can be enough to remove the protein from control by the cell's regulatory mechanisms.

One type of oncogene codes for a modified growth factor receptor that, unlike its normal proto-oncogene counterpart, continuously activates second messengers without having bound its growth factor (Fig. 9.17A), thus triggering cell division without cytokine signals. Another oncogene (Fig. 9.17B) causes a cell to secrete growth factors for which it has receptors, creating a simple feed-

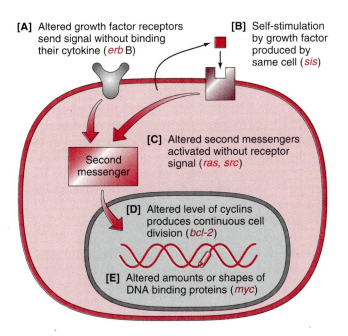

[A] Altered growth factor receptors send signal without binding their cytokine (*erb* B)

[B] Self-stimulation by growth factor produced by same cell (*sis*)

Second messenger

[C] Altered second messengers activated without receptor signal (*ras, src*)

[D] Altered level of cyclins produces continuous cell division (*bcl-2*)

[E] Altered amounts or shapes of DNA binding proteins (*myc*)

FIGURE 9.17
The cellular location and activities of the protein products of oncogenes, which lead the cell's escape from the regulation of cell division. Shown here are several ways in which oncogene products can alter the normal cell processes shown in Fig. 9.4. Examination of cancer cells reveals that one or more of these changes may have occurred. Examples of specific oncogenes are in parentheses.

back loop in which the cell stimulates itself to divide. A third type of oncogene codes for altered cellular second messenger molecules that carry "activate cell division" commands across the cytoplasm in the absence of any growth factor signal from outside the cell (Fig. 9.17C). Still other oncogenes alter the regulatory steps inside the nucleus, affecting the concentrations of cyclins (Fig. 9.17D) or of certain DNA binding proteins (Fig. 9.17E). In other cases, there may be an increase in the number of copies of the proto-oncogene in the DNA (gene amplification), causing it to function as an oncogene by producing abnormally high amounts of the normal protein. Chromosome rearrangement can result in a proto-oncogene being moved near a strong enhancer or being fused to a gene that is transcribed continuously or very frequently, as was found to be the case in a cancer called Burkitt's lymphoma.

Burkitt's lymphoma. The cancer called *Burkitt's lymphoma* affects B lymphocytes (B cells), a type of white blood cell involved in immunity (Chapter 12).

This cancer is associated with a gene called *myc*, whose product is a DNA-binding protein (Fig. 9.17E) that can immortalize a cell by preventing it from leaving the cell cycle and entering the G_0 resting phase (Fig. 9.3). The proto-oncogene counterpart is normally located on chromosome 8 in humans, but, in the transformed cells of Burkitt's lymphoma, sections of this chromosome, including the *myc* gene, have traded places with a small section of DNA from another chromosome, number 14, or 22, or 2. Such an exchange of DNA material is called a *translocation;* translocations are visible microscopically because the banding pattern on each chromosome is distinctive. In its new location, the *myc* gene is *deregulated,* that is, it has escaped from its normal control, so that it is expressed more often. The translocated *myc* gene is now located near the genes that code for different parts of the antibody molecule, a variable protein that is synthesized whenever the body is challenged by an infectious agent (Chapter 12). Because of its altered location near the antibody genes, any infectious challenge that stimulates transcription and translation of the antibody genes now also causes high levels of *myc* to be synthesized, sending the cell into cell division. The hypothesis of an association between antibody genes and the *myc* gene in Burkitt's lymphoma also offered an explanation of the epidemiological evidence that Burkitt's lymphoma occurs more often in patients who are infected with Epstein Barr virus and receive a second infectious challenge such as malaria.

Why should the *myc* gene be translocated, and why always to the region of the antibody genes? Cells that are differentiating to become B cells rearrange their antibody genes as part of this differentiation, which is why the body is able to generate such a huge number of different types of antibodies in response to the variety of bacteria and viruses it may encounter (Chapter 12). Since DNA translocation normally occurs in these cells at a very high rate, the frequency of abnormal translocations of DNA from other chromosomes is also increased. In this instance, the *myc* gene has not been altered but its chromosomal location has, resulting in its expression whenever the antibody genes are expressed.

Two-hit theory. Evidence suggests that the transformation of cells may require a combination of changes in several proto-oncogenes rather than a change in just one. For example, a viral oncogene named *ras* (after the rat sarcoma virus) was found to have perfect homology (identical nucleotide sequence) with a gene expressed in human bladder cancer. This gene could transform tissue culture cell lines that had already been immortalized (removed from having a finite limit to its number of cell divisions), but *ras* could not transform normal cells. When single oncogenes would not transform cell lines, scientists began testing these oncogenes in pairs. It was found that adding two oncogenes, *myc* and *ras*, to normal cells did transform them, while neither would do so alone. The *ras* proto-oncogene product is cyclic AMP, a cytoplasmic second messenger (Fig. 9.17C). Cyclic AMP is present in all animals, plants, and fungi, including yeast, meaning that it is highly conserved in the course of evolution and is thus presumed to be very important in all these organisms. If a cell has gone through the first step of transformation (immortalization by the *myc* protein), the presence of *ras* will complete the process, causing the cells to become unattached, lose their anchorage dependence, change shape from flat to rounded, and begin to secrete growth factors.

Transformation, in other words, seems to be a two-step process, a hypothesis called the *two-hit theory*. Under this theory, transformation to a cancerous cell requires two steps: first, immortalization, and second, the loss of anchorage dependence. Immortalization and the loss of anchorage dependence are hypothesized to be independent processes that may be triggered by separate events. Evidence from both epidemiology and cell biology now supports the two-hit theory.

Mutagenesis and Carcinogenesis

We have discussed viral oncogenes at length because the mechanisms are best understood when oncogenes are involved, but, in fact, chemicals, not viruses, are thought to be responsible for the majority of transformations. There is a clear correlation between the ability of an agent to act as a **mutagen** and induce mutation of DNA, and its ability to act as a carcinogen and induce cancer (see Box 9.1), although these terms are not synonymous. The fact that many oncogenes result from point mutations (single nucleotide changes) of proto-oncogenes suggests a mechanism by which

carcinogens might also produce the changes, independent of any viral infection. Transformation is initiated by a mutation in the DNA; the mutation or mutations can be caused by a virus, a chemical, or radiation.

In humans, there are on the order of 10^{16} cell divisions during the lifetime of the individual. Because of the limitations on the accuracy of DNA replication, at every cell division each gene has a 1 in 10^6 chance of being copied wrong. Since there are 10^{16} cell divisions, there should be a total spontaneous mutation rate of 10^{16} divided by 10^6, or about 10^{10} mutations per human lifetime.

However, only a very small fraction of these mutations lead to cancer. If single mutations were the primary cause of cancer, we would expect that the rate of incidence for new cancers would be the same for individuals of every age. That is clearly not the case for most cancers; the rate of incidence of new cancers increases with age, particularly in advanced age. What prevents the overwhelming majority of these mutations from initiating a cancer? One likely explanation is that the mutation must occur in a proto-oncogene in order to lead to cancer. It also appears that a single mutation is not enough, and that mutations may need to accumulate in more than one proto-oncogene in the same cell for that cell to become fully transformed.

Tumor initiators and tumor promoters. **Tumor initiators** are agents that begin the process of transformation. Initiators cause permanent damage in the DNA. Mutagens, including tobacco smoke, are tumor initiators. Most mutagens also require exposure at a later time to a **tumor promoter** before transformation is complete. Tumor promoters are not mutagenic by themselves. They can cause cells to go into cell division, and if some of the dividing cells contained mutations from an earlier exposure to an initiator, their chances of acquiring additional mutations leading to complete transformation are increased. Because the DNA damage from an initiator is permanent, a tumor promoter may have its effect years after exposure to the initiator. Tumor promoters include alcohol, phenobarbital, dioxin, saccharin, asbestos, and tobacco.

Tumor suppressor genes. Some of the rare, inherited forms of cancer have been associated with defects in genes other than proto-oncogenes, the tumor suppressor (or repressor) genes. The products of these genes normally inhibit cell division. If these genes are altered, their repressor activity may be removed. Alteration of a cell-division inhibitor leads to cell division (Fig. 9.7). Retinoblastoma and hereditary colon cancers are associated with defective tumor suppressors. More recently, a tumor suppressor gene, p53, has been found to be associated with many noninherited cancers. It is therefore likely that somatic mutations in the p53 gene contribute to transformation. (Somatic mutations are those that occur in the DNA of body cells rather than in the DNA of gametes as occurs in hereditary cancers.) Tumor cells continue to mutate and evolve; their mutation rate often seems higher than that for normal cells, possibly due to defects in checking proteins.

Immune surveillance.   A theory known as *immune surveillance* states that we would all get cancer all of the time if it were not for the immune system. There is increasing evidence for this theory: People who have weakened immune systems from any number of causes tend to develop cancers more frequently. This influence is striking in conditions that severely damage the immune system, such as AIDS (see Chapter 13), but the influence is also present in people whose immune systems are weakened less drastically from other causes, including chronic stress, sleep disorders, and so forth. Far more cells are mutated and transformed than ever develop into a cancerous tumor. A healthy and active immune system eliminates most of these cells as they arise. Any weakening of the immune system allows more transformed cells to grow and proliferate, and a higher rate of cancer is one of the results. People's immune systems also weaken with age, which is consistent with the finding that the incidence of new cancers increases with age.

THOUGHT QUESTIONS

1. What does the statement "Cancer is a disease of the genes, but it is not a genetic disease" mean? Is this statement always true, or are some cancers genetic diseases?
2. Recently the genetic defect associated with an inherited form of colon cancer was identified as a defect in a DNA checking protein. This finding was reported in the lay press as the discovery of "the colon cancer gene." Is this name misleading? To what extent can a defective repair mechanism be considered to be the same thing as a cause of a cancer?
3. Tobacco smoke contains chemicals that are tumor initiators and other chemicals that are tumor promoters. How does this combination contribute to the carcinogenicity of tobacco smoke?
4. Tobacco and alcohol act synergistically in increasing cancer risks. Can you explain this in terms of what is happening inside cells?

F. We Can Treat Many Cancers and Lower Our Risks for Many More

Surgery, Radiation, and Chemotherapy

Most of the present-day treatments of cancer involve one or more of three types of treatment: surgery, radiation, or chemotherapy. Surgery is limited to those cancers that produce visible tumors. Because single metastatic cells can lead to later recurrence, surgery is often combined with radiation or chemotherapy. In either radiation therapy or chemotherapy, the strategy is the same: Cancer cells are dividing cells; therefore agents that interfere with cell division should stop cancer cells. Radiation causes breaks in the DNA of dividing cells that are large enough that the cell cannot repair them and the cell cannot live with the damage. Chemotherapeutic drugs prevent DNA synthesis at several steps. Some of the drugs inhibit the synthesis of the nucleotides needed to build DNA; some substitute for certain nucleotides in newly synthesized DNA, preventing its further replication; and some inhibit an enzyme needed to unwind and rewind the double helix during its replication. Other chemotherapeutic drugs, some of which are natural plant products, prevent RNA synthesis or block mitosis. Still others act by damaging the DNA strands, thus preventing cell division and killing the cell.

There are tens of thousands of known chemical substances that have never been tested as possible carcinogens in animals. Animal testing is expensive and takes a long time to produce results; it will probably take many, many decades (and many billions of research dollars) to test all these substances. Clearly, we need a quick *screening method* that will tell us which substances are more likely to be carcinogenic; these substances can be tested first, while the testing of less likely carcinogens can wait.

Devised by Bruce Ames of Cornell University, the Ames test is a screening method that detects *mutagens* capable of bringing about a particular type of mutation in a culture of *Salmonella* bacteria. The bacteria used are from a strain called *his*⁻, unable to synthesize histidine, an amino acid required for the manufacture of bacterial proteins and hence for bacterial growth. Most bacteria are *his*⁺, meaning that they can make their own histidine from other materials. In the Ames test, *his*⁻ bacteria are grown in a medium containing just a small amount of histidine, which allows just enough growth to give mutations a chance to occur. Soon, however, the histidine is all used up, and the bacteria will die unless that have mutated from *his*⁻ to *his*⁺. A very low rate of spontaneous mutations results in only a few colonies capable of growth beyond this point, but any chemical that increases mutations will result in many more growing colonies. By counting these colonies, the Ames test can identify mutagens and can also distinguish between stronger and weaker mutagens.

Remember that the Ames test was designed as a *screening method* for carcinogens. The basis of the Ames test is the observation that many known carcinogens are also mutagenic. This is simply a statement of *correlation;* it does not necessarily indicate a causal relation between mutagenesis and carcinogenesis. Not every mutagen is a carcinogen, so the Ames test should not be regarded as a definitive result. The value of the Ames test is that it permits us to focus attention on chemicals that are mutagenic in bacteria, because experience teaches us that these chemicals are *more likely* to be carcinogenic in animals. The indiscriminate testing of animals using one substance after another would be wasteful and expensive and also raises ethical problems (see Chapter 2); the Ames test gives us an indication of which substances are worth testing *first,* using more expensive tests.

Bruce Ames, the originator of the Ames test, has also pointed out that nearly *any* substance can be made mutagenic in sufficiently high doses. We are surrounded with thousands of naturally occurring carcinogens (mostly weak ones), and yet we do not all get cancer from them. Perhaps, he argues, we should study our mechanisms of defense against these carcinogens rather than concentrating on merely identifying one carcinogen after another.

Side effects of radiation and chemotherapy. One drawback to radiation and chemotherapy is that they are nonspecific: Both kill any type of dividing cell. Hair follicles contain rapidly dividing cells, so hair loss frequently accompanies these treatments. Cells of the immune system are also rapidly dividing and so are killed. Hair follicle cells and immune cells are not killed off completely and so repopulate. Hair grows back and people regain their immune cells. During the time when people's

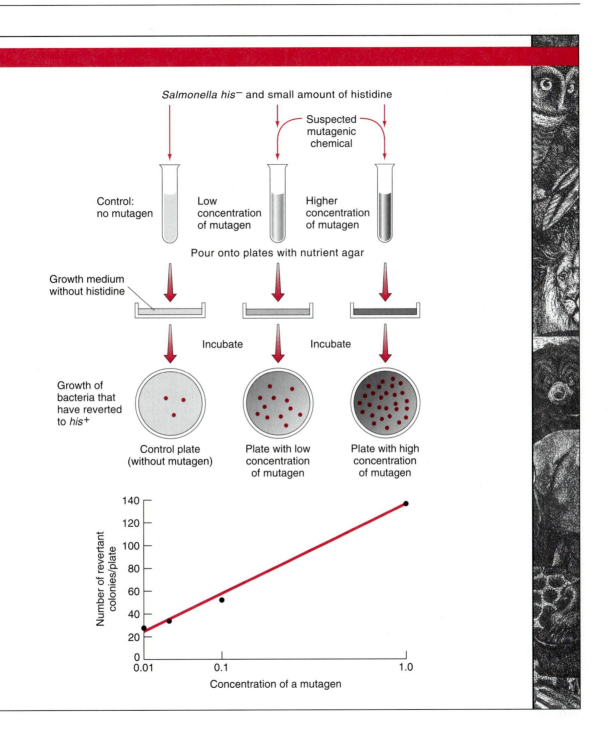

immune systems are compromised, they must be very careful to avoid situations that might expose them to infectious diseases. Both radiation and chemotherapy may destroy a specific type of immune cell, memory cells, which "remember" which diseases the person has been exposed to or vaccinated against (Chapter 12). If these memory cells are killed, even a person who has regained the ability to form new immune responses will have lost previous immunities and therefore may need to be revaccinated. Another drawback is that since both radiation and some chemotherapeutic drugs damage DNA, they increase the risk for the development of secondary cancers.

Chemotherapeutic drugs put a selective pressure on the population of transformed cells. As a result, any cells that become resistant to the drug quickly overgrow the drug-susceptible cells. Several drugs, each of which works by a different mechanism of action, are often used in combination to minimize the development of drug resistance.

Increased survival rates.

There has been an increase in the survival rates of many types of cancers. In most patients, leukemias, Hodgkin's lymphoma, and testicular cancer can now be cured.

For other cancers, it is not so clear what is meant by an increase in survival rate and how this relates to a "cure." When we try to evaluate the meaning of statistical terms like "survival rate," we need to know a lot about how the numbers were gathered and what definitions are being used for certain terms. What is often reported as survival rate refers to the number of persons with cancer still living 5 years after diagnosis compared to the total number of persons with cancer. For example, it is estimated that it takes 9 years for a breast cancer to develop, spread, and kill a person. If past methods of detection led to discovery of the cancer 7 years after its inception, very few people would have been alive 5 years after its discovery. Now, with better detection methods and better public education for breast self-examination, breast cancers are discovered earlier, and the 5-year survival rate looks much improved. Does this mean therapies have improved, or does it simply reflect the fact that people are now finding the cancers at 2 years into their development rather than at 7 years? It is not always easy to distinguish advances made through better or earlier diagnosis from advances made in the treatment of cancers after they have reached comparable stages of development.

New Cancer Treatments

Current research in cancer treatment is following several lines. New chemotherapeutic agents, both natural and artificial, continue to be sought and developed. New strategies are also being developed to make chemotherapeutic agents that more specifically target cancer cells and lessen the number of normal cells that are damaged.

Immunotherapies.

New treatments are also being developed based on boosting the immune system's ability to fight off tumors. If the immune system gets rid of many transformed cells, why can't we vaccinate people against cancer like we do against many infectious diseases? One reason is that the immune system can only act against cells it perceives to be *nonself*. The surface molecules of cancer cells are often the same as those on normal cells; they are just expressed in the wrong amounts or at the wrong times. The immune system cannot distinguish the latter two possibilities as differing from the normal; it can only detect new or different cell surface molecules. Some cancers, especially cancers initiated by chemical carcinogens, do have new or altered molecules at their surface. These molecules are called *tumor-associated antigens*. Unfortunately these are different in each person and even in two different chemically induced tumors in the same person. What new molecules will be present cannot be predicted, so vaccines cannot be developed against them. Some tumors induced by viruses do have common tumor-associated antigens and vaccines can be developed. Feline leukemia is the most successful example to date of a cancer that can be prevented by vaccination.

Other therapeutic strategies are being based on boosting the activity of the immune system with cytokines. Another approach is to make chemotherapeutic agents more selective by "addressing" the drugs to particular tissues by tagging the drugs with antibodies.

Experimental and unconventional therapies.

A good deal of cancer research is clinical research aimed at assessing the efficacy of new therapies. After extensive laboratory research, new agents or new ways of applying those agents are tested on patients who have given their informed consent (Chapter 2) to be part of these studies. Only by gathering data in properly designed and controlled clinical trials can the risks and benefits of new treatments be demonstrated.

Because cancer is greatly feared and because it is not always curable, people put their hopes in unproven remedies. Moss (1989) has examined the various unconventional cancer therapies and treatments that have been publicized in the last few decades. Some of these, such as laetrile, achieved a large and devoted following. The supporters of laetrile finally became so influential that the National Cancer Institute conducted careful clinical trials and announced in 1981 that laetrile had proven to be worthless.

Allocation of research funds. Research directed toward understanding the basic biology of cancer continues. Some people feel treatment cannot be rationally designed unless the underlying biology is known, while others feel that such basic understanding is not important. The latter group use arguments such as the following: We still do not know the basic biology underlying the disease polio, but development of a vaccine for its prevention has eliminated our need to know. The real dilemma is a problem in the allocation of resources. How much money should we spend on treating cancer patients, how much on improving methods of treatment, how much on laboratory research to discover more information on the causes of cancer, and how much on cancer prevention activities? How much funding should be aimed at particular types of cancer, such as breast cancer as compared to colon cancer? There are no clear answers here because we cannot accurately predict how well or how soon funds spent on certain activities (especially research) will translate into a reduction of cancer incidence rates or cancer deaths. Cancer prevention is clearly very cost-effective, but the cost-effectiveness of the other alternatives may be difficult to assess.

Cancer Management

Some people feel more research dollars should be spent on *cancer management* rather than cancer treatment. Cancer management includes the development of drugs or strategies to minimize the side effects of cancer treatments. Examples include cold-capping, a procedure in which a cold pack is applied to the scalp during chemotherapy so as to slow cell division in hair follicle cells, thus decreasing hair loss. Another possibility is the development of antinausea drugs. The possible medicinal use of marijuana to overcome nausea

and restore appetite in chemotherapy patients is being studied (see Chapter 11).

Cancer management also includes support groups and grief therapy. The aim of these approaches is to improve the quality of life—to treat the person, not the disease. Women who were in support groups after recurrent breast cancer lived longer than those who were not. Survival rates have been found in some cases to be influenced by mental attitude (Chapter 12). Patients who were optimistic, who were aggressive, or who were "determined fighters" had statistically longer survival rates and higher cure rates than those who were pessimistic or who resigned themselves early to their fate. There is a large body of psychological literature on "learned helplessness," a phenomenon in which a person or an animal experiences repeated stresses from which there is no escape and for which no remedy is available. Such individuals "learn" that there is nothing they can do to change anything, a lesson that they then apply in other areas of their lives. When such people get cancer, their learned helplessness results in a much lower survival rate and in a shorter life span.

Cancer Prevention

As we learn more about the causes of cancer, it appears that one of the most successful strategies may be cancer prevention—preventing cancer may be far easier than curing it.

Smoking remains a major cause of cancer. The Centers for Disease Control and Prevention state that cigarette smoke causes 30 times more lung cancer deaths than all regulated air pollutants combined. Exposure to secondhand smoke, for example as a result of living in a home with someone who smokes, causes the deaths from lung cancer of 3000 nonsmokers a year in the United States. This is in part because tobacco smoke contains both tumor initiators and tumor promoters, but also in part because tobacco smoke suppresses the immune system. Exposure to secondhand smoke is also responsible for nearly 300,000 infections per year in infants younger than 18 months.

Some dietary regimens have been associated with a decreased risk of cancer: lower total calories and low fat, for example. Food containing antioxidants may help; these include beta carotene and vitamins A, C, and E. The American Cancer Society has issued guidelines promoting fresh foods high in vitamins A and C, especially fresh vegetables of the plant family Cruciferae. Cruciferous vegetables include cabbage, radish, turnip, broccoli, cauliflower, kale, kohlrabi, mustard greens, and brussels sprouts.

High-fat diets may be an important risk factor in cancers of the colon and rectum, and in postmenopausal cancer of the breast. There is some evidence that high-fiber diets lower the risks of colon and rectal cancers.

Given the available evidence, the most important actions that average citizens can do to lower their cancer risks are these:

1. Don't smoke! Also, avoid secondhand smoke from poorly ventilated rooms where others smoke. *This is the single greatest change you can make to reduce your cancer risk, far outweighing all other possible measures.*

2. Follow a diet low in fats, high in fiber, and high in antioxidants (e.g., beta carotene, vitamins A, C).

3. Avoid occupational exposures to potential carcinogens; minimize exposure through appropriate use of safety equipment.

4. Avoid exposure to radioactive substances and X rays above necessary minimum levels; avoid needless exposure to ultraviolet light from the sun or tanning booths.

5. As you age, be sure to get checkups at regular intervals, including screening that will detect the common cancers in their earliest and most easily treated stages. If you are a woman, learn to practice breast self-examination.

THOUGHT QUESTIONS

1. Explain how the theory of evolution accounts for the development of cancer cells that are resistant to chemotherapy.
2. What characteristics would you look for in an ideal chemotherapeutic drug for the treatment of cancer?
3. Secondhand smoke is a cancer risk. Does this biological reality change the ethical debate in which individual freedom of choice must be balanced against the ethic of not doing harm to others?
4. Do you agree with the following statement? When someone's chances for survival are predicted to be very low, any and all treatments are justified. In other words, any treatment is good as long as it is not harmful. Try to apply this thinking to such unproven remedies as laetrile.
5. Is it ever ethically permissible to give up on treatment? Are there things other than treatments that can be done for a dying person? Do you think people who have terminal cancer should be told of their condition? Try to justify your answer. What ethical assumptions underlie your argument?

CHAPTER SUMMARY

All organisms are built of cells. Normal cells occasionally divide, but they always stop dividing when enough cells are present because of such phenomena as contact inhibition. Cells also differentiate as they divide, becoming more and more restricted in their potential to form different kinds of cells. Both cell division and cell differentiation are the result of differential expression of genes, and both are influenced by molecular signals (cytokines) secreted by neighboring cells. Cancer cells differ from normal cells in abnormal responses to these signals—they don't stop dividing (they are "immortal"), and they remain undifferentiated.

A few cancers have genetic causes, but most are caused by environmental factors. These factors include exposures to certain viruses and infective agents, dietary and other behavioral factors, and exposure to a long list of carcinogens, including ionizing radiation (radioactivity), ultraviolet light, tobacco smoke, and a variety of occupational carcinogens.

Cancerous tumors can be removed surgically, but other forms of cancer therapy target any cells that are dividing, destroying many healthy cells along with the cancer. New, more specific chemotherapies are being developed, including therapies based on boosting the body's immune system. We can best reduce our risks for cancer by avoiding tobacco smoke and other carcinogens (including ultraviolet light and ionizing radiations) and by following a low-fat, high-fiber diet rich in beta carotene and vitamins A and C.

KEY TERMS TO KNOW

anchorage dependence (p. 254)

cancers (p. 250)

carcinogens (p. 269)

cell cycle (p. 253)

cells (p. 250)

contact inhibition (p. 254)

determined (p. 257)

differentiation (pp. 252, 257)

doubling number (Hayflick limit) (p. 263)

doubling time (p. 254)

gene expression (p. 255)

growth factors (p. 255)

homeostasis (p. 253)

immortal (p. 264)

malignant (p. 266)

maternal effect gene (p. 260)

metastasize (p. 266)

mutagen (p. 276)

oncogene (pp. 268, 274)

organizer (p. 260)

potentiality (p. 257)

prospective studies (p. 272)

proto-oncogene (p. 274)

second messengers (p. 255)

stem cells (p. 264)

synergistic effect (p. 270)

tissues (p. 252)

transformation (p. 264)

tumor initiator (p. 276)

tumor promoter (p. 276)

tumors (p. 264)

CONNECTIONS TO OTHER CHAPTERS

Chapter 3   Cancers are caused by mutated genes or by changes that bring normal genes to abnormal locations in the genome.

Chapter 5   Several cancers have different incidence rates in different human populations.

Chapter 8   High-fat diets with low fiber content increase the risks for several cancers.

Chapter 12   Good mental outlook and immunological health can improve cancer survival rates.

Chapter 13   Certain otherwise rare cancers occur more frequently among AIDS patients.

Chapter 14   As cells or organisms become larger, the ratio of their surface area to their volume changes.

Chapter 15   Many cancer-fighting drugs are plant products.

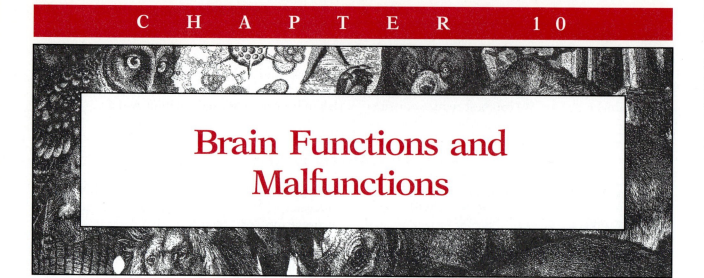

Brain Functions and Malfunctions

A man in his sixties finds it difficult to walk unless he shuffles his feet along the floor, just an inch or two at a time. Last week he tried to walk by lifting his feet, but he stumbled after only two quick but awkward steps. A woman in her seventies used to work crossword puzzles and solve math problems with ease, but she has forgotten how to do these tasks. Her friends, whom she no longer recognizes, say that she is not the same person that they knew just a few years earlier. A younger woman has lost all motivation to work, to keep herself clean, or just to go on living. "What for?" she asks, "none of it matters." These three people have three very different diseases: Parkinsonism, Alzheimer's disease, and depression. Could it be that all of their assorted symptoms are caused by chemical imbalances in the brain? In this chapter we will examine some of the brain's normal functions, including the ways in which nerve cells communicate with one another by secreting certain chemicals called neurotransmitters. A number of disease conditions, including those mentioned above, are caused by malfunctions involving neurotransmitters. We will also examine some of the biological processes involved in learning and memory formation and in the sleep–wakefulness cycle.

A. Brain Function Depends on the Activity of Brain Cells

The issues covered in this chapter all relate to the function of the nervous system and especially the brain. We therefore begin with some background information about the brain, the cells of the nervous system, and the mechanisms used by the nervous system to carry signals from one part of the body to another.

Structure of the Brain

The vertebrate nervous system consists of a **central nervous system,** which includes the *brain* and *spinal cord*, and a **peripheral nervous system,** which includes a series of *peripheral nerves* and several *special sense organs* such as the eyes and ears. The functions of the brain are studied by biologists and also by psychologists.

Central to the function of the entire nervous system is the brain itself, which receives information from various parts of the body through peripheral nerves. The brain is the center where most neural activity takes place, where most decisions are reached, and where most bodily activities are both directed and coordinated. Except for spinal reflexes (described later), nearly all activities that the body performs are carried out in response to signals originating in the brain and sent to the rest of the body through the peripheral nerves. We are aware of many of these signals as they occur; such signals are often said to represent "conscious" activity, although it is difficult to define *consciousness* in a manner that permits good experimental investigation.

The brain is an immense network of nerve cells and their fibers, yet it weighs less than 3 pounds (about 1.3 kilograms) in adult humans. The corresponding parts of human brains and the brains of other vertebrate animals have the same details of structure and of embryological development (Fig. 10.1). Such corresponding parts are said to be *homologous* to one another (Chapter 4), meaning that they share a common evolutionary heritage.

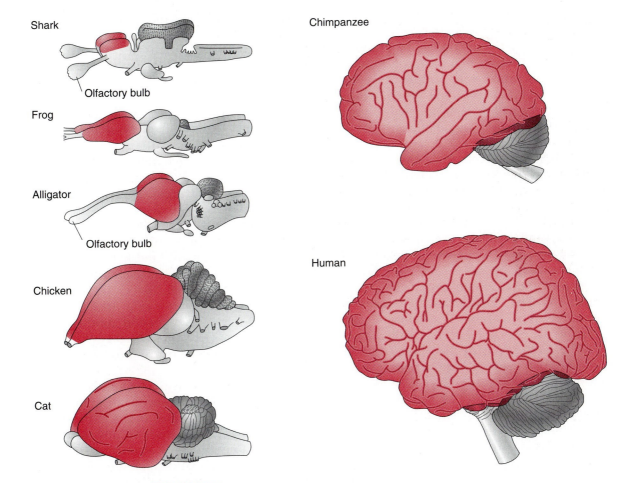

FIGURE 10.1

Correspondences among the brains of several vertebrate species. The cerebrum (part of the forebrain) is colored; the cerebellum (part of the hindbrain) is shaded gray. The olfactory bulbs also correspond in all these species; they are concealed beneath the greatly expanded cerebral hemispheres in chimpanzees and humans.

The folds and constrictions in the partially developed brains of embryos allow us to recognize several major brain divisions. These same divisions are also recognizable in various adult animal brains. The major regions of the brain are the *forebrain, midbrain,* and *hindbrain,* each composed of millions of nerve cells (neurons) and each containing a central cavity (Fig. 10.2). These divisions are similar in various vertebrate species, but their proportions differ: The midbrain and hindbrain make up a larger proportion of the total in primitive vertebrates, while the forebrain, especially the cerebral hemispheres, is larger in mammals, especially primates. Other changes in proportions are related to the senses that each species uses: Brain regions concerned with vision are enlarged in those species that rely upon vision, while species that rely on smell or hearing have enlarged brain regions devoted to those functions.

Forebrain.   The forebrain is divided into a paired portion, the *telencephalon,* which includes the cerebrum and olfactory bulbs, and an unpaired portion called the *diencephalon.* In early vertebrates, the forebrain was concerned with the sense of smell, and much of it still is. Emotional responses like anger, fear, and sexual response are in large measure governed by smell and are therefore also located in the forebrain, specifically in the diencephalon. The *hippocampus,* concerned with certain types of learning, is also located in the forebrain, as are a series of cavities called *ventricles.*

The *cerebrum* is divided into two halves, the right and left *cerebral hemispheres.* Thoughts and actions originate in the cerebral hemispheres, as do most of our "higher" functions of intellectual thought and reasoning. The cerebral hemispheres of mammals are much larger than in other vertebrates. This is particularly true in higher primates like ourselves, where the cerebral hemispheres make up the largest part of the brain. Conscious activity and higher thought originate in the highly folded (convoluted) surface layer of the cerebral hemispheres known as the *cerebral cortex.* Below the cortex lie many series of cellular interconnections (fiber tracts), which constitute the *cerebral white matter.* Deeper still lies a more densely colored *cerebral gray matter,* which includes several clumps of cell bodies known as **basal ganglia.** Several of these structures are involved in the disorders described later in this chapter.

Midbrain.   The midbrain includes a narrow passageway, the *cerebral aqueduct,* connecting the forebrain and hindbrain ventricles. Below the aqueduct lies the *ventral tegmental area,* which includes the brain's *positive reward centers (pleasure centers;* see Chapter 11) and also the *reticular formation,* important in keeping us awake and alert.

Hindbrain.   The hindbrain includes the *cerebellum* and the *medulla oblongata.* Functionally, the cerebellum is primarily concerned with balance and with the coordination of complex muscle movements. The medulla is concerned with such involuntary functions as breathing, functions which must continue even in sleep.

Blood–brain barrier.   The brain has few internal blood vessels; most of the arteries and veins that supply the brain run along the brain surface only. The cells deep in the interior must therefore receive most of their nutrition through the *cerebrospinal fluid,* which fills the brain's interior cavities (ventricles). This cerebrospinal fluid communicates with the blood supply across a thin membrane, the *tela choroidea.* Nutrients and other small molecules cross this membrane, but there is no direct flow of fluid from the blood to the cerebrospinal fluid; consequently, the tela choroidea and its parallel blood vessels are called the *blood–brain barrier* (Fig. 10.3). There is a separate tela in the forebrain and another in the hindbrain; a network of small blood vessels lies on the outer surface of each tela and follows many of its folds.

Studying the brain.   Our knowledge of the brain and its functions derives from many types of study. Anatomical studies begin with dissections of the brain and its parts, and also include microscopic examination of the brain and its cells. Comparison of the brains of various species reveals a great deal about evolution and adaptation. In general, animals with more complex behavior patterns have larger brains with larger and more highly folded cerebral hemispheres. The increasing size and complexity of the cerebral cortex are especially apparent in species similar to ourselves who rely primarily on learned behavior.

A variety of methods are used to study brain function. One technique involves implanting a recording electrode that measures electrical activity during various brain activities. Another method

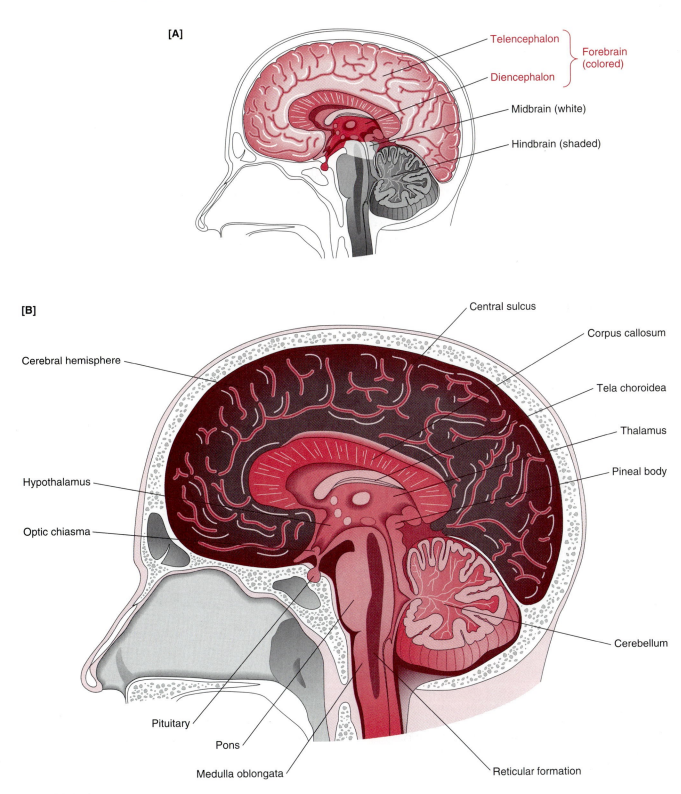

FIGURE 10.2
The human brain in median section. Only structures located in the midline plane are visible in this view; off-center structures such as the basal ganglia and hippocampus cannot be seen in this view.

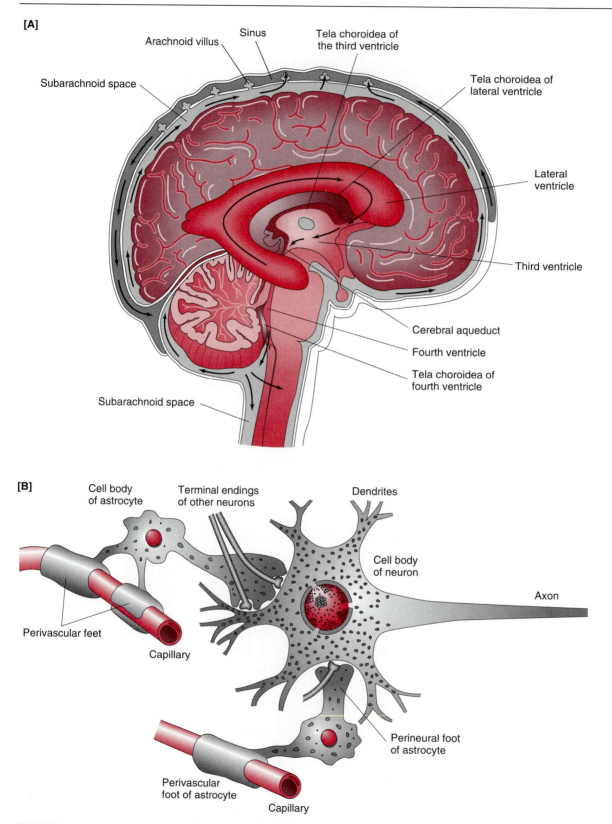

FIGURE 10.3   Components of the blood-brain barrier. [A] The ventricles of the brain and the flow of cerebrospinal fluid. Areas shown in color represent cavities filled with cerebrospinal fluid, which circulates in the directions shown by the arrows. [B] Neuroglial cells (astrocytes) nourishing the cell body of a neuron.

uses a stimulating electrode through which brain activity can be stimulated in experimental animals and the animals' resultant behavior observed. A related technique, *positron emission tomography* (PET), creates a computer-generated picture of various brain activities, such as glucose metabolism and blood flow, within the living brain. Another technique, called the *ablation* technique, is to destroy a portion of the brain (usually with an electric current) and to study the resulting changes in behavior.

Cells of the Nervous System

The functions of the nervous system, including the brain, are carried out largely by **neurons,** the cells that carry **nerve impulses.** Each neuron contains a cell body that includes a *nucleus* and surrounding cytoplasm. Of the cytoplasmic processes that extend out from this cell body, those that conduct impulses toward the cell body are called **dendrites,** while those that conduct impulses away from the cell body are called **axons.** The cytoplasm of each neuron has many RNA-rich clumps called *Nissl granules,* present in the cell body but not the axon (Fig. 10.4A).

Many axons, but not all, are surrounded by a series of special rolled-up *Schwann cells,* which form a structure called the **myelin sheath** (Fig. 10.4B,C). Each Schwann cell is wrapped around the axon in the manner of a jelly roll, and is separated from the next Schwann cell on that axon by an interruption known as the *node of Ranvier.* The myelin sheath acts as an insulator, preventing nerve impulses from spreading sideways from one axon to another rather than traveling the length of the axon. The disease called multiple sclerosis (MS) results from destruction of the myelin sheath by cells of the body's own immune system.

One important function of any neuron is to conduct a nerve impulse along its membrane surface. Each nerve cell conducts its impulse primarily in one direction: In *sensory neurons,* the axons are oriented so they conduct impulses toward the central nervous system, while in *motor neurons,* the axons are oriented to conduct impulses outwardly from the central nervous system to an *effector organ,* which is usually either a muscle or a gland. Nerve impulses must pass from one nerve cell to another across a gap known as a **synapse.**

Aggregations of neurons or their parts have distinctive names in the nervous system. Bundles of axons are called **nerves** throughout the peripheral nervous system or *tracts* within the brain and spinal cord. Clumps of cell bodies are called *ganglia* throughout the peripheral nervous system or *nuclei* within the central nervous system. An exception are the basal ganglia, which retain their traditional name even though they are located in the brain and are thus actually nuclei. Some neurons have cell bodies that do not belong to any such clumps, however.

In addition to nerve cells, the nervous system contains several types of nonneural cells called *neuroglia.* Neuroglial cells in the brain have cellular extensions that wrap around the neurons and other cellular extensions that wrap around small blood vessels. Through these extensions, the neuroglia carry nourishment from the small blood vessels to the nerve cells, forming another part of the blood–brain barrier (Fig. 10.3B). The neuroglia also provide structural support for the brain tissue and are closely related in origin to cells of the immune system (Chapter 12).

Nerve Impulses

Much of what we know about the nature of nerve impulses was learned using the giant axons of the squid (Fig. 10.5A), a distant relative of the octopus and the chambered nautilus. All of these animals belong to a group called the *Cephalopoda,* which is a class of the Phylum Mollusca (see Box 4.1, p. 102). The animals of this class need to react both quickly and forcefully to stimuli that could signal danger; they all have giant axons as part of their quick-response system. These axons are many times the diameter of typical axons in humans and so are ideal for studying nerve impulses, as we will see below.

Electrical potentials.   One way to study nerve impulses within a single neuron is by using a very sensitive voltmeter attached to needlelike probes (electrodes) that conduct electricity. Voltmeters read differences in the amount of electrical charge in contact with its two electrodes; a giant axon has a large enough diameter so that one of the electrodes can be placed inside the axon and the other electrode outside (Fig. 10.5B). Voltmeter readings show a difference in charge between the inside and outside of the axon, a difference called an *electrical*

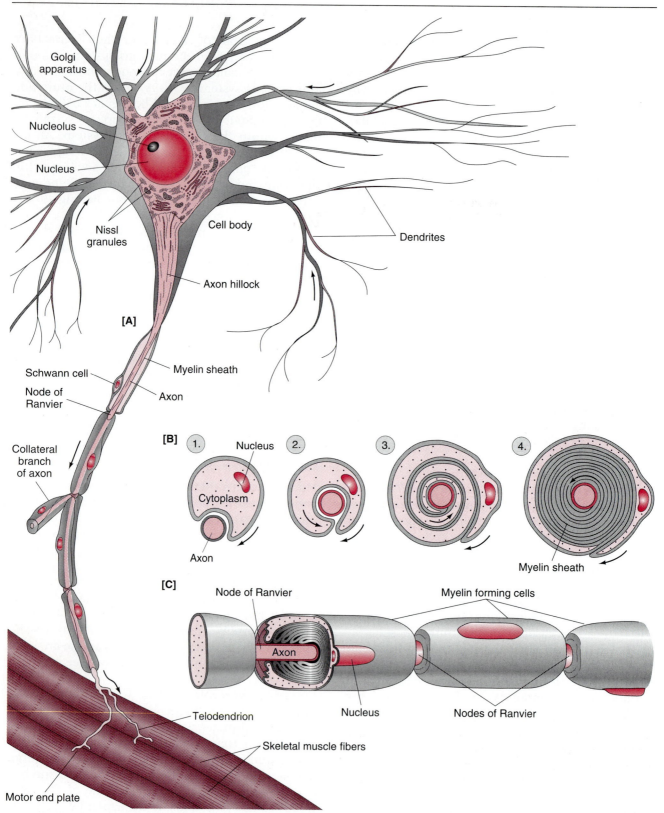

FIGURE 10.4   A neuron and its myelin sheath. [A] Structure of a typical motor neuron. [B] Stages in the formation of the myelin sheath. [C] Structure of the myelin sheath.

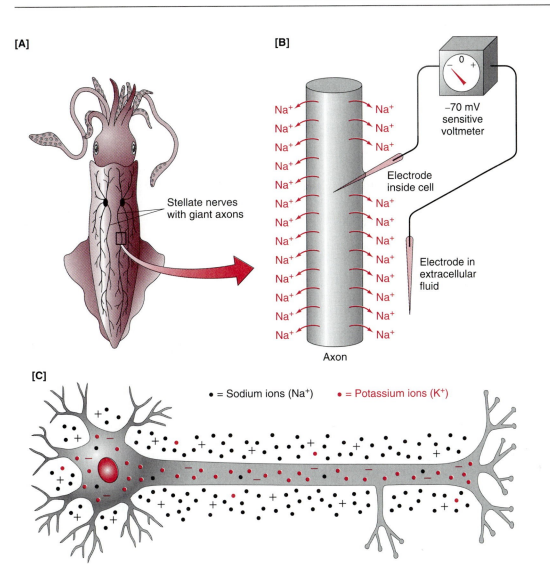

FIGURE 10.5
Resting potentials in the giant axons of the squid. [A] Location of the giant axons in the squid (*Loligo*). [B] Measurement of the resting potential (membrane polarization) in these giant axons, produced by the pumping of sodium ions (Na^+) outward across the cell membrane. [C] The distribution of sodium ions (Na^+) and potassium ions (K^+) responsible for the resting potential.

potential, measured in units called millivolts (mV). This electrical potential is produced by **sodium-potassium pumps,** composed of a series of membrane proteins engaged in the active transport of sodium ions (Na^+) and potassium ions (K^+) across the cell membrane. (Active transport is described in Chapter 8.) For every two potassium ions transported into the cell, three sodium ions are transported across the membrane toward the

outside, producing a buildup of sodium outside the cell and a buildup of potassium inside (Fig. 10.5C). The excess of positive charge outside the cell creates a potential difference of -70 mV across the cell membrane of the axon, meaning that the inside of the cell is negatively charged in comparison to the outside. We say that a membrane is *polarized* when there is an electrical potential across it. The electrical potential across an axon membrane is called a *resting potential* because it remains close to -70 mV as long as the neuron is not conducting an impulse.

Measuring nerve impulses.   The nerve impulse is a *wave of depolarization* traveling along the cell membrane. The easiest way to demonstrate this is by using a voltmeter with electrodes applied to different points along the length of the axon. When the nerve cell is not conducting an impulse, the two electrodes detect the same local concentration of ions (thus the same electrical potential), so that the voltmeter needle reads zero. As the nerve impulse passes, the needle deflects first one way, then the other (Fig. 10.6). Along the length of an axon, the nerve impulse travels at a relatively rapid rate that varies with the diameter of the fiber and with the presence or absence of a myelin sheath. Unmyelinated fibers generally have small diameters, between 0.3 and 1.3 micrometers (1 μm $= 10^{-6}$ m) and conduction velocities of 0.5 to 2.3 meters per second (abbreviated m/s). Myelinated fibers vary from 3 to 20 μm in diameter; conduction velocities vary from 3 m/s in the smallest fibers to 120 m/s in the largest.

Action potentials.   Nerve cells are *electrically excitable*; that is, they can be stimulated electrically. When the nerve cell membrane is sufficiently stimulated, channels through the membrane open and let sodium ions flow back in, locally depolarizing the membrane for a short time. Changes in voltage over time can be measured with electrodes (Fig. 10.5), producing graphs (Fig. 10.7). A slight *depolarization* of the membrane (a voltage change toward zero) or a *hyperpolarization* (an increase in the voltage difference) will be transmitted a very short distance to adjacent portions of the cell membrane; these effects decrease rapidly with distance and thus may disappear quickly. However, a depolarization that reaches or exceeds some **threshhold** will trigger nearby membrane channels to open (Fig. 10.7). The rapid inflow of sodium ions (Na^+) and outflow of potassium ions (K^+) results in a characteristic type of electrical discharge known as an **action potential,** or *spike*. The shape of this action potential can best be seen by connecting a recording electrode to an oscilloscope (whose display is similar to that of a television screen) instead of to a voltmeter. The action poten-

tial causes the adjacent portion of the cell membrane to depolarize and produce another action potential equal in strength to the first. In this way, the action potential rapidly spreads along the cell membrane with no reduction in its size or intensity. It is this propagation of the action potential that we recognize as a nerve impulse. Action potentials maintain their direction of travel because previously opened channels are prevented from reopening for a short period of time (the refractory period), during which the sodium-potassium pumps reestablish the resting electrical potential.

Neurotransmitters

Even though nerve cells are electrically excitable, the transfer of information from cell to cell (across a synapse) is generally chemical, not electrical. Any chemical substance that can stimulate or inhibit an action potential is called a **neurotransmitter.** The vast majority of synapses between nerve cells are chemical synapses in which a chemical neurotransmitter crosses the narrow space separating two adjacent cells (Fig. 10.8). In a chemical synapse, depolarization of the membrane in the presynaptic or transmitting cell (Fig. 10.8A) causes the release of neurotransmitters (Fig. 10.8B). The neurotransmitters released into the synapse bind to receptors on the next cell (the *postsynaptic* cell), where they alter the membrane potential (Fig. 10.8C).

Experiments demonstrating neurotransmitters. The concept of a chemical neurotransmitter was first suggested by the British physiologist Henry H. Dale but was first demonstrated by his German-American colleague Otto Loewi, with whom Dale shared the Nobel Prize in 1936. Dale and Loewi already knew that the *vagus nerve,* running down from the brain into the abdominal cavity, sends branches to the heart, which slow down the heartbeat. They also knew that the heart of a frog could be maintained for hours in a physiological salt solution (a solution containing various ions at concentrations close to those in the intact organism) and would keep beating, even if separated from the rest of the animal. Loewi removed the beating hearts of two frogs and placed them in salt solutions. One of his preparations contained nothing but the heart, but the other contained a

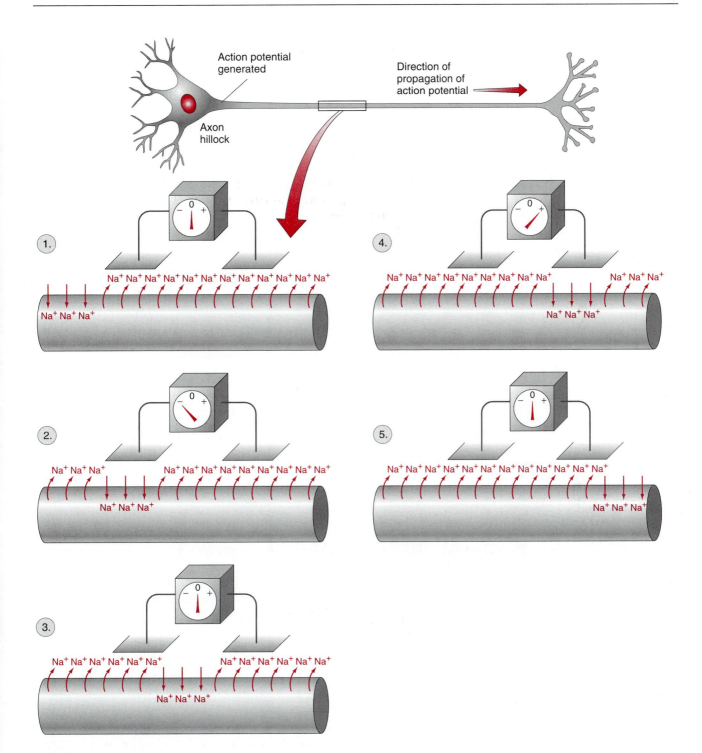

FIGURE 10.6
Detection of a nerve impulse using a sensitive voltmeter. The voltmeter shows differences between Na^+ ion concentrations at the two electrodes. The voltmeter reads zero when the concentrations are the same at the two electrodes, as is true before (step 1) or after an impulse passes by (step 5). As the impulse arrives at the electrode on the left (step 2), the concentration of Na^+ ions is lower on the left, so the voltmeter registers a negtive voltage difference. The situation reverses when the impulse arrives at the electrode on the right, so the voltmeter registers positive (step 4).

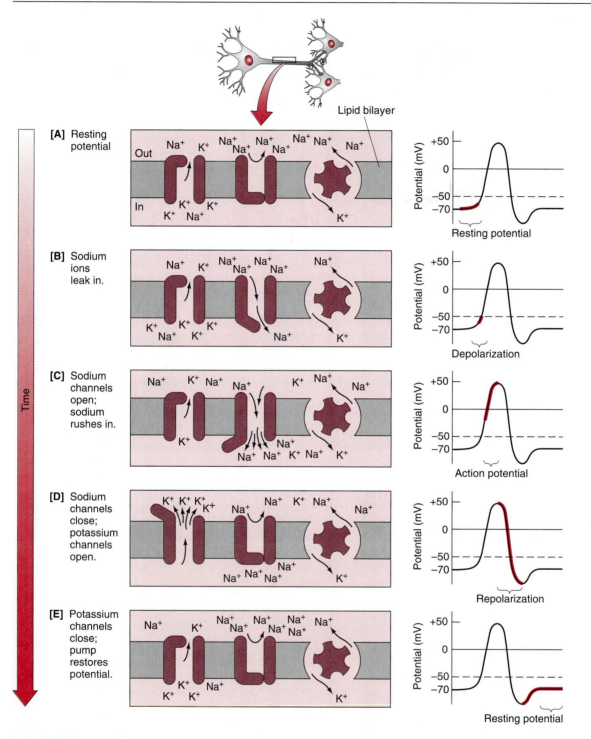

FIGURE 10.7   How an action potential is generated in a nerve cell. A sudden change in the ion balance outside and inside of a neuron's plasma membrane leads to an action potential. [A] The neuron is quiet, and the resting potential stays at −70 mV. [B] A stimulus has allowed some sodium ions to leak in, raising the potential (graph, right-hand panel), and partially depolarizing the cell. [C] When the potential passes a threshold of about −50 mV, the sodium channels open wide. Now the potential reverses and becomes positive (+50 mV, the top of the curve). This is called the *action potential*. [D] Potassium channels open as sodium channels close, allowing potassium ions to rush out. The potential begins to fall as positively charged ions leave the cell (right-hand panel). This phase is called *repolarization*. [E] The sodium-potassium pump restores the balance of sodium ions and potassium ions, and the resting potential returns.

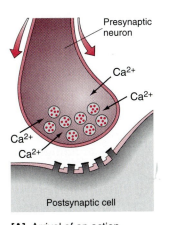

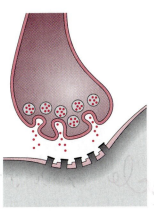

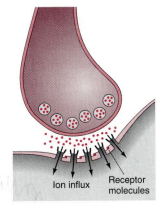

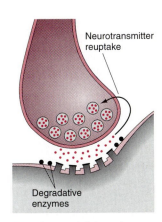

[A] Arrival of an action potential (heavy arrows) triggers the entry of calcium ions (Ca²⁺) into the cytoplasm of the presynaptic cell.

[B] The calcium triggers fusion of cytoplasmic vesicles with the plasma membrane of the presynaptic neuron, releasing the neurotransmitters inside them.

[C] The neurotransmitter binds to receptor molecules on the postsynaptic cell, opening channels that allow positive ions (Na⁺, K⁺, or Ca²⁺) to enter that cell's cytoplasm, depolarizing its membrane.

[D] After a brief time, the neurotransmitter is either recycled back into the cell that secreted it (reuptake), or else it is degraded by an enzyme like cholinesterase.

FIGURE 10.8

Structure of a chemical synapse and its function during neurotransmitter release and degradation. The synapse shown is an excitatory synapse; an inhibitory synapse would work identically except that in step [C] Cl⁻ ions would enter or K⁺ would leave the postsynaptic cell, increasing membrane polarity and making the neuron less likely to fire.

carefully preserved vagus nerve. When Loewi electrically stimulated the vagus nerve, the heart connected to this nerve slowed down. Loewi then used a dropper to take some of the fluid surrounding this heart and transfer it to the other heart, which no longer had any nerves leading to it. The second heart also slowed down, showing that some chemical had been present in the first preparation that could be transferred by dropper to the second, for this was the only connection between the two. The chemical was then isolated and identified as *acetylcholine*, the first neurotransmitter to be studied experimentally. Loewi's preparation also allowed the testing of various drugs or other substances that could block or otherwise modify the effect of acetylcholine. These drugs, in turn, could often be used to study whether a particular synapse used acetylcholine as a neurotransmitter.

Types of neurotransmitters. Over a dozen chemicals are now known that can function as neurotransmitters (Table 10.1). Most neurons in the peripheral nervous system use the neurotrans-

mitter acetylcholine; a few use norepinephrine. But the brain uses a much larger variety of neurotransmitters, including all those listed in Table 10.1. Each particular neuron secretes one type of neurotransmitter primarily or exclusively. Each cell responding to a neurotransmitter must have a specific receptor for that neurotransmitter.

Removal of neurotransmitters from the synapse. After a neurotransmitter evokes its response, there is always a mechanism that stops further transmission of the signal by removing loose (unbound) neurotransmitter molecules from the synapse (Fig. 10.8D). Many neurotransmitters are reabsorbed by the cell that secreted them, permitting the same molecules to be recycled and reused. This process, called *reuptake*, is typical of many synapses in the brain. Also, several neurotransmitters can be chemically degraded by enzymes. The amine neurotransmitters can be degraded by the enzyme *MAO* (monoamine oxidase). An enzyme, *cholinesterase*, breaks down acetylcholine molecules after they have stimulated

TABLE 10.1   NEUROTRANSMITTERS

Amine neurotransmitters

Epinephrine
Norepinephrine
Dopamine
Serotonin
Histamine

Amino acid neurotransmitters

Aspartic acid (aspartate)
Glutamic acid (glutamate)
Gamma-aminobutyric acid (GABA)
Glycine

Protein or peptide neurotransmitters (neuropeptides)

Somatostatin
Beta-endorphin
Leu-enkephalin
Met-enkephalin
Substance P

Ester neurotransmitter

Acetylcholine

Gas neurotransmitter

Nitric oxide (NO)

postsynaptic cells outside the brain. Any process interfering with the chemical breakdown or reuptake of neurotransmitters will result in continued presence of the neurotransmitter and thus excessive stimulation of the postsynaptic cell; likewise, an enhancement of chemical breakdown or reuptake results in decreased neurotransmission.

In addition to the synapses between nerve cells, neurons can also form synapses to other cell types, such as muscle cells and glands. Muscle cell membranes, like nerve cell membranes, are electrically excitable; they also contain specific receptors for neurotransmitters, especially acetylcholine. Neuromuscular junctions are thus synapses in which the muscle cell is the postsynaptic cell. Some pesticides work by blocking the action of cholinesterase at neuromuscular junctions and elsewhere. Any synapse stimulated by acetylcholine remains continually stimulated (because

the acetylcholine is never broken down), and the insect receiving the pesticide dies with most of its muscles in a state of rigid contraction. Since most animal nervous systems use acetylcholine as a neurotransmitter, such pesticides are toxic to all animals, great and small, including insects, pets, and humans. A toxic dose, however, depends on body size, so an amount that might kill most insects would only make your pet sick and might not noticeably affect you at all.

How the Brain Produces Movement

Voluntary movements are produced when the brain sends nerve impulses to the body's skeletal muscles. Because the muscles are attached to the skeleton, contraction of the muscles brings about movement of the skeleton and of the body as a whole. Most movements would be jerky and uncontrolled if only a single muscle were involved; controlled, steady movements require the simultaneous contraction of several muscles that pull in different directions to smooth and steady the movement.

Muscle cells contain several proteins. The two major proteins, *actin* and *myosin*, may be arranged in orderly bands, giving certain types of muscle fibers a *striated* (cross-banded) appearance common to both skeletal and cardiac types of muscle tissue. (Smooth muscle lacks this cross-banding because the actin and myosin fibers are arranged at random intervals.) Muscle cell membranes, like nerve cell membranes, have a series of sodium pumps that maintain an electrical potential by actively transporting positively charged sodium ions to the outside of the cell. When a motor neuron releases the neurotransmitter acetylcholine onto the surface of a skeletal muscle fiber, the acetylcholine binds to acetylcholine receptors, causing the muscle cell membrane to depolarize. This causes a release of calcium ions from interior vesicles; these calcium ions bind to a small protein, *troponin*. Troponin is normally bound to myosin, preventing the myosin from binding to actin. But troponin has a greater affinity for calcium ions, so the binding of calcium ions to troponin exposes the myosin and allows the myosin to bind to actin. Cross bridges between actin and myosin can now form, and the pull between these two proteins produces the contractions. Many filaments of actin and myosin slide along one another to produce forcible contraction of the muscle fiber as a whole. The coordinated contraction of many individual muscle fibers leads to controlled movements of the arms and legs, or of the body as a whole.

1. In the ablation technique, the function of a brain portion is investigated by destroying it and studying the behavioral defect produced. What are this technique's limitations? Could you study how a piano or automobile works by inserting a probe, destroying some local region, and studying the resulting defects? Could you study the function of a radio this way? Or a computer?
2. Suppose a scientist proposes the hypothesis that in a select group of hard-to-reach cells, a particular amino acid functions as a neurotransmitter. How would you test this hypothesis? What kinds of drugs or poisons would you look for to help you study the properties of this hypothesized neurotransmitter?

B. Some Diseases Are Related to Neurotransmitter Malfunction

Central to the science of neurobiology is the theory that all mental functions and dysfunctions, including mental illness, are the result of the actions and interactions of neurons and neurotransmitters. The diseases described here are all characterized by malfunctioning neurotransmitters; several are also characterized by the degeneration of particular groups of neurons.

Parkinsonism

This disease presents some of the clearest evidence for a malfunction in a neurotransmitter system resulting in a neurological disorder. The disease produces muscle tremors, including a distinctive "pill-rolling" movement of the thumb and forefinger. People with Parkinsonism have difficulty in initiating voluntary movements. This difficulty is not a true paralysis because patients who stop other activities and concentrate on their voluntary movements can temporarily improve their performance. The walking gait of Parkinson's patients is very characteristic: The body above the waist leans forward, while the feet shuffle slowly in small steps and are barely lifted from the ground.

Parkinsonism results from the degeneration of neurons in a bundle of darkly pigmented fibers called the *substantia nigra*. The degeneration of these fibers impairs the production of the neurotransmitter *dopamine*, depriving certain other brain structures of this neurotransmitter, including the cells of the basal ganglia that normally respond to dopamine. Evidence for the hypothesis that dopamine underproduction plays a large role in Parkinsonism comes from the effectiveness of the drug *L-dopa* (levodopa) in temporarily alleviating many of the symptoms of Parkinsonism. Since dopa is a precursor of dopamine, it is hypothesized that supplying L-dopa (a synthetic form of dopa) can increase dopamine production and thus relieve the symptoms of a dopamine deficiency. Unfortunately, this form of treatment increases dopamine production everywhere, not just to the portion of the brain that needs it. The excess of dopamine in other places may cause serious side effects, such as schizophrenia-like symptoms.

One promising form of therapy that has already been tested is the implantation of fetal tissue into the brains of Parkinson patients. The hypothesis is that the fetal tissue will grow and replace the missing or damaged cells. Adult tissue is unsuitable for this purpose because the brain loses its capacity to regenerate new nerve cells at an early age; even tissue from newborn babies or infants has much less regenerative capacity than fetal tissue. This is one reason why many people in the medical community welcomed the 1993 lifting of an earlier ban on the use of fetal tissues in medical therapy and in medical research. However, the use of fetal tissue raises objections from opponents of abortion because the tissue is obtained in most cases from aborted fetuses.

Huntington's Disease

Huntington's disease (formerly called Huntington's chorea) is a degenerative neurological disease whose genetic basis was discussed in Chapter 3. The disease is marked by uncontrollable spasms or twitches (choreic movements) of many muscles, beginning between ages 40 and 50. As the disease progresses, the spasms become more pronounced, and the patient gradually loses control of all motor functions and of mental processes. A slow death occurs within a few years of the disease's onset.

Huntington's disease results from a destruction of certain brain cells in one of the basal ganglia. Because many of the damaged cells normally inhibit the production and release of dopamine, one major effect of their destruction (and one sign of the disease) is an overproduction of dopamine. Huntington's disease is therefore in some ways a defect in the opposite direction from Parkinsonism. Just as the paucity of dopamine in

Parkinsonism leads to a deficiency of movement, the overproduction of dopamine in Huntington's disease leads to an excess of movement.

How does the destruction of cells in the basal ganglia result in an overproduction of dopamine? One answer may lie in another neurotransmitter, gamma-aminobutyric acid (GABA). GABA functions as a **feedback** mechanism in many neuronal pathways, inhibiting a neuron that has fired from firing again unless another stimulus, larger than the first, is received. GABA normally has this type of feedback inhibition effect on the dopamine-secreting pathways of the substantia nigra (Fig. 10.9). The destruction of the GABA-secreting neurons removes this inhibition and results in an overproduction of dopamine. Drugs known to inhibit the action of dopamine will temporarily reduce the symptoms of Huntington's disease.

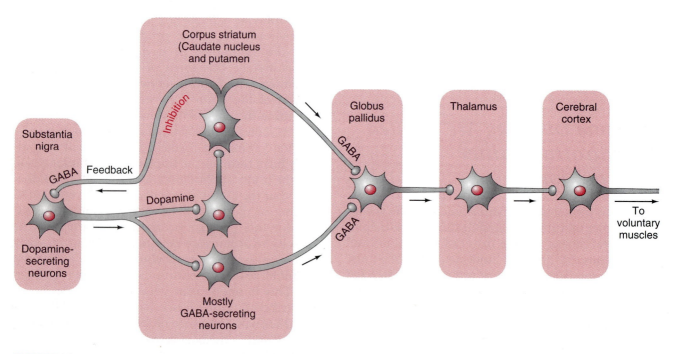

FIGURE 10.9

Feedback inhibition leading to dopamine overproduction. Shown is a neuronal circuit in which dopamine-secreting neurons in the substantia nigra are normally kept at a low level of discharge because GABA secreted by neurons originating in the corpus striatum slows their firing. In Huntington's disease, these GABA neurons degenerate, and their feedback inhibition is removed, so more dopamine is released, causing an excess of movement in the voluntary muscles. All structures here are parts of the forebrain; the corpus striatum and globus pallidus are among the basal ganglia. Arrows indicate the direction of nerve impulses (action potentials).

Alzheimer's Disease

Alzheimer's disease is a form of progressive mental deterioration in which there is memory loss and a loss of control of body functions that ends in complete dependency and death.

Clinical evidence suggests that nerve transmission across certain synapses that use acetylcholine is impaired in Alzheimer's patients, especially in the hippocampus and cerebral cortex. Drugs that inhibit cholinesterase can temporarily improve memory and other brain functions in Alzheimer's patients, although they cannot arrest the course of the disease. Inhibiting cholinesterase would amplify the effect of any existing acetylcholine. This fact, together with the gradual and progressive nature of the mental decline, has caused many workers to hypothesize that the disease is caused by a gradual loss of acetylcholine receptors in postsynaptic cells, while others believe that the primary defect is one involving the synthesis of acetylcholine itself. In 1993, a gene called *apo-e4* was identified to be present in a majority of patients with the most common form of Alzheimer's disease, but the significance of this finding is not yet clear.

Depression

Depression is a disorder marked by feelings of total helplessness, despair, and frequent thoughts of suicide. Many people suffering from depression attempt suicide, and some succeed. The smallest task, such as getting out of bed in the morning, can seem overwhelming. Everyone has unhappy or pessimistic feelings from time to time, but a person suffering from depression has these feelings nearly all the time and to a severe degree. Depression is about twice as prevalent among women as among men.

Reduced amounts of several neurotransmitters have been found in patients suffering from depression. In particular, the brains of depressed patients who commit suicide are found at autopsy to contain reduced amounts of serotonin.

A number of drugs are effective in treating the symptoms of depression. These drugs act in either of two ways: Some drugs act by blocking the action of the enzyme MAO, which degrades many neurotransmitters; other drugs, including Prozac, Elavil, and Tofranil, inhibit the reuptake of serotonin and several other neurotransmitters by presynaptic cells. In either case, the effect is the same: The neurotransmitter remains active for a longer time after it is secreted and results in a greater (or a more lasting) stimulus being passed on to the next cell. Drugs that inhibit serotonin synthesis can reverse the effects of antidepressant drugs, while drugs that inhibit several other neurotransmitters do not have this effect.

How might decreased amounts of a neurotransmitter bring about the symptoms of depression? One possible explanation is that, when something good happens, most people receive a pleasurable stimulus (called a *reinforcement*) in the form of a stimulation to the brain's positive reward centers, and this stimulation makes them more likely to repeat whatever behavior led to the reinforcement (Chapter 11). However, the reinforcement mechanism is not working in patients who are depressed, so they are never rewarded, nor do they learn to repeat whatever behavior led them to the sensation.

Neurotransmitters and Other Diseases

Neurotransmitters may also malfunction in several other diseases in addition to those already described. *Schizophrenia,* for example, is a disorder characterized by frequent delusions and auditory and/or visual hallucinations. Schizophrenia appears to result from an excess of the neurotransmitter dopamine; drugs that stimulate or mimic dopamine (like amphetamines or L-dopa) make schizophrenic symptoms worse, while drugs that block dopamine (such as chlorpromazine and haloperidol) generally lessen symptoms. Further evidence comes from the occurrence of schizophrenic side effects when drugs are used to treat Parkinsonism, and the occurrence of Parkinsonian side effects when drugs are used to treat schizophrenia.

Neurotransmitter activity may also be involved in *epilepsy,* a disorder marked by brain seizures, usually mild, characterized by uncontrolled electrical activity in the cerebral cortex. Both the cerebral cortex and the hippocampus (a deeper part of the forebrain) contain many feedback pathways involving GABA-secreting neurons, inhibiting a neuron that has fired from firing again.

An impairment of one or more of these GABA feedback pathways is hypothesized to allow a stimulated neuron to keep firing, perhaps causing an epileptic fit. Some of the drugs that block GABA synthesis or GABA receptors can bring on epileptiform seizures, thus lending support to the hypothesis. What is not fully explained is the temporary nature of the seizures, the nature of the much longer intervening periods, and the precipitating event or events that initiate seizures.

THOUGHT QUESTIONS

1. In what ways is Parkinsonism a neurological disorder? In what ways is it a psychiatric disorder? Answer the same questions for Alzheimer's disease, Huntington's disease, depression, and schizophrenia. What kinds of distinctions (if any) can you make between psychiatric and neurological disorders?

2. Do you think that people diagnosed with early-stage Alzheimer's disease have a right to end their own lives? Do they have a right to leave instructions directing others to terminate their lives at some future time when they themselves are unable to make such a decision? Does it make a difference how much of their cognitive functions remain? Does it make a difference whether they are in pain? From a deontological perspective, does one ever have the "right" to commit suicide? What is the source of this right?

3. Suppose you were investigating why depression occurs more often in women than in men. How might you test whether differences in upbringing and other cultural influences were at work? Would it be useful to study the prevalence of depression in different cultures? What methodological, ethical, or social problems would such a study face? Would you examine many people or few? What if *depression* were defined differently in each culture? Could you investigate depression by studying an outcome like suicide? What new ethical or social problems might arise from the results of such a study?

4. Is it ethical to give a drug that changes someone's personality? If a drug like Prozac is given to a patient, and the patient kills someone, can the doctor be held responsible?

C. Biological Rhythms Govern Sleep and Wakefulness

Although few people have had personal experience with the disorders examined in the last section, all of us experience sleep and wakefulness nearly every day. Sleep and wakefulness are distinguished by different patterns of electrical activity in the brain. These patterns change in a regular way, which we call a *biological rhythm*.

Sleep

Despite the fact that we each spend almost one-third of our existence in various forms of sleep, remarkably little is known about this particular mental state. At the end of each day, we usually have a strong urge to sleep; we can postpone this urge, but only to a limited extent. People deprived of sleep do not function well when awake (they make more mistakes, for example), and people who awake from a "good night's sleep" feel refreshed and alert.

The electrical activities of single cells take place across their membranes. The cumulative effect of ions flowing across the membranes of many cells can be measured by an *electroencephalogram* (EEG). Electrodes pasted onto the scalp detect different patterns of ion flow in the brain. EEGs show several levels or stages of sleep. Stage 1 is characterized by regular respiration, slowed heart rate, drifts in mental imagery (daydreaming), and low-voltage EEG wave patterns of about 7 to 10 Hz. (Hz stands for *Hertz*, a measure of frequency equivalent to cycles, or waves, per second.) Subjects aroused from stage 1 sleep will often say "I wasn't sleeping." Sleep stages 2, 3, and 4 also have characteristic EEG patterns (Fig. 10.10A). A further sleep stage is characterized by rapid eye movements (*REM*s) noticeable as movements of the eyeball beneath the closed eyelids of sleeping subjects, including cats and dogs as well as humans. About 80 percent of human subjects awakened during REM sleep tell of some dream they were having, often in vivid detail, while subjects awakened during other sleep phases do not remember any dreams.

During a typical night's sleep, an adult goes through about four or five sleep cycles. Each cycle

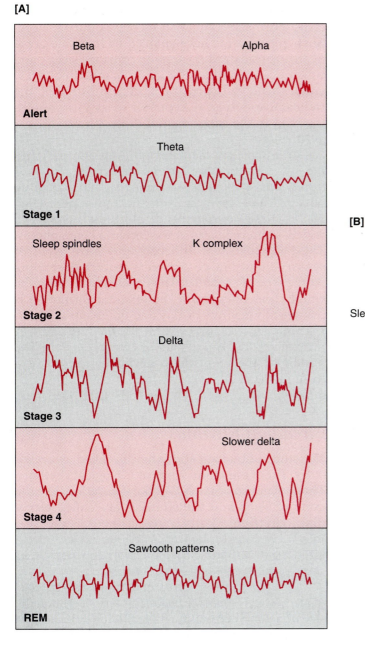

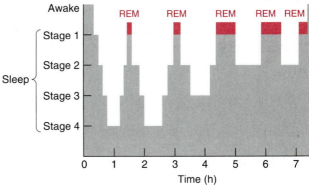

FIGURE 10.10

Characteristics of different stages of sleep. [A] Each sleep stage
has a characteristic pattern of brain waves, as revealed by an electroencephalogram (EEG). [B] A typical night's sleep consists of four to five
sleep cycles of 80 to 120 minutes' duration each. A complete cycle
goes through four stages and ends with a REM episode, but the later
cycles may not go through all four stages.

begins with non-REM stages 1, 2, 3, and 4, in order, and usually ends with a REM period, after which the next cycle begins. The proportion of REM sleep increases (while the proportion of non-REM sleep decreases) with each successive cycle. If subjects are allowed to awaken by themselves (with no alarm clock or other external stimulus), the awakening usually occurs following a REM phase or during stage 1 (Fig. 10.10B). Sleep cycle patterns vary widely according to the individual and the circumstances.

Evidence continues to mount that REM sleep is extremely important. Volunteers who are deprived of REM sleep begin to hallucinate (or dream while still awake) and show other problems after a day or two of deprivation, even if they have had normal or above-normal total hours of sleep, including normal or above-normal amounts of all the other sleep stages. If finally allowed to sleep as much as they wish, REM-deprived subjects sleep more than usual, and a higher than usual proportion of their sleep is REM sleep. People deprived of any of the non-REM phases don't show any abnormal symptoms.

Drugs can in some cases alter the natural occurrence of sleep. Caffeine, amphetamine, and other stimulants can interfere with the onset of sleep, though sensitivity to this effect varies with the individual. Alcohol and barbiturate drugs will bring on sleep more readily, especially in persons already tired, but this drug-induced sleep is less restful because it has longer stage 3 and stage 4 intervals and shorter REM episodes. Muscle relaxants have no effect on sleep except to overcome muscle tension that may be inhibiting the onset of sleep in some people. (Most drugs taken as aids to sleep are either barbiturates or muscle relaxants.) There is no totally safe "sleeping pill," and all drugs that alter sleep patterns usually have other side effects as well.

The role of neurotransmitters in the control of sleep and wakefulness is unclear. The rate of metabolism of neurotransmitters or certain other chemicals during sleep can be studied by labeling these chemicals radioactively. For example, studies in which animals are fed radioactive tryptophan (a chemical precursor from which serotonin is synthesized) show that the rate at which tryptophan is converted into serotonin increases during sleep. Likewise, studies have found that the rate of for-mation of the breakdown products (metabolites) of radioactive serotonin also increases during sleep. Such experiments have enabled us to locate areas of neurotransmitter activity during sleep and wakefulness. In these studies, as in other biological studies with radioactively labeled substances, the radioactive dosage is kept low so as to minimize the risks to the experimental subjects or to the experimenter.

An important mediator of sleep and wakefulness is a group of neurons called the *reticular formation,* or *reticular activating system,* radiating outward and upward from the brain stem (Fig. 10.2). These neurons send "alertness" signals to widely scattered parts of the cerebral hemispheres. These signals seem to accompany most types of sensory input but do not seem to vary depending on the type of stimulus. Their message seems to be simply "pay attention!" The reticular formation is usually more active by day and quiescent at night. Fortunately, it is also possible for the reticular formation to awaken us to an emergency in the middle of the night or to deviate in other ways from its usual 24-hour rhythm, as conditions warrant.

Circadian Rhythms

The study of biological rhythms is called *chronobiology.* There are many kinds of biological rhythms. Those of approximately 24 hours' duration, like the cycle of sleep and wakefulness, are called **circadian rhythms** (Latin *circa,* "about," and *die,* "day"). Various biological functions can be monitored on a 24-hour basis, and a majority of them show some recurrent circadian rhythm.

One place where circadian rhythms are controlled is in the *pineal body* (Fig. 10.2). About the size of a pencil eraser, the pineal body is located in the brain on the roof of the diencephalon. During times of darkness, the pineal body secretes a hormone called *melatonin,* which is not secreted during times of illumination. By illuminating various parts of the body separately from the rest, biologists have experimentally determined that the pineal body is sensitive to the light that it receives right through the skull and brain! If a laser (a highly focused beam of light) is focused on the pineal body through the head, and is turned on and off in a 24-hour rhythm, all parts of the body follow the established circadian rhythms, even when the rest of the body is in darkness. If, instead, the head is kept in darkness while other parts of the body are illuminated, the effect is the same as if the body and head were both in total darkness.

The source of circadian rhythms is an important question. If the rhythms are **endogenous** (internal in origin), how do they keep tuned to the 24-hour cycle of the world around us, and how can they adjust to different time zones, seasons, and work shifts? If, on the other hand, the rhythms are **exogenous** (external in origin), how do external clues regulate the body's cycles, and can these external clues be easily manipulated? The first serious attempts to answer these questions began with *isolation experiments,* such as those conducted in Mammoth Cave in Kentucky. In these experiments, volunteers who were kept from all sources of natural light were allowed to set their own daily routines. Nearly all individuals maintained fairly constant circadian rhythms in their sleep–wake cycles and also in body temperature, activity, and a variety of physiological measurements. Circadian rhythms were all maintained despite the constant environment and were rather uniform for each individual. Most significantly, the circadian rhythms maintained in these isolation experiments were generally slightly more than 24 hours, between 25 and 26 hours for most individuals. This drift from synchrony with the day–night cycle of the outside world showed that endogenous rhythm-keeping mechanisms existed.

How, then, does the external world exert its influence, causing most of us to maintain a 24-hour daily rhythm? Our current understanding is that the external environment provides us with certain time-related clues, or *zeitgebers* (German for "time-givers"). The most important zeitgeber is the rise and fall of melatonin in response to day–night cycles of light intensity. Our own activity and our bombardment with external stimuli (including social stimuli) during daylight hours are also important zeitgebers.

An important mediator of the 24-hour circadian rhythms is a group of neurons called the *suprachiasmatic nucleus.* Destruction of this nucleus abolishes all circadian rhythms. Under experimental conditions of continuous darkness or continuously dim lighting, the suprachiasmatic nucleus maintains an endogenous circadian rhythm with a cycle of slightly more than 24 hours in length. Under more natural conditions, the light received by the pineal body adjusts (or entrains) the rhythm of the suprachiasmatic nucleus to follow a 24-hour cycle, with a peak of activity in the late morning and with greatly reduced activity throughout the hours of darkness.

Gradual or slight disturbances in the natural 24-hour rhythms can occur as a result of short-distance travel, seasonal changes, and the semiannual change of clocks at the beginning and end of daylight savings time (or "fast time"). The effects of these changes are minimal in most cases. More drastic effects are felt in the phenomenon known as *jet lag,* which occurs when our 24-hour rhythms are disturbed as a result of travel through several successive time zones. The major symptom is fatigue (the rigors of travel—even north to south—add to this) plus a desire to sleep or remain awake at inappropriate times for a few days until the body readjusts (entrains) to the new cycle. There is a statistical increase in susceptibility to infection among people who have traveled long distances by air; while some of this may be due to other conditions of air travel, the increase is greater among people who have flown east–west than among those who have flown south–north, suggesting that interrupted circadian rhythms play a part. People who work occasional night shifts, so that their bodies are not able to adjust to any particular rhythm, also show an increase in fatigue-related events such as the number of accidents.

THOUGHT QUESTIONS

1. Disruption of circadian rhythms can suppress the immune system. Do you think that this might be the reason college health centers report that students get more illnesses during final examinations? What other factors might be at work? How would you attempt to distinguish among these factors? (Consider, for example, whether increased illnesses should be expected to continue after exams have ended, or whether the illness rate during the fall semester examination period should be the same as the illness rate during the spring semester examination period.)

2. Sleep deprivation, especially deprivation of REM sleep, makes a person hallucinate. It also

lowers a person's "psychological defenses" (including one's resolve to stand up for what one believes in), which explains why it is often used in political torture. Do you think it is ethical for police interrogators to obtain a confession from a sleep-deprived person? Do you think confessions obtained in this way should be allowed in a court of law?

D. Learning Is a Product of Brain Activity

The Neurobiology of Learning

When we change in response to the changing world around us, this is known as **adaptation.** Adaptation can take place through physiological, immunological, and neurological mechanisms and can be either conscious or unconscious. **Learning,** which consists of lasting changes in behavior or knowledge in response to experience, is an important form of adaptation that is generally assumed to be mediated by the nervous system. Even longer lasting than this is the type of evolutionary adaptation that is best explained as the result of natural selection (Chapter 4).

There are different types of learning that are distinguished by the types of information that are learned and by the neurological basis for the processing and storage of that information. **Declarative memory** or knowledge is mostly conscious remembrance of persons, places, things, and concepts, requiring the actions of the hippocampus and certain parts of the cerebral cortex. Memory of how to do things, **procedural memory** or knowledge, does not require the hippocampus, and is not necessarily conscious. People with hippocampal damage can still learn how to do new things, although they will not consciously recall that they can.

Procedural Learning

Procedural learning can be very simple and is more primitive in the sense that it occurs in simpler animals such as mollusks and insects. Procedural learning also appears earlier during development. Human infants at first learn procedurally: how to eat, how to move, how to respond to gravity. At about the age of 2, the age at which the number of brain cells has reached its maximum and a critical level of complexity of connections has been achieved, declarative learning begins. Although the exact age will vary with the individual, the order of occurrence of the stages does not.

The three simplest kinds of procedural learning are habituation, sensitization, and classical conditioning. These simple forms of learning are distinguished from higher learning in that they can be involuntary: The learner changes his or her behavior without showing any awareness of the learning process. Most animals, no matter how minimal their nervous systems, can learn in these simple ways.

When a stimulus is presented repeatedly, an animal may learn not to respond to it. This is known as **habituation.** Single-celled organisms can stop responding to signals as a result of changes in their membranes or receptors. Single nerve cells can cease responding due to changes in their receptors or to an increase in the threshold for generation of their action potentials. The behavioral responses of whole, multicelled organisms are also examples of habituation. When a stimulus does not result in harm, the organism learns it does not need to expend energy responding to the stimulus. Humans habituate to all kinds of signals: If you move to a new location, you see and hear many things that people who have lived there for a while have learned not to notice any more. After a time, you no longer see or hear them either. If something unusual happens, the habituation can be overcome. Habituation is thus context-specific to some extent.

Another type of simple learning, **sensitization,** is the opposite of habituation. Sensitization occurs when an intense and often aversive stimulus, such as a loud gunshot, increases subsequent responses to other stimuli. On the cellular level, for a prolonged period of time, nerve cells become more capable of generating an action potential, a change known as *long-term potentiation.* Several hypotheses have been suggested to account for long-term potentiation. One recent hypothesis involves a type of glutamate receptor known as *NMDA receptors* (*N*-methyl-D-aspartate receptors). The secretion of glutamate (a neurotransmitter) causes these receptors to respond with an increase in nitric oxide

(NO) by the postsynaptic cell. The nitric oxide then diffuses from the postsynaptic cell to the presynaptic cell, where it enhances the future release of glutamate. Glutamate secretion thus enhances future glutamate secretion in a positive feedback loop. Extreme overstimulation can lead to glutamate poisoning (characterized by convulsions), a condition made worse by the food additive MSG (monosodium glutamate).

A third type of simple learning is called **classical conditioning,** a change in which an organism learns to associate a stimulus with a particular response. Classical conditioning is also called *Pavlovian conditioning* because it was first demonstrated by the Russian physiologist Ivan Pavlov around 1900. Dogs salivate when they see or smell food. If a bell is rung each time food is presented, the dog will learn after a very few repetitions to salivate when it hears the bell ring, even if no food is present.

Humans can learn through classical conditioning, very often without being consciously aware of it. Fears sometimes become associated with various objects, colors, or smells when we are young because those stimuli were present when we were hurt in some way, or because they remind us of other unpleasant stimuli. One person we know grew up with a decades-long aversion to gelatin desserts. It seems that the kindergarten he attended always brought in such desserts on a tray, piled in cubes whose wiggling movements reminded him of certain caterpillars. Caterpillars are something that most children would avoid eating, so the shaking movements in this case conditioned an aversion to the desserts. Adults too become conditioned. In one case, several people became nauseated every time they saw the carpeting in a hospital. It turned out that the carpet was the same color as the chemotherapeutic drugs that these people had taken for treatment of their cancers, drugs that had made them sick to their stomachs. The people had become conditioned, associating the color of the carpet with the cause of their nausea.

Declarative Learning and Memory

In contrast to the three types of simple learning already described, more complex types of learning require the activity of the cerebral cortex. There is evidence that complexity of experience actually contributes to the size of the cortex. Rats raised in "enriched environments"—in large cages with other rats and with "toys"—develop a thicker and more elaborated cortex.

To become a **memory,** a piece of information must be *acquired, stored,* and *retrieved.* Many acquired sensory stimuli can be retrieved for only a short period of time. For example, you may be able to recall having heard a particular sound if someone asks you about the sound within a few minutes of the time you heard it, but not after that. Information that is quickly forgotten is said to be part of *short-term memory.*

Long-term storage requires structural changes within the brain. In order to be changed from short-term into *long-term memory,* a sensory input must be processed through part of the forebrain called the *hippocampus.* People who have suffered damage to the hippocampus are unable to form new long-term memories. They can recall things from before the time of the damage, indicating that the storage sites themselves are not in the hippocampus. Their recall of newly acquired information or experiences is limited to the short term; thus, long-term retention of a memory may also be said to require a "lack of forgetting." This is especially true of *declarative memories,* meaning memories that can be cognitively and consciously known.

The role of the hippocampus in turning short-term memories into long-term ones involves the making of new cellular connections with other parts of the brain. New "cell assemblies" are formed into loops of several neurons; the stimulation of any one of these neurons results in information transfer around the whole loop. The neurons already exist, but the loop is formed by the formation of synapses connecting each cell to the next cell in the loop. Interaction in these cell assemblies may need to continue for years before a memory is permanently stored.

As time passes after the acquisition, *memory consolidation* occurs. While long-term memories are forming, and even after they have been stored, they are organized and restructured based on even more recent experiences. Existing knowledge is constantly being reordered in the light of new knowledge. Memory consolidation relies on information processing, one innate aspect of which is the capacity to make generalizations from specific experiences. If we live in the city and have only walked in a forest one time, we will mentally picture all forests as being like the one we walked in. Further experience, either "in person" or acquired through seeing pictures or reading stories, will enable us to reformulate the initial generalization we made, replacing it with another generalization. In addition, we will still be able to recall some very specific aspects of the particular forest we walked in.

Emotional states can modify the process of memory consolidation: We are more apt to remember something that we would normally not consider worth remembering if we associate it with an event that had great emotional meaning for us (either positive or negative). People who lived through the assassination of John F. Kennedy in 1963 or the explosion of the space shuttle *Challenger* or through flood or hurricane disasters can often remember vivid details of where they were and what they were doing. The same is generally true of events with great personal meaning, like weddings, deaths, accidents, or other significant events.

Abstraction and Generalization

Memory consolidation also relies on the capacity to conceptualize, an extension of the ability to generalize. Researchers can demonstrate the capacity for generalization by using what are called *oddity problems*. Monkeys are shown three or four objects, all alike except one, and they are rewarded for picking the different one. If the first set is two toy trucks and a car, they must learn to pick the car. Presented with a totally different set, two oranges and an apple, they must learn to pick the apple. After many such sets, each different, have been presented, the monkeys develop a concept of "oddness" and pick the single object immediately. This type of learning involves the temporal lobe of the cortex. Much of human learning and memory is processed by the cerebral cortex, consolidating memories and forming concepts, although other parts of the brain are also involved.

THOUGHT QUESTIONS

1. Does the formulation of hypotheses in the scientific method rely upon our capacity to generalize? How can we be sure that our generalizations are accurate? Can we ever "know" how far a generalization can be applied?
2. Is science learned as declarative knowledge or as procedural knowledge? Is science taught as declarative knowledge or procedural knowledge? Is the way science is learned or taught different from the way(s) in which other subjects are learned or taught?

CHAPTER SUMMARY

The work of the brain is carried out by nerve cells (neurons). These neurons maintain differences in ion concentration across the cell membrane and thus a difference in electrical charge known as the resting potential. Neurons carry nerve impulses in the form of action potentials down the length of their axons. Neurons communicate with one another across spaces called synapses by means of such neurotransmitters as acetylcholine, norepinephrine, serotonin, dopamine, GABA, and others. Defects of dopamine neurotransmission may lead to Parkinson's disease, while excessive neurotransmission may produce the uncontrolled movements that characterize Huntington's disease. Defective acetylcholine transmission may characterize Alzheimer's disease, while a deficiency of serotonin is associated with depression.

Sleep is a natural brain function that follows definite stages, each with a distinctive EEG pattern. Dreams, which play an important role in sleep, coincide with periods of rapid eye movements (REMs). Many physiological processes follow a circadian rhythm that is set endogenously and fine-tuned by light acting on the pineal body.

Learning involves activity in the brain but does not require consciousness of the activity. Procedural knowledge can include habituation, sensitization, and classical conditioning, none of which require conscious awareness or involve either the hippocampus or the temporal region of the cerebral cortex. Declarative learning is conscious learning, requiring the hippocampus and cerebral cortex for stimulus processing, memory consolidation, and the formation of generalizations and abstractions.

KEY TERMS TO KNOW

action potential (spike) (p. 292)

adaptation (p. 304)

axons (p. 289)

basal ganglia (p. 286)

central nervous system (p. 284)

circadian rhythms (p. 302)

classical conditioning (p. 305)

declarative memory (p. 304)

dendrites (p. 289)

endogenous (p. 303)

exogenous (p. 303)

feedback (p. 298)

habituation (p. 304)

learning (p. 304)

memory (p. 305)

myelin sheath (p. 289)

nerve impulses (p. 289)

nerves (p. 289)

neurons (p. 289)

neurotransmitter (p. 292)

peripheral nervous system (p. 284)

procedural memory (p. 304)

sensitization (p. 304)

sodium-potassium pump (p. 291)

synapse (p. 289)

threshold (p. 292)

CONNECTIONS TO OTHER CHAPTERS

Chapter 3 Some brain disorders have a genetic basis.

Chapter 4 Similarities in the brains of different species have resulted from evolution.

Chapter 8 Neurotransmitters are made from dietary amino acids.

Chapter 11 Psychoactive drugs alter the functions of the brain, in some cases by affecting neurotransmission.

Chapter 12 Brain activity can influence our general health. Sleep deprivation, for example, can impair the immune system.

Chapter 14 Assistive devices can be used to help overcome or adapt to deficits in brain function.

Chapter 16 Brain disorders may arise from environmental pollution.

Chapter 16 Many drugs that influence brain activity are obtained from tropical rainforest plants.

The Use and Abuse of Drugs

We live in a society in which the consumption of drugs is commonplace. In fact, the United States is the number one drug-producing and the number one drug-using country in the world. Many of these drugs are prescribed for medical reasons: to fight bacterial infections, to fight cancer, or to regulate the body's physiological processes. Many other drugs are taken without medical supervision and for a wide variety of purposes. Many of these drugs are legal; some are not. There is said to be a "drug problem" in our society, which has led to a "war on drugs"; yet there does not seem to be societal agreement on the answers to some very basic questions: What is a drug? What are the various legitimate uses of drugs? What is addiction and what makes a drug addictive? In this chapter we will examine some recent research in biology and in related fields relevant to these questions. We will also examine the basic biological principles that underlie our understanding of how drugs work. An understanding of the respiratory, circulatory, and excretory systems will help us see how drugs enter and become distributed around the body, how long they stay in the body, and how they are eventually removed.

A. Drugs Are Chemicals that Alter Biological Processes

Definitions

The term **drug** can have many meanings, depending on the context in which the term is used. To biologists, a drug is *any chemical substance that alters the function of a living organism* other than by supplying energy or needed nutrients. In a medical context, a drug may be thought of as *any agent used to treat or prevent disease*. Those who work in the field of drug addiction define a **psychoactive drug** as *any chemical substance that alters consciousness, mood, or perception*. As with all definitions in science, these are open-ended; no one definition can cover every possible situation. Each of these definitions expresses a slightly different concept; thus each is correct within its contextual field.

This chapter emphasizes psychoactive drugs, those that alter consciousness, mood, or perception. All psychoactive drugs alter biological functions and are thus considered to be drugs on the basis of the first definition. Many, but not all, are also drugs by the second definition.

In order to understand how psychoactive drugs work, we need to know some general principles that apply to all types of drugs. The study of drugs, their properties, and their effects is called *pharmacology*. Pharmacological principles explain how drugs can be both effective and dangerous, how both concentration and time affect the activity of drugs, and how interactions between drugs can change drug activity.

Drug Activity in Organisms

The effects of any drug vary with the **dose** of the drug, meaning the amount given at one time. There is an *effective dose,* the amount "effective" in producing the desired change; in medicine, the effective dose is also called a *therapeutic dose.* For almost all drugs there is also a *toxic dose,* the amount at which the drug produces harmful effects, and a *lethal dose,* the amount that kills the organism. The more commonly used term, *overdose,* includes both toxic and lethal doses. Some drugs have a wide *margin of safety;* that is, the toxic dose is many hundreds or thousands of times more concentrated than the therapeutic dose. For other drugs, the toxic dose or even the lethal dose may be very close to the effective dose.

The exact amount that is effective, toxic, or lethal differs with the chemical structure of the drug. It also differs from one individual to the next depending on the size of the individual and many other physiological variables. We will discuss some of these variables later in the chapter.

Routes of drug entry into the body. Drugs can enter the body in many different ways. Drugs that are swallowed (taken orally) must be able to withstand the acidic environment of the stomach and then be absorbable by the cells of the intestinal lining in the same way that food substances are taken up (Chapter 8). The presence of food in the intestine can interfere with drug uptake.

Some drugs, for example anesthetics, can be produced in a gaseous form and can thus be inhaled. Inhaled drugs are taken into the blood by *diffusion.* There is a high density of blood capillaries in the lungs that run near the surface of the air pockets (alveoli); together these provide a tremendous surface area for carrying out their normal function, the exchange of oxygen and carbon dioxide (Box 11.1). Inhaled drugs rapidly diffuse into the blood because of this extensive surface.

Most substances that are inhaled are not gaseous but are inhaled instead as small particles. All forms of smoke, including smoke from cigarettes, marijuana, and the burning of fossil fuels (industrial smoke and automobile exhaust), are particulate. In order for a chemical to be absorbed from these particles, the particle must first adhere to the lung tissue. Various chemicals can be highly concentrated within a particle, even if the concentration is low when calculated as amount of chemical per volume of air. Thus an adherent particle may cause substantial damage to the lung tissue, inhibiting the normal functioning of the tissue and damaging the organism's health. Similarly, sniffing particles, such as cocaine powder, can cause local damage to the cells that line the nasal passages.

Cocaine, a product of the coca plant, *Erythroxylon coca,* native to the high Andes, exists in many forms and provides a good illustration of how the form of the drug and the mode of entry can affect the dose of drug actually received. Coca leaves were and are chewed by South American Indians, especially those living at high altitudes. The concentrations absorbed from the gut when coca leaves are chewed are quite low, because the cocaine is present in an ionized form. Ionized molecules carry a charge and thus do not readily diffuse across the nonpolar portion of the plasma membranes of the cells lining the gut (Chapter 8). In contrast, when cocaine is purified into powder form, although the cocaine itself is still in an ionized form, sniffing the powder increases both the rate of absorption and the amount absorbed into

BOX 11.1   GAS EXCHANGE IN THE LUNGS

The lungs are the body organ by which vertebrate land animals take up oxygen and give off carbon dioxide, a waste product of their metabolism (Chapter 8). This process has two parts: the mechanics of breathing and the exchange of gases.

Breathing is accomplished, not by the lungs themselves, but by the diaphragm, a muscular layer below the lungs. When the muscles of the diaphragm contract, they pull the diaphragm downward, expanding the chest cavity. The laws of physics tell us that the *pressure* of a gas depends on the number of gas molecules in a given volume; therefore, when the volume of the chest cavity becomes larger, the pressure inside it decreases. The air pressure inside the lungs is then less than the air pressure outside the body. Physics also tells us that gases flow from areas of higher pressure to areas of lower pressure, so air enters the lungs (inhalation). (Hiccups result when the downward movement of the diaphragm is more rapid than normal.) When the diaphragm relaxes, it returns to its higher position, decreasing the volume of the chest cavity and pushing air back out (exhalation). (The arrows in [A] indicate the up-and-down movement of the diaphragm.)

Once air has entered the lungs, gases are exchanged between the air sacs of the lung (alveoli) and the blood vessels of very small diameter (capillaries) that surround them. Gases move by diffusion (Chapter 8) across the cell membranes of the alveolar cells and the cells of the capillaries. Oxygen is in higher concentration in the inhaled air in the alveoli than it is in the capillaries, so oxygen diffuses *into* the blood. Carbon dioxide is in higher concentration in the blood than it is in the inhaled air, so it diffuses *out* of the blood into the alveoli, to be exhaled. The *rate* of diffusion depends on the *difference* in concentration and on the amount of *surface area* over which the diffusion can occur. The huge number of capillaries [B] and the compartmentalized, saclike structure of the alveoli [C] provide an enormous surface area, making diffusion very rapid. (The effects of compartmentalization on surface area are discussed in Chapters 9 and 14.) Any other substances that enter the lungs and that can diffuse across cell membranes will enter the blood by the same rapid mechanism. Particulate matter, such as the chemicals in smoke, can also adhere to the inner surfaces of the alveoli, delivering drug to the bloodstream.

When we breathe, dust, bacteria, and other particles enter our respiratory tract along with the air. Many of these particles are trapped on a sticky fluid (*mucus*) that coats the respiratory lining. The cells lining the upper respiratory tract have many hairlike projections called *cilia* on their surfaces. These cilia beat rhythmically in a coordinated way; the beating of the cilia moves the mucus and its trapped material upward, out of the respiratory tract. Tobacco smoke stops the beating of the cilia; inhaled bacteria and viruses consequently work their way down into the lungs instead of being eliminated. People who smoke therefore have a much higher incidence of respiratory tract infections and other infectious diseases of the lungs than do nonsmokers. The particulate matter, which can include cancer-causing chemicals in highly concentrated form, also works its way into the lungs and begins the process of transformation of normal cells into cancer cells (Chapter 9). Smokers therefore have higher rates of lung cancer than nonsmokers.

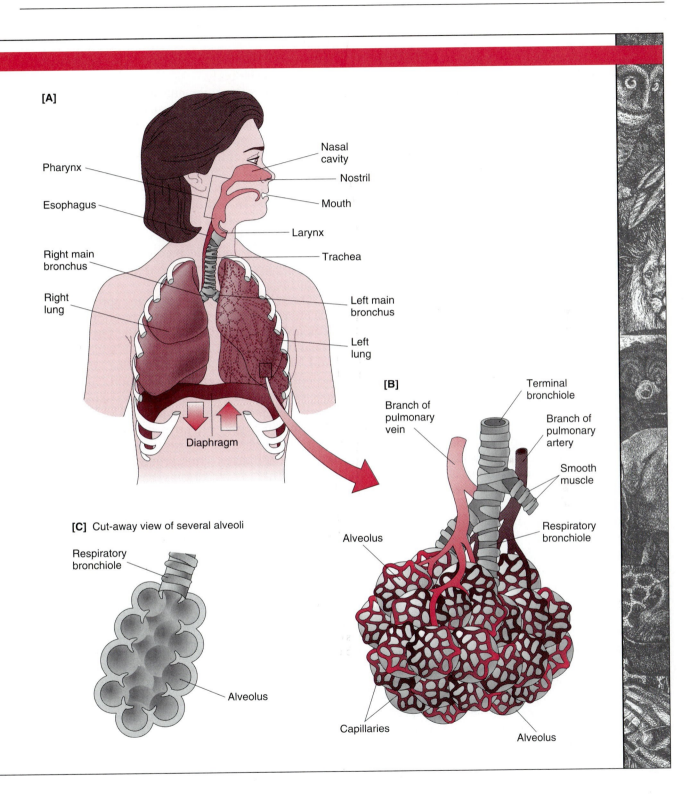

[A]

Pharynx

Esophagus

Right main
bronchus

Right
lung

Nasal
cavity

Nostril

Mouth

Larynx

Trachea

Left main
bronchus

Left
lung

Diaphragm

[B]

Branch of
pulmonary
vein

Terminal
bronchiole

Branch of
pulmonary
artery

Smooth
muscle

Respiratory
bronchiole

Alveolus

Capillaries

Alveolus

[C] Cut-away view of several alveoli

Respiratory
bronchiole

Alveolus

the blood vessels under the lining of the nasal passages. Cocaine can be further purified, resulting in the un-ionized form called *free-base* cocaine and the still more highly purified form known as *crack*. Un-ionized molecules are less polar than ionized molecules and are therefore more rapidly absorbed across cell membranes. In addition, free-base and crack cocaine are no longer powders. They are smoked rather than sniffed, further increasing the rate of absorption, since the surface area of the lungs over which the drug is absorbed is much larger than the surface area of the nasal passages. Each further purification and change in the mode of entry increases the dose and therefore the likelihood of reaching a toxic or lethal dose. The addictive potential (see below), while always great, also increases with each purification.

Drugs may also be administered by injection, that is, with a needle. In a drug injection, the needle can be placed in many locations, including into a muscle (intramuscular), under the skin (subcutaneous), or directly into a vein (intravenous, or IV). Each of these routes may be appropriate for different therapeutic drugs. For example, insulin, a drug given to control a sugar transport disease called diabetes, is usually given intramuscularly so that it will be released into the blood over a period of time rather than all at once; insulin cannot be given orally because it is a protein that would be digested in the gut.

Uptake from an intravenous injection is faster than it is from other routes, since entry is not dependent on absorption. When a drug must be absorbed from the gut, the nasal passages, or the lungs, only a portion of the drug is taken up; intravenous injection puts the drug directly into the bloodstream. Accidental overdoses can more easily arise from injection than from other routes of administration. The biological consequences of drug abuse became more severe after the invention of the hypodermic needle and syringe in 1853. All forms of drug injection can be dangerous if contaminated needles are shared because these practices can transmit infections such as hepatitis and AIDS from one user to another.

Distribution of drugs throughout the organism.
Although the concentration of a drug is initially highest at the local site of entry, the blood vessels (veins and arteries) soon distribute the substance throughout the body.

Transport through veins and arteries is one aspect of drug distribution; transport into the tissues is another. The smallest blood vessels, called *capillaries*, are the places where substances not injected intravenously enter and leave the bloodstream (Box 11.1 and Chapter 14). In most areas of the body, almost every tissue cell is within a few cells of a capillary. In addition to the lungs, areas of the body that have many blood capillaries, such as the liver, kidneys, and heart, will receive a higher dose of most drugs and will receive it faster than will those areas with fewer capillaries (e.g., skin, muscle, or fat).

One site that is very rich in capillaries is the placenta in pregnant females. The mother's blood does not circulate directly into the fetus, but maternal blood vessels in the placenta are close to the fetal blood supply (Fig. 11.1). The microscopic structure of the placenta varies among mammalian species; humans have a particularly close association between the maternal and fetal blood supplies, resulting in a very effective delivery of nutrients and other molecules from the mother's blood into the fetal capillaries. Thus, when a woman is carrying a fetus, many drugs that are transported throughout the mother's body will also be distributed to the fetus. In many cases, the doses that will be toxic or lethal for the fetus are much lower than the doses harmful to adult tissues. Drugs with only a small effect on the mother may, therefore, profoundly and permanently damage the fetus.

Various parts of the circulatory system differ in the molecules that they allow through. In order to act on the nerve cells of the central nervous system, psychoactive drugs must pass across the *blood–brain barrier*, which has several components (Chapter 10). Drugs that do not easily cross this blood–brain barrier will remain in the bloodstream while they are in the brain, even though they can leave the bloodstream in other locations of the body. Most antibiotics, for example, do not cross the blood–brain barrier, making it difficult to treat bacterial infections in the brain. In contrast, alcohol is particularly effective at crossing the blood–brain barrier and in causing both short-term and long-term damage to brain cells.

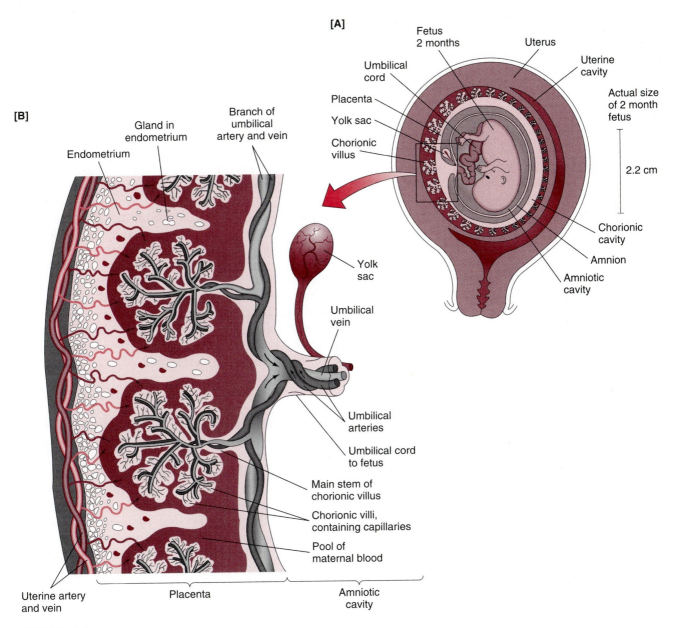

FIGURE 11.1
Maternal and fetal circulation in the placenta. [A] Pregnant uterus. [B] Placenta. Both these drawings show the relationship between the separate fetal circulatory system (shown in gray) and maternal circulatory system (shown in red) at 8 weeks of fetal development.

Drug Receptors and Drug Action on Cells

The activity of a drug depends on its molecular structure and on its dosage or concentration. In many cases, drug activity also depends on the existence of specific receptors for the drug.

Specific receptors for drugs.   After a drug has reached the site of its action, it acts on individual cells within the organism. The ability of a drug to act on a particular cell most often depends on the presence of a cellular **receptor** for that drug. Picture a receptor as being like a glove. For a drug to bind to that receptor, the drug must have the shape of a hand. A small hand will fit into a large glove, but there will be some hand size that has the "best fit." A large hand will not fit into a small glove. Substances that are structurally similar to a drug (analogues) and have the best fit to the receptor will have the greatest activity. Those that do not fit will have no activity.

Many chemical molecules can have right-handed or left-handed shapes depending on the directions in which the bonds between their atoms are formed. That is, two molecules composed of identical atoms can differ in their shape. Right-handed molecules are said to be mirror images of left-handed molecules. As an aid in visualizing this, hold your hands flat. Place them in front of you, with the fingers of both hands pointing up and with the thumbs of both hands pointing toward the left (so that the palm of your left hand is facing toward you and the palm of your right hand is facing away from you). Look at your right hand in a mirror. Does the image of your right hand in the mirror look like your left hand?

Now put your right hand into the right glove of a fitted pair, then try to put it into the left glove. Your right hand will fit only in the right glove of the pair. Similarly, biological receptors are specific for right-handed or left-handed molecules, not both.

Tissue locations of receptors correlate with locations of drug actions. Receptors can be located on the cell surface (plasma membrane receptors) or within the cytoplasm of the cell. Not all types of cells will have all types of receptors. The numbers of copies of a particular receptor molecule can also change over time. The only cells that can respond to a particular drug are those with receptors for that drug. Thus, whether or not a drug will be active in a particular tissue depends both on the ability of the drug to get to the tissue and on the presence of receptors for that drug on the cells of that tissue.

The binding of neurotransmitters to receptors on postsynaptic cells (Chapter 10) is equivalent to drug-receptor binding. In fact, many psychoactive drugs act as either agonists or antagonists of neu-

rotransmitters. An **agonist** is a substance that elicits a particular response or stimulates a receptor; an **antagonist** is a substance that inhibits a response. The interactions of the activities induced in all the receptor-bearing cells will produce the functional effect(s) on the organism as a whole.

Side Effects and Drug Interactions

Because drugs are taken by whole organisms, not by isolated cells, many effects are likely to be produced, especially for drugs that circulate throughout the body. Although a drug is usually given with an intent to produce a specific effect, drugs almost always have effects other than the intended ones; these other effects are called **side effects.** Side effects may be weak or strong and may vary from person to person. Side effects may also vary from beneficial or harmless to harmful or even lethal. Calling something a *side effect* simply means that it is not the main effect or the reason for which the drug was taken.

Types of interactions. When two or more drugs are taken, they may interact in various ways. The simplest interaction is called an **additive effect.** When two drugs elicit the same response, an additive interaction means that the total response elicited is equal to the sum of the responses produced by each drug individually (Fig. 11.2A). In a **synergistic interaction,** one drug *potentiates* the action of the other so that the total response is *greater* than the sum of the responses to each drug separately (Fig. 11.2B). An antagonistic interaction is one in which one drug inhibits the action of another so that the total response is *less* than the response to the two drugs individually (Fig. 11.2C).

Mechanisms of interactions. One mechanism for drug interactions is for one drug to change the threshold for response to the other drug. The **threshold** is a value (in this case, a drug dose) below which no effect is detectable. One drug may lower the threshold for response to the second drug, making the organism responsive to a lower concentration of the second drug. This could happen, for example, if one drug increased the number of receptors or the affinity of the cell receptors for the other drug. On the other hand, one drug may antagonize the action of a second drug by raising

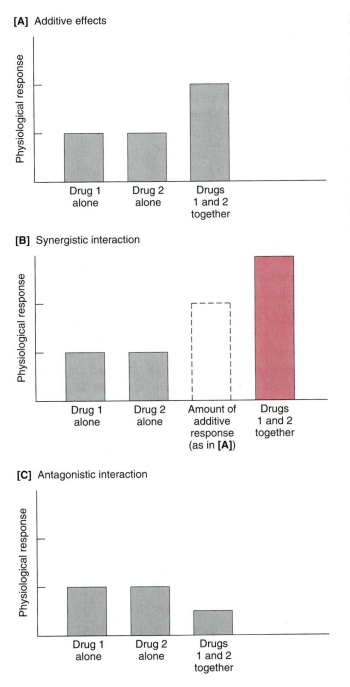

FIGURE 11.2
Drugs may interact in various ways.

the threshold for response to the second drug or decreasing the number or affinity of receptors for it. Again, this is similar to the actions of neurotransmitters (Chapter 10).

Frequency of drug interactions.   Drug interactions are an increasingly important problem, as many people take medication on a lifetime basis for control of chronic conditions such as high blood pressure. A recent study on an elderly population demonstrated that the probability of having some adverse interactions from medications was 75 percent if they were taking five different medicines and 100 percent if they were taking eight or more medications on a regular basis. Potentially harmful drug interactions are not solely a problem for elderly people, however. There can also be harmful interactions between various street drugs, between street drugs and medications, or between any of these and alcohol.

Elimination of Drugs from the Body

In contrast to food molecules, which may be stored (Chapter 8), most drugs begin to be broken down and/or removed from the body as soon as they enter. The active form of the drug may be chemically altered, in a process called *drug metabolism,* so that it is no longer active. This can occur very locally, as in the inactivation of neurotransmitters within synapses between nerve cells (Chapter 10), or it can occur in a central location such as the liver.

Metabolic elimination of alcohol.   Alcohol is metabolized by the cells of the liver. The metabolic breakdown of alcohol involves several steps, but the overall rate is limited by the amount of the proton carrier NAD available to be reduced to NADH (see Chapter 8) and by the levels of the enzyme *alcohol dehydrogenase.* The liver can metabolize 7 to 10 milliliters (ml) (about 0.25 to 0.33 fluid ounces) of alcohol per hour; if the rate of intake exceeds the rate of metabolism, intoxication results. As Table 11.1 shows, intake of most forms of alcoholic beverages will easily exceed this limit.

One step in the metabolic breakdown of alcohol involves the production of acetaldehyde, which is immediately broken down by another enzyme, *acetaldehyde dehydrogenase.* Some people become sick from small amounts of alcohol because they lack this enzyme, leading to a buildup of acetaldehyde, a toxic chemical. In some populations, the frequency of the lack of this enzyme is quite high; for example, as many as 50 percent of all

TABLE 11.1 BEVERAGES CONTAINING EQUIVALENT AMOUNTS OF ETHYL ALCOHOL (ETHANOL)			
BEVERAGE	ALCOHOL CONTENT (%) (VARIES SOMEWHAT)	AMOUNT OF BEVERAGE EQUIVALENT TO ONE DRINK	QUANTITY OF ETHANOL
Beer and ale	4	12-oz. can (350 ml)	1/2 oz. (15 ml)
Table wine	12–15	4-oz. glass (100 ml)	1/2 oz. (15 ml)
Dessert wine (sherry or fortified wine)	20	2/3 glass (2.5 oz., or 70 ml)	1/2 oz. (15 ml)
Distilled liquor (whiskey, vodka, etc.)	40	1.25 oz. (35 ml)	1/2 oz. (15 ml)
Liqueurs	40	1.25 oz. (35 ml)	1/2 oz. (15 ml)

individuals in Japan and China are affected. Disulfiram (Antabuse), a therapeutic drug used in the treatment of alcoholism, works by inhibiting acetaldehyde dehydrogenase in people with normal levels of this enzyme. People who take this drug and then drink alcohol become very sick.

Some drugs alter the metabolism of other drugs.

One way that drugs can interact is by altering the body's ability to metabolize other drugs. Barbiturates are eliminated from the body by enzymes contained in liver cells and the use of barbiturates causes more of these enzymes to be made. Because these same enzymes eliminate other drugs, barbiturate use will lower the body concentrations of other medications, including steroid hormones. Body concentrations of estradiol, a steroid in birth control pills, is decreased by barbiturates, and taking birth control pills and barbiturates at the same time may result in an unwanted pregnancy.

Drug excretion.

Drugs and their metabolites are eliminated from the body by the normal pathways of **excretion** (Box 11.2). Drugs are excreted primarily in the urine, but also in saliva, sweat, and air exhaled from the lungs. In nursing mothers, drugs may also be excreted into breast milk, with obvious consequences for the infant if the excreted drugs are still active, which is very often the case.

Drug half-lives.

The length of time required for the concentration of a drug (in its active form) to be decreased by half is called the **half-life** of the drug. Cocaine, for example, has a half-life of 5 to 15 minutes, meaning that, regardless of the mode of intake of the drug or of the initial starting dose, half of it will be gone in 5 to 15 minutes (Fig. 11.3A). Thus, a drug with a half-life of 10 minutes will remain active for a shorter time than a drug with a half-life of 10 hours (Fig. 11.3B), even when they started out at the same initial concentration in the blood. Drugs that are inactivated in a simple one-step process often have a shorter half-life than those whose inactivation pathways are more complex. For example, the half-life of alcohol is very short in most people, while marijuana remains for up to a week after its use.

Differences in drug half-lives have implications for drug dosage and interactions between drugs. Because drugs with long half-lives will remain in the body for a long time, a second dose will add to the remaining fraction of the first dose to produce a higher total concentration than would result from either dose separately (Fig. 11.3C). Drugs with long half-lives also have the potential to interact with other drugs taken a long time after the first drug. Thus it is not necessary that two drugs be taken at the same time for them to interact.

THOUGHT QUESTIONS

1. The physiological effects of a drug usually depend upon its concentration in body tissues. If two people, one of whom weighs 130 pounds (about 60 kilograms [kg]) and the other 200 pounds (about 90 kg), take the same *amount* of a drug (one beer each, for example, or two aspirins), will the concentration reached in their body tissues be the same? If other factors are equal, which person do you think will be more strongly affected by the drug?

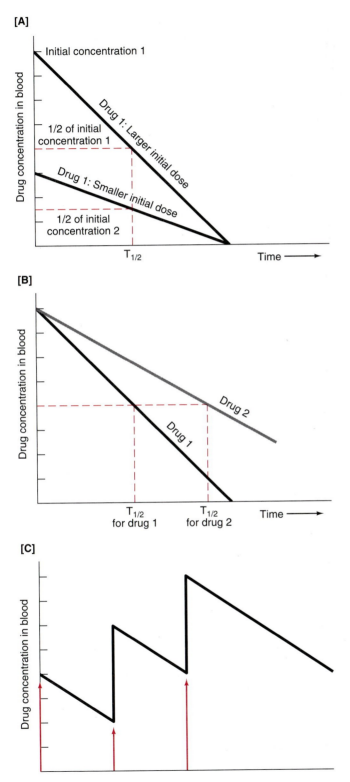

2. A drug administered by an intramuscular injection takes longer to spread through the body than a drug administered intravenously. Why?

3. How would you set up an experiment to monitor the rate at which a drug is delivered to various body tissues? Consider both animal and human test subjects.

4. Why do left-handed and right-handed forms of the same molecule frequently differ in their ability to act as drugs? When a difference of this kind affects the activity of a drug molecule, what does it indicate?

5. How long will it take for the blood concentration of a drug whose half-life is 20 hours to be reduced to one-eighth of its peak concentration?

6. Devise a model of the mechanism of the action of two drugs with their receptors that would account for additivity between the two drugs. How would you modify this model to account for antagonism? How would you modify it to account for synergism?

B. Psychoactive Drugs Affect the Mind

The brain coordinates the activities of the rest of the body (moving, sleeping, eating, breathing), but does it also produce activities we associate with the mind? Does biochemical activity in the brain produce perceptions, emotions, moods, and personality? Does it produce thoughts, dreams, and hopes? The central theory of neurobiology is that the mind and the brain are one and the same. As a framework that guides (and limits) research, the

FIGURE 11.3
Half-lives of drugs. The half-life of a drug ($T_{1/2}$) is the time required for its active concentration to be reduced by half. [A] For each drug, $T_{1/2}$ is the same regardless of its initial concentration. [B] Each drug has a characteristic half-life. Drug 2, with a longer $T_{1/2}$, stays active in the blood longer. [C] Taking a second dose of drug before the first dose is completely eliminated will result in a higher serum concentration, even if the two doses taken are equal.

BOX 11.2   THE EXCRETION OF URINE

The process of removal of material from the body is called excretion. Excretion is vital to maintaining the blood and tissues at the proper concentrations of many types of ions. For example, calcium (Ca^{2+}) is an important second messenger in muscle contraction, and potassium (K^+), sodium (Na^+), and chloride (Cl^-) all function in maintaining charge gradients across membranes and in propagation of nerve action potentials (Chapter 10). Levels of each of these is monitored and controlled by feedback loops which include the kidneys [A]. Urine is produced in three steps—filtration, reabsorption and secretion—which all occur in the nephrons of the kidneys [B], [C].

The first step, filtration, takes place in the many thousand glomeruli. Each glomerulus consists of a mesh of capillaries surrounded by the Bowman's capsule. The circulatory system brings blood to the kidney capillaries that surround the convoluted tubules [C]. All small molecular weight substances, including ions, amino acids, glucose, urea, and water, are filtered out of the blood and into the Bowman's capsule. Larger molecules, such as proteins, are retained in the blood.

The second step, reabsorption, takes place in the loops of kidney tubules, which are also surrounded by blood vessels [C]. Here many of the ions, amino acids, and glucose are reabsorbed by active transport (Chapter 8) via molecular pumps in the membranes of the nephron cells. Much water is then reabsorbed by osmosis.

The third step, secretion, takes place farther down the tubules. Larger molecular weight substances, including drugs and toxins, are secreted into the tubules. After these three stages, the liquid and its contents is called *urine*. Urine goes to the collecting ducts [C] and then from the kidneys to the bladder [A], [B], where it is stored for excretion.

Many drugs, or their metabolites, enter the urine at either the filtration or secretion steps and are excreted in the urine.

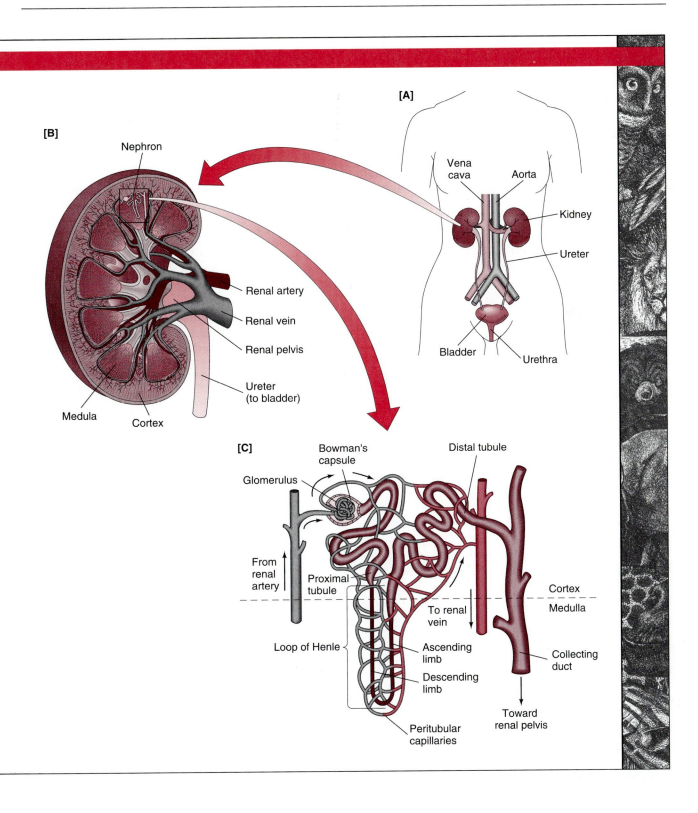

[A]

Vena cava
Aorta
Kidney
Ureter
Bladder
Urethra

[B]

Nephron
Renal artery
Renal vein
Renal pelvis
Ureter (to bladder)
Medula
Cortex

[C]

Bowman's capsule
Distal tubule
Glomerulus
From renal artery
Proximal tubule
Cortex
Medulla
To renal vein
Loop of Henle
Ascending limb
Descending limb
Collecting duct
Peritubular capillaries
Toward renal pelvis

assumption that mind equals brain is a good example of a research paradigm (Chapter 1). There is probably no theory in biology that is more controversial, both among biologists and between biologists and the public.

There is now considerable evidence in support of this theory. Electrical and biochemical activity have been measured in the brain during dreams and thought. Some diseases in which there is brain degeneration can be accompanied by changes in personality. Mental illnesses such as depression are associated with changes in brain neurotransmitters and can be treated with drugs (Chapter 10). Additional evidence for the "mind equals brain" theory comes from research on psychoactive drugs. Psychoactive drugs act directly on the nerve cells of the brain or central nervous system (CNS) to produce changes in consciousness, mood, or perception in highly drug-specific ways. Unlike *sensation,* which is the mere receipt of stimuli, *perception,* a higher-order brain function, includes the processing and interpretation of stimuli coming from the physical world. Both sensation and perception can be altered, sometimes permanently, by drugs. People may become paranoid due to permanent damage from drugs; that is, they perceive danger that is not present in their incoming sensations. They may also experience hallucinations, chemically induced changes in sensation.

Some psychoactive drugs (opiates, marijuana, and nicotine) work through specific receptors, and some (like amphetamines and hallucinogens) work because they are structurally similar to a neurotransmitter and bind to receptors for that neurotransmitter. Drugs that do not work via receptors (e.g., caffeine and alcohol) have a more generalized effect because they act on many types of cells.

Opiates and Opiate Receptors

Opiates are *narcotics,* that is, drugs that cause either a drowsy stupor or sleep. Most narcotics also cause some degree of euphoria (literally, "good feeling"), and most are highly addictive. In low doses, most narcotics can be used as painkillers (analgesics). High doses of narcotics can produce coma and death. Most of our common narcotics, including heroin, morphine, and codeine, are derived from the opium poppy (*Papaver somniferum*) and thus are known as *opiates.* The painkilling and

euphoria-inducing properties of opiates have been known for thousands of years in Asia, as has their ability to cause addiction. Pain reduction results from changes in the release of the neurotransmitters acetylcholine, norepinephrine, dopamine, and a pain-related substance called *substance P.* In people who are not in pain, opiates produce a euphoria by action on neurons in certain locations in the brain.

Opiates and similar chemicals (opioids) act on cells via one or more specific opiate receptors to produce their various effects (Fig. 11.4). Synthetic antagonist drugs have been made that block the binding of opiates to their receptors. These narcotic antagonists, of which naloxone (Narcan) and naltrexone are examples, are useful in the treatment of opiate overdoses. Some other synthetic drugs act as both agonists and antagonists: They bind to opiate receptors, thus blocking the effects of narcotics like morphine (antagonistic action), but their own binding produces some euphoric effect (agonistic action), sometimes leading to their abuse.

Opiate receptors are found on the membranes of many other types of cells in addition to neurons. Opiates consequently relax various muscles, including those of the colon. Consequently, morphine is an ingredient in some prescription antidiarrheal medications (Paregoric), and severe constipation is an effect of long-term opiate use.

Several neurobiologists hypothesized that opiate receptors would never have evolved unless they served some adaptive function rather than just allowing the organism to learn how to become addicted. This line of reasoning led to the hypothesis that opiate receptors must have some normal physiological molecules that could bind to them. The search for such molecules led to the discovery of *endorphins* and *enkephalins,* sometimes called the *endogenous opiates,* since they are produced within the body (endogenous), rather than being taken in from the outside (exogenous). These are peptides (short chains of amino acids), which have a molecular shape very similar to that of a portion of certain opiate molecules. The current theory is that the endogenous opiates act as inhibitory neurotransmitters, decreasing the activity of the neurons that normally signal pain and stress. These same endogenous opiates have other effects throughout the body, including actions on the immune system, which are discussed in Chapter 12.

[A] Cellular locations of psychoactive drug actions

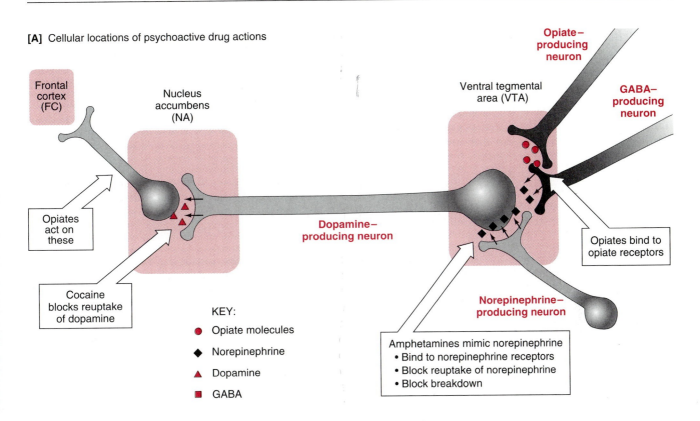

Frontal cortex (FC)

Nucleus accumbens (NA)

Ventral tegmental area (VTA)

Opiate–producing neuron

GABA–producing neuron

Opiates act on these

Opiates bind to opiate receptors

Cocaine blocks reuptake of dopamine

Dopamine–producing neuron

Norepinephrine–producing neuron

KEY:
- ● Opiate molecules
- ◆ Norepinephrine
- ▲ Dopamine
- ■ GABA

Amphetamines mimic norepinephrine
- Bind to norepinephrine receptors
- Block reuptake of norepinephrine
- Block breakdown

[B] Locations of psychoactive drug actions within the brain

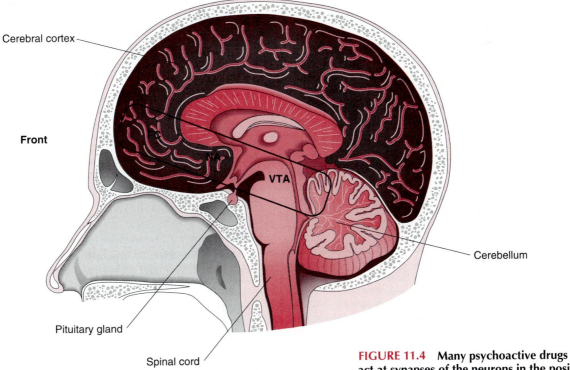

Cerebral cortex

Front

VTA

Cerebellum

Pituitary gland

Spinal cord

FIGURE 11.4 **Many psychoactive drugs act at synapses of the neurons in the positive reinforcing areas of the brain.**

Marijuana and THC Receptors

A number of psychoactive drugs are contained in marijuana smoke, the most active of which is Δ9-tetrahydrocannabinol (THC). There are receptors for THC in the parts of the brain that influence mood. Binding of THC to these receptors produces an altered sense of time, an enhanced feeling of closeness to other people, and an intensity of sensory stimuli. In higher doses, marijuana can also cause hallucinations. Unlike the opiate receptors for which endogenous brain chemicals have been found, endogenous substances that bind to THC receptors have not yet been found.

THC causes release of norepinephrine by the nerves in the median forebrain bundle, producing euphoric effects. There are also THC receptors on the cells of the hypothalamus (Chapter 10), a secretory part of the brain that regulates the steroid sex hormones (Chapter 6). Long-term marijuana use decreases testosterone levels and numbers of sperm in males and alters the menstrual cycle in females.

Nicotine and Nicotinic Receptors

Cigarette smoke contains over 1000 drugs, a large number of which are carcinogens (Chapter 9). The primary psychoactive drug among them is nicotine. In the brain, nicotine acts to stimulate the cerebral cortex, possibly by a direct effect on the cortical neurons, which have a series of nicotinic receptors. Nicotine also acts by stimulating nicotinic receptors on the neurons in the sympathetic ganglia, releasing the neurotransmitters acetylcholine, epinephrine, and norepinephrine. In the brain, norepinephrine produces increased awareness. In the rest of the body, these neurotransmitters produce a variety of physiological effects: increased heart rate and blood pressure, constriction of blood vessels, and changes in carbohydrate and fat metabolism.

Amphetamines: Agonists of Norepinephrine

Amphetamines are an example of a type of drug called *CNS stimulants*. All amphetamines are derivatives of ephedrine, a drug originally obtained from the *mah huang* plant (*Ephedra sinica*). CNS stimulants increase behavioral activity by *increasing* the activity of the reticular formation. The reticular formation is the portion of the brain that normally maintains a baseline level of neuronal activity in the brain as a whole, thus keeping the organism at a baseline level of wakefulness and awareness (Chapter 10). CNS stimulants have side effects on organs outside the brain. Because of their effects on judgment and their effects on other organ systems, such as the heart and diaphragm (Box 11.1), they are dangerous drugs, accounting for 40 percent of all drug-related trips to the emergency room and 50 percent of all sudden deaths due to drugs.

Amphetamines mimic the effects of the neurotransmitter *norepinephrine* by binding to norepinephrine receptors (described in Chapter 10). They can also indirectly increase norepinephrine activity by blocking its reuptake from the synapse and by inhibiting monoamine oxidase (MAO), the enzyme that normally breaks down norepinephrine. Either mechanism results in more norepinephrine remaining in the synapse to act on the postsynaptic cell (Fig. 11.4A). Prolonged use of high doses of amphetamines can induce a form of psychosis that includes aggressiveness, delusions, and hallucinations, possibly because the body has produced an oversupply of an enzyme involved in norepinephrine synthesis.

A controversial drug of this group is methylphenidate (Ritalin). In most human subjects, it has mild amphetaminelike effects similar to those described above. However, the drug has quite a different effect in children who have attention-deficit hyperactive disorder (ADHD, also called ADD)—it reduces their hyperactivity. The reason for this effect seems to be related to the fact that hyperactive children actually have a lower than normal function in the reticular formation, the area of the brain that keeps the rest of the brain alert. Such children may need to be constantly moving around to arouse their reticular formation. Ritalin's chemical stimulation of this area obviates the children's need for movement.

LSD: An Agonist of Serotonin

Lysergic acid diethylamide (LSD) is a derivative of the fungus *Claviceps purpurea*, which grows on rye. LSD, and the related compound *psilocybin* from mushrooms found in Central and South America, are structurally related to the neurotransmitter *serotonin* (Chapter 10). These drugs can activate nerve cells that normally respond to serotonin, leading to heightened sensory perception and hallucinations. *Altered perception* means a change in the awareness of stimuli that actually exist. Colors may become more bright, or sounds more clear. Perceptions of the sizes of objects and of speed or time may be altered. *Hallucination,* on the other hand, is the perception of things for which no outside physical stimuli were received. Hallucinations can be visual, auditory, olfactory, or cognitive. Heavy use of these serotonin agonists leads to permanent brain damage, with symptoms ranging from impairments of memory, attention span, and abstract thinking to severe, long-lasting psychotic reactions.

Caffeine: A General Cellular Stimulant

Not all psychoactive drugs produce their effects via action on CNS neurotransmitters and their receptors. Some produce their effects within cells. Caffeine, for example, works inside cells to increase their rate of metabolism by inhibiting the breakdown of the second messenger, cyclic AMP. Caffeine thus has a general stimulatory effect on cells throughout the body, including neurons in the brain. Like other CNS stimulants, caffeine increases the general level of awareness via action on the neurons of the reticular formation of the brain. In higher doses, or in more susceptible individuals, it can also produce insomnia (inability to sleep), anxiety, and irritability. It increases the heart rate, the respiratory rate, and the rate of excretion of urine by the kidney. It dilates peripheral blood vessels, but constricts the blood vessels of the CNS, which produces headaches in some people at high concentrations.

Caffeine can be derived from several types of plants, including the beans of the coffee plant (*Coffea arabica*), the leaves of the tea plant (*Thea sinensis* or *Camellia sinensis*), the seeds of the cocoa plant (*Theobroma cacao,* from which we get chocolate), and nuts from the kola tree (*Cola acuminata,* an African tree, the source of cola beverages). Table 11.2 shows the amounts of caffeine in different beverages and nonprescription medications.

TABLE 11.2	CAFFEINE CONTENT OF BEVERAGES AND NONPRESCRIPTION MEDICATIONS	
		APPROXIMATE CAFFEINE DOSE (mg)
Hot beverages, per 6 oz. cup (170 ml)		
Coffee, brewed		100–180 mg
Coffee, instant		100–120 mg
Tea, brewed from bag or leaves		35–90 mg
Cocoa		5–50 mg
Coffee, decaffeinated		2–4 mg
Caffeine-containing carbonated beverages, per 12 oz. can (350 ml)		35–60 mg
Stimulants, per tablet		
Vivarin, Caffedrine		200 mg
NoDoz		100 mg
Analgesics, per tablet		
Excedrin extra strength		65 mg
Midol maximum strength		60 mg
Anacin, Bromo Seltzer, Cope, Emprin		32 mg

Alcohol: A CNS Depressant

Alcohol belongs to a category of drugs called *CNS depressants* because they depress the functioning of the CNS by inhibiting transmission of signals in the reticular formation. Included in this category are barbiturates and tranquilizers. Because these drugs lower the general level of awareness, they are also called sedatives or hypnotics. Because they all affect the reticular formation, when two or more CNS depressants are taken together, the effect is stronger (either additively or synergistically) than when either one is used alone. The actions of barbiturates and tranquilizers are mediated through receptors on neurons in the brain, while alcohol produces a more generalized effect because it acts by making all cell membranes more fluid.

Alcohol is soluble in both water and fat and is thus readily able to pass through the plasma membrane of the cells forming the blood–brain barrier. The portions of the brain affected will depend on dose; the higher the dose the deeper into the brain the alcohol penetrates. Even low doses of alcohol can impair a person's response time, with devastating (often fatal) consequences if that person is driving a motor vehicle or boat. In the United States, alcohol is involved in over 60 percent of all motor vehicle fatalities and in over half of all drownings. Higher doses of alcohol suppress the reticular formation's stimulation of those portions of the brain involved in involuntary processes, such as the brainstem and the medulla oblongata. Depression of the medulla can result in cessation of breathing (respiratory arrest) and death.

The effects of alcohol on behavior can be predicted by the blood alcohol level (Table 11.3). Blood alcohol content is measured as the number of grams of alcohol in each 100 ml of blood. Thus 1 gram (g) of alcohol per 100 ml equals a 1 percent blood alcohol content and 100 mg equals a 0.1 percent level. Since alcohol is evenly distributed throughout the body, the actual blood alcohol content that results from drinking a given amount of alcohol will vary with the blood volume of the person, which is approximately proportional to the muscle weight of the person. A blood alcohol level of 0.1 is the level at which a person can be charged with driving while intoxicated (DWI) or operating (a motor vehicle) under the influence (OUI) of alcohol in most states of the United States, and several states have amended their laws to make the limit even lower. At a blood alcohol content of 0.05, the probability of being involved in an automobile accident is 2 to 3 times higher than it is for someone who has not been drinking.

THOUGHT QUESTIONS

1. Does the finding that many drugs act directly on the cells of the brain to alter perception, mood, and consciousness necessarily mean that there is no "mind" or "spiritual essence" apart from the physical entity of the brain?

2. Can a person accept the scientific findings in neuroscience without accepting the central theory that the mind and the brain are one? Can the central theory of neurobiology ever be proven beyond doubt?

3. How many eqivalent "drinks" are there in a bottle of wine, which is typically 750 ml? How many "drinks" are there in a quart of liquor? (1 ounce = 28.4 ml; 1 quart = 32 ounces)

C. Some Psychoactive Drugs Are Addictive

Some drugs that have uses as medicines, and many others that do not, are also used socially. All cultures have used at least some drugs, particularly psychoactive drugs, for nonmedical purposes that can be described as social. For example, in many cultures, wine is a common accompaniment to food, and wine is also used in many religious ceremonies.

Excessive or harmful social use of a drug is considered drug abuse or substance abuse. Drug abuse is a major problem that affects many thousands of people, their families, and often many other people as well. Most drugs that are abused socially are psychoactive drugs. These drugs have both direct effects on the user and indirect effects on other people due to the user's altered behavior.

If psychoactive drugs have been used by all cultures, why not sanction their use? In cultures where social use of psychoactive drugs has been sanctioned by tradition, the uses often have been highly ritualized or ceremonial, and the decision of when and how much drug to use is not left up to the individual. Psychoactive drugs such as alcohol impair a number of higher-order mental functions (Table 11.3). One of these is judgment, the very mental function needed for a person to be able to distinguish between "use" and "abuse." In addition, most, but not all, drugs that are abused are addictive.

Addiction is often defined as a compulsive "physiological and psychological" need for a substance, implying that there is both a biological basis and a mental basis for addiction. However, as more psychologists have accepted the neurobiology paradigm that "mind equals brain" and that all brain functions are biochemically based, the distinction between physiological and psychological addiction has become increasingly blurred. The above definition of addiction carries with it the assumption that all addictive drugs are psychoactive. The term *psychoactive drug* is not a synonym for *addictive drug*, however, because not all psychoactive drugs are physiologically addictive. Hallucinogens such as LSD, for example, are not known to be addictive. The use of addictive drugs

TABLE 11.3 BRAIN AND BEHAVIORAL EFFECTS OF ALCOHOL CONSUMPTION

BLOOD ALCOHOL LEVEL (%)*	NO. OF DRINKS†	EFFECTS
0.01–0.04	1–2	Slightly impaired judgment Lessening of inhibitions and restraints Alteration of mood
0.05–0.06	3–4	Disrupted judgment Impaired muscle coordination Lessening of mental function

0.07–0.10	5–6	Deeper areas of cortex affected:
		    Reduced reaction time
		    Exaggerated emotions
		    Talkativeness or social withdrawal
		    Mental impairment
		    Visual impairment

0.11–0.16	7–8	Cerebellum affected:
		    Staggering
		    Slurred speech
		    Blurred vision
		    Greater impairment of judgment, coordination, and mental function

0.17–0.20	9–10	Midbrain affected:
		    Inability to walk or do simple tasks
		    Double vision
		    Outbursts of emotion

0.21–0.39	11–15	Lower brain affected:
		    Stupor and confusion
		    Increased potential for violence
		    Noncomprehension of events

0.40–0.50	16–25	Activity of lower brain centers severely depressed:
		    Loss of consciousness
		    Shock

0.51	26	Failure of brain to regulate heart and breathing: Coma and death

*Blood alcohol content is given for a 150-pound person; in general, persons weighing less will experience higher blood alcohol levels after consuming the same amounts.
†Based on equivalent amounts of alcohol: 12 ounces of beer, 4 ounces of wine, 1.25 ounces of liquor (See Table 11.1)

by people in the United States is shown in Fig. 11.5.

Dependence and Withdrawal

Addictive drugs cause a physiological **dependence,** meaning that the person can no longer function normally without the drug. Once a person has become dependent on a drug, cessation of drug taking produces the biological symptoms of **withdrawal.** The length of time required for the development of dependence varies with the drug, as does the severity of withdrawal. Dependence on morphine and related drugs develops very quickly; withdrawal begins within 48 hours of the last dose and lasts about 10 days. People are very ill during

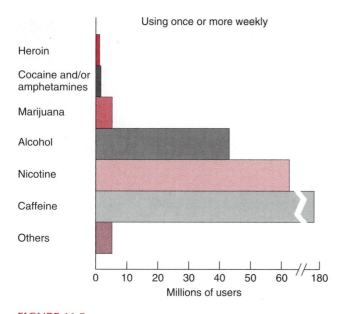

FIGURE 11.5
The use of addictive drugs in the United States in 1991.
People are listed as "users" if they used the drug at least
once a week (or, for heroin, at least once in the past year).

this time, but withdrawal is rarely fatal. With-drawal from alcohol dependence is physically much more severe and can sometimes be fatal.

Psychological dependence.   One form of addiction is psychological dependence. In this form of addiction, the physiology of the brain has become dependent on the drug in such a way that the person can no longer function normally without the drug. The sensations felt during withdrawal tend to be the opposite of the sensations produced by the original drug taking. For example, when one has become dependent on depressant drugs, anxiety and agitation may result when they are withdrawn. People who have become dependent on painkillers will feel pain when the drug is withdrawn, long after the original, biological source of the pain has gone away, and these pain sensations often lead to resumption of the drug.

Cocaine, caffeine, nicotine, and marijuana all produce psychological dependence. Withdrawal from these drugs is typified by symptoms that affect the physiology of the brain rather than the physiology of the entire body, and because of this they were originally thought to be nonaddictive. In light of the neurobiology paradigm, it is now

known that all four are addictive. In fact, if we define level of addiction as the degree of difficulty of getting through withdrawal without returning to the drug, nicotine must be considered as one of the most highly addictive drugs known.

One aspect of psychological drug dependence in humans involves the context—the places and people—in which the drugs are taken. *Conditioned withdrawal syndrome* refers to the fact that visual cues associated with drug taking (for example, seeing one of these people or places) can bring on the physiological symptoms of withdrawal in drug-dependent people even when the body is not actually in withdrawal, possibly leading them to seek the drug again. For this reason, many drug recovery programs recommend that recovering addicts stay away from specific people and locations associated with drug use.

Brain Reward Centers and Drug-Seeking Behaviors

Neuroscientists have discovered both negative and positive reward, or reinforcement, systems in the brain. Certain things make us feel good (positive reward, or **positive reinforcement**). Stimulation of the nerves in the positive reinforcement system leads to a repetition of the behavior. Basic biological functions like eating and sexual activity are repeated because they activate these nerves. Throughout our lives we learn other experiences that stimulate these centers. Our ability to derive pleasure from certain experiences, such as the feeling of "a job well done" or the feelings evoked by a beautiful painting or a sunset, is *learned* to a large extent; our families, religions, cultures, and other influences operate from birth to teach us to view certain experiences as positive and certain experiences as negative.

Two theories of drug addiction.   One theory of drug addiction is that people use drugs to escape from some kind of pain, either physical pain or psychological pain, caused by many possible factors. This theory suggests that drug-taking behavior results from the attempt to inhibit or avoid the negative reward and to avoid withdrawal. The removal of an unpleasant stimulus is called **negative reinforcement.**

A newer theory suggests that drug addiction results from stimulation of the positive reward centers. Of the psychoactive drugs, the ones that cause addiction are those that stimulate the positive reward system in the brain. They directly stimulate these centers, neurochemically producing the positive sensation. This theory is supported by the findings that addiction may occur very rapidly, well before physiological dependence has begun. Both theories may be valid for different drugs or different individuals, meaning that there may be more than one mechanism for addiction.

Evidence from behavioral experiments in rats.

Much of the research in the area of drug-seeking behavior has been done with rats. Rats fitted with tubes so that they can give themselves drugs, either into their bodies or directly into specific parts of their brains, rapidly learn to self-administer addictive drugs. They increase the frequency of self-administration if allowed, while they do not self-administer nonaddictive drugs. Although physiological dependence does develop, these experiments suggest that dependence is the result of addiction rather than its cause.

A part of the brainstem known as the *ventral tegmental area* (Fig. 11.4B) is thought to be the positive reward (reinforcing) center. This has been demonstrated in several types of experiments. If electrodes are placed into the ventral tegmental area, rats will activate the electrodes, electrically stimulating that area of the brain; they quickly learn to repeat the behavior, giving themselves repeated electrical stimulation, preferring it even over food. The same type of experiment can be done on rhesus monkeys, with the same results: They will refuse food if their choice is between eating (even after starvation) and stimulating their ventral tegmental areas electrically. Electrodes placed in other areas do not produce repeated self-stimulation, and destruction of the ventral tegmental area of the brain stops the behavior.

In another type of rat experiment, tubes are placed into an area of the brain so that by pushing a lever the rat can administer drug directly to the area. By their action on neurons or their receptors, psychoactive drugs stimulate nerve impulses in the brain, producing the same effects as elicited by electrical stimulation of those same neurons. The negative reward centers are in the periventricular areas of the brain, while the positive reward center, as previously stated, is in the ventral tegmental area. The nerves of the ventral tegmental area make synaptic connections to the nerves of another brain area, the nucleus accumbens, which are involved in the processing or interpretation of the signal (Fig. 11.4A). Self-administration of drug to the negative reward center is not reinforcing and thus is not repeated, while self-administration to the ventral tegmental area or the nucleus accumbens is.

When rats are able to take drugs by a more normal route, rather than directly into their brains, the results are similar: They repeat the drug taking if the drug is addictive and do not when the drug is nonaddictive. Measurements of changes in electrical activity in different parts of the CNS have shown that activity increases in the ventral tegmental area (and generally nowhere else) following administration of an addictive drug.

The ventral tegmental area of the brain includes the reticular formation described above, so that many of the drugs that act on the reticular formation also act on the positive reinforcing area of the ventral tegmental area. Amphetamines indirectly stimulate the neurons of the ventral tegmental area (Fig. 11.4B), elevating mood. For this reason they have been successful in the treatment of depression. Cocaine acts on brain cells of the ventral tegmental area that secrete dopamine. Most researchers think that the euphoria produced by cocaine is due to its effects on the dopamine-secreting cells because the euphoria can be stopped with drugs that block dopamine receptors on postsynaptic cells (Fig. 11.4A). Opiates, marijuana, caffeine, and alcohol all produce ventral tegmental self-reinforcing effects.

Not all psychoactive drugs are addictive. Hallucinogens, for example, do not produce repeated self-administration by rats and thus are not considered addictive under this theory. Nicotine does not initially produce self-administration in rats; in fact, initially it is strongly aversive. However, after a rat has been exposed to nicotine several times, it begins to self-administer the drug and this becomes a strongly persistent behavior, that is, a behavior that is hard to break.

Conditioned learning in drug addiction.

Addiction is both a biological response and a learned

behavioral response in which the behavior being learned is the drug-seeking and drug-taking behavior. The sensation produced by the drug does not need to be learned—it is a direct chemical stimulation of the nerves—but the behaviors that resulted in the sensation become learned. This is a type of procedural learning called *operant conditioning,* in which behavior is learned as a result of its consequences. Drug seeking and drug taking can be learned by operant conditioning if taking the drug is usually associated with stimulation of the positive reward centers in the brain. Being in certain places or being with people with whom drug taking has occurred provide strong learned cues that bring on a physical sensation of craving for the drug. (See the section on conditioned learning in Chapter 10.) Thinking of those places or longing for those people will bring on craving for the drug and can also bring on the sensations produced by the drug itself. Seeing or thinking about aspects of the drug taking itself is often sufficient to bring on these feelings. A person dependent on cocaine reported that the sight of someone wearing a gold watch would bring on the sensations of a cocaine high, since gold watches were one of the items he had often stolen to purchase his cocaine. The contribution of the "drug culture" to the reinforcement of drug-taking behavior is enormous: Many of the symbols and rituals associated with the culture become contextual cues. Attempts to block and reverse drug dependence must take into account these learned associations, which are called *conditioned place preferences.* Rehabilitation of drug addicts is usually very difficult because of the strength of these learned associations. However, the chances of successful rehabilitation are increased if the addict can be helped to develop an aversion to the old behaviors while learning to substitute new behaviors.

Effects of long-term use. One of the effects of long-term use of psychoactive drugs is that they erase the ability of the ventral tegmental nerves to respond to the normal positive signals; appreciation of a good meal, enjoyment of the company of friends, happiness from helping others may all disappear. We tend to interpret positive experiences as "pleasure"; thus the positive reinforcement centers have sometimes been called the *pleasure centers.*

Drugs compete with the normal neurotransmitters in these brain centers. Long-term use of addictive drugs decreases the number of receptors on the nerve cells (see below) so that these centers are only triggered in the drug-abusing person by taking drugs, and no longer by the experiences that used to be pleasurable.

Drug Tolerance

One of the biological effects of addictive drugs is that they produce **tolerance,** which is also called *homeostatic compensation* (i.e., the body adjusting to new conditions). This means that the same dose of drug exerts a decreased effect when administered repeatedly. It also means that in order to produce the same effect, greater amounts of the drug must be taken. Some people have suggested that addicted people be given "all the drugs they want" to keep them off the street. There is a simple biological reason why such an approach would not work, and that is that drug tolerance develops. A person needs more and more drug to produce the original psychoactive effect. Higher doses affect other body systems and begin to produce negative mental states such as hostility and paranoia. The increasing doses may often become toxic or lethal.

Drug tolerance can be shown on a graph as a shift of the drug's dose-response curve to the right (see Fig. 11.6). While the graph shows that tolerance has occurred, it does not tell us anything about the mechanism(s) that produced it. There are two broad categories of mechanisms: metabolic and cellular. Metabolic tolerance occurs when the body produces an increased amount of the enzymes involved in the breakdown of the drug. Tolerance to barbiturates develops at least in part because they stimulate the synthesis of the liver enzymes responsible for their elimination, so their rate of elimination increases with repeated use.

Cellular tolerance results from changes in the receptors for the drug, principally the receptors on the nerve cells in the brain. Either a drug-induced decrease in the number of receptors or an increase in their response threshold results in tolerance. Heroin produces these changes in the brain within a week or two of daily use. If use is more frequent

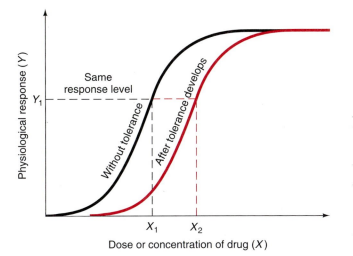

FIGURE 11.6
Drug tolerance causes a shift of the dose–response curve. Each curve shows the physiological effects (responses) at increasing drug concentrations (doses). Compare the two dose–response curves shown here. For a given response, the dose (X_1) that would produce that response can be determined from the curve on the left. Notice that a higher dose (X_2) is required to produce the same response after drug tolerance develops (curve on the right).

than once a day, a higher level of tolerance develops; that is, even fewer receptors are present on the nerve cells so an even higher dose is required to produce an effect.

Some drugs can cause permanent damage to receptors, and therefore tolerance to these drugs becomes permanent. For most drugs, however, tolerance is not permanent but disappears gradually with time. The time period varies greatly from one drug to another.

1. If a person has never learned to feel pleasure from daily activities, will they be more attracted to the "artificial" pleasure offered by drugs? Can this suggest anything to us about drug prevention strategies? One slogan says that "hugs are better than drugs." Does this slogan correlate with neurobiological findings? Can this idea be applied in drug prevention programs?

2. Since psychoactive drugs work directly on the brain cells, are individuals exempted from re-

sponsibility for their own drug use? Apply this reasoning to alcohol and tobacco as well as illegal drugs.

D. Drug Abuse Impairs Health

Drug Effects on the Health of Drug Users

In addition to addiction, the use of psychoactive drugs has many effects on the health of the individual drug user. Drug use that negatively affects the health of individuals or society is called **drug abuse,** or **substance abuse.** In the United States, deaths from drug abuse increased from 6500 in 1979 to 10,000 in 1988. The majority of these deaths were due to heroin and cocaine, with the remainder mostly the result of the misuse of legal drugs such as alcohol (Wysowski et al., 1993).

Alcohol. Among the most harmful of drugs, particularly in light of the frequency of its use, is alcohol. While each of the other types of CNS depressants are designated as controlled substances, subject to fines and/or imprisonment for possession or sale in the United States and in many other countries, alcohol is not. The reasons for the difference are certainly not biological, as alcohol (ethanol) is a powerful depressant comparable to the others in terms of both short-term and long-term risks to health.

Most of the biological effects of alcohol on the body are **acute effects,** reversible in a matter of hours or days. Over time, however, permanent damage results from the **chronic effects.** Alcohol-induced biochemical imbalances permanently damage tissues such as the brain, liver, and heart and other muscle tissue, resulting in dementia, cirrhosis and other liver diseases, and cardiovascular disease, respectively. Tissue damage may in turn contribute to altered uptake of and sensitivity to medically necessary drugs, including antibiotics, antidiabetic drugs, and other medications. Alcohol in the gut also destroys certain vitamins and interferes with the absorption of others. This is why vitamin deficiencies rarely seen in industrial countries occur in those countries among alcoholics (Chapter 8).

Alcohol depresses the immune system, leaving alcoholics very susceptible to infectious disease, including tuberculosis. Since alcoholics frequently substitute alcohol for food, they are often malnourished in either total calories or in micronutrients, which can further depress the immune system (Chapter 8). Alcohol consumption is also associated with an increased risk for cancer (Chapter 9).

Caffeine.   Caffeine, which is even more widely used than alcohol, can have significant negative effects on health. Consumption of more than 10 cups of coffee a day causes chromosome damage (which can lead to birth defects), respiratory difficulties, and heart and circulatory problems.

Tobacco.   Tobacco is carcinogenic in any form. Smoking tobacco is correlated with lung cancer, while chewing tobacco is correlated with cancer in the mouth, a form of cancer that often metastasizes to other parts of the body (Chapter 9). Tobacco and alcohol together produce a synergistic increase in the number of cancer deaths, beyond what either would produce without the other (Chapter 9). Since nicotine also suppresses the immune system, as well as the action of the cilia in the respiratory tract, smokers have a high incidence of other lung diseases, including emphysema and respiratory tract infections such as pneumonia and bronchitis. Passive smoke is related to an increase in sudden infant death syndrome as well as to the incidence of pneumonia and bronchitis in the first year of an infant's life. Since an infant's immune system is only partially developed, passive smoke can be much more damaging to infants than to adults.

Marijuana.   The present-day controversy over the use and abuse of marijuana has led to a large amount of research on its effects. Drug tolerance does develop, as does physiological dependency in some heavy users. Although marijuana has a few possible medicinal uses, it is also known to have several biologically adverse effects. Possible medicinal uses include the reduction of eyeball pressure in glaucoma patients, the stimulation of appetite and suppression of nausea in cancer patients un-

dergoing chemotherapy, and the stimulation of appetite in AIDS patients. The U.S. Congress has approved the medical use of marijuana for only a small number of individual patients by special legislation.

The harmful effects of marijuana are much better understood scientifically. When smoked in cigarette form, much of the particulate matter in marijuana smoke stays in the lungs and builds up to form tar. Marijuana smoke produces more tar per weight of plant material than does tobacco smoke, and the tar is equally carcinogenic. It also inhibits the immune cells that clear debris from the lungs and protect against airborne infectious bacteria and viruses. All forms of marijuana alter the production of reproductive hormones, decreasing the production of sperm in men and ovulation in women. Men who use marijuana over long periods of time often develop fatty enlargement of the breasts (gynecomastia).

Designer drugs.   The term *designer drugs* refers to those drugs that are slight structural alterations of existing drugs. MPTP and MPPP are two designer derivatives of Demerol (meperidine), itself a derivative of opium. These two have psychoactivity similar to other opiates, but are also potent neurotoxins (nerve cell poisons). They particularly destroy the nerve cells in the substantia nigra, the area of the brain that controls movement, causing movement defects similar to those found in Parkinson's disease, a condition that mostly affects elderly people (see Chapter 10). In 1985, 400 cases of Parkinsonism in young people were found to be due to MPTP. After these cases, these drugs were added to the list of controlled substances, so use dropped. Designer drugs are often made to get around laws that ban particular drugs by name. New designer drugs may be legal until laws are rewritten to cover them.

Over-the-counter drugs.   Over-the-counter (OTC) drugs are those that may legally be sold without a prescription. Although people tend to view OTC drugs as "safe" because a prescription is not required, there are actually thousands of deaths from OTC drugs each year and over one million cases of drug poisoning, generally from overdose. Aspirin is second only to barbiturates as the drug most frequently used in suicides and suicide attempts.

Several OTC preparations contain combinations of many drugs. Most of the OTC sleeping pills are combinations of aspirin and antihistamines. Cold remedies often have many ingredients, and the liquid ones contain high concentrations (25 percent) of alcohol. These combination drugs are marketed by urging people to "cover the bases" by treating all the symptoms. There is the unstated assumption in such advertising that the taking of unneeded drugs is harmless; there is no mention of possible negative effects of any of the ingredients.

Drug contaminants and additives. Although prescription drugs and over-the-counter drugs are not "safe" unless taken as directed (and even then are not without their side effects), the consumer can at least be assured that the product contains the drug it claims to contain and that the quantities of it are standardized from one batch to the next. Although no medication is sold as an unmixed, pure compound, the purchaser can be assured that harmful impurities are not present. None of these assurances exist, however, for the purchaser of street drugs. Many impurities are present from the chemical synthesis itself. Many others are deliberately added. Strychnine is sometimes added to LSD ("white acid"), supposedly to sensitize the nerves to the drug. It is also sometimes added to marijuana without the knowledge of the purchaser. By acting on the nerve cells, strychnine causes abnormal muscle contractions (convulsions) and is therefore sometimes used as rat poison. There is no specific antidote for strychnine, so it cannot be counteracted once it is taken, and it can cause permanent nerve damage in the brain and elsewhere.

Steroids. Although most commonly abused drugs are psychoactive drugs, not all drugs that are abused belong to this category. Anabolic steroids are abused instead because of their hormonal effects on physical development. Certain athletes of both sexes (bodybuilders, weight lifters, swimmers, runners, football players, and others) have used these drugs because they cause an increase in muscle mass, resulting in a bulkier and more powerful physique. These drugs are not addictive, but their many dangerous side effects include damage to the reproductive organs and the circulatory system, especially the heart, as well as increased hair in some places and premature baldness in others. Deaths have occurred among amateur and professional athletes from the abuse of steroids.

Drug Effects on Embryonic and Fetal Development

Aside from the biological effects on the person taking a drug, there are many effects on developing embryos. Some can result from damaged gametes and thus can be the result of either maternal or paternal drug taking. Many more drugs affect fetuses in utero.

Caffeine. Some studies on rats show that caffeine intake comparable to 12 to 24 cups of coffee per day resulted in offspring with missing toes, while a dose comparable to as little as 2 cups per day delayed skeletal development.

One retrospective study on the effects of caffeine intake on pregnancy found that of sixteen pregnant women whose estimated daily intake was 600 mg (milligrams) or more of caffeine (8 cups), fifteen had miscarriages, stillbirths, or premature births. In this study, caffeine intake of men was also examined. In a subgroup of thirteen fathers with a daily intake of 600 mg, while the mother's intake was less than 400 mg, only five of the births were normal. All of the births in which both the parents consumed less than 300 to 450 mg (4 to 6 cups) were normal. Any paternal effects were presumably caused by chromosome damage prior to conception. Another retrospective study on pregnant women by Infante-Rivard et al. (1993) showed that an amount of caffeine equivalent to as little as half a cup a day increased the frequency of miscarriages; this study also demonstrated that some of the effects of caffeine might occur *before* pregnancy. However, Eskenazi (1993) points out that other studies of caffeine have reached inconsistent and inconclusive results for a number of reasons. Among these reasons are the difficulty of measuring caffeine intake when people use different brewing methods and cup sizes, the lack of control for noncaffeine ingredients in caffeinated beverages, and a variety of other methodological differences such as inconsistencies in the number of people and the types of beverages studied. Another difficulty with retrospective experimental designs is that it is often impossible for a person to reliably recall their drug intake, particularly so for the so-called "soft" drugs, such as caffeine.

Nicotine. Nicotine damages the placenta, increasing the likelihood for miscarriages, premature births, and damage to the fetus. Nicotine crosses the placenta very quickly and remains in the fetal circulation longer than it does in the mother's bloodstream. Nicotine causes oxygen deprivation in the fetus, as do carbon monoxide and cyanide from cigarette smoke. Since oxygen is required as the terminal electron acceptor in the production of ATP (Chapter 8), cells deprived of oxygen will have less ATP and be less able to carry out cell synthesis functions necessary to produce new cells in the growing fetus. Oxygen deprivation is made worse by nicotine-induced damage to the blood vessels, including those of the placenta.

Alcohol. Alcohol that is consumed by a pregnant woman will be quickly distributed into the blood of the fetus at the same concentration as is present in the mother's blood, causing severe and permanent mental and physical birth defects called **fetal alcohol syndrome.** The prevalence of fetal alcohol syndrome in the United States in 1983 was 1 to 3 affected children per 1000 total births and 23 to 29 per 1000 births to alcohol-abusing mothers. The period during which the fetus is most sensitive to damage from alcohol is the first month, often before pregnancy is recognized. There is also evidence that alcohol abuse by women before conception correlates with decreased fetal growth, even when the mother abstains during pregnancy itself. There are few data on the fetal effects of heavy alcohol consumption on the part of the father. Alcohol is toxic to sperm and five or more drinks daily decrease the number of sperm produced.

Drug combinations. Alcohol, caffeine, and nicotine all increase the blood levels of the neurotransmitter acetylcholine, lowering placental blood flow. The effects increase with the dose of the drug (a positive dose-response correlation) and also with its duration in the body. Because the fetus lacks the enzymes for breaking down either alcohol or caffeine, the concentrations of these drugs stay higher longer in the fetus than in the maternal circulation. Because there is a higher incidence of smoking among people who abuse alcohol, the interactions of these drugs are also significant. Marijuana also crosses the placenta and is correlated with low birth weight and prematurity. Barbiturates readily cross the placenta, and barbiturate use by pregnant women can cause birth defects. The combination of marijuana or barbiturates with any of the other drugs mentioned increases the risks to the fetus.

Persistence of drugs after birth. Drugs passed on to the fetus in utero may stay present in the child for prolonged periods. Phencyclidine (PCP), known as angel dust, may still be present in the blood of a 5-year-old child of a PCP-using mother. Because children's brains and immune systems continue to develop after they are born, the presence of toxic drugs will interfere with brain and immune system development in young children.

THOUGHT QUESTIONS

1. Divide a piece of paper into three columns. In the first column, list all the characteristics you can think of to describe addictive drugs; in the second, list the characteristics of psychoactive drugs; in the third, list the characteristics of drugs of abuse. Also list specific drugs in each column. Are all addictive drugs drugs of abuse? Are all drugs of abuse addictive? Are all psychoactive drugs addictive? Are all psychoactive drugs drugs of abuse?

2. Many factors contribute to making a drug dangerous. In what ways are street drugs more dangerous than a chemically similar drug obtained from a licensed manufacturer?

3. Does the American cultural expectation of a quick fix for life's pains contribute to our tremendous use of legal drugs? Does our widespread and somewhat casual use of legal drugs contribute to drug abuse?

4. Read the package inserts for some drugs that you have purchased over the counter. Do these inserts indicate that there are any potential negative effects? What kind of warnings do the package labels contain? Study the advertisements for these medications. Do the advertisements and the labels give you the same impression of the products?

E. Drug Use and Abuse Have Effects on Populations

Recall that drug abuse has been defined as drug use that negatively affects the health of individuals *or society*. We have examined the effects of drugs on cells and on organisms, but an ecological perspective on biology teaches us that actions in cells and organisms will produce actions in populations.

Drug Abuse as a Public Health Problem

There are different concepts of how to protect society from the consequences of drug use by some members of that society. One concept views drug addiction as a crime and drug abuse as a law enforcement problem. Another concept, referred to as *harm reduction*, is based on a view of addiction as a disease and treats drug abuse as a public health problem.

Public health measures based on harm reduction. Many European countries have followed harm reduction strategies. Their drug-abuse rates (and their crime rates) are much lower than in the United States, which has followed the crime concept for the most part. In England, for example, heroin can legally be prescribed to those who are addicted. The approach seems to work in several ways: Harm to the addicts (from the dangers of overdose or impure formulations) is minimized. Their incentive to commit crimes or prostitution to pay for drugs or to recruit others into becoming addicts is taken away. (Many drug dealers are addicts who recruit others so that they will have a steady supply of customers, allowing them to use the profits from the drug trade to support their own habit.) Under this approach, heroin-use rates in England have dropped, while rates of heroin use in the United States have risen. The illegal heroin trade has withered in England because its profits have been removed, and new cases of addiction are rare because the drug is available only to persons registered as already addicted. Most promising of all is the fact that about 25 percent of heroin addicts spontaneously give up the habit on their own.

The approach in several other European countries, including Germany, Switzerland, and the Netherlands, allows drug users freedom from arrest if they follow a few simple rules (staying in certain locations, for example), while government funds are spent more in public health campaigns aimed at education, prevention, and rehabilitation. While the unlicensed selling of drugs is illegal, the criminal justice system is used very little in the attempts to minimize the harm done by addictive drugs, either to the addicts themselves or to society as a whole.

Public health measures aimed at prevention. Another public health measure is education. Where prevention of addiction has been tried, it is cheaper and more successful than rehabilitation. Education as to the risks of drug use does decrease drug use (Fig. 11.7). European countries that emphasize education and prevention have experienced much lower drug abuse rates than the rest of the industrialized world.

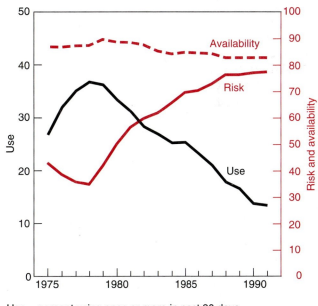

Use = percent using once or more in past 30 days

Risk = percent saying great risk of harm in regular use

Availability = percent saying fairly easy or very easy to get

FIGURE 11.7
Perception of risk influences the use of marijuana. The availability of marijuana has remained uniform over the past 20 years. Use has decreased in inverse proportion to perception of risk.

Social Attitudes Regarding Drug Use

Where is the boundary between the rights of the individual and the rights of the group in relation to drug use? Many people are inclined to leave this matter up to the individual if there is "no harm to society." However, in many if not all forms of drug abuse, there is clearly a harm to society. Bodily harm done to others while under the influence of any psychoactive drug and the commission of crimes to pay for drugs are two obvious effects of drugs on populations. In some instances, once the societal effects of individual drug abuse have been documented, laws have been passed to limit these effects. The U.S. public has accepted laws designed to protect the non-drug-using citizen from the harmful effects of others' use (or abuse) but has not accepted laws that are perceived as infringement on individual rights. For example, laws designed to protect others from exposure to secondhand smoke have been successful at limiting where or when people may smoke, while attempts to pass laws prohibiting individuals from smoking at all have not been successful. A similar approach is being tried on college campuses in educational efforts to raise awareness about the secondhand effects of binge drinking.

Legalization of drugs?

Occasional critics on both the political left and right have suggested legalizing various dangerous drugs, taxing them, and treating them as public health problems the way we treat alcohol and tobacco, with heavy reliance on education and prevention programs. Few people who work in the field of drug addiction favor this approach because these drugs are truly dangerous and addiction is easy to establish and very difficult to break. Legalization without the structures that are part of harm reduction policies would, they feel, result in increases in health risks.

The harm done by alcohol shows that a drug need not be illegal to cause considerable social harm. In addition to its negative effects on the user's health, alcohol abuse also causes harm to others. Alcohol use currently causes deaths in motor-vehicle accidents, boating accidents, drownings, and many other causes of accidental injury, including industrial accidents. It also is responsible for much employee absenteeism, job loss, and school failure. Alcohol is frequently a factor in ac-

quaintance rape (also called "date rape"), child neglect, child abuse, spouse abuse, divorce, and suicide. Alcohol and other psychoactive drugs can also do great harm in safety-sensitive occupations such as commercial transportation (airline pilots, air traffic controllers, railroad engineers), power plant operations, the nuclear industry, and much of the military. We tend to hear about the increases in crime when alcohol use was illegal during the Prohibition on alcoholic beverages in the period from 1920 to 1933 in the United States, but it is also true that the incidence of alcohol-related accidents and disease was greatly *decreased* during Prohibition.

Social policy.

In Chapter 2 we outlined steps for arriving at policy decisions on societal issues that are influenced by science. Have these methods been followed on the issues presented by drug use and abuse? Nicotine addiction causes far more deaths (from lung cancer) than all other drugs combined, yet it is legal! Alcohol abuse ruins more families and careers than do illegal drugs, and causes more fatal accidents, yet it is legal in most places. Marijuana users, on the other hand, cause far less harm to others, yet the substance is illegal in most places. Caffeine is not regarded by most people as a drug and is readily available even to children, yet it is certainly addictive. The inconsistencies go on and on. Clearly, decisions as to which drugs should be legal and which should be illegal are not always made on scientific criteria.

THOUGHT QUESTIONS

1. Is information about the biological effects of drugs an effective prevention or a deterrent against drug use? See if you can find any published information about the effectiveness of various educational programs.
2. What criteria could be used to distinguish between the use and abuse of caffeine? Can these same criteria be applied to other drugs?
3. What ethical considerations govern the use of animals in testing drugs for safety and effectiveness? What about the use of human volunteers?
4. Pick a drug for discussion that is currently illegal or legally controlled. What experimental findings would cause you to change your mind (in either direction) regarding its legalization or wider availability?

CHAPTER SUMMARY

A drug is a chemical substance that produces one or more biological effects (usually several). These effects depend on the molecular structure of the drug, on its dosage or concentration, and in many cases on the bodily location of specific receptors for the drug. There are many medicinal uses for drugs, and drugs have been used both medically and socially by all cultures at least since the beginnings of recorded history.

All drugs, whether used medically or nonmedically, must enter the body by some route and be distributed around the body, usually by the circulatory system. Drugs have their effects on cells either by direct action on the cell membranes or by stimulating receptor molecules to alter one or more cellular functions. Actions on cells in different tissues produce different physiological effects.

There are always other effects in addition to that for which the drug was taken, and these side effects are every bit as real as the intended effect. Psychoactive drugs cross the blood–brain barrier and act directly on the nerve cells of the CNS. Those that are addictive stimulate activity in the positive reward center of the brain. The possible side effects of psychoactive drugs include permanent damage to the brain cells and interference with normal physiological functioning of other organ systems, impairing the health of drug users. Many drugs can cross the placental barrier and cause damage to a fetus in utero; other drugs can be transmitted to infants through breast milk. The use of addictive drugs has a tremendous cost to society.

KEY TERMS TO KNOW

acute effects (p. 329)

additive effect (p. 314)

agonist (p. 314)

antagonist (p. 314)

chronic effects (p. 329)

dependence (p. 325)

dose (p. 309)

drug (p. 308)

drug abuse (substance abuse) (p. 329)

excretion (p. 316)

fetal alcohol syndrome (p. 332)

half-life (p. 316)

negative reinforcement (p. 326)

positive reinforcement (p. 326)

psychoactive drug (p. 308)

receptor (p. 314)

side effects (p. 314)

synergistic interaction (p. 314)

threshold (p. 314)

tolerance (p. 328)

withdrawal (p. 325)

CONNECTIONS TO OTHER CHAPTERS

Chapter 2 Attempts to limit the effects of drug abuse raise numerous ethical issues.

Chapter 6 Drug use can affect the physiological regulation of sex hormones.

Chapter 8 Drugs interfere with nutrient pathways at many levels.

Chapter 9 Drugs such as tobacco and marijuana contain many cancer-causing agents.

Chapter 10 Drugs interfere with the normal processes of the brain.

Chapter 12 The brain and the endogenous opiates interact with the immune system.

Chapter 13 Drug use by injection is a major risk factor in the transmission of HIV infection and AIDS.

Chapter 15 Most psychoactive drugs are plant products or are derived from plant products.

The Mind–Body Connection

Whatdo we mean by health, and how do we achieve it? What makes one person healthy and someone else chronically ill? Certainly there are many answers to these questions, and we have touched on some of them in previous chapters. A person's genetic heritage plays a role in some diseases (Chapter 3), as does what a person eats (Chapter 8). How and where a person lives are important because they determine the person's exposure to infectious microorganisms (Chapter 5) and to hazardous chemicals. Our immune system helps to remove damaged tissue and repair the body, preventing some diseases before we know we have been exposed and bringing us back to health after we have been sick. Genetics, nutrition, and exposure to chemicals and microorganisms all affect the functioning of the immune system. It is often the case, however, that some people in a particular area will get sick, while their relatives in the same area with about the same exposure, diet, and genetic background may not. In this chapter we will examine the theory that a person's mental and emotional states are factors in health or disease, and that these exert their effects because they interact with the immune system.

A. The Mind and the Body Interact

What Is Psychoneuroimmunology?

In the chapters of this book we have covered some biological fields in which theories have been debated, tested, and modified for 150 years—a long time in the science of biology. In this chapter we will be discussing a very new field of study, **psychoneuroimmunology.** This new subject area is built upon a *central organizing theory*: the premise that the mind, through the action of the nerves, affects the functioning of the immune system and therefore affects human health. Each part of the name contributes to the overall meaning: *psycho,* from the Greek word *psyche,* meaning "the mind"; *neuro,* referring to the nerves and the brain; and *immunology,* the study of the immune system. Psychoneuroimmunology is the study of how these body systems, which were previously assumed to be independent of one another, interact with and influence each other. Psychoneuroimmunology is thus a new paradigm since it embodies both a new theory and a new field of study (see Chapter 1).

Central to the paradigm are large and important issues with which people have struggled for millennia: What do we mean by **health**? Why do we get sick? How we answer these questions will help to define what areas of investigation are valid with regard to the cure and prevention of disease and/or the promotion of health.

Historical Antecedents

No theories arise spontaneously, and neither did psychoneuroimmunology. Its roots lie in ancient observations that personality, emotional state, and attitude influence when and if people get sick and how sick they get. Doctors in China, India, and (later) Greece rejected supernatural forces (the gods, evil spirits, or magic) in favor of natural (biological) forces as the explanations for both health and disease. Each of these traditions maintained that there was a life force or life spirit, called *qi* (or *ch'i*) by the Chinese, *prana* by the Indians, and *pneuma* by the Greeks, which was present in living organisms for the length of time that they were alive. A person was healthy when the life forces were balanced and unhealthy when the forces were out of balance. A person's mental and emotional states could alter the balance or imbalance of the life force.

The Greek physician and teacher Hippocrates advocated the "rational" study of diseases. Since it was believed that nature followed a rational course, it followed that diseases had natural causes and that those causes should be discernible and knowable entirely by the mind. Hippocrates taught that the body contained four fluids, or *humors* (from the same Greek word that gives us *humid*). Each of the four humors corresponded to a personality type: Sanguine people were rich in blood, cholic people in yellow bile, melancholic people in black bile, and phlegmatic people in phlegm (we still use some of these words). Each personality type also predisposed people to a further excess of the corresponding humor, creating diseases, also classified into four types according to which humor was present in excess. A person with an excess of black bile, for example, would have a "hot" personality and would be prone to "hot" diseases accompanied by fevers. People who had an even balance of all four humors enjoyed good health and were thus said to be in "good humor."

As European science developed in the 1600s, the criteria for "knowability" changed. In 1616 an English physician, William Harvey, described the circulation of the blood on the basis of dissection and observation, not purely on rational thought. At about the same time, Galileo invented the thermometer, and this was used by another Italian, Santorio, to demonstrate that people said to have an excess of black bile were no more hot than other people. Not only did these two discoveries undermine the notion of the four humors, but they ushered in an era in which hypothesis testing was added to the standards of "knowing" about human health. Although the scientific method has vastly increased our knowledge of health and disease, a negative aspect has been that things that could not be seen or in some way quantified have come to be viewed as irrelevant to the explanation of health and disease.

Christianity viewed humans as unchanging reflections of God and therefore beyond the scope of study by the scientific method. The French philosopher René Descartes offered a way around this problem by positing that the mind was separate from the body: The mind was the seat of the spiritual essence, and hence belonged to the realm of the church, and the body was a purely physical essence, and hence suitable for scientific study. This split of mind from body, or *Cartesian dualism* as it is sometimes called, had a profound effect on the study of biology and medicine (and on some church doctrines). As medicine strived to become less of an art and more of a science, the split became wider. It is just this split, however, that psychoneuroimmunology postulates is artificial. In seeking to rejoin the mind to the body, scientists working in this field have found new ways to detect and quantify connections between the "mind" and the rest of the body. There is an underlying assumption here, as there was in Chapter 11, that all of the functions of the "mind" (thoughts, emotions, hopes, and dreams) can be studied in terms of brain biochemistry. Some psychoneuroimmunologists take this assumption one step further, postulating that the mind is more than the brain; it is the integrated, inseparable network that includes the nervous system, the endocrine system, and the immune system. Other scientists in this field would be more likely to say that the psyche (or mind or spirit) affects the brain and the body in ways that can be studied by biology. Think about these distinctions as you proceed through this chapter.

THOUGHT QUESTIONS

1. Think about the difference between the following two statements: (a) The mind is the neuroendocrine/immune system and (b) The mind can be studied by studying the neuroendocrine/immune system. Are both scientific statements? Would both allow the formulation of falsifiable hypotheses?

2. Can a person accept the data produced by psychoneuroimmunologists without taking a stand on which of the assumptions stated in thought question 1 is correct?

B. The Immune System Maintains Health

In order to understand how the central theory of psychoneuroimmunology can be tested, we need to become familiar with some of the biology of the immune system, the nervous system, and the endocrine system. These three systems in the body have communication as their primary function.

A Sense Organ for Noncognitive Stimuli

Cognitive stimuli are those that are processed by the neurons of the brain (Chapter 10). Cognitive stimuli enter our bodies through the senses we normally think of: hearing, vision, taste, smell, and touch (see Box 12.1). There are many other sensory inputs that occur constantly of which we are not consciously aware. These *noncognitive stimuli* are regulated by a combination of the nervous and endocrine systems. Our breathing movements (which continue even in sleep), the secretions of our glands, and the contraction and dilation of our blood vessels are all activities that our bodies monitor and respond to but of which we are not usually consciously aware. To these "noncognitive" activities, scientists working in the field of psychoneuroimmunology would add the functioning of the immune system.

A healthy multicelled organism can be viewed as an ecosystem of cells in which the parts are in dynamic equilibrium, or homeostasis (see Box 1.1). The immune system is the sense organ that detects whether or not this homeostasis exists and attempts to bring the organism back to this state if it does not exist. We can talk about homeostasis in many physiological contexts: temperature regulation in warm-blooded organisms, for example. Immunological homeostasis is a more general state, suggesting that the cells have a way of asking, "Are we all together?" "Are we in harmony?" As we will see below, the immune network is aided in this task by its ability to both send and receive chemical messages from the nervous and endocrine systems.

In the past, the immune system was often described using the metaphor of a vast military establishment fending off foreign invaders. Newer knowledge of immune functions has led to a new metaphor, that of the immune system as being a communication network that carries on a "conversation" throughout the organism, checking to make sure that all the parts of the organism are contributing. In this new view, the central function of the immune system is the maintenance of *self*, meaning the aggregate of cells forming a cooperative unit that we recognize as a multicellular organism.

The view that the immune system functions to maintain multicellularity is supported by observations that the complexity of the immune system has evolved in parallel with the complexity of multicelled organisms. Even simple multicellular animals like sponges have immune systems. Cells called amoebocytes travel throughout the sponge, engulfing dead cells or other *nonself* cells. (Cells and molecules that are not contributing to *self* are called *nonself*: Thus, dead, damaged, or cancerous cells, as well as some outside molecules, are nonself.) Since a sponge does not have differentiated tissues, overgrowth by cells that have become cancerous (Chapter 9) is not likely to disrupt the functioning of the whole organism. Its ability to discriminate self from nonself does not need to be very precise. In more complex organisms with specialized tissues, however, overgrowth by cancer cells disrupts the function of the tissue and thus of the organism as a whole. Since cancer cells disrupt the cellular ecosystem, they are nonself. The difference between a cancerous cell (nonself) and a normal cell (self) can be very small compared to the sum of their surface molecules. Immune cells specific enough to recognize such subtle differences have evolved only in vertebrates. Vertebrates have retained the more primitive type of immune system and another type of system has been added to it, as we will see shortly.

The Lymphatic Circulation

The immune system is very diffuse, consisting of mobile cells that travel throughout the body and often are not confined to specific locations. This diffuse organization is in contrast to the nervous system, which is organized in distinct nerve pathways, and the endocrine system, in which cells are associated in distinct entities called *glands*. Immune cells "crawl" through the spaces between the cells in tissues and are transported longer distances via the blood and a second circulatory system called the **lymphatic circulation** (Fig. 12.1, p. 342).

In the spaces between cells in all tissues is a water-based liquid called *lymph*. (All cells must be constantly in contact with water, both inside and outside the cell, as it is the repulsion of lipid molecules by water that keeps cell membranes intact.) Immune cells called *macrophages* move in the lymph, cleaning up any dead or damaged cells in a manner very similar to the way amoebocytes of more primitive animals get rid of nonself cells. Lymph flows from the intercellular spaces into *open-ended* lymphatic capillaries and then into larger collecting vessels, the lymphatics (Fig. 12.1). The lymphatic circulation is thus more akin to the open (one-way) circulatory systems of some insects and crustaceans than it is to the blood circulation, which is a closed system (Chapter 14).

In humans there is no pump to move fluids through the lymphatic system the way the heart moves fluids through the blood. Muscle contractions and movements of the individual provide what little push this system gets.

The lymphatic circulation not only helps remove wastes and damaged cells from tissues, but it solves the problem of how the immune system can monitor all of the cells and molecules in all of the tissues of the body. This is accomplished by macrophages bringing molecules to centralized locations (the lymph nodes, tonsils, and adenoids) for checking by other immune cells called *lymphocytes*. In an active immune response, ten times the normal number of lymphocytes enter the nodes, causing the nodes to "swell." Our lymph nodes are what we commonly refer to as "swollen glands" when we are sick. The fact that they get larger during sickness indicates that an immune response is occurring.

In the mid-twentieth century, the function of tonsils and adenoids as tissues of the immune system was not known. Tonsils and adenoids were routinely removed from children who had repeated respiratory or middle ear infections since the swelling of these tissues during infections can make children's breathing difficult or block the eustachian tube, which connects the back of the throat to the middle ear. Fortunately, there are backup systems in the immune tissues so that removal rarely had serious consequences, but today tonsils and adenoids are left in place unless the blockage is extreme.

Development of Immune Cells

The immune system's capacity to distinguish self from nonself prevents the immune system from reacting against self while allowing it to eliminate nonself. This distinction is made in an ingenious way. Individual immune cells have highly specific binding characteristics. In the jargon of immunology, we say that each immune cell can "recognize" only one **antigen,** an antigen being any molecule that is detected by the immune system. The population of immune cells as a whole, however, can recognize a vast diversity of antigens. This combination of individual cell specificity, along with population diversity, is what allows the immune system to distinguish between self and nonself.

The success of living organisms depends on their ability to detect and respond to changes in their internal and external environments. The detectors for environmental cues vary in the type of stimulus to which they respond. Eyes contain photodetectors that respond to light. Ears contain hair cells that respond when their hair is vibrated by sound waves; different hair cells have hairs of different length and thickness that are made to vibrate by different pitches (sound waves of different frequencies). Humans can respond to more than twenty different physical or chemical signals, using specialized receptor cells located in five different types of sense organs. Some sense organs respond to very different types of stimuli; the ear, for example, responds to both sound and to body position. Other sensory modes are detected by cells in more than one type of sense organ; chemical signals are detected by cells both on the tongue and in the nose (see the accompanying table.)

Other animals can detect other physical signals that we don't detect because they have specialized receptor cells of different types than ours. Invertebrate eyes contain cells capable of responding to the degree of polarization of light. Some types of fish can detect electrical fields, while some types of birds can respond to magnetic fields with the help of symbiotic iron-carrying bacteria in their ears.

In animals with simple nervous systems, detection of stimuli and integration of multiple stimuli may all take place in the region of the detector; that is, it does not require a central nervous system. In animals with more complex nervous systems, signals from detectors of different modalities are processed in an interconnected way, usually in the central nervous system. Those signals of which we are often consciously aware are called *cognitive stimuli*, while those of which we are not usually aware are *noncognitive stimuli*. The stimuli listed in the table are cognitive stimuli. Our bodies can also detect and respond to many noncognitive stimuli, including the concentrations of various ions, blood pressure, blood oxygen levels, and many more.

White blood cells. Cells of the immune system are primarily white blood cells. They are called that even though many spend as much time in the lymph as they do in the blood, because under the microscope they look clear or white by comparison to the oxygen-carrying red blood cells. Among the various types of white blood cells are the macrophages and the lymphocytes, mentioned above. There are two types of lymphocytes: the **B lymphocytes** (B cells), which make blood proteins called *antibodies* (explained below), and the **T lymphocytes** (T cells), some of which kill infected cells directly and some of which help other immune responses. Another type of white blood cell with immune functions is the neutrophil, which is important in removing bacteria.

New blood cells are continuously produced throughout the individual's life. In this, immune cells differ strikingly from nerve cells, which lose their capacity to divide further when the individual is still a child. Both white and red blood cells are produced in the bone marrow, the porous interior of the major bones, and in the spleen (Fig. 12.1). New blood cells are produced only to replace

SENSORY STIMULI TO WHICH HUMANS CAN RESPOND

SENSORY MODE	SENSING SYSTEM	QUALITY SENSED	DETECTOR CELLS
Light	Retina of eye	Brightness	Rods
		Color	Cones
		Contrast	Cones
		Motion	Cones
		Size	Cones
Touch	Skin	Light touch	Meissner's cells
		Vibration	Pacinian cells
		Temperature	Temperature detectors
Chemical	Tongue		Taste bud cells at:
		Sweet	Tip of tongue
		Sour	Sides of tongue
		Salty	Tip and sides
		Bitter	Back of tongue
Chemical	Nose	Floral	Different detector cells
		Fruity	in lining of nose
		Musky	
		Sharp	
Sound	Inner ear	Pitch	Hair cells
		Tone	
		Loudness	
Position	Inner ear	Gravity	Macula cells
		Side-to-side	Vestibular cells

blood cells that have been lost through injury or that have reached the end of their life span. Both the types of cells and the numbers of each type that are produced are tightly regulated. These regulatory signals arrive in the form of **cytokines,** chemical messengers that deliver information between cells and control both cell division and cell differentiation, as well as many other cellular processes (Chapter 9).

Differentiation of blood cells. Blood cells differentiate in several steps, each step involving a cell division (Chapter 9). Of the offspring cells, some become white blood cells that circulate throughout the body, while others remain in the bone marrow as undifferentiated or partially differentiated *stem cells.* Thus the bone marrow maintains a supply of cells with the ability to replenish the blood cells throughout the lifetime of the individual. Scientists are now studying how to transplant these less differentiated cells from one individual to another. Such a procedure offers the possibility of establishing blood cells and an immune system in individuals lacking them, or in individuals in

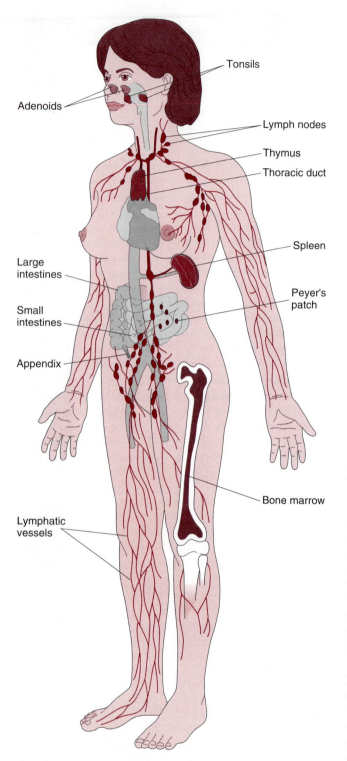

FIGURE 12.1
Macroscopically visible locations of the immune system.

Labels on figure:
Adenoids
Tonsils
Lymph nodes
Thymus
Thoracic duct
Spleen
Peyer's patch
Large intestines
Small intestines
Appendix
Bone marrow
Lymphatic vessels

whom immune cells have been killed off, for example, as a result of cancer therapy (Chapter 9).

The differentiation of many types of immune cells occurs completely in the spleen and bone marrow. Other immune cells may differentiate in the lining of the gut or in the lower layers of the skin. The skin and the gut lining are thus important organs of the immune system.

The diversity of antigen receptors.   Each B or T cell has receptors that can bind to only one specific antigen, yet the immune system as a whole is able to recognize over 10^{11} (100,000,000,000) different antigens. The human genome, as large as it is, is not large enough to have a separate gene for each of 10^{11} specific antigen receptors. How is it possible to have receptors for this many different antigens?

Something unique happens to the DNA of B and T cells during their differentiation. The gene regions coding for their antigen receptors are actually multiple regions containing far more DNA than needed to code for proteins the size of B or T cell antigen receptors (Chapter 3). In each cell, the DNA rearranges, and some portions are cut out of the DNA, making the rearrangements permanent in that cell and its offspring cells. Each mature T cell or B cell is thus able to synthesize a unique protein that functions as a receptor for only one specific antigen.

These DNA rearrangements occur randomly in each developing T or B cell so that the populations of T and B cells as a whole will contain cells that can bind to over 10^{11} different antigens. Not only are the rearrangements random, but they take place *independently* of the organism being exposed to any particular antigen. *Even before* being infected by a particular type of virus, for example, an individual already has a small number of lymphocytes that can bind to that virus.

Population selection.   Immature T lymphocytes develop in the bone marrow, then travel through the blood to an organ called the *thymus* (Fig. 12.1) to receive the cytokine signals that stimulate their final differentiation into T cells. The developing T cells are tested and sorted by matching their antigen receptors against the self antigens on the surfaces of the accessory cells of the thymus (Fig. 12.2). In a process called **population selection,**

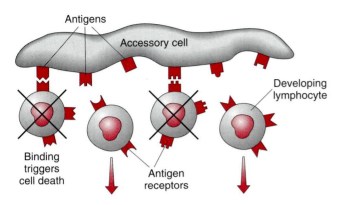

FIGURE 12.2
Lymphocyte population selection. Self cells have characteristic antigens on their surfaces. Developing lymphocytes encounter these antigens on accessory cells. Any developing lymphocyte whose antigen receptors bind to accessory cell antigens is eliminated. Lymphocytes whose receptors do not bind develop to maturity. The immune system's capability of discriminating self from nonself thus results from the selection of a population of responding cells from those that arise randomly.

those T cells whose receptors match an accessory-cell antigen would be capable of reacting against self. These cells are eliminated, an important process that usually protects you against reacting immunologically against your own cells. Cells with receptors that do not match self-antigens go on to become fully mature T lymphocytes.

Thus, while any one cell can only match one antigen, a population of cells is selected that will not react with self but that will have the potential of reacting with any other antigen encountered later (Fig. 12.2). The capability of the immune system for telling the difference between self and nonself is thus a characteristic of the *population* of cells, not a characteristic of any single cell.

B cell differentiation.   The specific antigen receptors on B lymphocytes are cell surface-bound antibody proteins. Like each of the T cells, each B lymphocyte goes through a process of gene rearrangement to produce a final DNA code for its antigen receptor prior to any exposure to antigen. Also like T cells, the population of developing B cells undergoes selection to eliminate any B cell whose surface antibody could bind to self molecules. The population of B cells left after selection thus includes only B cells that can react with non-

self antigens. The reaction of a B cell upon encountering its specific antigen is to secrete antigen-specific proteins called **antibodies,** which then circulate in the blood or in secretions such as saliva and tears. Population selection thus also ensures that only those antibodies that are against nonself molecules will be secreted.

Acquiring Specific Immunity

A person is not born with specific immunity, but is born with the capacity to acquire it. This capacity is the result of the two processes, explained above, that occur before a person is ever exposed to a particular antigen: production of the vast numbers of different antigen receptors by DNA rearrangements and population selection of lymphocytes.

After undergoing these processes during their differentiation, lymphocytes are released into the bloodstream, where some circulate at all times, ready to go into action when needed. When some nonself antigen is detected (referred to as an *immune challenge*), additional white blood cells (lymphocytes and other types) are called in from the bone marrow. These white blood cells are carried by the bloodstream to the lymph nodes, tonsils, and adenoids (Fig. 12.1), where they encounter nonself antigens brought from the tissues by the lymphatic circulation.

Production of antigen-specific clones.   Most lymphocytes will not be able to bind any given antigen and so will continue on their way. A small number of lymphocytes will be able to bind to that antigen. Those lymphocytes that can bind do so, which triggers them to begin to divide (Fig. 12.3). This cell division of a small selected number of cells produces an increased subpopulation of identical cells, called a **clone,** able to recognize that antigen; this clone of cells then fights off the antigen. If you are exposed to a particular virus (such as influenza or flu virus), for example, a subpopulation of lymphocytes which recognize that flu virus will develop and kill off the virus.

Immunological memory.   Antigen recognition and cell division take some time, so the first time a person encounters a particular flu virus, the virus has time to make that person sick before the immune system fights it off. In other words, on the

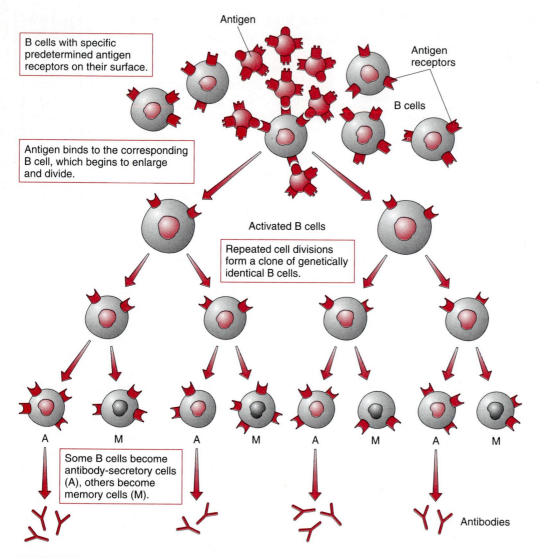

FIGURE 12.3
Clones of antigen-specific lymphocytes.

first exposure, the immune system may not block the virus fast enough to prevent the disease, but then, as specific immunity develops, it is able to stop the virus, thereby ending the disease. The clone of specific lymphocytes produced in that first encounter remains in the body as *memory cells,* sometimes for the lifetime of the individual. The second time the person is exposed to that same flu virus, this subpopulation of memory cells is ready to stop the virus more quickly than on the first encounter, in most cases quickly enough to prevent

the illness. The response is also greater because the number of cells in the clone is greater than it was before the first exposure to that antigen. *This accelerated response on the second or succeeding exposures to the same antigen is called* **specific immunity**.

Lymphocytes also encounter self molecules in the lymph nodes, but remember that in a fully functional immune system the lymphocytes that could have bound to self molecules have been eliminated from the population. If self molecules become altered, lymphocytes will exist that can bind and thus remove cells bearing the altered molecules; transformed (precancerous) and damaged cells are removed in this way.

Immunization.   Remember that we are not born with specific immunity, but with the potential to develop it. We develop specific immunity only to those things to which we are exposed in our individual lifetimes, and so one person's "immune repertoire" will not be the same as another's.

This is the basis for **vaccination,** also called **immunization.** A person is given molecules from various disease-causing bacteria or viruses, but in a form that will not cause disease. The person's immune system responds to this artificial challenge and establishes a subpopulation (clone) of memory cells, which stand ready to protect the person from later exposure to the real bacteria or virus. Immunization, working as it does *with* the disease-preventive powers of the body, has become one of our most effective ways of preventing many infectious diseases.

Mechanisms for Removal of Antigens

Many antigens recognized by the immune system are parts of whole bacteria, viruses, or cancer cells. It is these whole cells or organisms that need to be removed from the body, and there are several mechanisms by which the activated immune system does so.

Cell-mediated mechanisms.   Some T lymphocytes, a type called *cytotoxic T cells*, can kill cells directly, removing cancer cells or cells infected with virus (Fig. 12.4A). A particular T lymphocyte is able to kill only a target cell whose antigen matches the antigen receptors on the T cell.

Antibody-mediated mechanisms.   B lymphocytes do not kill antigens directly. Instead they make and secrete antigen-specific proteins (antibodies), which then circulate in the body fluids. Specific antibody molecules bind to their matching antigens on the bacteria or virus. This antigen-antibody complex can then combine with other blood proteins called *complement*. Antibody and complement together can break apart bacterial cell membranes and can inactivate viruses that are not inside cells (Fig. 12.4B). In addition, antibodies and complement can also coat bacteria, allowing the bacteria to be engulfed and killed by other white blood cells called *neutrophils*. Unlike lymphocytes, neutrophils do not have antigen receptors, so they cannot bind to most bacteria directly. Instead, they have receptors that bind to one end of antibody molecules. As a result, once a bacterium is coated with antibodies, the other ends of the antibody molecules can be bound by a neutrophil, which will then take up the bacterium and digest it (Fig. 12.4C).

Some antibodies work not by killing bacteria but by preventing bacterial adherence to the host (Fig. 12.4D). Most bacteria cannot initiate disease without adhering to the host; this is especially true of respiratory viruses and oral bacteria. If antibody binds to these organisms, it can block their adherence, preventing the disease. Antibody present in the mucous linings of the respiratory tract and the oral cavity is especially important in blocking the adherence of organisms trying to gain entry through those routes, causing them to pass harmlessly through the body and to be excreted as waste.

Other antigens, such as toxins excreted by bacteria, are soluble molecules, not parts of bacterial cells. Toxins can become inactivated by having specific antibody bind to them (Fig. 12.4E). Many of our most successful vaccines actually stimulate the production of antitoxins, that is, antibodies against toxins. The lethal results of diphtheria, for example, result from the action of a bacterial toxin. Immunity conferred by diphtheria vaccine produces an antitoxin that prevents the disease.

Helper T cells.   Both cytotoxic T lymphocyte responses and responses involving antibodies are made stronger ("helped") by another class of antigen-specific T lymphocytes called *helper T cells*. Helper T cells do not get rid of antigen; instead, they secrete a cytokine called *interleukin-2*, which boosts the strength of the responses of cytotoxic T cells and B lymphocytes to antigens. Without these cells, the immune response of the cytotoxic T cells and the B cells is often not strong enough to prevent disease. This is the situation when HIV infection results in AIDS (Chapter 13).

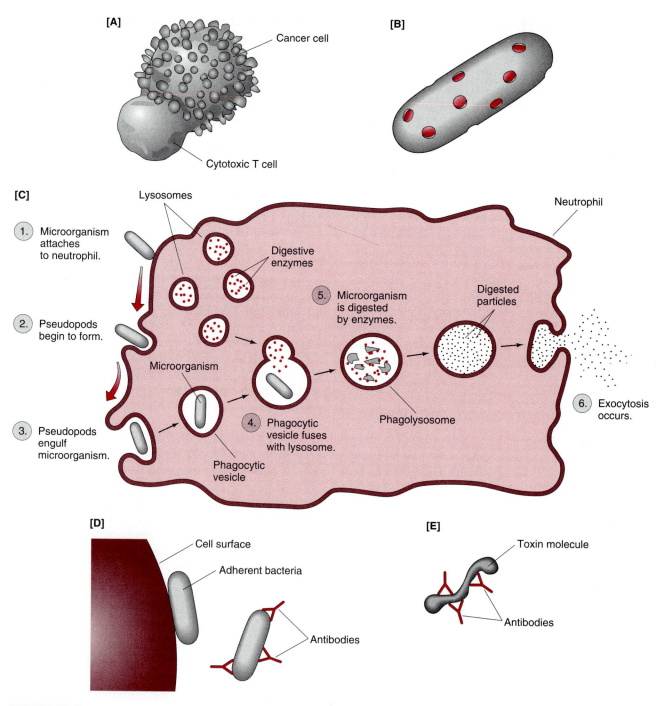

[A] Cancer cell

Cytotoxic T cell

[B]

[C]

Lysosomes

Neutrophil

1. Microorganism attaches to neutrophil.

Digestive enzymes

5. Microorganism is digested by enzymes.

Digested particles

2. Pseudopods begin to form.

Microorganism

3. Pseudopods engulf microorganism.

4. Phagocytic vesicle fuses with lysosome.

Phagolysosome

6. Exocytosis occurs.

Phagocytic vesicle

[D]

Cell surface

Adherent bacteria

Antibodies

[E]

Toxin molecule

Antibodies

FIGURE 12.4

Five ways the immune system can eliminate nonself antigens. [A] Antigen-specific cytotoxic T lymphocytes killing a cancer cell. [B] Lysing of bacteria by antibody and complement. [C] Phagocytosis (engulfing) of antibody-coated bacteria by neutrophil white blood cells. [D] Antibody blocking of bacterial adherence to host cells. [E] Antibody inhibition of toxin activity.

Feedback Loops Turn Off an Immune Response

Once an immune response has been activated, it does its job of removing antigens, whether the antigens are soluble molecules, like toxins, or are molecules that are part of bacteria, virus, or cancer cells. As long as the nonself antigen is present, the immune response stays activated. Once the antigen is no longer present, it would be inefficient for the process to continue in high gear. As in many other physiological systems, **feedback loops** exist in the immune system to shut off a response when it is no longer needed. Antibody feeds back to suppress B lymphocyte production of more antibody. Cytotoxic T cell activity is also suppressed by feedback systems.

Feedback within the lymphocyte populations.

Activation and suppression are not on–off phenomena. Remember that we are dealing with subpopulations of lymphocytic cells. At any given instant, some of the lymphocytes able to react to a given antigen are being stimulated and others are being suppressed. Measured on a population basis, when the activity curve is going up (Fig. 12.5A), more cells are being activated than are being suppressed. This stage is called *activation*. At the plateau (Fig. 12.5A), an **equilibrium** exists, with the same numbers of cells being activated as are being suppressed. The same individual cell is not being both activated and suppressed; rather, the number of cells in the antigen-specific subpopulation that is being activated is equal to the number being suppressed. As the activity curve slopes back down, more cells are being suppressed than are activated, a stage called *suppression*. When the curve achieves a new base-line level, another equilibrium state has been reached.

It is important to realize that when the population activity has returned to base line, it does *not* mean that nothing is happening. A low number of cells are being activated and suppressed, keeping the immune system "tuned up" and ready for its next response.

Feedback is antigen-specific.

Activation of an immune response is antigen-specific; that is, not all of the B cells or T cells are activated, only those that match the antigen. Feedback inhibition is also

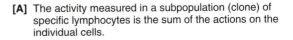
[A] The activity measured in a subpopulation (clone) of specific lymphocytes is the sum of the actions on the individual cells.

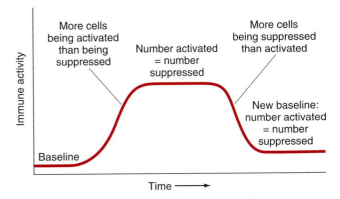

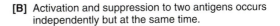
[B] Activation and suppression to two antigens occurs independently but at the same time.

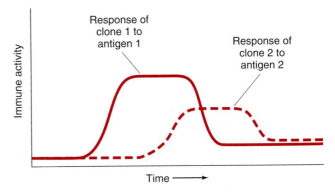

FIGURE 12.5
Activation and suppression of immune responses. [A] The activity measured in a subpopulation (clone) of specific lymphocytes is the sum of the actions on the individual cells. [B] Activation and suppression to two antigens occurs independently but at the same time.

antigen-specific; the entire immune system is not shut off, only those cells of the subpopulation that had been reacting to that antigen.

Since the activation and later suppression of each immune response are antigen-specific, several immune responses can be occurring at one time and be at different stages of the response. As shown in Fig. 12.5B, a response to one antigen may be in the suppression phase, while response to another antigen is just beginning. The maximum response to an antigen depends on the number of cells in that lymphocyte clone; clones with more cells produce bigger responses. The maximal response may be different to different antigens (Fig. 12.5B).

Natural Immunity

If we have to *acquire* our own immunity, why aren't we killed by our first exposure to a bacteria when we are very young? There are many answers to this important question.

Maternal antibody.   First of all, we do temporarily have some specific immunity transferred from our mothers, as some antibodies can cross the placenta and enter the fetal circulation (Fig. 11.1), where they can protect the infant for six months or so after birth. Other antibodies can be passed from mother to child in breast milk, particularly in the first week or two of breast-feeding. Remember that these antibodies will protect the infant against things to which the mother has acquired immunity, which may or may not be the antigens to which the infant is exposed. This is a type of **passive immunity,** meaning antigen-specific immunity transferred from another individual. Immunity can also be passively transferred from one adult to another by transfusing blood containing specific antibody. Passive immunity is only temporary. Antibody proteins, like all proteins in the body, are constantly being degraded, and the recipient of antibody has not acquired the antigen-specific B cells to produce more.

Antigen-nonspecific activities of the immune system.   There are other parts of the immune system that function without antigen specificity and that operate even on our first exposure to some, but not all, new antigens. These parts of the immune system are collectively called **natural immunity.** Natural immunity is inborn, not acquired. It does not involve cell division of antigen-specific subpopulations (clones) of lymphocytes; therefore, natural immunity, which is antigen-nonspecific, is not stronger on the second exposure to the same antigen.

The blood proteins called complement are able to bind to some types of bacteria (causing lysis) or to inactivate some types of virus without the aid of antibodies. Viruses nonspecifically induce lymphocytes to secrete *interferon*, a molecule that then prevents replication of other virus strains as well as the virus strain that induced its secretion. Cells capable of engulfing particles (macrophages and neutrophils) can engulf some types of bacteria and fungi without antibodies and thus seem to be homologous to the nonspecific immune system of invertebrates. In evolution, these cells have been retained, while antigen-specific immunity has been added. Other cells, the *NK lymphocytes,* or *natural killer cells,* act without a prior immune reponse to kill tumor cells in a manner similar to that of cytotoxic T cells.

Inflammation and Healing

Another important function of the immune system is to promote the growth and repair that occur after injury, whether the injury is due to microorganisms or due to physical damage to tissues.

Inflammation.   As described above, cells and cytokines of the immune system are mobilized to get rid of microorganisms or transformed cells. At a physiological level, this process is called **inflammation.** In the first century A.D., the Roman physician Celsus described the "four cardinal signs of inflammation": *rubor* (redness), *calor* (heat), *dolor* (pain), and *tumor* (swelling). If you have ever had a scraped knee or a splinter in your finger, you no doubt have experienced these cardinal signs. The injured area becomes red, hot, sore, and slightly swollen.

The fact that inflammation is carried out by cells of the immune system was discovered by Elie (or Ilya) Metchnikoff, a Russian scientist, in the very early 1900s. He observed that when he placed a thorn into a starfish, mobile cells within the starfish were able to push the thorn out and repair the hole. In Metchnikoff's view, these cells were maintaining the "harmony" or "integrity" of the organism. Although Metchnikoff received a Nobel Prize for his work in 1908, the importance of his discoveries was not fully recognized until recently, when his view of the function of the immune system regained attention.

Chemotaxis.   The cells that Metchnikoff observed in the starfish were the amoebocytic cells, the primitive form of nonspecific, or natural, immunity. These are the same types of cells that are at work in inflammation in humans. When tissue is wounded, the damaged cells change the acidity (pH) of the local fluids, activating various chemical factors. Some of these factors attract macrophages and neutrophils to the area. These cells can discern concentration gradients of certain chemicals and move in response to those gradients, a behavioral response called *chemotaxis*. If the wound includes bacteria, products of the bacteria also provide chemotactic signals to call in nonspecific immune cells.

Effects on blood vessels.

Other chemical mediators constrict the blood vessels beyond the site of the wound, causing blood to build up in the capillaries close to the wound. These changes in blood flow result in the redness, heat, and swelling of inflammation. Other chemical factors increase the permeability of the capillaries in the local area. Fluid escapes from the slightly permeabilized bloodstream into the intercellular area, and the immune cells are able to crawl through the capillary walls and into the tissue. There they remove the damaged tissue and bacteria and secrete the cytokines called *growth factors* that are the first step in repair. These growth factors stimulate cell division, providing offspring cells to replace the cells lost in the wound. Thus, when we talk about a wound **healing,** we are referring to a process coordinated by immune cells.

Fever.

In addition to local effects on blood vessels, inflammation produces effects throughout the organism. One such effect, namely *fever*, is what we often recognize as the symptom of having an infection. Some viruses and bacteria induce macrophages to secrete several cytokines (including interleukin-1), which induce fever by acting on part of the brain called the hypothalamus (see Chapter 10). The increased body temperature inhibits the growth of bacteria and also enhances the immune response to the bacteria. This is one of many examples now known in which products of the immune system act on the cells of the nervous system. Chemical messengers produced by immune cells during inflammation also act on the endocrine system, leading to the synthesis of several hormones that influence carbohydrate metabolism and immune responses.

Harmful Immune Responses

Our metaphorical view of the immune system as our defender against disease may lead us to assume that the immune system is always protective. There are many situations when it is not. These abnormal reactions are generally against a specific antigen or a small number of antigens. The remainder of the antibodies and clones of immune cells function normally against other antigens.

Autoimmune diseases.

Autoimmune diseases result when the immune system begins to make an immune response to self, resulting in both antibodies and cytotoxic T cells that react with antigens in the body's own tissues. The immune cells and/or antibodies then try to rid the body of these antigens as if they were nonself, resulting in damage to the body's own tissues. Although the mechanisms that produce tissue damage are known for some autoimmune diseases, the factors that trigger autoimmunity are unknown.

Multiple sclerosis is an autoimmune disease in which cytotoxic T cell clones, which are specific for an antigen on the insulating (myelin) sheath, develop around nerve cells in the brain (Chapter 10). These cytotoxic T cells migrate to the brain, where they kill the cells bearing their antigen. The ensuing damage to the nerve sheaths causes a variety of problems, depending on exactly which nerves have been affected.

In insulin-dependent diabetes mellitus (IDDM), both autoreactive T cell clones (those capable of reacting against self) and B cell clones that produce autoantibody (antibody to self) develop. These destroy the cells in the pancreas that produce the hormone insulin. Since insulin controls the cellular uptake of glucose, its absence produces severe consequences throughout the body.

Allergies.

People who suffer from allergies do so because their immune systems react atypically to some antigens from which the host does not need protection (pollen or dust mites, for example), but not to all antigens. The atypical response produces a special type of antibody called IgE, specific for these substances, which are called *allergens*. IgE binds to certain white blood cells called *mast cells*. When the person later encounters the same allergen, the allergen binds to the IgE on the mast cells, triggering the explosive release of histamine. Histamine is one of the chemicals that plays a positive role in the first stages of inflammation, dilating blood vessels to allow the entrance of immune cells into the tissue. In cases of allergic reaction, larger amounts of histamine are suddenly released (Fig. 12.6), producing the various symptoms of allergy. Whether an allergic response produces runny eyes, sneezing, or shortness of breath depends on the tissue in which the mast cells were triggered (the eyes, nasal lining, or the lungs). Because the symptoms are produced by histamine, antihistamine medications stop the symptoms by blocking the binding of histamine to cells in the blood vessels. Antihistamines do not prevent the immune response or the release of histamine by the mast cells.

Because allergy is an antigen-specific immune response, it shows memory and a greater response on the next exposure, which is why people's allergies can worsen over time. Although there are thousands of different substances that are allergenic in some people, each person is usually allergic to only a few. The severity of allergy to one substance will not predict the severity of allergy to some other substance, since each is a separate, antigen-specific response. We are not yet able to predict who will become allergic or what they will become allergic to, but there does seem to be some

inherited component, because allergies do run in families.

Plasticity of the Immune System

Plasticity is a word used in biology to denote things that are changeable. Almost every biological system is plastic to some degree, but the immune system is probably among the most plastic. Some or all parts of the immune system can be either inhibited or strengthened.

Immunological tolerance.   When inhibition of immune responses is antigen-specific, it is called immunological **tolerance.** For example, people who suffer from allergies can often be desensitized, that is, made immunologically tolerant to those particular antigens. The desensitization procedure consists of giving the person repeated small doses of the substance he or she is allergic to. The procedure must be carried out very carefully because giving the wrong dose, either too much or too little, will make the allergy worse, not better. Because both allergy and its desensitization are antigen-specific, the procedure must be carried out for each separate allergen. Induced tolerance to one antigen does not change the ability of the individual to react to other antigens.

Immunosuppression.   Other factors can suppress the functioning of all or parts of the immune system in antigen-nonspecific ways. Many environmental pollutants and other chemicals suppress

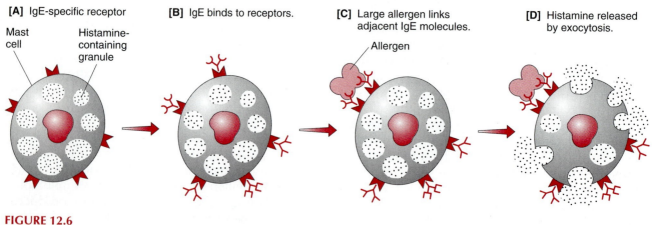

[A] IgE-specific receptor

Mast cell   Histamine-containing granule

[B] IgE binds to receptors.

[C] Large allergen links adjacent IgE molecules.

Allergen

[D] Histamine released by exocytosis.

FIGURE 12.6
Allergic release of histamine.

the immune system and are thus said to be **immunosuppressive.** Other immunosuppressive factors include alcohol, cocaine, heroin, and malnutrition. Since immunosuppression inhibits the workings of all B cells, or all T cells, rather than just antigen-specific clones, immunosuppression is generally correlated with an increase in disease.

On the other hand, in autoimmunity, where the immune system has turned against itself, suppression of immune reactivity may be needed to prevent tissue damage. Because immunosuppression is nonspecific, however, it also increases susceptibility to infectious disease, and the risk–benefit ratio of performing immunosuppressive procedures must be carefully assessed in each case.

Immune potentiation.   Antigen-nonspecific immune enhancement, usually called immune potentiation or **immunopotentiation,** is also possible. Tumor necrosis factor (TNF), a cytokine produced by macrophages in response to some bacterial infections, can cause lysis of tumor cells. TNF was discovered in the 1970s by Dr. Lloyd Olds as the result of observations made in the 1890s by Dr. William Coley. Coley had noted that some patients had a spontaneous remission from sarcoma, a type of cancer, if they had also had erysipelas, a severe infection with streptococcal bacteria. (Now such infections would be treated with antibiotics and would not progress to the point that would stimulate much TNF production.) Olds found that when tumor-bearing mice were injected with Coley's streptococcal toxins, their tumors regressed, and the active factor, TNF, was eventually isolated. Clinical applications for TNF in cancer therapy are now being investigated.

The importance of stress as a factor in regulating the immune system (immunomodulation) is now being extensively studied as part of the psychoneuroimmunology paradigm. We will examine psychologically produced immunosuppression and immunopotentiation in more detail after we examine the workings of the neuroendocrine system.

THOUGHT QUESTIONS

1. Since the beginnings of immunology at the turn of the century, its language has often been very military. The immune cells are said to "protect us from invaders" or to "kill off foreign antigens." Locate an immunology textbook or an article on immunology from the popular press. Can you find examples of military language in these accounts?

2. We need words to convey what we imagine the immune system to be doing, based on experimentation and hypothesis testing. Do you think that the new imagery of the immune system as a communications system has taken hold at this time because we are in the "information age," or because new discoveries brought about a need for new terminology? To what extent are scientific terms metaphors for reality and to what extent are they models? Can the words we choose cloud our view or prevent us from being open-minded about new hypotheses?

3. Make a list of the thoughts that come to your mind when the word *self* is used in a nonimmunological context. Make a second list for your thoughts about what *self* means in immunology. Is there any overlap between your two lists?

4. When people get a bacterial or viral infection, they often get a fever. Why? Why do people sometimes get a fever after a vaccination?

C. *The Neuroendocrine System Consists of Neurons and Endocrine Glands*

The *endocrine glands* are a series of organs that secrete chemical products directly into the bloodstream. (In contrast, glands that release their secretions into the digestive tract or through the skin are called *exocrine glands.*) These secreted chemicals alter the function of target organs and thus can be said to carry messages from one organ to another. The chemicals secreted by endocrine glands are called **hormones.** Because hormones are distributed throughout the body by the bloodstream, the target or receiving cells can be far removed from the organs that secreted these hormones.

It was once common for the endocrine glands to be described as an endocrine system because the actions of these glands were thought to be separate from the other known communication system of the body, the nervous system. In the last few decades, biologists have discovered that several hormones originally thought to be secreted only by cells of the endocrine glands are also secreted by brain cells. Other endocrine secretions are chemically related to the neurotransmitter substances (Chapter 10) originally thought to be secreted only by the cells of the nervous system. Today, the endocrine and the nervous systems are considered to be so completely intertwined that many scientists now refer to them collectively as the **neuroendocrine system.**

The Autonomic Nervous System

All vertebrate nervous systems can be divided into a *central nervous system* (CNS), consisting of a brain and spinal cord (see Chapter 10), and a *peripheral nervous system*. The peripheral nervous system connects the CNS to the more distant parts of the organism (the sensory receptors, muscles, glands, and organs) and is itself comprised of the somatic nervous system and the autonomic nervous system. The somatic nervous system, largely under conscious, voluntary control, carries signals to and from the skeletal (voluntary) muscles, skin, and tendons. The **autonomic nervous system** has been considered, at least by Western scientists, to be largely involuntary, carrying signals to and from the gut, blood vessels, heart, and various glands, and thus regulating the internal environment (*milieu intérieur*) of the body. *Autonomic* literally means "self-governing" or "self-regulating," reflecting the fact that the autonomic nervous system can work by itself without any input from the centers of conscious awareness. The autonomic nervous system can thus regulate body functions even while we are distracted or asleep.

The sympathetic and parasympathetic nervous systems.   The autonomic nervous system consists of two functionally separate divisions, called **sympathetic** and **parasympathetic** (Fig. 12.7). Functionally, the two divisions of the autonomic nervous system have opposite effects. In general, the nerve cells (neurons) of the sympathetic system ready the organism for heightened activity, while the parasympathetic neurons do the opposite. The neurons of the sympathetic and parasympathetic

divisions differ from one another both anatomically and chemically. In a sympathetic nerve pathway, the first neuron runs a short distance from the spinal chord to a ganglion (collection of cell bodies) close to the spinal cord, where it makes a synapse onto a second neuron. The second neuron is long, running from the ganglion to the target organ, and secretes *norepinephrine* as its principal neurotransmitter (see Chapter 10). In a parasympathetic pathway, the first neuron is very long, running from the spinal chord to a ganglion close to, or on, the target organ. The second neuron, after the synapse in the ganglion, is usually very short. The neurons of the parasympathetic division secrete the neurotransmitter *acetylcholine.* Tissues and organs throughout the body receive nerve endings and neurotransmitters from both of these divisions of the autonomic nervous system.

Fight or flight.   Imagine that you are crossing a street. You hear a loud noise! You turn your head suddenly, and a large truck is heading right at you! Your heart begins to pound faster, your sweating increases, your voluntary muscles are stimulated (and their threshold for action is lowered), your breathing speeds up, and your digestive organs stop digesting your last meal. (You may even feel nauseous in extreme cases.) All these are the results of stimulation of different organs and tissues by the sympathetic division and its neurotransmitter, norepinephrine (also called *noradrenalin*). In general terms, the sympathetic nervous system prepares the body for **fight or flight,** reactions that require large amounts of energy and oxygen for voluntary muscle contraction.

Rest and ruminate.   Now imagine having finished a sumptuous candlelight dinner with your favorite food, elegant service, and quiet music playing softly in the background. You are relaxing in a comfortable chair or sofa with your favorite drink in your hand and wonderful company nearby. Just thinking about this scene (or reading this paragraph to yourself slowly and calmly) may relax you, cause your heartbeat and your breathing to slow down, your sweating to stop, your voluntary muscles to relax (and to raise their threshold for action), and your blood to be diverted to your digestive organs, which are now digesting the sumptuous meal. These are all the effects of the

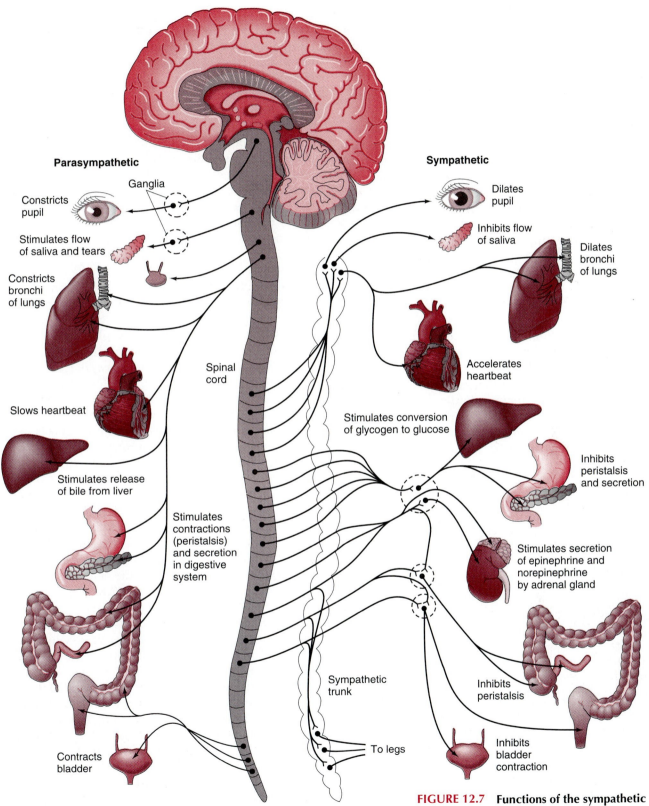

Parasympathetic

Constricts pupil

Ganglia

Stimulates flow of saliva and tears

Constricts bronchi of lungs

Slows heartbeat

Stimulates release of bile from liver

Spinal cord

Stimulates contractions (peristalsis) and secretion in digestive system

Contracts bladder

Sympathetic

Dilates pupil

Inhibits flow of saliva

Dilates bronchi of lungs

Accelerates heartbeat

Stimulates conversion of glycogen to glucose

Inhibits peristalsis and secretion

Stimulates secretion of epinephrine and norepinephrine by adrenal gland

Sympathetic trunk

To legs

Inhibits peristalsis

Inhibits bladder contraction

FIGURE 12.7 Functions of the sympathetic and parasympathetic nervous systems.

parasympathetic division and its neurotransmitter, acetylcholine. The parasympathetic division prepares the body to *rest and ruminate,* activities that use less oxygen while replenishing the body's store of energy supplies.

Regulation by the hypothalamus.

During rest or when the organism is not receiving much sensory input, the parasympathetic system is predominant (Fig. 12.7). Saliva, bile, and stomach enzyme secretion are stimulated, as is peristalsis (rhythmic muscle contraction) in the stomach and intestine, while the pupils of the eye constrict, the bronchi of the lungs constrict, and the heart rate slows. During moments of physical exertion or emergency, a part of the brain called the *hypothalamus* signals the sympathetic system to dominate, stimulating dilation of the pupils and bronchi, conversion of the storage molecule glycogen to glucose for conversion to energy (Chapter 8), acceleration of the heartbeat, and slowing of the digestive processes. When the stimuli that produced this sympathetic response are no longer present, the parasympathetic system predominates once again. In extreme cases, this switch can be rapid, producing a *rebound effect,* as when a person feels woozy or faint after an emergency situation is over.

Emotional stimuli trigger the autonomic nervous system.

As you may have been able to demonstrate to yourself as you imagined the scenes described above, the actual frightening or relaxing situation need not be present. The fight-or-flight response can be triggered by just thinking of tense situations or by watching a frightening movie. Similarly, a rest-and-ruminate response can be brought about just by relaxing comfortably and imagining a relaxing, pleasurable situation. The triggers for these responses can thus originate completely within the brain of the person imagining them. We say that the responses can be triggered by *emotional stimuli.*

The Stress Response

Research on the fight-or-flight response by Dr. Hans Selye showed that it is the first step of a larger series of physiological reactions termed the *general adaptation syndrome.* These reactions are produced by chemicals secreted by the nervous system and also by the endocrine and immune systems. The process begins when some stimulus or force, called a *stressor,* causes the body to deviate at least temporarily from its normal state of balance, **homeostasis.** The body's *response* to this deviation from homeostasis is called **stress** or the **stress response** (Fig. 12.8). Stress consists of physiological and immunological changes that allow our bodies to fight off or remove ourselves from stressors and return to homeostasis, and thus stress can be a useful response. However, when stress persists too long, it can become harmful, causing disease or even death.

Alarm.

Alarm, the first stage of the stress response, includes the fight-or-flight response. The hypothalamus at the base of the brain stimulates the sympathetic neurons to secrete norepinephrine, stimulating an endocrine gland called the adrenal gland to secrete epinephrine (also called adrenalin). Epinephrine and norepinephrine together bring about the physiological changes known as flight or fight. In addition, sympathetic neurons release norepinephrine directly into the lymphoid organs (the spleen, thymus, and lymph nodes), stimulating these organs to release their store of lymphocytes into the bloodstream. As the lymphocytes are released, the organs in which they were stored decrease in size. The hypothalamus also secretes a hormone called adrenocorticotropic hormone (ACTH), which stimulates the adrenal gland to produce corticotropin-releasing hormone (CRH), which in turn stimulates *steroid hormones* (cortisol) to be released from the cells of the outer layer of the adrenal gland (Fig. 12.8).

Resistance.

If the stress continues, the animal enters the second stage, resistance, in which the body mobilizes its resources to overcome the stressor and attempts to regain homeostasis. New stores of steroid hormones are synthesized, keeping blood levels of these hormones elevated. If the stressor is a disease or injury, inflammation begins and phagocytic cells are chemically attracted to the area. Epinephrine acts on the heart to increase the heart rate (the number of contractions per minute) and cardiac output (the strength of the contractions). Norepinephrine increases the flow of blood to the heart and muscles for possible increased activity, while constricting other vessels, diminishing the flow of blood to the gut, skin, and kidneys (Fig. 12.8). Cortisol suppresses lymphocyte activity, thereby decreasing the cytokine interleukin-2; it also suppresses inflammation and the inflammatory cytokine interleukin-1.

Exhaustion. If the stressor is not successfully overcome, the adaptation syndrome reaches its third phase, exhaustion. The steroids made in the resistance phase are used up and the animal is unable to make more. During the exhaustion phase, the action of the sympathetic nerves tapers off, while the endocrine organs take over. Adrenal hormones stimulate another endocrine gland, the pituitary gland, to secrete endorphins and enkephalins. These are chemically related to opioid drugs (Chapter 11) and alter the activity of neurons and of immune cells.

The pituitary was once called "the master gland" because its hormone secretions controlled the activity of many other endocrine glands, but it is now known that the pituitary itself is under the control of the brain via the hypothalamus. Hypothalamic cytokines also act on other cells within the brain itself, changing the activity level of neurons, and this change affects behavior, heat production, and many other functions.

Some stress responses are due to actual physical danger, but stress responses can also result from the demands of work or school (e.g., deadlines or examinations), or the actions of the people in our lives. Whether these demands are real or perceived, they can have the same biological consequences: the physiological changes that characterize the stress response.

Effects of stress on health. To what extent does the stress response influence human health? The

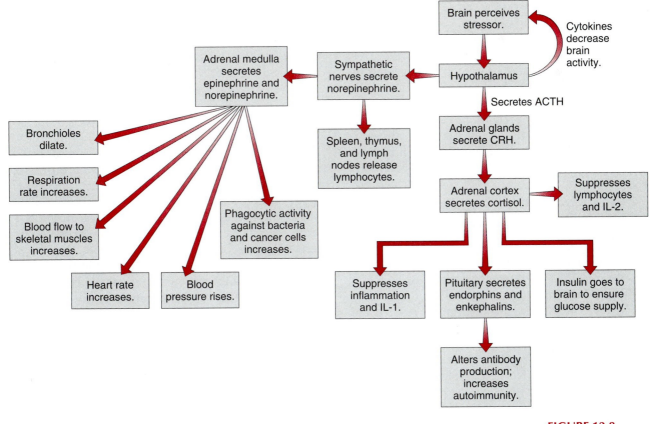

FIGURE 12.8
The stress response.

answer is, to a great extent. Stress (i.e., the stress response) is an important risk factor in heart disease, and there is considerable evidence that people exposed to chronically high levels of stress are statistically more likely to become ill with infectious diseases, to remain ill for longer periods, and to suffer more severe consequences, even death. Depression is much more common in people subject to chronic stress. Among cancer patients, those who have better coping skills for dealing with stress (as measured by psychological tests) have higher survival rates, compared with patients who have the same forms of cancer but poorer coping skills.

The Relaxation Response

When actual physical danger has passed, the stress response will abate. During a stress response, the adrenal and pituitary glands mediate the response of the sympathetic nervous system. This response will gradually be reversed by the actions of the parasympathetic nervous system. This reversal is called the **relaxation response.**

It is sometimes more difficult to get our mentally induced stresses to abate. Since the stress response can be mentally induced, it has been hypothesized that the relaxation response should be mentally inducible as well. Some cultures and some religions have been more open to this idea than others. Many practices that are aimed at evoking this relaxation response, including Yoga, transcendental meditation, and others, originated in Asian traditions. Hindu yogis have learned how to consciously control the actions of their autonomic nervous systems and to bring about levels of activity even below the normal levels for the resting state. Measurements made by Western scientists have shown that these yogis are able to lower their blood pressure, breathing, oxygen consumption, heart rate, and metabolic rates.

Other cultures, including many Western cultures, have been less open to the idea that the relaxation response can be controlled. The traditional definition of the autonomic nervous system emphasized that it governed involuntary functions over which we do not have conscious control and had the unintended effect of discouraging research on any possible interactions of the autonomic nervous system with our emotions and other conscious body states. Recently, however, some of the methods for conscious control of autonomic processes are being borrowed from other cultures. Some athletes have learned meditation, while others have learned a related technique called *imaging.* In imaging, one invokes specific mental imagery in order to put one's body as well as mind in a certain state of relaxed determination to succeed in sport. Western medicine has begun to use similar techniques to help cancer patients in fighting their cancers (described further in a later section on mental imaging).

Less conscious strategies can also produce the relaxation response. Studies using measurements of blood pressure and other physiological indicators have shown that contact with pets can reduce stress and bring about relaxation. Older people who keep pets have also been shown to live longer than those who do not, even when comparison is made between people of comparable health status initially. Heart-attack victims are much less likely to have a second heart attack if they care for a pet. Studies such as these show a statistical correlation between two factors. From such data by themselves we cannot assign a causal relation between the two; that is, we can not say that pets caused the increased longevity or improved health.

The Placebo Effect

In clinical trials testing new drugs, a common experimental design is for one group of people to receive the test drug and for another group, the *control group,* to receive a **placebo,** a preparation that is similarly colored and flavored but that does not contain the test ingredient. To be in such a study, people must have given their informed consent (Chapter 2); that is, each person must have been informed as to the nature of the test being conducted and then signed a written consent form. Many studies are conducted *double blind;* that is, neither the subject nor the experimenter knows who is receiving the placebo and who is receiving the actual drug. For the drug to be considered effective, there must be a statistically significant difference in outcome between the group receiving the experimental drug and the control group receiving the placebo.

Studies like this were initially designed to demonstrate whether particular drugs were or were not effective. They have also shown, over and over again, that the people who receive the placebo in such tests have a significant change from the base-line values of whatever parameters are being measured. They also experience many "side effects," although they have not received a drug. In experiments on pain perception, people who are given placebo painkillers will very often experience a reduction in pain. People who have purchased street drugs will often feel the reaction they seek even when the drugs are in such low concentration that no real effect could be produced.

Such **placebo effects** have often been treated as an annoyance in research; that is, they make it more difficult to demonstrate the "real" effects of test compounds. The existence of such effects, however, is additional evidence that the mind can bring about physiological changes in the body, in addition to the effects of the stress response and the relaxation response.

1. Is adrenalin (epinephrine) a hormone or a neurotransmitter?
2. How does thinking about an annoying or threatening event produce stress?

D. The Neuroendocrine System Interacts with the Immune System

We have discussed how the nervous system and the endocrine system communicate with each other. In recent decades, it has also become apparent that both of these systems also communicate with the immune system. The emerging model is one that sees the nervous, endocrine, and immune systems not as separate entities but as one interacting communications network. Extending the metaphor, we might say that the nervous system provides the hardwiring of the system, since nerve cells have distinct locations and make direct connection with one another via synaptic junctions. The immune cells, by analogy, are mobile synapses, responding and transmitting messages at the same time they themselves are moving, the original cellular phones. Endocrine glands could be thought of as more like microwave transmission towers, having a fixed location themselves but sending out their messages chemically rather than over wires (nerves).

The above is intended to aid the student in conceptualizing how the immune system works and should not be taken too literally. Mental images of this kind can, however, provide a basis for discussion and can suggest hypotheses to test. When used this way and shared by a number of scientists, an image becomes a *model* (Chapter 1). Psychoneuroimmunology proceeds by testing hypotheses suggested by the model of a system of interactive chemical communications linking the immune, nervous, and endocrine systems.

Unlike literary images and metaphors, which are unconstrained, those in science must be tested against reality. Models in science gradually become more limited, or more well defined, as data are gathered from observation and experimentation. Models are not descriptions of reality but are evolving concepts, consistent with the body of data known at that time. The psychoneuroimmunology model as it exists today is already more limited than the communications network metaphor just described. The "hardwired" nerves are not completely immobile because synaptic connections between cells are formed and lost throughout life. There are also known to be limits to the mobility of immune cells. Because the model is relatively new, its limitations are now in the process of being identified. Scientific explanations are always tentative, but nowhere is this more so than in the case of new models that have not yet stood the test of time. The reader is strongly encouraged to think critically about the information presented here. Where are the gaps? What still needs to be done?

Psychoneuroimmunology has an appealing central model, but models or theories that are appealing or that fit with our common sense do not necessarily stand up to scientific scrutiny. Is there any scientific evidence that supports the hypothesis that the nervous, endocrine, and immune systems are interconnected? Yes. The first line of evidence comes from the discovery that these systems use the same cytokines for cellular communication. In addition, there is both structural and functional evidence, gathered from *in vitro* studies ("in glass," that is, studies in test tubes), and *in vivo* studies ("in life," that is, studies in animals and humans). We will examine each of these in turn.

Shared Cytokines

The term **cytokine** includes all molecules that are secreted by cells for the purpose of communication with other cells. The ability of any cell to respond to a particular cytokine is dependent on whether the cell has receptors for that cytokine. Therefore, for the cytokines secreted by the nervous, endocrine, or immune systems to have an effect on the other systems, one must ask if there are receptors for them on cells of the other systems. The answer is yes, there are. Candace Pert and Solomon Snyder discovered in the early 1980s that there were receptors on immune cells for endorphins, cytokines produced by nerve cells. Since that time, receptors have been found for many other cytokines that interconnect the nervous, immune, and endocrine systems. B cells have receptors for many neurotransmitters, including norepinephrine and enkephalins. The presence on immune cells of receptors for these neurotransmitters suggests that immune cells could respond to these transmitters, as does the finding that the number of these receptors per cell increases during immune activation. Lymphocytes possess receptors for several endocrine cytokines induced during stress (ACTH and steroid hormones) or in pain (beta-endorphin). Neurons in the hypothalamus have receptors for the immune cell cytokine interleukin-1. The glial cells that feed the neurons in the brain have interleukin-2 receptors (Chapter 13), as do some endocrine cells.

Nerve Endings in Immune Organs

Evidence of another sort came from studies done by David Felten on the innervation of the organs of the immune system. Neurons of the sympathetic nervous system were found to terminate in the immune organs, such as the spleen, thymus, lymph nodes, bone marrow, and lymphoid tissue in the gut. The sympathetic nerve cells release norepinephrine, and the immune cells in these organs have receptors for norepinephrine.

These studies used a technique called immunohistochemistry. Immune organs from animals are frozen and sliced very thin (*histology*) and then are stained with antibodies (*immuno-*) covalently bound to enzymes (*chemistry*). Such techniques borrow the exquisite antigen specificity of the immune system. Antibodies that recognize some molecule you want to detect are produced artificially in cell cultures. The antibody used by Felten's group recognizes and binds to an enzyme used in the synthesis of norepinephrine but does not bind to other chemicals. Thus, the antibodies give a specificity to the technique, just as they do during a natural immune response in the body. The assay enzyme then allows detection by producing a color change in the location where the antibody has bound. Some typical results are shown in Fig. 12.9. While most nerve cells terminate in synapses with other nerve cells or on muscle cells (see Chapter 10), Felten's immunohistochemical studies showed nerve cells terminating and releasing norepinephrine in proximity to the cells of the immune system.

Studies of Cytokine Functions

Demonstrating that a receptor is present, or even that both the receptor and its cytokine are present together in a tissue, does not by itself show that any effect follows the binding of the cytokine to its receptor. For that, experiments known as *func-*

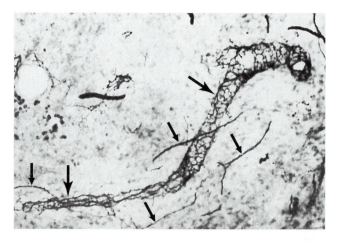

FIGURE 12.9
Immunohistochemical staining showing the presence of the neurotransmitter norepinephrine in the rat spleen. Large arrows: blood vessel. Small arrows: nerve fibers synthesizing norepinephrine.

tional assays (techniques used to measure a response) need to be done.

Immune cytokines increase hormone secretion.

If secretion of one cytokine can be shown to happen after administration of another cytokine, this is evidence of a functional connection between the two. Such studies can also give us clues as to the way in which an effect can be triggered, which we call the *mechanism* (or *mechanisms,* since more than one trigger is often possible).

Functional assays can also tell us the order in which events happen. For example, during acute bacterial infection, the secretion of adrenal and pituitary hormones increases. Functional studies showed that the effect is not a direct one. If pituitary cells are stimulated with bacterial molecules in vitro they do not secrete these hormones. When immune cells are exposed to these bacterial products, however, they secrete cytokines. If immune cell cytokines (including interleukin-1, interleukin-2, or tumor necrosis factor) are administered to animals, the level of pituitary hormone in the blood increases.

Neuroendocrine effects on immune cells.

Among the cytokines secreted by the brain cells are the enkephalins and endorphins. When enkephalins are given to living rats, immune responses are altered, including antibody responses and cell-mediated responses (Fig. 12.8). Interestingly, low doses of enkephalin increase antibody production, while high doses suppress it. Low doses of enkephalins also increase some destructive aspects of the immune response, including both allergic and autoimmune responses, while high doses of enkephalins depress those responses.

The glucocorticoid steroid hormones (cortisol, cortisone, corticosterone, and several related compounds secreted by the adrenal cortex) can have inhibitory effects on the immune system. Steroid hormones given to animals decrease the numbers of cells in lymphoid organs and suppress interleukin-1 and interleukin-2 secretion (Fig. 12.8). Steroids also increase the susceptibility of the animals to disease and activate latent infections (infectious organisms that have been present but have not brought about disease now do so).

A chemically similar compound, *hydrocortisone,* is used medicinally as an *anti-inflammatory* agent. Remember that normal inflammation is the healing phase of the immune response. In insect bites, severe poison ivy, athletic injuries, and rheumatoid arthritis, the swelling and pain are the result of the inflammatory response. In these situations, the annoying symptoms are actually an indication that the immune system is at work. Thus, a person who chooses to take an anti-inflammatory drug is choosing to suppress the healing processes of the immune system in order to suppress the negative symptoms of inflammation. The symptoms may be so severe that immunosuppression is needed, but long-term immunosuppression by corticosteroids will likely have adverse consequences on other aspects of health.

Stress and the Immune System

Results such as those just mentioned must be extrapolated with caution because the intact living body is more complex than any experimental system. Another type of study, therefore, looks at changes in cytokine and immune functions after real-life events. Rather than giving the cytokine to the experimental subjects, they are instead exposed to stress.

As was discussed above, stress is a series of physiological events involving both the nervous and endocrine systems. That stress also involves the immune system has been shown in many studies on animals and humans. Short-lasting stress may actually strengthen aspects of the immune system. Engulfment of bacteria by white blood cells in mice is increased by short-term stress brought on by conflict (Fig. 12.8). The ability of the natural killer cells (described in an earlier section of this chapter) to kill tumor cells can be increased by restraining rats on a single day so as to increase their stress response. However, several days of restraint-induced stress causes a decrease in natural killer cell activity.

The production of interleukin-2 is decreased during stress. This cytokine, secreted by helper T cells, is necessary for a full-strength antigen-specific immune response. So decreased levels of interleukin-2 decrease the response of antibody-producing B cells and cytotoxic T cells (Fig. 12.8).

An anti-inflammatory feedback loop.

The anti-inflammatory activity of the steroid hormones may

provide a natural feedback loop to keep the extent of inflammation within physiological bounds. *Addison's disease* is a disease that results from insufficient secretion of these hormones by the adrenal gland. People with Addison's disease have such low levels of these anti-inflammatory cytokines that they suffer severe inflammation when they get bacterial infections and need to be given anti-inflammatory drugs. The opposite problem occurs in *Cushing's syndrome*, a disease in which persistently high levels of steroid hormones result in immunosuppression. People with Cushing's syndrome get frequent bacterial infections, and, since the normal inflammatory response is lacking, the infections can be fatal.

Higher levels of one of these hormones, corticosterone, cause the death of pre-T cells that are developing in the thymus. Daily fluctuations (circadian rhythms, Chapter 10) in the plasma levels of corticosterone also correlate with the circadian rhythm of the numbers of B and T cells circulating in the blood. Stress can, however, suppress immune function even in rats whose adrenal glands (the source of corticosterone) have been removed, demonstrating that there must be other pathways as well.

Effects of stress on health.

The biochemical events of stress can be started by psychological factors, and prolonged stress can suppress the immune system. There have been many demonstrations of immune suppression in people undergoing various types of stress. It follows that long-term psychological stressors might produce conditions in which disease can develop. In one experimental design, blood samples are taken from young, basically healthy students during exams, and various immune parameters are compared to base-line levels measured in blood samples from the same students one month prior to exams. In this type of experiment, each person serves as his or her own control, minimizing differences due to factors other than the tension of exam situations (see below). Exam periods brought on an increase in adrenal cortex hormones and a decrease in natural killer cell activity. There were also changes in the numbers of T cells and a decrease in the ratio of helper T cells to suppressor T cells. Since helper T cells stimulate immune responses and suppressor T cells inhibit them, a decrease in the ratio of

the two means less help and more suppression, for an overall decrease in T cell responsiveness. Such short-term stress was correlated with an increase in disease, primarily upper respiratory tract infections. Many college health centers report an increase in student admissions for infectious diseases during exam periods.

Another study examined the immune functions and health status of men who had separated or divorced within the previous year. The experimental design was different than the one used with the medical students. One group of people, the divorced men, were compared to people in a control group of married men. Not only were the immune systems of the divorced men impaired, they experienced a greater number of illnesses than did the controls. Comparisons were also made between those men who had not initiated the separation or divorce and those who had. Those who had not initiated the break were significantly more immunosuppressed and had more illnesses than those who did. Studies like this employ statistical methods for determining if the differences between groups are greater than could have been predicted by chance only.

Other researchers found that elderly people who had been caring for a spouse with Alzheimer's disease demonstrated a decrease in three different measures of immune function. The elderly people undergoing prolonged stress got sick more often than those in the control group. The immune systems of these caregivers stayed depressed after the death of the spouse.

Several studies have showed decreased immune function in people with clinical depression. A prospective study showed that people who were depressed had a higher incidence of cancer 17 years later. A *prospective* experimental design is one in which a group of people are entered into a study, physical exams and/or psychological exams are given (depending on the purpose of the study), and then their outcomes are monitored at various later times. One strength of a prospective study is that no one knows ahead of time who will be sick and who will not; base-line data are taken before the outcomes are known. One weakness is that the percentage of people in any particular group who will get a particular disease may be very low, so that the number of people entering the study must be very large. Many people will leave the study for unrelated reasons. Many other factors can influence the outcome; to some extent this can be corrected by statistical methods, but only those factors that have been identified can be factored out.

What mechanisms could bring about cancer after clinical depression? Immune activity is compromised, and other functions are also impaired, including levels of an enzyme that repairs damaged DNA. Breaks and misreadings of DNA occur rather frequently, but normally several "proofreading" mechanisms check the DNA and repair most of the mistakes; mistakes that remain uncorrected are capable of transforming cells. The suppressed immune system then fails to remove these transformed cells before they have become established as cancer (see Chapter 9).

Individual Variation in the Stress Response

There are many types of stressors. The effect of stressors on health will be highly variable, since the effects will be modified by many additional factors. For example, of the people exposed to infectious mononucleosis, prevalent in college-age populations, not everyone becomes sick, and of those who do, some become much sicker than others. It is important whether the stress occurs before or after the immune challenge started. The severity and duration of the stress are also important. Genetic factors have some role; in animal studies, different strains of mice (each genetically inbred so as to minimize genetic variation within the strain) respond differently to stressors. Psychological factors are just as important. If an animal is able to establish coping behaviors, the effect of the stress period on the immune system will be lessened.

Personality profiles and life events have some bearing on disease susceptibility and disease progression in humans as well. Testing methods have been developed for quantifying the psychosocial impacts of life events (Table 12.1). The use of such methods has indicated that certain life events can increase the probability that cancer will develop, although the results have been highly variable from study to study.

We cannot predict how much immunosuppression will be sufficient to result in disease in a given individual. Both the degree and duration of suppression that result in disease will likely be different for different people. Some studies suggest that coping styles can help regulate the degree of impact that stressful events will have on individual health. When psychological tests are given to matched sets of cancer patients (with the same kind of cancer in comparable stages of the disease), those with more optimistic or aggressive personalities show higher survival rates than those who are more easily resigned to their fate. Other factors, such as environmental pollutants, drugs, alcohol, and malnutrition, may also weaken the immune system. If a person's immune system is already weakened by one or more of these factors, the additional immunosuppressive effects of stress are more likely to result in disease.

Conditioned Learning in the Immune System

If brain cells can interact with immune cells and psychological factors can influence the onset and progression of disease, could a person learn how to control their own immune response? As odd as this idea may sound, evidence in support of it is accumulating.

Conditioned immunosuppression.   Classical or Pavlovian conditioning is a type of unconscious learning in which the organism learns to associate one stimulus with another. After such conditioning, presentation of the second stimulus will bring on the physical effects of the first stimulus (see Chapter 10). Robert Ader, a psychiatrist, and Nicholas Cohen, an immunologist, worked together to try to explain why some of Ader's mice had been dying unexpectedly in his studies on a drug called cyclophosphamide. This drug suppresses the immune system and in fact is given to recipients of organ transplants so that their immune systems do not reject their transplants. Ader's mice had been receiving cyclophosphamide along with saccharin in their drinking water. Later the mice received only the saccharin, but their immune systems again became suppressed although saccharin itself has no direct effect on the immune system. What Ader and Cohen demonstrated in several controlled studies is that the dying mice had been conditioned. The mice had learned to

TABLE 12.1   STRESSFUL LIFE EVENTS

The relative stressfulness of each event is indicated by the numbers at the left. The most stressful life event was assigned a value of 100, and other events were assigned lower values in proportion to their effects on stress. Divorce, for example, caused stress in 73% as many individuals experiencing this event as was true for death of a spouse.

Stressfulness	Life Event	Stressfulness	Life Event
100	Death of spouse	29	Change in responsibilities at work
73	Divorce	29	Son or daughter leaving home
65	Marital separation	29	Trouble with in-laws
63	Jail term	28	Outstanding personal achievement
63	Death of close family member (except spouse)	26	Spouse begins or stops work
53	Major personal injury or illness	26	Begin or end school
50	Getting married	25	Change in living conditions
47	Fired at work	24	Change in personal habits (self or family)
45	Marital reconcilation	23	Trouble with boss
45	Retirement	20	Change in work hours or conditions
44	Change in health of family member (not self)	20	Change in residence
40	Pregnancy	20	Change in schools
39	Sex difficulties	19	Change in recreation
39	Gain of new family member	19	Change in church activities
39	Business readjustment	18	Change in social activities
38	Change in financial state	17	Mortgage or loan less than $40,000
37	Death of close friend	16	Change in sleeping habits
36	Change to different occupation	15	Change in number of family get-togethers
35	Change in number of arguments with spouse	13	Change in eating habits
31	Mortgage over $40,000	13	Vacation
30	Foreclosure of mortgage or loan	12	Christmas
		11	Minor violations of the law

associate the immunosuppressant chemical, cyclophosphamide, with the saccharin. Later the effect of the first stimulus, immunosuppression, could be produced by giving them only the second stimulus, the saccharin water without cyclophosphamide.

These experiments demonstrating conditioned immunosuppression have been repeated with several other paired stimuli. Such conditioning has been shown to improve the health of mice with autoimmune disease; preliminary results suggest that conditioned immunosupression is also useful in the treatment of people with autoimmune disease.

Conditioned immunopotentiation.   If animals are challenged with antigen paired with another stimulus, a conditioned immunopotentiation can be demonstrated. In a normal immune response, a second exposure to the same antigen would produce an increase in specific immunity to that antigen. In these experiments, however, the animals later have an increase in their antigen-specific immune response when they are given only the paired stimulus without the antigen.

Voluntary Control of the Immune System

Other work has shown that people can learn to voluntarily regulate many of the physiological processes mediated by the autonomic nervous system. People can learn to regulate the temperature of their hands, their blood pressure, their heart rate, and their galvanic skin resistance (resistance to electrical conductivity, which is a measure of the amount of sweat on the skin; Chapter 14). Since the immune system communicates with the autonomic system, these findings raised the hypothesis that parameters of the immune system may also be subject to voluntary control. Several studies have shown that this is possible using different self-regulation procedures, including relaxation, mental imaging, biofeedback, and emotional support. Voluntary potentiation of several immune parameters has been shown, including white blood cell engulfment of bacteria, antibody production, lymphocyte reactivity, and natural killer cell activity. Women with metastatic breast cancer, all of whom were receiving medical treatment of their cancers, survived longer if they were part of support groups than if they were not.

Mental imaging. In one study, subjects were asked to make mental images of their neutrophils becoming more adherent, a cellular process that might make neutrophils more efficient at getting to a disease site. When the adherence of their neutrophils was measured after several sessions of practicing such imaging, it was significantly increased in comparison to the neutrophils from people who had simply relaxed without forming the mental image of their neutrophils. Adherence was also measured in neutrophils from a third group who made mental images but did not have the practice sessions. Neutrophils from this group did not show an increase in adherence, showing that effective imaging takes a period of training.

Biofeedback. In *biofeedback*, measurement devices are placed on people so that they can monitor the results of their self-regulation (Chapter 14). Biofeedback has proved to be effective for some people in the management of chronic pain and migraine headaches. Although it is too soon for there to be conclusive data on whether voluntary regulation of immunity will translate into improved health, preliminary studies suggest that it will. People with HIV infection and AIDS have remained healthier when they have used these techniques.

Studies on populations. If the mind and the emotions can influence the immune system, and the immune system helps fight off many diseases, how far can the disease-fighting process be controlled by the mind? In recent years, statistical evidence has been accumulating from studies in both the United States and China to show that dying patients can exercise control over their disease processes to the extent that they actually influence the time of their death.

Large-scale studies are often done on entire populations by studying death certificates and comparable records. When the date of death is examined for a large number of patients in a population at large, several interesting regularities appear. For instance, the overall mortality rate is lower for a period of several days to either side of each person's birthday, and this is compensated by an increasing mortality rate about a week or two later. Such data make it appear that dying patients are eager to survive to reach their birthdays and that they can postpone the inevitable by as much as a week or two. Other studies have shown similar statistical effects demonstrating the ability of people to postpone the time of their death until after holidays or family events (e.g., weddings) of special importance to them. In China, this effect even has a name: the Harvest Moon phenomenon. The Harvest Moon Festival is a traditional family celebration in which the oldest and most respected woman in each family is expected to prepare a large feast to celebrate with her entire family. Studies of death certificates show that older Chinese women have a reduced mortality rate around the time of this festival. It is difficult to determine whether this effect is primarily the result of the activities in which the matriarch engages, the increased esteem or importance that she receives, her desire not to disappoint others, or simply the anticipation of the big event. Studies on Jewish populations have shown a similar decline in mortality around the time of Passover.

A complex interaction between expectation and mortality has also surfaced in a recent study in China. Traditional Chinese astrology divides the calendar into 12-year cycles. Each year is represented by a different animal, and people born in that year are said to be under control of that animal's influence, with which certain diseases are associated. In this study, elderly Chinese patients were surveyed to see whether or not their disease matched the predictions of Chinese astrology. Patients with a disease that matched their astrological year were then compared with patients having the same disease but a different astrological year. The patients whose diseases matched their astrological year experienced higher mortality, and this effect was proportional to the patient's belief in traditional Chinese astrology. Presumably, patients who believed that they had the disease that was fated for them in the stars more willingly gave up the struggle and resigned themselves to an earlier death. Although these phenomena are well documented, the mechanism(s) by which they are produced are not known.

Changing Theories of Disease Causation

Since the mid-nineteenth century, Western medicine has tended to view disease as having external causes. The ascendancy of this view can be traced back to the work of Louis Pasteur, who championed the theory of "specific etiology" as part of the germ theory of disease. In this theory, each disease has one specific and identifiable cause. This theory led to much highly successful research that associated single species of microorganisms (bacteria, viruses, and parasites) with specific diseases. Such research ultimately produced vaccines and antibiotics, which have successfully controlled many infectious diseases.

Working at the same time as Pasteur, another French scientist, Claude Bernard, questioned what it meant for a microorganism to "cause" a disease. He observed that there were very great differences in individual response to microorganisms; some people got sick and even died, while other people who were also exposed did not get sick. Bernard, a physiologist, developed an alternative theory, that of the *milieu intérieur*, or inner environment, as being an equal determinant in whether or not a person became sick. While a microorganism might be a *necessary* cause of a disease, it was not a *sufficient* cause. A microbe was like a seed that would only sprout into disease if *le terrain*, the "soil," allowed it.

The past century of research in immunology and more recently in psychoneuroimmunology suggests that Bernard's theory had seized upon some truth that Pasteur's theory had omitted. Even the diseases for which an infectious agent is known are not caused by the microorganism alone, but rather by the outcome of a complex process in which the microorganism disturbs homeostasis while host mechanisms attempt to restore it. Bernard's theory is compatible with a view of the neuro-endocrine-immune communication network as a sensory organ by which deviations from homeostasis are detected and corrected.

One of the primary differences between the two theories is that one is essentially linear (one cause produces one effect), and the other is not. Most biological systems are nonlinear. Statements of the kind "X is the cause of Y" often fail to tell the whole story. If X and Y are interacting in a feedback loop, Y may just as accurately be called a cause of X. The effects are bidirectional; that is, X and Y influence each other. Moreover, each cause may have multiple effects; if we have found one, there may be many others as well. When one feedback loop interacts with another loop, changes in part of one loop will affect the other components of that loop and all the other loops it interacts with.

Today, many of the diseases that are still without effective cures do not appear to have simple, single causes. Scientists are examining other medical traditions, searching to widen our conceptual framework of the causes of disease. Psychoneuroimmunology borrows from Chinese, Indian, and other Asian traditions that view health as the balance of life forces. The psychoneuroimmunology paradigm also uses a *functional* model of the body, a model that regards the body as an entity that is in a constant state of change. Health is the state in which these forces are in balance, in dynamic equilibrium (homeostasis); disease is the state in which they are not. African traditions in which a person's health and well-being are seen to be influenced by the social environment in which the person lives coincide with the psychoneuroimmunology view that mental and emotional factors can affect health and disease. Widening one's point of view and being open to new ideas are integral parts of science. New hypotheses are formed by the melding of ideas and then must be followed by the hard, and often slow, work of hypothesis testing.

THOUGHT QUESTIONS

1. Given that chronic or severe stress generally weakens the immune system, how might such an apparently harmful relationship have evolved?
2. Is stress always harmful?

3. List the types of experiments done in psychoneuroimmunology. Do the results of any of these falsify the hypothesis that the mind and the body interact? Do any of these results prove that the mind and the body interact?

4. Is the Cartesian concept of the mind the same thing as the brain? Is *mind* simply the name we give to the *workings* (or functions) of the brain?

5. Is the statement "The mind is nothing more than the actions of the brain" a testable hypothesis? How might it be falsified?

CHAPTER SUMMARY

The immune system is a system that works to detect and correct deviations from homeostasis within the organism. Immunity is antigen-specific and acquired by the individual. Permanent rearrangement in the DNA that codes for antigen receptors produces the tremendous diversity of the binding sites of these receptors. A population of cells is then selected that can detect and react to nonself but not to self. Products of the immune response (antibody proteins and cytotoxic T cells) rid the body of the specific antigen that induced their production. The immune system retains a memory of the encounter so that it can react faster and more strongly to subsequent exposures to that antigen.

The immune system is not an independent system, but interacts with the nervous and endocrine systems in an integrated and multidirectional way. Communication among these three systems is mediated by chemicals called cytokines. Factors that interrupt this communication network will prevent the restoration of homeostasis within the organism, producing what we call disease. Mental states can affect the functioning of the endocrine and immune systems and can either increase or decrease disease-fighting activity and tissue regeneration and repair.

KEY TERMS TO KNOW

antibodies (p. 343)

antigen (p. 339)

autonomic nervous system (p. 352)

B lymphocytes (p. 340)

clone (p. 343)

cytokine (pp. 341, 358)

equilibrium (p. 347)

feedback loops (p. 347)

fight-or-flight response (p. 352)

healing (p. 349)

health (p. 337)

homeostasis (p. 354)

hormones (p. 351)

immunization (vaccination) (p. 345)

immunopotentiation (p. 351)

immunosuppression (p. 351)

inflammation (p. 348)

lymphatic circulation (p. 339)

natural immunity (p. 348)

neuroendocrine system (p. 352)

parasympathetic neurons (p. 352)

passive immunity (p. 348)

placebo (p. 356)

placebo effect (p. 357)

population selection (p. 342)

psychoneuroimmunology (p. 336)

relaxation response (p. 356)

specific immunity (p. 344)

stress (stress response) (p. 354)

sympathetic neurons (p. 352)

T lymphocytes (p. 340)

tolerance (p. 350)

CONNECTIONS TO OTHER CHAPTERS

Chapter 1   Psychoneuroimmunology is a good example of a paradigm.

Chapter 3   The variety of antigens recognized by the immune system is based on the rearrangements of just a few genes, resulting in a great variety of proteins.

Chapter 5   The ability to form an immune response to any particular antigen depends on cell surface proteins. These proteins are coded by genes that are highly polymorphic; gene frequencies for these genes vary among populations.

Chapter 7   The genes that allow kin recognition in many species are also used by the immune system to recognize self.

Chapter 8 Poor nutrition suppresses the immune system.

Chapter 9 Suppressed immune function greatly increases risks for cancer.

Chapter 10 The brain can greatly affect many immune functions.

Chapter 11 Many drugs suppress the immune system. Many drugs mimic some of the actions of the autonomic nervous system.

Chapter 13 Suppression of the immune system is characteristic of AIDS.

AIDS and HIV

The disease known as AIDS first received public attention in 1981. It quickly became one of the most feared and widely discussed diseases of our time. As we saw in Chapter 12, the immune system has several ways of protecting the body from disease. When people have AIDS, their immune systems no longer function properly, so they are at great risk for conditions that would almost never affect a healthy person. The spread of AIDS has been accompanied by the spread of misconceptions concerning the disease. In this chapter, we will summarize what is known about this dreaded disease and also address certain misconceptions.

A. AIDS Is an Immune System Deficiency

The acronym **AIDS** stands for *a*cquired *i*mmunodeficiency *s*yndrome. *Acquired* means that the illness is not genetically inherited as the result of a defective DNA message. *Immunodeficiency* means that some part or parts of the immune system are not functional, and *syndrome* means that a wide range of symptoms are associated with the condition. Since the immune system normally functions to protect the body from disease-causing microorganisms, people whose immune systems are deficient can become seriously ill.

The Immune System

The body has many ways of protecting itself from diseases that are caused by bacteria, viruses, or other microorganisms. The skin protects the body's surface against entry. The mouth, vagina, and many other potential entry points are coated with mucous secretions that continually wash away adherent bacteria, inhibit bacterial growth, and promote healing.

A much more specific type of protection is afforded by the immune system. The immune system is capable of distinguishing between *persistent* molecules (those that the immune system has been educated to consider *self;* Chapter 12) and new or transient molecules (those that the immune system has not been educated to consider self). The latter may be molecules from outside the body, as in an infection, or they may be new molecules that have appeared internally, such as molecules made by cancer cells (Chapter 9). In a state of health, the immune system does not respond to molecules that are recognized as self. Molecules determined to be *not self* will trigger an immune response that results in the inactivation or destruction of the molecules or of the cells bearing these molecules.

Many of the actions of the immune system depend upon cells called *lymphocytes,* a type of white blood cell that circulates throughout the body via the bloodstream and the lymphatic circulation. *B lymphocytes* (B cells) react to nonself molecules by making and secreting soluble proteins called *antibodies*. Antibodies protect against disease by neutralizing the disease-causing microorganism or its products or by identifying the microorganism as a target for engulfment by other white blood cells. *T lymphocytes* (T cells) do not secrete antibodies, but instead react to nonself molecules by increasing the number of specific *cytotoxic* ("cell killer") T lymphocytes. T cells protect against diseases by killing cells infected by viruses or cells that have become cancerous. Any of the above actions can end the course of the disease and prevent its recurrence (see Fig. 12.5). A properly functioning immune system thus keeps infections in check and prevents minor infections from becoming major life-threatening events.

There are other types of T cells in addition to cytotoxic T cells. **Cytotoxic T cells** bear a surface molecule called CD8, so they are also called **CD8 T cells.** Another type of T cell, the **helper T cells,** carry a surface molecule called CD4, and thus are also known as **CD4 T cells.** It is these helper T cells that provide the connection between the two main types of immune response by secreting a molecule called *interleukin-2* or *IL-2* (Fig. 13.1). B cells and cytotoxic T cells that have become partially activated by exposure to nonself molecules will not become fully activated (to produce antibody, or to become cytotoxic) until they bind interleukin-2. The helper T cell, which produces this interleukin-2, is thus central to both antibody and cellular immune responses (Fig. 13.1).

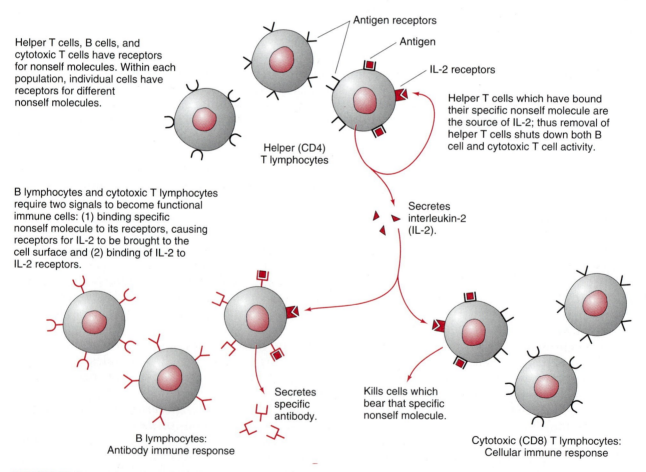

Antigen receptors

Antigen

IL-2 receptors

Helper T cells, B cells, and cytotoxic T cells have receptors for nonself molecules. Within each population, individual cells have receptors for different nonself molecules.

Helper (CD4) T lymphocytes

Helper T cells which have bound their specific nonself molecule are the source of IL-2; thus removal of helper T cells shuts down both B cell and cytotoxic T cell activity.

B lymphocytes and cytotoxic T lymphocytes require two signals to become functional immune cells: (1) binding specific nonself molecule to its receptors, causing receptors for IL-2 to be brought to the cell surface and (2) binding of IL-2 to IL-2 receptors.

Secretes interleukin-2 (IL-2).

Secretes specific antibody.

Kills cells which bear that specific nonself molecule.

B lymphocytes: Antibody immune response

Cytotoxic (CD8) T lymphocytes: Cellular immune response

FIGURE 13.1
Interactions between lymphocytes and their cytokines.

Immunodeficiency

An **immunodeficiency** is an absence of one or more of the normal functions of the immune system. Immunodeficiencies can be either inherited or acquired. One type of inherited immunodeficiency, the severe combined immune deficiency syndrome (SCIDS), caused by a lack of the enzyme adenosine deaminase (ADA), is discussed in Chapter 3.

There are many ways to acquire immunodeficiencies. Malnutrition, alcohol, and drugs such as cocaine and marijuana can depress the functioning of the immune system. Psychological stress and depression can decrease immune functions (Chapter 12). A variety of pollutants, including cigarette smoke, can also decrease immune function. Many microorganisms cause immunodeficiencies, although no others are as severe as AIDS. People who acquire immunodeficiencies by any of these means get sick more often than people with healthy immune systems, and their illnesses are more severe.

Many immunodeficiencies are temporary: If the inhibiting factor is removed, the immune system will recover. In addition, in cases in which only a portion of the immune system has been affected, the unaffected parts of the immune system may provide a backup. The immunodeficiency of AIDS, on the other hand, is long-lasting; the immune system does not recover, and the disease is fatal. The hallmark symptom of AIDS is a decrease in the number of CD4 T cells in the blood, which decreases the availability of interleukin-2 and thus weakens both the antibody and the cytotoxic T cell-mediated immune responses (Fig. 13.1). When both the antibody- and cell-mediated responses are impaired, as in AIDS, the body becomes extremely susceptible to infection and even the most minor infection can quickly become life-threatening.

B. AIDS Is Caused by a Virus Called HIV

We now know that AIDS is an infectious disease caused by a virus known as HIV (human immunodeficiency virus). However, when the syndrome first began to appear in the United States at the very end of 1980, the cause, and even the fact that it was an immunodeficiency, was not known. In this section we will trace the steps that led to the identification of this immunodeficiency and of its causative agent. How were hypotheses suggested? How were these hypotheses tested? What types of evidence are necessary to call something the "cause" of a disease? Does such evidence rule out other hypotheses?

Discovery of the Connection between HIV and AIDS

Unusual pneumonia cases.   In June of 1981, five cases of pneumonia were reported from the city of San Francisco, in which a microorganism called *Pneumocystis carinii* had been identified. The report appeared in *Morbidity and Mortality Weekly Report* (*MMWR*), a publication of the Centers for Disease Control and Prevention (CDC), which tallies all cases of sickness (morbidity) and death (mortality) in the United States due to certain kinds of diseases called *reportable diseases*. They are called "reportable" diseases because a physician seeing a patient with such a disease is legally obligated to report this fact to the CDC so that accurate data can be collected and published. (The numbers of cases are reported; patients' names are not.) Reportable diseases listed in *MMWR* include most of the serious contagious diseases caused by microorganisms. Each issue of *MMWR* also contains articles written by alert clinicians who have observed patterns of disease that are "unusual" for a particular geographic area or season, which are too frequent, or which are occurring in an age group that does not usually get the disease. Tallies of reportable illnesses will often be the first indication of an unusual spread of a known disease, while reports of unusual illnesses can be the first clue of a new disease. Having such information collected and published by the CDC allows identification of trends that might not be visible to a single observer. The study of this type of statistical information about the occurrence and spread of diseases in whole populations is called **epidemiology.**

Pneumonia, a disease characterized by fluid in the lungs, can be caused by many different microorganisms, but is generally associated with bacteria and viruses. The five cases of *Pneumocystis* pneumonia reported in *MMWR* were quite unusual, since *Pneumocystis* is a parasite that is neither a bacteria nor a virus. All five cases were from a single geographic area and close together in time, and thus represented what epidemiologists call a *case cluster*.

Unusual cancer cases.

Later in the summer of 1981, a dermatologist in New York City, Dr. Alvin Friedman-Kien, noticed an unusual cancer, called *Kaposi's sarcoma*, among many of his young homosexual male patients, a finding that he reported in *MMWR* (July 1981). Kaposi's sarcoma, a cancer of the cells lining the walls of blood vessels, results in red or purple raised patches on the skin. Kaposi's sarcoma was rare in the United States and had previously been found only in elderly men of Italian or Eastern European Jewish descent. The Kaposi's sarcoma seen in the reported cluster was far more aggressive than that seen in elderly men, meaning that it spread much faster, and it was present in the internal organs, not just on the skin. Aggressive Kaposi's sarcoma was also known to occur in kidney transplant patients. Transplant recipients have to take medication to suppress the activity of their immune systems, which would otherwise reject the tissue transplanted from another person; this fact suggested the hypothesis that the Kaposi's sarcoma took on its aggressive form when the immune system was suppressed.

The immunodeficiency hypothesis.

Were the Kaposi's cases in any way related to the unusual pneumonia cases? Did immunodeficiency underlie both the unusual pneumonias and the unusual cancers? One of the *P. carinii* pneumonia patients was found to have a severely decreased number of helper T cells. As we have seen above, helper T cells are one of the central types of cytokine-secreting cells in the immune system; a person lacking helper T cells thus has a suppressed immune system (Fig. 13.1). Other pneumonia and Kaposi's patients were also found to have decreased helper T cells. This evidence fit the hypothesis that these two very different outcomes (a cancer and a pneumonia) formed a syndrome resulting from the same underlying mechanism, namely immunodeficiency. This syndrome was given the name AIDS.

However, because so many things can cause immunodeficiencies, it was not known whether this syndrome was a disease caused by a microorganism or whether some other factor was causative. Reported cases accumulated quickly: 87 in the first 6 months of 1981, 365 in the first 6 months of 1982, and 1215 in the first 6 months of 1983.

Evidence for and against lifestyle hypotheses.

Epidemiologists gathered information from patients, trying to establish any common links between the cases: Had they all been exposed to the same chemical agent? Did they all live in the same geographical area or in the same household? Were the patients known to one another? For a while, since the initial cases occurred in homosexual men, the search was for common "lifestyle factors," on the assumption that there was such a thing as a "homosexual lifestyle," one characterized by some drug or dietary factor that was shared by most or all homosexual men. Several researchers hypothesized that the immunodeficiency was due to an overload of the immune system by chronic exposure to nonself molecules via promiscuous sexual activity. Others doubted these hypotheses because the effects seemed to be specifically targeted on one type of cell, the CD4 or helper T cell, so they searched instead for infectious microorganisms that homed in on this type of cell and that might be transmitted by sexual contact.

The viral hypothesis.

Was the infectious microorganism a bacterium, a fungus, a protozoan, or a virus? Support for the hypothesis of a viral agent came when some cases of AIDS were reported among hemophiliacs. People with hemophilia lack the genes that code for certain blood proteins necessary for forming blood clots after an injury. Hemophilia can be life-threatening because a minor cut or scrape can cause the person to bleed to death. One preventive measure is to give hemophiliacs blood-clotting proteins from other people. This works well, but only temporarily; transferred clotting agents, like all proteins in the body, are eventually broken down by protein-degrading enzymes. In the case of hemophiliacs, new protein is not made, so the clotting factors must be supplied repeatedly in order to maintain the proper concentrations. These clotting factors are obtained from blood pooled from many donors. Before it is given to another person, the blood plasma (the clear, protein-rich fluid in which blood cells are suspended) is filtered to remove bacteria and fungi. The pore size of the filters, however, is usually large enough so that virus particles can pass through. AIDS could be an infectious disease, and it was reasoned that since hemophiliacs receiving filtered blood plasma were contracting AIDS, the infectious agent could be a virus.

Viruses that specifically infect human T cells. A few patients were found to be infected with previously known viruses (including Epstein-Barr virus and cytomegalovirus), but these viruses do not act specifically on T cells. Scientists reasoned that, since AIDS seemed to be a new disease, its cause would not turn out to be a previously well-known virus. One laboratory that was already studying viruses at the time was the National Cancer Institute's Laboratory for Tumor Cell Biology, headed by Robert Gallo. Gallo has said that it was the occurrence of AIDS in hemophiliacs that convinced him that a virus must be involved. Gallo's laboratory was studying *retroviruses*, a type of virus whose genetic information is in the form of RNA that is copied to make DNA. The first human retrovirus to be isolated was human T cell leukemia virus, HTLV-I, which was isolated in 1979.

As its name implies, this virus causes a form of leukemia, a blood cell cancer, but it also causes a mild immunodeficiency in some patients. It is known that the immune system can remove some transformed (cancerous) cells as they arise (see Chapter 9), so that suppressed immunity might result in the growth of cancers. However, it is also known that some cancers produce immunosuppressive factors. Therefore, it could not be said whether the immunodeficiency seen in some HTLV-I patients was caused by the virus or was the result of the cancer or some other factor(s). Nevertheless, it was known that HTLV-1 was transmitted from person to person by sexual contact and that it specifically attacked T cells, which fit the pattern seen for AIDS.

Evidence from reverse transcriptase. Meanwhile, in France, evidence of a retrovirus was found in 1983 in the tissues of a patient with the chronic lymphadenopathy (swollen lymph nodes) that often precedes AIDS. (The lymph nodes are the structures that enlarge when the body is fighting an infection; people often refer to them as "swollen glands.") The evidence, reported by Luc Montagnier and his co-workers at the Pasteur Institute, was the detection in these tissues of *reverse transcriptase*, an enzyme used by retroviruses to produce DNA from RNA (see below). Reverse transcriptase is common to all retroviruses, and methods were already available for detecting it. Although Montagnier's group had not yet found the virus, just one of its enzymes, they named it lymphadenopathy-associated virus, or LAV.

A new human T cell virus. Gallo's group was also isolating retroviruses from people with AIDS. Could these retroviruses from AIDS be the same virus as LAV, HTLV-I, or HTLV-II (another human retrovirus discovered by Gallo's group in 1983)? Although the presence of any retrovirus can be detected by finding reverse transcriptase (as Montagnier's group had done), identification of a specific retrovirus requires an antibody or nucleic acid sequence for each separate strain of virus. Antibodies or nucleic acid probes are very useful for identification purposes because their binding is highly specific, but to produce them you must first have large quantities of whatever molecules you want to identify: purified virus, in this case. Viruses cannot replicate outside of a host cell, but these host cells may be grown in the laboratory rather than in an animal. Viruses are thus obtained by growing them in laboratory culture in whatever kind of cell is their normal host. In working with HTLV-I and HTLV-II, Gallo's laboratory had developed a method for growing human T cells in the laboratory, and with it they had discovered the T cell-derived growth factor that is now called interleukin-2 (described earlier). HTLV-I and HTLV-II were grown in cultured cells, and antibodies were made to each virus (see Box 13.1). These antibodies were used to show that the new retrovirus from AIDS patients was neither HTLV-I or HTLV-II, and Gallo's group named it HTLV-III.

Identification and detection of HIV. To determine whether HTLV-III was the virus found in AIDS patients, an antibody would have to be made against it. As before, this required growing large

BOX 13.1   METHODS FOR IDENTIFICATION OF HIV OR OTHER SPECIFIC VIRUSES

[A] HIV is first grown in human T cells in a laboratory dish, to produce large amounts of HIV.

[B] HIV is used to immunize an animal, which will make antibodies (Ab) that bind only to HIV. The animal will not become infected with HIV because its cells do not have human CD4 molecules.

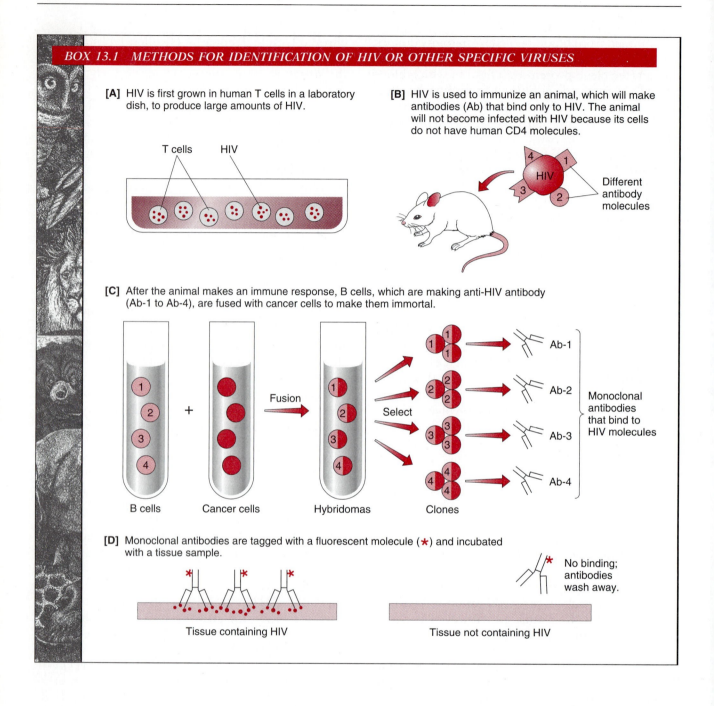

[C] After the animal makes an immune response, B cells, which are making anti-HIV antibody (Ab-1 to Ab-4), are fused with cancer cells to make them immortal.

[D] Monoclonal antibodies are tagged with a fluorescent molecule (✱) and incubated with a tissue sample.

quantities of the virus. However, when the new retrovirus was put into human T cells growing in the laboratory, it killed the cells. Michael Popovic in Gallo's lab found a type of T cell in which the new retrovirus could be grown without killing the host cells. This T cell line was derived from a human T cell leukemia (blood cell cancer, Chapter 9).

Once the method for growing quantities of HTLV-III had been developed, antibodies were made that were specific for HTLV-III, as had already been done for HTLV-I and HTLV-II (see Box 13.1). These antibodies were then used to test viruses isolated from three groups: a control group that consisted of healthy heterosexuals, a group of AIDS patients, and a group of patients with "AIDS-related complex." (*AIDS related complex*, or ARC, was a term used to describe cases of HIV infection having less severe symptoms than AIDS; at that time ARC was assumed to be an early stage of AIDS.) The virus isolated from AIDS patients and from some ARC patients was identified as being HTLV-III. Some viruses were isolated from the healthy control subjects, but these were not HTLV-III.

HTLV-III-specific antibodies were also quickly put to use to detect the new virus in the blood of AIDS patients and controls. The virus was found in 80 to 100 percent of AIDS patients, in varying percentages of people in the risk groups (see below), and only rarely in healthy individuals outside the risk groups. These results were strong evidence that this new retrovirus, HTLV-III, was the cause of AIDS. HTLV-III and LAV were later determined to be two strains of the same virus. The International Committee on the Taxonomy of Viruses studied the naming problem and decided in 1986 that both HTLV-III and LAV should be renamed as **human immunodeficiency virus,** or **HIV.**

Establishing Cause and Effect

Koch's postulates.   How do we know that HIV is the cause of AIDS? There is a set of rules that has traditionally been used to identify a microorganism as the cause of a particular disease. These rules were formulated in the late 1800s by Robert Koch, a German physician, and have come to be known as **Koch's postulates:**

> First, the microorganism suspected as the causative agent must be present in all (or nearly all) animals or people with the disease.
>
> Second, the microorganism must not be present in undiseased animals.
>
> Third, the microorganism must be isolated from a diseased animal and grown in pure culture (that is, a culture containing no other contaminating microorganisms).
>
> Fourth, the isolated microorganism must be injected into a healthy animal, the original disease must be reproduced in that animal, and the microorganism must be found growing in the infected animal and be reisolated from it in pure culture.

Koch used these rules to show that the bacterium *Bacillus anthracis* was the cause of anthrax, a fatal disease in sheep that was decimating European herds in the 1870s. Koch later used his postulates to identify a bacterium now known as *Mycobacterium tuberculosis* as the causative agent for human tuberculosis. Koch received a Nobel Prize for his demonstration of the causes of anthrax and tuberculosis.

Limitations to Koch's postulates.   While these rules are straightforward to state, they are difficult to fulfill. Every animal is host to many bacteria, so finding one species that is present only in diseased animals is not easy. Koch and his contemporaries developed bacterial culture media and techniques that enabled them to grow pure cultures of certain bacteria, but the growing of many other bacterial species (such as those killed by exposure to air) required technology that did not exist in Koch's time. In studying anthrax, a disease of sheep, the fourth postulate—requiring production of the disease in an experimentally infected animal—was straightforward. In Koch's later studies on tuberculosis, ethical considerations dictated that the fourth postulate could not be fulfilled by infecting a healthy human, so an animal was used instead as a model, or experimental patient.

Despite the difficulties associated with meeting the requirements of Koch's postulates, they have been very useful. About a dozen or so infectious diseases are controllable by vaccination, including rabies, poliomyelitis, whooping cough, tetanus, and diphtheria; the infectious microorganisms responsible for these diseases and others were identified on the basis of Koch's postulates. A microorganism that has been shown to cause a disease is called a *pathogen*.

There are other diseases for which these postulates have not been demonstrated. As we have seen with human retroviruses, difficulties in growing the infectious agent in culture became even more acute when viral, rather than bacterial, diseases were studied. Because many bacteria and viruses that infect one species do not infect other species, it is not always possible to find animal models for human diseases, as Koch did for tuberculosis. Further, even the best animal models can never reproduce all of the aspects of a human infection. Some diseases may be caused by the interactions of more than one species of bacteria or virus; in such cases, Koch's last postulate would not work because the organisms isolated in pure cultures would no longer have the same effect. In many diseases there are healthy carriers (people who are infected and can transmit the infection to someone else, but who do not become ill themselves), so the second postulate is not met. Other diseases, such as some cancers, may be multifactorial, meaning that no single factor is causative by itself. Nevertheless, Koch's postulates are the standards by which scientists establish cause and effect for most *infectious* diseases, meaning those diseases that can spread from one patient to another.

Does HIV fulfill Koch's postulates as the cause of AIDS?

Because Koch's postulates are the accepted standard of proof for asserting cause and effect in infectious disease, to say that HIV *causes* AIDS carries with it the implication to many in the scientific community that Koch's postulates have been fulfilled. This implication has been contested by some scientists, most notably by the virologist Peter Duesburg, who maintains that three out of the four postulates have not been met. The presence of the virus itself has been difficult or impossible to demonstrate in some people. The basis for considering someone to be HIV-infected is the presence in their blood of antibodies to the virus, not the virus itself. Duesburg maintains that Koch's first postulate has therefore not been fulfilled. This particular point has become less controversial since newer, more sensitive techniques (like the polymerase chain reaction described in Chapter 3) have been used to detect the virus in more people. The third postulate has also been difficult to satisfy; Luc Montagnier demonstrated that HIV could only be made to replicate in culture if there were other infectious agents present. The fourth postulate has not been fulfilled since only two nonhuman animals have been found in which HIV will grow (chimpanzees and macaque monkeys), and when these animals are infected with HIV, they do not develop AIDS.

Although many scientists agree that Koch's postulates have not been entirely fulfilled (and perhaps cannot be), the failure to fulfill them does not prove that the suspected agent does *not* cause the disease. Gallo has pointed out that Koch's postulates cannot even be strictly applied to the diseases that Koch himself studied. Many people, for example, can be healthy carriers of the tuberculosis bacterium, and so the second rule is not always met.

Criteria other than Koch's postulates.

Because of this limitation of Koch's postulates, several scientists have suggested other criteria for establishing causality, particularly of viral diseases.

The criteria used by Gallo for stating that HIV is the sole cause of AIDS are as follows:

1. HIV or antibody to HIV is found in the vast majority of persons with AIDS.

2. HIV is found in a high percentage of people with ARC.

3. HIV is a new virus and AIDS is a new disease.

4. Wherever HIV is found, AIDS develops; where there is no HIV, there is no AIDS.

5. People who received transfusions of blood contaminated with HIV developed AIDS.

6. HIV infects CD4 T lymphocytes, a cell type depleted in AIDS.

7. On autopsy, HIV is found in the brains of people who have died of AIDS, and dementia (loss of brain cell function) is a symptom of AIDS.

Necessary causes and sufficient causes.

It is not always easy to identify cause and effect. There are scientists who do not agree that Gallo's criteria are sufficient to establish cause. There are others who say that Koch's rules are the correct criteria, but that, in the case of HIV and AIDS, the criteria have not been met, particularly rule 4. At present, the vast majority of scientists agree that HIV is a *necessary cause* of AIDS; that is, someone who is not infected with HIV will not get AIDS. However, not all scientists agree that HIV is the sole or *sufficient* cause of AIDS (that is, no other factors are required) because there is a great difference in the course of the infection among various persons with AIDS.

There are several lessons to be learned here. Hypotheses about cause and effect of diseases are debated both in written articles and at scientific meetings. Many additional hypotheses are suggested and tested, and the number of scientists involved is often (as in this case) very large. Eventually a consensus develops as to the explanation which best fits the data. Consensus is never 100 percent agreement, but public policy and public health decisions must be made nevertheless. The fact that there are scientists who do not agree with the consensus does not mean that the consensus is wrong or that it is right. Scientists who disagree with a particular consensus generally do not dispute the data (professional honesty in data presentation is both expected and assumed); rather, they disagree in most cases over data interpretation. The gradual accumulation of additional data may settle a given controversy (e.g., in the case of HIV and AIDS), but at any time new data may arise that require a change in the consensus explanation. Scientific debate and research must always be open to new possibilities because the findings of science are always tentative or provisional.

Controversy has arisen even among those who agree with the consensus position. Scientific research is a human endeavor and therefore has all of the strengths and failings of other forms of human endeavor. Along with the quest for new knowledge, jealousy, politics, and professional competition can all come into play. The search for the cause of AIDS is no exception. This search has been extremely well-documented by several authors. The most notorious professional controversy in this field developed between French and American researchers. They agreed on the conclusion that HIV was the important virus, but they each claimed that they had been first to discover it (see Box 1.2, pp. 12–13). The same events may receive very different accounts in the writings of authors with different points of view or in different situations, or even by the same author at different times.

Saying that a particular virus causes a disease, especially a disease with such a diffuse group of symptoms as AIDS, really tells us little or nothing by itself. All of the "how" questions remain. How does the virus infect? How do cellular effects progress to clinical symptoms? How can the disease be prevented or stopped? How is the infection transmitted? How contagious is it? How likely is it that HIV infection will become AIDS? How do people cope with such a disease? It is interesting that the articles written by scientists, either for other scientists or for the public, are almost entirely devoted to answering the first three questions, while the literature written by nonscientists is much

more concerned with the last four. We will examine each of these questions later in this chapter.

Viruses and the HIV Virus

Viruses: living or nonliving?   **Viruses** are bits of either DNA or RNA that cannot replicate by themselves but can replicate inside a cell (called the *host cell*) using the biochemical machinery of the host. Viruses thus blur the distinction between *living* and *nonliving*. Biologists define a living organism as one that can replicate itself, which viruses cannot; yet once inside a host, viruses can cause the host to replicate the virus, something that is not a characteristic of any known nonliving thing.

Human cells, animal cells, plant cells, and bacteria can all serve as hosts to viruses. For each virus, however, there are usually only certain species that can serve as its host, and within that host only certain types of cells can be host cells. This restricted host range is called *host specificity* and is one basis for distinguishing different types of viruses.

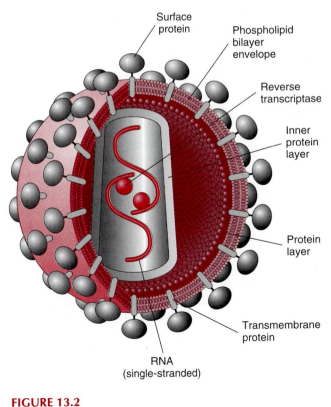

Surface protein

Phospholipid bilayer envelope

Reverse transcriptase

Inner protein layer

Protein layer

Transmembrane protein

RNA (single-stranded)

FIGURE 13.2
The structure of HIV.

A particular virus has either DNA or RNA, and the nucleic acid can be either single-stranded or double-stranded, distinctions that are used in viral classification. Viral genomes vary in size; some have only enough nucleic acid to code for 3 to 10 proteins, while others contain code for 100 to 200 or more proteins. HIV, the virus that causes AIDS, is at the small end of this size range. Virus particles consist of the nucleic acid, an outer protein shell, and, in some viruses, including HIV, a lipid bilayer membrane called the *envelope* that also contains some viral proteins (Fig. 13.2).

The viral life cycle.

The growth cycle of a virus includes many stages (Fig. 13.3). To enter a cell, the virus must attach to some molecule on the host cell surface. The first step in any infection is adherence of the microorganism to the host; this explains the restricted host range or host specificity, because each type of virus is able to bind only to specific host cell molecules, which are usually membrane proteins. The species of virus or bacteria that can adhere to dog or cat cells, for example, usually cannot adhere to human cells, which explains why we usually cannot catch diseases from our pets.

After a virus attaches to a host cell, the viral nucleic acid, sometimes with some viral proteins, crosses the cell's plasma membrane into the cytoplasm, usually with the help of energy derived from the host cell. Viruses whose nucleic acid is DNA may insert their DNA into the host DNA, or, for other DNA viruses, the viral genome may remain separate. In either case, many copies of the virus are made using the host's molecular machinery for DNA replication. Some types of viruses whose nucleic acid is RNA may replicate their RNA genome directly, while others (the retroviruses) convert their RNA to DNA. Viral DNA can then insert into the host genome, and many copies are made using the host's machinery, in this case the machinery for transcription.

The final stage of the viral life cycle consists of the release of infective viruses from the cell by the rupturing (also called *lysis*) of the cell or by virus budding out through the host cell membrane. By escaping from the host cells, the virus can infect other cells and repeat its life cycle. Some lytic or budding viruses can remain infectious during a cell-free period, but many types of virus rapidly lose infectivity when they are outside of a host cell.

The HIV virus.

The genomic RNA of HIV is surrounded by a protein shell, which is surrounded by a lipid membrane or envelope. Some proteins float in this envelope, and it is these proteins that adhere to human helper T lymphocytes. The T cell protein to which the virus adheres is the CD4 protein found only on a few types of human cells: helper T cells and some macrophages, a type of white blood cell that can engulf microorganisms. Helper T cells, macrophages, and cells related to macrophages are thus the only cells that become directly infected by the HIV virus. To a large extent, HIV follows the typical retroviral life cycle shown in Fig. 13.3. HIV is an enveloped virus whose genome consists of two copies of a single strand of RNA (Fig. 13.2). Once attached to the host cell, HIV enters by fusion of its envelope with the host's plasma membrane. Viral lipid and envelope proteins thus end up in the host cell plasma membrane, and the viral particle is released into the host cell's interior (Fig. 13.3).

As we have already mentioned, HIV is a retrovirus, a type of virus whose genetic information is in the form of RNA. In order to make multiple copies of the virus, the RNA must first be used to make DNA; since this is the reverse of the usual DNA-to-RNA process of *transcription* (Chapter 3), the RNA-to-DNA process is called **reverse transcription.** After entry into the cell, the first step in HIV replication is the activation of the viral enzyme *reverse transcriptase* that uses the viral RNA as a template (or pattern) from which to synthesize complementary DNA. The first DNA strand is, in turn, the template for synthesis of the second strand of DNA. The viral RNA is broken down, or degraded, and the now double-stranded DNA is incorporated into the host's DNA.

The viral DNA can now be transcribed, using host enzymes, into many copies of messenger RNA (mRNA) and viral genomes, each comparable to the original viral RNA. The viral genomic RNA can be enclosed in a coat of newly synthesized viral proteins, ready to infect another cell. Some viral mRNA is translated into viral envelope proteins that are inserted along with host cell membrane proteins into some of the host cell's internal membranes (the *rough endoplasmic reticulum,* described in Chapter 3). From here, viral membrane proteins are transported via the same pathway by which host membrane proteins are

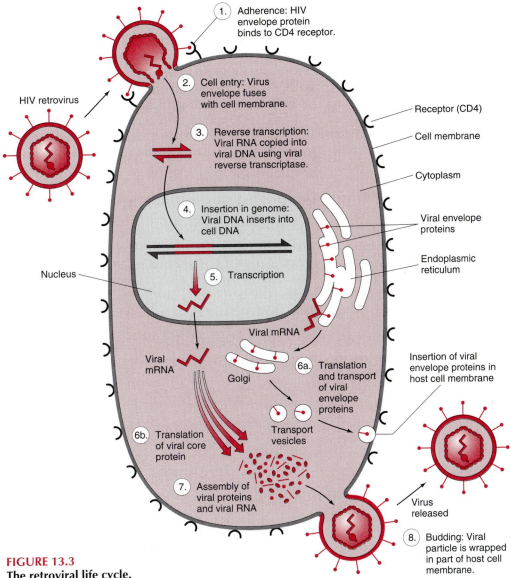

FIGURE 13.3
The retroviral life cycle.

transported: first to the *Golgi apparatus,* then via transport vesicles to the plasma membrane, where multiple copies of viral envelope proteins build up on the surface of the host cell (Fig. 13.3). The viral RNA genome with its protein coat joins a portion of the plasma membrane containing viral envelope proteins; when the virus buds out, it carries along a piece of the host cell membrane, which becomes the viral envelope.

The HIV can escape from the host cell by budding out into new cell-free enveloped virus particles or by causing lysis of the host CD4 cell,

further contributing to the immunodeficiency. Viremia, the presence of cell-free viruses in the blood, occurs early in the infection (before antibody is detectable), then decreases and is not high again until the very last stages of the disease. Free viruses can infect more CD4 cells.

The HIV envelope and genome contain all of the molecular determinants that make the virus *virulent* (able to cause disease), *infective* (able to enter a cell), *cytopathic* (able to kill or inactivate the cell), and *cell-specific* (entering only certain types of cells). The HIV genome also contains some *regulatory genes*, meaning genes that turn other genes "off" or "on." Regulatory genes are redundant in some viruses; that is, the genes could be lost and the virus would still be virulent. In HIV, all of the regulatory genes seem to be essential for the viral life cycle; thus they may be targets for drugs or vaccines because blocking any essential step of the viral life cycle would inhibit the whole cycle.

There are currently two types of HIV known: HIV-1 and HIV-2. Both have the same routes of transmission (see section D), but the rate of transmission of HIV-1 is five to ten times higher than the rate for HIV-2. People infected with HIV-1 are three to eight times more likely to lose immune function, have a decrease in CD4 T cells, and progress to AIDS than people infected with HIV-2. As we shall see, there is some evidence that infection with HIV-2 affords some protection against infection with the more devastating strain HIV-1.

Evolution of Virulence

Virulence is the ability of a pathogen to overcome host defenses, thereby causing serious illness or death. From an evolutionary standpoint, virulence poses a severe problem for the pathogen: If it kills its host, it deprives itself of a suitable habitat and food supply. Clearly, a pathogen that causes minimal harm to its host is assured a longer time span for itself and its offspring to continue living in the same place than a pathogen that kills the host.

Although the AIDS epidemic presumably began in the late 1970s, HIV is related to several other viruses that infect higher primates such as monkeys and apes without causing AIDS. These viruses are nonvirulent—they spread from one host to another without causing serious illness or death. Thus they have been around long enough—thousands of years at the very least—to have evolved a type of symbiotic relationship with their hosts (see Chapters 8 and 15).

Evolutionary biologists who study bacteria and viruses believe that the evolution of virulence is related to the pathogen's *fitness*, meaning its capacity to perpetuate itself into future generations (Chapter 4). The fitness of virulence depends in large measure on the likelihood of transmission of the pathogen. When a new strain originates, it must compete with the older, nonvirulent strains. A virulent pathogen can proliferate rapidly in a host, but

if it spreads from host to host at a slow rate, it may kill off its hosts before being able to colonize new ones; such a virulent strain will actually be less fit and will soon die out. Concomitantly, a virulent strain that also spreads rapidly from host to host will soon outcompete its nonvirulent relatives within the host population. If the process is rapid enough, an epidemic occurs. Under this hypothesis, we suspect that by 1981 increases in intravenous drug abuse and increases in the numbers of people having sex with multiple partners had given the HIV pathogen new opportunities to spread—rapidly enough to cause an epidemic.

In the long run, natural selection among hosts favors the evolution of host defenses against the pathogen, including both physiological or chemical defenses. These changes in the host reduce microbial virulence directly, and a change to host behaviors less conducive to the pathogen's spread will also slow transmission. Under these conditions of slowed transmission the less virulent strains once again become relatively more fit than the virulent strains, and the cycle repeats.

THOUGHT QUESTIONS

1. How might scientists decide whether or not two strains of virus are actually the same?

2. What are some of the reasons that evidence of the types called for by Koch's postulates may be impossible to obtain for some infectious diseases?

3. What do you feel the response of scientists should properly be to dissenters such as Peter Duesberg? Do dissenters serve a useful role in the process of science? If so, what role?

4. Why might the writings of scientists and nonscientists address different concerns? To what extent could these several concerns overlap or interact with one another?

5. If a viral RNA from HIV contains the base sequence AAUGCA, what would be the base sequence on the first strand of DNA produced by reverse transcription? What would be the sequence of the second DNA strand transcribed from the first one? (You may need to review material from Chapter 3 in order to answer this question.)

6. What are the differences between the terms *virulent* and *infective*? How does the term *cyto-*

pathic differ from *cell-specific?* Could a virus be virulent without being infective? Virulent without being cytopathic? Virulent without being cell-specific?

C. HIV Infection Progresses in Certain Patterns, Often Leading to AIDS

Events in Infected CD4 Cells

We have already seen that HIV virus binds to cells that have a molecule called CD4 on their surfaces. It then enters these cells, replicates, and goes on to infect more cells. What does it do to the cells it infects?

When HIV infects cells carrying CD4, it can kill or inactivate them. There are several possible mechanisms by which this depletion of T cells and other CD4 cells may occur; these mechanisms, described below, include the fusion of cells into larger masses, the direct killing of cells, and indirect processes that also result in cell death. We don't really know which of these several mechanisms results in the most cell loss. The result is the same: Losing CD4 cells in any of these ways results in immune deficiency (Fig. 13.1). Knowing the mechanism, however, will be critical for developing strategies for treatment of the disease.

Cell fusion.   The envelope of the HIV virus carries a protein that binds to the CD4 protein on the surface of human T cells (Fig. 13.3). During the infection process, some of the viral protein begins to appear on the plasma membrane of the infected T cells. This same viral protein, now carried by the infected T cell, can bind to CD4 on other, uninfected T cells. When this binding occurs, instead of triggering fusion between the viral envelope and the host plasma membrane, it triggers fusion of the infected and uninfected host cells, spreading the infection from one cell to another. At least in the laboratory, this can happen over and over until many dozens of T cells have been fused into a multinucleated "giant cell" (also called a *syncytium*), which can no longer carry out the immunological activities of the normal CD4 (or

helper) T lymphocytes. This inactivation of the CD4 T lymphocytes may contribute to the immunodeficiency; people whose viral isolates form syncytia when grown in the laboratory have a much more rapid progression of AIDS.

Cell lysis.   In a test tube, HIV can directly kill the cells it has entered by causing lysis. The importance of this direct killing process in human hosts has been questioned because there is so little free virus in the blood during the time when the T cells are disappearing. It has now been found that much more virus is hidden in the lymph nodes where it replicates at all stages of the infection and where it may cause cell lysis.

Cell suicide.   HIV may also trigger the T cell to kill itself when the T cell responds to another infection; under these conditions, the T cells commit a type of cellular suicide called *apoptosis*, in which their DNA breaks up into small fragments. When a healthy immune system is activated by a pathogen, the CD4 T cells begin to synthesize DNA and to divide; in fact, such activation is used as a standard procedure to get T cells to grow in the laboratory. Jean-Claude Ameisen of the Pasteur Institute has found that, when laboratory-grown CD4 T cells isolated from HIV-infected people are activated by other pathogens, instead of dividing, the CD4 cells die by apoptosis. It is not yet known whether this also occurs in the human body, although there has been one case in which an HIV-positive individual received a smallpox vaccination (which triggers an immune response) and very rapidly succumbed to AIDS. This is one reason why it may be difficult to develop a vaccine against AIDS, because the very T cells whose division any vaccine would stimulate would also die by apoptosis.

Indirect inactivation.   Evidence is accumulating that HIV can also lead to CD4 depletion by indirect mechanisms. Jay Levy at the University of California School of Medicine in San Francisco found that there were HIV isolates from AIDS patients that did not kill CD4 cells in the test tube (*noncytopathic* strains), despite the fact that these patients had depleted numbers of CD4 T cells. More recently, he found that noncytopathic HIV strains still led to the depletion of human T cells from mice into which they had been transplanted. One hypothesis is that depletion of CD4 cells could result indirectly if HIV caused production of the wrong cytokines or inhibited the production of cytokines needed for T cell growth.

Progression from HIV Infection to AIDS

Three stages of HIV progression.   A *productive infection* is one that involves more and more cells until the entire host is affected. How does infection at the cellular level progress to become a disease of the organism? The answer to this question is different for each infectious disease. We have seen how HIV enters cells and replicates within them. The progression from HIV infection to AIDS follows three stages: the initial infection, an asymptomatic phase, and a third phase called disease progression (Fig. 13.4).

The initial, or acute, infection stimulates both an antibody immune response, in which antibodies to HIV are produced, and a cell-mediated immune response, in which there is an increase of CD8 cytotoxic T cells that can kill cells containing HIV virus. Because the cells that contain virus are the helper T cells and the macrophages, elimination of infected cells also produces immune deficiency. The initial infection may be accompanied by flulike symptoms—fever, swollen lymph nodes, and fatigue—which, because they are similar to the symptoms for many other diseases, are often not diagnosed as an acute HIV infection.

After the initial phase, the person enters the asymptomatic phase. The levels of free virus in the blood decrease, although it is not known what brings about this decrease. The asymptomatic phase can last for a few months to several years. It is uncertain what factors influence the duration of this phase or what triggers the onset of the third phase. The antibodies that are synthesized in the acute phase can initially neutralize the virus, that is, prevent it from infecting more cells. Neutralization is accomplished by the antibody binding to the virus, thus blocking portions of the virus that are critical for the infective process (see Fig. 12.5). HIV are probably replicating and CD4 cells are probably being eliminated at slow rates in this phase, but the immune system's backup capacity is still sufficient to keep the infection under control so the person does not feel ill.

In the third phase, the levels of virus in the blood increase once more, while the numbers of CD4 T cells in the blood decrease. The virus mutates to forms that no longer match the antibodies produced in the acute phase. The antibodies thus cannot bind to the virus and therefore no longer neutralize the virus. This rapid mutation has evolved in the viruses as a way of evading the immune systems of their hosts.

The CD8 cells that develop in the initial phase kill or inactivate cells that contain HIV, the CD4 T cells, the very cells that are necessary to sustain a protective immune response. As we have seen, the virus itself may kill CD4 cells; the host's own im-

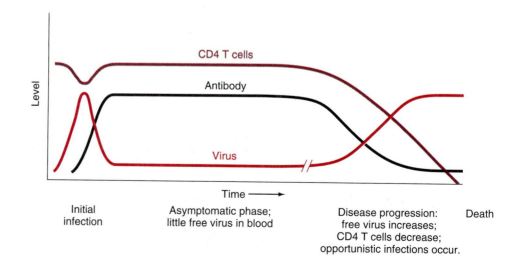

FIGURE 13.4
The course of HIV infection.

mune system in the form of CD8 T cells is now also turned against the CD4 cells. Death generally follows when the level of CD4 cells falls very low in the third phase.

Other cells bearing CD4 may also be infected and possibly inactivated by HIV. As mentioned previously, macrophages are white blood cells that engulf pathogens and damaged cells or molecules, thereby removing them. Macrophages and the cytokines that they produce also help induce T cells to initiate CD4 and CD8 T cell responses. Lung tissue contains tissue macrophages, and the brain contains certain macrophagelike cells called *microglia;* both these cell types can bear the CD4 molecule and can thus be infected and killed by HIV. Elimination of macrophage cells contributes to the risk for opportunistic infections, which may, in turn, lead to the patient's death.

Opportunistic infections.   Opportunistic infections are infections resulting from microorganisms that are always present in the person or the environment but that are normally kept in check by a healthy immune system, so that they seldom cause illness. Immunodeficiency, however, impairs the body's ability to fend off these organisms, allowing them to cause illness that may in some cases be serious. The occurrence of opportunistic infections is one of the primary symptoms of AIDS. The probability of transmission of one of these opportunistic infections from a person with AIDS to someone else with a healthy immune system is very low because the healthy immune system of the exposed person will stop the opportunist.

The particular opportunistic infections that develop are different in different areas of the world. In the United States, typical opportunistic infections that accompany AIDS are pneumocystis pneumonia, caused by the organism *Pneumocystis carinii,* and a fungal infection caused by an organism called *Toxoplasmodium;* both of these infections are very rare in people who have normal immune systems. Another fungus called *Candida* (a yeast) causes infections of the mouth, esophagus, or vagina; these infections may occur in the absence of AIDS, and are nearly always mild in form, but *Candida* infections in people with AIDS are much more severe. In Africa, the more common opportunistic infection accompanying AIDS

is tuberculosis, a bacterial infection caused by *Mycobacterium tuberculosis.* (Robert Koch's work with this disease and its pathogen was described earlier in this chapter.) Of 6.5 million people currently HIV-infected in Africa, 3.5 million also have tuberculosis. Tuberculosis is a disease in which there are active periods and periods of remission; HIV infection increases the frequency of reactivation of tuberculosis and also the mortality rate.

Other symptoms.   In addition to the opportunistic infections and the reduction in the numbers of CD4 T lymphocytes, AIDS patients may also suffer from low levels of interleukin-2, high fevers, night sweats, general weakness, mental deterioration (dementia), and severe weight loss, though these last two symptoms may not develop for a long time. Dementia may be related to elimination of macrophagelike microglia cells from the brain. Another type of CD4-bearing cell are the chromaffin cells of the gut, which play a role in absorption of nutrients; elimination of these cells may be related to the weight loss.

AIDS-related complex.   Many people infected with HIV develop an *AIDS-related complex (ARC),* a set of milder symptoms than AIDS. Originally it was thought that ARC was a pre-AIDS condition and that everyone who had ARC would end up with AIDS. The CDC did not initially require reporting of ARC, assuming that these cases would later be reported as AIDS cases, resulting in underestimates of HIV infection rates. As time passed since the initial description of the disease, researchers noticed that several people died while still showing only the symptoms of ARC, not AIDS. Other people with ARC have lived a long time without dying and without having AIDS. Distinctions are no longer made between ARC and other categories of HIV infection; they are all simply called *HIV infection.*

Variations in disease progression.   Does everyone infected with HIV get AIDS? Does everyone with AIDS die from the disease? Despite the tremendous amount of information on this topic, we do not have definitive answers to these questions. There is certainly a lot of individual variation in the speed with which HIV infection progresses to disease. There are some people who have been

infected with HIV for years and have yet to develop AIDS, including certain prostitutes in several parts of Africa. Several recent studies have shown that these so-called *nonprogressive* HIV infections are often characterized by a very low amount of the virus, but it is not clear whether this reflects a low infective dose initially or an immune system that has successfully kept the viral population low. It may be that the virus is *latent,* or dormant, in some people and that eventually every infected person will come out of the asymptomatic phase and become ill. Many other types of viruses, following an initial mild illness, go into latency, and may cause further disease at a later time. There are many HIV cases where the time between infection and AIDS is known, and a latency period of 2 to 10 years has been documented in some people. A long-term study of HIV-infected homosexual men in San Francisco showed that after 12 years, 65 percent had progressed to AIDS, but 35 percent had not. It may yet turn out that the progression from HIV infection to AIDS is not inevitable. Certainly, avoidance of other immunosuppressive factors, including drugs, alcohol, and stress, can help maintain health (see Chapters 11 and 12).

One group of researchers followed several dozen professional sex workers (prostitutes) in west Africa who were infected with the less virulent strain HIV-2. A significant finding of this study was that HIV-2 infection seems to offer these women a certain degree of protection against the more virulent strain HIV-1 (Travers et al., 1995; Cohen, 1995). Compared to other prostitutes, those in the study group had higher rates of infection for other sexually transmitted diseases (STDs), falsifying the hypothesis that the lower HIV-1 rates were simply the result of safer sex practices.

At least one case has been reported of an infant infected with HIV at birth who subsequently became HIV-negative. The HIV-1 infection in this infant was confirmed by two different PCR tests (see below) taken 32 days apart, and genetic markers confirmed that the DNA in these two tests were from the same strain, making it unlikely that a laboratory mistake had been made. The infant proved to be HIV-negative at the age of 12 months and has remained so ever since; at the age of 5, the child continues to enjoy good health and normal development.

Tests for HIV Infection

How can people tell if they are infected with HIV? When a person becomes infected with HIV, B lymphocytes respond to the virus. This response takes a couple of weeks or months and results in the production of antibodies to HIV in the blood. Development of specific antibodies is called **seroconversion,** and once the antibodies have developed, the person is said to be *HIV-positive* (HIV+). As discussed earlier, the virus itself is very difficult to detect, so the basis of HIV tests is the presence of antibodies to HIV.

There are two common tests for HIV, the *ELISA* test and the *Western blot* test, as described in Box 13.2. For every diagnostic test there is a frequency of **false positives,** test results that are positive when the person does not really have the condition that the test is designed to detect. In addition to false positives, every type of test also has some frequency of **false negatives,** test results that are negative when the person really has the condition being tested for. The frequency of false negatives determines the **sensitivity** of the test, while the frequency of false positives determines the **specificity** of the test.

Every diagnostic test must be thoroughly tried on samples from thousands of individuals whose actual status is known before the test can be sold. These trials must be conducted blind; that is, at the time they are running the trials, the person doing the testing cannot know whether the test samples came from persons infected with HIV or not. Afterward, the code is cracked, and the true infection status (known beforehand but concealed from the researchers) is compared to the status revealed by the test. In this way the frequency of false results can be quantified (and minimized to the extent possible). The reliability of a given test depends both on its sensitivity and its specificity. The more sensitive a test, the less often it will miss a truly positive result; the more specific a test, the fewer cases that are truly negative will be reported as positive (Box 13.3, p. 386).

In the case of HIV testing, the ELISA test is done first. The sensitivity of the ELISA test is high—less than 1 percent false negatives—but it is not very specific—as high as 2 to 3 percent false positives. For this reason, the Western blot test is usually conducted following a positive ELISA result, because the Western blot test rarely gives false positive results. Why not use the Western blot as the initial test? The reason is that Western blots are more costly and technically more difficult, particularly for large numbers of tests. Even the ELISA test is too costly for widespread use in many countries.

Both the ELISA and Western blot tests are based on the detection of an antibody specific for HIV in a patient's blood plasma. A second generation of tests based on the polymerase chain reaction (PCR) may replace both these antibody-based tests. The main advantage of PCR tests is that they detect viral RNA rather than antibody, so they become positive soon after infection, not weeks or months later, as for antibody tests.

A Vaccine against AIDS?

Vaccines.   A highly successful strategy for prevention of many infectious diseases has been vaccination. A vaccination is really a controlled exposure of a person to molecules similar or identical to those carried by the infectious organism. The material to which the person is exposed is called the **vaccine,** named after the fact that the first successful vaccine (against smallpox) used the *vaccinia* virus from the sores of infected cows (Latin *vacca*). Exposure to a vaccine stimulates the immune system to make an immune response to the molecules, and vaccination is therefore also called *immunization*. The infectious agent itself is not used so that the person is not given the disease. The vaccine may be another microorganism, closely related to the infectious microorganism but nonvirulent to humans (as when *vaccinia* from cows is used to protect against smallpox), or it may be the infectious microorganism itself treated in such a way to *attenuate* it (make it nonvirulent). Older vaccines used whole microorganisms, but today some molecule(s) vital to the organism's life cycle or to its ability to cause disease are more frequently used instead.

Several laboratories are attempting to develop vaccines that would prevent HIV infection (preexposure immunization) or would prevent the progression of HIV infection to AIDS (postexposure immunization). Postexposure immunization is called *immunostimulation* by some authors, who prefer to reserve the term *vaccination* for preventive strategies.

Biological barriers to successful vaccines against AIDS.   The many roadblocks to the development of a successful AIDS vaccine have been reviewed several times, most recently by Letvin (1993). These roadblocks include variation of the virus, a lack of knowledge about which immune responses are protective against HIV, and a lack of animal models in which to test trial vaccines. There is a gradual change in the viral RNA and DNA sequences over time, including changes that occur within a single patient. Is it possible to develop one vaccine that could stimulate a protective immune response in every person vaccinated and that would continue to protect infected people as the viral nucleic acid sequences changed? The answer right now is maybe: Maybe there are some sequences that do not change very much or for which changes have no effect on recognition by the immune system. The latter is possible because the immune system actually recognizes protein *shapes*, not sequences of nucleic acids or amino acids. A change in nucleic acid sequence will change the amino acids put into the protein during its synthesis, but some changes may not alter the shape of the completed and folded protein. If the shape has not changed, the immune system may still be able to recognize the altered protein.

Not all immune responses against HIV are protective, as can be seen by the fact that HIV-infected people form an immune response to HIV but still eventually get AIDS. Proteins involved in parts of the viral life cycle are being targeted for vaccine development, but it is not known if these will stimulate protective responses. Stimulating an immune response by vaccination may actually trigger progression to AIDS in someone already infected with HIV, as happened with the person mentioned earlier who rapidly progressed to AIDS after a smallpox vaccination.

Assuming that protective responses can be stimulated, it is likely that the materials used for preexposure immunizations will need to be different from the materials used for postexposure immunizations. Preexposure vaccines for some diseases prevent adhesion to or entry of the

The ELISA and the Western blot test use immunological techniques to detect HIV-specific antibodies in a person's blood and thus are called immunodiagnostic techniques. In the ELISA (*Enzyme-Linked ImmunoSorbent Assay*), laboratory-grown HIV are immobilized onto a surface, generally small wells made of a plastic that is especially designed to tightly bind protein molecules (1). The rest of the plastic is coated with other proteins to block any nonspecific binding of proteins used later in the assay. The immobilized virus particles then act as binding sites for specific antibodies when they are exposed to blood plasma (or serum) from the person being tested. If that person's plasma contains antibodies whose specific binding sites match molecules on the virus, the antibodies bind to the immobilized virus (2). There will be many other antibody molecules in the person's blood that do not match any HIV molecule; these will not bind. The wells are then rinsed, removing any unattached antibody molecules. The viral molecules and any specific, anti-HIV antibodies bound to them are so tightly attached that they do not wash away. Anti-HIV antibodies are then detected by a second antibody to which an enzyme has been attached (3). The binding sites on the enzyme-linked antibody match amino acid sequences on the constant portion of human antibody molecules, so they bind, not to the plastic or to the HIV, but to any human antibodies present. Again, any unbound antibody is washed away, then enzyme substrate is added (4). Enzymes are proteins that catalyze biochemical reactions; the enzymes used in these tests are ones that will split a substrate molecule causing a change in color in the media. A well in which the medium has changed color (5) is thus a well in which the blood plasma used in the test contained antibody specific for HIV, an initial bit of evidence that the person is HIV-positive.

If an ELISA suggests the presence of anti-HIV antibodies, the Western blot test is done. The viral proteins are separated by a technique called *electrophoresis*. Proteins are solubilized with detergents and put onto a gel. The gel is mostly water, but also contains other molecules that give it stiffness (see Chapter 14) and slightly retard the movement of molecules through the gel. An electrical field is applied to the gel; the solubilized proteins have a uniform negative charge and so move in the electric field toward the positive electrode (1). The separated proteins are then transferred out of the gel onto special paper that has a high affinity for proteins. The transfer again uses the fact that proteins move in an applied electric field, but this time the field is applied perpendicular to the plane of the gel so the proteins leave the gel (2). The paper is then exposed to blood plasma from the person being tested. Specific antibodies bind, but in this case they bind not to the whole virus but to some individual virus protein. The plasma may contain specific antibodies that bind to some viral proteins but may lack antibodies to other viral proteins (3). Unbound antibody is rinsed away and bound antibody is detected, as in the ELISA, with enzyme-linked antibody that binds to all human antibodies (4). If the enzyme-linked antibody finds human antibody to bind to, it will catalyze a color change when the substrate is added. Dark bands will appear where the test person's blood plasma contained antibody specific for that viral protein; where there was no specific antibody, no band will appear (5). If antibody to specific proteins is present, the person has tested seropositive.

Neither of the immunodiagnostic tests tests for virus itself. Newer tests are being developed that do test for the virus; these tests are based on a technology called polymerase chain reaction (Chapter 3).

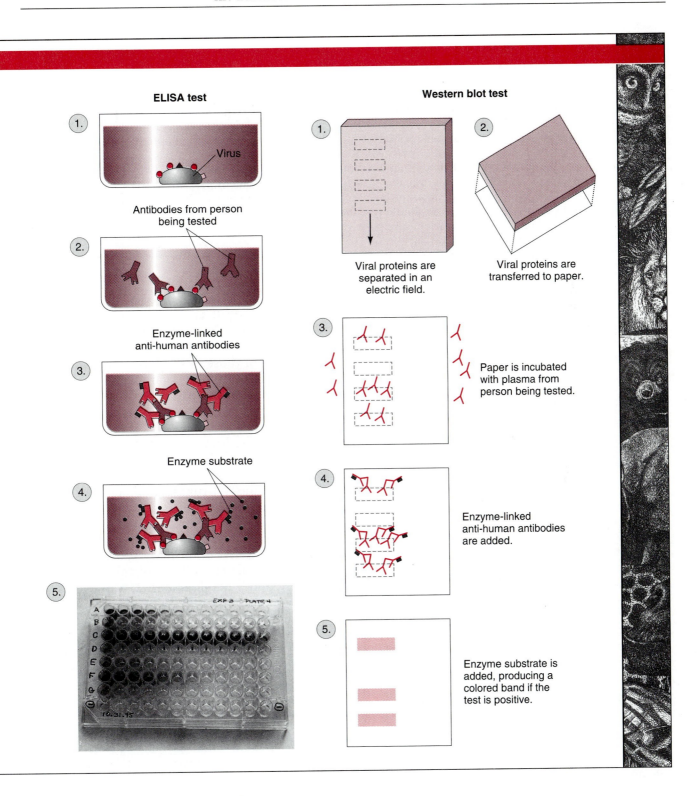

ELISA test

1.

Virus

Antibodies from person being tested

2.

Enzyme-linked
anti-human antibodies

3.

Enzyme substrate

4.

5.

Western blot test

1.

Viral proteins are
separated in an
electric field.

2.

Viral proteins are
transferred to paper.

3.

Paper is incubated
with plasma from
person being tested.

4.

Enzyme-linked
anti-human antibodies
are added.

5.

Enzyme substrate is
added, producing a
colored band if the
test is positive.

pathogen into host cells, while a postexposure vaccine will probably have to target some different molecule(s) involved in the progression of the disease. There are two variants or phenotypes of HIV, those that cause syncytium formation and those that do not, and it is the nonsyncytium variants that are more often found in HIV-infected but healthy people. The nonsyncytium variants may be more infectious, that is, more likely to be the form transmitted to another person. This form of the virus may be the best target for preexposure vaccines. The syncytium-inducing variants are more frequently isolated from individuals who show a rapid progression from HIV infection to AIDS.

The lack of animal models is a significant problem. The effects on each step in an immune response can be studied in vitro, but protection from disease can only be evaluated in an animal that gets the disease. Ethical considerations would bar the early testing of vaccines on human volunteers in any disease known to have a high percentage of fatalities or for which there is no known cure (See Chapter 2).

Drug Therapy for People with AIDS

Biological barriers to the development of antiviral drugs.   Very few drugs are helpful against viral diseases. Antibiotics, which are highly effective against bacterial diseases, do not work against viruses. Most antibiotics are designed to work either on bacterial cell walls or on bacterial ribosomes (small cell structures used in protein synthesis, as described in Chapter 3). This approach works well against the bacteria without causing harm to the host because host cells lack cell walls and bacterial ribosomes are structurally different from those of the host. Since viruses do not have much of their own machinery but use the cell machinery of the host, blocking the virus may also cause harm to the host. As detailed knowledge becomes available about the few enzymes HIV does have, new drugs may be developed to target these enzymes specifically.

Antiviral drugs.   Drug development has focused primarily on preventing or delaying the progression from the asymptomatic stage into AIDS. One drug that has some effect is *zidovudine*, or *ZDV* (trade name Retrovir, formerly known as azidothimidine, or AZT). Like several other drugs currently under development, ZDV prevents viral replication by inhibiting reverse transcriptase (Fig. 13.5), thus preventing the virus from making the DNA that it needs to complete its infective cycle (Fig. 13.3). This drug has shown promise in prolonging the asymptomatic phase (and delaying the late phase) in people whose CD4 T cell counts remain above 400 cells per milliliter of blood. It also reduces the rate of HIV transmission from pregnant women to their babies during birth. ZDV works best against the nonsyncytium HIV phenotype. Another drug, *didanosine* (Videx, formerly called dideoxyinosine, or DDI), also delays the progression of AIDS, but has frequent side effects such as peripheral nerve pain (neuritis) in 34 percent of patients and pancreatitis in 10 percent.

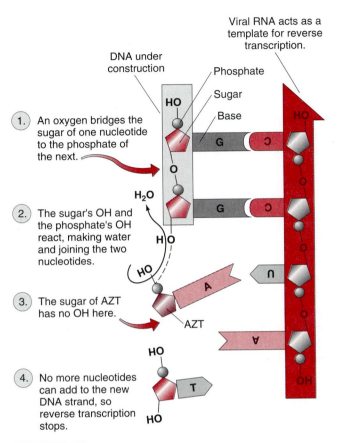

1. An oxygen bridges the sugar of one nucleotide to the phosphate of the next.

2. The sugar's OH and the phosphate's OH react, making water and joining the two nucleotides.

3. The sugar of AZT has no OH here.

4. No more nucleotides can add to the new DNA strand, so reverse transcription stops.

DNA under construction

Viral RNA acts as a template for reverse transcription.

Phosphate

Sugar

Base

AZT

FIGURE 13.5
How the drug zidovudine (ZDV, or AZT) interrupts the replication of the HIV virus.

Combination therapy. In bacterial diseases it is sometimes necessary to use more than one antibiotic, since bacteria can become resistant to one antibiotic. The rule for *combination therapy* is to pick two agents that work by different mechanisms, since resistance to one antibiotic that has one mechanism of action often confers resistance to other antibiotics that work by the same mechanism. Resistance to antiviral drugs is a potential problem as well, especially for therapy against a virus such as HIV, whose genome changes substantially over time. Since there are so few viral enzymes to choose from, combination therapy against HIV has used two or more agents with the same mechanism of action, that is, two or more reverse transcriptase inhibitors. Human trials with zidovudine in combination with didanosine have produced greater prolongation of the asymptomatic phase than ZDV alone.

Therapy and prevention of infections in persons with AIDS. Nearly all of the people who die as a result of AIDS actually die because of the opportunistic infections that accompany immunodeficiency. There are therapeutic drugs available for treatment of many of these infections, such as a combination of the drugs trimethoprim and sulfamethoxazole for *Pneumocystis* pneumonia, for example.

Equally important is prevention of other infections in people with AIDS. During the phase of CD4 T cell depletion, people are very susceptible to infectious diseases carried by people who are not infected by HIV. A cold or the flu can have grave consequences in an immunodeficient person. Bacteria picked up from food can be equally hazardous. Food preparation for people with AIDS must be meticulous.

THOUGHT QUESTIONS

1. What lifestyle choices could be made by a person to decrease the chances of becoming immunodeficient? Would those choices also be important for an HIV-positive person?
2. In the ELISA test explained in Box 13.2, how many of the enzyme-linked antibody molecules will bind if there are no specific, anti-HIV antibodies in the blood plasma being tested?
3. A positive control is an experimental control designed to ensure that conditions expected to give a positive result do in fact give such a result. A negative control is designed to ensure negative results when negative results are expected. For the tests explained in Box 13.2, what would you include in these tests as a positive control? What would you include as a negative control?
4. What steps in the ELISA test determine its specificity for HIV? What steps determine its sensitivity?
5. Could an ELISA test be used to test for antibodies specific to other viruses besides HIV? What step(s) would need to be changed, and how would you change them?
6. If a vaccine against AIDS were developed, how would you go about testing it? Remember that AIDS only develops in people, so animals cannot reliably be used as subjects. Would your test

have a control group? How would you ensure that the conduct of the test was ethical?

7. What side effects would you imagine that a drug like zidovudine might have, if any? Why?

8. In addition to biological barriers, are there non-biological barriers to the development of an AIDS vaccine or antiviral drug? What might they be? How might they be overcome? Should biologists be involved in addressing any nonbiological barriers, or is that task more properly left to nonbiologists?

D. Knowledge of AIDS Transmission Can Help You to Avoid AIDS Risks

Routes of Transmission

How is HIV virus passed from one person to another? The general term for passage of a pathogen from one individual to another is **transmission.** Transmission of some pathogens is very complex, involving other hosts. In these cases the pathogen has different phases, and these phases have different host specificities. The malarial parasite is an example (see Chapter 5). Other pathogens can be transmitted via contaminated water (e.g., cholera bacteria or poliovirus) or via water droplets such as from a sneeze (e.g., influenza virus).

HIV does not have any other animal hosts and does not remain infective in water or in air. It is passed only in certain body fluids: blood (including menstrual blood), semen, and vaginal fluids. For transmission to occur, these fluids must come in contact with the rectal mucosa or with the bloodstream via breaks in the mucous membranes or skin of another person. (HIV must make contact with cells bearing the CD4 molecule.) Transmission can occur from mother to fetus, and a few cases are known in which HIV has been transmitted via breast milk. There are low numbers of HIV particles in the saliva or tears from 1 to 2 percent of HIV-infected people (Barr et al., 1992). Saliva contains antiviral activity, and there have been no known cases in which HIV has been transmitted

via saliva, including human bites, or via tears. HIV is not found in feces or urine.

The percentage of AIDS cases in the United States transmitted by various routes is shown in Fig. 13.6. This figure represents cumulative AIDS cases between 1981 and 1990; therefore, it includes a higher percentage of cases transmitted by blood transfusion than is the case now that blood products are screened for the presence of HIV before they are given to a recipient.

Various behaviors or activities have been grouped into *risk categories*, based on what is known about transmission routes (Bartlett and Finkbeiner, 1991). It is very important to notice that **risk** (the probability of transmission) is now categorized by *specific behaviors* and not by population groups. The following four categories are based on the risk of transmission from an infected person to an uninfected person. In making choices about these behaviors, we must assume that every person whose HIV status is unknown to us is possibly HIV infected and is a potential source of transmission.

Category I: high-risk behaviors. These are behaviors during which transmission is very likely, accounting for over 99 percent of all cases of AIDS transmission.

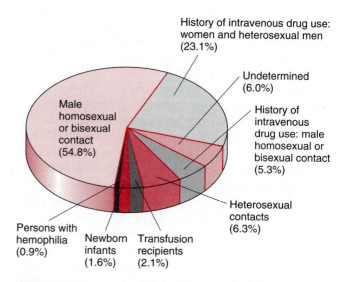

FIGURE 13.6
Routes of AIDS transmission in the United States. Data based on 1990 statistics from the Centers for Disease Control.

1. Behaviors in which the passage of blood, vaginal fluids, or semen is very likely, such as anal or vaginal intercourse with an infected person without the protection of a condom (unsafe sex).

2. Intravenous drug use in which needles or syringes are shared with an infected person.

3. Transmission from an infected mother to her child before or at birth.

Of all these behaviors, anal intercourse is considered the riskiest because semen contains an enzyme (collagenase) that breaks down the lining of the rectum and exposes blood vessels, a form of injury to which the vagina is much more resistant. In addition, vaginal intercourse is not as likely to result in infection as anal intercourse because the cells of the rectal mucosa are receptive to HIV infection (Adachi et al., 1987; Levy, 1988), while the intact epithelial lining of the vagina is a significant barrier (Padian, 1987). Nevertheless, the risk of HIV infection during vaginal intercourse is greater for male-to-female transmission than for female-to-male transmission.

Category II: likely risk behaviors.   This category includes routes by which documented HIV transmission has occurred, but with lower frequency.

1. Anal or vaginal intercourse using a condom (safer sex, not safe sex). Condom use is not completely safe. Condoms fail as birth control for about 10 percent of couples who use them; this means they can also fail as HIV prevention. Although the rate of failure seems to be low (less than 5 percent) for condoms *used properly,* many users do not exercise proper care in putting condoms on or taking them off, so estimates of failure rates *in general use* can be as high as 20 percent. Reliable data of this kind are difficult to obtain.

2. Breast-feeding (transmission to a baby from an infected mother or to a woman from an infected baby).

3. Receiving a blood transfusion or organ transplant. This risk was high prior to 1987, but is now considered very low in the United States because of careful screening of blood products and donated organs.

4. Artificial insemination. As with blood transfusion, the risks are now low when donated semen

has been tested for HIV.

5. Infection of health care professionals by needle-stick injuries.

6. Dental care by an infected worker. (There is a single cluster of six cases involving only one dentist; no other cases are known in which an infected health care worker has transmitted virus to a patient.)

7. Oral sex. (This is included only on the basis of anecdotal evidence; there are no cases in which transmission by this route has been documented.)

Category III: low-risk behaviors.   This category includes routes that are biologically plausible, based on knowing the presence of virus in body fluids, but no cases have been confirmed.

1. Sharing toothbrushes or razors or other implements that may be contaminated with blood.

2. Being tattooed or body pierced to produce ornamental scars or for jewelry.

3. Transmission via tears or saliva (as in deep kissing with exchange of saliva).

Category IV: no-risk behaviors.   Included are transmission routes considered not biologically possible for HIV.

1. Shaking hands.
2. Sharing a toilet.
3. Sharing eating utensils.
4. Being sneezed on.
5. Working in the same room.
6. Transmission via pets.
7. Close-mouthed kissing (kissing with no exchange of saliva or blood).
8. Transmission by mosquitoes or other insects (Box 13.4).

Communicability

Another question people ask about HIV is, "How contagious is it?" meaning, "If I am exposed, how likely is it that I will become infected?" The term used by the medical community to mean the likelihood of transmission after exposure to HIV is *communicability.* (*Contagious* simply means "capable of being transmitted"; it does not refer to probability of transmission.) The concept of communicability, or likelihood of transmission, is

BOX 13.4   CAN MOSQUITOES TRANSMIT AIDS?

Two frequently asked questions are, "Why don't mosquitoes transmit HIV?" and "How do we know that they don't?" Epidemiological evidence from studies examining the distribution of AIDS cases among the U.S. population shows that the frequency of AIDS and of the number of unexplained cases is no higher in areas that are mosquito infected than in other areas. In Africa, the people with the disease are mostly babies and sexually active young adults; mosquitoes do not bite these groups any more frequently than they bite other people. On all continents, children who are not yet sexually active often get mosquito bites, but they don't have AIDS unless they are born with it. Laboratory experiments have shown that although HIV and HIV-infected cells may be taken up by mosquitoes that bite infected people, it is not transmitted this way to other people. Several factors are involved: HIV cannot replicate in mosquitoes or survive long in their bodies (because mosquitoes are not a host for HIV), and the amount of blood ingested (3 to 4 microliters, millionths of a liter) is too small to contain an infective dose of HIV or HIV-infected cells. Mosquito saliva may also have substances that inhibit the virus. (Antiviral substances are known to be present in human saliva.) Because of the many biological factors involved in the transmission of a disease by an insect, it is highly unlikely that a single mutation in either the mosquito or the virus would significantly alter this situation.

Booth and colleagues at the Institute for Tropical Medicine in North Miami, Florida, proposed in 1986–1987 that the high rate of AIDS in the town of Belle Glade, Florida, could be attributed to the squalor and crowding of its people (conditions favorable to the transmission of any infectious disease) and to the mosquitoes breeding in nearby swampy lands, "where 100 insect bites a day are not unusual." The U.S. Centers for Disease Control studied this situation further and falsified Booth's hypothesis. The CDC concluded that the high incidence of AIDS in Belle Glade was attributable to sexual contact and shared needles, not insects. It is conceivable that the presence of malaria or some other mosquito-borne infection might make it easier for an HIV infection to develop into full-blown AIDS (after all, tuberculosis and certain viral infections have this effect), but there is no firm evidence that this occurs in human populations.

Similar questions arise in the case of other blood-sucking animals such as bedbugs (which are insects) and ticks (which are more closely related to spiders). There is one account of an African tick, which is endemic in the same parts of Africa where AIDS is, that carries enough blood and live virus to make transmission theoretically possible. However, as in the case of mosquitoes, children below the age of sexual activity do get bitten in significant numbers but do not get AIDS. Transmission by ticks and bedbugs is at most exceedingly rare.

directly related to the concept of risk. **High-risk behaviors** increase the likelihood of transmission.

There are at least two ways to address the question of communicability; one is an epidemiological approach and the other is a microbiological approach. The epidemiological approach compares the number of encounters to the number of infections throughout the population or within certain population subgroups. Many diseases, such as influenza or hepatitis, are very communicable; that is, there is a high probability that exposure will lead to infection and disease. It is difficult to designate the probabilities for HIV transmission in

comparison to other diseases because there is a period of weeks or months before antibody develops, and there is often a period of years between infection and disease symptoms, during which people may not know they are HIV-infected. The number of encounters with HIV is sometimes not known for an individual, and is less known for all of the individuals comprising a population.

For some routes of transmission the rate is more accurately known. For example, for transmission from mother to child, there is a 30 to 35 percent chance of an infected mother passing the virus to the fetus, meaning that 30 to 35 percent of the babies born to HIV-positive mothers are born infected, but these rates of transmission from mother to child can be reduced with the drug ZDV. In general, it can be said that HIV is much less communicable than a virus such as hepatitis (another virus spread by contact with contaminated blood), but that the communicability varies with the risk behavior.

The microbiological approach to determining communicability is to quantitate (measure numerically) what is called the *infective dose,* the number of pathogenic particles that must be transferred to result in a productive infection. This value is not known for HIV. The number of HIV particles in general is higher in semen than in vaginal fluids, but this can be very different at different stages of infection. Very early, in the weeks or months before antibody develops, and also very late, when the CD4 T cell count is low and the antibody concentration has dropped, the number of HIV particles in blood (including menstrual blood) and genital fluids is much higher than at other stages. From accidents in which health care workers have been exposed to infected blood, it is known that there is a higher probability of infection when a person has been splashed with large quantities of blood onto open sores in the skin and a lower probability when they have been pricked by a needle. For every 250 reported needle sticks, there has been one transmission. Viral load is certainly a factor, but only one of many factors. The precise infectious dose is not known and probably varies from one person to another. For example, persons with open genital sores due to other sexually transmitted diseases (STDs) are more likely to become infected than people without this additional factor.

Transmission of other diseases from an HIV-negative person to an HIV-positive person is much more likely than transmission of HIV from an HIV-positive person to an HIV-negative person. Since people with AIDS have severely impaired immune systems, their risk of catching diseases from other people is very high. When they become ill, they get much sicker than a person with a healthy immune system would.

Susceptibility versus High Risk

What is the difference between the terms *susceptibility* and *high risk*? To examine this question, let us go back and look further at how knowledge of AIDS and HIV developed. The fact that early cases were reported among hemophiliacs suggested an infectious cause for AIDS. As more cases were reported, the affected individuals seemed to fall into five groups, which became known as "the five H's": homosexual males, hemophiliacs, heroin addicts, Haitians, and hookers. From an epidemiological perspective, these categories may have served a useful purpose at the time, but they quickly become imprinted in the minds of scientists and the public, allowing a complacency on the part of people who were not in these groups. To epidemiologists, "the five H's" was merely a convenient term to identify clusters of reported cases of a mysterious, new syndrome. Such identification did not in itself imply anything about cause and effect or transmission, but was useful in suggesting hypotheses that could be tested.

High-risk groups. To epidemiologists, the term **high-risk group,** as applied in connection with a particular disease, simply means that there is a higher *frequency* of the disease among members of that group. Use of the term implies nothing about the possible reasons for the increased frequency, which may stem from increased exposure or increased susceptibility or both. *Exposure* to a disease means coming in contact with the disease agent. Increased exposure can sometimes be due to shared behaviors, but there are many other possible explanations. A disease may have a higher frequency in a certain group of people if all the people in that group came from the same geographic location so that they were exposed to the same toxic chemical, or if they all ate food from the same source and so were exposed to the same foodborne pathogen. It does not mean that these people are more susceptible; anyone else exposed to the same factors would also have become sick. **Susceptibility** to a disease means the ability to contract that disease *if exposed.* Humans are susceptible to HIV and most other animals are not. Susceptibility to disease can vary from one individual to another, and it can be genetically and/or environmentally influenced (malnutrition, for example, may make a person more susceptible to many infectious diseases).

Several misperceptions about AIDS have resulted from the early identification of specific high-risk groups. There is a lesson here on the power of language to influence both public and private behavior. First, as mentioned above, it allowed people not in the high-risk groups to assume they were not susceptible. In addition, it allowed people to assume that every person within a high-risk group was equally likely to be a carrier of disease (and that people outside of these groups were unlikely to carry the disease). Haitians, in particular, suffered adverse consequences by being classified as "high risk." In efforts to screen blood donors, before the nature of the disease was known and before tests were available for screening blood, all Haitians were barred from donating blood in the United States. In the resultant hysteria, some Haitians were evicted from their homes and lost their jobs, Haiti's tourist trade collapsed, and Haitian dictator Jean-Claude Duvalier's state police rounded up and incarcerated homosexuals in Haiti. Haiti's ambassador to Washington wrote a letter published in the *New England Journal of Medicine* deploring the damage done by North American semantic carelessness. As he pointed out, being from a certain country does not contribute to disease in the way that socially acquired characteristics do (multiple sex partners, intravenous drugs).

The fifth *H*, hookers, always seemed problematic because the other risk categories were predominantly or exclusively male. Why did so few women contract the disease at first? If women could contract the disease, why only prostitutes? There was a period of time when women were thought of as "carriers" even though they were dying of AIDS themselves. Now it appears that women are just as susceptible as men to AIDS infection, but that the epidemic in the United States began among homosexual men and spread only slowly to women.

Population frequency of disease. Once mechanisms of transmission are known, it becomes more appropriate to focus on high-risk behaviors than on high-risk groups. The frequency of infection within a discernible group of people can sometimes, however, play a role in an individual's risk. People within a high-risk group are at risk to the extent that they engage in high-risk behaviors. Their risk may be increased to the extent that their partners in high-risk behaviors are also members

of a group in which the frequency of infection is high. A higher population frequency of a disease increases risk by increasing the chance of encountering an infected person, not by altering any individual's susceptibility. (Remember that membership in a group, either a group with a high frequency of infected individuals or one with a low frequency, does not tell you whether a particular individual is or is not infected.)

As we have seen, the lower frequency of HIV infection among females in the United States led many people to assume that women were less susceptible. Therefore, when women began to be sick, there was a further misconception that the virus must have mutated to become more virulent, and if it could mutate once, it could mutate again, and heterosexuals would be susceptible. Women and heterosexual men have always been susceptible to HIV infection, as amply demonstrated by the pattern of the infection in Africa, where the number of men and women infected has been about equal. The pattern of infection in the United States has changed over time, but it is possible to explain all of the changes on the basis of frequency of HIV in various subpopulations, not on changes in virulence of the virus.

Susceptibility to HIV. We still do not know definitely whether one category of people is more susceptible to HIV than others. Although transmission by vaginal intercourse is not as likely as it is by anal intercourse, this does not mean that vaginal sex is safe, only that the number of infections per number of encounters is lower. It also does not mean that women are less susceptible than men. It appears that, if there are breaks in the vaginal epithelium (for example, as a result of other STDs), women are just as likely to be infected as men. So again, the risk is related to particular practices, not to differences in susceptibility, and these practices carry comparable risks for all groups of people. For example, data collected in both the United States and Africa seem to show that receptive anal sex is just as risky for females as for males.

Other factors that increase risk. Epidemiology shows that there is a positive correlation in both sexes between the rate of HIV infection and the number of sexual partners and also a positive correlation between the infection rate and the frequency of previous infections with other kinds of STDs. Sexual transmission can occur either vaginally or anally, and it can occur from male to female, female to male, or male to male. (Female to female transmission, though possible, is statistically rare.)

The use of other drugs besides injectable drugs also increases the probability of infection, particularly among women. In New York City, 32 percent of female crack cocaine users were HIV-positive, compared to 6 percent of other women. It has not been shown that crack is a cofactor (a factor that increases biological risk), but the culture that goes along with crack use is often one of a high incidence of sexual activity and of sexually transmitted diseases. In any locale, including college campuses, the impaired judgment that accompanies a drug or alcohol high cannot be discounted as a factor working against sexual abstinence or the practice of "safer sex."

Refusal skills.   There is another aspect to "risk," and that is the risk resulting from a particular individual's inability to say "no" to high-risk behaviors. Economic and cultural factors can put severe limitations on an individual's "refusal skills." In some cultures, for example, women may not be able to insist that their male sexual partners use a condom. Education about the risks of HIV and AIDS must do much more than provide people with information about transmission routes, as we will discuss shortly.

Public Health and Public Policy

While *medicine* deals with individual cases of disease, **public health** deals with populations and seeks to minimize the levels of particular diseases in those populations. Public health measures are usually organized efforts. Many of these efforts require legislation and most require funding, so they are usually carried out by governments or by large, often international, organizations. Many nongovernmental organizations (NGOs) have been crucial in educating the public about AIDS and in caring for persons with AIDS and their families.

History of public health responses to disease. In each nation in which the AIDS epidemic has spread, the governmental response was molded by the unique history and social customs of each nation. Some nations sought to restrict the immigration of HIV-infected people; others did not. Some jurisdictions segregated certain types of AIDS patients, while others did not. Hospital care and medical insurance for AIDS patients varied greatly from one country to another. Some nations instituted needle exchange programs for drug addicts; others did not.

The response of the U.S. government to AIDS has been very different from past responses to other diseases. During a cholera epidemic in New York City in 1832, alcoholics and poor Irishmen were rounded up and detained based on the mistaken beliefs that the disease (which is now known to be a waterborne, bacterial disease) resulted partly from intemperance and that all poor Irish were alcoholics. In 1916 during a polio epidemic, the New York City government carried out a house-to-house roundup and forceable quarantine of children thought to have polio. In the period from 1918 to 1920, 18,000 prostitutes were put into detention facilities until they were determined to be noninfectious for venereal disease. Many more examples exist.

The difference in the public response to AIDS can be attributed partly to its nature of transmission (that is, it is not communicable by casual contact), partly to the political organization of an "AIDS community," and partly to changes in civil rights laws during the period since 1950. Before 1950, U.S. law gave only weak support to individual rights. The Supreme Court decision in *Brown* v. *Board of Education*, as well as the resultant Civil Rights Act of 1964 and the Voting Rights Act of 1965, have given much greater strength to the rights of individuals in the face of discrimination by the many. More recently, the Rehabilitation Act of 1973 mandated that employers receiving federal funds cannot discriminate against someone with a handicapping condition who is otherwise qualified, and a disease that does not endanger others is considered a handicapping condition. This was upheld in *School Board of Nassau County* v. *Arline* (1987), in which a person with tuberculosis won the right to continue work. A bioethical principle known as the *harm principle* provides a moral limit on the exercise of freedom of individuals when others may be injured. In the case of AIDS, a person could morally be prevented from deliberately spreading the disease but could not be prevented from working or attending school or living in a particular place.

HIV testing. The issue of mandatory testing for HIV antibodies remains very controversial. There are issues of confidentiality involved, since many people fear (correctly or incorrectly) they would be discriminated against if they were identified as being HIV-positive. At the same time, medical professionals would like to know the HIV status of their patients both as a way to provide better care for the infected person and as a way to better protect health care workers against unintentional infection. Current laws prohibit testing most individuals without the individual's consent. The federal government, however, does test all people who apply for the Peace Corps, the Job Corps, and the military; Foreign Service Officers and their spouses; and applicants for immigration. The 1992 International Conference on AIDS, which was to have been held in the United States, was moved to Amsterdam to protest the U.S. policy of denying entry to any HIV-positive person; that policy has since been relaxed. Fourteen states screen all prisoners and six states segregate those who are HIV-positive.

Another issue related to HIV testing involves the notification of sexual partners. Public health officials have handled other sexually transmitted diseases by notifying and testing sexual partners of all infected persons, then their sexual partners, and so on. Some doctors have argued against this practice in the case of AIDS among women, which is often diagnosed during prenatal testing when they are pregnant. Many HIV-positive women (and their unborn fetuses) become victims of domestic violence when their HIV status is reported to their partner (North & Rothenberg, 1993).

Guidelines for handling blood. People likely to come in contact with blood—e.g., dentists and surgeons as well as their auxiliary workers, sports coaches and trainers, and security personnel—are now required to follow a series of guidelines known in the United States as the *Universal Precautions for Blood-Borne Pathogens.* The term *universal* refers to the fact that all blood must be handled as if it were infected, since the true presence or absence of pathogens in a particular person's blood is not known. The guidelines are meant to limit transmission of any blood-borne pathogen (of which hepatitis is much more communicable and much more frequent than HIV). They include mechanisms for cleaning blood spills, reporting accidents and injuries in which workers have come into contact with blood, and for education of workers as to the risks involved in handling blood. The U.S. guidelines were developed by the CDC, and the Occupational Safety and Health Administration (OSHA) has been charged by the federal government with monitoring the compliance of employers with these guidelines. Every college and university, for example, must have an infection-control plan.

The risk of transmission of AIDS through blood transfusion depends a good deal on the methods of blood donor selection. In the United States, blood donors are volunteers. Blood is screened for antibody to HIV, but blood donors are not. Blood donors are educated as to the behaviors that transmit AIDS, and they are given a chance to place a tag anonymously on their donated blood if they have been involved in high-risk behaviors, to decrease the risk of using blood from a person during the period of acute infection before the development of antibody. In the United States, the risk of transmission from a blood transfusion is presently estimated to be 1 in 225,000.

In countries that rely on professional donors, the safety of the blood supply is much less assured than in countries that rely on voluntary donations. In India, 30 to 50 percent of blood donors are professionals who sell their blood an average of 3.5 times per week. In one city in India, 200 professional donors were screened, and 86 percent were found to be HIV-positive. Often, economic factors play a large role in people's ability to modify their behavior.

Access to health care. Another aspect of public policy is access to health care. In the United States, the CDC is the body which develops the criteria that define AIDS. The criteria have been changed as new information has become known, but the wording of the definition is important because people with AIDS are eligible for some types of care from the government that other people are not.

AIDS was originally defined as a set of symptoms including particular opportunistic infections. The proposed new CDC definition includes all those persons who are HIV-positive and whose CD4 T cell count is below 200.

Many studies have shown that HIV-positive women have a poorer *prognosis* (prediction of outcome) than HIV-positive men, both in the United States and elsewhere in the world. This is probably a result of women generally having poorer access to medical care for the infections that accompany AIDS. The care of people with AIDS has put a strain on public health monies and the time that public health workers and other medical personnel have to spend on other diseases. In some countries of Africa, more than a quarter of all 1992 public health expenditures were for AIDS: Zimbabwe, 26.5 percent; Malawi, 35.3 percent; Tanzania, 40.6 percent; and Rwanda, 63.5 percent.

Educational campaigns. Educational campaigns aimed at increasing AIDS awareness have had conflicting approaches. Some organizations distributed free condoms and promoted their use, while others emphasized abstinence. U.S. Surgeon General C. Everett Koop stated that the only safe sex is a faithfully monogamous relationship with a faithfully monogamous uninfected partner, and the next best thing is the use of a condom.

Education about HIV and its transmission changed the behavior of homosexual men so that the rates of infection within this group began to decline. This subgroup is generally well educated and has provided many models for educational efforts to reach other groups. Because the factors guiding people's private behavior often differ greatly from one group to another (based on language, income, geography, religion, and cultural background), educational campaigns designed for one group will often need to be modified for each different locale and target group.

Information is not the same thing as education. Giving people information about how HIV is spread may not help unless the reasons underlying their high-risk behaviors are addressed. In any sexually transmitted disease, the motivations of people having mutually voluntary sex will be different from the motivations of commercial sex workers (prostitutes) and vastly different from the motivations of people such as street children having sex for survival. Many teenagers and young adults engage in sexual activity (often including high-risk activity), and those who do not are frequently subjected to very strong peer pressure to conform. Education often includes strategies for raising self-esteem and providing support for those who are trying to avoid high-risk behaviors.

Worldwide Patterns of Infection

As of 1992, the worldwide total of HIV infected adults was 10 to 12 million. It has now become a *pandemic*, a worldwide epidemic. There are estimated to be 1 million HIV-infected children worldwide, and in many countries AIDS has reversed the hard-won decreases in the infant mortality rate. In addition, there are 10 million children orphaned by the death of parents from AIDS. In the United States a cumulative total of 335,000 people have been diagnosed with AIDS (1993); as many as 2 million may be HIV-infected.

HIV compared to other infections. How do these numbers compare to the numbers of deaths from other diseases? The answer varies from one country to another. In the United States there are more than 10 times as many deaths per year due to heart disease and 5 times as many deaths due to cancers as there are deaths due to HIV. Persons with AIDS are predominantly young people, while the frequency of heart disease and cancer are greater in older people. Death from other infectious diseases remains very high in many countries of the world. Two million children per year die of measles, for example.

How do these numbers compare to the incidence of other diseases that are transmitted by sexual contact? In the United States and worldwide, the incidence of AIDS infection is lower than that of many other sexually transmitted diseases. Each year in the United States, 5 to 10 million people are infected with chlamydia, and there are 120,000 new cases of syphilis, with incidence rates of the latter rising among teenagers and young adults. Also, about 31 million people carry type II herpes simplex (the most common genital herpes virus), and 500,000 new cases are reported to the

CDC annually. Most of the sexually transmitted diseases other than AIDS are not fatal (untreated syphilis is an exception), but they have other serious consequences, which include (for different diseases) sterility, paralysis, arthritis, and chronic pain in adults as well as severe disease in newborns when transmitted from the mother. Since chlamydia, gonorrhea, and syphilis are bacterial diseases, most can be successfully treated with antibiotics if treatment is started early enough. In contrast, there is no drug to eradicate herpes since it is a virus. Antibiotics put a strong selective pressure on bacterial populations, selecting for mutated variants that are resistant to the antibiotic. Many strains of bacteria are now resistant to many different antibiotics, severely limiting the options for treatment of diseases caused by these strains.

People who have contracted any other sexually transmitted disease have already engaged in behavior that puts them at risk for HIV infection. Since properly used condoms can, in general, prevent the transmission of sexually transmitted diseases as well as that of HIV, someone who has contracted a sexually transmitted disease has probably not used a condom during intercourse. Moreover, they have further increased their risk for HIV infection because the presence of open genital sores from a sexually transmitted disease greatly increases the probability that contact with HIV will result in HIV infection.

Worldwide HIV incidence. Obviously, AIDS-related statistics vary according to the way in which AIDS is defined. The World Health Organization (WHO) criteria for diagnosing someone with AIDS are very different from the criteria set by the CDC in the United States. The WHO criteria are based entirely on clinical description (symptoms), not on HIV status or T cell counts. The WHO criteria are used in countries where monetary or technical considerations make testing for HIV infection impossible, although clinical diagnosis is confirmed in wealthy countries by testing for antibody to HIV (Box 13.2). Worldwide surveillance of numbers of AIDS cases is therefore not the same as surveillance of HIV infection, which must be estimated from the numbers of AIDS cases. Many people with AIDS are difficult to distinguish from people who are immunodeficient from other causes, such as undernourishment, and so may or may not be counted as AIDS cases.

Africa has the highest prevalence of AIDS (the numbers of infected people per total population), and in most areas the rate is still increasing. In the areas with the highest prevalence, such as Uganda, the percentage of the population that is infected has stabilized at around 20 percent. Infection rates stabilize when the rates of new infections are balanced by the death rate (the number of deaths per 100,000 population, Chapter 6) from the disease. For AIDS to result in a net decline in the total population in countries with high birth rates, it has been estimated that an infection rate of 50 percent would be required.

The area with the greatest rate of increase in HIV infections is Asia, with Thailand being the most severe. In both Asia and Africa the rate of infection is about equal among men and women, while in North America, South America, and Europe there is still a greater prevalence among men. In the United States, a majority of infected women are between the ages of 15 and 25, and 80 percent of newly infected women are either intravenous drug users or sexual partners of intravenous drug users.

What does the future hold? In the absence of any vaccines or drugs, prevention remains the best hope for control of AIDS. Effective prevention programs will require that research scientists, medical professionals, and educators work together, rather than in isolation. Common language must be found in which these groups can communicate with each other and with the people they serve. Education cannot be unidirectional: Professionals educating the people on the street must also learn from those people.

As it was put by Jonathan Mann in his opening remarks to the VIII International Conference on AIDS in 1992:

We have seen important success in basic and applied research, yet that research in isolation from concern about access to its achievements has severely limited its impact on lives of people with HIV. . . . If we believe that the entire problem of AIDS is really only about a virus, then we really only need a virucide or a vaccine. Yet if AIDS is deeply, fundamentally about people and society—and if societal inequity and discrimination fuel the spread of the pandemic—then to be effective against AIDS, we would have to address these issues.

1. What advice would you give to college students about the best ways to avoid getting AIDS? How might you modify the advice for different groups of people?

2. What factors are involved in determining how much money gets spent on medical research on any particular disease?

3. Is medical research disease-specific? In other words, will the knowledge gained from studying one disease be applicable only to that disease?

CHAPTER SUMMARY

The antibodies and cytotoxic T lymphocytes of the immune system protect the body from disease, but they are only produced in the presence of inter-leukin-2 secreted by the CD4 helper T cells. When some part of the immune system doesn't work, an immunodeficiency results. HIV is the virus that causes the immunodeficiency of AIDS by destroying CD4 T cells and other cells displaying CD4 molecules. Immunodeficiency results in sickness. In addition, microorganisms that are normally kept in check by healthy immune systems can cause serious and possibly lethal opportunistic infections when the immune system is compromised by AIDS.

HIV testing can detect antibodies to HIV. This virus is transmitted most often by sexual intercourse, by sharing intravenous drug needles, and by other less common means in which blood, semen, or vaginal fluids are exchanged between people. Since 1981 AIDS has spread to become a worldwide pandemic. There are currently no cures. Prevention is key, but will depend on many groups of people listening to each other, learning from each other, and working together.

KEY TERMS TO KNOW

AIDS (acquired immuno-deficiency syndrome) (p. 367)
cytotoxic (CD8) T cells (p. 368)
epidemiology (p. 369)
false negatives (p. 382)
false positives (p. 382)
helper (CD4) T cells (p. 368)

high-risk behaviors (p. 390)
high-risk group (p. 391)
human immunodefi-ciency virus (HIV) (p. 373)
immunodeficiency (p. 369)
Koch's postulates (p. 373)

opportunistic infections (p. 381)
public health (p. 393)
reverse transcription (p. 376)
risk (p. 388)
sensitivity (p. 382)
seroconversion (p. 382)
specificity (p. 382)

susceptibility (p. 391)
transmission (p. 388)
vaccine (p. 383)
virulence (p. 378)
viruses (p. 375)

CONNECTIONS TO OTHER CHAPTERS

Chapter 3   HIV is a retrovirus that produces DNA from RNA.
Chapter 4   Virulence follows a long-term pattern in its evolution.
Chapter 6   Condoms are useful for both contraception and preventing the transmission of AIDS.
Chapter 8   A decline in appetite (and therefore a decline in nutritional status) often occurs in the late stages of AIDS.
Chapter 9   A healthy immune system targets many cancer cells. Kaposi's sarcoma is an otherwise rare cancer that occurs more frequently among AIDS patients.
Chapter 10   Dementia from loss of brain cells is one of the late-developing symptoms of AIDS.
Chapter 11   The abuse of injectible drugs is a major risk factor for AIDS.
Chapter 12   The immune system that normally protects us from infection is greatly impaired by HIV infection.

Bioengineering

T he topics covered in this chapter all deal with the ways in which biology and engineering, two quite disparate disciplines, have been combined. The combination has provided a new way of looking at living organisms, at the kinds of materials of which they are made, and at the way size and shape affect their function. Bioengineering has produced technologies for measuring and interpreting the electrical fluxes of living tissues (thus often helping in the diagnosis of disease), for taking over when human tissues or organs have failed, and for expanding the abilities of both able and disabled persons. Many applications of bioengineering to medicine involve disease therapies that serve as examples of euthenics (Chapter 3) and are thus quite different from the approaches commonly followed by the fields covered in Chapters 8 through 13. In addition, bioengineering applications of lessons learned from ecology have led to the development of biological systems approaches to cleaning up pollution.

Throughout this book we have tried to show that various paths of inquiry can be taken to examine the same question. In many cases, scientists working from different perspectives, such as a cellular or a population approach to the same question, were not working together. By contrast, in bioengineering, biological strategies and engineering approaches are intentionally combined, and people with vastly different training, background, and world view work together.

How does an interdisciplinary approach differ from specialized work, in which biologists do biology, engineers do engineering, sociologists do sociology, and so forth? In the words of Lucien Gerardin, an engineer who fostered collaboration with biologists:

It is scarcely an exaggeration to say that the true specialist of the future will know everything about nothing. We have not yet reached this situation but it is already difficult to see how very specialised knowledge taken by itself can be usefully applied to human society.

True progress can come only from a composite view; invention consists of connecting things never before connected, and yet one must be aware of their existence before one can connect them. If specialists from different disciplines were brought together this would provide the initial conditions for some effective meeting of different points of view. Instead of one mind trying to grasp everything there would be an association of minds, all pooling their knowledge. (Gerardin, 1968:7–8)

Gerardin called what would result from such collaborations "cross-road sciences." He anticipated that the effort would be entirely different from specialized work, and would produce results that would be more than simply the sum of the parts. In his view the specialized sciences are "analytical and represent collections of factual knowledge"; while the crossroad sciences would be "synthetic and represent concepts relating these facts to one another."

Bioengineering is just such a crossroad discipline. As Gerardin implied, there are obstacles to combining different disciplines. One of the obstacles is the lack of a common language. As we saw in Chapter 1, every research field is constructed around a concept or model of how the world, or some part of the world, works. Hidden within each central concept are *assumptions* that are acceptable to the people within the field but may not be to people in other fields. Often, assumptions are taught as "givens" in a particular field and are so deeply embedded that a person working in the field may not realize the extent to which the assumptions have or have not been tested. Workers in one field may not be conscious enough of these assumptions to articulate them to someone in a different field, and they may not even realize that they need to be articulated. Before the benefits of the synthesis of any disciplines can be realized, there must be a commitment to learn each other's languages and each other's paradigms. There must also be the tolerance to see that another point of view and another way of doing things have value. There must be the trust that the hard work of synthesis will result in something valuable.

Will the issues and problems that we face today be more easily solved by specialized approaches or by interdisciplinary efforts? If there have been obstacles to combining biology and engineering, will there be even greater obstacles to combining even more disparate points of view? What can we learn from bioengineering about the potential benefits from and obstacles to crossroad or interdisciplinary collaborations?

A. Form and Function Concern Engineers and Biologists

Biological organisms live in a physical world, subject to the same physical forces as are nonbiological systems. As Steven Vogel of Duke University has put it, "Much of the design of organisms reflects the inescapable properties of the physical world in which life has evolved, with consequences deriving from both constraints and opportunities." Seen from a physics or engineering perspective, living things have developed what Vogel terms a set of "alternative technologies," which are different from the technologies devised by humankind. The method by which the technologies have been generated is also very different. Biological technologies have developed by natural selection, that is, without prior calculation and without anticipation of future forces. They are not "designed" in the rational, deliberate sense that human technologies are. Biological technologies persist when they are successful in meeting the physical forces to which they are subjected, that is, when they have contributed to the fitness of an organism.

Scaling

To a great extent, the physical forces acting on an organism depend on its *size*. **Scaling** is an engineering approach that compares how different *dimensions* (measurable quantities) of things change in proportion to one another as their size changes. A bridge that is twice as long as another but that has the same shape will weigh far more than twice as much—in fact, it will weigh 8 times as much ($2 \times 2 \times 2$). The same principle applies to plants and to animals (Fig.14.1). The effects of scaling alter many of the relationships between size and the structural and metabolic characteristics of cells, tissues, organisms, and ecosystems.

Surface area versus volume.   If you compare organisms of different sizes, their various dimensions do not all change in the same proportion. If a series of related animals maintained the same *shape* as they varied in size, the volume would be proportional to length times width times height ($L \times W \times H$), or to L^3. Because surface area increases in proportion to length times width (L^2), surface area does not increase as fast as volume. The *ratio* of surface area to volume thus *decreases* as length increases (Fig. 14.2). Through evolution, organisms have adjusted in various ways to compensate for this.

Every organism has an internal environment that it must maintain. The contact between the internal environment and the external environment

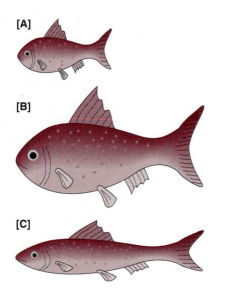

FIGURE 14.1

The scaling of the weight of objects of the same and different shapes. Weight, the effect of gravity on an object, is proportional to its volume, and volume is proportional to the length times the height times the width. If fish B is twice the length (*L*), twice the height (*H*), and twice the width (*W*) of fish A, the weight of fish B will be eight times that of fish A, regardless of the actual values of *L, H,* and *W*.

$$\frac{2L \times 2W \times 2H \text{ for fish B}}{L \times W \times H \text{ for fish A}} = 8$$

Now look at fish C. It is twice the length of fish A, while the height and width are the same as for fish A. In this case the weight is twice as great as that of fish A, but fish C is not the same shape as fish A. Changes in size must be compensated for by changes in shape.

occurs over the *surface* of the organism. The organism's mass and its basal metabolic rate (the rate at which an organism burns fuel to supply its functioning when it is at rest) both depend on the *volume*. If we are dealing with a warm-blooded animal such as a mammal or a bird, the amount of body tissue generating heat will also vary in proportion to L^3. Body heat, however, is lost across the surface of the animal, and heat loss will therefore vary in proportion to L^2. An animal twice as big as another will generate 8 times the body heat ($L^3 = 2 \times 2 \times 2$) internally, but only experience 4 times the heat loss ($L^2 = 2 \times 2$) across its body surface. This makes a larger warm-blooded animal more able to keep warm in a cold climate. In fact, this is the basis for *Bergmann's rule* (Chapter 5), which says that a geographically variable species of mam-

mals or birds will tend to have larger body sizes in cold climates and smaller body sizes in warm climates.

Another strategy resulting from evolution is to maintain a constant ratio of surface area by subdividing the organism into compartments (Fig. 14.2). Evolution from single cells into multicelled organisms (Chapter 9) accomplished just that.

Allometry.   The major method of adapting to the consequences of scaling, however, has been *not to maintain constant proportions,* but to change shape in various ways as size increases. As size increases, increase in any dimension that maintains a constant ratio to size is called *isometry;* increase in any dimension that changes out of proportion to size is called **allometry.** Organisms that grow isometrically maintain the same shape as they get larger; the *Nautilus* (Box 4.1), whose shell grows in the shape of a logarithmic spiral, is an example. Most organisms, however, grow allometrically, meaning that they change shape as they get larger (Fig. 14.1).

An animal that is twice as big as another has 8 times the living tissue to nourish, but only 4 times the intestinal surface area across which it can absorb nourishment if it maintains the same shape throughout. (It would therefore be only half as efficient, since $\frac{4}{8} = \frac{1}{2}$.) In each group of animals that varies in size, the larger animals usually differ allometrically from the smaller ones, having more elongate and usually more twisted digestive tracts, plus other adaptations (like internal folds or fingerlike villi) that increase the surface area of the intestine.

Weight is approximately proportional to the cube of length or width. However, the weight-bearing capacity of limbs and other structural supports is proportional (for a given shape) to their cross-sectional area, which varies (like other areas) in proportion to the length squared (L^2). This means that an animal that doubled in size would weigh 8 times as much, but would only have 4 times the supporting strength in its limbs *if* it maintained constant shape. (If a mouse were enlarged to the size of an elephant, it wouldn't be able to stand up without breaking its bones.) The usual solution to this dilemma is *not to maintain the same shape,* so that large animals (like elephants and brontosaurs) have proportionately thicker legs. Very

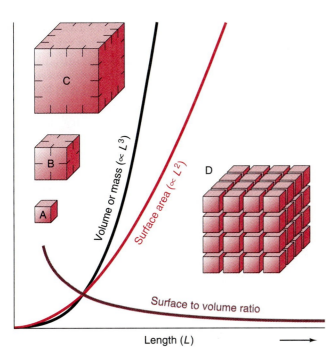

FIGURE 14.2
Allometric change. When the magnitude of organisms change, their different dimensions do not change equally (isometrically), but instead change unequally (allometrically). For example, as the length of a cube increases, its volume increases faster than its surface area. Compare box B to box A. Compare box C to box A. As size increases, the ratio of surface area to volume *decreases*. This ratio can be restored to the same values as for box A by subdividing it into smaller cubes (see D). The symbol $\propto$ means "is proportional to."

large land animals have also evolved more cylindrical limbs and a more vertical (instead of a crouching) stance.

We have seen that in colder climates, animals tend to be larger because of the advantage conferred by the smaller ratio of surface area to volume. There are limits, however, to the size range over which a disproportionate change in this ratio is biologically possible. If a cow used food energy at the same rate, relative to its mass, as a mouse, its internal temperature would be so great that steaks would come precooked. Metabolic rate, therefore, changes allometrically, not isometrically, with volume. The metabolic rate of larger organisms is actually slower and their food needs are therefore smaller, in proportion to their mass, than the metabolic rate of smaller animals. In scaling terms,

metabolic rate has been found to be proportional to mass according to

$$\text{Metabolic rate} \propto m^{3/4}$$

where the symbol $\propto$ means "is proportional to." A larger animal needs more food than another animal half its mass (the exponent is positive), but it does not need twice as much food (the exponent is less than 1). An isometric relation between metabolic rate and mass would be written as

$$\text{Metabolic rate} \propto m^{1}$$

Scaling can also predict the population density or numbers of organisms per unit area that can be supported by an ecosystem (see also Chapter 16). Larger animals need more food, so there must be fewer of them per square mile than smaller animals if food supply is to keep pace with food use. This is an *inverse proportion*, meaning that one variable (population density) decreases as another variable (animal mass) increases. Inverse proportions are represented by negative exponents: m^{-1}. As we have just seen, however, the food needs of animals are not proportional to their mass, but to $m^{3/4}$. Scaling studies have found that population density is proportional to $m^{-3/4}$.

$$\text{Population density} \propto m^{-3/4}$$

Scaling theory hypothesizes that the constraints imposed by physics result in these proportions. Scaling principles successfully predict the allometric change of many biological attributes of adult organisms with respect to their body mass. These include running or flying speed, heartbeat or wingbeat frequency, hearing frequency, skeletal mass, and bone cross section.

Design Elements

In any mechanical or biomechanical system, any force that tends to deform an object is called a **stress.** The stresses on all organisms and structures are of the same four basic types, as illustrated in Fig. 14.3. The strength of both living and nonliving structures is described in terms of their resistance to stress. Materials like brick, which resist compression, are said to have *compressive strength*, while materials that resist tension are said to have *tensile strength*.

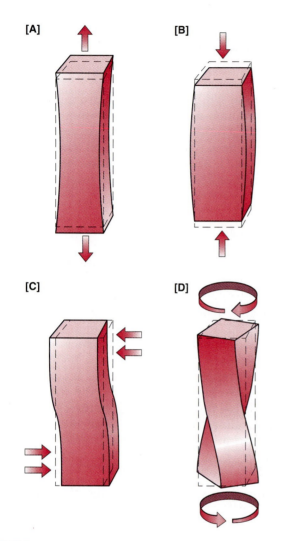

FIGURE 14.3
Types of stresses. Stress, in the physical sense of the word, is force per unit of cross-sectional area. The type of stress depends on the directions in which the forces operating on an object are applied, indicated by the arrows. The dotted lines for each bar show the position of the bar before the stress was applied. The solid lines show its position as the result of the force. [A] *Tensile stress* is the result of two forces working perpendicular to the cross-sectional area and in opposite directions away from each other; in other words, *pull.* [B] *Compressive stress* is the result of two forces working perpendicular to the cross-sectional area in opposite directions but toward each other; in other words, *push.* [C] *Shear stress* is the result of two forces working in opposite directions but along lines of action that do not coincide; in other words, *slide.* [D] *Torsional stress* is the result of two rotational forces applied in opposite directions; in other words, *twist.*

Human-built structures do not always resist stress in the same ways as biological materials: Curved surfaces and acute angles are much more common in biological technologies, while flat surfaces and right angles are more common in human technologies. Load-bearing elements in living systems are most often cylindrical and hollow, whereas load-bearing elements in human structures are often flat: walls, floors, and beams.

Strength is related to cross-sectional shape. Tall buildings and organisms are subject to both shear stresses and compressive forces. For the same amount of material, I beams offer the maximum resistance to shear stresses, and consequently are used commonly in construction. As shown in Fig. 14.4, hollow cylinders have somewhat less shear strength than an I beam (Fig. 14.4C,D), but they resist shear in any direction. For the same amount of material, the cross-sectional shape that gives the strongest support against compressive forces is a hollow cylinder. Hollow cylinders thus offer the best combination of strengths when a variety of stresses can be expected, as is the case for living organisms. The load-bearing structures of both plants and animals are hollow cylinders, although examples of I beams can also be found in biology (the lower jaws of many mammals, for example).

A tree must resist vertical compressive force from its own weight and horizontal shear force from the wind in almost any direction. Seen in cross section, a tree trunk is a series of concentric hollow cylinders (Fig. 14.4E). Note that *hollow* means "not solid," rather than "empty"; in fact, the cylinders in trees are filled with liquid. Liquids have great compressive strength in all directions. On a microscopic scale, the cells that give wood its strength are also vertically oriented hollow cylinders filled with fluid.

The leg bones of many animals are hollow cylinders (Fig. 14.4F). Bone tissues are also composed of a series of concentric cylinders on a microscopic scale. Bone gains additional strength from the fact that the spaces are filled with cells, which are basically liquid phospholipids and water. In addition, since bone is a living material, it can respond to new stresses by growing new crosspieces. In general, biological structures are much more responsive to changes in stresses than non-

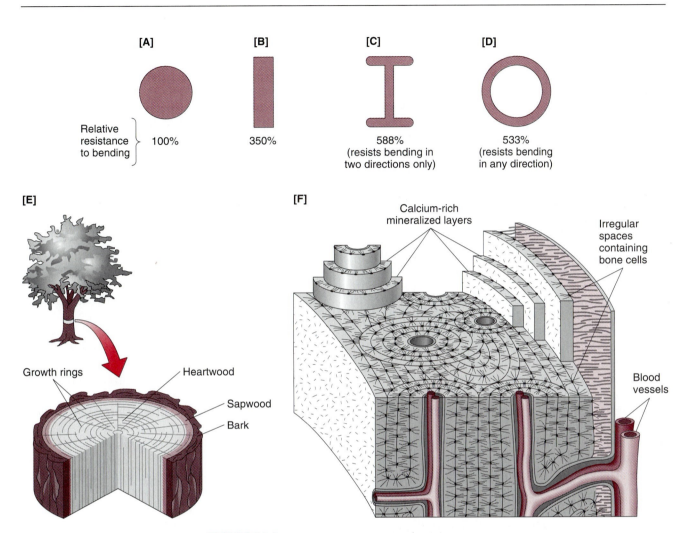

FIGURE 14.4

Relative strength of different shapes. When the same amount of material is made into vertical beams of different cross-sectional shapes, they have different resistance to shear stress or bending. If we define the resistance to bending of a cylindrical solid [A] as 100%, then the rectangular beam [B] has 350%, or 3.5 times, the resistance to bending as the solid cylinder. The I beam [C] has 588% of the resistance to bending as the solid cylinder, but resists only in two directions. The hollow cylinder [D] has 533% of the resistance to bending as the solid cylinder (somewhat less than the I beam), but it offers this resistance in all directions equally. The hollow cylinder also has the greatest compressive strength. Both tree trunks [E] and bones [F] are constructed of hollow cylinders in both overall shape and on a microscopic scale.

biological structures since they can change their composition in response to new forces.

Biological shapes result from evolution. As the preceding examples show, the structure and form of organisms frequently follows sound engineering principles, including scaling according to size and strength derived from shape. How do we explain

this in biological terms? In Chapter 4, we described the pre-Darwinian views of William Paley, who held that organisms had been consciously designed by God. Perfection is easily explained by Paley's hypothesis, but limitations ("the best that could be done with available materials") cannot.

The theory of evolution suggests a better explanation. Organisms vary in their construction in every generation, and those that conform to the best engineering principles have the best chance of surviving and reproducing when confronted with the changing physical forces of the world. Those individuals that are suboptimal in design survive less often. For example, following a late winter storm, an ornithologist named H. C. Bumpus examined hundreds of sparrows that fell out of the trees, half frozen, across the campus and environs of Brown University in Providence, Rhode Island. Bumpus and his students nursed back to health as many birds as they could. Bumpus also measured the body length, wing length, and seven other measurements of all the birds, and compared the measurements of the birds that survived with those of the birds that perished. In each case, the birds with extreme values (those with both the longest and the shortest wings, the longest and the shortest body lengths, and so forth) were among those that died, while the survivors were in most cases close to the average value in each trait.

Selection of this kind keeps each species close to the best possible engineering design that it is capable of realizing, within the limitations of its genome. Notice that this is not always the same as the best possible design of all. Natural selection will work toward the maximization of fitness, but maximum fitness is not always the same as maximum mechanical efficiency. There are also certain design limitations in living things that might not constrain an engineer designing a nonliving system. For example, the body parts of most animals must maintain contact with a blood supply, sensory nerves, and so forth. This precludes, for example, parts that would spin like a wheel. Parts of animals must also be capable of growing, and in many cases must be capable of repair in case of injury, limitations that would not constrain the human engineer.

Biomaterials

The materials of which living organisms are composed are very different from the materials from which human cultures have built buildings, roads, and machines, despite the fact that the stresses on all materials are of the same four basic types, as described above. The strength of materials, like the strength of design elements, is described in terms of the forces to which they are resistant; materials that have tensile strength are able to resist tensile stress, and so forth.

Biomaterials with tensile strength. Biological materials that have high tensile strength include *silk*, a protein made by spiders, silk moths, and other members of the arthropod phylum; *collagen*, a protein found in a pure form in the tendons of vertebrate animals; *chitin*, a polymeric sugar filling the role of tendons in arthropods such as crustaceans, insects, and spiders; and *cellulose*, a polysaccharide that provides tensile strength in plants. Table 14.1 compares the tensile strength of biomaterials with that of nonbiological materials; you can see that they compare very favorably.

Gels. Response to compression, shear, and torsion is accomplished in a variety of ways. While nonbiological materials are *stiff*, or rigid, and must be strong enough to resist forces, biomaterials tend to be *pliant*, giving with forces rather than resisting them. Some organisms, particularly smaller ones that do not have as much weight of their own to support, can get all of their compressive strength from water, without the need for solid support. Liquids do not resist shear stresses and torsion stresses very well, so various biological polymers are included in the water to resist these forces. The "jelly" of jellyfish and sea anemones is collagen fibers sparsely distributed in water. The collagen fibers are loosely cross-linked to form a *gel*, a flexi-

TABLE 14.1 TENSILE STRENGTH OF MATERIALS	
MATERIAL	TENSILE STRENGTH (Newtons per square centimeter)
Arterial wall	200
Brick	500
Concrete	500
Pine wood (dry, measured in the direction of the grain)	10,000
Collagen	10,000
Bone (fresh)	up to 10,300
Keratin (human hair)	20,000
Mild (non-high tension) steel	40,000
Nylon thread	100,000
Spider silk	200,000

ble semisolid that resists shear and torsion stresses. Similarly, animal cells contain cross-linked filaments of the cytoskeleton, made of a protein called *actin,* which form the cytoplasm into a gel.

A similar shock-absorbing function is also carried out by the polysaccharides, including such polymers as carrageenans and alginates found in seaweeds that receive harsh buffeting from ocean waves and currents. These polymers form gels that trap great amounts of water; a typical gel is only 2 percent solid material. These polymers also serve as molecular sieves, which filter molecules of different size and charge. Gel-forming molecules, particularly the ones derived from algae and higher plants, are now used commercially to form gels that thicken jams, jellies, and ice cream.

Pliant materials.   Pliant materials deform in response to a stress. They differ in the rapidity with which they return to their original shape. Some biomaterials "snap back" rapidly and are put to use as energy storage devices. The protein *resilin,* for example, is found in the hinge of insect wings. Muscle energy is used to pull the wing down; some of this energy is stored as a downward deformation of the resilin hinges. When the muscle contraction ends, the energy stored in the resilin enables it to return to its original position, helping bring the wing back up. *Elastin,* a protein in our skin and blood vessel walls, serves the same purpose, springing skin and vessels back to their original position after deformation by a stress. Other pliant materials recover their shape more slowly. We are shorter in the evening than we are when we first wake up; compression on cartilage all day causes it to become thinner, but as we sleep (in a horizontal position, relieving the compression), cartilage regains its original shape.

Composites.   Nonbiological materials rely heavily on metals; several types of metals may be together in a material, but they are alloyed in such a way that the properties of the material are the same at all locations and in all directions. Biomaterials, in contrast, have very little metallic content and are more often **composites.** In a composite, small areas of one material are embedded in a different background material called the matrix. The embedded material and the matrix each retain

their own properties, so that at a macro level, the material can have the distinctive properties of both of the micromaterials in the composite.

In composites, strength is often a property of a fibrous protein portion, while resiliency is a property of the matrix. In cartilage, the toughness comes from collagen fibers, while resilience comes from the elastin-containing matrix. The matrix can be strengthened if the function of the material requires more toughness and less resilience. For example, certain bivalve mollusks (especially mussels of the genus *Mytilis*) cling to rocks with the aid of tough composite protein threads called *byssus threads.* Composites can be further strengthened by mineralization, in which case they are called *biological ceramics;* tooth enamel and mollusk shells are examples.

The external skeleton of arthropods is composed of *cuticle,* a composite made of chitin fibers in a proteinaceous matrix. Unlike pure chitin, which gives tensile strength when it connects insect muscles to the exoskeleton, the composite provides rigidity in all directions. In larger arthropods, the cuticle composite is further strengthened by being calcified; that is, calcium carbonate ($CaCO_3$) crystals are formed within it. *Cartilage* is a composite of collagen plus hyaluronic acid, a special amino acid that does not occur in most proteins. Cartilage is found in organisms in situations where shear strength and torsion strength are more important than compressive strength, so it is not calcified. Cartilage is one of many examples of pliant composite biomaterials. *Keratin* is a composite of two proteins forming animal hair (including wool, alpaca, and mohair) as well as horn and bird feathers. It is strong while still pliant.

Layered composites.   Both toughness and resiliency can be changed if materials are layered. In insect cuticle, the chitin fibers are laid down in layers, each with a specific orientation, rather like wood fibers in plywood. Outer layers of the cuticle need to be tougher, and so the matrix proteins have been modified to make them tougher. Biological ceramics are similarly strengthened by their layered morphology. *Bone* is a composite of collagen in layers called *lamellae* in a matrix of other proteins, the whole strengthened by crystals of calcium phosphate called hydroxyapatite, or

$Ca_5(PO_4)_3(OH)$; the arrangement of materials in bone was discussed earlier (see also Fig. 14.4). The calcium carbonate crystals existing within mollusk shells can be different shapes (columns; flat tablets; long, thin crystals; or aggregates of unoriented crystals). These are then layered parallel or perpendicular to each other (Fig. 14.5); a similar layering occurs in plant cell walls. What we call mother-of-pearl is a layered composite of calcium carbonate crystals in a matrix of conchiolin, a glue-like protein. Light reflecting off the different layers give the shell its rainbow of color and swirly appearance. Thus these composites are beautiful in at least two ways: the beauty of a material exquisitely matched to its function and the beauty of intricate color and pattern.

Characteristics change with forces. Some composite biomaterials change their characteris-

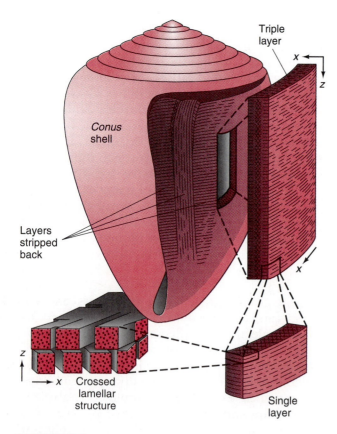

FIGURE 14.5
Many biological ceramics are layered composite materials. Shown here are the layers of shell in a mollusk (the whelk, *Conus*).

tics as physical conditions change. Slugs, for example, move along mucus trails. They are able to do this because when the mucus is first secreted under the front of their foot, it is a soft solid and resists shear, so that the slug is able to push against it. When a certain level of shear stress is applied, however, the mucus turns to liquid, flowing toward the following end of the slug while the slug itself glides forward. When the slug is no longer pushing on that portion of the mucus, it changes back from its liquid state to its more solid state, and a more posterior portion of the slug then pushes against that section of mucus. Several waves of solid–liquid–solid transitions thus occur under the slug.

Materials also may change over longer time periods. A study of rabbit leg tendon found that there was a small increase in stiffness and strength of the tendon material with exercise. Changes in the physical characteristics of the material are due, in part, to the ability of the organism to alter the cross-linking between molecules in the collagen.

Adhesion

How to get things to stick together, but only when you want them to, is one of the physical challenges faced by living organisms. Biological **adhesion** does not always need "sticky" substances, as we think of glue.

Capillary adhesion. Highly effective adhesion results when all available nooks and crannies are filled by some substance. Once two very smooth surfaces have been brought very close together, they are difficult to pull apart; anyone who has ever tried to pry apart clean glass microscope slides will appreciate this. Biological surfaces, however, are almost never that smooth. Many land organisms rely on a very thin layer of water for adhesion. When a thin layer of water separates two fairly smooth hydrophilic, or water-absorbing, surfaces, surface tension operates to keep the surfaces from pulling apart. As illustrated in Fig. 14.6, pulling the surfaces apart would require the creation of new surfaces (air–water interfaces), which would be opposed by the cohesion (surface tension) of the water. It may seem that this would not be a very large force, but it is large enough to keep tree frogs attached to vertical panes of glass or to trees. Cap-

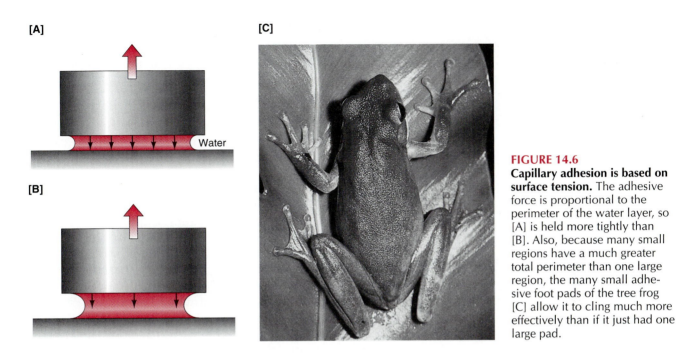

FIGURE 14.6
Capillary adhesion is based on surface tension. The adhesive force is proportional to the perimeter of the water layer, so [A] is held more tightly than [B]. Also, because many small regions have a much greater total perimeter than one large region, the many small adhesive foot pads of the tree frog [C] allow it to cling much more effectively than if it just had one large pad.

illary adhesion is also the force that keeps contact lenses on our eyes.

Suction cups. Other organisms attach themselves by biological suction cups. After pressing the flexible surface of their tube feet against the rigid surface of a rock, an octopus or starfish then slightly retracts the center of the tube foot, reducing the pressure in the small space between the foot and the rock. Because the pressure inside is then lower than the pressure outside, the air or water pressure outside will push the organism against the rock with enough force to keep it attached, provided no air or water leaks into the space to equalize the pressure (Fig. 14.7). Unlike capillary adhesion, adhesion by suction cups works underwater as well as in air.

Underwater glue. Other organisms may use secretions that work like glue. Mussels anchor themselves to the rock by means of the byssus threads previously mentioned. These threads are secured to the rock by an adhesive that the mussels secrete. This adhesive is remarkable for its ability to work under water, and its composition is being investigated by the navy for possible applications in underwater ship repair or in medicine, for such uses as securing artificial hips to bone.

Implantable Devices

In addition to using an engineering approach to study how biological systems have solved engineering problems, bioengineers have designed various devices to substitute for some of the functions of natural tissues and organs, a form of euthenics (Chapter 3). Some of these devices are made to be put inside a living organism. The major engineering problems that must be overcome in the design of such *implantable devices* are adhesion and **biocompatibility.** The materials used must be stable in the biological environment and must exert minimal effect on the surrounding tissue.

Many properties are important to ensure biocompatibility. The materials must resist corrosion or leaching by biological fluids over very long periods of time. The materials must be smooth at a microscopic level, since any surface roughness can promote the formation of blood clots. The materials must not trigger a response by the immune system (Chapter 12) because immune cells and their secretions could damage tissue around the device if they try to isolate it from the body. The material must not be carcinogenic. The physical chemistry of the surface can be engineered to either prevent or promote adhesion. In some cases, cell adhesion is needed to "heal" the implanted

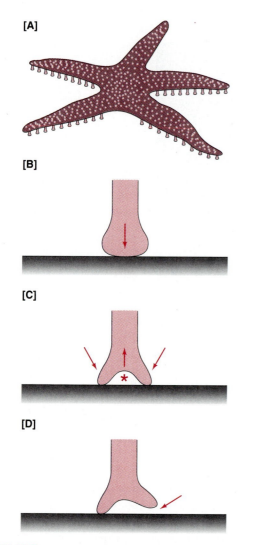

FIGURE 14.7
Suction cups. [A] A starfish has hundreds of water-filled tube feet. [B] By pumping water, it presses one of its tube feet against a rock. [C] It then retracts the center of the attached zone, creating a space [*], where the pressure is less than the pressure outside the space. External air or water pressure, indicated by the multiple arrows, push the tube foot against the rock, keeping it attached. [D] The attachment fails if the seal around the edge of the space fails. Since the starfish has many hundreds of tube feet, detachment of one will not result in detachment of the whole organism from the rock before that foot can reattach.

material into the tissue, but in other cases, such as the inside of an artificial blood vessel, adhesion is not desirable, as cells should travel through the vessel, not adhere to it. The material must stand up to sterilization before it is implanted, because the development of an infection around a device generally means that the device must be removed. In addition to biocompatability, the expense, weight, strength, and fatigue life of the materials are important. *Fatigue life* is the resistance to breakdown after repeated cycles of stress and unstress, which can be more relevant in biological applications than strength measured under constant conditions.

Artificial hip joints (Fig. 14.8) utilize three types of biocompatible materials. The ball section is made of cobalt-chromium or titanium alloys. To form the joint, the ball is fitted into a socket of high-density polyethylene plastic that replaces the shock-absorbing function of the cartilage that lined the original joint. The ball and socket are cemented with polymethylmethacrylate (PMMA) into surgically prepared sites in the top of the leg bone and in the hip bone, respectively. The PMMA does not "glue" the implant but rather acts like mortar between bricks, absorbing the compressive forces generated by gravity and by movement. Over time in some people, the PMMA can crack, so methods for cementless hip implants are being explored, methods that induce the ingrowth of natural bone into the implant. The angle of placement of the ball section into the top of the leg bone is also important; it is this angle that changes the mechanics of how stress is absorbed and spread through the leg, rather than being concentrated on the contact point. A hip joint, for example, is subjected to repetitive weighting and unweighting (fatigue stress) as the person walks.

Some implantable devices, such as artificial hips and artificial heart valves, are purely mechanical devices, while others may employ electronics. None of these implantable devices are perfect substitutes for the real thing. Body parts are not as easily replaceable as engine parts. Developing ways to prevent disease or injury is as important as developing ways to repair their consequences.

Assistive Devices

Another avenue in which biologists and engineers collaborate is in the design and the development of *assistive devices*. These devices replace a function in individuals lacking that function. A person may

The design of different modes of data entry into computers allows people with different physical abilities to express themselves. Both computer hardware and software have been adapted for use by people with impaired control over their hand movements, such as occurs in diseases like cerebral palsy. Input to the computer can be made by eye gaze, by puffs of air from the mouth, or by gesture. Voice synthesizers can speak for people who are unable to speak. Voice synthesizers have also been incorporated into several devices for people with visual impairments, from relatively simple devices such as talking thermometers to more complex systems such as computers fitted with optical scanners that scan books and read the text aloud to blind people. Software that enlarges on-screen text makes the text readable for visually impaired persons.

Improved designs of artificial limbs and wheelchairs can offer some increase in mobility and comfort. Advances have been made in the construction of these devices from new materials that are lightweight. Advances have also been made in customized wheelchair designs, tailored to the size and shape of the individual or designed for specific purposes like racing or climbing stairs. New designs for wheelchairs reduce the tendency toward pressure sores that result from impaired blood circulation due to prolonged sitting.

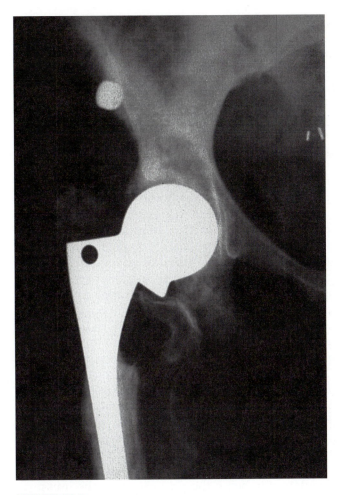

FIGURE 14.8
An artificial hip joint.

genetically lack a function. Even functions that are genetically possible may not be present because of environmental influences on gene expression, or they could be lost later because of factors such as accident or aging. The ability to biologically repair or regenerate a lost tissue and its function is limited.

Assistive devices are also called *adaptive technologies*, a term denoting technologies that extend people's abilities. One of the U.S. National Institutes of Health, the National Institute on Disabilities and Rehabilitation Research, funds research in this area. Another place to contact for further information is the National Information Center for Children and Youth with Disabilities [(800) 999-5599 in the United States].

THOUGHT QUESTIONS

1. Throughout this text we have mentioned various strategies that have evolved for increasing surface area in larger organisms. What are some of those strategies?

2. In the story *Gulliver's Travels*, which was written by Jonathan Swift in 1726, Gulliver is captured by the Lilliputians. The Lilliputians are people-shaped but only one-twelfth as tall as Gulliver. The Lilliputians calculated that since Gulliver was twelve times as big as themselves, he would need 12^3, or 1728, times as much food. Were they correct? Why or why not?

3. Why would capillary adhesion not work under water?

4. In what ways are body parts not equivalent to engine parts?

5. Why is an alarm clock that vibrates marketed as an "Assistive Device for the Deaf" rather than as

a vibrating alarm clock? Do assistive devices only assist disabled people?

6. Who else besides biologists and engineers should contribute to the design of assistive devices?

7. Gerardin called specialized sciences "analytical" and crossroad sciences "synthetic." List as many synonyms as you can think of for *analytical* and as many as you can think of for *synthetic*. Do you agree with Gerardin's statement?

B. Engineering Principles Describe Flow and Pressure in Biological Fluids

Living organisms are in constant contact with fluids, both internally and externally. For aquatic organisms, the fluid will be liquid (water). For terrestrial organisms, internal fluids are liquid and external fluids are more often gases. Important parameters for biological systems include the fluid pressure (the force per unit area with which the fluid presses against the walls of its container), flow rate (volume moved in a set length of time), and fluid volume. To a great extent, biological fluids act the same as similar fluids in nonbiological systems.

Blood as a Fluid

Blood flow.   *Hemodynamics* is the application of physical principles to the flow of blood through closed-loop circulatory systems, the type found in all vertebrate animals (Box 14.1). At a molecular level, the circulatory system is not closed; it is constantly losing water through excretion and leakage into the interstitial fluids. The kidneys return much of the lost water to the blood, thus acting to maintain total blood volume and ion concentrations. Despite this, the circulation can be treated for our present purposes as a closed system at the organismal level. The blood moves through this closed system because it is pumped by the heart. **Blood flow** is the volume of blood flowing through a vessel or group of vessels in a set length of time. Since flow is thus a rate parameter, its units are a proportion, usually stated as milliliters per minute

(ml/min). The rate of flow through a tubular vessel is determined by pressure gradients (see below) and by the diameter of the tube.

Blood pressure.   **Blood pressure** is the force per unit area with which the blood pushes against the vessel walls. Because blood is in tubes, blood pressure also propels the blood forward. Blood pressure, like other kinds of pressure, is measured in millimeters of mercury (abbreviated *mmHg*), meaning the pressure equivalent to the downward pressure exerted by a column of mercury of the specified height. When the heart contracts, sending a spurt of blood into the large artery called the aorta, the arterial pressure increases. The pulsing of the heart produces pressure waves, in which the maximum arterial pressure is called the *systolic pressure*, and the minimum arterial pressure is called the *diastolic pressure*. A blood pressure reading of 120/80 (read "120 over 80"—a so-called normal value) means that the maximum pressure on the artery walls when the heart first contracts is 120 mmHg, and the minimum between heartbeats is 80 mmHg. Pressure waves ensure that there is always a difference in pressure (a pressure gradient) between any point and another point farther downstream. Flow occurs in pressure gradients, with the direction of flow being from an area of higher pressure to an area of lower pressure, and the flow rate being the greatest when the pressure difference is the greatest.

Elasticity of blood vessels.   Normally when the heart beats (contracts), pushing blood into the aorta, the walls of the aorta expand, absorbing some of this sudden change in pressure, and storing the energy in the elastic biomaterial of which the aorta is made (Fig. 14.9). As the surge is finishing, the elasticity of the aorta pulls the walls back in, a recoil that helps push the blood along. When the aorta material has become less elastic due to disease (*arteriosclerosis,* or hardening of the arteries), it can no longer expand, in which case the increased blood from the heartbeat results in increased blood pressure rather than in increased vessel diameter. Increased blood pressure and lack of the boost from arterial elasticity forces the heart to work harder; over a long period of time this may result in an enlargement, and then failure, of the heart.

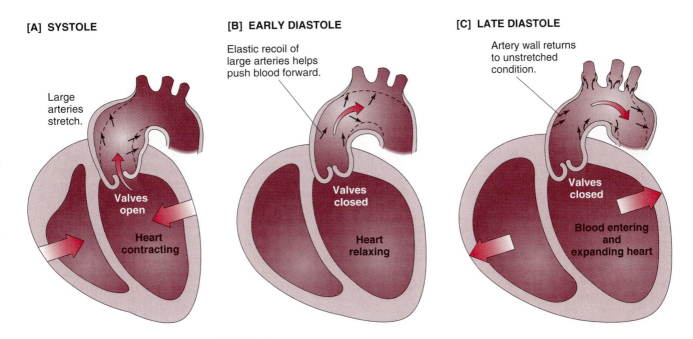

[A] SYSTOLE

Large arteries stretch.

Valves open

Heart contracting

[B] EARLY DIASTOLE

Elastic recoil of large arteries helps push blood forward.

Valves closed

Heart relaxing

[C] LATE DIASTOLE

Artery wall returns to unstretched condition.

Valves closed

Blood entering and expanding heart

FIGURE 14.9

Arterial elasticity relieves blood pressure and helps to move the blood along. Pressure is a physical stress (See Fig. 14.3) with the dimensions of force per unit area. [A] When the heart contracts and blood enters the aorta, the pressure on the arterial walls increases. Some of the pressure is relieved by the stretching of the wall, increasing the diameter of the aorta. [B] Between heart contractions, the elasticity of the aorta returns it to its resting diameter, exerting pressure on the blood. Like water, blood is a noncompressible fluid (it cannot be compressed to relieve the pressure), so the increase in pressure causes the blood to flow. Valves in the heart and at the base of the aorta prevent the blood from flowing backward. [C] The stretching and elastic return of the aorta is repeated along its length, pushing the blood along.

Decreased vessel diameter increases blood pressure. Arterial blood pressure is also increased if the inner diameter of the artery becomes smaller (Fig. 14.10, p. 414). Such is the case in *atherosclerosis*, a blockage of blood vessels due to the buildup of fatty deposits, the leading cause of *coronary heart disease*. Atherosclerosis of the vessels supplying the heart's muscle tissue may lead to a heart attack and damage to the heart tissue.

Regulation of blood flow and pressure. The body's remarkable capacity for maintaining a steady hemodynamic state is possible because of a number of sensing systems and feedback loops. There are pressure sensors (*baroreceptors*) in the artery walls, which initiate a nerve impulse when the artery wall is stretched by increased pressure. The nerve impulse travels to the *medulla oblongata*

of the brain (Chapter 10), where the cardio-inhibitory center is stimulated, reducing cardiac output (volume pumped) and dilating very small arteries (arterioles) to relieve pressure. Response is rapid; the force of gravity can cause pressure changes when we make sudden movements, but these centers compensate to maintain constant pressure. There are also arterial *chemoreceptors* that respond to changes in the blood concentration of oxygen and carbon dioxide. Somewhat slower responses can be achieved via hormones that can act directly on the vessel smooth muscle, resulting in decreased vessel diameter, or indirectly via action on the kidney and change in blood volume. Emotional state also affects blood pressure: Rage or fear stimulates the sympathetic nervous system and constricts the diameter of the arterioles, increasing pressure; depression, grief, and loneliness decrease sympathetic stimulation, decreasing

BOX 14.1   THE HUMAN CIRCULATORY SYSTEM

The veins collect blood from the body's organs and carry this blood to the heart. Blood from everywhere except the lungs enters the right atrium and is pumped into the right ventricle and then out through the pulmonary arteries to the lungs. In the lungs, the blood receives oxygen, while carbon dioxide diffuses away and is exhaled (Chapter 11). Oxygenated blood from the lungs returns to the left atrium of the heart and is pumped to the left ventricle. The left ventricle has an extremely thick, muscular wall whose contractions pump the blood via the arteries to all body regions, including the head, arms, internal organs, and legs. At each destination, arteries branch to form smaller arterioles, eventually forming thin-walled capillaries whose diameter is scarcely larger than the blood cells passing through them. Gases, nutrients, and metabolic waste products are exchanged with the surrounding tissues across the thin capillary walls. The capillaries converge to form larger vessels and eventually veins, which carry the blood back to the heart. The arteries, capillaries, and veins form a continuous closed loop through which the blood is moved by the pumping of the heart and by lesser forces such as the contraction of nearby muscles and the elastic rebound of arteries (Fig. 14.9).

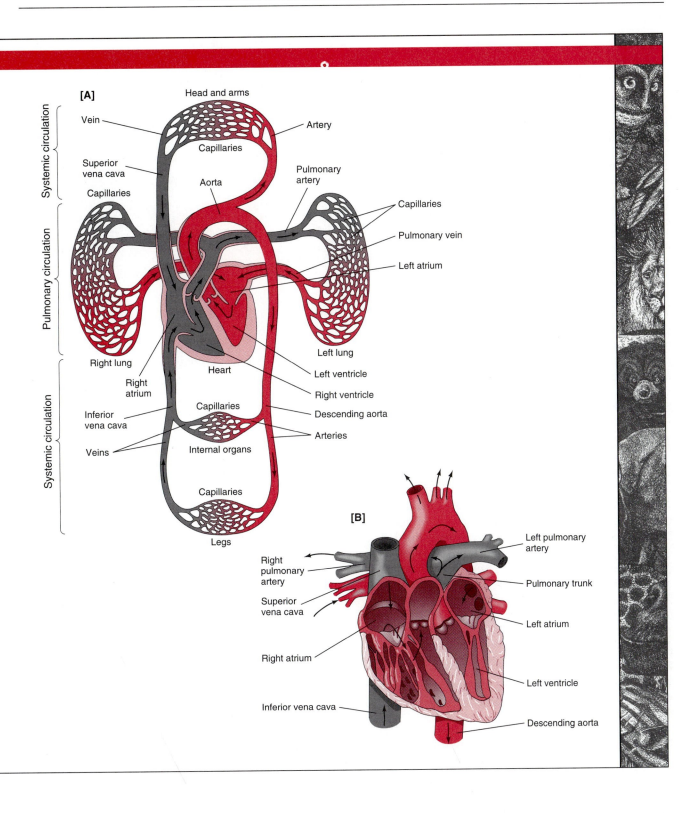

[A]   **[B]**

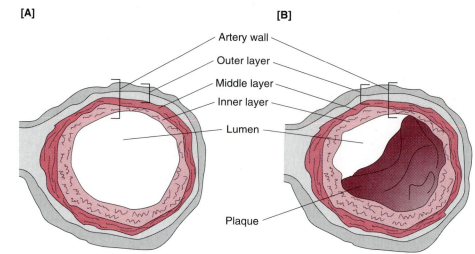

Artery wall
Outer layer
Middle layer
Inner layer
Lumen
Plaque

FIGURE 14.10
Atherosclerotic plaque reduces the effective diameter of an artery. Although the outer diameter of the artery has not changed, the inner diameter, or lumen, through which the blood can flow has become smaller by the formation of lipid deposits on the inner wall of the vessel. These deposits may also become calcified, reducing the vessel elasticity, further increasing the pressure.

blood pressure (Chapter 12). In summary, while the blood pressure in a person can be analyzed in much the same way as water pressure in pipes, there is an important difference in the complexity and responsiveness of the biological versus nonbiological systems, with biological systems able to constantly alter fluid pressure, flow, and volume.

Plant Fluids

Osmosis and turgor.   Plants, like animals, are influenced by fluid physics, but have met these challenges in very different ways. Like all cell membranes, plant cell membranes stay intact only because they are surrounded by water, a fluid. The lipid bilayer membrane is impermeable to ions, yet it is permeable to water; this results in a movement of water called **osmosis** (Box 14.2). Plant cells are also surrounded by a rigid cell wall made of cellulose. Plant cells take up water, which pushes against this cell wall, producing **turgor** (Box 14.2), a result of fluid pressure that makes plant cells stiff. It is thus water pressure, rather than a skeleton, which keeps nonwoody plants upright. In woody plants, a rigid biomaterial called *lignin* provides increased support.

Water transport.   A plant also needs to move water from its roots to all of its cells. There are several hypotheses for how this transport is accomplished, but the one that best fits existing data is the *transpiration-pull theory*, the idea that water is pulled up through a plant rather than being pushed up

from its roots. Although water pressure does build up in the roots of plants, it is not of sufficient magnitude to push water against gravity to the heights achieved by many plants. When a water transport vessel (*xylem*) is cut, water continues to be pulled upward above the cut, but none is pushed into the cut from below. The rate of water movement is greatest when the plant is in sunlight, a time when water loss by evaporation from the leaves (**transpiration**) is also greatest.

Transpiration is controlled by openings called *stomates*, or *stomata*, in the undersides of leaves (Fig. 14.11). When these are open, gaseous exchange with the atmosphere occurs: Carbon dioxide (CO_2), needed for photosynthesis (Chapter 15), is taken up, and oxygen, a by-product of photosynthesis, is given off. However, when the stomates are open, water is also given off, since the water concentration of the air is lower than that of the xylem.

Water molecules are very polar, with the oxygen atoms partially negative and the hydrogen atoms partially positive. The oxygen of one water molecule forms weak attractive bonds, called *hydrogen bonds*, with one of the hydrogen atoms of an adjacent water molecule. Water, being strongly hydrogen-bonded, has a high amount of *cohesion*. As water molecules are lost through the stomates, cohesion pulls the next water molecules up to take their place. A tension (tensile stress) is thus put on the thin column of water in the xylem, pulling the whole column upward (Fig. 14.12, p. 418). The opening and closing of the stomates is accomplished by *guard cells,* which change shape in response to turgor (Fig. 14.11).

Rapid movement in plants.

Regulation via turgor pressure is the key to how plants like the Venus flytrap (Box 15.1) and the sensitive *Mimosa* (Fig.

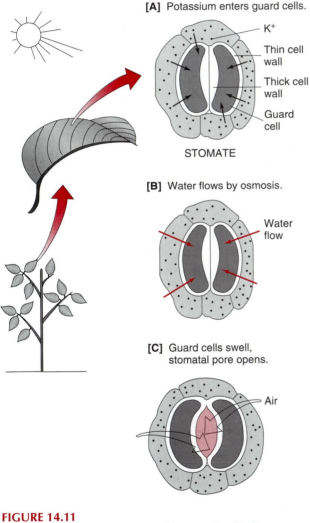

[A] Potassium enters guard cells.

K⁺

Thin cell wall

Thick cell wall

Guard cell

STOMATE

[B] Water flows by osmosis.

Water flow

[C] Guard cells swell, stomatal pore opens.

Air

FIGURE 14.11

Stomates are opened and closed by guard cells that are sensitive to changes in turgor. [A] When turgor is low, which is also when water supplies are low, the guard cells are closed. When turgor increases in the adjoining cells, ion channels in the guard cell membranes open, allowing K⁺ and Cl⁻ to enter the guard cells. [B] Water molecules flow by osmosis, which increases the turgor in the guard cells, resulting in a change in shape. [C] In the new shape, a space between the two guard cells opens and allows the exchange of gases, including CO_2, O_2, and water vapor.

14.13, p. 419) are capable of rapid movements without nerves or muscle cells. In these plants, touch causes ion channels to open, causing an outflow of ions and a hypertonic situation. Water flows out in response, and turgor drops. In plant cells, the cytoplasm of one cell is connected to the cytoplasm of an adjoining cell at small openings called *plasmodesmata.* The changes of ion concentration are thus passed along to the next cell, where ion channels are stimulated, passing along the change in turgor. When turgor drops in the cells where the leaflet joins the stem, the leaflets collapse toward each other (Fig. 14.13). When changes in turgor have propagated to the base of the leaf stem, which may take more than one touch, the whole leaf droops. Thus, in plants, fluid flow is accomplished without a pump, and movement is accomplished without contractile muscle fibers.

THOUGHT QUESTIONS

1. How does blood flow compare to water flow in a plumbing system? How does it compare to water flow in a forced hot water heating system?
2. List all the different stimuli that the human body can sense. (Before beginning, think about what you have learned in this and other chapters.)
3. How can plants move without nerves and muscles?
4. Do plants have sense organs?

C. Electrical Activity Occurs in Living Organisms

Because ions are charged particles, and because biological membranes are impermeable to them, membranes can act to allow a separation of charge from one side of the membrane to the other. The lipid part of the membrane's bilayer thus acts as an *insulator* separating the charges. Anytime there is a difference in charge on two sides of an insulator, there is stored potential energy, measured as a **voltage** between the two sides (Chapter 10). If the opportunity arises, by the opening of a hole in the membrane (a membrane channel), this difference in charge will result in

BOX 14.2   OSMOSIS

Ions cannot cross cell membranes because it is energetically unfavorable for a charged particle (ion) to move through a nonpolar region such as the interior of the lipid bilayer. Water molecules are polar but not charged; they are not ions. Water molecules are small enough so that they can pass through membranes in spite of their polar character. The interior (cytoplasm) of a cell is water that contains many ions. On its exterior side, a cell is also in water that contains ions. If the concentration of ions is not the same on both sides of the membrane, physical forces will tend to equalize the concentrations. Since ions cannot cross the membrane but water can, water will move toward the side with the higher ion concentration. (As shown in [A], the side with the higher ion concentration can also be said to have the lower water concentration, so water moves from an area of high water concentration to an area of low water concentration.) This movement is called *osmosis*, and the force producing this movement is called *os-*

motic pressure. Earlier, we said that fluids *flow* in pressure gradients; the same is true here: water flows because of a gradient in osmotic pressure.

Osmotic pressure is especially important in fluid flow in plants. At a cellular level [B], plant cells have a vacuole that can store water. When the vacuole is full, the water pushes the plant cell against the cellulose cell wall, a phenomenon called turgor pressure. Turgor pressure thus acts to oppose osmotic pressure, preventing any more water from entering the cell and keeping the cell membrane at equilibrium [C]. Seen at the level of the whole plant, it is turgor that keeps the plant stem vertical.

In a *hypertonic solution*, that is, when the salt concentration outside the cell is higher than that inside the cell, water leaves the cell by osmosis, emptying the vacuole and reducing the turgor pressure [D]. Since turgor keeps the plant stem vertical, a loss of turgor pressure causes the plant to *wilt*.

---

ion flow, just as we have seen previously for pressure gradients resulting in fluid flow. Voltages are constantly changing as the cells use them to do work, for example, depolarizing the membrane to begin and propagate an action potential in a nerve cell (Chapter 10) or to begin contraction in a muscle cell (Chapter 10). Voltages also change when the cell rebuilds the gradients by pumping ions back across the membrane. This is true in any organism—plant, animal, or procaryote—since all have membranes that act as insulators and that contain ion channels.

Bioelectrical Activity as a Diagnostic Tool

Electrical activity at a cellular level can be detected on an organismal level, and is used diagnostically in medicine. Changes in electrical activity are correlated with many types of diseases and are often detectable before other disease symptoms are. Electrical activity in the heart is monitored by a diagnostic test called an *electrocardiogram*, which is abbreviated either ECG or EKG. Electrical activity in the muscles is monitored by *electromyography*

[A] Osmotic pressure in a
hypotonic medium

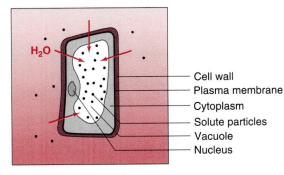

H₂O

Cell wall
Plasma membrane
Cytoplasm
Solute particles
Vacuole
Nucleus

Because of dissolved materials,
water concentration is lower
inside cell, so water moves
into cell.

[B] Turgor pressure in a
hypotonic medium

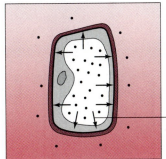

Turgor
pressure

Turgor pressure inside cell
is high; cell becomes turgid.

[C] Equilibrium in an
isotonic medium

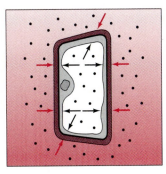

Osmotic pressure equals
turgor pressure.

[D] Hypertonic medium

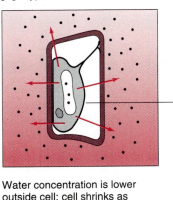

Plasma
membrane
of shrunken
cell

Water concentration is lower
outside cell; cell shrinks as
water flows out.

(EMG). Electrical activity in the whole brain is measured by an *electroencephalogram* (EEG), which shows sleep patterns (Chapter 10), epilepsy, and many neurological disturbances.

Evoked potentials. Electrodes can also be placed over particular areas of the brain and electrical activity measured as the organism is given various sensory stimuli. *Evoked potentials* are the electrical activity resulting from something like sound. Measurements of evoked potentials are now used to test hearing in infants too young to be

able to indicate in other ways whether or not they have heard a sound stimulus. Such a measurement does not tell you what the infant perceives in response to that sound, only that sound has resulted in some electrical activity in the brain, indicating that the sensory apparatus in the ear and the auditory nerve are intact and functional.

Skin conductivity. So-called lie detectors work somewhat differently than the previously mentioned tests. Electrodes placed on the skin are used to measure changes in the *conductivity* of the skin

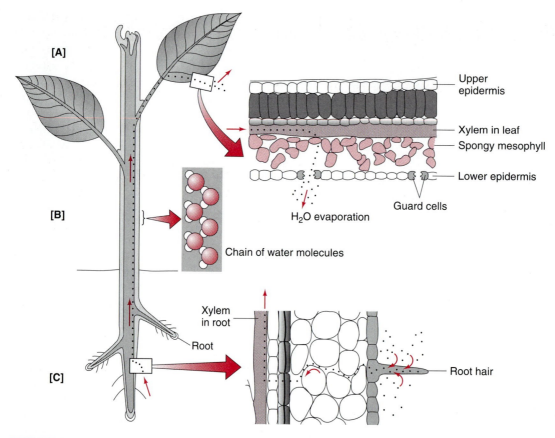

[A]

Upper epidermis

Xylem in leaf

Spongy mesophyll

Lower epidermis

Guard cells

H₂O evaporation

[B]

Chain of water molecules

Xylem in root

Root

[C]

Root hair

FIGURE 14.12

The transpiration-pull theory. Transpiration creates a tensile stress that acts to pull water up through a plant. [A] When water is lost by transpiration (evaporation through the leaves), water molecules move in to take their place. [B] Strong cohesive forces from hydrogen bonding hold the water molecules together, so that evaporation pulls a whole chain of water molecules upward. [C] Water is taken up through the root hairs by osmosis, and it enters the xylem.

surface, a measure of how well electrical charge travels. Because sweating increases the electrical conductivity of the skin surface, measurements of conductivity can detect very small increases in sweating caused by nervousness.

Biofeedback

Many of the techniques used to monitor electrical activity can be used not only by clinicians, but by patients themselves. As covered in Chapter 12, physical, psychological, and emotional stress can change brain activity, heart activity, and other parameters such as muscle tension and skin conductivity. A person can learn to calm physiological

stress responses and other functions such as heart rate, once thought to be involuntary, by exerting mental control in a process called **biofeedback.** Monitoring devices help people learn biofeedback: By watching their own EEGs, EKGs, EMGs, or skin conductivity on an oscilloscope, people can learn to successfully change them. EEG biofeedback can help overcome some types of depression, chronic pain, and insomnia (inability to sleep). Tension headaches and stress-related muscle spasms can be lessened by EMG biofeedback. Other physical parameters, such as skin temperature, reflecting changes in blood flow, can also be used for biofeedback in treatment of such conditions as migraine headaches.

FIGURE 14.13
Controlled changes in turgor cause rapid and selective movement in some plants, such as this *Mimosa*.

THOUGHT QUESTIONS

1. What uses do plants make of electrical activity from ion flow?
2. What are the advantages of diagnostic tests like EEGs and EKGs? What are the disadvantages?

D. Bioengineering Can Help Solve Environmental Problems

Bioremediation of Oil Spills

In the ecological cycling of chemicals, some organisms (primarily plants and photosynthetic bacteria) use energy to build simple, inorganic molecules into complex, organic molecules. Other organisms are uniquely suited to breaking down complex molecules, deriving their energy from the process. The latter are mostly decomposers, a part of the producer–consumer–decomposer loop that keeps matter cycling through the biosphere (Chapters 15 and 16). Decomposers are primarily fungi and bacteria.

Bioremediation is an approach to engineering in which the natural decomposition activities of living organisms are put to work cleaning up contaminated soil and water. There is a distinction between the terms *biodegradation* and *bioremedia-* *tion*. **Biodegradation** refers to the natural decomposition processes that go on without human intervention; **bioremediation,** in contrast, implies the manipulation of biodegradative processes by humans. Both have been used in the cleanup of oil and chemical spills and in wastewater treatment.

Oil and water. When there has been an oil spill at sea, little oil dissolves in the water. Oil is a mixture of many different chemical compounds, the exact mixture depending on where the oil came from and whether or not it has been refined. Most of the compounds are *nonpolar*, built almost entirely of atoms such as carbon and hydrogen that do not attract electrons and that form bonds over which electrons are evenly dispersed. Water, on the other hand, is very *polar* (see also Chapter 8), meaning that there is a partial separation of positive and negative charges. Other molecules that are polar are likely to be soluble in water, while those that are nonpolar, such as the constituents of oil, are less soluble in water.

Because crude oil is not very soluble in water and because it is less dense than water, it floats on top, gradually spreading across the surface and forming a slick. Although slicks may be only a few molecules thick, they interfere with organisms that need to absorb oxygen at the water surface. Evaporation removes the smaller, lighter-weight components of the slick; the components left behind may then be more dense than water and sink, eventually inhibiting bottom-dwelling organisms. Oil slicks are also broken up into small aggregates of oil molecules by the action of the waves and the wind, a process called *dispersal*. These aggregates, or the slick itself, can be driven ashore onto beaches, where they interfere with more living things. Some of the compounds in oil are directly toxic to organisms; other organisms are killed because the oil coats and blocks the surfaces over which they exchange gases with the air or water. Air trapped in the feathers or fur of water birds and mammals gives them natural insulation and buoyancy that is destroyed if they become coated with oil (Fig. 14.14). Biological membranes are stable structures only because the molecules making up these membranes are suspended in water (Chapter 8). Contact with a nonpolar substance such as oil can seriously disrupt the integrity of cell membranes.

FIGURE 14.14
Oil-coated seabirds. Such birds die unless the oil is removed by detergents. Many die even if the oil is removed because the detergents also cause harm.

Evaporation, sinking, or dispersal of oil may make it disappear from view, but the oil molecules are unchanged. Likewise, many of the mechanical means employed for spill cleanup may remove the oil but do not chemically change it. Changing the molecules entails chemical reactions called *degradation* at which decomposer organisms are very effective.

Oil-degrading microorganisms. Oil-degrading bacteria and fungi are found in all types of water habitats, both fresh and salt, and include representatives of over seventy different genera. No one species can degrade all of the different molecular compounds in oil. Even when bacteria are present that are genetically capable of producing oil-degrading enzymes, the rate and extent of biodegradation will depend on many other factors. Such factors include the temperature, the amount of oxygen, and the availability of other nutrients.

Since many of these environmental factors are unpredictable, the success of biodegradation at any given site will also be unpredictable.

Probably the most critical factor for biodegradation is the availability of nutrients such as nitrogen, phosphorus, and iron. These elements are necessary for microbial synthesis of proteins and nucleic acids, and as enzyme cofactors, much as they are in other organisms (see Chapter 8).

Bioremediation by nutrient enrichment. Bioremediation consists of attempts to enhance biodegradation. One strategy for bioremediation is nutrient enrichment, which, in concept, is much like soil fertilization (Chapter 15). It assumes that there are already oil-degrading microorganisms present that are adapted to the local conditions. Nutrient shortages will prevent these bacteria from growing fast enough and to a great enough density (number of bacteria per liter of water) to overcome a sudden release of large quantities of oil. Supplying nutrients supports a greater growth of degradative bacteria. The concept has been tested in several nearshore spills and was extensively tested after the *Exxon Valdez* spill in March of 1989 by the Environmental Protection Agency (EPA), in conjunction with Exxon and the state of Alaska. In these tests, the rate of biodegradation on a 110-mile stretch of beach treated with fertilizers was accelerated two- to fourfold for a period of at least 30 days. A second application of fertilizer after 3 to 5 weeks accelerated the rate even more. The fertilizers were applied in ways that resulted in the nutrients being concentrated at the oil–water interface, the very area where the bacteria need to do their work. Bacterial activity appeared to be increased down to depths of at least 50 centimeters, of significance on beaches where oil has seeped into the ground. Exxon has also used nutrient enrichment to successfully accelerate cleanup of a subsurface spill in a wildlife refuge on Prall's Island in New Jersey. Algal blooms and other adverse effects did not occur.

Bioremediation by introduction of bacteria. The bacterial species indigenous to a spill area may not have the right enzymes to degrade the compounds present in the type of oil in that spill. Each bacterial species is able to make specific enzymes that can degrade specific shapes of molecules and not others. Even when the indigenous species are active, they may not be present in sufficient numbers to degrade the oil at a fast enough rate to be effective. Hence, introduction of bacteria into a spill site can be advantageous. The bacteria may be a mixture of indigenous species and species chosen for their ability to degrade the compounds present; an introduced bacterial population containing many species will be able to digest more of the oil components than would any single species. In other cases, the indigenous species may be collected, grown to great numbers in a laboratory or on-site, and then reintroduced, usually along with nutrients.

Since enzymes are proteins, the enzymes that an organism can make are those for which it has genetic material. Some people advocate using genetic engineering to make bacteria that can make a greater range of enzymes. The first bacterial species to be given a patent was one that had been engineered to degrade oil. This approach may have some feasibility in spills involving unusual oil compounds for which there are no naturally occurring biodegrading bacteria. It may also have potential for remediation of biohazardous wastes other than oil.

Although the concept of seeding of bacteria sounds plausible, we have not yet perfected the process. Two species of commercially available bacteria were seeded onto beaches polluted by the *Exxon Valdez,* but there was no significant enhancement of biodegradation compared to that on unseeded beaches. Any introduced bacteria, whether engineered or isolated from nature, would, of course, need to be genetically stable (having a low mutation rate), be nonpathogenic to humans or other animals and plants, and be unable to produce toxic metabolic intermediates from the oil compounds. All of these conditions appear to be met by the bacteria studied so far. Bacteria could potentially grow to proportions great enough to upset the ecological balance of other organisms, but they would likely die off as soon as the oil was used up. Initial experience suggests that a greater problem may be getting new bacteria to become established long enough to degrade the oil.

Risk-benefit ratios of bioremediation. Before bioremediation can be considered practical, it must be shown to be more effective and/or less hazardous than other remediation methods. Effectiveness has been shown for some nearshore spills; it has yet to be demonstrated for spills in the open ocean. It may be difficult to keep bacteria in contact with the oil for long periods of time in the ocean. It is also much more difficult to control and monitor conditions at sea than conditions nearshore, in order to conduct the experiments to establish efficacy.

Other remediation methods have their hazards, so bioremediation may compare favorably. Hot water or steam is sometimes sprayed onto beaches to disperse oil. Oil is not soluble in hot water, but the heat makes the oil less viscous (runnier) so it becomes possible for the force of the water to remove the oil from the beach. This may also wash the oil back out to sea, where it will need to be collected, separated from the water, and disposed of, or may force it deeper into a porous beach. The heat kills many organisms, although only during the period of time in which the heat is being applied. Some chemical dispersants have also been used; however, the ones used initially were detergents that were found to be toxic to the beach organisms the method was attempting to protect. Newer dispersants have been developed that are nontoxic. In contrast, nutrients for bacterial bioremediation were applied to the beaches polluted by the *Exxon Valdez* spill from small boats offshore, minimizing the mechanical damage from foot and machine traffic on the beach.

Marsh areas are especially sensitive to mechanical damage. The few studies that have been done so far suggest that marshes may be good candidates for bioremediation of oils on the surface, although the efficacy of bioremediation for cleaning oil that has seeped into the ground is limited by the low oxygen levels in marsh mud.

Future research. Bioremediation is unlikely to become a first response; when oil threatens the shore, it is preferable to prevent it from reaching the shore, if possible. Where oil is pooled or in tar blobs on beaches, bacterial action is limited; in these cases some mechanical cleanup must be done prior to bioremediation. Many important areas still need to be addressed, as the following needs list from the Office of Technology Assessment (1993:26) suggests:

Better understanding of environmental parameters governing the rate and extent of biodegradation in different habitats

Improved methods for enhancing the growth and activity of petroleum-degrading bacteria

Field validation of laboratory work

Investigation of what can be done to degrade the more recalcitrant components of petroleum, for example, certain high molecular weight hydrocarbons, such as asphaltene

Increased knowledge of the microbiology of communities of microorganisms involved in biodegradation

Better understanding of the genetics of regulation of biodegradation

Bioremediation of Wastewater

Biological solutions to engineering problems include the design and operation of wastewater treatment facilities, all of which depend at their core on the activity of microorganisms.

Wastewater.   *Wastewater* is water that has been used, for whatever purpose. Sewage is water that contains human wastes; if not treated, sewage is a major route for the spread of infectious diseases. Ingestion of fecally contaminated water spreads bacterial diseases such as shigella, typhoid fever, and cholera, and viral diseases such as hepatitis A. Swimming or wading in contaminated water spreads parasitic diseases such as schistosomiasis. Microorganisms and pollutants in fecally contaminated water are concentrated by filter-feeding animals, and so mollusks collected from polluted waters may be highly infectious or poisonous to people who eat them.

Gray water is water that has been used for things like showers or washing clothes, where it has not contacted human wastes. In virtually all homes and other buildings, gray water joins sewage in common drains. In some cities, industrial effluents and rainwater from storm drains also join sewage drains. Thus, chemical pollutants and fecal pollutants are mixed together and all water must be treated as wastewater, even though the particulate, or solid, content of the wastewater may be very low (about 0.03 percent).

Septic tanks and leach beds.   Streams and soil contain bacteria capable of biodegrading waste materials; thus, for the very low population densities of times past, natural biodegradation was sufficient. Even today, in many areas where soil is suitable and population density is not too great, the wastes from each home can be collected and adequately treated by septic tanks and leach beds (Fig. 14.15). The treated water goes back into the ground, not back into the home.

Wastewater lagoons.   On a slightly larger scale, fecal material can be separated from wastewater in a shallow lagoon in which wastewater can be contained for about 30 days. This is sufficient time for solids to settle and for sunlight, air, and microorganisms to kill off the bacteria and viruses that came from humans or animals via feces, blood, or other means, including many microorganisms that cause human disease. Wastewater lagoons are actually complex ecosystems. The growth of algae is fostered, because algae metabolize by photosynthesis and use carbon dioxide and produce oxygen in the process. Oxygen produced by the algae keeps the lagoon aerated, allowing the growth of aerobic bacteria that digest organic matter and kill fecal bacteria.

Three-stage wastewater treatment.   Any of these simple systems work well if they are not overloaded. Proper sanitation and wastewater treatment have probably played a role equal to that of antibiotics in the widespread decrease in infectious disease. A number of factors can overload these systems: excess rainwater, increased agricultural use, an increase in population densities, or an increase in water usage per person. Centralized municipal water treatment plants have become increasingly sophisticated but ultimately still rely on bacterial biodegradation. Water is treated in three stages (Box 14.3). Primary treatment consists of the mechanical separation of particles by skimming off floating material, by filtering, and by allowing heavier particles to settle in an undisturbed tank. Secondary treatment is largely bacterial digestion of organic materials. Tertiary treatment consists of fine filtration to remove microscopic bacteria and many dissolved ions.

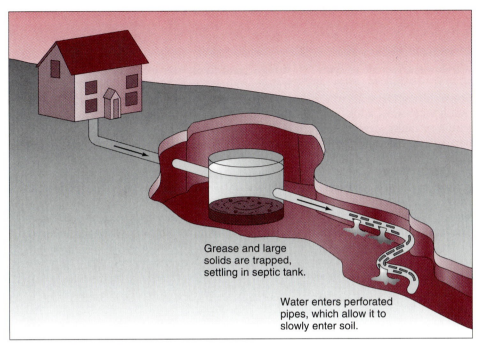

Grease and large solids are trapped, settling in septic tank.

Water enters perforated pipes, which allow it to slowly enter soil.

FIGURE 14.15
Septic tanks and leach beds. The soil minerals act as filters to trap some pollutants, while soil bacteria biodegrade materials that can be decomposed. This type of system can only be effective when the wastewater input does not exceed the biological capacity of the soil bacteria. Putting things down the drain that would kill the soil bacteria would also make the system nonfunctional.

Indicator bacteria. Tests for water purity consist of looking for the incidence of **indicator species** of bacteria. These are bacteria that are so consistently present in human feces (and absent in other sources) that their presence in water indicates contamination of that water with human wastes. They also must remain alive for as long as pathogenic bacteria might remain alive, so that when the indicator species is gone from treated water, it may be safely assumed that any pathogens that might have been present are also gone. The most common indicator species are the *fecal coliform bacteria* of the species *Escherichia coli*. Other coliform bacteria live in soil and thus are not good indicator species. Under normal circumstances, most strains of *E. coli* are not pathogenic, although they can sometimes cause diarrhea and urinary tract infections; thus the numbers of coliforms present are not usually important in themselves, but rather as an indication of the level of fecal contamination. A new and dangerous strain of *E. coli*, called O517, has caused several deaths from improperly cooked ground beef, but the outbreaks have been unrelated to drinking water supplies.

Coliforms in water samples are quantitated by the *most probable number* method after growth on special media, as illustrated in Fig. 14.16, p. 426. The acceptable *coliform count* depends on the water being tested. The U.S. Environmental Protection Agency Drinking Water Standards set the acceptable levels, as well as the minimum number of samples that must be tested. For example, treated water from secondary wastewater treatment facilities can be discharged into small bodies of water if the count is at or below 23 per 100 ml (= 23 per deciliter) and is acceptable for release into the ocean, where more dilution can occur, at counts of up to 240 per 100 ml. Water discharged from tertiary treatment facilities can be used to irrigate food crops if the coliform count is less than 2.2 per 100 ml and can be used for irrigation of landscaped or recreational areas if the count is as high as 23 per 100 ml.

Treatment by marshlands. Some towns are experimenting with using marshlands to treat wastewater. Since many natural wetlands have been destroyed by pollution or development, towns have created new wetlands. Wastewater goes through primary treatment and then is pumped into these marshes. In marshes, plants remove nitrogen and phosphorous, degrade organic

BOX 14.3   THREE-STAGE WASTEWATER TREATMENT

Primary wastewater treatment has three components: [A] skimmers to remove floating oil and grease, [B] filtration through screens to remove large debris, such as stones, sticks and rags, and [C] sedimentation tanks, much like a giant septic tank, where suspended solids are allowed to settle out, forming sludge. Settling may be enhanced by the addition of chemicals called *flocculents*, which aggregate particles, making them larger so that they settle faster.

The part that does not settle goes on to secondary wastewater treatment, in which aerobic bacteria remove up to 90% of the organic wastes by one of two methods, either activated sludge treatment or trickling filters. In activated

sludge treatment, the effluent from primary treatment goes into [D] giant aeration tanks in which air or oxygen is bubbled through it. Especially important at this stage are bacteria called *Zoogloea*, which not only biodegrade organic material in the sludge to carbon dioxide and water, but also act as a flocculent, aggregating suspended particles. [E] Trickling filters are not actually filters at all, but are rocks or plastic that provide a large surface area on which aerobic microorganisms grow, forming a gelatinous surface layer and biodegrading the organic material that passes over it. In either case, some of the remaining material settles out in another set of [F] settling tanks.

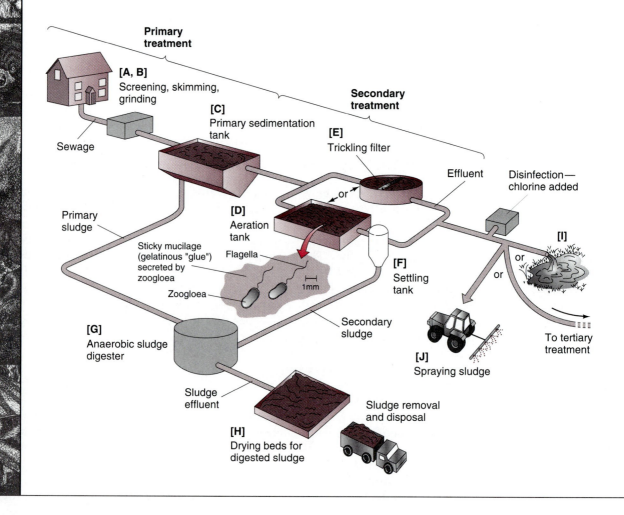

Sludge from primary and secondary treatment undergoes further digestion by [G] anaerobic bacteria. Some bacteria degrade the organic matter to organic acids and carbon dioxide. Methane-producing bacteria digest the organic acids and release methane gas (CH_4), which is used as fuel for heating the digester and running the power equipment of the treatment plant. Undigested sludge is [H] dried and incinerated or put in landfills, although it contains valuable nutrients (about one-fifth the amount in commercial lawn fertilizer) and soil conditioners, much the same as compost, and is sometimes used as such.

All municipalities in the United States must have at least primary and secondary wastewater treatment. Where this is the last step, the secondary treatment effluent is [I] disinfected with chlorine and discharged into some body of wa-

ter or [J] sprayed onto fields designated for that purpose.

In other cities, the secondary effluent goes on to tertiary treatment, where the remaining organic material is removed by [K] fine activated carbon filters. There also remain high amounts of nitrogen and phosphorous. Phosphorous, sodium, chloride, and other ions are chemically or electrically precipitated out in [L] desalination tanks, and water is chlorinated before being discharged. Nitrogen can be removed by [M] denitrifying bacteria, which convert nitrogen compounds to gaseous ammonia, which is evaporated into the air. [N] Chemical pollutants such as pesticides and herbicides are removed by specific means. Tertiary treatment, although very costly, results in water that is once again fit for drinking.

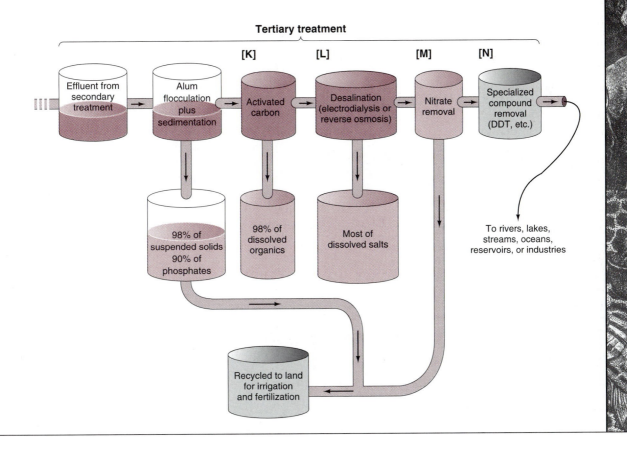

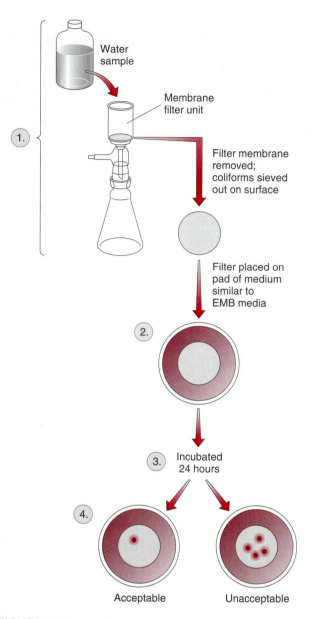

FIGURE 14.16
The most probable number method for quantitating fecal coliform bacteria in water samples. (1) A 100-ml sample of water is filtered through a filter with pores small enough to trap all the bacteria from the water. (2) The filter is incubated on top of nutrient medium called EMB or lauryl tryptose medium, selective media that only support the growth of coliform bacteria. (3) The plate is incubated 24 hours. (4) The number of bacterial colonies is then counted, and each colony is presumed to have grown from a single bacterium. The number of colonies is thus equal to the number of bacteria in 100 ml of water. Counts of 4 or greater are unacceptable in drinking water.

(sewage) wastes, and even filter out toxic chemicals. Other small-scale wastewater treatment is done in greenhouses. In tanks containing plants such as water hyacinths and cattails, algae and microorganisms decompose the organic materials, which are then used as nutrients by the plants. The water then passes to other tanks containing snails and zooplankton, which consume the algae and microorganisms; the zooplankton are then eaten by fish, and the water is further purified by marshes. Such systems work well, but will only work if the quantities of wastewater do not exceed the capacity of the treatment ecosystem. The demonstrated success of such small-scale treatment facilities serves to emphasize the important role that reduction of wastewater, and/or its separation into different catch mechanisms, could play.

Water Treatment

Flocculation and filtration.   Treated wastewater is not at this time recycled immediately into the drinking water supplies, but in most cases it could be. Municipal drinking water supplies instead are treated in separate water treatment plants. Incoming water is stored in reservoirs to allow the particulates to settle. Particles of dirt, particularly of clay, as well as bacteria and viruses that are too fine to settle, are removed by flocculation with aluminum potassium sulfate (alum), a process that was known to reduce the incidence of cholera long before it was known that cholera is caused by bacteria. Water is then passed over beds of sand or diatomaceous earth (Fig. 14.17), which adsorb microorganisms to their surfaces, and is then disinfected by chlorination. As in wastewater treatment, coliform counts are used as the test of water purity.

Filtration is a necessary step in water purification. Some viruses and the cysts of a pathogenic protozoan called *Giardia lamblia* are not killed by subsequent chlorination treatments that kill the indicator species. They are, however, removed by the flocculation and filtration processes.

Chlorination and fluoridation.   Chlorine (Cl_2) binds to the organisms it inactivates, which in turn use up the chlorine. Because chlorine is very reactive, it has some toxicity to other organisms, including humans, if its concentrations are too

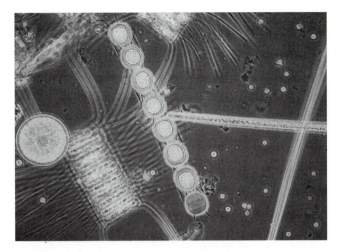

FIGURE 14.17
Diatoms are single-celled aquatic organisms. Their elaborate boxlike cases are made of silica, another example of a biological ceramic. Where these organisms have died in huge numbers, they can be mined from the ground as diatomaceous earth. Because each box is solitary, all together they have an enormous surface area. This, in conjunction with the high free surface energy of silica, gives diatomaceous earth a great capacity for adsorption; hence its use in water purification in water treatment plants and in swimming pools. It is also used as a mild abrasive in some toothpastes.

high. The risk of damage from chlorine is currently considered acceptable, on balance with the benefit (disinfection) obtained from its use. Chlorine levels must, however, be carefully monitored, to be sure that effective but nontoxic levels are maintained. Some water treatment facilities have begun to use ozone as a disinfectant, producing the ozone electrically on site. This is highly effective, although very expensive.

Many water treatment facilities also add fluoride (Fl^-) to the water. The fluoride ion can substitute for phosphate in the hydroxyapatite crystal that makes up tooth enamel. The resultant fluorapatite-containing biological ceramic is less soluble than ordinary enamel in the acidic pH produced by the bacteria that cause dental cavities. Fluoride is most effective for children during their tooth-forming years, but it has decay-preventing benefits in adults as well, because enamel is constantly being remodeled even in adults. Fluoride is also effective via other methods of application, where it is not added to municipal water supplies.

THOUGHT QUESTIONS

1. In what ways could temperature affect the efficiency of biodegradation of an oil spill?
2. Another attempt at bioremediation occurred when beaches in Spain were contaminated with oil from the tanker *Aragon*. The spraying of fertilizer on the beaches did not yield such promising results as it did in the case of the *Exxon Valdez* spill. What factors could explain why nutrient enrichment that works in one site might not work in another?
3. Since the availability of oxygen affects the rate of biodegradation, would you guess that most degradative enzymatic processes are aerobic or anaerobic?
4. In what ways is wastewater treatment efficient? In what ways is it inefficient? Would it increase efficiency to separate gray water from wastewater? Could this easily be done?

CHAPTER SUMMARY

Biologists and engineers use different vocabularies to express their ideas, and may have different philosophies regarding what are important questions to answer. Where the two disciplines have found ways to overcome these differences, the outcome has been fruitful. Study of the materials and designs that living things have developed has led engineers to new designs and composite materials. The degree of complexity in biological organisms far surpasses that in nonliving things, as living things can change and

respond to their environments in ways that nonliving things cannot. Nevertheless, engineering approaches have been valuable in describing how organisms function.

Bioengineering has resulted in the invention of devices for monitoring biological processes, for replacing damaged body parts, and for extending human abilities. Bioengineering has also suggested new ways to approach environmental problems such as wastewater treatment and oil spills.

KEY TERMS TO KNOW

adhesion (p. 406)
allometry (p. 400)
biocompatability (p. 407)
biodegradation (p. 419)

biofeedback (p. 418)
bioremediation (p. 419)
blood flow (p. 410)
blood pressure (p. 410)

composites (p. 405)
indicator species (p. 423)
osmosis (p. 414)
scaling (p. 399)

stress (p. 401)
transpiration (p. 414)
turgor (p. 414)
voltage (p. 415)

CONNECTIONS TO OTHER CHAPTERS

Chapter 3   The range of possible sizes or shapes that an organism can be and the range of its possible responses is constrained by its genome.

Chapter 3   Most implantable devices are euthenic measures.

Chapter 4   Evolution acts to make organisms conform to the sizes, shapes, and responses that would be predicted by physics.

Chapter 8   Biodegradative bacteria use oil or other pollutants as their source of nutrition.

Chapter 9   The evolution of multicellularity is a response to the physical limitations set by the surface/volume ratio.

Chapter 10   Measurement of the electrical activity of the brain can tell us about brain function and malfunction.

Chapter 12   The body's physical processes can be altered by the mind through biofeedback.

Chapter 15   Bioremediation by nutrient enrichment, like soil fertilization, supplies rate-limiting nutrients.

Chapter 16   Scaling equations predict the amount of food and space that a species will require.

Plants and Crop Production

Plants are essential to human life and to the well-being of most other organisms on Earth. The oxygen that we breathe is produced by plants. The food that we eat comes from plants, either directly or indirectly via animals that eat the plants. Without plants, most other forms of life would soon die out.

Plants are also the world's richest energy source. We all use plant energy as food. Wood is the most commonly used household fuel for much of the world's population, and fossil fuels such as coal and petroleum are the products of plants and other photosynthetic organisms of past geological ages. On a worldwide basis, plant photosynthesis (primary production) yields about 6×10^{17} kilocalories per year (abbreviated kcal/yr), or the equivalent of a sugar cube 5 kilometers (km), or 3 miles, on each side. The maximum rate of energy use by humans is estimated (on the basis of measurements in New York City) at about 10^8 kcal/yr per person. If all humans used energy at this rate, then plant life on Earth would support a maximum of 6 billion people, even if all the energy produced by plants were used by our species alone. With a 1993 world population estimated at 5.6 billion people (Chapter 6), we are rapidly outgrowing the world's chief energy resource. Population control (Chapter 6) and energy conservation (Chapter 16) are among the solutions to this problem. Another solution consists of growing more edible plants more efficiently. This requires a knowledge of how plants work to capture light energy and how we can make them work even better. Plants, in other words, are key to solving the world's energy crisis as well as the world's food shortages.

A. Plants Capture the Sun's Energy and Make Food

Plants capture radiant energy from the sun and use this energy to make energy-rich compounds, especially carbohydrates, which they store for their own use as energy sources at a later time. When humans and other animals eat plants, they harvest some of this energy.

Photosynthesis

Plants and energy. Plants are essential to other organisms as an energy source because plants can convert energy from the sun into forms of energy that other organisms can use. Organisms that can make all of their own organic (carbon-containing) molecules from simpler molecules are called **autotrophs.** Nearly all plants use light energy from the sun as their energy source, making them *phototrophs*. Because they are both autotrophs and phototrophs, we can call them photoautotrophs. Most of the species in all the other kingdoms are **heterotrophs,** organisms that cannot make their own organic compounds from inorganic materials and that are therefore absolutely dependent on the organic compounds made by plants and other autotrophs. The energy source used by most heterotrophs is chemical bond energy, making them *chemotrophs;* because they are both heterotrophs and chemotrophs, we can call them chemo-heterotrophs.

From a human-centered perspective, most autotrophs are also called *producer* organisms because they produce compounds usable by other organisms, including ourselves. Most heterotrophs are *consumers* that must obtain their energy by eating other organisms. Plants are the major source of food energy for other organisms, including both primary consumers (those that eat plants directly) and secondary and higher-order consumers that eat other consumers. When organisms die, their complex organic molecules are broken down by other heterotrophs called *decomposer* organisms; the breakdown products can then be recycled and used by other living things.

Energy enters the biological world as sunlight and flows through producer, consumer, and decomposer organisms in turn (Fig. 15.1). At each step, energy can be converted to different forms, for example, from light energy to ion gradient energy to chemical bond energy. At every energy transfer, some energy is lost by being converted into forms that are not usable by organisms. Most of this unusable energy escapes as heat. Because of the loss of heat, the process would run down and stop altogether unless new energy were continually supplied. In the living world, this new energy is

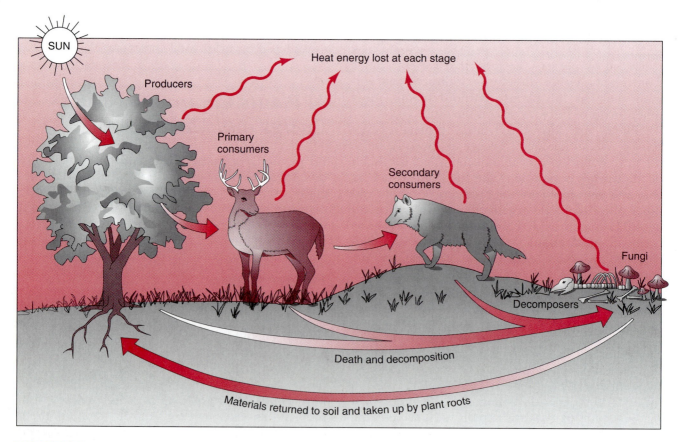

FIGURE 15.1

Energy flow through biological systems. Energy enters as sunlight. Producer organisms convert the sunlight to chemical bond energy usable by other organisms, the consumers. Energy locked in the chemical bonds of dead producers and consumers is released by decomposer organisms as heat. Energy is also lost as heat at each step, so more energy must continuously enter the system for the process to continue.

provided in the form of sunlight (Fig. 15.1). Plants are essential to the global energy flow because they are the principal means by which solar energy is captured and changed into forms that organisms can use.

Energy and pigments.   **Photosynthesis** is the process by which plants gather energy from sunlight and use this energy to make carbohydrates (e.g., sugars and starches; Chapter 8) from atmospheric CO_2 and water. This overall process of photosynthesis can be summarized by the following equation:

$$6CO_2 + 6H_2O \longrightarrow C_6H_{12}O_6 + 6O_2$$

carbon
dioxide   water   glucose   oxygen

The capture of light energy for photosynthesis takes place in certain light-sensitive molecules (pigments), of which **chlorophyll** is the most important. Because these molecules absorb some wavelengths of light and reflect others, we see the reflected wavelengths as colors. Chlorophyll absorbs blue and red light and reflects green light, which is why so much of the living world looks green (Fig. 15.2). In addition to chlorophyll, photo-

synthetic organisms (including plants) possess various other pigments, such as carotenes and xanthophylls, which absorb light of other colors and pass the energy on to chlorophyll. These plant pigments are useful to the plant because they enable the plant to take advantage of light energy of different wavelengths. Many of these *accessory pigments* occur only in certain groups of photosynthetic organisms and are used to identify these groups and reconstruct their evolution. For example, similarity between the pigments of green algae, bryophytes, and vascular plants has been used to argue that bryophytes and vascular plants probably evolved from green algae (see Appendix, p. 492).

In temperate climates, chlorophyll breaks down in the autumn and is no longer made. In the absence of chlorophyll, the other plant pigments become apparent. These pigments reflect other colors of light, resulting in the fantastic rainbow of leaf colors in the fall foliage (see Fig. 1.1, p. 4).

In all photosynthesizing organisms that have eucaryotic cell structure (Chapter 4), the photosynthetic pigments are contained in cellular organelles known as **chloroplasts** (Fig. 15.3A). Within each chloroplast are stacks of flattened membrane vesicles called *thylakoids* (Fig. 15.3B).

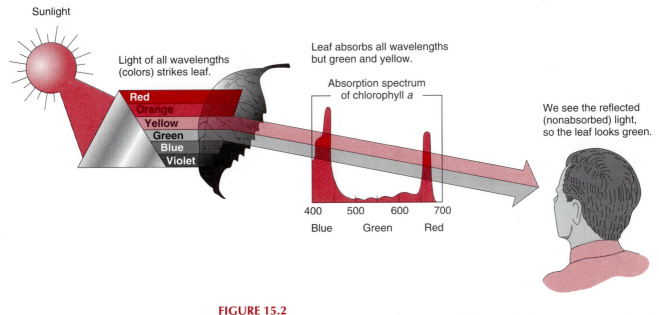

FIGURE 15.2
The colors we perceive are due to the wavelengths of visible light that are reflected from objects.

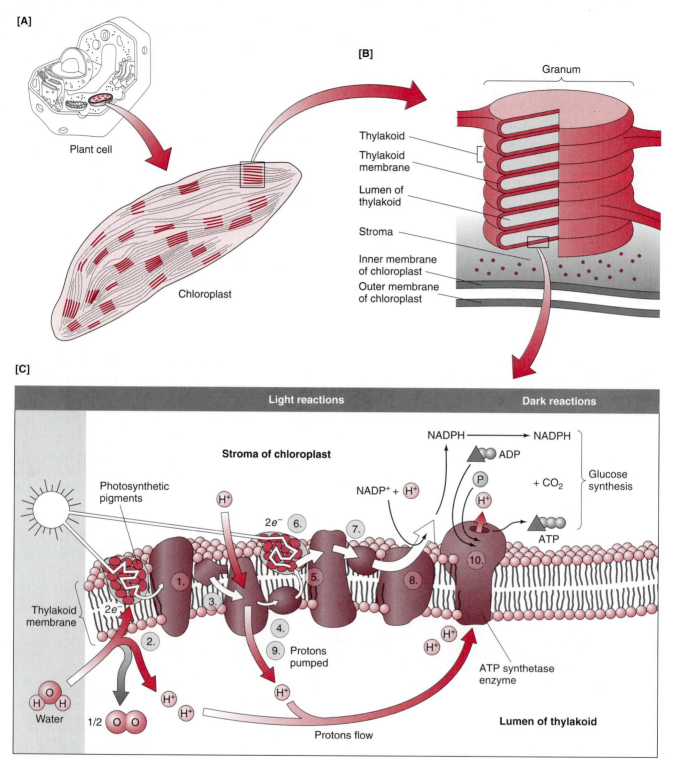

FIGURE 15.3 **Photosynthesis: location and reactions.** [A] Chloroplasts are organelles in plant cells, primarily leaves. [B] In chloroplasts are stacks of membrane vesicles called thylakoids, containing the light-absorbing pigments. [C] In the light reactions, pigment molecules absorb light energy (1), used to break water into hydrogen and oxygen (2). Oxygen diffuses away. Each hydrogen is also split into an electron (e^-) and a proton, also called a hydrogen ion (H^+). Hydrogen ions are released to one side of interior membranes of thylakoids. Electrons are passed along a chain of carrier proteins (3, 4, 5, 7, 8). Another set of light-absorbing pigments (6) is associated with carrier protein 5, and again converts light energy into energy carried by electrons. Carrier protein 3 is a proton pump; electron energy is used to move more hydrogen ions into the thylakoid interior, adding to the hydrogen ion gradient across the membranes (9). This gradient is then used to power synthesis of energy-rich molecule ATP (10). During dark reactions, energy stored in NADPH and ATP is used to synthesize glucose from CO_2.

The remainder of the chloroplast, which is outside the thylakoids, is known as the *stroma*. Certain procaryotic organisms carry out photosynthesis without chloroplasts, but we will concentrate on plant photosynthesis using chloroplasts.

Photosynthesis: light reactions. Photosynthesis takes place in stages. The first stage consists of the reactions that take place in the thylakoids; these require light energy and so are called *light reactions* (Fig. 15.3C). Light energy is captured by the photosynthetic pigments, and the captured energy is used to split water molecules into hydrogen and oxygen. The oxygen is released to the atmosphere, where it is useful to oxygen-dependent organisms, including humans. Each hydrogen molecule is further split into a hydrogen ion (also called a proton, or H^+) and an electron (e^-). The electrons move down a chain of electron carrier proteins in the thylakoid membrane and are ultimately delivered to a molecule called NADP, forming NADPH. NADP is an energy-carrying molecule and is made in part of niacin, a vitamin described in Chapter 8.

The hydrogen ions are pumped into the interior space of the thylakoids, thus forming an ion gradient. As we have seen for other ion gradients, the hydrogen ion gradient is a way of storing energy. The stored energy is then used to power the synthesis of the energy-rich molecule ATP (Fig. 15.3C). ATP is used by the cell for all its biological activities.

In summary, in the light reactions of photosynthesis, captured light energy is converted into the high-energy phosphate bonds of ATP and the bonds in NADPH. This stored, light-derived energy is later used in the dark reactions that form the remainder of photosynthesis. Some of the light energy is transformed into heat. Plants prevent overheating by evaporating water from their leaves, a process called *transpiration* (Chapter 14). Most of the water absorbed by any plant is eventually lost through transpiration.

Photosynthesis: dark reactions. The ATP and NADPH created in the thylakoids of the chloroplasts are not released. Instead, they move into the stroma of the chloroplasts, where the ATP supplies energy and the NADPH provides hydrogen for the synthesis of glucose ($C_6H_{12}O_6$) from CO_2, the source of the carbon and oxygen atoms. Because the reactions that use ATP and NADPH do not directly require light, they are called the *dark reactions* (Fig. 15.3C). The net outcome of the dark reactions is that atmospheric carbon dioxide is "fixed," or incorporated, into plant organic material, principally sugars such as glucose.

Under most conditions, glucose is immediately used as an energy source in metabolism or converted into other carbohydrates such as sucrose, fructose, or starch for long-term energy storage. If the energy is later needed by the plant at a time when photosynthesis is not possible or is in a nonphotosynthetic part of the plant, storage compounds can be converted back into glucose and broken down to supply energy. Most storage products are carbohydrates such as sucrose (table sugar), found in sugar cane and sugar beets, and starch, found in potatoes and cereal grains. In seeds, many plants (including corn, palm, and most nuts) use oils as storage products instead of or in addition to carbohydrates; the energy stored in seeds is used when the seed germinates. Growth of the new plant depends on energy from the seeds until enough new leaves are produced to carry on photosynthesis for themselves.

Tissue Specialization in Plants

In many algae and other simple plants, each cell carries out its own photosynthesis, absorbs its own nutrients, and gets rid of its own waste products. As was true for the evolution of animals, the evolution of plants on land has been characterized by an increasing specialization of parts, each for a different function (Chapter 9). Higher plants have groups of similar cells organized into **tissues** and groups of tissues organized into **organs** like leaves and roots. For example, each leaf in a higher plant is an organ, while each cell layer within the leaf constitutes a tissue. Algae are photosynthesizing organisms that lack such specialized tissues.

The simplest plants containing separate types of tissues are the mosses, liverworts, and hornworts, often grouped as Bryophyta (see Appendix, p. 491). In these and all higher plants, the diploid stages develop from fertilized eggs (zygotes) that are surrounded by sterile, nonreproductive cells, a characteristic that distinguishes such plants from algae.

Vascular versus nonvascular plants. The most important group of plants are the **vascular plants,** including all plants which have conducting tissues (vascular tissues) that transport fluids, generally through tubular cells surrounded by rigid cell walls. Vascular plants contain two types of conducting tissue: *xylem,* which usually conducts water and minerals upward, and *phloem,* which conducts a water solution of photosynthetic products (mostly sugars) in both directions but more often downward. The existence of these vascular tissues in the *stem* allows the other parts of the plant to specialize (Fig. 15.4). Underground *roots,* which also contain vascular tissue, can receive photosynthetic products from chlorophyll-containing tissues above and need not carry out photosynthesis themselves. The roots also anchor the plant in the soil, and the vascular tissues conduct water and dissolved nutrients from the roots to the aboveground parts, where the water is needed for both photosynthesis and transpiration. Vascular tissues have rigid cell walls which provide the support that allows many vascular plants to grow tall. We build most of our houses out of lumber made from these sturdy vascular tissues.

Nonvascular plants such as mosses and liverworts never have true roots or deep underground parts because all parts of the plant carry out photosynthesis. They can never grow very tall because they lack vascular tissue that would conduct fluids and because they lack the roots and stems that provide anchorage and support in vascular plants.

Photosynthesis in most nonvascular plants takes place throughout the plant. In most vascular plants, however, photosynthesis takes place only in specialized organs called *leaves.* The middle tissue layer of each leaf features many chloroplast-rich cells and air spaces that permit the diffusion of CO_2 into the photosynthesizing cells and O_2 out of them (Fig. 15.4).

Flowering plants. The most highly evolved vascular plants reproduce with the aid of *seeds,* which are small diploid embryos, surrounded by protective layers and capable of being dispersed to new locations away from the parent plant. Seed plants also contain true, branching roots and leaves with multiple veins. The largest and most diverse, as well as ecologically dominant, group of seed plants are the *angiosperms* or flowering plants (An-

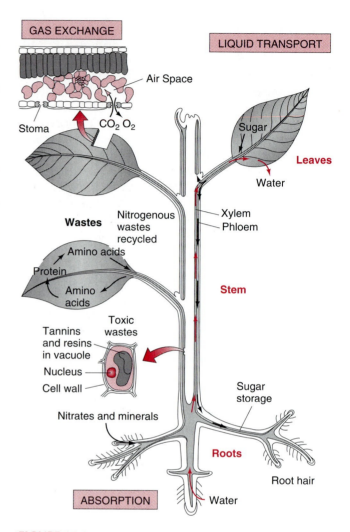

FIGURE 15.4

The parts of vascular plants are specialized for different functions. Leaves carry out photosynthesis and gas exchange; roots absorb water and dissolved minerals; vascular bundles in the stems connecting these parts distribute materials from place to place throughout the plant.

giospermae or Anthophyta). Seeds in this largest group of plants develop within elaborate reproductive structures called *flowers* (Fig. 15.5). An egg with a haploid set of chromosomes is produced by the female part of the flower. The haploid structures produced by the male part of the same flower (or a different flower or plant of the same species) are called *pollen.* Pollination is the introduction of the pollen onto the female structures of the flower (Fig. 15.5). A pollen tube grows from the pollen grain to the ovary where fertilization occurs, resulting in a diploid zygote. Many flowers are

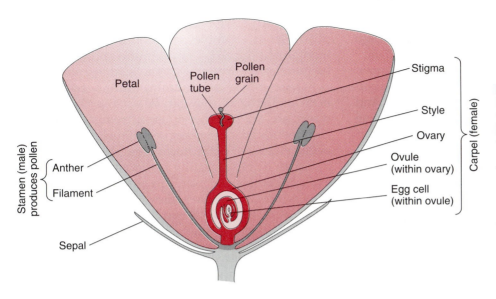

FIGURE 15.5
Diagrammatic structure of a "complete" flower, containing both male and female parts together. After fertilization, the ripened ovary or ovaries will form a fruit. Variations in the number and structure of the parts shown here are among the most useful characters in plant classification. Some plant species have separate male flowers with undeveloped female parts and female flowers with undeveloped male parts.

pollinated by wind, but a much larger number are pollinated by insects. The relations between flowering plants and the insects that pollinate them are often quite elaborate and are the key to much of the diversity and also the evolutionary success of both flowering plants and insects (Chapter 16).

After fertilization, the structures surrounding the zygote mature into a seed. The seed-bearing structures in a flower ripen into *fruits.* In addition to the seeds themselves, fruits often contain tissues that attract various animals by means of conspicuous colors, fragrant odors, carbohydrate nutrients, or a combination of these. Animals that eat the fruits disperse the seeds, often carrying them far from the parent plant. Among the fruits that humans eat are many that we commonly recognize as fruits (e.g., apples, peaches, melons) and others that we do not usually regard as fruits (e.g., nuts, grains, cucumbers, tomatoes, peppers, eggplant).

THOUGHT QUESTIONS

1. What is the source of the energy that becomes ATP in the light reactions of photosynthesis? Is this the same or different from the source of energy for ATP production in mitochondria (Chapter 8)? Do plants have mitochondria?
2. Do animals have any molecules that can absorb light energy? Where might you expect to find them?

B. Plants Make Many Other Useful Products

Plant Products of Use to Humans

Humans and other animals find many ways of benefiting from the use of plant parts. Most importantly, plant products are eaten as foods, that is, as sources of energy and other important nutrients (see Chapter 8). Many of these foods are the carbohydrates (and occasionally oils) that plants have stored for their own use as energy sources at a later time. Some human and animal foods are the fruits by which plant seeds are dispersed, as described earlier.

Although the principal use of plant products is as food, people also use plants as sources of beverages, medicines, flavorings, fragrances, dyes, poisons, or simply as decoration. Beverages include coffee, tea, cola, an assortment of fermented beverages (beer and wine among them), and an even greater assortment of juices. Many plant parts are used as spices, fragrances, or flavorings, from barks like cinnamon and roots like ginger to flowers and flower parts like cloves and vanilla. Often, an essential oil or other ingredient is squeezed or extracted from the appropriate plant part and used in concentrated form. Some of these extracts are used as fragrances or food ingredients; others are used as poisons to help capture and subdue large animals. For example, roots containing rotenone are used by native South Americans to help capture fish by temporarily paralyzing them; rotenone is also used as a pesticide. Manioc, a tropical root, is used as the source of both an arrow poison and tapioca, a food that also serves as a thickening agent.

Wood is another commercially important plant product used the world over as a fuel and as a building material in the form of lumber. The history of human civilization would have been very different without spears, axes, hoes, boats, and many other objects made predominantly of wood. Paper and paper products are also made largely from wood.

One of the most important uses of plant products is in the field of medicine. Our modern arsenal of medicines is still derived largely from plant products, many of them tropical: digitalis and other heart medications from the purple foxglove (*Digitalis purpurea*) and related species, the blood-pressure-lowering drug reserpine from the Indian snakeroot (*Rauwolfia serpentina*), the neurotransmitter blocking drugs atropine and belladonna from the nightshade plant (*Atropa belladonna*), the sedative scopolamine from the Jimson weed or thorn apple (several species of *Datura*), codeine (a cough suppressant) and morphine (a painkiller) from opium poppies (*Papaver somniferum*), the antimalarial drug quinine from the bark of the cinchona tree (*Cinchona ledgeriana*), and many others. Just in this decade, two new anticancer drugs, vinblastine and vincristine, were discovered in the rose periwinkle (*Catharanthus roseus*), a rainforest plant native to Madagascar, and another anticancer agent, taxol, was isolated from the Pacific yew (*Taxus brevifolia*).

Nitrogen Requirements

How plants use nitrogen.   The carbohydrates produced by photosynthesis contain carbon, hydrogen, and oxygen only. Many other plant products contain other elements, notably nitrogen. Because proteins are made from amino acids containing nitrogen, the synthesis of proteins requires significant quantities of nitrogen. The simplest amino acids can be made from compounds in the Krebs cycle (Chapter 8) by the addition of an amino group (NH_2). The amino group can be supplied from soluble ammonium compounds (containing the NH_4^+ ion). Amino groups can also be transferred from one amino acid to another.

Once nitrogen-containing amino acids have been synthesized, they can be used as the starting materials for all the other nitrogen-containing compounds that the plant needs, including nucleic acids, vitamins, plant hormones, chlorophyll, and a variety of bitter-tasting compounds called *alkaloids*. Alkaloids are useful to plants because, even in very small amounts, they discourage animals from eating the plants, both because of their bitter taste and because they often mimic or interfere with the functions of neurotransmitters in these plant predators (Chapter 10). One hypothesis for the evolution of alkaloids is that, after they formed as waste products or as by-products of other processes, they conferred a selective advantage upon the plants that possessed them in comparison to those that did not. Over many generations, plants with alkaloids survived in greater numbers because similar plants without alkaloids or with lower alkaloid levels were more often eaten by plant-eating animals (*herbivores*). A similar type of selection brought about the evolution of alkaloids of increasing bitterness or toxicity, or alkaloids that repelled a wider variety of potential herbivore species.

Nitrate: a limiting nutrient.   Plants need several chemical elements in order to make the biological molecules they need. Carbon can be obtained by photosynthetic plants from atmospheric carbon dioxide, and water can serve as the source for hydrogen and oxygen as well as dissolved ions. The source of nitrogen, however, is a bit more complex.

Most plants get their nitrogen from the soil as dissolved nitrates (NO_3^- ions), which they then convert first into nitrites (NO_2^- ions) and then into ammonia (NH_3) as it is needed for amino acid synthesis. A plant does not accumulate any more ammonia than it immediately needs, because excess ammonia reacts with water to form ammonium hydroxide (NH_4OH), which is strongly damaging to most organic tissues.

Many plant species are limited in their occurrence by the availability of dissolved nitrates. A nutrient whose absence makes a species unable to grow in a particular place is called a **limiting nu-**

trient. If the amount of the limiting nutrient were increased (by the use of fertilizers, for example), more growth would take place, assuming that other nutrients were present in nonlimiting quantities. For many plants in many places, nitrates are a limiting nutrient. This is why nitrates are among the most important fertilizers used in agriculture. In many places, crop yields can be dramatically increased by adding nitrates to the soil.

The nitrogen cycle. Nitrogen moves through the world's ecosystems, changing from one molecular form to another, forming a loop called the *nitrogen cycle* (Fig. 15.6). Plants and other living organisms are an important part of this cycle: Chemical end products released by one organism are used in biochemical reactions by other organisms. Some of these organisms can incorporate (fix) atmospheric nitrogen into molecular forms that other organisms can use, while other organisms release nitrogen as a gas to the atmosphere. In each ecosystem, living organisms are thus united with each other and with their physical surroundings (including the atmosphere) by the nitrogen cycle.

The nitrogen cycle is a good example of a *nutrient cycle;* all such cycles are closed loops where materials such as nitrogen are exchanged among producers, consumers, decomposers, and their physical surroundings, including the atmosphere. The materials never leave the ecosystem. This

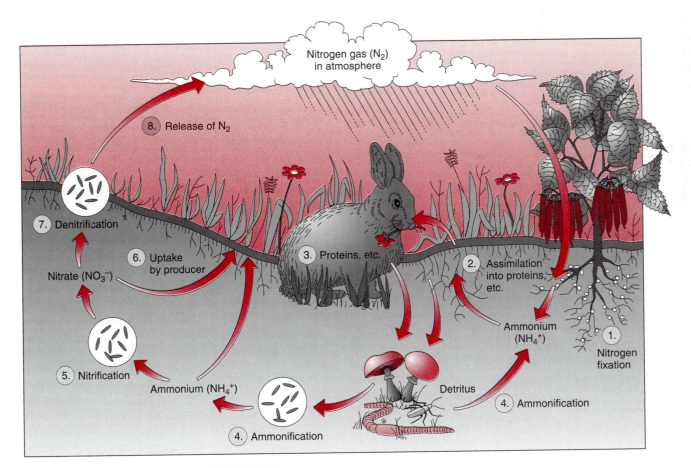

FIGURE 15.6

The nitrogen cycle. Nitrogen is abundant in the atmosphere. It is fixed into biologically usable forms (1), assimilated by plants and other producers (2, 6), and made into proteins and other compounds that are used by consumer organisms (3). Nitrogen-containing compounds are then broken down by decomposer organisms (4, 5, 7) and returned to the atmosphere (8). The organisms and the atmosphere form a closed ecosystem.

contrasts with the flow of energy in the same systems, which is "open" in the sense that heat energy is always lost and sunlight must be constantly supplied to keep the cycle going (Fig. 15.1).

The most abundant source of nitrogen is atmospheric nitrogen gas (N_2). Plants need nitrogen but, like other eucaryotic organisms, they are unable to make the enzymes necessary to use nitrogen from the atmosphere. They are therefore dependent upon certain procaryotic microorganisms that have the ability to fix atmospheric nitrogen into ammonia and its compounds.

Most nitrogen-fixing organisms are blue-green bacteria (Cyanobacteria) such as *Nostoc*. In addition, several kinds of heterotrophic nitrogen-fixing bacteria, such as *Rhizobium, Azotobacter, Klebsiella,* and *Clostridia,* can incorporate atmospheric nitrogen into ammonia (NH_3) or ammonium compounds (Fig. 15.6, step 1). Ammonia is the form of nitrogen used in the synthesis of amino acids, which are assimilated into proteins (Fig. 15.6, steps 2 and 3). Ammonia can also be freed from organic material in the soil by decomposer organisms called *ammonifying bacteria* (Fig. 15.6, step 4).

Most plants, however, cannot take up ammonia; they can only take up nitrogen as nitrite (NO_2^-) ions or as nitrate (NO_3^-) ions. Ammonium compounds produced by nitrogen-fixing or ammonifying bacteria in the soil can be converted into nitrites by still other bacteria, such as *Nitrosomonas,* which obtain their energy from this conversion process. The nitrites are then converted into nitrates by *Nitrobacter,* another bacterium (Fig. 15.6, step 5). These bacteria are *autotrophs,* as are plants, but, unlike plants, they derive their energy from energy-rich inorganic chemical bonds rather than from sunlight. (Thus they are are a subtype of autotroph called *chemoautotrophs,* while photosynthetic plants and bacteria are a subtype called *photoautotrophs.*) Plants can absorb both nitrates and nitrites (Fig. 15.6, step 6) and convert them into NH_3 to be used in synthesis of proteins and nucleic acids. Certain bacteria can also convert NO_3^- into nitrogen (N_2), which they release into the atmosphere (Fig. 15.6, steps 7, 8).

Adaptations by which plants obtain nitrogen.
Plants have evolved a variety of ways to obtain their nitrogen. Most plants simply absorb nitrogen compounds (mostly nitrates) from the soil. Nitrates are often a limiting nutrient to such plants, meaning that they can only grow in soils in which nitrate supplies are adequate.

Plants that grow in nitrogen-poor soils need to have other ways of coping. Some plants harbor nitrogen-fixing microorganisms that can fix atmospheric nitrogen. Among these are the plants that ensure the availability of nitrogen-fixing microorganisms by growing *root nodules* not far below the soil surface. These root nodules actively attract the growth of nearby soil bacteria, chiefly *Rhizobium,* some of which are capable of fixing atmospheric nitrogen to the benefit of the plant (Fig. 15.7). Nitrogen-fixing bacteria in root nodules also solve the ammonium uptake problem because they deliver ammonia directly to the vascular tissues of the plant. The root nodule serves as a culture chamber for these microorganisms. The nitrogen-fixing reactions use a great deal of energy, which the plants supply to the bacteria.

The relationship between the plant and the bacteria is called **mutualism,** an interaction in which both species are benefited. Mutualism is a type of **symbiosis,** a close association between organisms of two species. In many cases, symbiotic organisms are intimately associated; some have become so permanently interdependent that neither species can complete its life cycle without the other.

Plants of the family Leguminosae, including beans, peanuts, peas, locusts, and alfalfa, are all capable of adding nitrogen compounds to the soil because of their symbiotic association with nitrogen-fixing *Rhizobium* bacteria. This is the basis for the practice of *crop rotation,* in which a field that has been depleted of nitrates for several years (by the growing of plants that absorb large quantities of nitrogen compounds from the soil) is planted with one of these legumes for a year. When the legume is plowed under, the nitrogen compounds that it and its symbiotic bacteria have produced are returned to the soil in biologically available form, reducing or eliminating the need for the addition of nitrate-containing fertilizers.

Rice plants growing in paddies obtain their nitrogen in a different way, which includes a more complex symbiosis that does not use root nodules. In these paddies, nitrogen-fixing blue-green cyanobacteria live on water ferns and release ni-

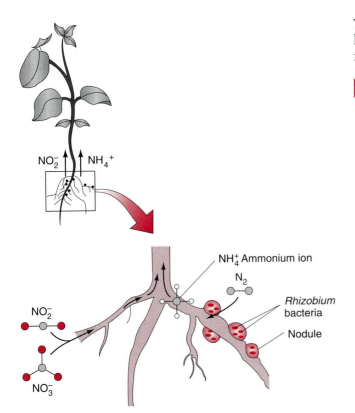

FIGURE 15.7
Nitrogen fixation by symbiotic *Rhizobium* bacteria. The photograph shows root nodules that contain *Rhizobium*.

trogen compounds to the surrounding water. The nitrogen compounds contribute to the growth of the rice.

One of the most unusual ways in which plants can obtain nitrogen is by eating animals. Most carnivorous plants live in nitrogen-poor habitats. They have evolved various adaptations to trap and kill small animals (mostly insects) and derive the nitrogen from the digested proteins (Box 15.1).

THOUGHT QUESTIONS

1. Consider the value judgments that may be hidden in a term such as *useful* (as in "useful product"). Are plants that are useful to humans more valuable than plants that are not? Is utility the only criterion for making something valuable? Is economic value the only way of measuring value? What might some of the other ways be? How would you measure them?
2. What are some ways in which large plants (e.g., trees) are used by other plants? What are some ways in which plants are used by animals? List as many possible uses as you can.

C. Crop Yields Can Be Increased by Overcoming Various Limiting Factors

Of the almost 250,000 species of plants known, some 80,000 are edible by humans. Of these, however, only about 30 form the major crop plants. The cultivation of plants as food for humans was probably *the* critical step in the development of *agriculture* (Chapter 4), an achievement that led also to the establishment of human civilization. Most important in their influence on the development of civilization were the cereal grains (wheat, rice, corn, oats, and others). Today, wheat, rice, corn, and potatoes provide more of the world's food than all other crops combined.

Hunger, malnutrition, and starvation are endemic in many parts of the world, and there are pockets of poverty in even the wealthiest countries (Chapter 8). Rapid increases in the world's population (Chapter 6) have intensified these problems. One approach to addressing these problems is to increase crop production, but there are very few paths available for doing this. One is to increase the amount of land under cultivation; this approach has its geographic limits as well as very high biological costs in terms of the destruction or displacement of natural ecosystems (see Chapter 16). Another possibility is to increase the yields or

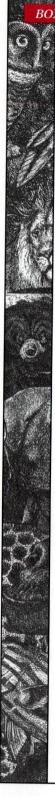

BOX 15.1   CARNIVOROUS PLANTS

Plants that have evolved to live in nitrogen-poor soils have a number of different ways of coping. Some plants can absorb the ammonium ions (NH_4^+) contained in decaying leaf litter, but they do so by exchanging the NH_4^+ ions for hydrogen ions (H^+), thus making the soil more acidic. Microbial decay of leaves releases even more acid, eventually killing all those plant and microbial species that are not *acid-tolerant*. The bacteria that convert ammonium to nitrites and then to nitrates are not acid-tolerant and thus are inhibited in this environment. The habitat that results is an *acid bog* with high organic content but few nitrates.

One rather unusual solution to the problem of obtaining nitrogen in an acid bog is to digest animal proteins. Few plants, however, can be so fortunate as to have an animal die and leave its carcass in the soil just within reach of their root system. How, then, to obtain animal protein? *Carnivorous plants* obtain their nitrogen from attracting and catching insects and other small animals and digesting their proteins. Most carnivorous plants live in such nitrogen-poor habitats as acid bogs, where moisture and insects are both usually abundant.

A variety of mechanisms have evolved in a number of plants to trap the insects, digest their proteins, and absorb the resulting amino acids; one such evolutionary path, for the pitcher plant, is shown here. The insect traps can either be passive or active. Passive mechanisms can include pitfalls (as in pitcher plants), sticky materials that form a passive flypaper, and traps whose funnel-shaped entrances are (like lobster traps) easy to enter but very hard to escape from. Active mechanisms can include trapdoors, active flypapers (such as the sundew, whose hairs bend to enclose and further hold their victim), and the unique closing mechanism of the Venus flytrap.

PASSIVE TRAPS

The tropical pitcher plant, *Nepenthes*

The pitcher plant, *Sarracenia*. Insects, attracted to this plant by its odor, become trapped when they fall inside, where enzymes can break down their proteins.

Steps by which pitcher plants may have evolved

Sarracenia psittacium, a pitcher plant with one trap inside another. The entrance works like a lobster trap because insects entering the chamber have difficulty finding the opening by which they entered. When they crawl down the long tube, the hairs inside form a second trap that allows them to crawl deeper in only one direction.

External view

Cutaway view

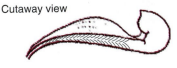

(box continues)

BOX 15.1   (continued)

ACTIVE TRAPS

The sundew plant, *Drosera*, which forms an active flypaper whose sticky hairs bend to enclose and further trap their victim

A Venus flytrap, *Dionaea*

A sundew plant closing over a trapped insect

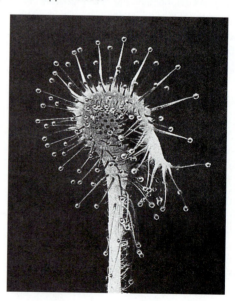

the nutrient content of crop species, through such techniques as soil improvement, using fertilizers and advanced irrigation techniques; pest control; and the use of different plants or new genetic strains of plants. Each of these methods seeks to increase yield in a different way: by supplying a limiting nutrient; by supplying water (which is often limiting in the same way); by controlling natural enemies of the crop species; or by developing new crop plants, such as drought-resistant strains, pest-resistant strains, or strains with diminished nutrient requirements capable of growth on marginal or poor soils. In any event, increased crop yields are not likely to solve all problems related to human hunger and malnutrition. Many experts believe that present supplies of food would be adequate to feed all the world's populations (and perhaps more), if only the food were more equitably distributed. The unequal distribution of food supplies involves political and economic questions that go far beyond biology.

Fertilizers

In those locations where nitrogen or some other nutrient is a limiting factor for the growth of plants, crop yields can often be dramatically increased by adding the appropriate fertilizer to the soil. A **fertilizer** is a substance that supplies organic or inorganic nutrients needed by plants. In the largest number of locations, nitrogen is the limiting nutrient for many plant species, but phosphorus (P), in the form of phosphates, can also be an important limiting nutrient, especially in the case of ornamental flowers or crops whose edible portion is a seed, flower, or fruit. Phosphorus often signals the plant to begin setting flowers, seeds, and fruit. Potassium (K^+) is often important as well. It is used by plants to control the opening and closing of stomates, and it may be scarce (and therefore limiting) in certain soils.

Organic and inorganic fertilizers. Fertilizers can be either organic, including various manures or composts (see p. 445), or inorganic. All organic fertilizers contain their nutrients in the form of complex molecules that break down slowly. The slow breakdown gives these fertilizers the advantage of *slow release*, a process which provides a steady supply of soluble nutrients that is more likely to be absorbed by growing plants than to be washed away in the manner of many inorganic fertilizers. One disadvantage of organic fertilizers is their high bulk and cost of transportation, a feature that often limits their use to the immediate locality of their production. In China, human and animal wastes have been used as fertilizers for centuries. This practice provides needed nutrients to crops, but it also has the undesirable effect of spreading parasites (especially flukes) and other infectious diseases.

Crop rotation. Crop rotation, described earlier in this chapter, provides organic fertilizers without need of transportation whenever plants that accumulate nitrogen are plowed under. The alternation of cereal grains with soybeans or alfalfa is a form of crop rotation widely practiced in many large regions of North America and Asia. Farmers using organic practices employ crop rotation, sometimes in combination with the use of various animal and plant wastes as fertilizer. The benefit of crop rotation for maintaining the fertility of the soil was recognized in ancient Rome.

Phosphorous sources. Nutrients are also slowly released from fish meals and bone meals, which are dried, powdered bone. These fertilizers supply phosphorus in the form of the mineral hydroxyapatite, an inorganic crystal that contains both calcium and phosphates. The phosphates provided by these fertilizers are insoluble; they are converted to soluble form only slowly, often by the very plants that use them. (The slow solubility is a disadvantage at first, especially if the practice is begun in soil that is nutritionally deficient; chemical fertilizers may be necessary during the initial phase-in period.) Fish or fish meals also contain additional phosphorous in the form of nucleic acids from the nuclei and organic phospholipids of the membranes of the cells of the dead animal. Native Americans throughout the eastern woodlands traditionally grew their corn in mounds, under which they customarily buried a fish that provided both phosphate and nitrogen as fertilizer for the corn.

Chemical fertilizers. Inorganic chemical fertilizers usually take the form of nitrate or phosphate salts sold as powders or granules to be applied to fields, usually by mechanical equipment. When they were first introduced, chemical fertilizers were cheap and readily available. The benefits to crop production were immediately evident, and, during the twentieth century, the use of chemical fertilizers increased greatly in countries that could afford them, principally North America, Europe, and Australia.

The problems associated with the prolonged use of inorganic chemical fertilizers did not become evident until they had been in general use for several decades, in accordance with what some call the *law of unintended consequences.* This has also been stated as the first law of ecology: You can never change just one thing.

Excessive use of inorganic fertilizers can kill off the soil organisms required for maintaining soil fertility, thus requiring the use of even more fertilizer for subsequent crops. Compared to organic fertilizers, inorganic minerals are relatively expensive to produce or extract, and most of them must be transported over great distances. Cheap supplies that formerly existed near places in which fertilizers were in demand have in most cases been depleted, and many are now mined far away from the places where they are used. For example, large quantities of nitrate minerals are mined in Chile for use as fertilizers in countries in the northern hemisphere thousands of miles from Chile. Fossil fuels are used in most of the mining and transportation industries as well as by the tractors that are typically necessary to spread these fertilizers on fields. Most chemical fertilizers are costly to transport to where they are needed, and in most of the developing nations of the world they are prohibitively expensive. (Organic fertilizers can also be expensive to transport on a per-mile basis, but they are more often locally available.)

Runoff.   Additional costs associated with the use of fertilizers include the problems of runoff and stream pollution. Runoff of fertilizers from agricultural fields causes excessive growth of algae (algal blooms) in many freshwater ponds and streams. Fertilizers are concentrated nutrient sources and are not solely nutrients for the crop plants for which they are intended. When fertilizers enter bodies of water, they supply nutrients to algae in the water. The growth of algae had previously been held in check by the lack of some limiting nutrient. Once that nutrient is present, the growth of these algae can be so rapid that they use up all the oxygen in the water, causing the death of many fish. The decomposition of dead fish releases more nutrients, and this further accelerates the growth of the algae.

Eutrophication.   Even when algal growth takes place more slowly, it can cause the filling-in of a pond or lake, a process called **eutrophication.** Eutrophication is a type of **ecological succession,** a situation where one organic community sets up the proper environmental conditions for a group of organisms that could not have lived in the original environment. The new group of organisms becomes established and eventually replaces the original community with another community of species that is suited to the changed conditions. This succession may then repeat several times, until a *climax community* is reached in which no further change occurs for long periods. In the case of eutrophication, the algae make the habitat suitable for an intermediate community of floating vegetation such as *Sphagnum* moss. These create a quaking bog, which in turn serves as a habitat for moisture-loving shrubs and eventually for trees (Fig. 15.8). The climax community differs from place to place, but is commonly some kind of forest. This type of ecological succession occurs naturally, particularly in small lakes or ponds that do not have a significant flow-through of water, but it is greatly accelerated by runoff that contains fertilizers.

The problems of runoff and eutrophication can originate from animal manures (e.g., from dairy farms) or from the use of inorganic chemical fertilizers, especially those applied to the soil in soluble form, as most nitrates are. The amount of rainfall and its seasonal distribution can greatly influence runoff problems. Runoff problems can sometimes be minimized by applying fertilizers at the proper time of year, just before the nutrients are most needed by the plants and are most likely to be absorbed.

Soil Improvement and Conservation

Soil is the loose material derived from weathered rocks and supplemented with organic material from decaying biological organisms. This organic material supports the growth of plants as well as

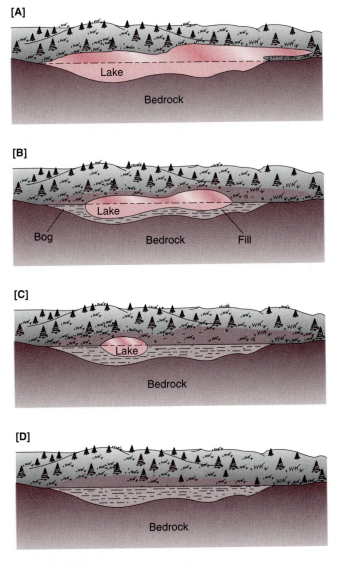

FIGURE 15.8
Eutrophication: an ecological succession that can be accelerated by fertilizer runoff. The lake [A] shrinks as it is filled in with decomposing material, initially as a bog [B, C] and finally as forest [D].

many bacterial, protist, and animal species. Soil is thus the product of both biological and geological processes (Fig. 15.9).

Humus.   Organic material in the soil is broken down by bacteria and fungi, which function as decomposers. The partially decomposed organic matter is called *humus*, and it serves several important functions in the soil. It binds the particles of weathered rock together to form small aggregates, which give the soil its structure. The small holes between particles help hold water and oxygen and provide space for the growth of minute root hairs, extensions of single cells on the roots of plants. While the larger roots provide the anchorage for the plant and in some plants can reach deeply into the ground in search of water, it is the root hairs that provide the enormous surface area for absorption of nutrients, just as the microvilli of the intestine provide the surface area for absorption of nutrients from mammalian guts (Chapter 8). Because root hairs are so delicate, they cannot push into soil, but must grow into already existing spaces in the soil structure.

Humus has an overall negative charge, which helps hold positively charged nutrient ions such as potassium (K^+), calcium (Ca^+), and ammonium (NH_4^+) in the topsoil, where they are available to plants' roots. Inorganic fertilizers can provide chemical nutrients, but they do not contribute to humus and soil texture as organic fertilizers can. Plowing under organic matter helps build humus. Fields can be planted with grasses rather than being left bare for the winter. This both prevents erosion and adds to humus when the grass is plowed under in the spring.

Humus can be produced by *composting*. Organic material (leaves, grass clippings, food waste) is layered with manure or soil. In a matter of months the decomposer bacteria from the soil turn the organic matter into humus. Currently less than 1 percent of organic wastes are composted in the United States, while much more composting is done in Europe. Many U.S. municipalities encourage composting, either by teaching people how to do it in their own backyards or by having centralized composting facilities where people can bring their yard wastes. It is estimated that solid waste in U.S. landfills could be reduced by 20 percent if organic material were composted.

A nonrenewable resource.   Topsoil is eroding faster than it is being formed on approximately one-third of the world's croplands. Soil that is lost can be regenerated by biological processes, but only very slowly—1 inch of topsoil takes between 200 and 1000 years to renew, depending on the climate and other factors. Groundwater pollution, excess fertilizer, and/or runoff from highway salt,

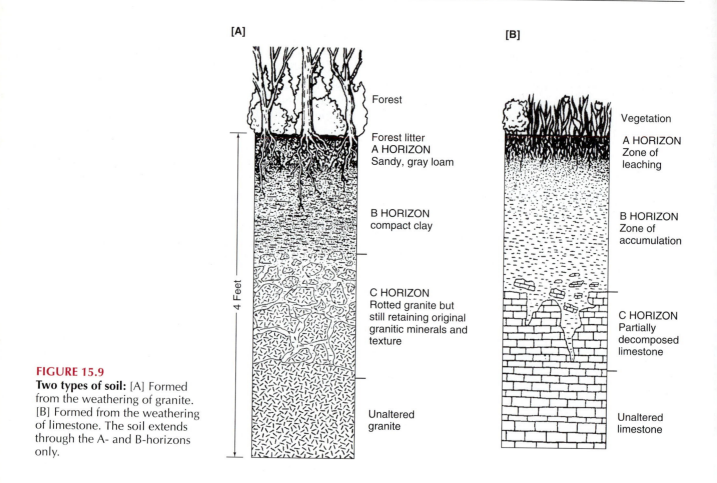

FIGURE 15.9
Two types of soil: [A] Formed from the weathering of granite. [B] Formed from the weathering of limestone. The soil extends through the A- and B-horizons only.

can kill the soil microorganisms and stop the production of topsoil. Because soil cannot be regenerated within a time period that is relevant within a human lifetime, it must be treated in the same way as a nonrenewable resource. In view of soil as a nonrenewable resource, farming and other land uses are best done in ways that are *sustainable*. Sustainable practices are those that lead to no net loss of a resource. Important practices in sustainable agriculture include crop rotation, use of organic, humus-building fertilizers, and the prevention of erosion.

Irrigation and Hydroponics

The **irrigation** of crops is the process of adding water, a vital nutrient for the growth of plants. As we have seen, plants need water to supply the hydrogen used in photosynthesis. They also need water for transport of all their other essential

nutrients, as well as the products of their synthesis reactions. About 90 percent of water absorbed by a plant is evaporated through its leaves during transpiration, dissipating excess heat. Also, the biophysical strength of plants is based partly on the properties of water (Chapter 14).

Irrigation is expensive in regions where water is scarce and where irrigation is therefore most needed. Traditional methods of spray irrigation lose much of the water to evaporation. Newer methods include *drip irrigation*, in which irrigation tubes (pipes) with tiny holes are laid at or below ground level. The tiny holes, spaced every few inches, are designed to leak or drip water slowly into the soil, providing moisture but minimizing evaporation. All irrigation systems, whether drip or traditional, require a large initial investment in pipes or ditches, pumping stations, and the like, plus a supply of fresh or desalinized water (seawater from which the salts have been removed).

Freshwater sources can become the subject of political disputes between neighboring governments, such as those between Israel and Jordan or between California and Arizona.

Hydroponics

Although plants could not live without water, they can live without soil. Some plants, including epiphytes (Chapter 16), grow naturally without soil. The growing of crop plants without soil is called **hydroponics.** In places where there is very little soil, or where the soil is unsuitable, hydroponics offers a possible alternative agricultural method. In a hydroponic system, the plants are grown with their roots immersed in tanks through which water carrying dissolved nutrients is allowed to flow. The water is recirculated (and therefore conserved), and its dissolved materials are frequently monitored and adjusted.

Advantages of hydroponics. A well-managed hydroponic system can produce yields in excess of more traditional soil-based systems. Since these systems are generally inside greenhouses, the produce that results is free of blemishes. Other advantages include the elimination of much of the labor that was traditionally concentrated on soil care, including tilling, planting, fumigation, and irrigation. Water is provided to plants directly and more efficiently, and is recycled so as to minimize its use. This is extremely important in arid climates, where sunshine may abound but where water is scarce. Instead of traditional fertilizers, which can diffuse beyond the reach of plant roots, mineral nutrients are added to water directly, where the excess not taken up by the plants remains available and can be recycled. This way, use of nutrient supplements is minimized, and runoff problems are also minimized. Disease and pest control can be handled easily by adding the necessary chemicals to the water directly in soluble form, with far less danger of contamination spreading to local water supplies or domestic animals and humans.

Disadvantages of hydroponics. Many of the disadvantages of a hydroponic system involve cost: the initial construction cost, which is high; the high cost of maintaining equipment and greenhouse facilities; the costs of nutrients; and the salaries of highly trained personnel for constant monitoring to prevent nutritional problems (such as nutrient depletion) and waterborne diseases. Hydroponic systems may be in locations that are distant from the Equator, where the provision of artificial light and heat becomes a cost factor.

Plants are subject to disease, as are all living organisms; waterborne diseases are those in which the disease-causing organism (pathogen) can live in water. Even if the pathogen does not grow well in water, its ability to survive in water will enable it to be transmitted from plant to plant via this route. Because most hydroponic systems concentrate on a single plant species (a practice that is known as **monoculture;** see the section below on monoculture and pests), diseases can spread rapidly throughout the colony. Disease organisms, whether they are involved in human disease or in plant disease, are very often specific for one or a few host species. When a pest or a pathogen encounters a monoculture, it can spread very quickly because of the proximity of other individuals of the same host species.

For these reasons, hydroponics is easier to justify economically for plants that are profitable in small quantities (e.g., pharmaceutical plants or certain vegetables) rather than for staple crops such as cereal grains. Hydroponic methods are potentially useful where either growing space or soil are limited, as in proposed space stations. On Earth, hydroponics works best in warm climates such as Israel or southern Italy, where light energy is abundant and artificial heating is unnecessary, or in countries such as Japan, where people are willing to pay extra for produce that is "picture-perfect."

Chemical Pest Control

Each year, 30 percent or more of many crops are destroyed by insects and other crop pests. In addition to insects and their larvae, which damage the growing plants, other crop pests, including various rodents, damage many crops in the field and also consume stored grain supplies. Fungi also destroy up to 25 percent of stored crops. Nutrients that are consumed by pests are not available to the plant or to humans for food.

Monoculture and pests. Some of the farming practices that have increased crop yields over the last few centuries have also increased the susceptibility of crops to damage by pests. For example, large mechanized farms (farms that are heavily dependent on the use of machinery) planted with a single crop constitute enormous pure stands (monocultures), often thousands of hectares in size. (The *hectare*, or metric unit of land area, is an area 100 meters (m) by 100 m, equivalent to about 2.47 acres.) Monocultures are especially suitable for mechanized agriculture and for several other economies of scale; as a result, monoculture is extremely profitable. However, monocultures are not natural ecosystems and thus lack some of the natural ecosystem controls on pest populations (e.g., natural predators). Monocultures make it easier for a pest species to spread rapidly, especially if neighboring farms are also planted with the same crop. If neighboring fields were planted with different crops, the spread of pest species would be interrupted or made more difficult. Also, planting the same crop year after year facilitates the continued survival of pest species from year to year in the form of eggs or larval stages. In contrast, planting multiple crops in rotation interrupts the life cycle of a pest that depends on one particular crop species for its propagation.

Pesticides. Chemical pesticides were used in ancient times. (The Romans dusted sulfur, which we now know acts as a fungicide, on their grapes.) Enormous increases in the use of chemical pesticides occurred during the late nineteenth and twentieth centuries. Arsenic and copper compounds were widely used in the nineteenth century, but an increasing number of twentieth-century pesticides have been derived from petroleum. During the first several decades of their use, chemical pesticides greatly reduced the level of crop damage due to pests, and they continue to be used in many countries because they increase crop yields.

For much of the twentieth century, economic pressure encouraged the use of chemical pesticides, both for crop treatment and for postharvest treatment with fungicides. The postharvest treatments have given many farm products longer shelf lives, allowing transportation across longer distances. As a result, people in the industrialized world have come to expect perfect, blemish-free produce in every store and at most any time of the year, even for crops that do not grow at all in their local area.

A classic example of a chemical pesticide is *DDT*. First developed in the late 1930s, DDT was found to kill large numbers of different insect species. During World War II, DDT was sprayed on soldiers to control body lice and similar pests. Its effectiveness in insect control was followed by extensive spraying campaigns on crops following the war, first in the United States and Europe, and then in other countries. Because it killed so many species of insects, DDT led to reduced crop damage and a greatly increased food supply in many countries. In addition, DDT and similar chemicals controlled mosquitoes and other disease-carrying insects, resulting in reduced disease levels; disease reduction and increased food supplies both contributed to population increases in many countries (Chapter 6). The use of DDT marked the time of greatest optimism in the use of chemical pesticides. This era was brought to a close by the discovery of DDT's toxic effects on nontarget species, findings that were convincingly made public by Rachel Carson (1962). The term *nontarget species* is an apt one for a phenomenon that is another example of the operation of the law of unintended consequences: If the target species is the one you intended to kill, the nontarget species are those that are killed unintentionally. Often, unintentional effects are not immediately detected, not because of ill-will, but because it can be difficult to predict where, when, and how the unintended effects will show up.

Consequences of pesticide use. There are many problems associated with the use of chemical pesticides such as DDT. The pesticides themselves are generally expensive; most of them are petroleum derivatives, and a great deal of energy is used in their extraction and further synthesis. Attempts to control pests with chemical pesticides have in several cases brought about increased levels of pest-related devastation several decades later. Pesticides like DDT are toxic to a wide variety of harmful and beneficial species alike. They may kill so many of the target species' natural enemies that the population size of the target species subsequently increases (after a time delay) above its earlier levels. Another problem with frequent pesticide use is that the target species develop pesticide-resistant mutations so that the pests no longer respond to the spraying. Over 400 insect species, for example, are now DDT-resistant. Widespread use of the same pesticide year after year favors the evolution of pesticide-resistant mutant strains. Once they originate, these resistant strains of pest species spread rapidly because of selection by the pesticide itself.

Biomagnification.   DDT and many other long-lasting insecticides that are applied to crops become concentrated in the bodies of the pests that eat those crops, and then further concentrated (and thus more toxic) in the beneficial species that eat the pests, a principle known as **biomagnification** (Box 15.2). As with fertilizers, there is also a runoff problem in which the pesticide finds its way into groundwater supplies and then ultimately into streams and lakes. Pesticides may contaminate drinking water supplies in this way, and they may also poison the fish in our lakes and streams. Long-lasting pesticides like DDT also accumulate and are concentrated in fish and in the species of birds that eat the fish (Box 15.2). In particular, DDT and other chemical pesticides interfere with calcium metabolism, causing a thinning of the egg shells among owls, hawks, bald eagles, and other consumer species in many ecosystems. For this and other reasons, DDT is now banned in most of the industrialized countries, though it is still used in some parts of the third world.

Pesticides and neurotransmitters.   Many pesticides kill insect pests by blocking the function of their neurotransmitters (Chapter 10). Because some insect neurotransmitters also occur in humans and other vertebrates, these pesticides can also block neurotransmitter function in humans. Although the recommended concentrations properly used for pest control would not kill a person, they can cause permanent injury. The concentrated forms in which the pesticides are manufactured, transported, and stored are more dangerous and may be lethal; high concentrations can also result from biomagnification. Unfortunately, the banning of DDT use in many countries has resulted in the development and use of other chemicals that are even more toxic to nontarget species, including humans.

Integrated Pest Management

Integrated pest management (IPM) is a newer approach to crop pest management, one that uses a combination of techniques. The term *management* is meant to convey the intent to keep pest populations under control, so that they stay below the levels at which they cause economic harm. Total pest eradication is in most cases viewed as a goal that can only be achieved at an unacceptably high cost (including the cost to the environment or to society as a whole) or which cannot be achieved at any cost. The term *integrated* means that all available tools are used in a mix of strategies that includes chemical controls (such as pesticides), biological controls (such as the maintaining of a population of the pest's natural enemies), cultural control (such as public education), and regulatory control (such as public policy legislation). Integrated pest management avoids or reduces most of the risks of chemical pesticides. One must be aware, however, that a zero level of pest damage to crops can never be achieved with these methods. Integrated pest management requires the monitoring of pest populations to assess the possible damage that they may do and to allow the application of no more pesticide than is necessary to control pest populations below acceptable limits. Because integrated pest management relies more on biological controls than previous techniques, it requires a good working knowledge of the ecology of the pest species, especially a knowledge of its natural enemies and the other factors that control its numbers.

Economic impact level.   An important feature of any integrated pest management program is the concept of an economically acceptable level of the pest population, which is generally expressed as an **economic impact level (EIL).** The economic impact level is the threshold level above which corrective action must be taken. Pest populations are constantly monitored, and as long as the populations stay below the economic impact level, they are left alone, and the cost of countermeasures is saved. The cost of corrective measures includes the cost of expendable materials such as pesticides, the cost of using and maintaining necessary equipment, the cost of labor, and costs to the environment (including cleanup). In order for

BOX 15.2   TROPHIC PYRAMIDS AND BIOMAGNIFICATION

Producer organisms (such as plants) supply food energy to the primary consumers that eat them. The primary consumers, in their turn, supply food energy to secondary consumers, and so on. In each conversion, much of the food energy is lost in the form of heat, so that 10,000 kilocalories (kcal) of sunlight provided to the grass yields only 1000 kcal of grass energy to the cows that eat the grass, who in turn provide only 100 kcal of energy to the people that eat beef steak. These relations can be represented in a *trophic pyramid,* also called a *food pyramid,* as shown in illustration [A].

Similar food pyramids could be drawn in proportion to *biomass* (quantity of biological tissues) or numbers of individual organisms rather than energy; most such diagrams would have basically the same pyramidal shape.

If a long-lasting pesticide (or a heavy metal ion) were added to the plants at the base of this pyramid, most of the pesticide would be passed along and concentrated into a smaller and smaller amount of biological tissue with each successive conversion. The concentration at each step would be the total quantity of the pesticide divided by the total biomass. With a reduction in biomass at each successive step, but with little reduction in the quantity of the pesticide, the *concentration* of the pesticide increases with each successive step in the trophic pyramid. This is the phenomenon of biomagnification, as exemplified by such long-lasting pesticides as DDT. Illustration [B] shows the biomagnification pyramid specifically for DDT in Long Island Sound, New York, around 1970. The DDT concentration was measured as 0.000003 ppm in the water, but increased to 25 ppm in the tissues of fish-eating birds such as ospreys, an increase of more than 8 million times.

---

integrated pest management to become widely adopted, it must result in a net savings most of the time. As compared with "calendar" spraying (spraying at a particular time of year, without any regard to the level of the pest population or the need to spray), integrated pest management saves costs in chemicals and equipment, but there are some costs in monitoring pest populations and in the use of biological controls.

Introduction of predator species. Planting crops in smaller, separated patches instead of pure stands is one way in which the spread of pest species can be controlled without the use of chemical pesticides. Planting seasons may be modified so as to interrupt the life cycles of the pests. The most important techniques in integrated pest management, however, are those that take advantage of the natural enemies that keep the pest species in

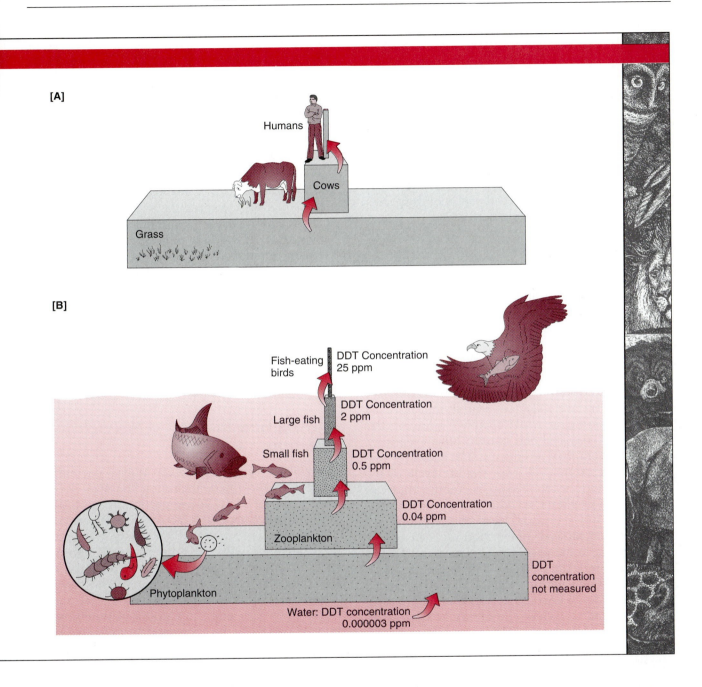

check. Rather than focusing on killing the pest directly, an effort is made to identify a predator that specifically targets the crop pest. If such a species can be identified, then any measure that encourages the growth, development, and proliferation of the predators will keep the pests in check. For example, the bacterium *Bacillus thuringiensis* (BT) can attack a wide variety of larval insect pests and prevent them from destructively feeding on crops; therefore many plant growers use BT on their crops. Planting a noncrop species that ensures the continuance of a population of predator species can control the population levels of the targeted pest. Integrated pest management can most often be accomplished at a much lower cost than the application of chemical pesticides, especially because a predator species need not be applied repeatedly since it will reproduce naturally on its own, something that a chemical pesticide never does. In addition, because the predator is often specific for the pest, damage to nontarget species occurs much less often.

For example, cotton has long been a crop of commercial importance in the southern United States, India, Egypt, and elsewhere. Traditional pests included the boll weevil and the pink bollworm. Spraying with chemical pesticides initially reduced the levels of these pests, but, by the 1960s, pesticide resistance had developed in both species. Despite increased spraying, pest populations continued to increase. Worse yet, the chemical sprays destroyed many of the pests' natural enemies, such as the spined soldier bug, and the destruction of the natural predators allowed other pest species, such as the tobacco budworm (previously unimportant as a pest of cotton), to become significant pests—in some cases more devastating than the traditional ones.

In both Texas and Peru, integrated pest management techniques have been successfully used to control cotton pests. Soldier bugs and other natural predators are collected, reared, and released on the cotton fields, while chemical spraying has been greatly reduced and is used only selectively, though not eliminated entirely. The planting season is timed so as to disrupt the life cycle of the bollworm moth; when the moths emerge, they can find no cotton plants on which to lay their eggs. Stalks and other unused parts of the plants are shredded and plowed under soon after each harvest, denying to the pest insects places to hide and overwinter until the next growing season. In some places, corn and wheat are interplanted with the cotton in order to encourage the growth of natural predators and to reduce the ability of the cotton pests to spread from one field to the next.

Alfalfa is another important crop on which integrated pest management techniques have successfully been used. As a nitrogen-rich legume, alfalfa is useful in crop rotation schedules, and its high-quality protein is valued as an animal feed for many domestic animal species. In the United States, alfalfa ranks fourth (behind corn, cotton, and soybeans, and ahead of wheat) in the land area devoted to its cultivation. The principal pests of alfalfa are two related species of alfalfa weevils (genus *Hypera*). At least nine natural enemies of these weevils have been identified, most of them

wasps that parasitize either the weevil larvae or other stages of the weevil life cycle. The weevil's life cycle can also be disrupted by harvesting alfalfa early. Alfalfa pests were formerly controlled with chemical pesticides, but this practice was sharply curtailed when one such pesticide, heptachlor, showed up in the milk produced by cows that had eaten the treated alfalfa.

Use of pheromones.   Also part of integrated pest management is the spraying of *pheromones*, chemicals normally used in animal communication (Chapter 7). The insect pheromone *glossyplure* is used by female pink bollworms to attract their mates. Spraying this pheromone on cotton fields confuses the male insects and interferes with their ability to locate the females, resulting in a natural birth control that is very specific to the pink bollworm and that has no effect on other species. It is, moreover, a chemical to which the bollworm can never develop a natural resistance without impairing its own ability to mate.

Altering Genetic Strains through Selection

Many plant characteristics are at least partially genetically determined, including the size, texture, and sweetness of the edible portion. Also under genetic influence are many factors that determine the *hardiness* of crop plants, their drought resistance, their rate of growth under different soil conditions, their dependence on artificial fertilizers, and their resistance to various pests and plant diseases. Therefore, the yield, both in terms of the amount of crop per acre and the amount of nutrition per unit of crop, can be increased by selective plant breeding.

Artificial selection.   Selective breeding is also called **artificial selection.** As carried out by both animal and plant breeders, the practice was already well known in Charles Darwin's time, and served as a model for his theory of natural selection (see Chapter 4). Darwin realized that great changes in agriculturally important plants and animals had been made within his own lifetime by British animal and plant breeders. These breeders chose the individuals of the species that best exemplified the trait they desired. They allowed these individuals to mate, while preventing mating among individuals that did not have the desired trait.

Artificial selection among crop species can be used to change almost any trait in one direction or the other. A closely related wild species may offer a desired trait, such as a nutritionally more complete protein, in which case the wild species may be crossed with the crop plant as a first step toward the production of a nutritionally superior strain. If this makes you wonder how the concept of crossing members of different species can be reconciled with the biological species definition given in Chapter 4, remember that the definition refers to populations (not individuals) that do not *naturally* interbreed. It is often possible to get individuals under domesticated conditions to do what would not happen in the natural population, for example, by dusting pollen artificially from a cultivated species onto a wild relative.

Figure 15.10 shows the results of 50 years of selection to produce corn plants with high or low oil content, or high or low protein content. However, attempts to change only one trait at a time without taking into account the interconnectedness of various components of fitness can often result in the production of an inferior strain. For example, it does no good to select for corn plants with larger kernels or larger ears unless the stalks and root systems are capable of supporting them and unless the plants are sufficiently drought-resistant and disease-resistant to survive under field conditions. Modern agricultural practices include the selection for several traits at once, resulting in harmonious combinations of traits that are well adapted to function together as a whole.

By selectively planting crops with desirable traits and by avoiding the use of genetic strains with less desirable traits, scientists have increased crop yields in many nations dramatically in the last few hundred years, and the seeds of these genetic strains command a high price. Around the world, nations that have achieved the most efficient agricultural production (high dollar yields of major crops per worker-day or per dollar invested) have generally become wealthy, while those countries that have the least efficient agricultural production are generally among the poorest. Thus there is a very high correlation between the affluence of a nation as a whole and the efficiency of its agricultural

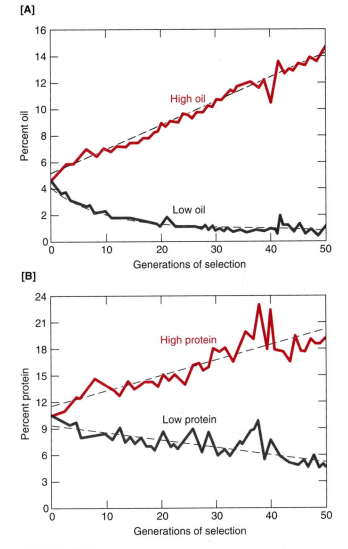

FIGURE 15.10

The results of 50 years of selection on the oil content [A] and protein content [B] of corn *(Zea mays).* By selectively breeding only those plants that had the highest or lowest protein or oil content in each generation, botanists have changed the inherited characteristics of each strain. One result is the development of more nutritious foods for humans and domestic animals.

production. The development of new genetic strains, each suited to a particular climate and soil type, is among the most important components of agricultural efficiency, rivaling even the mechanization of agricultural work. These strains greatly increased crop yields in industrialized countries during the 1950s and 1960s. In the United States, crop production has doubled on the same or less land since the 1940s.

The green revolution.   In the 1960s and 1970s, an effort was made to export many new and improved genetic strains of plants from North America and Europe to other parts of the world. This effort, loosely termed the "green revolution," was aimed at improving both the agricultural yields and the nutritional content of crops. For example, some agricultural scientists developed a more nutritious variety of corn, high in lysine, an amino acid in which corn is usually deficient. High-lysine corn provides more complete protein for human nutrition (Chapter 8). Genetically improved crop strains are, however, as subject to the law of unintended consequences as are other biological interventions. For example, many of these new strains grew well using mechanized agriculture but had never been tested using third world agricultural practices, and many of them produced disastrously low yields when cultivated under third world conditions. The promised green revolution never materialized in the third world to the extent that it had in the United States. The worst legacy of this experience is that a whole generation of third world farmers are now enduringly suspicious of the agricultural experts who had tried—and are still trying—to help them.

Altering Plant Strains through Genetic Engineering

The kinds of changes to a species that can be accomplished by artificial selection are limited by the *natural range of variation* that already exists within a species or its close relatives. Genetic engineering offers a newer method for customizing food crops by giving them genetic traits that they normally lack, including traits that do not exist in the species. These may include the ability to live in nitrogen-poor soils and other marginal habitats, the ability to fix atmospheric nitrogen and make their own nitrates, or the ability to provide a more complete type of plant protein for human consumption.

Some of the more general aspects of genetic engineering were described in Chapter 3, to which the reader is referred for descriptions of the methods generally used.

Gene identification.   Genetic engineering depends first upon the identification of the genetic trait or traits that are judged to be desirable. One goal, for example, might be the change of a plant protein to make it more nutritionally complete as a protein source for humans (see Chapter 8). This might be achieved by adding some particular amino acid that the protein normally does not contain in sufficient quantity, a goal that has already been accomplished in some corn varieties by artificial selection.

In order to bring about such a change through genetic engineering, the gene responsible for the protein in question would have to be identified and its nucleic acid sequence determined. Once the DNA sequence of the gene was known, its sequence could be altered so as to add the DNA code for the missing amino acid, then the altered gene could be reinserted into the reproductive cells of the plant. Future progeny of the genetically altered plant would then carry the altered gene. Genetic engineering of this kind is anticipated for the near future.

Genetic engineering is not simple. Whether or not the genetically altered plant actually made the protein would depend on whether the altered gene had been inserted into a portion of the genome that was actually transcribed into mRNA. Inserting a gene does absolutely no good unless that gene is transcribed and translated (Chapter 3), which means that the gene must be located "downstream" from the signals that control gene expression (Chapter 9). One way of achieving this control is to combine a new or altered gene with a DNA fragment that contains its controlling sequences (Fig. 9.7, p. 257). The altered mRNA must be translated into a protein whose altered amino acid sequence is not immediately degraded by the cell. If the protein is going to be eaten, the new protein must also be palatable.

Plant viruses insert new genes.   Scientists can use viruses to insert a new or altered gene into a plant genome. First they use restriction enzymes to make "sticky ends" on the DNA fragment and on the viral DNA (see Chapter 3). The sticky ends then bind the gene to the viral DNA, after which scientists make use of the ability of viruses to incorporate into the host genome (see Chapters 3 and 13). The virus enters the plant cells and adds its DNA and the new gene to the plant's DNA.

One virus used in genetic engineering experiments in plants is the *tobacco mosaic virus* (TMV). Like most viruses, the tobacco mosaic virus is restricted in its choice of hosts. Tobacco mosaic virus is restricted to tobacco plants, which makes it very unlikely that it could spread new traits to other species. Tobacco mosaic virus has the advantage that both its biology and that of its host, the tobacco plant (*Nicotiana tabaccum*), have been intensively studied for decades. The goal would not be to make a better tobacco plant, but to use the tobacco plant as a biological factory to produce some medicine, or a similar plant product, even small quantities of which would be valuable. In cases in which it doesn't matter what plant is used in producing a particular compound, the tobacco plant is a logical choice, because methods for its cultivation are well known. The new medicine would need to be purified to remove the nicotine and other tobacco plant molecules, but this might not be any more expensive than purifying the medicine from its original plant source. Further, the tobacco plant may grow in places where the original plant source will not, or it may grow faster. Success in such endeavors thus might encourage tobacco farmers to grow more plants for pharmaceuticals and fewer for cigarettes. Especially if the pharmaceutical plants became more profitable, tobacco farmers would have reason to switch away from growing the crop currently the leading cause of lung cancer and other diseases such as emphysema.

Bacterial plasmids insert new genes.

Since tobacco mosaic virus does not enter cells of other species besides the tobacco plant, this virus is unsuitable for changing most crop species. In cases in which TMV cannot be used, genetic engineers have experimented principally with *Agrobacterium tumefaciens*, a bacterial species that normally induces tumor formation in a variety of plant species. DNA fragments known as *plasmids* can carry new genes into host species, using methods described more fully in Chapter 3. In the normal life cycle of *Agrobacterium*, plasmids serve to introduce bacterial genes into the cells of the host plants, which cause tumors to form. The plasmids used in genetic engineering are modified so that they are still capable of introducing bacterial genes

into the host plants, but they lack the gene that causes the tumor formation, so these plasmids no longer induce tumors. The *Agrobacterium* plasmid has the further advantage that many different plant species can serve as hosts to *Agrobacterium* and can incorporate the plasmid; several other plasmids are useful only in one or two host species.

Transgenic plants.

Genetic engineering using *Agrobacterium* was first achieved in 1983. Since that date, over a dozen plant species have received genetically engineered genes using *Agrobacterium* plasmids. In all, **transgenic** plants (meaning those with genes derived from another species) have been produced in over twenty species, including tomatoes, potatoes, carrots, alfalfa, peas, cotton, and sugar beets. For example, transgenic cotton plants have been developed that produce a natural insecticide, a toxin that is derived from the bacterium *B. thuringiensis*. These plants are thus resistant to most pest species, including cotton boll weevils, without any need for insecticide spraying or application of BT.

Yet another use for genetic engineering in plants is the introduction of genes for herbicide tolerance. If a crop plant is given a gene that allows it to resist (or tolerate) a particular herbicide, then the herbicide can be used as a weed killer to control the presence of other plant species that might compete with the crop for water and other limited resources. Herbicides are subject to many of the same dangers and pollution problems as pesticides, and those that are not biodegradable can poison both soil and water.

Ethical questions.

Genetic engineering also has its potential risks, most of which are poorly understood and quantitatively uncertain. Growing crop plants with inserted genes requires the release of genetically engineered plants into the environment. What would occur if these plants got out of control? What if they also inadvertently acquired some undesirable characteristic? What if, in addition to this undesirable characteristic, they also acquired the ability to spread faster and outbreed competing genetic strains? The answers to these questions generally vary from species to species, making it important for the questions to be raised (and the research conducted to answer them) again and again for each new instance. For example, the use of the *Agrobacterium* plasmid has been criticized because the original form of this plasmid allows the bacteria to stimulate plant tumor formation. Suppose that a genetically altered strain of a plant containing an *Agrobacterium* plasmid managed to reacquire the gene for plant tumor formation, either by mutation or (more likely) by genetic recombination with wild strains of *Agrobacterium;* the plant might then grow tumors. The probability of such an event must be carefully estimated if reliable risk-benefit ratios (see Chapter 2) are to be obtained and if defensive measures against such mutant strains are to be planned in advance. Potential economic benefits from genetically engineered crops are generally easy to estimate; risks often are much more uncertain.

A few critics, notably Rifkin (1983), have consistently opposed all biotechnology, especially transgenic research, on the grounds that the dangers are unknown and potentially very great. We are, these critics say, attempting to alter nature by going considerably beyond the bounds that nature intended. Rifkin goes so far as to question whether *any* transgenic research can be ethical. In the terminology explained in Chapter 2, Rifkin might be described as a deontologist who believes that any transplantation of a gene from one species to another is inherently unethical, a stance that prevents the experimental measurement of certain risks (among other things). One possible answer to such criticisms—one grounded in natural law ethics (Chapter 2)—is to point out that rearranged genomes happen all the time in nature without drastic consequences. Gene transfer (introgression) between related plant species happens fairly often in higher plants, and cross-breeding has been transferring genes between domestic plant strains of the same species for centuries.

Fox (1986, 1992) argues more as a utilitarian who simply believes that the risks are great and should be very carefully evaluated before we proceed. The risks (or potential dangers) of biotechnology do exist, and some of them have been listed earlier. Only in a few cases, however, do we have reliable estimates of any probability of their occurrence.

The results of several studies of field performance of transgenic tomato plants were reported by Kung and Arntzen (1989) as follows:

1. Field tests were conducted with local community support.

2. Tests show that the genetically altered plants did not adversely affect the test site environment.

3. Introduced traits did not reduce yields or adversely affect growth.

4. Field performance was at least as good as, if not better than, performance in greenhouse tests.

Test results such as these are reassuring, but the possible hazards of the transgenic plants or the transgenes escaping from cultivation was never specifically assessed. Criticisms of genetic engineering, including those by Michael W. Fox and Jeremy Rifkin, are not likely to be satisfactorily answered unless such risks are also assessed and unless far more test data become available.

The possible benefits of the genetic engineering of crops are immense, although the dollar costs may also be immense. The biological dangers are not fully known. Given the law of unintended consequences, the difficulties of trying to assess risks should not be underestimated. The risks may prove to be minimal, or they may prove to be significant, and the dangers may prove to be easily controlled even in worst-case scenarios—we will never know unless we undertake the relevant investigations for each species. Only if we have investigated the possible dangers will we be able to assess the possible risks. That is one reason why it may be desirable to proceed with testing, and why the evaluation of risks would be welcomed by many people.

Marketing the results. In 1994, the Flavr-Savr tomato was introduced into U.S. markets as the world's first commercially available genetically engineered crop. The gene that was introduced into this tomato variety allows the tomatoes to stay on the vines longer and ripen there. Claims made on behalf of the Flavr-Savr tomato are that it tastes better and has a longer shelf life. Some opposition has surfaced among restaurants that refuse to carry any genetically engineered foods. Only time and experience will tell whether consumers will accept genetically altered foods. For better or worse, our experience with the Flavr-Savr tomato may form attitudes that will be carried over to other genetically engineered foods, attitudes that will undoubtedly persist for many decades to come.

1. The runoff of fertilizers from agricultural fields often produces algal blooms. Can you explain why this would be so? (What limits algal growth under normal conditions?) Would the problem be greater with organic fertilizers or with inorganic ones? Why do you think so?

2. Compare the volume occupied by the same weight of commercially available potting soil or of sand. Compare the amount of water that can be held by equal weights of soil and sand. Now compare those results to the amount of water that can be held by soil from your area. Does the soil in your area contain a lot of humus? What ways can you think of to improve the quality of your soil?

3. Can you see any evidence of erosion in the area where you live? What natural processes or human activities may contribute to soil erosion?

4. Once the toxic effects of DDT on nontarget species received widespread publicity, agricultural use of the chemical was banned in the United States and in many European countries. The United Nations considered imposing a worldwide ban, but this effort was stopped by many third world nations insisting they needed the pesticide to help control both crop pests and mosquitoes. Do you think DDT should have been banned in countries like the United States? Do you think the ban should have been extended worldwide?

5. What do you think would happen if DDT were allowed to be used in limited amounts? How might the limits be enforced? What would be done if a farmer found that he or she could kill more pests (and thus increase crop yields) by using more DDT and causing potential future harm to the environment? How intrusive would enforcement agencies need to be? Is a total ban more practical than a limited ban? Would rationing work? Would an "agricultural prescription" system (similar to medical prescriptions for drugs) work?

6. Once pest resistance to a pesticide arises, it can quickly spread through any populations that are exposed to that pesticide. Explain this fact using your knowledge of genetic mutations and natural selection.

7. Experimenters attempting to alter a genetic strain by selecting for one trait sometimes end up changing not only that trait but some other trait along with it. Use your understanding of genetics to explain why this would be so.

8. Jeremy Rifkin and other critics of genetic engineering have argued that the escape of a genetically engineered strain from cultivation would be a chaotic event whose consequences are inherently unknowable. Do you agree or disagree? Is there any way of planning for such events? If a genetically altered strain of plants (say, tomatoes that stayed on the vine longer to develop better flavor or color) were found growing outside of cultivated areas, what should our response be?

C H A P T E R   S U M M A R Y

Plants make carbohydrates, which they store for their own later use. The synthesis of carbohydrates is carried out by the process of photosynthesis, using water and atmospheric carbon dioxide as raw materials and sunlight as an energy source. Proteins, nucleic acids, vitamins, and alkaloids require nitrogen for their synthesis. Most plants get their nitrogen from dissolved nitrates, a requirement that may limit the occurrence of many plant species to soils that contain adequate nitrogen. Some plants are able to make use of microorganisms that can fix atmospheric nitrogen and incorporate the nitrogen into a form that the plants can use.

Crop yields can be increased by supplying limiting nutrients via fertilizers or via soil improvement, by supplying water, by controlling pests that compete with the plant or with humans for the energy produced by plants, and by altering the traits of the plants either by selective breeding or by genetic engineering. Integrated pest management and genetic engineering can be used to reduce our dependence on chemically produced pesticides and fertilizers.

KEY TERMS TO KNOW

artificial selection (p. 452)

autotrophs (p. 429)

biomagnification (p. 449)

chlorophyll (p. 431)

chloroplasts (p. 431)

ecological succession (p. 444)

economic impact level (EIL) (p. 449)

eutrophication (p. 444)

fertilizer (p. 443)

heterotrophs (p. 429)

hydroponics (p. 447)

integrated pest management (IPM) (p. 449)

irrigation (p. 446)

limiting nutrient (p. 436)

monoculture (p. 447)

mutualism (p. 438)

organs (p. 433)

photosynthesis (p. 431)

symbiosis (p. 438)

tissues (p. 433)

transgenic (p. 455)

vascular plants (p. 434)

CONNECTIONS TO OTHER CHAPTERS

Chapter 2   Genetic engineering of crops raises ethical issues. Pest control and fertilizer uses have both costs and benefits.

Chapter 3   Genetic engineering techniques can be used on crop species as well as other species.

Chapter 4   Plants have adapted to their environments in the course of evolution. Artificial selection can be compared to natural selection.

Chapter 6   Feeding the world's growing population will require increased crop yields and more nutritious crops.

Chapter 8   Undernutrition and malnutrition affect human health. Plants are the source of most nutrients that we need.

Chapter 9   Several anticancer drugs are plant products. The plasmid used in plant genetic engineering was originally a tumor-inducing plasmid.

Chapter 11   Most drugs are plant products and several are psychoactive in humans.

Chapter 14   Many artificial materials are patterned after plant tissues. Water flow in plants is a biophysical process.

Chapter 16   Clearing more land for agriculture threatens biodiversity and destroys ecosystems.

Biodiversity, Extinction, and Endangered Habitats

I n the shadow of trees over 60 meters tall (more than 200 feet, or as high as a 17-story building), workers use bulldozers and other heavy equipment to clear a 100-m-wide (312-foot-wide) path through a tropical forest. They are building a new road that will bring commerce and communications to the people of the region and will enable them to send their agricultural products, crafts, and minerals to markets in faraway countries. For each kilometer of roadway, they are destroying 10 hectares (10 ha, or about 24.7 acres) of tropical rainforest. The building of the road brings many high-paying jobs to the workers who build it, and the road will also open up new lands for agriculture and human settlement, a process that will destroy even more forest. The trees are important in themselves, and also because they provide a set of environmental conditions (that is, **habitat**) to thousands of species.

The number and variety of species are called biological diversity or, simply, **biodiversity.** Biodiversity is measured most easily by the number of distinct species present. More broadly, biodiversity also includes genetic diversity within species and also ecological diversity within habitats or ecosystems. In this chapter, we will consider the importance of biodiversity, the conditions that support biodiversity, and the many threats to biodiversity.

Recall from Chapter 4 that species are reproductively isolated groups of interbreeding natural populations. Most new species originate by a process of geographical speciation, in which reproductive isolation evolves during a period of geographic separation. The processes that give rise to new species increase biodiversity, while the processes that result in the extinction of species decrease biodiversity.

A. Biodiversity Means Variety among Living Species

Cataloging and Categorizing Biological Diversity

A wealth of species. There are nearly 1,500,000 species of organisms currently known to science, and a bit more than half are insects (Fig. 16.1). Vascular plants make up approximately 250,000 species, so that two-thirds of all species on Earth are either vascular plants or insects.

Our knowledge is far from complete. It is estimated that the total number of species is actually between 5 and 30 million. Such estimates are extrapolations from the few studies in which an effort was made to identify *every* species in a given area. In one such study, between 4000 and 5000 species were found in a *single gram* of sand. In another study, 5000 species were found in a gram of forest soil, and these 5000 were almost completely

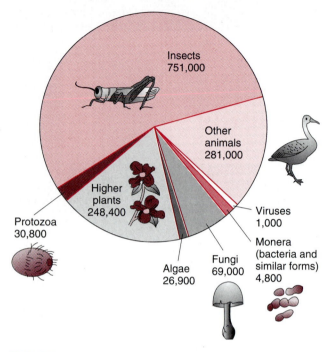

All organisms: total species, 1,413,000

Insects
751,000

Other animals
281,000

Higher plants
248,400

Protozoa
30,800

Viruses
1,000

Monera
(bacteria and similar forms)
4,800

Fungi
69,000

Algae
26,900

FIGURE 16.1
Major groups of species currently known to science as of 1992.

different from the species found in the gram of sand. The more thoroughly we look, the more species we find.

Despite our incomplete knowledge of the extent of biodiversity, we have begun to use what we do know to test hypotheses about the factors that contribute to the richness of diversity. Present levels of biodiversity are the result of many processes. Above all, the process of speciation increases biodiversity and the forces that produce extinction decrease biodiversity. Since these are both evolutionary processes, the study of biodiversity depends on an understanding of evolution. By looking at present-day biodiversity in different places, we can test certain theories that give a better understanding of evolutionary and ecological processes.

Communities and ecosystems. Biological diversity implies more than just the number of species. It implies that those species are living together and interacting with one another in well-integrated **communities.** Each species has its

own **niche,** meaning its way of life or its own role in the life of the community. Each species (or its products) provides part of the niche for many other species in its community. An increase or decrease in the population size of any one of the species is therefore likely to have consequences for all the others. A community plus the physical environment surrounding it and interacting with it is called an **ecosystem.** In addition to living species, ecosystems include the soil, water, rocks, and atmosphere in which those species live. The largest ecosystem of all, that of the entire planet, is called the **biosphere.**

When a new species originates (Chapter 4), it must not only become genetically distinct, it must also find its niche and fit into the ecosystem in which it finds itself. Otherwise, its populations will decrease in that ecosystem and it may undergo extinction. If, on the other hand, a species finds a niche that allows it to integrate into the ecosystem, its population will increase and the ecosystem will become more complex due to its presence. *Biodiversity is thus a measure of ecosystem complexity.*

Energy and biodiversity. One of the major determinants of biodiversity is latitude. Tropical ecosystems contain a much richer diversity of species and genera than temperate-zone ecosystems. The richest diversity on land occurs in tropical rainforests, while the richest marine ecosystems are those of warm-water coral reefs (Color Fig. L). Large taxonomic groups, including many entire families and several orders, are confined to tropical ecosystems, and nearly every major group reaches its maximum diversity in the tropics. In marked contrast are the Arctic and Antarctic ecosystems, which are relatively sparse in biodiversity and in ecological complexity.

Why should this be true? One of the great theoretical problems in evolutionary biology is the reasons for greater species diversity in the tropics. One possible explanation is provided by the **energy-stability-area (ESA) theory** of biodiversity. Places with greater amounts of solar energy and more continuous levels of solar energy have more species. More solar energy falls on the tropics, making them energy-rich areas, and species diversity also reaches its maximum there. Ecologists have hypothesized that each biological population requires a certain minimum amount of energy to maintain a population size capable of reproducing itself. Photosynthesizing plants capture this energy directly from sunlight; most other species obtain the energy from their food (Chapter 15). For an equal amount of nutrients (an important proviso), greater quantities of **biomass** (mass of living things) will grow in areas that are the hottest (most solar energy input) and the most humid (a consistent supply of water for photosynthesis). Individuals in a high-energy locale can more easily find all that they need to sustain themselves within a small area, while individuals in a low-energy locale may need to search over a wider area. For this reason, according to the ESA theory, energy-rich habitats can support a larger number of different species living together within a smaller space.

Climate stability. Stable climate on a yearly and on a geologic time scale allows speciation, each species becoming uniquely suited to fill a niche. When climate is stable, there are many niches in a particular area, each niche differing slightly in the amount and type of energy that is available. In the areas where seasonal changes are less extreme, multiple species will specialize to fill different niches in other ways: living at different heights in a vertically stratified forest, occupying different kinds of microhabitats, exploiting different food resources, or attracting different species of pollinators.

Habitats at higher latitudes have less stable climatic conditions than do the tropics. They have been subject to more disruption by geological processes such as ice ages, and they have had a higher rate of past extinctions, resulting in a lower biodiversity than in the more stable habitats. On a shorter time scale, climatic conditions in temperate regions produce a much wider variety of changes in weather (temperature, rainfall, humidity) within the span of a year than do tropical regions. The organisms that live in temperate regions must be able to adapt to these changing conditions. A species adapted to living all year long in New England or Scandinavia may also be able to live in many other places where the climatic conditions of some particular New England season persist year-round. Because temperate regions experience changing conditions throughout the year, species adapted to the living under such

varying conditions typically occupy large geographic ranges. Energy limitations dictate that there will be fewer species when each has a wider range.

The effect of area on biodiversity. Area is also an important factor in biodiversity. The greater the area, the more species will be present, assuming other conditions to be the same. For a wide variety of habitats, the relation between species diversity and area is expressed by the equation $S = CA^z$, where S is the number of species, C is a constant peculiar to a particular group of organisms and to a particular region (and which is discovered by cataloging those organisms), A is the area of the island or habitat, and z is an exponent that may vary. For a wide variety of situations, the exponent z has values close to 0.3, meaning that the number of species approximately doubles for each tenfold increase in area.

Size of organisms. Within a particular area, the number of species is greater for those types of organisms that are small. For organisms that are one-tenth the size of some other type of organism, there are 1000 times the number of species. This is partly dependent on energy requirements, which are subject to allometric scaling of the type discussed in Chapter 14. As the length of an organism decreases, its volume decreases at a greater rate. Its energy requirements are proportional to its volume and so also decrease at a greater rate than the length.

An additional reason has to do with the environment as it is experienced on the scale of a particular organism. The potholes and molehills seen by large organisms may appear as enormous craters and mountains to smaller organisms. The surface area experienced by a smaller organism is much greater than the surface area experienced by a larger organism. (This is a biological application of a new branch of mathematics called *fractals*.) At the scale of the small organism, smaller differences in other conditions also become important. Thus, more niches are able to be identified by small organisms than by large ones, and if these small niches are stable, more species will arise to fill them. For these reasons there are more than twice as many species of insects as of other, relatively larger, animals (see Fig. 16.1).

How Biodiversity Affects Our Lives

Hunters and gatherers. The study of biological diversity is as old as our need for food, clothing, shelter, and medicines, because all of these things, as well as tools and weapons, are made from the millions of other species that inhabit our planet. Before the development of agriculture, our ancestors roamed the forests and grasslands in search of these items. People who still live this way do so by *hunting and gathering.* Simply to survive, hunters and gatherers must become thoroughly familiar with the many species that surround them. They must learn which species to eat, which to avoid, and which have medicinal value for various ailments. Familiarity with each species is important if one is to learn where to find it and how to collect or capture it. In one part of New Guinea, the evolutionary biologist Ernst Mayr found that the natives distinguished and named as many species of birds as he did. The techniques of identification that he knew as a university-trained ornithologist were common knowledge among these hunters and gatherers, and they could also identify plants, rocks, insects, and four-footed animals that he could not.

Effects of agriculture. The development of agriculture caused people to pay much closer attention to the detailed characteristics of only a few dozen species at a time, and less attention to the myriad other species living around them. Still, most agricultural peoples supplement their diets by occasional hunting, and their medicines are almost always obtained from plants growing in the wild. Our modern arsenal of medicines is still derived largely from plant products, many of them tropical (see Chapter 15). When knowledge of medicinal plants was discovered or rediscovered during the Renaissance, it was often noticed that only certain species of plants had medicinal properties. Species with medicinal properties were often hard to distinguish from closely related nonmedicinal species, yet it was crucial that correct identifications be made. This realization was one of the important stimuli to the development of systematic botany as well as pharmacology; our modern concept of species grew in part from these attempts to distinguish closely related plants.

The value of preserving unknown species. The *study* of biological diversity is still an important endeavor for people who live close to nature as well as for scientists who seek to understand nature; it is also important as the prelude to any efforts to protect and preserve biodiversity. The *preservation* of biological diversity is important for other reasons, of which three broad types may be distinguished. First there are the reasons related to the benefits of particular—but unknown—species that can only be preserved—given our ignorance as to which they are—by preserving all. Preserving species that might become the sources of important drugs is an example of this type of reasoning. We know that many plants have yielded important drugs; other plants have yielded important foods, dyestuffs, paper, and rubber. It is therefore likely that species now living but poorly known possess a wealth of new possibilities for these and like products; therefore we must preserve them all for the sake of those that may someday prove useful.

A second reason for preserving biodiversity pertains to the wild relatives of our domesticated species. Many botanists (and some zoologists) have argued that the store of genetic variation, and therefore the possible number of genetic traits from which to choose, is greatly reduced in each of our domestic species, and is much greater in their wild relatives. It is therefore in our long-range best interests to preserve the wild relatives of all domestic species and varieties, in order that newly discovered desirable properties (or properties that *become* desirable) can be bred into domestic stocks from their wild relatives. For example, corn or maize (*Zea mays*) is one of the world's valuable domesticated species of plants, but the domesticated variety is an annual plant that must be replanted each year, an operation that involves both considerable labor and expense. In the 1970s, however, a wild relative named *Zea diploperennis* was discovered growing in the Mexican state of Jalisco, confined to a small mountain tract, just days before the land was scheduled to be cleared and the species wiped out. The Jalisco species was found to be resistant to a number of diseases that afflict the domestic varieties. Best of all, unlike all other species and varieties of corn in the world, the newly discovered species grew as a perennial, meaning that an individual plant would be able to produce its corn year after year without replanting. If some of these genetic traits could be introduced into domestic corn, either by breeding or by genetic engineering, it could represent billions of dollars worth of savings for the farmers of all corn-producing regions. Had the Jalisco corn not been discovered in time, an important genetic reserve for this important domestic species would have been lost forever. This is just one instance; similar arguments can be given for the preservation of the genetic resources of other species in zoos, botanical reserves, and gene banks, but the most cost-effective way to preserve these genetic resources is to promote the survival of the wild species or varieties in their natural habitats.

Preserving ecosystem stability. A third argument for the preservation of biological diversity depends more directly on the diversity and multiplicity of species. *There are no ecosystems that are comprised of only one or a few species.* Recall that a community is a group of species whose needs are interdependent. Stable communities are stable in part for the reason that materials are recycled: many producer, consumer, and decomposer species (Chapter 15) are all present. A small group of species is much less likely to form a complete and stable community. Multiple species of each kind ensure that the ecosystem is less easily disrupted. For example, many species of decomposers will ensure that a greater variety of biological materials can be decomposed and recycled. Multiple prey species will ensure a more stable food supply for each species of predator, for they can easily survive a scarcity of one prey species by switching to other species for their food.

Many ecosystems are unstable in the sense that removal of just one *keystone species* will cause the balance among dozens of other species to collapse, so that the disappearance of one species will cause further extinctions and lead to other drastic changes in the ecosystem. Other ecosystems, such as tropical rainforests, are thought to be more stable than this, a consequence of the large number and variety of species. In a typical rainforest, there are hundreds of species of trees, with no single species comprising more than 5 percent or so of the total. There are also hundreds of bird species, thousands of insect species, and a large diversity of other animals and plants, some of them illustrated in the color figures following this chapter.

The health of animals is promoted by the variety of plants available for them to eat or to climb or to nest in, while the health and well-being of many of the plants depends on the variety of animals that can pollinate them, disperse their seeds, or fertilize the ground near their roots with their remains. While it is obvious that the survival of any of these species depends on the survival of the ecosystem as a whole, it is equally true that the stability of the ecosystem depends on the variety of species that are present.

Paradoxically, some tropical ecosystems are also very fragile, despite the number of species present. Water is most efficiently used if it is cycled in an ecosystem. The evaporation and transpiration of water from rainforests contributes to the cloud cover which produces the rains that sustain the rainforest ecosystem. Removal of the trees from a rainforest disrupts the water cycle and prevents the rainforest from reestablishing itself. In addition, the soils underlying many rainforests are very poor in nutrients. Most of the nutrients, rather than entering the soil, remain in the organic matter, such as the leaf litter that accumulates on the forest floor, and these nutrients are recycled within the living community of the forest. The trees are both the source of forest litter and the means of preventing its erosion. If the trees are removed, the nutrients are irrevocably lost, and the nutrient-poor soil will never support much growth of any kind.

On a global scale, the stability of the Earth's atmosphere depends on its major ecosystems. For example, photosynthesis by plants, especially rainforest plants, helps limit the buildup of carbon dioxide that contributes to global warming. The same plants are also the principal source of the oxygen in our atmosphere. The health of our atmosphere thus depends on the continued health of major tropical ecosystems.

THOUGHT QUESTIONS

1. Why would the perennial growth habit in the corn from Jalisco be considered a valuable trait? Under what agricultural conditions (and in what nations) would this trait be especially valuable?

2. How easily could the genes for perennial growth be identified? How easily could these genes be introduced from the Jalisco corn into the domestic varieties? (You will probably need to review parts of Chapters 3 and 15 in order to answer this question.)

3. What would be the effects of clearing 100 square kilometers (100 km^2) from a small rainforest 5000 km^2 in size (about 2000 square miles, approximately the size of Delaware)? What would be the effect of clearing 1000 km^2 of this same forest?

4. In what way is the atmosphere part of the biosphere?

B. Extinction Reduces Biodiversity

Extinction in the Past

The time interval from approximately 200 to 65 million years ago is known as the Age of Reptiles, or Mesozoic era (Fig. 4.7, p. 105). Of all the species that lived during this time and have been preserved in the fossil record, none are still alive today—they are all considered **extinct.**

The reasons behind the extinction of a species (or a group of species) are complex and are not always fully understood. In some cases, the fossil record shows that a competing group of species appeared on the scene shortly before extinction occurred. In other cases, no specific cause can be identified.

Types of extinction.   A **lineage** is an unbroken series of species arranged in ancestor-to-descendent sequence. If we had a complete record of the history of life on this planet, each lineage would extend back in time to the common origin of all earthly life. Working forward from earlier times, each lineage extends either to a species alive today or to one that has become extinct.

Two important kinds of extinction can be distinguished. **True extinction** means that an entire lineage has died out without issue—no living descendents exist. *Trilobites* (Fig. 16.2) are an example of a group of organisms that, according to current theories, have undergone true extinction. **Pseudoextinction,** also called *phyletic transformation,* is the kind of change that occurs when a species undergoes change into something recognizable as a different species. The ancestral horse, *Hyracotherium* (formerly called Eohippus), is extinct in this sense: There are none alive today, but their descendents, the modern horses, are still alive. Many of the traits of *Hyracotherium,* and the genes that contributed to those traits, persist among modern horses. In some cases, we do not know whether an extinct group has undergone true extinction or only pseudoextinction—the evidence does not permit us a clear choice. For example, the dinosaurs are no longer alive, but if living birds are their descendents, as many scientists believe, then the dinosaurs are only pseudoextinct.

Mathematical models of extinction.   Has extinction occurred randomly? Do the chances of extinction remain constant over time and space? Several biologists of recent decades have hypothesized that extinction occurs at random over vast time periods. It is worth examining random extinctions as a mathematical model in which the same probability of extinction applies to many taxa over a long time. The mathematical model gives us a way to test various hypotheses (or to search for causes) by comparing *real* (and presumably nonrandom) extinctions with theoretical predictions based on a model of random extinctions. In other words, whatever our opinion on the randomness of extinction, investigation of the theoretical model of random extinction is a good *research strategy* that might help us discover either randomness or nonrandomness in the real world. Several studies have compared extinction in the fossil record of particular animal groups with the random model and have found, for example, that many early invertebrate groups suffered most of their extinction early in their history rather than at a constant rate through time.

Departures from randomness can be of two types: situations in which extinction is much less likely than random models predict, and situations in which extinction is much more likely. We will first examine situations in which the probability of extinction is reduced.

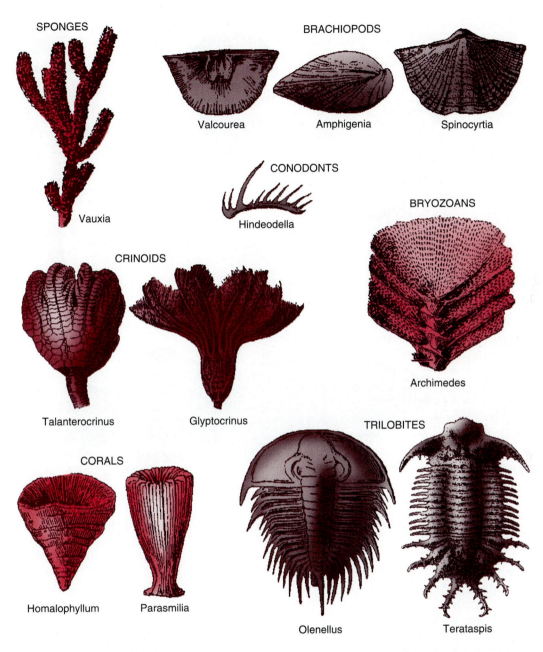

SPONGES

Vauxia

BRACHIOPODS

Valcourea   Amphigenia   Spinocyrtia

CONODONTS

Hindeodella

BRYOZOANS

Archimedes

CRINOIDS

Talanterocrinus   Glyptocrinus

CORALS

Homalophyllum   Parasmilia

TRILOBITES

Olenellus   Terataspis

FIGURE 16.2
Paleozoic animal life. The groups shown here were all abundant during Paleozoic times, and all suffered considerable extinction at the end of the Permian period. The trilobites (belonging to the phylum Arthropoda) and conodonts (probably phylum Chordata) are completely extinct; the other groups have some living members.

Living fossils. Are there ways to make extinction less likely, or to avoid extinction entirely? To answer this question, we can examine the several species and genera known as *living fossils*. These are species or genera that have survived for many millions of years without extinction and often with very little morphological change. Several of these "living fossils" are described below; the time

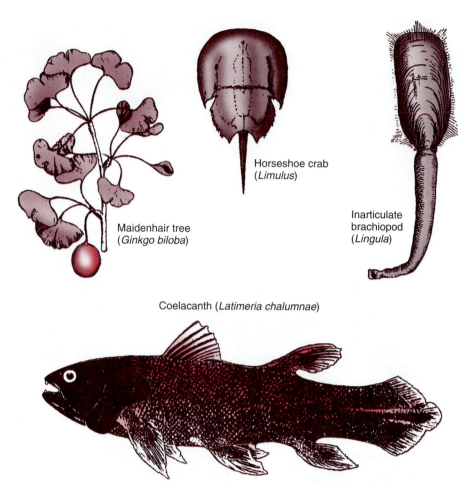

Horseshoe crab
(*Limulus*)

Inarticulate
brachiopod
(*Lingula*)

Maidenhair tree
(*Ginkgo biloba*)

Coelacanth (*Latimeria chalumnae*)

FIGURE 16.3
Living fossils, species that have
avoided extinction for long periods
of time.

periods in which their ancestors lived are shown in
Fig. 4.7. They include:

Psilophyton, a primitive vascular plant (Chapter 15) closely resembling the earliest land plants of the Silurian period

Ginkgo, an ornamental plant, native to China, which closely resembles its Mesozoic ancestors and which is planted in many urban areas around the world because it tolerates urban pollution (Fig. 16.3)

Methanomonas, a primitive, bacteria-like organism placed in a separate kingdom (the Archaebacteria) because of its primitive type of metabolism and its unique nucleic acid sequences (Chapter 4, also Appendix, p. 491)

Lingula, an inarticulate brachiopod (phylum Brachiopoda, see Appendix, p. 493) with a wormlike body enclosed in a two-valve shell that has no hinge and with a feeding structure (called a lophophore) that strains suspended particles from the water (Fig. 16.3)

Neopilina, a deep-water mollusk (phylum Mollusca, see Appendix, p. 493) with a low-domed conical shell resembling that of the extinct genus *Pilina,* one of the most primitive mollusks

Limulus, the horseshoe crab (phylum Arthropoda, class Merostomata), which closely resembles its Paleozoic ancestors (Fig. 16.3)

Latimeria, a large, rare Indian Ocean fish belonging to a group (the coelacanths) that otherwise became extinct during the Mesozoic era (Fig. 16.3)

Sphenodon, a lizardlike reptile confined to the northern island of New Zealand, the only living remnant of an order (the Rhynchocephalia) that flourished before the dinosaurs did

Several characteristics are shared by all these different living fossils: They all have locally large populations (therefore a sufficiently large gene pool to allow adequate genetic variation); they are all adapted to dependably persistent habitats (like deep ocean waters); they do not depend on a narrow range of food species (some of them will eat anything within a certain size range); and their reproductive stages (pollen, spores, or larvae) are dispersed mechanically by wind or ocean currents rather than by other species. If there are any secrets to long-term survival, these species have surely stumbled upon them.

Mass extinctions.   At certain times, hundreds of species have become extinct within a geologically short interval of time. One such mass extinction occurred at the end of the Cretaceous period, marking the end of the Mesozoic era; another, even larger, mass extinction occurred at the end of the Permian period. (The geologic time periods are shown in Fig. 4.7).

Mass extinctions, like living fossils, are examples of departures from the models of random extinction governed by probability. Across hundreds of taxa, these random extinction models predict extinction rates that do not vary greatly over time. In contrast, living fossils result from long spans of time without extinctions, while mass extinctions are episodes with a greater-than-expected number of extinctions occurring in a short time.

Mass extinctions devastate biodiversity. Over half of the families and 85 percent of the genera became extinct in the last 5 million years of the Permian period, and much higher rates occurred in some phyla. Although many more species and genera perished in this mass extinction than in any other, the Permian event has attracted less attention than other mass extinctions because nearly all the species were unfamiliar types of organisms (for example, crinoids and brachiopods; Fig. 16.2) that lived in underwater habitats such as shallow inland seas.

The mass extinction at the end of the Cretaceous period has attracted the most attention because many well-known animals became extinct at this time: dinosaurs (the reptilian orders Saurischia and Ornithischia), flying reptiles (order Pterosauria), several types of marine reptiles (orders Sauropterygia, Ichthyopterygia, and others),

ammonoids (an extinct group of mollusks, mostly with large, coiled shells), and several groups of fish, plants, and other organisms (Fig. 16.4). Any theory that proposes to deal with this mass extinction must at minimum explain the extinctions that occurred among many types of organisms on land as well as in the seas. The several hypotheses that have been proposed for mass extinctions are summarized in Box 16.1. Some of the proposed mechanisms require a very brief period during which all extinctions suddenly took place; other mechanisms would be more likely to take place over an extended period of time or at different times on different continents.

David Raup and John Sepkoski, who have studied extinction rates in a number of fossil groups, suggest that episodes of increased extinction have recurred periodically, approximately every 26 million years since the mid-Cretaceous. The late Cretaceous extinction of the dinosaurs and ammonoids was just one of the more drastic in a whole series of such recurrent extinction episodes. The possibility that mass extinctions may recur periodically has given rise to such hypotheses as that of a companion star with a long-period orbit deflecting other bodies from their normal orbits, causing some of them to fall to Earth as meteors, wreaking widespread devastation upon impact.

The asteroid impact hypothesis.   Of the various hypotheses attempting to account for the late Cretaceous extinctions, the one that has attracted the most attention in recent years is the asteroid impact hypothesis that was first suggested by Luis and Walter Alvarez. According to this hypothesis, the Earth collided with an asteroid of an estimated 10 km diameter, or with several asteroids, the combined mass of which was comparable. The force of collision spewed large amounts of debris into the atmosphere, darkening the skies for several years before the finer particulates settled. The reduced level of photosynthesis led to a massive decline in plant life of all kinds, and this led to massive starvation of plant eaters (herbivores) first and meat eaters (carnivores) subsequently. The postulated mass extinction would have occurred very suddenly under this hypothesis.

One interesting test of the Alvarez hypothesis is in the occurrence of the rare earth element *iridium*

FIGURE 16.4
Types of organisms that perished in the extinctions at the end of the Cretaceous period.
[A] Ammonoids. [B] Artist's view of various Mesozoic reptiles.

(Ir). The Earth's crust contains very little of this element, but most asteroids contain a lot more. Debris thrown into the atmosphere by an asteroid collision would presumably contain large amounts of iridium, and this material would be carried by atmospheric currents all over the globe. A search of sedimentary deposits that span the boundary between the Cretaceous and Tertiary periods shows that several do in fact show a dramatic increase in the abundance of iridium briefly and precisely at this boundary. This *iridium anomaly* offers strong support for the Alvarez hypothesis.

Note: These explanations are not necessarily mutually exclusive. Mass extinctions may be best explained by a combination of hypotheses.

CLIMATIC EXPLANATIONS

1. Warming or cooling of the Earth as a whole.
2. Changes in seasonal fluctuations of climate.
3. Climate changes brought about by changes in ocean currents.

GEOLOGICAL EXPLANATIONS

4. Changes brought about by increased mountain building.
5. Changes brought about by the changing positions of continents (plate tectonics), including changes that brought about the disappearance of shallow inland seas.
6. The "Arctic Ocean hypothesis," according to which the Arctic Ocean became land-locked, isolated, and therefore much colder, and was subsequently reconnected with the Pacific Ocean (across the Bering Strait), producing changes in ocean currents and weather patterns.

BIOLOGICAL EXPLANATIONS

7. Extinction of many cycads and other non-flowering plants, causing subsequent extinctions of plant eaters (herbivores) and later of the carnivores that fed upon the herbivores.

8. Ecological changes brought about by the increasing cooperation of insects and flowering plants (angiosperms).
9. In marine habitats, the expansion of the planktonic Foraminifera, bivalves, and bottom-feeding predators, including many Crustacea, which in turn brought about changes in marine ecosystems.
10. Extinction by human agency.

EXTRATERRESTRIAL EXPLANATIONS

11. Changes in Earth's orbit, including possible changes that were brought about by an undiscovered planet X.
12. Increases in cosmic radiation, causing an increase in mutation rates on Earth.
13. Explosion of a nearby supernova.
14. Collision of the Earth with a large single object such as an asteroid (e.g., Alvarez and co-workers' hypothesis for the late Cretaceous mass extinction).
15. Collision of the Earth with many smaller objects (e.g., comets) thrown out of their normal positions by an undiscovered planet X or by a small companion star (sometimes called Nemesis) periodically passing through an envelope of cometary orbits (the Oort Cloud, thought to surround the entire solar system).

However, no asteroid body has ever been recovered.

An asteroid of this size would be expected to leave an immense crater, even if the asteroid itself was disintegrated by the impact. The intense heat of the impact would produce heat-shocked quartz in many types of rocks. Also, large blocks thrown aside by the impact would form secondary craters surrounding the main crater. To date, several such secondary craters have been found along Mexico's Yucatan peninsula, and heat-shocked quartz has been found both in Mexico and in Haiti. Several candidate locations for the primary impact site have been suggested, including one underwater site called Chicxulub, in the Caribbean Sea off the coast of Yucatan.

Pleistocene extinctions.   Many extinctions took place during the last 2 million years, an interval that includes the Pleistocene epoch and its several glacial episodes. A number of species of large mammals, reptiles, and flightless birds became extinct during this time, and several hypotheses have been advanced as explanations. The most obvious hypothesis attributes the extinctions to changes in climate and on the advance and retreat of Pleistocene glaciers. However, careful examination of the fossil record by Martin and Klein (1984) shows that this hypothesis fails to explain the timing of the extinctions, most of which did not coincide with the extremes of temperature.

According to a second hypothesis, widespread glaciation brought about a decline in sea levels, an emergence of land bridges (including those across Panama, the Bering Strait, and the English Channel), and a consequent introduction of new species from one landmass to another. The newly introduced species competed with other species already present, and many extinctions ensued among the species that could not adjust to the altered conditions. Paleontologist G. G. Simpson used this hypothesis to explain many of the extinctions that followed the emergence of the land bridge across Panama, connecting South America (previously an island continent) with North America. In the exchange of animals that followed the establishment of this land bridge, many species became extinct. Most of these extinctions, however, happened early in the Pleistocene, leaving a large group of later extinctions still to be explained.

The human role in extinctions. By comparing species that became extinct in the last 50,000 years with others that did not become extinct, an interesting pattern emerges: Only large, conspicuous species of mammals, reptiles, and flightless birds became extinct, while smaller animals (including rodents, bats, and small birds) or marine animals suffered very little or no extinction. This pattern suggests yet another hypothesis: The activities of humans, including both hunting and alterations of habitat, may have played a large role in these extinctions. Circumstantial evidence favors the hypothesis of extinction by human agency: In those cases where the time of first human arrival can be dated (e.g., Madagascar, New Zealand, or certain Pacific islands), the evidence shows the accumulation of animal bones beginning with human arrival, and most of the extinctions took place soon afterward, within several hundred years of the arrival of humans on each new continent or island group (Fig. 16.5). Moreover, the species that became extinct were always those that humans would be apt to hunt, mostly large herbivores, while species that would have been difficult to hunt, or too small to be worth hunting, usually survived.

Species Threatened with Extinction Today

Several vertebrate species have become extinct within historic time. The dodo (a large, flightless,

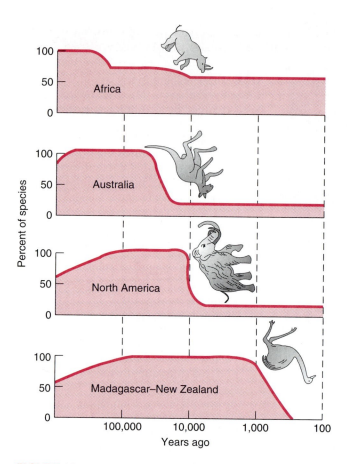

FIGURE 16.5
The timing of extinctions in various parts of the world followed soon after humans arrived in those places. In Africa, on the other hand, extinctions took place more gradually because humans had been present for a much longer period of time.

pigeonlike bird) and the passenger pigeon are two famous examples, but many other species of birds, many plants, and numerous species of insects also became extinct. Hawaii was home to some 50 species of land birds at the time of its European discovery in 1778. Human activities since that time have caused the extinction of one-third of these species, and the investigations of archaeological sites show that an additional 35 to 55 species had been hunted to extinction by the native Hawaiians before the arrival of Captain Cook. (Olson and James, 1991). On New Zealand, the giant moa (a flightless, ostrichlike bird) was hunted to the brink of extinction by the Maoris, the indigenous people of that island nation. Much the same thing happened on Madagascar, where an even larger flightless bird, the elephant bird, had become extinct long before European colonists arrived. Most recently, and on a larger scale, it is now estimated that more than 100,000 plant and animal species became extinct during the decade of the 1980s.

Endangered species.

A species threatened with extinction is called an **endangered species.** Official lists, which differ from one another, are maintained by a variety of governments and international organizations and are based on a variety of criteria. The United States, for example, maintains a list of endangered species in the United States as part of the Endangered Species Act. International lists are maintained by such organizations as the International Union for Conservation of Nature and Natural Resources (IUCN).

One indication that a species may be endangered is a dwindling in numbers. The extinction of a species is nearly always preceded by its becoming rare, and rarity may thus be the prelude to extinction. However, the only type of rarity qualifying as the harbinger of extinction, therefore qualifying a species as endangered, is the kind in which the entire species is represented by only one or a few populations, all of which are small. Once the population size falls below a certain minimum, several processes work to hasten extinction. One of these is *genetic drift* (Chapter 4), a phenomenon in which gene frequencies in small populations change erratically and not necessarily adaptively, thus hastening extinction. A second phenomenon is the increased occurrence of matings between related individuals, leading to a condition called **inbreeding depression,** in which homozygous recessive traits occur more frequently than they normally would in non-inbred populations. Since many homozygous recessive traits are harmful (Chapter 3), their occurrence reduces the fitness of the population as a whole and hastens its extinction. A third phenomenon is that environmental fluctuations (due to changing seasons, weather phenomena, and so forth) demand different genotypes at different times. Large populations that contain many different genotypes are thus more likely to survive; small populations lacking sufficient genetic diversity are more susceptible to extinction.

Extinction of a niche.

Other indications that a species is threatened are the disappearance (or impending disappearance) of its habitat, or the disappearance of another species on which its niche depends. A remarkable example of the latter is the tambalacoque tree (*Calvaria*) of the island of Mauritius in the Indian Ocean. Although these trees are long-lived, none of their seeds were ever observed to germinate, even when planted. A botanist who studied these seeds noticed that they had a very hard outer husk that mechanically prevented the seed within from breaking through. When this outer husk was damaged slightly by abrading the seeds before they were planted, they successfully germinated. The same effect was obtained by feeding these seeds to turkeys, a bird about the size of the dodo. The dodo was a type of bird that had lived on the island of Mauritius before becoming extinct in the 1800s. We now believe that the dodos fed on the seeds of this tree, and the hard outer husk was an adaptation that permitted the seeds to survive the digestive action of the dodo's stomach. (Seed-eating birds often intentionally swallow pebbles, and these pebbles work like pulverizing machines in their muscular stomachs to abrade such things as tough seeds.) Once the dodos became extinct, no new *Calvaria* trees were able to grow. New *Calvaria* trees are now growing with the aid of humans, saving them from the threat of certain extinction.

Species currently in danger.

The list of endangered species is long and growing. Among mammals, the list includes giant pandas, gorillas, orangutans, elephants, manatees, caribou, timber wolves, and dozens of less familiar species. Endangered bird species include bald eagles and whooping cranes. There are also a large number of endangered fish, amphibians, reptiles, insects, and plants. The International Council for Bird Preservation estimates that close to 2000 species of birds have already been driven to extinction in the last 2000 years, mostly by human agency, and that another 11 percent (1029 of the 9040 cataloged species) are endangered.

Some species, of course, are endangered because of indiscriminate hunting, fishing, or poaching. These are mostly large and conspicuous organisms. However, a much larger threat to biodiversity lies in the destruction of natural habitats, a process that threatens thousands of species at once, including those that are inconspicuous and poorly known. Among these small and inconspicuous organisms, the number of endangered species is certainly many more than those on any official list, and many of these species will undoubtedly become extinct before they have even been discovered and adequately described.

THOUGHT QUESTIONS

1. Of the various hypotheses listed in Box 16.1, which would predict rapid extinction (say, in less than 500 years), and which would predict a more gradual extinction over a longer period of time? Which hypotheses would account for both marine and terrestrial extinctions?
2. Experts still disagree as to whether the great extinctions of the past occurred suddenly or gradually. If an article appeared that claimed new evidence for either of these hypotheses, how would you go about evaluating this evidence?
3. The presence or absence of dinosaur fossils is used to determine whether a certain bed of rock is Cretaceous or Tertiary. How would this practice influence research on the question of whether dinosaur extinction came about gradually (at different times in different places) or suddenly (simultaneously everywhere)?
4. In 1995, the United States Department of the Interior proposed to downgrade the legal status of bald eagles (*Haliaeetus leucocephalus*) from "endangered" to "threatened." If you were asked to decide this issue, what evidence would you seek? Why is the U.S. government involved in this matter? Bald eagles also live in Canada and part of northern Mexico. Does the U.S. government have any authority to declare the bald eagle "endangered" or "threatened" in those places?

C. Some Entire Habitats Are Threatened

Among the most serious threats that any species faces is the destruction of its habitat. Of all the endangered species in the world, an estimated 73

percent are endangered because of the destruction of their habitats. Habitat destruction threatens many species at once, often thousands at a time. One ecologist has identified eighteen areas of the world as "hot spots" (Fig. 16.6) where threatened habitats put many thousands of species at the risk of extinction simultaneously (Myers, 1988, 1990).

Because of the latitudinal gradient in species diversity, the tropics contain many more species than the temperate or polar regions. Wilson (1992) estimates that one Peruvian farmer clearing land to grow food for his family will cut down more kinds of trees than are native to all of Europe.

What impact does habitat destruction have on human lives? We will examine two specific types of habitat destruction: the destruction of tropical rainforests and desertification.

Tropical Rainforest Destruction

Ecosystems that are different in locale but are similar in type are grouped together into **biomes;** examples of biomes include tundra, grassland, desert, and rainforest. Each biome thus groups together areas of similar climate, similar habitats, and similar assortments of species on various continents (Fig. 16.7).

The tropical rainforest biome. **Tropical rainforests** (see the color figures at the end of the chapter) are a biome that encompasses habitats of low latitude (from about 15°S to 25°N), continually warm temperatures, and year-round high precipitation. Average temperatures are usually around 25°C, with only small fluctuations (22° to 27°C) and with seasonal extremes no lower than 20°C. Precipitation is high throughout the year, with annual totals of 1800 millimeters (1800 mm, about 71 inches) or more. Although there is seasonal variation, no month averages less than 60 mm (2.36 inches) of rainfall. Under these conditions, humidity stays moderate to high at all times. The most conspicuous vegetation consists of tall trees, 30 m to 60 m (about 100 to 200 feet) in height, up to the height of a seventeen-story building. Their leafy tops make a *continuous canopy* through which tree-living animals can roam widely without ever coming down to lower levels (Fig. 16.8). The tallest of trees may protrude above this level and are called *emergents*. The taller trees

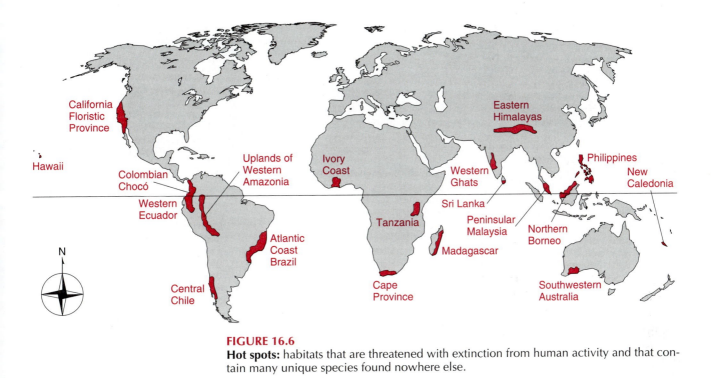

FIGURE 16.6

Hot spots: habitats that are threatened with extinction from human activity and that contain many unique species found nowhere else.

are often buttressed at their bases with a variety of woody supports, some of which give the base of the tree a star-shaped or fluted cross section (shown in Fig. 16.8A and in Color Fig. D) instead of a circular one.

Lateritic soils. The continuously high rainfall washes soluble minerals from the upper layers of many rainforest soils, a process called *leaching.* The high rainfall leaches away even partially soluble minerals, leaving the soil nutrient-poor. What remains behind after extensive leaching is a dark or reddish type of very hard, mineral-poor soil called *laterite.* Most attempts to grow crops on lateritic soils have been very disappointing because of their low nutrient content and frequent mudslides. The amount of rainfall is also usually too high for most crops. Because the forest maintains its own rain clouds, rainfall can become too low if too much rainforest is destroyed (see below).

Ecological diversity. Rainforests have a high diversity of plant and animal species, several of which are shown in Fig. 16.9, p. 476, and in the color figures at the end of this chapter. Some biologists have estimated that more than half of the world's species occur in the rainforests and nowhere else. For example, Costa Rica, a nation about the size of West Virginia, has more species of birds than the whole United States and Canada combined!

Many plants specialize to take advantage of the well-lit conditions on the forest's edge or in temporary clearings created by fallen trees, but much of the continuous rainforest receives little direct sunlight below the canopy (see Fig. 16.8B), and most plants need to be shade-tolerant as a result. Variety among trees makes for a variety of habitats for **epiphytes** (plants that use other plants instead of the ground for support, such as orchids perched high in the trees) and for many other species of both plants and animals. Animals and plants specialize to occupy various habitats within the forest floor, the canopy, along tree trunks, or in the clearings created by river banks, landslides, or fallen trees.

The great diversity of small-scale habitats within a rainforest contributes to the maintenance of diversity of species. One method that ecologists use to study this diversity is by fogging the trees above them with a biodegradable insecticide and studying all the insects that fall into the cone-shaped traps below (Fig. 16.10, p. 477).

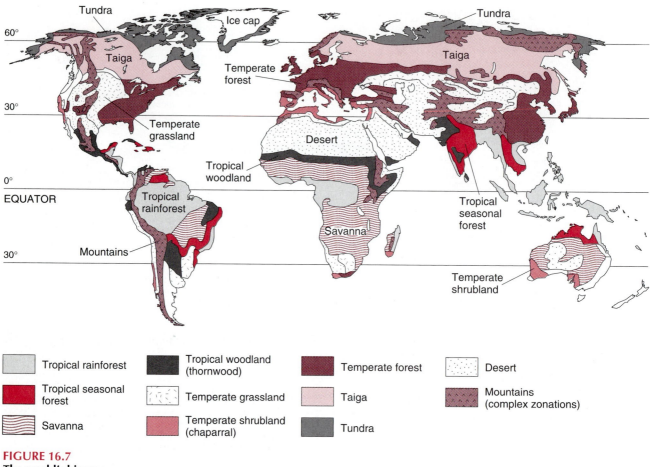

FIGURE 16.7
The world's biomes.

One result of this great diversity is the large number of interactions that occur among all the many species present, including predation, competition, and varying degrees of mutualism. Most tropical plants have elaborate mechanisms to ensure seed dispersal by animals. For example, shrubs of the genus *Piper* have fruits that are often visited by bats, who disperse the seeds.

Complex interactions among species. As an example of the complex interrelationships among tropical species in the rainforests, consider the figs (*Ficus*), an extremely successful group of tropical shrubs and trees that often reach great heights. The edible part of a fig is an aggregate fruit or receptacle (Fig. 16.11), which is unusual in that several dozens of flowers are contained within it. Many of these flowers are home to the tiny fig wasps (family Agaontidae) that pollinate the plants. Figs produced in the cooler months bear *winter receptacles* containing mostly neuter (sterile) flowers in which a female wasp lays her eggs before she dies. The wasps develop throughout these months and emerge in the spring. The male wasps emerge first, wingless and nearly blind. They move around inside the fig searching for female wasps, still in the pupa (cocoon) stage. The male then chews his way into the female pupa and inseminates her before she emerges; he then dies without ever leaving the fig. The female wasp emerges later; as she leaves the fig receptacle, she picks up pollen from the male flowers located near the exit. She then flies around in search of fresh spring-season figs, a second type of receptacle containing both female and sterile flowers. The female wasp enters the fig, then roams around inside and

[A] Layers in the rainforest

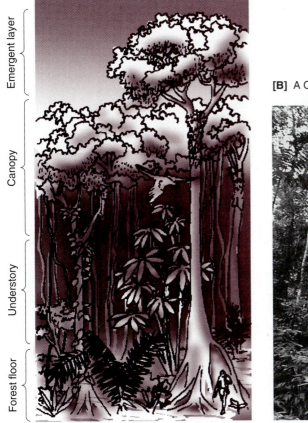

[B] A Costa Rican rainforest

FIGURE 16.8
Tropical ranforests. [A] Profile view of a rainforest, showing vertical stratification. [B] The interior of a Costa Rican rainforest. Notice the marked differences in light levels from the canopy above to the understory below.

lays her eggs by the hundreds, all the while pollinating the female flowers that she visits. The tiny new wasps that emerge then repeat the process and produce a second generation of wasps. Female wasps emerging late in the year lay their eggs in a third type of receptacle, containing only sterile flowers. The wasps emerging from this third generation lay their eggs in the winter receptacles, completing the yearly cycle. Three types of receptacles are thus home to three generations of fig wasps each year, with male flowers appearing only in the first (winter) type of receptacle and female flowers (and thus seeds) only in the second. All three types of receptacles contain sterile flowers, which alone support the development of new fig wasps. The wasps develop only within these sterile

flowers, which they also use as food. The figs are pollinated only by the wasps, each species of figs generally supporting its own species of wasps.

The story does not end there, for the seeds of the figs will not grow if they fall beneath the tree that bore them. This is a general phenomenon: Established trees have such an overwhelming advantage when they compete with their offspring that the offspring have little or no chance if they fall near the parent. The seeds must therefore be dispersed far from the parent tree if they are to succeed, and must have both moisture and organic nutrients in order to germinate. In some species of figs, the seeds germinate first as epiphytes upon the branches of other trees, then later grow roots reaching down into the ground; the seeds of these

[A]

[C]

[B]

FIGURE 16.9
The rainforests of Costa Rica are home to many plant species. [A] *Palicouria.* [B] *Spathophyllum,* showing leaves damaged by the feeding activities of herbivores. [C] *Piper* inflorescences. [D] A variety of different leaf types.

[D]

species must find their way to aboveground perches in order to germinate properly.

The service of dispersing the fig seeds is performed by a variety of animal species, different species of animals scattering the seeds of different figs in different areas. In parts of Indonesia and Malaysia, the best dispersers of fig seeds are orangutans (*Pongo pygmaeus;* Fig. 16.12). These fairly large apes practice *quadrumanual clambering,* a form of locomotion through the trees in which the orangutan's weight may hang from a support above, often using the feet as well as the hands.

FIGURE 16.10
These conical traps are set to catch insects that fall from above when the trees are sprayed with an insecticide.

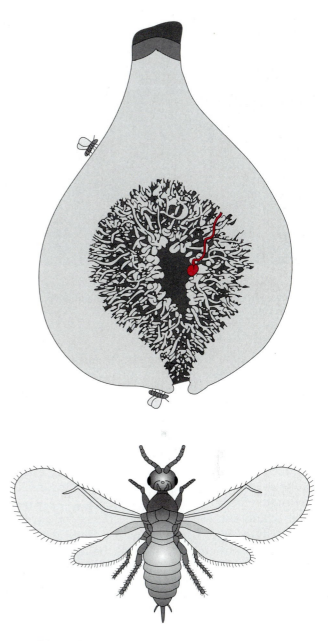

Quadrumanual clambering is very inefficient energetically for a large ape. An orangutan, if it is to avoid starving, must therefore eat enough calories in any one place to sustain it on its high-energy journey to the next place, which may be miles—and days—away. Wild orangutans are nearly always hungry and are continually wandering in search of energy-rich foods, the most preferred of which are usually figs. Captive orangutans will often gorge themselves into obesity. The wide-roaming habits of orangutans are ideal for the figs, for the orangutans will consume hundreds of figs, containing tens of thousands of seeds, then wander for miles through the forest. When the orangutan defecates amidst the branches, it leaves behind hundreds of fig seeds, together with a supply of moist, nutrient-rich fertilizer that helps the seeds sprout and establish themselves. The process only needs to work once every several years in order to replenish the supply of fig trees in a forest.

The orangutans must also cope with the fact that the fig trees flower and bear fruit at different times of the year. Since there are many species of figs, each fruiting in a different seasonal pattern, a resourceful orangutan can usually manage to find mature figs in almost every month of the year. The skill of the orangutans in finding fig trees in the proper state of ripeness is an interesting lesson in why intelligence was a selectively favored trait in apes more generally, including human ancestors. Using their costly and inefficient means of locomotion, an orangutan would surely starve to death if it

FIGURE 16.11
Cross section through a fig (*Ficus*), showing a female fig wasp (*Blastophaga*) ready to enter. An enlarged view of the female wasp is shown below; the male has a similar body but is wingless. One of the many flowers inside the fig is shown in color.

simply searched at random for figs, which constitute only 5 percent or less of the trees in the forest. Even this 5 percent figure is misleading, because only a small fraction of the figs are bearing fruit at

FIGURE 16.12
The orangutan *(Pongo pygmaeus),* **showing quadrumanual clambering.**

any given time. The orangutan therefore needs to know *where* these fruit-bearing figs are located and how to reach them by the most efficient route. This means that there is a selective advantage for a good spatial memory—a mental map of the forest covering many square miles. On its way to a fruiting fig tree, a forebear of today's orangutans would have stopped to examine other fig trees (as long as they were not too far out of the way), inspecting each to estimate when it would next be ready to bear fruit. The orangutan with the best chance of survival would have been the one whose intelligence allowed it to remember where all the fig trees were, when each was last visited, how far along its figs were at the time, and when the time would be ripe to visit each particular tree again. The orangutan who could remember this information for only a few fig trees might not survive, and the ape that could remember two dozen fig trees might survive just barely, but the one who could remember a hundred fig trees in this way would be guaranteed a steady food supply all year round.

The interrelatedness of the lives of figs, wasps, and orangutans (and goats, birds, monkeys, humans, and other seed-dispersing species in the places where the orangutans do not live) shows how complex the relations may be among the species of the rainforest. Destroy a few fig trees, and many orangutans may starve. The removal of a few orangutans will decrease the ability of the remaining fig trees to disperse their seeds, which also diminishes their ability to provide homes for fig wasps, or food for the insect-eating species that feed on the wasps. The clearing of a portion of the rainforest thus has consequences far beyond the portion actually cleared, for it diminishes the health of the whole ecosystem for miles around.

Ecological succession.   Many tropical botanists have remarked at how slow the rainforest is to reclaim an area that humans have cleared and then abandoned. Certain *pioneer species* will grow in such clearings, and these are generally the same shade-intolerant species that are adapted to take advantage of the natural clearings created by tree falls. These species take advantage of the open sunlight while they can, only to be replaced at a later stage by the more shade-tolerant species. In this way, entire species assemblages (*communities*) are replaced by a **succession** of other communities that take over, one after another, until a **climax community** emerges that is not replaced. Although a few pioneer species will colonize an area in a few months or years, it may take centuries for the rainforest to again reach its climax community and regain its former canopy height and species density. Because the small-scale conditions may be different than in previous successions, tropical forests, even when they regrow, may not reach the same climax community as the forest that was replaced.

Deforestation.   Around the world, rainforests are being destroyed at an alarming rate. One expert on tropical rainforests estimates that they are disappearing at the rate of 150,000 square kilometers per year (150,000 km²/yr, or 410 km²/day, an area equivalent to the size of Manhattan island every 3.5 hours), and that at least 40 percent of the world's rainforests have already been destroyed! (Jacobs, 1988). Among the contributing factors that have been identified, Jacobs (1988) lists the following:

1. "Systematic discrimination by Man Against Everything that is not Man." Human arrogance, in other words, gives rise to an ethic in which "man is free to help himself to everything in nature" and the needs of other species do not count.

2. Growth of human populations.

3. The rights of poor farmers to earn a meager living, even if they do so by "reclaiming" the poor soil beneath rainforests for purposes of cultivation. In places where land reform breaks up large holdings into smaller plots managed by numerous small farmers, one effect is a more thorough exploitation of the land and its resources.

4. Progress and development in third world countries brought about by individuals striving for a better life. This is especially noticeable in Brazil, home of the world's largest rainforest in its Amazon basin.

5. Free market forces, which encourage short-term economic exploitation of the rainforest, including both the trees and the minerals beneath the soil.

6. Technology, including bulldozers, which make possible the destruction of rainforests on an unprecedented scale.

7. Economic growth and industrial development, especially by big corporations and governments in the third world countries in which rainforests occur.

8. A longstanding hatred of the rainforest by many agricultural people living nearby, something akin to the "pioneer spirit" of Americans who subdued the frontier in the period from about 1750 to 1900. This attitude is usually not held by hunters and gatherers whose lives depend upon the forest, but by the often more numerous farming communities living in the clearings or on the periphery of the forests.

Tropical deforestation has many causes (Fig. 16.13 and Color Fig. E). Some rainforests are cleared because of the mineral wealth (definitely nonrenewable) that lies beneath them. In other cases, it is the trees themselves that are being harvested for timber. Trees are renewable given a sufficient length of time, but many logging operations destroy the trees much faster than they can grow back, so that the forest could not long sustain continued logging on such a scale. Much more rainforest destruction is carried out in the name of agriculture. The land reclaimed from the rainforests is put to agricultural use, as grazing land for cattle ranchers or as farmland for the planting of various crops, either for local human consumption or for commerce. In rainforest regions with lateritic soils, agricultural use of the land quickly

[A]

[B]

FIGURE 16.13 **Tropical deforestation.** [A] Trees cleared to make way for agriculture. [B] Timber ready for transport. These operations destroy many hectares of forest at a time; the results include a loss of topsoil and extensive erosion, often leaving the land unsuited for agriculture or human habitation.

exhausts the low nutrient content of the soil. Problems with drainage and erosion also arise in many places where the soil holds water poorly.

Since rainforest vegetation (through leaf transpiration) contributes to the rain clouds that maintain the rainy climate, any large-scale clearing of the rainforest is bound to alter the delicately balanced water cycle. Unlike the small clearings created by natural treefalls, large areas cleared from rainforests tend to change the regional patterns of precipitation, usually for the worse (see the next section, on desertification). With reduced rainfall, the land supports less vegetation, either in crops or in the surrounding forest. Thousands of animal species retreat into what is left of the forest itself, where they become more crowded and more vulnerable to extinction. Within a few years, the crop yields often fall drastically. The use of these lands as ranges for grazing is not particularly profitable either, and it also accelerates the trend toward desertification in many places. Many rainforests are cleared by burning, a process that adds tons of carbon dioxide and smoke to the Earth's atmosphere. The loss of rainforest photosynthesis lessens our planet's capacity to cope with this increase in carbon dioxide and contributes to a trend toward global warming.

Tropical rainforests occupy a vital position in the biosphere. Of all the photosynthesis that plants perform on our planet, almost half of it takes place in tropical rainforests. Most important, this photosynthesis is the principal source of atmospheric oxygen and the principal safeguard against a global increase in carbon dioxide. Unfortunately, human activities are already producing carbon dioxide at a rate so fast that global photosynthesis rates (Chapter 15) cannot keep up.

Tropical rainforests are thus an important global resource whose continued health benefits the entire planet, in addition to the countries in which the rainforests occur.

Deforestation is a problem at all latitudes, not just in the tropics. Norway, Russia, Canada, and the United States all have vast forests that consume carbon dioxide and contribute oxygen and water to the atmosphere. These forests are harvested for their timber and cleared for agricultural and other uses. Timber resources can be renewable if properly managed, especially if only a portion of the trees are cut in any one location. The clear-cutting of large tracts of land is far more damaging to forest ecosystems and to the global atmosphere that all forests support.

Desertification

Climatic zones. The growth of deserts takes place in many areas, but the process is more striking in Africa. The northern half of Africa is occupied by the world's largest desert, the Sahara. Global patterns of atmospheric convection (Fig. 16.14) conspire to rob this large region of all its moisture on a continuing basis: Prevailing high air

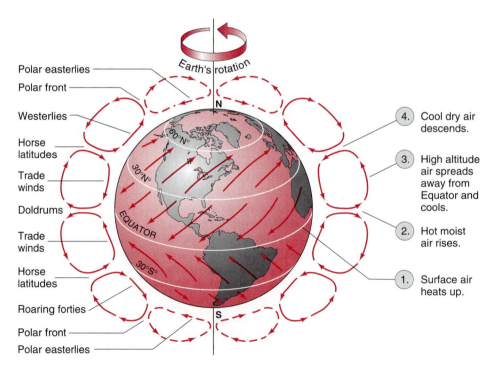

FIGURE 16.14
Global patterns of prevailing winds. Notice that winds blow away in both directions from latitudes 30°S and 30°N, carrying moisture away from these latitudes and creating desert regions.

Polar easterlies
Polar front
Westerlies
Horse latitudes
Trade winds
Doldrums
Trade winds
Horse latitudes
Roaring forties
Polar front
Polar easterlies

Earth's rotation

4. Cool dry air descends.
3. High altitude air spreads away from Equator and cools.
2. Hot moist air rises.
1. Surface air heats up.

pressure creates winds that favor evaporation of whatever moisture there is and the transport of this moisture away to greener pastures.

To the south of the Sahara, much of Africa is characterized by a series of climatic zones arranged very much in parallel bands of varying moisture (Fig. 16.15): tropical rainforest along the south coast; then, heading north, a patchy mosaic of forests interrupted by more open land; then a more open woodland with scattered trees and shrubs only, giving way to a *savanna* (tropical grassland) farther inland; then a type of dry pastureland, the Sahel region; and finally the desert. Each band has more rainfall than the one to its north, and less than the band to its south. The overall pattern has local exceptions where the land is mountainous, but in general it prevails across most of Africa north of the Equator. Each of these climatic bands supports a distinctive kind of vegetation and a distinctive kind of culture among its human inhabitants.

The climatic zones listed above are not static, but are slowly changing. The Sahara is very slowly advancing southward by a process called **desertification.** The other vegetational bands are moving southward with it. A similar process has taken place in other directions, for the Sahara now reaches westward to the shores of the Atlantic and eastward to the shores of the Red Sea and beyond into the Arabian Peninsula. Archaeological excavations throughout this part of Africa confirm that the land was much more wet and the vegetation lush within the last few thousand years, as the bones (and cave paintings) of hippopotamus and crocodiles attest.

Factors leading to desertification. How does desertification take place? At least within the Sahel, an important factor is the overgrazing of pasture lands by flocks of domestic animals: goats, cattle, camels, sheep, and other species. The result is a loss of vegetative cover. Without the many plant roots that hold the soil and its moisture, the precious topsoil blows away. The land, which can no longer support plant life, becomes a desert.

Another important process takes place farther south, where rainforests and thorn forests are cleared for agricultural use. The type of shifting cultivation practiced in these locations is called **slash-and-burn agriculture** because of the means

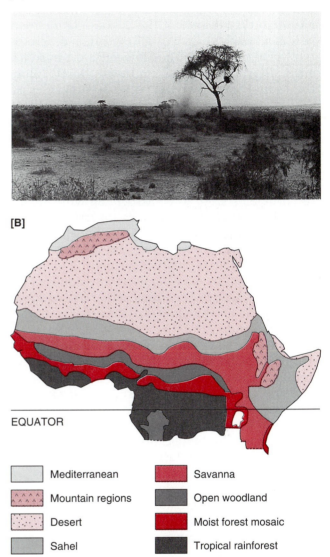

[A]

[B]

EQUATOR

Mediterranean	Savanna
Mountain regions	Open woodland
Desert	Moist forest mosaic
Sahel	Tropical rainforest

FIGURE 16.15 **African vegetational zones.** [A] The Sahel, with a dust storm visible in the distance. [B] Distribution of vegetational zones in Africa north of the Equator.

used to clear the forests (Color Fig. E). After several decades of agricultural use, the land is abandoned and new land is cleared. There are several unfortunate consequences of this method of agriculture: The abandoned land is never totally reclaimed by forest ecosystems; the agricultural land has much less ability to retain moisture than the forests it replaces; and the dry pasture of the Sahel replaces much of the abandoned fields.

Desertification around the world.   The problem of desertification is not limited to Africa, although the advance of the Sahara claims more new land each year than all the other deserts of the world combined. The Mojave Desert in southern California and the Great Indian Desert (along the India-Pakistan border) are two other deserts that are advancing on adjacent agricultural land and pasture land. The situation in India may be broadly similar to that in much of Africa. The situation in the western United States is a bit different because desertification is only in its early stages in most places and because most of the problems seem to be associated with the use of underground water reserves for irrigation and for domestic use in cities like Los Angeles. Farmers and ranchers throughout the western United States use underground water deposits (*aquifers*) for irrigation. In many cases, these aquifers are either shrinking or becoming saltier as ocean waters (e.g., from the Gulf of California) encroach further inland.

The immediate effects of desertification are the loss of cropland and rangeland, an effect that is felt keenly but locally. In a few cases, there is also increasing conflict over water rights, for example, between California and Arizona over the use of the Colorado River, or between Turkey and Syria over the use of the Euphrates. Much more serious, however, are the long-term consequences. With reduced vegetational cover, the ground retains less moisture. This means that the air above can become drier, and rain clouds are far less likely to form. The absence of rain clouds results in reduced rainfall, which in turn accelerates the process of desertification.

Prospects for reversing desertification.   Can desertification be arrested or turned back? Yes, but only very slowly, very expensively, and with a concerted effort over many years. Israel has had great success in "making the desert bloom," turning desert and scrubland (like the Sahel) into agricultural land. One key to this process is irrigation, an expensive undertaking in a dry climate. Many of the world's desert regions occur in poor nations that do not have the financial resources to repeat Israel's successful experiment in desert reclamation.

Valuing Habitat

Although we have examined only two of the world's many biomes, some of the conclusions we have drawn pertain to other types of ecosystems as well. In particular, habitat destruction threatens many ecosystems, whether they be coastal wetlands, pine forests, or the African Sahel.

Ways of assessing value.   Philosophers often distinguish between *intrinsic value*, the value that something has as an end in itself, and *instrumental value*, the value that something has as a means to some other end. Dollar bills, for example, have no intrinsic value; they are valued only because of what we can buy with them, and they would be useless in a society that did not accept them in trade. We will soon examine the instrumental value of various habitats as places where valuable resources can be obtained. Before we do so, let us also point out that many people also value other living species, and entire ecosystems, as having a high intrinsic value. The habitat that sustains living ecosystems is likewise valued intrinsically by many people. Also, in the view of many people, no species has a right to destroy another species or to deprive it of its habitat or its means for continued existence.

Habitat destruction and ethics.   Habitat destruction can take many forms, including the clear-cutting of forests and the draining of swamps. In some cases, the destruction takes place to permit the building of housing tracts or shopping centers. In other cases, land is cleared for agricultural use. In still others, extractive industries like mining or logging simply exploit the land on a one-time basis for its mineral wealth or its standing crop of trees. Many of the social, political, and economic forces identified above as impinging on tropical rainforests often threaten the destruction of other habitats as well.

When we pause to consider what the many cases of habitat destruction have in common, we soon realize that the same ethical issues recur in case after case: How important are natural communities? How important are their habitats? How important is it to leave nature undisturbed? How important is the possible use of land for agriculture or some other human endeavor? To what extent does the answer to the previous question depend on the quality of the soil, the economics of the country in question, or other factors?

All of these are basically ethical questions, or parts of a larger, all-embracing ethical question: Is it better to preserve a particular ecosystem in its "natural" state, or is it better to convert the area into agricultural or similar use? Viewed one nation at a time, the forces that push toward one alternative or the other are generally weighted against the natural ecosystem: the pressure of human population, the need for land and food, the need for income, the need for economic development. Measured against all these forces is the value of undisturbed wilderness, a value that is not always obvious to those living nearest to it.

When we consider the worldwide ecosystem of the Earth as a whole, however, the balance seems to shift in the other direction, in favor of preserving the natural environment. The advance of the Sahara, or the destruction of rainforests, threaten the planet with consequences far worse (or far greater) than the continuation of poverty and underdevelopment in any one country. The case for Brazil can easily be argued in these terms: The preservation of the Amazon rainforest is best for the planet as a whole, and the economic best interests of Brazil would be viewed as secondary if the good of the planet were given priority. Perhaps this makes sense to North American environmentalists and philosophers, but it is certain not to be a very popular attitude in Brazil! Moreover, it is the Brazilians who are largely in control of their rainforest, and they are right to resent any suggestion that they sacrifice the well-being of their nation's economy for the "greater good" of a global environment that the wealthier nations of the North have already started to destroy. A similar argument can be directed against the industrial nations of North America and Europe: A reduction of resource consumption by these nations would reduce pollution, reduce the trend toward global warming, and benefit the planet as a whole.

It is easy, on utilitarian principles, to argue that the good of the planet should take precedence over the economic well-being of any single nation. However, on just about any principle of fairness, it is just as easy for the Brazilians to argue that they should not bear the entire burden for a sacrifice that benefits the whole world. If the world benefits from the Brazilian rainforest, then the world should somehow pay to maintain it in its natural state of affairs. If rainforests offer such good protection against global warming, then all nations should contribute to their maintenance, perhaps in proportion to the amount of carbon dioxide that they generate.

Sustainable use.   It does not take long to realize that many forms of habitat destruction are driven by very shortsighted goals. As an example, the harvesting of slow-growing trees brings only short-term gains, and only to a small number of people (those in a single industry, sometimes only a single company), but the damage that it causes both to the local economy and to the biosphere as a whole may be irreversible. The same may be said of cutting down the rainforest for the planting of those crops that grow poorly on lateritic soils. What we need instead are **sustainable uses** of forests, uses that allow people to derive profit from maintaining the forest instead of from destroying it. Most nations that contain rainforests are nonindustrialized, and many of them are also poor. If they are able to achieve a sustainable economic use of the rainforest, they will have an economic incentive to maintain the forest rather than to destroy it. It is therefore in the long-term best interests of all nations and all people to help tropical nations develop such sustainable uses.

Examples of sustainable use include ecotourism (discussed below), and the gathering of small amounts of high-income rainforest products such as pharmaceutical plants. For example, the anticancer drugs vinblastine and vincristine, derived from a rainforest plant (the rose periwinkle of Madagascar), are responsible for sales of $180 million a year. In 1991, the pharmaceutical firm of Merck and Company entered into a $1 million agreement with Costa Rica's National Institute of Biodiversity. Scientists working for the Institute were to identify as many rainforest plants as they could (the total number is estimated to be 12,000), extract samples from them, and send the more promising ones to Merck for tests of their medicinal value. Merck's $1 million investment should be compared to their annual sales of close to $15 billion in 1994. In 1990, Merck sold $735 million worth of just one drug, Mevacor (lovastatin), a cholesterol-lowering drug derived from a soil fungus. If even one new drug brings in a small fraction of this amount, Merck will recoup their original investment many times over.

Certain kinds of rainforest agriculture can be sustainable, but others are not, and much remains to be learned from experimentation and frequent reevaluation. Most promising are the attempts at mixed uses of rainforest habitats, allowing tall trees, shorter trees and shrubs, and smaller plants to persist side by side. Coffee, vanilla, cocoa, cashews, bananas, and a variety of spices are potential candidates for such experimental attempts. Many uses of rainforest plants are traditional, but they could benefit economically from improvements in harvesting, transport, and marketing. Tropical plants that could easily be marketed more widely include amaranth (a nutritious and drought-resistant grain), fruits such as durians and mangosteens, and the winged bean (*Psophocarpus tetragonolobus*) of New Guinea, a fast-growing plant that produces spinachlike leaves, young seed pods that resemble green beans, and mature seeds similar to soybeans, all without the use of fertilizers. Hundreds of other fruits are grown and eaten in the tropics but are only rarely exported. Botanists and agricultural scientists can often assist in identifying which plants might profitably be grown or harvested in a given region without harm to the environment, and which could be marketed internationally to bring much-needed income and provide local people with an economic incentive to maintain the forest ecosystem. Agencies responsible for economic development in tropical nations need to find new ways to exploit the rainforests without destroying them—to develop rainforest ecosystems into sustainable resources for both local and worldwide benefit. Such agencies will need help from biologists familiar with tropical species and tropical ecosystems. Likewise, forestry practices in developed countries need to be managed so as to ensure a continued supply of trees. This means logging at a rate no greater than the rate of regrowth. Reforestation occurs readily in well-managed northern forests, but areas are too often clear-cut and a diverse forest is replanted with a single species of tree, a form of monoculture (Chapter 15). While this practice may maintain a supply of one species of trees, it does *not* maintain habitat.

Ecotourism.

If Brazilians and other tropical nations are to preserve the rainforest instead of destroying it, they must have economic incentives to do so. One type of economic incentive is the small but growing market for ecologically based tourism, also called nature tourism or green tourism. This type of tourism seeks to develop tourist destinations so that there is minimal or no impact on the natural environment. In Costa Rica, ecotourism is now that country's second largest source of foreign income, behind coffee and ahead of bananas. Before the 1994 civil war, ecotourism was also a major source of revenue in Rwanda, thanks largely to the presence of mountain gorillas in the Virunga region, along the border with Zaire. By comparison, the rainforests of Brazil afford a largely untapped tourist resource. Of course, a minimum of land would have to be set aside for airports, roads, and hotels. Beyond this initial investment, however, ecotourism would provide economic incentives to leaving the rest of the rainforest untouched. Ecotourism not only provides a country with income, it also gives that country an economic incentive to preserve its own natural heritage for the benefit of all.

In the search for ways to stop the destruction of ecosystems, no solution will work if the rich and poor nations continue at odds with one another. If these nations continue in their battles over short-range economic interests, the planetary ecosystem will undoubtedly be the loser.

THOUGHT QUESTIONS

1. Do you think undisturbed habitats have intrinsic value, or only instrumental value? In other words, is habitat valuable as an end in itself, or only because of the uses to which it might be put?
2. How much habitat destruction do you think takes place at the hands of wealthy people, and how much at the hands of poor people? (The answer may differ from country to country.) What would it take to secure the cooperation of both rich and poor people in an effort to halt desertification or rainforest destruction? Could you easily appeal to both rich and poor together, or

would it be easier to appeal to the two groups separately?

3. E. O. Wilson has estimated that one Peruvian farmer clearing land to grow food will cut down more species of trees than are native to all of Europe. Do you think there may have been greater numbers of trees in Europe before human populations grew to their present density? More species of trees?

4. Are there examples of habitat destruction occurring near where you live or go to college? What are they? What factors contribute to the destruction?

D. Pollution Threatens Much of Life on Earth

You have probably heard of toxic dumps or of the *Exxon Valdez* oil spill (Chapter 14). Toxic dumps and oil spills are two of the many kinds of pollution that threaten our environment. The original meaning of the verb *pollute* was to "contaminate," or "make dirty." Today, **pollution** may be defined as anything that is present in the wrong quantities or concentrations, in the wrong place, or at the wrong time. Although there is room for people to disagree about acceptable quantities, usually there is general agreement that pollution exists when it affects human health or kills other organisms. Crude oil from oil spills can kill thousands of aquatic birds, mammals, fish, and other organisms. Lead, cadmium, and other heavy metals can cause brain damage and other neurological defects. Toxic dumps can poison people and raise cancer rates. In some of the worst cases, numerous unidentified chemicals were dumped together into the same toxic waste site, forming a "witch's brew" that underwent further and often unpredictable chemical reactions to produce additional hazards. In the United States and certain other countries, it is now illegal to dispose of many chemicals except by government-approved methods.

Every society throws things away as garbage, and the things thrown away can tell us a good deal about a society. Garbage dumps (solid waste disposal sites) are among the largest human-made structures on Earth. The study of garbage (sometimes called "garbageology") is a small but growing research field in anthropology. By studying the remains of our throwaway culture, garbageologists can interpret the kinds of things we value, the ways in which we live, and the changes that have occurred in our culture over time.

When we flush our toilets or send our garbage to a landfill, we are polluting, by the definition given above, because we are contributing to the accumulation of solid and liquid wastes. When you drive your car, you are contributing to the pollution of the air that we all must breathe. All of us contribute to pollution in many different ways. Even our breathing releases carbon dioxide into the atmosphere.

Does this mean that every act mentioned in the previous paragraph is an immoral act? Certainly not. In order to live, we must breathe, eat, urinate, and defecate. Nearly every food product requires us to throw away inedible parts (skin, bones, pits, rinds, shells), packaging materials, or unfinished remains. To go on living, we must continue to pollute in certain minimal ways. So why all the fuss?

Detecting and Measuring Pollution

Pollution in most cases is a problem of quantities, and sometimes also of location. Clearly, you don't want your garbage to accumulate in your living room. Suppose, for the moment, that garbage disposal services were not available, and you had to dispose of your household garbage yourself, as many people in rural places still do. You could perhaps bury it in the backyard, or just toss it away. If you only tossed away bones, rinds, and other **biodegradable** materials (things that can be broken down by bacteria, fungi, and other *decomposer* organisms), then you might be able to dispose of your own garbage in this way. Up to a point, that is. Your backyard may have enough decomposer organisms to break down and recycle your own personal wastes, or perhaps even your family's wastes. Of course, this depends in part on the size of your family, and also on the size of your backyard. Clearly, your backyard would not be able to accommodate the garbage produced by an entire town or city.

Pollution: a problem of quantities. Pollution—whether it is residential or industrial—is thus a problem of quantities. Just about every known pollutant is harmless in some sufficiently small quantity. Pollutants become bothersome or toxic as quantities increase. In most cases, measuring pollution means measuring quantities, for pollution is a matter of quantities.

How do we know whether or not pollution exists, or to what extent? If pollution is known to occur in a particular place, how does one measure the extent to which it has been cleaned up? Most of the specific answers depend upon the measuring of particular chemical substances. First, a particular chemical pollutant or breakdown product must be identified as an indicator of the pollution in question. Second, the concentration of this particular chemical must be measured repeatedly at various places and times.

Solving pollution problems usually requires that quantities of various pollutants be monitored frequently. Regulatory agencies must measure the quantities of pollutants. Responsible companies must measure these same quantities themselves if they wish to avoid fines. Companies that perform environmental monitoring or that help to write environmental impact statements need people who are skilled in chemical analysis, especially quantitative analysis. The decision as to which chemical should be measured is largely a matter for chemists and also for physicians familiar with the health hazards caused by the pollutant. The special field of *toxicology* deals with the damage done to human (or animal) health by poisons, including environmental pollutants. Students interested in careers related to pollution control may therefore wish to obtain quantitative skills, not only in biology but also in physics, chemistry, and statistics.

Air pollution. *Air pollution* affects the air that we breathe, both indoors and outdoors. Automobile exhausts and industrial plants are major sources of outdoor pollution. Many outdoor air pollutants are oxides of carbon (carbon monoxide and carbon dioxide), oxides of nitrogen, and oxides of sulfur released as combustion products. Certain other gases, such as chlorine, benzene, and hydrogen sulfide, are also released by some industrial processes. Outdoor air pollution can also include certain bacterial pathogens, mold spores, and allergens such as pollen.

Air pollutants can become trapped in the ventilation systems of buildings and cause *indoor air pollution*. Indoor air pollution includes many of the same components as outdoor air pollution (including pollen), plus additional pollutants such as additional bacteria, benzene, or formaldehyde formed by the chemical decomposition of materials used in building construction. Buildings with poorly designed ventilation systems can impair the health of the people working in them, a phenomenon called "sick building syndrome" by some people. This may be one of the causes for a 70 percent increase in the incidence of allergies in the United States over the last two decades. Studies in England show that people working in older buildings with open windows as a source of ventilation suffer fewer respiratory illnesses than people working in newer, self-contained office buildings with recirculated air. Secondhand cigarette smoke is another form of indoor air pollution. A major study of this form of pollution in England showed that cancer rates among nonsmoking married women were much higher if their husband smoked than if he did not, presumably because these women were breathing the polluted indoor air containing the secondhand smoke (Chapter 9). Campaigns to banish smoking from restaurants and other public places are motivated by such health considerations.

Pollution prevention. Personal awareness is a crucial component in the prevention of pollution. Unless we realize how we are polluting, it is unlikely that we will take any corrective action to pollute less. It is therefore important for each of us to inform ourselves about the disposal of our garbage and our industrial waste, about the emissions from our cars and from nearby (and faraway) smokestacks, and the cleanliness of our beaches, playgrounds, drinking supplies, and foods.

Because pollution is a consequence of the ways in which we live and work, there are certain things we can each do to reduce the amount of pollution that we cause. Most of these measures also have further benefits, such as the saving of money or contributing to human health. For example, carpooling saves money, while bicycling to work saves even more money and often contributes to health and fitness as well. Recycling saves money, too, and we can all recycle such items as paper, bottles, and cans. Making an aluminum soda can from recycled materials requires only a small fraction of the electricity needed to make the same can from aluminum ore. Already, many industries are responding to market forces by using recycled materials in their products and by advertising their use of biodegradable or other "Earth-friendly" materials. This book, for example, is one of many printed on recycled paper.

Indicator species. More general indicators of pollution are also needed, especially pollution from hazards that have not yet been clearly identified. A number of environmentally sensitive species have been suggested as possible general indicators of pollution. These *indicator species* are sometimes also called "canary species," in reference to the coal miners' practice of taking canaries with them into the mines. Because the canaries were extremely sensitive to methane and other dangerous gases present in coal mines, the health of the canary reassured the miners and the sickness or death of a canary was always viewed by the miners as a danger signal. Frogs and other amphibians are used as canary species in freshwater habitats. Dolphins have occasionally been suggested as canary species for marine habitats because environmental pollution can stress the immune systems of these marine mammals (see Chapter 12) and raise their rate of infectious diseases. The grounding of marine mammals on beaches may also be a stress-related phenomenon that reflects marine pollution.

A Case in Point: Acid Rain

One of the most widespread and best understood pollution problems is that of acid rain. (We should really speak of "acid deposition," because much of the problem occurs as acid snow, acid fog, and dry deposition of acid dust or condensation.) The acidity of a substance is measured on a standard pH scale, where the lower the value, the higher the acidity (the more hydrogen ions). Distilled water has a pH of 7, most body tissues (such as blood) have pH values between 7.35 and 7.45, and ordinary rain has pH values between 5.5 and 6.0.

Rainwater of pH 5 or below is considered acid rain; values as low as 2.7 have been measured on occasion. A small change in pH can have an enormous effect on living organisms (Fig. 16.16).

Chemical tests of acid rain show that most of the acidity is due to either sulfuric acid (H_2SO_4) or nitric acid (HNO_3). Acid rain across the eastern United States and the Scandinavian countries (Norway, Sweden, etc.) is primarily of the sulfuric acid variety, while the acid content of acid rain in western United States locations like Denver, Colorado, is more than 50 percent nitric acid, derived primarily from the nitric oxides produced by automobile exhausts.

Sulfuric acid pollution begins when materials containing sulfur are burned in air, forming sulfur dioxide (SO_2). The sulfur dioxide combines with additional oxygen to make sulfur trioxide, which then combines with water to make sulfuric acid. Sulfur is a common impurity in many coal deposits and is also present in many of the ores from which lead, zinc, nickel, and certain other minerals are commonly obtained. In many parts of the United States, Canada, England, and Germany, the mining of metals from sulfur ores generates sulfur dioxide that ends up as acid rain in such downwind locations as the northeastern United States or Sweden. The burning of high-sulfur coal in electrical power plants is an even larger source of sulfur dioxide pollution that eventually falls as acid rain. This type of acid rain is a political as well as an environmental problem because the governments of New York or of Sweden are relatively powerless to control pollution that originates in Illinois, Indiana, or Germany. The United States and Canada have accused one another of being major sources of cross-border acid rain pollution.

Acid rain erodes and slowly dissolves marble and limestone statues and buildings (Fig. 16.16C), including those of the famous Parthenon in Athens, Greece. In localities where the rock formations are predominantly limestone and other carbonate rocks, acid rain is neutralized as it runs through these rocks or as it percolates through the soils derived from the weathering of these rocks. However, granitic rocks, such as those that predominate in New England, northern New York State, or Scandinavia, have little or no capacity to neutralize acid rain. As a result, acid deposition in

[A]

[B]

FIGURE 16.16
Some of the effects of acid rain. [A] Trees killed by acid rain in North Carolina. [B] Trout taken from the same lake in Ontario in 1979 at pH 5.4 (above) and 1982 at pH 5.1 (below). [C] Stone sculptures dissolving in acid rain.

those areas accumulates in ponds and lakes until it kills the fish. The Adirondacks of New York State contain hundreds of lakes and ponds that once teemed with fish, but whose fish populations have completely disappeared as a result of lake acidity.

Costs and Benefits

We have said that ethical decisions about pollution are decisions about quantities because the prob-

[C]

lem of pollution is, in large part, a problem of quantities. Expanding on this, we can say that the determination of both costs and benefits associated with pollution control depends upon the measurement of quantities. Whether costs are measured in dollars, in lives lost, or in the quality of human health, what counts is that they are measured. The same is true of benefits, which may be measured in dollars, in lives saved, or in the quality of human health or enjoyment.

One general problem is that costs are often easier to identify or to measure than benefits, and are subject to much less uncertainty. The costs of curtailing or modifying a manufacturing process can easily be measured in terms of costs for new equipment, costs to operate the equipment, and either increased wages (if additional employees are needed) or reduced employment (if an activity is curtailed). These costs are fairly certain and are easily measured. The benefits to the ecosystem are less certain and are harder to measure: If a particular set of changes is implemented, will the ecosystem recover? What levels will the pollutants reach at some distance from the source? How many lives will be saved or how much disease prevented as a result? What dollar value should be placed on the prevention of disease, on the bird or fish population, or on the health of the habitat as a whole?

Translating "quality" of health or enjoyment into something that can be quantified is a relatively new and important specialty within economics. One way of measuring these intangibles (certainly not the only way, but definitely one of the easiest ways) is by measuring the average dollar amount that people are willing to pay to obtain them. To take one small example, the value of a better neighborhood can be measured by the additional amount that people are willing to pay in order to live in it, compared with some other neighborhood. (This assumes, of course, that the people being studied have the means to make such a choice.) The value of a clean environment can be measured by the amounts of money that people are willing to forego in order to live in it, work in it, or visit it.

Attempts to measure environmental quality can be misunderstood if people look at the issue too narrowly. For example, if an indicator species is used to monitor environmental quality, many people will wrongly assume that the value of the indicator species is the only benefit that is worth considering. Salmon have a certain value as a commercial food species, but they have a far greater importance as a general indicator of pollution or of the health of river ecosystems as a whole. If salmon populations are used to measure pollution or pollution abatement in freshwater ecosystems, it is the value of the entire ecosystem that should be counted as a benefit, not just the commercial value of the salmon fishery. Likewise, bald eagles and spotted owls (neither of which have commercial value) may be used to indicate the general health of ecosystems, and it is the health of these entire ecosystems that should be counted as a benefit, not just the value we place on one or two indicator species. The death of the canary in the mine means much more to the miners than the price of a replacement bird.

THOUGHT QUESTIONS

1. How can one persuade legislators in one jurisdiction to spend money on antipollution measures if most of the damage occurs in other jurisdictions? For example, if acid rain in Norway and Sweden originates mostly in Germany, how can the people in Norway and Sweden influence legislators to change German laws and/or industrial practices?

2. Materials that cannot easily be recycled are generally taken to places specifically set aside as *dumps*. If such facilities are needed, where should they be located? Because dumps are generally unsightly, unclean (sometimes toxic), and subject to noisy traffic, people generally don't want to live near them. The motivating desire to have these facilities somewhere else is often symbolized by the phrase "not in my backyard" (NIMBY). One result is that dumps are often located wherever people have the least political clout. What further social problems arise from the widespread application of the NIMBY principle? Is there any socially responsible way to locate an undesirable facility? What is the best way to reduce the need for such facilities in the first place?

CHAPTER SUMMARY

The number and variety of species are referred to as biodiversity. Speciation increases biodiversity, while extinction decreases it. Maximum biodiversity occurs in tropical rainforests and in coral reefs. Habitats support the interdependent lives of biological communities of species. When conditions change, one community may be replaced by another in a process called succession. Destruction of these habitats threatens the survival of all the species that live there. Since a majority of the world's photosynthesis and oxygen production occurs in rainforests, their destruction is a worldwide threat to the atmosphere.

Pollution is a complex matter; solving pollution problems means much more than allocating blame. In most cases, it means measuring quantities, for pollution is a matter of quantities. Ethical decisions about pollution are also decisions about quantities because neither costs nor benefits can be measured without measuring all the quantities involved.

KEY TERMS TO KNOW

biodegradable (p. 485)
biodiversity (p. 459)
biomass (p. 461)
biomes (p. 472)
biosphere (p. 460)
climax community
  (p. 478)
communities (p. 460)

desertification (p. 481)
ecosystem (p. 460)
endangered species
  (p. 471)
energy-stability-area
  (ESA) theory (p. 460)
epiphytes (p. 473)
extinct (p. 464)

habitat (p. 459)
inbreeding depression
  (p. 471)
lineage (p. 464)
niche (p. 460)
pollution (p. 485)
pseudoextinction
  (p. 464)

slash-and-burn agricul-
  ture (p. 481)
succession (p. 478)
sustainable uses
  (p. 483)
tropical rainforests
  (p. 472)
true extinction (p. 464)

CONNNECTIONS TO OTHER CHAPTERS

Chapter 2   Pollution and habitat destruction violate several ethical injunctions, including the principle of doing no harm to others.

Chapter 3   Genetic diversity is an important component of biodiversity.

Chapter 4   Species arise through evolution.

Chapter 5   Gene frequencies change in populations.

Chapter 6   Carrying capacity is one factor that regulates population size.

Chapter 12   Pollution stresses the immune systems of many species.

Chapter 14   Cleaning up pollution is an important bioengineering problem.

Chapter 15   Plants are energy producers in every ecosystem and thus support many other species.

Color Figure A Batesian mimicry between a model species, the monarch butterfly (left) and its mimic, the viceroy (right).

Color Figure C Several woody vines (lianas) are visible in this photo, along with the aerial roots (visible along the tall trunk) of epiphytic plants growing high above. Notice also the differing light levels beneath the canopy.
[Photo courtesy of Sharon Kinsman.]

Color Figure B Geographic variation and mimicry in African butterflies. Left column: three unpalatable species *(Danaus chrysippus, Amauris crawshayi, Amauris niavius)*. Middle column: the African swallowtail butterfly *(Papilio dardanus)*, showing three different mimics, each occurring in the same geographic locality as the corresponding model shown to its left, and the nonmimetic form (top) from Madagascar, where no suitable model occurs. Right column, upper pair: *Hypolimnas misippus*, a species in which the male (top) is nonmimetic, while the female is a mimic. Right column, lower pair: *Hypolimnas dubius wahlbergi*, a species that mimics different species of *Amauris* (left column) in different localities.
[From W. Wickler, Mimicry in Plants and Animals. *McGraw-Hill, 1968, p. 29, Figure 4b.*]

Color Figure D Trees in tropical rain-forests often have buttress supports that give the base of their trunks star-like cross-sectional shapes.
[Photo by Richard Thom/Visuals Unlimited.]

Color Figure E Slash-and-burn agriculture quickly depletes the soil of its nutrients, requiring the fields to be abandoned and new fields to be cleared by burning more forest. *[Photo by G. Prance/Visuals Unlimited.]*

Color Figure F These slopes in Costa Rica were once forested, but the harvesting of trees had increased the frequency of landslides and the rate of erosion. *[Photo courtesy of Sharon Kinsman.]*

Color Figure G Orchids and other epiphytes growing on a tree in Central America. *[Photo by Max and Bea Hunn / Visuals Unlimited.]*

Color Figure H This *Guzmania nicaraguensis* (fam. Bromeliacae) has yellow flowers surrounded by modified leaves called *bracts,* whose bright red color attracts the hummingbirds that pollinate this species. Bright red colors are common in bird-pollinated plants. *[Photo courtesy of Sharon Kinsman.]*

Color Figure J This harlequin frog *(Atelopus)* in Costa Rica shows warning coloration as a deterrent to predators. *[Photo courtesy of Sharon Kinsman.]*

Color Figure I The poison arrow frog, *Dendrobates,* which was photographed in Costa Rica. The skin of all frogs in the family Dendrobatidae contain distasteful secretions that act to deter predators, and the red color of the frog serves to warn those predators. In South America, another species in this family is used by the Choco Indians of Columbia, who rub their blowgun darts on the frog skins. The skin secretions contain curare, a poison that inhibits the neurotransmitter acetylcholine; sufficient amounts of this poison can paralyze the muscles into which the dart lands. *[Photo courtesy of Sharon Kinsman.]*

Color Figure K Among the many species of birds that inhabit the rainforest are these five species of South American hummingbirds: l. Violet-tailed sylph *(Aglaiocercus coelestis);* 2. Crimson topaz *(Topaza pella);* 3. Wire-crested thorntail *(Popelairia popelairii);* 4. Tufted coquette *(Lophornis ornata);* 5. Booted rackettail *(Ocreatus underwoodii).* *[From Hopson & Wessells, p. 787. Figure 42-13.]*

Color Figure L Coral reef diversity in the Red Sea, Egypt. [Photo by Hal Beral / Visuals Unlimited.]

Color Figure M A bronzy hermit hummingbird visiting a passion flower in Costa Rica.
[Photo by Michael Fogden.]

Color Figure N Squirrel monkey *(Saimiri sciurea)* in Costa Rica. [Photo by Michael Fogden.]

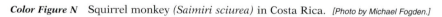

A Six-Kingdom Classification of Organisms

Classifications reflect the relatedness of different groups of organisms, as shaped by their evolution. Since the course of evolution is not fully known in every detail, classifications are very elaborate hypotheses, subject to continual revision, about the common descent of the species included in each group. Because of the complexity of the hypotheses reflected in classification schemes, classifications frequently differ from one another. Algae, for example, are sometimes listed as plants and sometimes as protists. Some biologists now group kingdoms III through VI into a "Domain Eucarya." See Box 4.3 for the distinctions between procaryotic and eucaryotic cells. Not included in this classification are the viruses, which many experts do not consider to be organisms because of their inability to carry out life functions independently of their hosts.

I. KINGDOM ARCHAEA

Phylum ARCHAEBACTERIA. A small group of simple, bacteria-like organisms with procaryotic cells. Nucleic acid sequences set these organisms apart from all other kingdoms.

> **METHANOGENS.** Methane-producing organisms.
> **EXTREME THERMOPHILES.** Organisms adapted to living in hot springs and similar environments, with high optimal growing temperatures.
> **EXTREME HALOPHILES.** Organisms adapted to extremely salty environments.

II. KINGDOM MONERA, OR PROKARYOTAE

This group includes the vast majority of procaryotic organisms, with nucleic acid sequences showing homology (similarity by common descent) to the nucleic acids of eucaryotic organisms.

Phylum BACTERIA (or "EUBACTERIA"). Procaryotic Monera that lack chlorophyll *a* and do not release oxygen.
Phylum CYANOBACTERIA. Procaryotic Monera, structurally similar to bacteria, but possessing chlorophyll *a* and releasing oxygen during photosynthesis.

III. KINGDOM PROTISTA

Eucaryotic unicells lacking plastids or other features of kingdoms IV through VI. *NOTE:* This definition excludes the algae, included below in the plant kingdom. Many biologists define the Protista somewhat differently to include the algae.

Phylum SARCODINA. Protists that move by extending pseudopods. Subphylum Rhizopodia includes protists with lobe-shaped pseudopods; subphylum Actinopodia includes protists with long, needle-shaped pseudopods. Some experts list these as two separate phyla.
Phylum MASTIGOPHORA (FLAGELLATA). Protists that move with the help of one to several whiplike flagella.
Phylum CILIATA. Protists that move by the beating of numerous hairlike cilia.
Phylum SPOROZOA. Nonmotile protists that reproduce using spores, including the species that cause malaria (Chapter 5).

IV. KINGDOM PLANTAE

Eucaryotic organisms possessing plastids. Most are multicellular and nearly all carry out photosynthesis using at least one other pigment in addition to chlorophyll *a*. Cell walls often present (Chapters 8, 15).

Subkingdom Thallophyta. Eucaryotic algae: Fertilized eggs (zygotes) not surrounded by sterile nonreproductive tissue.

Divisions include:

CHRYSOPHYTA. Golden-yellow algae and diatoms (Fig. 14.17).
PYRROPHYTA (DINOPHYTA). Dinoflagellates.
XANTHOPHYTA. Yellow-green algae.
EUGLENOPHYTA. *Euglena* and its relatives.
RHODOPHYTA. Red algae.
PHAEOPHYTA. Brown algae.
CHLOROPHYTA. Green algae.

Subkingdom Embryophyta. Plants in which zygotes are contained within an *embryo* in which they are surrounded by sterile nonreproductive tissue.

Bryophyta. Nonvascular embryophytes, including mosses and liverworts (16,000 species).
No specialized vascular tissue for conducting water throughout the plant; therefore, must live in moist areas.
No specialized support tissues; therefore, most are short.
Sperm must "swim" to reach the egg; therefore, fertilization requires moist conditions.
Haploid and diploid phases occur together; haploid phase is more prominent and diploid plant never grows independently.

Divisions include:

BRYOPHYTA. Mosses.
HEPATOPHYTA. Liverworts and hornworts.

Tracheophyta. Vascular plants (250,000 species).
Specialized tissue (**xylem**) carries water and minerals throughout the plant; other specialized tissue (**phloem**) carries the products of photosynthetic reactions throughout the plant (Chapter 14 and Fig. 15.4).
Diploid phase is more prominent than haploid phase.

Divisions of Tracheophyta without seeds:

PSILOPHYTA. Psilophytes (p. 466).
LEPIDOPHYTA. Club mosses and other lepidophytes.
ARTHOPHYTA, or SPHENOPSIDA. Horsetails and their relatives.
PTEROPHYTA. Ferns.

Divisions of Tracheophyta possessing seeds:

PTERIDOSPERMOPHYTA. Seed ferns.
CYCADOPHYTA. Cycads.
GINKGOPHYTA. Ginkgos (Figure 16.3).
CONIFEROPHYTA. Pines, spruces, and other cone-bearing plants (conifers).
GNETOPHYTA. *Gnetum, Ephedra,* and *Welwitschia.*

ANGIOSPERMAE, or ANTHOPHYTA. Flowering plants, the largest and most successful group (Figs. 14.13, 15.5, 16.9, 16.11, Box 15.1, and Color Figs. C, D, G, H, and M).

V. KINGDOM FUNGI, OR MYCOTA

Eucaryotic organisms possessing cell walls but no plastids, usually reproducing by spores, and carrying out absorptive nutrition.

Subkingdom MYXOMYCOTA. Slime molds (Fig. 9.1). Life cycle alternates between motile unicells and creeping multicellular or multinucleated aggregates that form fruiting bodies.
Subkingdom EUMYCOTA. True fungi, usually possessing branched threadlike absorptive filaments (hyphae).

VI. KINGDOM ANIMALIA

Multicellular eucaryotic organisms lacking plastids and developing from a hollow ball of cells (blastula, see Fig. 9.8).

Phylum PORIFERA. Sponges (Fig. 9.2). Animals possessing specialized cells but no organized tissues.
Phylum MESOZOA. A small group of species, containing organisms with only a few cells each, without organized tissues.
Phylum CNIDARIA. The coelenterates, or cnidarians, a diverse group possessing stinging cells, and including corals, anemones, hydroids, and jellyfish (Figs. 9.2, 16.2, and Color Fig. L). Two tissue layers (endoderm and ectoderm) present.
Phylum CTENOPHORA. "Comb jellies," animals with a "biradial" symmetry, like that of a two-armed pinwheel.
Phylum PLATYHELMINTHES. Flatworms, including planarias, parasitic flukes, and parasitic tapeworms. Tissues formed by three germ layers (ectoderm, mesoderm, endoderm) but no body cavity present.
Phylum RHYNCHOCOELA, or NEMERTEA. Ribbon worms; similar to flatworms but with a protrusible head structure (proboscis).
Phylum NEMATODA. Roundworms, including both free-living and parasitic species, second in abundance on Earth only to the arthropods. This and the next few phyla possess a body cavity called a *pseudocoel,* bordered by endoderm as well as mesoderm.

Phylum KINORHYNCHA. Parasitic worms related to roundworms.

Phylum GASTROTRICHA. Small worm-like animals related to roundworms.

Phylum GORDIACEA. Elongated "horsehair worms."

Phylum ROTIFERA. Microscopic aquatic organisms containing a ring of cilia that beat in a circular pattern resembling a rotating wheel.

Phylum ACANTHOCEPHALA. A group of parasitic worms with hook-studded heads.

Phylum ENTOPROCTA. A small group of filter-feeders (animals that strain small particles from the water).

NOTE: Phyla listed from this point on have a true body cavity (*coelom*) lined with mesoderm throughout.

Phylum PHORONIDA. A small group of filter-feeding wormlike animals. This and the next two groups feed by means of a cilia-covered structure known as a lophophore.

Phylum BRYOZOA. Small "moss animals," usually living in colonies, their numerous small tentacles giving them a mosslike appearance (Fig. 16.2).

Phylum BRACHIOPODA. Filter-feeding animals with a shell composed of two unequal parts and a stalk that attaches the adults to a fixed location (Figs. 16.2 and 16.3).

Phylum MOLLUSCA. A large and diverse group of animals possessing a cavity lined with a layer of cells (the mantle) that usually secretes some kind of an inflexible shell of calcium carbonate (Fig. 14.5).

Class MONOPLACOPHORA. *Neopilina* (p. 466) and its relatives; uncoiled mollusks with a one-part shell but repeated internal organs.

Class APLACOPHORA. Wormlike mollusks (solenogasters) with no shells.

Class POLYPLACOPHORA. Flattened mollusks (chitons) with multipart shells, feeding on the algae that they scrape from rocky surfaces.

Class GASTROPODA. Snails, with a one-part (univalve) shell that is most often coiled asymmetrically.

Class SCAPHOPODA. Small, symmetrical, tusklike shells.

Class BIVALVIA, or PELECYPODA. Clams, oysters, and other mollusks with two-part (bivalve) shells.

Class CEPHALOPODA. Ammonoids, nautiloids, squids, and octopus (see Box 4.1 and Fig. 16.4).

Phylum ANNELIDA. Segmented worms, including earthworms and sandworms.

Phylum SIPUNCULIDA. Peanut-shaped worms.

Phylum ECHIURIDA. A small wormlike group.

Phylum PENTASTOMIDA. Small parasitic worms whose head and claws give the appearance of a five-branched head.

Phylum TARDIGRADA. Tiny aquatic "water bears," segmented animals with clawed appendages.

Phylum ONYCHOPHORA. Segmented, wormlike organisms with clawed appendages.

Phylum ARTHROPODA. Animals with jointed legs, the largest phylum of all. An external skeleton composed of chitin and strengthened in many species with calcium carbonate (Chapter 14), permitting hingelike movements between more rigid parts.

Subphylum TRILOBITOMORPHA. Extinct trilobites (Fig. 16.2).

Subphylum CRUSTACEA. Crabs, lobsters, shrimp, barnacles, and so forth.

Subphylum CHELICERATA. Arthropods in which the first pair of appendages beyond the mouth are seizing structures called *chelicerae*, and usually with four pairs of walking legs.

Class PYCNOGONIDA. Sea spiders.

Class MEROSTOMATA. Horseshoe crabs. (Fig. 16.3).

Class ARACHNIDA. Scorpions, spiders, mites, and so forth.

Subphylum UNIRAMA. Arthropods in which the first pair of appendages beyond the mouth are a pair of food-tearing mandibles.

Class CHILOPODA. Centipedes.

Class DIPLOPODA. Millipedes.

Class PAUROPODA.

Class SYMPHYLA.

Class INSECTA. Insects, with three pairs of walking legs and with a body divided into head, thorax, and abdomen (Figs. 4.3, 4.4, 7.8, 16.1, and Color Figs. A, B).

Phylum CHAETOGNATHA. Arrow worms.

Phylum ECHINODERMATA. Crinoids, starfish, sea urchins, and so forth, possessing a water-vascular system, numerous tubelike feet, and in many cases a five-part radial symmetry (Figs. 7.4, 14.7, and 16.2).

Phylum HEMICHORDATA. Acorn worms, pterobranchs, and graptolites, related to the Chordata but not sharing all chordate characteristics.

Phylum CHORDATA. Animals with a notochord (a stiff, flexible rodlike structure), a dorsal hollow nerve cord, and gill slits, each at some stage of life.

Subphylum UROCHORDATA. Sea squirts (tunicates), salps, and their relatives, with actively swimming larval stages and generally with nonmotile filter-feeding adults.

Subphylum CEPHALOCHORDATA. Sea lancets such as amphioxus, with motile filter-feeding adult stages.

Subphylum VERTEBRATA. Animals with a backbone.

Class AGNATHA. Jawless fish.

Class PLACODERMI. Extinct, heavily armored fish with jaws.

Class CHONDRICHTHYES. Sharks and their relatives, with skeletons made of cartilage instead of bone.

Class OSTEICHTHYES. Bony fish with advanced types of scales (Fig. 7.3B and Color Fig. L).

Class AMPHIBIA. Frogs, toads, salamanders, and their relatives, reproducing by eggs laid in water (Figs. 4.10, 4.16, and Color Figs. I, J).

Class REPTILIA. Snakes, turtles, dinosaurs, lizards, crocodiles, and related animals, reproducing by eggs surrounded by a saclike structure (amnion) and laid on land in most cases.

Class AVES. Feathered vertebrates (birds) (Figs. 4.2, 7.2, 7.3A,D, 7.5, 7.6A, 7.9A, and Color Figs. K, M).

Class MAMMALIA. Mammals, including all species possessing hair and feeding milk to their young (Figs 4.1, 4.12, 4.13, 7.3C, 7.6B, 7.9B,C, 7.11, 7.12, and Color Fig. N).

Bibliography

Chapter 1

Achinstein, P. 1983. *The Concept of Evidence.* New York: Oxford University Press.

Bigelow, J., and R. Pargetter. 1990. *Science and Necessity.* New York: Cambridge University Press.

Brody, B. A., and R. E. Grandy, eds. 1989. *Readings in the Philosophy of Science,* 2nd ed. Englewood Cliffs, N.J.: Prentice-Hall.

Cann, R. L., M. Stoneking, and A. C. Wilson. 1987. Mitochondrial DNA and human evolution. *Nature,* 325: 31–36.

Davis, B. D., ed. 1991. *The Genetic Revolution.* Baltimore: Johns Hopkins University Press.

Depew, D. J., and B. H. Weber, eds. 1985. *Evolution at a Crossroads: The New Biology and the New Philosophy of Science.* Cambridge, Mass.: MIT Press.

Doyal, L. 1986. *Empiricism, Explanation, and Rationality.* London: Routledge & K. Paul.

Earman, J., ed. 1983. *Testing Scientific Theories.* Minneapolis: University of Minnesota Press.

Fuller, S. 1993. *Philosophy of Science and its Discontents,* 2nd ed. New York: Guilford.

Gutting, G. 1980. *Paradigms and Revolutions: Appraisals and Applications of Thomas Kuhn's Philosophy of Science.* Notre Dame, Ind.: University of Notre Dame Press.

Hacking, I. 1981. *Scientific Revolutions.* New York: Oxford University Press.

Harre, R. 1985. *The Philosophies of Science.* New York: Oxford University Press.

Hesse, M. B. 1980. *Revolutions and Reconstructions in the Philosophy of Science.* Bloomington, Ind.: Indiana University Press.

Hull, D. L. 1988. *Science as a Process: An Evolutionary Account of the Social and Conceptual Development of Science.* Chicago: University of Chicago Press.

Kitcher, P., and W. C. Salmon, eds. 1989. *Scientific Explanation.* Minneapolis: University of Minnesota Press.

Kuhn, T. S. 1970. *The Structure of Scientific Revolutions,* 2nd ed. Chicago: University of Chicago Press.

Laudan, L. 1984. *Science and Values: The Aims of Science and their Role in Scientific Debate.* Berkeley: University of California Press.

Lederberg, J., and E. L. Lederberg. 1952. Replica plating and indirect selection of bacterial mutants. *J. Bacteriol.,* 63: 399–406.

Merton, R. K. 1970. *Science, Technology, and Society in Seventeenth Century England.* New York: H. Fertig.

Nitecki, M. H., and D. V. Nitecki. 1992. *History and Evolution.* Albany, N.Y.: State University of New York Press.

Popper, K. R. 1959. *The Logic of Scientific Discovery.* New York: Basic Books.

Simon, H. A. 1977. *Models of Discovery and Other Topics in the Methods of Science.* Boston: D. Reidel.

Simpson, G. G. 1980. *Why and How: Some Problems and Methods in Historical Biology.* New York: Pergamon.

Watson, J. D. 1968. *The Double Helix.* New York: Atheneum.

Chapter 2

Beauchamp, T. L., and L. Walters, eds. 1978. *Contemporary Issues in Bioethics.* Belmont, Calif.: Wadsworth.

Cohen, C. 1986. The case for the use of animals in biomedical research. *New Eng. J. Med.,* 315: 865–870.

Copp, D., and D. Zimmerman, eds. 1984. *Morality, Reason, and Truth.* Totowa, N.J.: Rowman & Allanheld.

Dawkins, M. S. 1991. *Ethics in Research on Animal Behaviour.* London: Academic Press.

Edel, M., and A. Edel. 1959. *Anthropology and Ethics.* Springfield, Ill.: Charles C Thomas.

Fox, M. W. 1986a. *The Case for Animal Experimentation. An Evolutionary and Ethical Perspective.* Berkeley, Calif.: University of California Press.

Fox, M. W. 1986b. *Returning to Eden. Animal Rights and Human Responsibility.* Malabar, Fla.: Robert E. Krieger.

Haan, N., E. Aerts, and B. A.B. Cooper. 1985. *On Moral Grounds. The Search for Practical Morality.* New York: New York University Press.

Haber, J. G., ed. 1994. *Absolutism and its Consequentialist Critics.* Lanham, Md.: Rowman and Littlefield.

Hardin, R. 1988. *Morality within the Limits of Reason.* Chicago: University of Chicago Press.

Hargrove, E. C., ed. 1992. *The Animal Rights/ Environmental Ethics Debate.* Albany, N.Y.: State University of New York Press.

Kagan, S. 1989. *The Limits of Morality.* Oxford, U.K.: Clarendon Press.

Mappes, T. A., and J. S. Zembaty. 1992. *Social Ethics. Morality and Social*

Policy, 4th ed. New York: McGraw-Hill.

Miller, H. B., and W. H. Williams, eds. 1983. *Ethics and Animals.* Clifton, N.J.: Humana Press.

Newkirk, I. 1992. *Free the Animals!* Chicago: Noble Press.

Phillips, M. T., and J. A. Sechzer. 1989. *Animal Research and Ethical Conflict.* New York: Springer-Verlag.

Preece, R., and L. Chamberlain. 1993. *Animal Welfare and Human Values.* Waterloo, Ont.: Wilfrid Laurier University Press.

Rawls, J. 1972. *A Theory of Justice.* Cambridge, Mass.: Belknap Press, Harvard University Press.

Regan, T. 1982. *All that Dwell Therein. Animal Rights and Environmental Ethics.* Berkeley: University of California Press.

Rodd, R. 1990. *Biology, Ethics, and Animals.* Oxford: Clarendon Press.

Rollin, B. E. 1992. *Animal Rights and Human Morality*, revised ed. Buffalo, N.Y.: Prometheus Books.

Scheffler, S., ed. 1988. *Consequentialism and its critics.* Oxford, U.K.: Oxford Univertsity Press.

Singer, P., ed. 1985. *In Defense of Animals.* New York: Basil Blackwell.

Singer, P., ed. 1991. *A Companion to Ethics.* Cambridge, Mass.: Basil Blackwell.

Sterba, J. S., ed. 1994. *Morality in Practice*, 4th ed. Belmont, Calif.: Wadsworth Publishing Co.

U.S. Congress, Office of Technology Assessment. 1993. *Biomedical Ethics in U. S. Public Policy—Background Paper*, OTA-BR-BBS-105. Washington D.C.: U. S. Government Printing Office.

Williams, J., ed. 1991. *Animal Rights and Welfare.* New York: H. W. Wilson.

Chapter 3

Anderson, W. F. 1992. Human gene therapy. *Science*, 256: 808–813.

Avery, O. T., C. M. MacLeod, and M. McCarty. 1944. Studies on the chemical nature of the substance inducing transformation of pneumococcal types. Induction of transformation by a desoxyribonucleic acid fraction isolated from pneumococcus Type III. *J. Experimental Med.*, 79: 137–158.

Bishop, J. E., and M. Waldholz. 1990. *Genome.* New York: Simon and Schuster.

Bory, C., et al. 1991. Effect of polyethylene glycol-modified adenosine deaminase (PEG-ADA) therapy in two ADA-deficient children: measurement of erythrocyte deoxyadenosine triphosphate as a useful tool. *Adv. Exp. Med. Biol.*, 309A: 173–176.

Botstein, D., et al. 1980. Construction of a genetic linkage map in man using restriction fragment polymorphisms. *Am. J. Human Genet.*, 32: 314–331.

Brennan, J. R. 1985. *Patterns of Human Heredity.* Englewood Cliffs, N.J.: Prentice-Hall.

Cox, G. A., et al. 1993. Overexpression of dystrophin in transgenic *mdx* mice eliminates dystrophic symptoms without toxicity. *Nature*, 364: 725–729.

D'Alton, M. E., and A. H. DeCherney. 1993. Prenatal diagnosis. *New Engl. J. Med.*, 328: 114–120.

Davies, K. E., ed. 1986. *Human Genetic Diseases, A Practical Approach.* Washington, D.C.: IRL Press.

Davis, B. D., ed. 1991. *The Genetic Revolution.* Baltimore: Johns Hopkins University Press.

Edlin, G. 1990. *Human Genetics. A Modern Synthesis.* Boston: Jones and Bartlett.

Elmer-Dewitt, P. 1993. Cloning: Where do we draw the line? *Time*, Nov. 8, 1993, pp. 65–70.

Ferrari, G., *et al* . 1992. Transfer of the ADA gene into human ADA-deficient T lymphocytes reconstitutes specific immune functions. *Blood*, 80: 1120–1124.

Gelehrter, T. D. 1990. *Principles of Medical Genetics.* Baltimore: Williams and Wilkins.

Goldsby, R. 1971. *Race and Races.* New York: Macmillan.

Gostin, L., ed. 1990. *Surrogate Motherhood. Politics and Privacy.* Bloomington: Indiana University Press.

Gould, S. J. 1981. *The Mismeasure of Man.* New York: W. W. Norton.

Grompe, M., et al. 1991. Gene therapy in man & mice: adenosine deaminase deficiency, ornithine transcarbamylase deficiency, and Duchenne muscular dystrophy. *Adv. Exp. Med. Biol.*, 309B: 51–56.

Gusella, J. F., et al. 1983. A polymorphic DNA marker genetically linked to Huntington's disease. *Nature*, 306: 234–238.

Haller, M. H. 1963. *Eugenics: Hereditarian Attitudes in American Thought.* New Brunswick, N.J.: Rutgers University Press.

Holtzman, N. A. 1989. *Proceed with Caution.* Baltimore: Johns Hopkins Press.

Hubbard, R., and E. Wald. 1993. *Exploding the Gene Myth.* Boston: Deacon Press.

Jordan, I. K. 1991. Ethical issues in the genetic study of deafness. *In* Genetics of Hearing Impairment; R. J. Ruben, T. R. van de Water, and K. P. Steel, eds., *Ann. N.Y. Acad. Sci.*, 630: 236–239.

Kamin, L. J. 1974. *The Science and Politics of IQ.* Potomac, Md.: L. Erlbaum Associates.

Levitan. M. 1988. *Textbook of Human Genetics*, 3rd ed. New York: Oxford University Press.

Lewin, B. 1990. *Genes IV.* New York: Oxford University Press.

Lewontin, R. C., S. Rose, and L. J. Kamin. 1984. *Not in Our Genes.* New York: Pantheon.

Lusher, J. M., et al. 1993. Recombinant factor VIII for the treatment of previously untreated patients with hemophilia A. *New Engl. J. Med.*, 328: 453–459.

Marx, J. L. 1989. The cystic fibrosis gene is found. *Science*, 245: 923–925.

Muller, H. J. 1961. Human evolution by voluntary choice of germ plasm. *Science*, 134: 643–649.

Muller, H. J. 1965. Means and aims in human genetic development. *In* T. M. Sonneborn, ed. *The Control of Human Heredity and Evolution.* New York: Macmillan, pp. 100–122.

Pierce, B. A. 1990. *The Family Genetic Sourcebook.* New York: Wiley.

Ridley, M. 1993. A boy or a girl? Is it possible to load the dice? *Smithsonian* (June 1993): 113–123.

Roberts, L. 1987. Who owns the human genome? *Science*, 237: 358–361.

Sayre, A. 1975. *Rosalind Franklin and DNA.* New York: W. W. Norton.

Shannon, T. A. 1988. *Surrogate Motherhood: The Ethics of Using Human Beings.* New York: Crossroad.

Sinclair, A. H., et al. 1990. A gene from the human sex-determining region encodes a protein with homology to a conserved DNA-binding motif. *Nature*, 346: 240–244.

Singer, M., and P. Berg. 1991. *Genes and Genomes. A Changing Perspective.* Mill Valley, Calif.: University Science Books.

Suzuki, D., and P. Knudston. 1989. *Genethics: The Clash Between the New Genetics and Human Values.* Cambridge, Mass.: Harvard University Press.

United States, House of Representatives, Committee on Energy and Commerce. 1988. *OTA Report on the*

Human Genome Project. Washington, D.C.: U. S. Government Printing Office.

Vogel, F., and A. G. Motulsky. 1986. *Human Genetics: Problems and Approaches,* 2nd ed. Berlin: Springer-Verlag.

Watson, J. D. 1968. *The Double Helix.* New York: Atheneum.

Watson, J. D. 1990. Statement to Congress. *In* The Human Genome Project. Hearing before the Subcommittee on Energy Research and Development of the Committee on Energy and Natural Resources, U. S. Senate, 101st Congress, 2nd session, S. Hrg. 101–894.

Wilfond, B. S., and N. Frost. 1990. The cystic fibrosis gene: Medical and social implications for heterozyote detection. *JAMA,* 263:2777–2783.

Whitfield, L. S., R. Lovell-Badge, and P. N. Goodfellow. 1993. Rapid sequence evolution of the mammalian sex-determining gene SRY. *Nature,* 364: 713–715.

Chapter 4

Bowler, P. J. 1990. *Charles Darwin: The Man and His Influence.* Oxford, U.K.: Blackwell.

Briggs, D., and S. M. Walters. 1984. *Plant Variation and Evolution.* Cambridge: Cambridge University Press.

Darwin, C. R. 1859. *On the Origin of Species by Means of Natural Selection, or the Preservation of the Favoured Races in the Struggle for Life.* London: John Murray.

Darwin, C. R. 1871. *The Descent of Man, and Selection in Relation to Sex.* 2 vols. London: John Murray.

Darwin, C. R. 1992. *The Darwin CD-ROM.* San Francisco: Lightbinders.

Desmond, A. J., and J. Moore. 1992. *Darwin.* New York: Warner Books.

Dobzhansky, Th.G. 1970. *Genetics of the Evolutionary Process.* New York: Columbia University Press.

Eldredge, N. 1982. *The Monkey Business: A Scientist Looks at Creationism.* New York: Washington Square Press.

Futuyma, D. J. 1986. *Evolutionary Biology,* 2nd ed. Sunderland, Mass.: Sinauer Associates.

Gish, D. T. 1978. *Evolution: The Fossils Say No!* San Diego, Calif.: Creation-Life Publishers.

Kitcher, P. 1982. *Abusing Science: The Case against Creationism.* Cambridge, Mass.: MIT Press.

Leakey, M. G., et al. 1995. New four-million-year-old hominid species from Kanapoi and Allia Bay, Kenya. *Nature,* 376: 565–571.

Mayr, E. 1963. *Animal Species and Evolution.* Cambridge, Mass.: Belknap Press, Harvard University Press.

Mayr, E. 1991. *One Long Argument: Charles Darwin and the Genesis of Modern Evolutionary Thought.* Cambridge, Mass.: Harvard University Press.

Mayr, E., and P. D. Ashlock. 1991. *Principles of Systematic Zoology.* New York: McGraw-Hill.

McGowan, C. 1984. *In The Beginning: A Scientist Shows Why the Creationists Are Wrong.* Buffalo, N.Y.: Prometheus Books.

Minkoff, E. C. 1987. *The Case for Evolution.* Great Neck, N.Y.: Todd & Honeywell.

Minkoff, E. C. 1983. *Evolutionary Biology.* Reading, Mass.: Addison Wesley.

Morris, H. M., ed. 1974. *Scientific Creationism.* San Diego, Calif.: Creation-Life Publishers.

National Academy of Sciences. 1984. *Science and Creationism.* Washington, D.C.: National Academy Press.

Numbers, R. L. 1992. *The Creationists.* New York: Knopf.

Oparin, A. I. 1938. *The Origin of Life.* Translated by S. Morgulis. New York: Macmillan; reprinted 1953, New York: Dover.

Ospovat, D. 1981. *The Development of Darwin's Theory.* Cambridge, U.K.: Cambridge University Press.

Richards, R. J. 1992. *The Meaning of Evolution.* Chicago: University of Chicago Press.

Schopf, J. W. 1974. Paleobiology of the Precambrian: The age of blue-green algae. *Evol. Biology* 7: 1–43.

Strahler, A. N. 1987. *Science and Earth History: The Evolution/Creation Controversy.* Buffalo, N.Y.: Prometheus Books.

Weiner, J. 1994. *The Beak of the Finch. The Story of Evolution in Our Time.* New York: Knopf.

White, M. J. D. 1978. *Modes of Speciation.* San Francisco: W. H. Freeman.

Chapter 5

Beutler, E. 1992. The molecular biology of G-6-PD variants and other red cell enzyme defects. *Ann. Rev. Medicine,* 43: 47–59.

Cann, R. L., M. Stoneking, and A. C. Wilson. 1987. Mitochondrial DNA and human evolution. *Nature,* 325: 31–36.

Cavalli-Sforza, L. L. et al. 1988. Reconstruction of human evolution: Bringing together genetic, archaeo-logical, and linguistic data. *Proc. Natl. Acad. Sci.,* 85: 6002–6006.

Cavalli-Sforza, L. L., P. Menozzi, and A. Piazza. 1994. *The History and Geography of Human Genes.* Princeton, N.J.: Princeton University Press.

Chifu, Y., et al. 1992. Beta-thalassemia major resulting from a compound heterozygosity for the beta-globin gene mutation: further evidence of multiple origin and migration of the thalassemia gene. *Hum. Genet.,* 89(3): 343–346.

Dickerson, R. E., and I. Geis. 1983. *Hemoglobin: Structure, Function, Evolution, and Pathology.* Menlo Park, Calif.: Benjamin/Cummings.

Gould, S. J. 1981. *The Mismeasure of Man.* New York: W. W. Norton.

Herrnstein, R., and C. Murray. 1994. *The Bell Curve.* New York: Free Press.

Hershkovitz, I, et al. 1991. Possible congenital hemolytic anemia in pre-historic coastal inhabitants of Israel. *Am. J. Phys. Anthropol.,* 85: 7–13.

Hunt, E. 1995. The role of intelligence in modern society. *Am. Scientist,* 83: 356–368.

Kamin, L. J. 1974. *The Science and Politics of IQ.* Potomac, Md.: L. Erlbaum Associates.

Kay, A. C., et al. 1992. The origin of glucose-6-phosphate dehydrogenase (G-6-PD) polymorphism in African-Americans. *Am. J. Hum. Genet.,* 50(2): 394–398.

Lang, C. C., et al. 1995. Attenuation of isoproterenol-mediated vasodilation in blacks. *New Engl. J. Med.,* 333: 155–160.

Lewontin, R. C., S. Rose, and L. J. Kamin. 1984. *Not in Our Genes.* New York: Pantheon.

Lillyquist, M. J. 1985. *Sunlight and Health.* New York: Dodd, Mead.

Livingstone, F. B. 1985. *Frequencies of Hemoglobin Variants, Thalassemia, the Glucose-6-Phosphate Dehydrogenase Deficiency, G-6-PD Variants and Ovalocytosis in Human Populations.* New York: Oxford University Press.

McKusick, V. A. 1983. *Mendelian Inheritance in Man.* Baltimore: Johns Hopkins University Press.

Meindl, R. S. 1987. Hypothesis: A selective advantage for cystic fibrosis heterozygotes. *Am. J. Phys. Anthropol.,* 74: 39–45.

Osborne, R. T., C. E. Noble, and N. Weyl, eds. 1978. *Human Variation. The Biopsychology of Age, Race, and Sex.* New York: Academic Press.

Papiha, S. S., et al. 1991. Population variation in molecular polymorphisms of the short arm of the human X chromosome. *Am. J. Phys. Anthropol.*, 85: 329–334.

Rickards, O., et al. 1992. Genetic structure of the population of Sicily. *Am. J. Phys. Anthropol.*, 87: 395–406.

Roberts, D. F., and G. F. DeStefano, eds. 1986. *Genetic Variation and Its Maintenance.* Cambridge, U.K.: Cambridge University Press.

Rosatelli, M. C. et al. 1992. Molecular screening and fetal diagnosis of beta-thalassemia in the Italian population. *Hum. Genet.*, 89: 585–589.

Roth, E. F., et al. 1983. Glucose-6-phosphate dehydrogenase deficiency inhibits *in vitro* growth of *Plasmodium falciparum. Proc. Natl. Acad. Sci.*, 809: 298–299.

Rowell, L. B. 1986. *Human Circulation Regulation during Physical Stress.* New York: Oxford University Press.

Saha, N., and J. S.H. Tay. 1992. Origin of the Koreans: A population genetic study. *Am. J. Phys. Anthropol.*, 88: 27–36.

Shanklin, E. 1994. *Anthropology and Race.* Belmont, Calif.: Wadsworth Publishing Co.

Tills, D., A. C. Kopec, and R. E. Tills. 1983. *The Distribution of the Human Blood Groups and Other Polymorphisms*, Supplement I. Oxford, U.K.: Oxford University Press.

Walter, H., H. Matsumoto, and G. F. DeStefano. 1991. *Gm* and *Km* allotypes in four Sardinian population samples. *Am. J. Phys. Anthropol.*, 86: 45–50.

Warren, K. S., and A. A. F. Mahmoud. 1984. *Tropical and Geographical Medicine.* New York: McGraw-Hill.

Weatherall, D. J. 1993. The treatment of thalassemia—slow progress and new dilemmas. *New Engl. J. Med.*, 329: 877–879.

Weatherall, D. J., and J. B. Clegg. 1981. *The Thalassemia Syndromes.* Oxford, U.K.: Blackwell Scientific Publications.

Williams, W. J., E. Beutler, A. J. Erslev, and M. A. Lichtman. 1990. *Hematology*, 4th ed. New York: McGraw-Hill.

Winter, W. P., ed. 1986. *Hemoglobin Variants in Human Populations*, 2 vols. Boca Raton, Fla.: CRC Press.

Chapter 6

Allen, M. C., P. K. Donohue, and A. E. Dusman. 1993. The limit of viability—neonatal outcome of infants born at 22 to 25 weeks' gestation. *New Engl. J. Med.*, 329: 1597–1601.

Alonso, W. 1987. *Population in an Interacting World.* Cambridge, Mass.: Harvard University Press.

Bromwich, P., and T. Parsons. 1984. *Contraception. The Facts.* New York: Oxford University Press.

Brown, L. R. 1991. *The World Watch Reader on Global Environmental Issues.* New York: W. W. Norton.

Bulatao, R. A., and R. D. Lee, eds. 1983. *Determinants of Fertility in Developing Countries*, 2 vols. New York: Academic Press.

Byer, C. O., L. W. Shainberg, and K. L. Jones. 1988. *Dimensions of Human Sexuality*, 2nd ed. Dubuque, Iowa: Wm. C. Brown.

Chesler, E. 1992. *Woman of Valor: Margaret Sanger and the Birth Control Movement in America.* New York: Simon and Schuster.

Chesnais, J.-C. 1992. *The Demographic Transition. Stages, Patterns, and Economic Implications.* Oxford: Clarendon Press.

Donaldson, P. J., and A. O. Tsui. 1990. The international family planning movement. *Population Bull.*, 45(3): 1–47.

Ehrlich, P. R., and A. H. Ehrlich. 1990. *The Population Explosion.* New York: Simon and Schuster.

Hausknecht, R. U. 1995. Methotrexate and misoprostol to terminate early pregnancy. *New Eng. J. Med.*, 333: 537–540.

Knodel, J. 1977. Breast-feeding and population growth. *Science*, 198: 1111–1115.

Lappé, F. M., and R. Schurman. 1990. *Taking Population Seriously.* San Francisco: Institute for Food and Development Policy.

Maloy, K. 1992. *Birth or Abortion? Private Struggles in a Political World.* New York: Plenum.

Malthus, T. R. 1976 [1798]: *An Essay on the Principle of Population.* P. Appleman, ed. New York: W. W. Norton.

Mastroianni, L., Jr., P. J. Donaldson, and T. T. Kane, eds. 1990. *Developing New Contraceptives.* Washington, D.C.: National Academy Press.

Morowitz, H. J., and J. S. Trefil. 1992. *The Facts of Life. Science and the Abortion Controversy.* New York: Oxford University Press.

Noonan, J. T., Jr. 1965. *Contraception: A History of its Treatment by Catholic Theologians and Canonists.* Cambridge, Mass.: Harvard University Press.

Peng, X. 1991. *Demographic Transition in China. Fertility Trends Since the 1950s.* Oxford: Clarendon Press.

Petersen, W. 1975. *Population*, 3rd ed. New York: Macmillan.

Petersen, W. 1979. *Malthus.* Cambridge, Mass.: Harvard University Press.

Riddle, J. M. 1992. *Contraception and Abortion from the Ancient World to the Renaissance.* Cambridge, Mass.: Harvard University Press.

Salvatore, D., ed. 1988. *World Population Trends and their Impact on Economic Development.* New York: Greenwood Press.

Simon, J. L. 1986. *Theory of Population and Economic Growth.* New York: Oxford University Press.

Song, J., and J. Yu. 1988. *Population System Control.* New York: Springer-Verlag.

Spallone, P. 1989. *Beyond Conception. The New Politics of Reproduction.* Granby, Mass.: Bergin & Garvey.

Tien, H. Y., et al. 1992. China's demographic dilemmas. *Population Bull.*, 47(1): 1–44.

United Nations. 1991. *Consequences of Rapid Population Growth in Developing Countries.* New York: Taylor and Francis.

United Nations, Department of International Economic and Social Affairs. 1991. *Measuring the Dynamics of Contraceptive Use.* New York: United Nations.

United Nations. 1992. World Population Monitoring 1991, with Special Emphasis on Age Structure. *Population Studies no. 126.* New York: United Nations.

United Nations. 1993. *The Sex and Age Distributions of the World Populations*, 1992 rev.. New York: United Nations.

Weeks, J. R. 1989. *Population: An Introduction to Concepts and Issues*, 4th ed. Belmont, Calif.: Wadsworth.

Chapter 7

Alexander, R. D., and P. W. Sherman. 1977. Local mate competition and parental investment patterns in the social insects. *Science*, 196: 494–450.

Axelrod, R., and W. D. Hamilton. 1981. The evolution of cooperation. *Science*, 211: 1390–1396.

Boorman, S. A., and P. R. Levitt. 1980. *The Genetics of Altruism.* New York: Academic Press.

Bradbury, J. W., and M. B. Andersson, eds. 1987. *Sexual Selection: Testing the Alternatives.* New York: Wiley.

Clutton-Brock, T. H., ed. 1988. *Reproductive Success.* Chicago: University of Chicago Press.

Clutton-Brock, T. H. 1991. *The Evolution of Parental Care.* Princeton N.J.: Princeton University Press.

Cronin, H. 1991. *The Ant and the Peacock. Altruism and Sexual Selection from Darwin to Today.* Cambridge, U.K.: Cambridge University Press.

Dahlberg, F., ed. 1981. *Woman the Gatherer*. New Haven: Yale University Press.

Daly, M., and M. Wilson. 1983. *Sex, Evolution, and Behavior*, 2nd ed. Belmont, Calif.: Wadsworth.

Devor, E. J., and C. R. Cloninger. 1989. Genetics of alcoholism. *Ann. Rev. Genet.*, 23: 19–36.

Eibl-Eibesfeldt, I. 1989. *Human Ethology*. New York: Aldine de Gruyter.

Emlen, S. T., and L. W. Oring. 1977. Ecology, sexual selection, and the evolution of mating systems. *Science*, 197: 215–233.

Erikson, E. H. 1980. *Identity and the Life Cycle*. New York: W. W. Norton.

Gould, J. L., and C. G. Gould. 1989. *Sexual Selection*. New York: Scientific American Library.

Harlow, H. F. 1959. Love in infant monkeys. *Sci. Amer.*, 200: 68–74.

Harlow, H. F., and C. Mears. 1979. *The Human Model: Primate Perspectives*. Washington: V. H. Winston.

Hausfater, G., and S. B. Hrdy, eds. 1984. *Infanticide. Comparative and Evolutionary Perspectives*. New York: Aldine.

Hrdy, S. B. 1977. *The Langurs of Abu: Female and Male Strategies of Reproduction*. Cambridge, Mass.: Harvard University Press.

Hrdy, S. B. 1981. *The Woman That Never Evolved*. Cambridge, Mass.: Harvard University Press.

Hrdy, S. B. 1986. Empathy, polyandry, and the myth of the coy female. *In* R. Bleier, ed. *Feminist Approaches to Science*. New York, Pergamon, pp. 119–146.

Ito, Y., J. L. Brown, and J. Kikkawa. 1987. *Animal Societies. Theories and Facts*. Tokyo: Japan Scientific Societies Press.

Jolly, A. 1985. *The Evolution of Primate Behavior*, 2nd ed. New York: Macmillan.

Lee, R., and I. DeVore, eds. 1968. *Man the Hunter*. Chicago: Aldine.

LeVay, S. 1993. *The Sexual Brain*. Cambridge, Mass.: MIT Press.

Lewontin, R. C., S. Rose, and L. J. Kamin. 1984. *Not in Our Genes*. New York: Pantheon.

Lorenz, K. 1966. *On Aggression*. New York: Harcourt, Brace & World.

Lorenz, K. 1981. *The Foundations of Ethology*. New York: Springer-Verlag.

Pereira, M. E., and L. A. Fairbanks, eds. 1993. *Juvenile Primates. Life History, Development, and Behavior*. New York: Oxford University Press.

Rasa, A. E., C. Vogel, and E. Voland, eds. 1989. *The Sociobiology of Sexual and Reproductive Strategies*. London: Chapman and Hall.

Rivers, P. C., ed. 1987. *Alcohol and Addictive Behavior*. Lincoln, Neb.: University of Nebraska Press.

Ruse, M. 1979. *Sociobiology: Sense or Nonsense?* Dordrecht, Holland: D. Reidel.

Sahlins, M. 1976. *The Use and Abuse of Biology: An Anthropological Critique of Sociobiology*. Ann Arbor, Mich.: University of Michigan Press.

Trivers, R. I. 1985. *Social Evolution*. Menlo Park, Calif.: Benjamin Cummings.

Trivers, R. I., and H. Hare. 1976. Haplodiploidy and the evolution of the social insects. *Science*, 191: 249–263.

Wilson, E. O. 1975. *Sociobiology, the New Synthesis*. Cambridge, Mass.: Belknap Press, Harvard University Press.

Wynne-Edwards, V. C. 1962. *Animal Dispersion in Relation to Social Behavior*. Edinburgh: Oliver and Boyd.

Chapter 8

Brownell, K. D., and J. P. Foreyt, eds. 1986. *Handbook of Eating Disorders*. New York: Basic Books.

Bruch, H. 1979. *The Golden Cage. The Enigma of Anorexia Nervosa*. New York: Vintage.

Carper, J. 1988. *The Food Pharmacy*. Toronto: Bantam.

Chernin, K. 1985. *The Hungry Self. Women, Eating, and Identity*. New York: Times Books.

Committee on Diet and Health, National Research Council. 1989. *Diet and Health. Implications for Reducing Chronic Disease Risk*. Washington, D.C.: National Academy Press.

Gershwin, M. E. 1985. *Nutrition and Immunity*. Orlando, Fla.: Academic Press.

Gibney, M. J. 1986. *Nutrition, Diet, and Health*. New York: Cambridge University Press.

Gordon, R. A. 1990. *Anorexia and Bulimia. Anatomy of a Social Epidemic*. Cambridge, Mass.: Basil Blackwell.

Graham, G. G. 1993. Starvation in the modern world. *New Engl. J. Med.*, 328: 1058–1061.

Grobstein, C., et al. 1982. *Diet, Nutrition, and Cancer*. Washington, D.C.: National Academy Press.

Hsu, L. K. J., 1990. *Eating Disorders*. New York: Guilford.

Kates, R. W. 1993. Ending deaths from famine. *New Engl. J. Med.*, 328: 1055–1057.

Keil, J. E., et al. 1993. Mortality rates and risk factors for coronary disease in black as compared with white men and women. *New Engl. J. Med.*, 329: 73–78.

Khare, R. S., and M. S. A. Rao, eds. 1986. *Food, Society, and Culture*. Durham, N. C.: Carolina Academic Press.

Lappé, F. M. 1975. *Diet for a Small Planet*. New York: Ballantine.

Lloyd-Still, J. D. 1976. *Malnutrition and Intellectual Development*. Littleton, Mass.: Publishing Sciences Group.

McArdle, W. D., F. I. Katch, and V. L. Katch. 1986. *Exercise Physiology: Energy, Nutrition, and Human Performance*, 2nd ed. Philadelphia: Lea & Febiger.

National Research Council, Committee on Diet, Nutrition, and Cancer. 1982. *Diet, Nutrition, and Cancer*. Washington, D.C.: National Academy Press.

Reid, I. R., et al. 1993. Effect of calcium supplementation on bone loss in postmenopausal women. *New Engl. J. Med.*, 328: 460–464.

Schusky, E. L. 1989. *Culture and Agriculture*. New York: Bergin and Garvey.

Selvini-Palazzoli, M. 1985. *Self-Starvation. From Individual to Family Therapy in the Treatment of Anorexia Nervosa*. New York: Jason Aronson.

Sharman, A., J. Theophano, K. Curtis, and E. Messer. 1991. *Diet and Domestic Life in Society*. Philadelphia: Temple University Press.

Simoons, F. J. 1991. *Food in China. A Cultural and Historical Inquiry*. Boston: CRC Press.

Skolnick, A. A. 1993. "Female athlete triad" risk for women. *JAMA*, 270: 921–923.

Stein, Z., M. Susser, G. Saenger, and F. Marolla. 1975. *Famine and Human Development: The Dutch Hunger Winter of 1944–1945*. New York: Oxford University Press.

Stunkard, A. J., and E. Stellar. 1984. *Eating and Its Disorders*. New York: Raven.

Thomas, P. R. 1991. *Improving America's Diet and Health*. Washington, D.C.: National Academy Press.

Watson, R. R., ed. 1984. *Nutrition, Disease Resistance and Immune Function*. New York: Marcel Dekker.

Williams, S. R. 1985. *Nutrition and Diet Therapy*, 5th ed. St. Louis: Times Mirror/Mosby.

Chapter 9

Ames, B. N., and L. S. Gold. 1992. Animal cancer tests and cancer prevention. *J. Natl. Cancer Inst. Monogr.*, 12: 125–132.

Baquet, C. R., et al. 1991. Socioeconomic factors and cancer incidence among blacks and whites. *J. Natl. Cancer Inst.*, 83: 551–557.

British Society for Cell Biology. 1989. *The Cell Cycle.* Supplement to *J. Cell Sci.* 12. Cambridge, U.K.: Company of Biologists.

Cooper, G. M. 1992. *Elements of Human Cancer.* Boston: Jones and Bartlett.

Furth, M., and M. Greaves, eds. 1989. *Molecular Diagnostics of Human Cancer.* Cold Spring Harbor, N.Y.: Cold Spring Harbor Laboratory.

Greenwald, P. 1992. Keynote address: Cancer prevention. *J. Natl. Cancer Inst. Monogr.,* 12: 9–14.

Hayflick, L. 1994. *How and Why We Age.* New York: Ballantine Books.

Henderson, M. M. 1992. International differences in diet and cancer incidence. *J. Natl. Cancer Inst. Monogr.,* 12: 59–63.

Hirayama, T. 1992. Life-style and cancer: From epidemiological evidence to public behavior change to mortality reduction of target cancers. *J. Natl. Cancer Inst. Monogr.,* 12: 65–74.

Howe, G. R., et al. 1991. A cohort study of fat intake and risk of breast cancer. *J. Natl. Cancer Inst.,* 83: 336–340.

Jordan, V. C., M. K. Lababidi, and S. Langan-Fahey. 1991. Suppression of mouse mammary tumorigenesis by long-term Tamoxifen therapy. *J. Natl. Cancer Inst.,* 83: 492–496.

Kahane, D. H. 1990. *No Less a Woman: Ten Women Shatter the Myths about Breast Cancer.* New York: Prentice-Hall.

LaFond, R. E., ed. 1988. *Cancer: The Outlaw Cell.* Washington, D.C.: American Chemical Society.

Levy, S. M. 1985. *Behavior and Cancer.* San Francisco: Jossey-Bass.

Love, S. M. 1990. *Doctor Susan Love's Breast Book.* Reading, Mass.: Addison-Wesley.

Moss, R. W. 1989. *The Cancer Industry. Unraveling the Politics.* New York: Paragon House.

National Cancer Institute. 1989. *Cancer Statistics Review, 1973–1986.* Bethesda, Md.: U. S. Department of Health and Human Services, National Institutes of Health publication no. 89–2789.

National Research Council, Committee on Diet, Nutrition and Cancer. 1982. *Diet, Nutrition and Cancer.* Washington, D.C.: National Academy Press.

Page, H., and A. J. Asire. 1985. *Cancer Rates and Risks,* 3rd ed. Bethesda

Md.: National Institutes of Health publication DHHS-NIH-85–691.

Pickle, L. W., et al. 1990. *Atlas of U. S. Cancer Mortality among Nonwhites: 1950–1980.* Bethesda, Md.: National Institutes of Health publication no. 90–1582.

Riggan, W. B., et al. 1983. *U.S. Cancer Mortality Rates and Trends, 1950–1979,* 3 vols. Bethesda, Md.: National Cancer Institute.

Ruddon, R. W. 1981. *Cancer Biology.* New York: Oxford University Press.

Spemann, H. 1938. *Embryonic Development and Induction.* New Haven: Yale University Press.

Stellman, S. D., ed. 1986. *Women and Cancer.* New York: Haworth Press.

U.S. Department of Health and Human Services, Public Health Service. 1985. *The Health Consequences of Smoking: Cancer and Chronic Lung Disease in the Workplace: A Report of the Surgeon General.* Rockville. Md.: U. S. Public Health Service document no. DHHS-PHS-85–50207.

Wynder, E. L. 1991. Primary prevention of cancer: planning and policy considerations. *J. Natl. Cancer Inst.,* 83: 475–478.

Chapter 10

Ackerman, S. 1992. *Discovering the Brain.* Washington, D.C.: National Academy Press.

Birchwood, M. J., S. E. Hallett, and M. C. Preston. 1989. *Schizophrenia: An Integrated Approach to Research and Treatment.* New York: New York University Press.

Bloom, F. E., A. Lazerson, and L. Hofstadter. 1985. *Brain, Mind, and Behavior.* New York: W. H. Freeman.

Bradford, H. F. 1986. *Chemical Neurobiology. An Introduction to Neurochemistry.* New York: W. H. Freeman.

Cooper, J. R., F. E. Bloom, and R. H. Roth. 1991. *The Biochemical Basis of Neuropharmacology,* 6th ed. New York: Oxford University Press.

DeGroot, J. 1991. *Correlative Neuroanatomy,* 21st ed. Norwalk, Conn.: Appleton & Lange.

Feldman, R. S., and L. F. Quenzer. 1984. *Fundamentals of Neuropsychopharmacology.* Sunderland, Mass.: Sinauer Associates.

Gazzaniga, M. S. 1988. *Mind Matters: How Mind and Brain Interact to Create our Conscious Lives.* Boston: Houghton Mifflin.

Gormezano, I., and E. A. Wasserman, eds. 1992. *Learning and Memory: The Behavioral and Biological Substrates.*

Hillsdale, N.J.: Lawrence Erlbaum Associates.

Gottesman, I. I. 1991. *Schizophrenia Genesis. The Origins of Madness.* New York: W. H. Freeman.

Hall, Z. W. 1992. *An Introduction to Molecular Neurobiology.* Sunderland, Mass.: Sinauer Associates.

Hamilton, L. W., and C. R. Timmons. 1990. *Principles of Behavioral Pharmacology.* Englewood Cliffs, N.J.: Prentice-Hall.

Hefti, F., and W. J. Weiner, eds. 1988. *Progress in Parkinson Research.* New York: Plenum.

Hobson, J. A. 1988. *The Dreaming Brain.* New York: Basic Books.

Jack, D. C. 1991. *Silencing the Self. Women and Depression.* Cambridge, Mass.: Harvard University Press.

Johnson, G. 1991. *In the Palaces of Memory. How We Build the Worlds inside Our Heads.* New York: Knopf.

Kandel, E. L., and J. H. Schwartz. 1985. *Principles of Neural Science,* 2nd ed. New York: Elsevier.

Kleitman, N. 1963. *Sleep and Wakefulness.* Chicago: University of Chicago Press.

Lister, R. G., and H. J. Weingartner. 1991. *Perspectives on Cognitive Neuroscience.* New York: Oxford University Press.

Martinez, J. L., and R. P. Kesner, eds. 1991. *Learning and Memory. A Biological View,* 2nd ed. New York: Academic Press.

Marx, J. 1993. Alzheimer's pathology begins to yield its secrets. *Science,* 259: 457–458.

Moncade, S., and A. Higgs. 1993. The l-arginine-nitric oxide pathway. *New Engl. J. Med.,* 329: 2002–2012.

Moorcroft, W. H. 1989. *Sleep, Dreaming, and Sleep Disorders.* New York: University Press of America.

Noback, C. R., and R. J. Demarest. 1981. *The Human Nervous System. Basic Principles of Neurobiology,* 3rd ed. New York: McGraw-Hill.

Nolen-Hoeksema, S. 1990. *Sex Differences in Depression.* Stanford, Calif.: Stanford University Press.

Petri, H. L., and M. Mishkin. 1994. Behaviorism, cognitivism, and the neuropsychology of memory. *Am. Scientist,* 82: 30–37.

Sacks, O. 1985. *The Man Who Mistook His Wife for a Hat, and Other Clinical Tales.* New York: Summit.

Scheibel, A. B., A. F. Wechsler, and M. A.B. Brazier. 1986. *The Biological Substrates of Alzheimer's Disease.* New York: Academic Press.

Siegel, G., et al., eds. 1989. *Basic Neurochemistry*, 4th ed. New York: Raven Press.

Soloman, P. R. et al., eds. 1989. *Memory: Interdisciplinary Approaches*. New York: Springer-Verlag.

Weinberger, N. M., J. L. McGaugh, and G. Lynch, eds. 1985. *Memory Systems of the Brain*. New York: Guilford.

Young, D. 1989. *Nerve Cells and Animal Behavior*. Cambridge: Cambridge University Press.

Chapter 11

Bennett, G., C. Vourakis, and D. S. Woolf., eds. 1983. *Substance Abuse. Pharmacologic, Developmental and Clinical Perspectives*. New York: Wiley.

Bowman, W. C., and M. J. Rand. 1980. *Textbook of Pharmacology*, 2nd ed. Oxford, U.K: Blackwell Scientific Publications.

Engel, J., et al., eds. 1987. *Brain Reward Systems and Abuse*. New York: Raven Press.

Eskenazi, B. 1993. Caffeine during pregnancy: Grounds for concern? *JAMA*, 270: 2973–2974.

Goldstein, A. 1994. *Addiction. From Biology to Drug Policy*. New York: W. H. Freeman.

Hamilton, L. W., and C. R. Timmons. 1990. *Principles of Behavioral Pharmacology*. Englewood Cliffs, N.J.: Prentice-Hall.

Hollister, L. E. 1968. *Chemical Psychoses. LSD and Related Drugs*. Springfield, Ill.: Charles C Thomas.

Infante-Rivard, C., et al. 1993. Fetal loss associated with caffeine intake before and during pregnancy. *JAMA*, 270: 2940–2943.

Iversen, L. L., S. D. Iversen, and S. H. Snyder. 1987. *Handbook of Psychopharmacology*, vol. 19. *New Directions in Behavioral Pharmacology*. New York: Plenum.

Jacobs, B. I. 1984. *Hallucinogens: Neurochemical, Behavioral and Clinical Perspectives*. New York: Raven.

Kalivas, P. W. and H. H. Samson, eds. 1992. The neurobiology of drug and alcohol addiction. *Ann. N.Y. Acad. Sci.*, 654: 1–545.

Kane, J. M., & J. A. Liebermann. 1992. *Adverse Effects of Psychotropic Drugs*. New York: Guilford.

Kirsch, M. M. 1986. *Designer Drugs*. Minneapolis: CompCare Publications.

McKim, W. A. 1986. *Drugs and Behavior: An Introduction to Behavioral Pharmacology*. Englewood Cliffs, N.J.: Prentice-Hall.

Pasternak, G. W., ed. 1988. *The Opiate Receptors*. Clifton, N.J.: Humana Press.

Physician's Desk Reference. 1993. Oradell, N.J: Medical Economics Company, Inc.

Sanberg, P. and R. T. M. Krema. 1986. *Over-the-Counter Drugs. The Encyclopedia of Psychoactive Drugs*. New York: Chelsea House.

Segal, B. 1988. *Drugs and Behavior. Cause, Effects and Treatment*. New York: Gardner.

Snyder, S. H. 1989. *Brainstorming: The Science and Politics of Opiate Research*. Cambridge, Mass.: Harvard University Press.

Stimmel, B. 1991. *The Facts about Drug Use*. Yonkers, N.Y.: Consumers Reports Books.

Watson, R. R. 1992. *Drugs of Abuse and Neurobiology*. Boca Raton, Fla.: CRC Press.

Weil, A., and W. Rosen. 1983. *Chocolate to Morphine. Understanding Mind-Active Drugs*. Boston: Houghton Mifflin.

Wysowski, D. K., et al. 1993. Mortality attributed to misuse of psychoactive drugs, 1979–1988. *Public Health Rep.*, 108: 565–570.

Young, D. 1989. *Nerve Cells and Animal Behavior*. Cambridge, U.K: Cambridge University Press.

Chapter 12

Ader, R., D. Felten, and N. Cohen. 1991. *Psychoneuroimmunology*. San Diego: Academic Press.

Ader, R., and N. Cohen. 1993. Psychoneuroimmunology: Conditioning and stress. *Ann. Rev. Psychol.*, 44: 53–85.

Asterita, M. 1985. *The Physiology of Stress, with Special Reference to the Neuroendocrine System*. New York: Human Science Press.

Batuman, O., et al. 1990. Effects of repeated stress on T cell numbers and function in rats. *Brain Behavior and Immunity*, 4: 105–117.

Benson, H. 1975. *The Relaxation Response*. New York: Morrow.

Cooper, C. L. 1984. *Psychosocial Stress and Cancer*. Chichester, U.K.:Wiley.

Dantzer, R. and K. W. Kelley. 1989. Stress and immunity: An integrated view of the relationships between the brain and the immune system. *Life Sci.*, 44: 1995–2008.

Felten, D. L., et al. 1987. Noradrenergic sympathetic neural interactions with the immune system: Structure and function. *Immunol. Rev.*, 100: 225–260.

Hall, H., L. Minnes, and K. Olness. 1993. The psychophysiology of voluntary immunomodulation. *Int. J. Neurosci.*, 69: 221–234.

Holmes, T. H., and R. H. Rahe. 1967. The social readjustment rating scale. *J. Psychosomatic Research* 2: 213–218.

Kiecolt-Glaser, J. K., et al. 1991. Spousal caregivers of dementia victims: Longitudinal changes in immunity and health. *Psychosomatic Med.* 53: 345–362.

McCubbin, J. A., P. G. Kaufman, and C. B. Nemeroff, eds. 1991. *Stress, Neuropeptides, and Systemic Disease*. New York: Academic Press.

Nauts, H. C. 1989. Bacteria and Cancer—Antagonisms and Benefits, *Cancer Surv.*, 8: 713–723.

O'Dorision, M. S. and A. Panerai, eds. 1990. Neuropeptides and immunopeptides, messengers in a neuroimmune axis. *Ann. N.Y. Acad. Sci.*, 594:1–503.

Pert, C., et al. 1985. Neuropeptides and their receptors: A psychosomatic network. *J. Immunol.*, 135: 820s–826s.

Phillips, D. P., and D. G. Smith. 1990. Postponement of death until symbolically meaningful occasions. *JAMA*, 263: 1947–1951.

Plaut, M. 1987. Lymphocyte hormone receptors. *Ann. Rev. Immunol.*, 5: 621–669.

Sandi, C., J. Borrell, and C. Guaza. 1992. Behavioral factors in stress-induced immunomodulation. *Behav. Brain Res.*, 48: 95–98.

Selye, H., ed. 1980. *Selye's Guide to Stress Research*, vol. 1. New York: Scientific and Academic Editions.

Selye, H., ed. 1983. *Selye's Guide to Stress Research*, vol. 2. New York: Scientific and Academic Editions.

Sibinga, N. E. S., et al. 1988. Opioid peptides and opioid receptors in cells of the immune system. *Ann. Rev. Immunol.*, 6: 219–249.

Spector, N. H. 1980. The central state of the hypothalamus in health and disease. *In* P. Morgane and J. Panksepp, eds., *Physiology of the Hypothalamus*, pp. 453–517. New York: Marcel Dekker.

Spiegel, D., et al. 1989. Effect of psychosocial treatment on survival of patients with metastatic breast cancer. *Lancet*, 888–891.

Ursin, H., E. Baade, and S. Levine. 1978. *Psychobiology of Stress*. New York: Academic Press.

Chapter 13

Adachi, A., et al. 1987. Productive, persistent infection of human colorectal cell lines with human immunodeficiency virus. *J. Virology,* 61: 209–213.

Adams, J. 1989. *AIDS: The HIV Myth.* New York: St. Martin's.

AIDS 1993: The unanswered questions. *Science,* 260: 1219–1293.

Baltimore, D. 1995. Lessons from people with nonprogressive HIV infection. *New Engl. J. Med.* 332: 259–260.

Barr, C. E., et al. 1992. Recovery of infectious HIV-1 from whole saliva. *J. Am. Dent. Assoc.,* 123: 37–45.

Bartlett, J. G. and A. K. Finkbeiner. 1991. *The Guide to Living with HIV Infection.* Baltimore: Johns Hopkins University Press.

Blattner, W., R. C. Gallo, and H. M. Temin. 1988. HIV Causes AIDS. *Science,* 242: 514–517.

Booth, W. 1987. AIDS and insects. *Science,* 237: 355–356.

Bryson, Y. J. 1995. Clearance of HIV infection in a perinatally infected infant. *New Engl. J. Med.* 332: 833–838.

Cao, Y., et al. 1995. Virologic and immunologic characterization of long-term survivors of human immunodeficiency virus type 1 infection. *New Engl. J. Med.* 332: 201–208

Clarke, L. K. and M. Potts. 1988. *The AIDS Reader.* Boston: Branden.

Cohen, J. 1995. Can one type of HIV protect against another type? *Science* 268:1566.

Connor, E. M. 1994. Reduction of maternal-infant transmission of human immunodeficiency virus type 1 with zidovudine treatment. *New Engl. J. Med.* 331: 1773–1180.

Duesberg, P. H. 1988. Human immunodeficiency virus and acquired immunodeficiency syndrome: Correlation but not causation. *Proc. Natl. Acad. Sci. USA,* 86: 755–764.

Fitzgibbon, J. E., et al. 1993. Transmission from one child to another of the Human Immunodeficiency Virus Type 1 with a zidovudine-resistance mutation. *New Engl. J. Med.,* 329: 1835–1841.

Frumpkin, L. R. and J. M. Leonard. 1987. *Questions and Answers on AIDS.* Oradell, N.J.: Medical Economics Books.

Grmek, M. D. 1990. *History of AIDS. Emergence and Origin of a Modern Pandemic.* Princeton, N.J.: Princeton University Press.

International Conference on AIDS, VIII. 1992. *Conference Summary Report.* Amsterdam: CONGREX Holland B. V.

Keeling, R. P. 1986. *AIDS on the College Campus. American College Health Association Report.* Rockville, Md.: ACHA.

Kirp, D. L., and R. Bayer, eds. 1992. *AIDS in the Industrialized Democracies.* New Brunswick, N.J.: Rutgers University Press.

Koop, C. E. 1986. *Surgeon General's Report on Acquired Immune Deficiency Syndrome.* Washington, D.C.: U. S. Department of Health and Human Services.

Letvin, N. L. 1993. Vaccines against human immunodeficiency virus—progress and prospects. *New Engl. J. Med.,* 329: 1400–1405.

Levy, J. A. 1988. The transmission of AIDS: The case of the infected cell. *JAMA,* 259: 3036–3037.

Levy, J. A. 1993. Pathogenesis of human immunodeficiency virus infection. *Microbiol. Rev.,* 57: 183–289.

Liebowitch, J. 1985. *A Strange Virus of Unknown Origin.* New York: Ballantine Books.

Nelkin, D., et al., eds. 1991. *A Disease of Society: Cultural and Institutional Responses to AIDS.* New York: Cambridge University Press.

North, R. L., and K. H. Rothenberg. 1993. Partner notification and the threat of domestic violence against women with HIV infection. *New Engl. J. Med.,* 329: 1194–1196.

Padian, N., et al. 1987. Male-to-female transmission of human immunodeficiency virus. *JAMA,* 258: 788–790.

Pantaleo, G., et al. 1995. Studies in subjects with long-term nonprogressive human immunodeficiency virus infection. *New Engl. J. Med.* 332: 209–216.

Root-Bernstein, R. S. 1993. *Rethinking AIDS. The Tragic Cost of Premature Consensus.* New York: The Free Press.

Shilts, R. 1987. *And the Band Played On: Politics, People, and the AIDS Epidemic.* New York: St. Martin's Press.

Travers, K., et al. 1995. Natural protection against HIV-1 infection provided by HIV-2. *Science* 268: 1612–1615.

Chapter 14

Bement, A. L., ed. 1971. *Biomaterials. Bioengineering Applied to Materials for Hard and Soft Tissue Replacement.* Seattle: University of Washington Press.

Calder, W. A. 1984. *Size, Function and Life History.* Cambridge, Mass.: Harvard University Press.

Fry, J. C., et al., eds. 1992. *Microbial Control of Pollution.* New York: Cambridge University Press.

Gerardin, L. 1968. *Bionics.* New York: McGraw-Hill.

Mayor, M. B. and J. Collier. 1994. The technology of hip replacement. *Scientific American Science and Medicine,* May/June 1994.

Metcalf, H. J. 1980. *Topics in Classical Biophysics.* Englewood Cliffs, N.J.: Prentice-Hall, Inc.

Miller, G. T. 1992. *Living in the Environment,* 7th ed. Belmont, Calif.: Wadsworth Publishing Co.

Profio, A. E. 1993. *Biomedical Engineering.* New York: Wiley.

Tortora, G. J., B. R. Funke, and C. L. Case. 1989. *Microbiology,* 3rd ed. Redwood City, Calif.: Benjamin Cummings.

U.S. Congress, Office of Technology Assessment. 1991. *Bioremediation for Marine Oil Spills Background Paper, OTA-BP-O-70.* Washington, D.C.: U. S. Govt. Printing Office.

Vincent, J. F. V. 1982. *Structural Biomaterials.* New York: Wiley.

Vogel, S. 1988. *Life's Devices. The Physical World of Animals and Plants.* Princeton, N.J.: Princeton University Press.

Chapter 15

Alscher, R. G., and J. R. Cumming. 1990. *Stress Responses in Plants: Adaptation and Acclimation Mechanisms.* New York: Wiley-Liss.

Beevers, L. 1976. *Metabolism in Plants.* London: Edward Arnold.

Benton, J. J. 1983. *A Guide for the Hydroponic and Soilless Culture Grower.* Portland, Ore.: Timber Press.

Bottrell, D. R. 1979. *Integrated Pest Management.* Washington, D.C.: Superintendent of Documents, U. S. Government Printing Office.

Bruening, G. 1987. *Tailoring Genes for Crop Improvement: An Agricultural Perspective.* New York: Plenum.

Briggs, D., and S. M. Walters. 1984. *Plant Variation and Evolution.* 2nd ed. Cambridge, U.K.: Cambridge University Press.

Carson, R. 1962. *Silent Spring.* Boston: Houghton Mifflin.

Chrispeels, M. J., and D. E. Sadava. 1994. *Plants, Genes, and Agriculture.* Boston: Jones and Bartlett.

Crawley, M. J., ed. 1986. *Plant Ecology.* Oxford: Blackwell Scientific Publications.

Davies, J, and W. S. Resnikoff. 1992. *Milestones in Biotechnology: Classic Papers on Genetic Engineering.* Boston: Butterworth-Heinemann.

Dixon, R. O. D., and C. T. Wheeler. 1986. *Fixation in Plants*. Glasgow: Blackie.

Dover, M. J. 1985. *A Better Mousetrap: Improving Pest Management for Agriculture*. Washington, D.C.: World Resources Institute.

Fitter, A. H., and R. K. M. Hay. 1981. *Environmental Physiology of Plants*. New York: Academic Press.

Fox, M. W. 1986. *Agricide: The Hidden Crisis that Affects Us All*. New York: Schocken Books.

Fox, M. W. 1992. *Superpigs and Wondercorn. The Brave New World of Biotechnology and Where It All May Lead*. New York: Lyons & Burford.

Grierson, D., and S. N. Covey. 1988. *Plant Molecular Biology*, 2nd ed. Glasgow: Blackie.

Harlan, J. R. 1992. *Crops and Man*, 2nd ed. Madison, Wisc.: American Society of Agronomy.

Hedin, P. A., ed. 1985. *Bioregulators for Pest Control*. Washington, D.C.: American Chemical Society.

Horn, D. J. 1988. *Ecological Approach to Pest Management*. New York: Guilford Press.

Huffaker, C. B., and R. L. Rabb, eds. 1984. *Ecological Entomology*. New York: John Wiley.

Jeffrey, D. W. 1987. *Soil-Plant Relationships. An Ecological Approach*. Portland, Ore.: Timber Press.

Koshland, D. E. 1986. *Biotechnology. The Renewable Frontier*. Washington, D.C.: American Association for the Advancement of Science.

Kung, S. D., and C. J. Arntzen. 1989. *Plant Biotechnology*. Boston: Butterworths.

Kung, S. D., and R. Wu, eds. 1993. *Transgenic Plants*, 2 vols. New York: Academic Press.

Levin, M. A., and H. S. Strauss. 1991. *Risk Assessment in Genetic Engineering*. New York: McGraw-Hill.

Marco, G. J., R. M. Hollingworth, and W. Durham, eds. 1987. *Silent Spring Revisited*. Washington, D.C.: American Chemical Society.

Miller, G. T. 1992. *Living in the Environment*. Belmont, Calif.: Wadsworth.

Rifkin, J. 1984. Algeny: *A New Word—A New World*. New York: Penguin.

Salunkhe, D. K., and S. S. Deshpande, eds. 1991. *Foods of Plant Origin*. New York: Van Nostrand Reinhold.

Slack, A. 1980. *Carnivorous Plants*. Cambridge, Mass.: MIT Press.

Chapter 16
Almeda, F., and C. M. Pringle. 1988. *Tropical Rainforests: Diversity and Conservation*. San Francisco: California Academy of Sciences and Pacific Division, American Association for the Advancement of Science.

Alvarez, L. W., et al. 1980. Extraterrestrial cause for the Cretaceous–Tertiary extinction. *Science*, 208: 1095–1108.

Anderson, A. B., ed. 1990. *Alternatives to Deforestation: Steps toward Sustainable Use of the Amazon Rain Forest*. New York: Columbia University Press.

Blockhus, J. M., et al., eds. 1992. *Conserving Biological Diversity in Managed Tropical Forests*. Cambridge, U.K.: International Union for Conservation of Nature and Natural Resources.

Boo, E. 1990. *Ecotourism: The Potentials and Pitfalls*. Washington: World Wildlife Fund.

Boyle, R. H., and R. A. Boyle. 1983. *Acid Rain*. New York: Nick Lyons Books.

Brown, L. R. 1991. *The World Watch Reader on Global Environmental Issues*. New York: W. W. Norton.

Broad, R., and J. Cavanagh. 1993. *Plundering Paradise. The Struggle for the Environment in the Philippines*. Berkeley, Calif.: University of California Press.

Collins, M. 1990. *The Last Rain Forests: A World Conservation Atlas*. New York: Oxford University Press.

Donovan, S. K., ed. 1989. *Mass Extinctions: Processes and Evidence*. New York: Columbia University Press.

Ehrlich. P. R., and A. H. Ehrlich. 1981. *Extinction. The Causes and Consequences of the Disappearance of Species*. New York: Random House.

Erwin, D. H. 1990. The end-Permian mass extinction. *Ann. Rev. Ecol. Systematics*, 21: 69–91.

Erwin, D. H. 1993. *The Great Paleozoic Crisis: Life and Death in the Permian*. New York: Columbia University Press.

Gentry, A. H., ed. 1992. *Four Neotropical Rainforests*. New Haven: Yale University Press.

Goldfarb, T. D., ed. 1991. *Taking Sides. Clashing Views on Controversial Environmental Issues*, 4th ed. Guilford, Conn.: Dushkin Publishing Group.

Hurst, P. 1990. *Rainforest Politics: Ecological Destruction in South-East Asia*. London: Zed Books.

IUCN Conservation Monitoring Centre. 1986. *IUCN Red List of Threatened Animals*. Gland, Switzerland: International Union for Conservation of Nature and Natural Resources.

Jacobs, M. 1988. *The Tropical Rain Forest. A First Encounter*. New York: Springer-Verlag.

Johnson, R. W., et al., eds. 1987. *The Chemistry of Acid Rain*. Washington, D.C.: American Chemical Society.

Lucas, G., and H. Synge. 1980. *IUCN Plant Red Data Book*. Gland, Switzerland: International Union for Conservation of Nature and Natural Resources.

Martin, P. S., and R. G. Klein, eds. 1984. *Quaternary Extinctions*. Tucson: University of Arizona Press.

Mayr, E., and P. D. Ashlock. 1991. *Principles of Systematic Zoology*, 2nd ed. New York: McGraw-Hill.

Myers, N. 1988. Threatened biotas: "Hot spots" in tropical forests. *Environmentalist*, 8: 187–208.

Myers, N. 1990. The biodiversity challenge: Expanded hot-spots analysis. *Environmentalist*, 10: 243–256.

National Research Council, Committee on Atmospheric Transport and Chemical Transformation in Acid Precipitation. 1983. *Acid Deposition. Atmospheric Processes in Eastern North America*. Washington, D.C.: National Academy Press.

National Research Council, Committee on Monitoring and Assessment of Trends in Acid Deposition. 1986. *Acid Deposition. Long-Term Trends*. Washington, D.C.: National Academy Press.

Nitecki, M. H., ed. 1984. *Extinctions*. Chicago: University of Chicago Press.

Olson, S. L., and H. F. James. 1991. Descriptions of thirty-two new species of birds from the Hawaiian Islands. Part 1: Non-Passeriformes. Part 2: Passeriformes. *Ornithol. Monogr.*, 45: 1–88; 46: 1–88.

Pounds, J. A., and M. L. Crump. 1994. Amphibian declines and climate disturbance: The case of the golden toad and the harlequin frog. *Conservation Biol.*, 8: 72–85.

Raup, D. M. 1991. *Extinction: Bad Genes or Bad Luck?* New York: W. W. Norton.

Signor, P. W. 1990. The geologic history of diversity. *Ann. Rev. Ecol. Systematics*, 21: 509–539.

Terborgh, J. 1992a. *Diversity and the Tropical Rain Forest*. New York: Scientific American Library.

Terborgh, J. 1992b. Maintenance of diversity in tropical forests. *Biotropica*, 24: 283–292.

Thornback, J., and M. Jenkins. 1982. *The IUCN Mammal Red Data Book*.

Gland, Switzerland: International Union for Conservation of Nature and Natural Resources.

Wells, S. M., R. M. Pyle, and N. M. Collins. 1983. *The IUCN Invertebrate Red Data Book*. Gland, Switzerland: International Union for Conservation of Nature and Natural Resources.

Whitmore, T. C. 1991. *An Introduction to Tropical Rain Forests*. Oxford: Clarendon.

Whitmore, T. C., and J. A. Sayer, eds. 1992. *Tropical Deforestation and Species Extinction*. London: Chapman and Hall.

Wilson, E. O., ed. 1988. *Biodiversity*. Washington, D.C.: National Academy Press.

Wilson, E. O., 1992. *The Diversity of Life*. New York: W. W. Norton.

World Conservation Monitoring Centre. 1992. *Global Biodiversity. Status of the Earth: Living Resources*. New York: Chapman and Hall.

Yanarella, E. J., and R. H. Ihara, eds. 1985. *The Acid Rain Debate. Scientific, Economic, and Political Dimensions*. Boulder, Colo.: Westview Press.

Glossary

Abortion: Expulsion or removal of a fetus from the womb prematurely.

Acquired characteristics: Physiological or other changes developed during the lifetime of an individual.

Action potential (spike): A large reversal of polarization in a nerve cell membrane, resulting in a nerve impulse.

Active transport: Use of energy to transport a substance from an area where it is in lower concentration to an area where it is in higher concentration.

Acute effects: Short-term effects, such as those of a drug that disappear once the drug is cleared from the body.

Adaptation: (1) Any trait that increases fitness, or which increases the ability of a population to persist in a particular environment. (2) A physiological change in response to a stimulus that better prepares the body to withstand or react to similar stimuli.

Additive effect: A physiological response produced by two drugs given together that is the same as the sum of the effects of each drug given separately.

Adhesion: Attractive force between two substances, such as between water and a solid surface, or between two cells.

Age pyramid: A diagram that represents the age distribution of a population by a stack of rectangles, each proportional in size to the percentage of individuals in a particular age group.

Agonist: A drug that stimulates a particular receptor or that has a stated effect.

AIDS (Acquired ImmunoDeficiency Syndrome): Impairment of most parts of the immune system resulting from infection with human immunodeficiency virus, accompanied by opportunistic infections or rare cancers and leading, in most cases, to death.

Allele: One of the alternative forms of a gene.

Allen's rule: In any warm-blooded species, populations living in warmer climates tend to have longer and thinner protruding parts (legs, ears, tails, etc.), while the same parts tend to be shorter and thicker in colder climates.

Allometry: A change of proportion with increasing size, such as growth in which shape changes as size increases.

Altruism: Any act that increases another individual's fitness but lowers or endangers one's own fitness.

Amniocentesis: A procedure in which a sample of amniotic fluid is withdrawn from within the uterus during pregnancy in order to permit testing for enzymes or for fetal chromosomes.

Analogy: Resemblance resulting from similar evolutionary adaptation, as in wings of similar shape made of different materials.

Anchorage dependence: Inability of a normal eucaryotic cell to divide unless it is attached to a substrate.

Anemia: A disease symptom in which the blood has a reduced ability to carry oxygen.

Anisogamy: A condition in which gametes (eggs and sperm) differ in size and other characteristics.

Anorexia nervosa: A psychological eating disorder characterized by self-imposed starvation.

Antagonist: A drug that inhibits another or that inhibits a particular receptor.

Antibodies: Proteins, secreted by lymphocytes during an immune response, which bind specifically to the type of molecule that induced their secretion, thus helping protect the body from disease.

Anticodon: A three-nucleotide sequence in a transfer RNA molecule that pairs with a messenger RNA codon.

Antigen: Any molecule or part of a cell that can provoke a response by the immune system.

Antioxidant: A substance that prevents oxidation of a molecule by an oxidizing agent.

Arithmetic increase: Population increase in which a constant number of individuals is added in each time interval.

Artificial insemination: The introduction of sperm into a female's reproductive tract other than by sexual intercourse.

Artificial selection: Consistent differences in the contributions of different genotypes to future generations, brought about among domestic species by human activity.

Asexual reproduction: Reproduction (i.e., increase in the number of individuals) without the recombination of genes.

Assistive devices: Devices that replace or enhance a function, especially in individuals lacking that function.

Atherosclerosis: Deposits of fat, cellular debris, and calcium on the interior walls of the arteries.

ATP: Adenosine triphosphate, a molecule that stores energy in its chemical bonds.

Autonomic nervous system: Part of the peripheral nervous system that regulates "involuntary" physiological processes of the body; consists of the parasympathetic and the sympathetic divisions.

Autotroph: An organism capable of making its own energy-rich organic compounds from inorganic compounds, in most cases using sunlight.

Axon: An extension of a nerve cell that carries an impulse away from the nerve cell body.

B lymphocyte: A type of lymphocyte that makes antibodies.

Bacteriophage: A type of virus that infects bacteria.

Balanced polymorphism: A stable type of polymorphism caused by superior fitness of the heterozygous condition.

Basal ganglia (basal nuclei): A series of clusters of nerve cells deep within the forebrain.

Bergmann's rule: In any warm-blooded species, populations living in colder climates tend to have larger body sizes, compared to smaller body sizes in warmer climates.

Biocompatibility: Ability of a material to be stable in contact with living tissues while also exerting a minimal effect on those tissues.

Biodegradable: Capable of being broken down into simpler materials by decomposer organisms.

Biodegradation: Breaking down a chemical by biological action.

Biodiversity: Diversity among biological species, their genes, and their communities.

Biofeedback: Monitoring one's own physiological activity as an means of learning how to modify this activity, e.g., to reduce stress.

Biology: The scientific study of living systems.

Biomagnification: The increasing concentration of pollutants as one proceeds up the food energy pyramid from one trophic level to the next.

Biomass: The total quantity of living matter of a particular species or group of species.

Biome: A group of similar ecosystems in various locations around the world.

Bioremediation: Human manipulation of naturally occurring biodegradation processes, as in the cleanup of environmental pollution.

Biosphere: The ecosystem that includes the whole Earth and its atmosphere.

Birth control: Any measure intended to prevent unwanted births or to reduce the birth rate.

Birth rate (*B*): The number of births in a given time period divided by the number of individuals in a population at the beginning of that period.

Blood flow: Volume of blood moving past a selected point in a given time interval.

Blood pressure: Force exerted by blood per unit area.

Bulimia: A psychological eating disorder characterized by an overeating binge, followed by self-induced vomiting or laxative abuse.

Calorie: See *Kilocalories*.

Cancer: A group of diseases characterized by DNA mutations in which some cells divide without regard to the growth control signals of other cells.

Carbohydrates: Polar molecules used by organisms as energy sources and consisting of carbon, hydrogen, and oxygen, with hydrogen and oxygen atoms in a 2 to 1 ratio.

Carcinogen: A physical, chemical, or viral agent that induces cancer; its action is called carcinogenesis.

Carrying capacity (*K*): The maximum population size that can persist in a given environment.

Catalyst: A chemical that speeds up the rate of a chemical reaction but is returned to its original form at the end of the reaction and can thus be reused.

Cell cycle: The process by which a cell divides into two cells.

Cell: A unit of living systems that maintains an efficient ratio of surface area to volume; can be either free-living or part of a multicelled organism.

Central males: Males that exert dominance collectively and that are the chief defenders of a baboon troop or similar social group.

Central nervous system: The brain and spinal cord.

Chemical digestion: The use of enzymes and chemical reactions to break down food into molecules absorbable by cells.

Chlorophyll: A green pigment molecule that traps light in the light reactions of photosynthesis.

Chloroplasts: Photosynthetic organelles containing chlorophyll.

Cholesterol: A lipid with a multiringed structure, found in the cell membranes of most animal cells.

Chorionic villus sampling: A procedure in which a tissue sample is taken from part of the placenta for genetic testing.

Chromosomes: Elongated structures that contain DNA and usually protein as well; in animal and plant cells, the chromosomes are contained in the nucleus.

Chronic effects: Lasting or life-long effects.

Circadian rhythm: A biological change whose pattern repeats approximately every 24 hours.

Classical conditioning: A form of learning in which one stimulus (the conditioned stimulus) that repeatedly precedes or accompanies another (the unconditioned stimulus) becomes capable of evoking the response originally elicited only by the unconditioned stimulus.

Classification: An arrangement of larger groups of species that are subdivided into smaller groups on the basis of some organizing principle or theory.

Climax community: A stable community that no longer experiences succession.

Cline: A gradual geographic variation of a trait within a species.

Clone: The genetically identical cells or organisms derived from a single cell or individual by asexual reproduction or cell division.

Codominance: The ability of two alleles to produce their phenotypic effects simultaneously.

Codon: A coding unit of three successive nucleotides in a messenger RNA molecule that together determine an amino acid.

Coenzyme: A nonprotein substance needed for an enzyme to function.

Communicability: The probability that a disease-causing microorganism will be transferred from one individual to another, either directly or indirectly.

Community: A group of species that interact in such a way that a change in the population of one species has consequences for many other species in the community.

Complete protein: A protein that contains all of the amino acids considered essential for human nutrition.

Composites: Materials made of two or more separate materials combined, e.g., in layers, so that each material retains and contributes its own properties.

Concentration gradient: A situation in which the concentration of a substance is different in different locations or on opposite sides of a membrane.

Conditioning: See *Classical conditioning.*

Contact inhibition: The inability of a cell to divide if it is touching other cells.

Contraceptive: Any method that prevents conception (fertilization).

Control group: In an experiment, a group used for comparison. For example, if animals are experimentally exposed to a drug, then a control group might consist of similar animals not exposed to the drug but treated the same in every other way.

Convergence: Independent evolution of similar adaptations in unrelated lineages.

Creationist: One who insists on the biblical account of creation.

Crossing-over: The rearrangement of linked genes when chromosomes break and recombine.

Cytokines: Chemicals that carry information from one cell to another but which have no nutritional value or enzymatic activity of their own.

Cytoplasm: The portion of a cell outside the nucleus but within the plasma membrane.

Cytotoxic (CD8) T lymphocytes: Cells of the immune system that react specifically to a nonself molecule, becoming activated to kill cells bearing that molecule.

Data: Information gathered so as to permit the testing of hypotheses.

Death control: Any measure that reduces the death rate.

Death rate (*D*): The number of deaths in a given time period divided by the number of individuals in the population at the beginning of that period.

Declarative memory: Conscious remembrance of persons, places, things, and concepts, requiring the actions of the hippocampus and the temporal regions of the brain.

Deduction: Logically valid reasoning of the "if . . . then" form.

Demographic momentum: A temporary population increase that can be predicted in a population that has more prereproductive members and fewer postreproductive members than a population with a stable age distribution would have.

Demographic transition: An orderly series of changes in population structure in which the death rate decreases before a similar decrease occurs in the birth rate, resulting in a population increase during the transition period.

Demographics: The mathematical study of populations.

Dendrites: Nerve cell processes that receive signals and respond by conducting impulses toward the nerve cell body.

Deontological ethics: Ethics in which the rightness or wrongness of an act is judged without reference to its consequences.

Dependence: Inability to carry out normal physiological functions without a particular drug.

Desertification: The processes whereby habitats dry up and are replaced by an advancing desert.

Determined: A state of development in which the fate (future identity) of a cell's progeny is predictable.

Differentiation: The process of becoming different; a restriction on the set of future possibilities for a cell's progeny.

Diffusion: A process in which molecules move randomly from an area of high concentration to an area of low concentration until they are evenly distributed.

Diploid: Possessing chromosomes and genes in pairs, as in all somatic cells.

Dispersal: Spreading of a population to new localities.

DNA: Deoxyribonucleic acid, a nucleic acid containing deoxyribose sugar and usually occurring as two complementary strands arranged in a double helix.

Dominant: A trait that is expressed in the phenotype of heterozygotes; an allele that expresses its phenotype even when only one copy of the allele is present.

Dose: The amount of a drug given at one time.

Doubling number (Hayflick limit): The maximum number of times that a cell can divide before it dies.

Doubling time: The time required for the number of individual units in a population to double.

Drug: Any chemical substance that alters the function of a living organism other than by supplying energy or needed nutrients.

Drug abuse (substance abuse): Excessive use of a drug, or use of which causes harm to the individual or society.

Ecological succession: See *Succession.*

Economic impact level (EIL): The smallest population level of a pest species that reduces crop yields by an unacceptable amount.

Ecosystem: A biological community interacting with its physical environment.

Effective dose: The amount of a drug required to produce a desired effect.

Egg: The female gamete, larger than the sperm and nonmotile.

Egoism: An ethical system in which the rightness or wrongness of an act is judged by whether or not it benefits the individual making the decision.

Electrolyte: Charged particles in solution.

Emulsification: Breaking nonpolar substances into small droplets and giving them a polar surface to prevent their coming together and fusing again.

Endangered species: A species threatened with extinction.

Endocytosis: Bringing a particle into a cell by surrounding it with cell membrane.

Endogenous: Originating within the body.

Endosymbiosis: A theory that explains the origin of eucaryotic cells from large procaryotic cells that engulfed and maintained smaller procaryotic cells inside of the larger cells.

Energy-stability-area (ESA) theory: A theory that relates the biodiversity of an ecosystem to the amount of sunlight received, the fluctuating availability of sunlight, and the area required by individuals of each species to meet their needs.

Enzyme: A chemical substance (nearly always a protein) that speeds up a chemical reaction without getting used up in the reaction; a biological catalyst.

Epidemiology: The study of the frequency and patterns of disease in populations.

Epiphyte: A plant that lives upon and derives support, but not nutrition, from another plant.

Equilibrium: A state in which two or more processes are balanced in such a way that no net change occurs, as when certain molecules are produced and destroyed at equal rates.

Estrogen: A hormone that stimulates development of female sex organs prior to reproductive age and the growth of an ovarian follicle each month during the reproductive years.

Estrous period: A period in which a nonhuman female mammal is sexually receptive and sexually active.

Ethical relativism: The principle that different moral codes are appropriate for different societies.

Ethics: The study of moral rules and moral codes.

Eucaryotic: Made of cells that contain various organelles bounded by membranes, including nuclei surrounded by a nuclear envelope; chromosomes are usually multiple and contain protein as well as nucleic acids; cytoskeleton composed of structural and/or contractile fibers of protein.

Eugenics: An attempt to change gene frequencies through selection or changes in fitness. Raising the fitness of desired genotypes is called "positive" eugenics; lowering the fitness of undesired genotypes is called "negative" eugenics.

Euphenics: Measures designed to alter phenotypes (producing phenocopies) without changing genotypes.

Eupsychics: Social and educational measures that accommodate people with differences.

Eusocial: A form of social organization characterized by overlapping generations (parents coexist with offspring), strictly delimited subgroups (castes), and cooperative care of eggs and young larvae.

Euthenics: Measures designed to assist people to overcome some of the consequences of their phenotypes. Wheelchairs and eyeglasses are examples.

Eutrophication: An ecological succession in which a lake becomes filled with vegetation and eventually disappears.

Evolutionarily stable strategy (ESS): Any behavior or strategy that outcompetes all others and therefore increases in frequency within the population.

Exogenous: Originating outside the body.

Experiment: An artificially contrived situation in which hypotheses are tested by comparison to some known condition called the experimental control condition.

Experimental sciences: Sciences that rely primarily on experiments.

Exponential growth: A form of geometric growth without any limit, according to the equation
$$dN/dT = r\,N.$$

Extinct: No longer having any living members.

Extinction: Termination of a lineage without any descendents.

Fairness: The principle that all individuals in similar circumstances should receive similar treatment.

False negative: A negative test result in a sample that actually has the condition being tested for; indicates a lack of sensitivity of the test.

False positive: A positive test result in a sample that does not actually have the condition being tested for; indicates a lack of specificity of the test.

Falsifiable: Capable of being proved false by experience.

Feedback: Any process in which a later step modifies or regulates an earlier step in the process.

Females: Individuals who produce large gametes (eggs).

Fertilizer: Any substance artificially furnished to promote the growth of crops.

Fetal alcohol syndrome: Permanent brain damage and mental retardation, accompanied by abnormal facial features, caused by fetal exposure to alcohol while in the uterus.

Fiber: Material that cannot be digested but is needed to keep material moving through the digestive system at an optimal rate.

Fight–flight response: A set of physiological reactions brought about by the sympathetic nervous system in response to physical or mental alarm stimuli.

Fitness: The ability of a particular individual or genotype to contribute genes to future generations, as measured by the relative number of viable offspring of that genotype in the next generation.

Fossils: The remains or other evidence of life forms of past geological ages.

Founder effect: A type of genetic drift in which the gene frequencies of a population result from the restricted variation present among a small number of founders of that population.

Frame-shift mutation: A mutation caused by the addition or deletion of one or a few base pairs.

Game theory: A mathematical theory used by biologists to describe the evolution of certain structural and behavioral adaptations by processes comparable to game-playing strategies.

Gametes: Reproductive cells (eggs or sperm), containing one copy of each chromosome.

Gene: A portion of DNA that determines a single protein or polypeptide. In earlier use, a hereditary particle.

Gene expression: Transcription and translation of a gene to its protein product.

Gene frequency: The frequency of an allele in a population; the fraction of a population's gametes that carry a particular allele.

Gene therapy: Introduction of genetically engineered cells into an individual for the purpose of curing a disease or a genetic defect.

Genetic drift: Changes in gene frequencies in small to moderate populations as the result of random processes.

Genetic engineering: Direct and purposeful alteration of an individual genotype.

Genetics: The study of heredity, including genes and hereditary traits.

Genome: The total genetic makeup of an individual, including its entire DNA sequence.

Genotype: The hereditary makeup of an organism as revealed by studying its offspring.

Geographic isolation: Geographic separation of populations by an extrinsic barrier such as a mountain range or an uninhabitable region.

Geometric increase or growth: Increase in which a constant fraction (r) of the existing population is added during each successive time interval.

Germ cells: Gametes (egg or sperm cells), or their precursor cells.

Gloger's rule: In any warm-blooded species, populations living in warm, moist climates tend to be darkly colored or black; populations living in warm, arid climates tend to have red, yellow, brown, or tan colors; and populations living in cold, moist climates tend to be pale or white in color.

Group selection: Selection that operates by differences in fitness between social groups.

Growth factors: Chemical messengers which signal cells to divide.

Habitat: The place and environmental conditions in which an organism lives.

Habituation: A form of learning in which an organism learns not to react to a stimulus that is repeated without consequence.

Half-life: For a drug, the time that it takes for the level of the drug in the body to be reduced by half.

Haplodiploidy: A form of sex determination characteristic of the insect order Hymenoptera, in which males have one copy of each chromosome (haploidy) while females have a pair of each type of chromosome (diploidy).

Haploid: Containing only unpaired chromosomes, as in gametes or procaryotic organisms.

Hardy-Weinberg principle: In a large, random-mating population in which selection, migration, and unbalanced mutation do not occur, gene frequencies tend to remain stable from each generation to the next.

Healing: Repair of tissue brought about by immune cells (macrophages and neutrophils) and their cytokines (growth factors).

Health: The ability of an organism to maintain homeostasis or to return to homeostasis after disease or injury.

Helper (CD4) T lymphocytes: Cells of the immune system that react specifically to a nonself molecule by secreting interleukin-2, a cytokine needed for full activity of either B cells or CD8 T cells.

Hemoglobin: The oxygen-carrying protein in red blood cells.

Hereditarian: A position attributing biological differences to heredity rather than to environment.

Heterotroph: An organism not capable of manufacturing its own energy-rich organic compounds and therefore dependent on eating other organisms or their parts to obtain those compounds.

Heterozygous: Possessing two different alleles of the same gene in a genotype.

High blood pressure (hypertension): A condition in which the pressure inside the blood vessels is always above the optimal pressure range for that species.

High-density lipoproteins (HDLs): Proteins that carry lipids away from tissues via the bloodstream; often referred to as "good cholesterol."

High-risk behaviors: Behaviors or actions that increase the probability of undesirable outcomes, such as the transmission of a disease.

High-risk group: A subpopulation of people who share some characteristic (behavioral, geographic, nutritional, or other) and who have a higher frequency of a particular disease than the general population.

Homeostasis: Ability of a complex system (such as a living organism) to maintain conditions within narrow limits. Also, the resulting state of

dynamic equilibrium, in which changes in one direction are offset by other changes that bring the system back to its original state.

Homology: Shared similarity resulting from common ancestry.

Homozygous: Possessing two like alleles of the same gene in a genotype.

Hormone: A chemical messenger, transported through the blood, that affects the activity of the cells in a target tissue.

Human Immunodeficiency Virus (HIV): The virus that causes AIDS, presumably by infecting and inactivating cells of the immune system which bear a molecule called CD4.

Hydroponics: The practice of growing plants without soil.

Hypothesis: A suggested explanation that can be tested.

Immortal: A property of transformed cells that relieves them from having a limit on the number of times they can divide.

Immunity: See *Specific immunity.*

Immunization (vaccination): Artificial exposure to an antigen that evokes a protective immune response against a potential disease-causing antigen similar in structure to the antigen in the vaccine.

Immunodeficiency: A decreased activity of some part of the immune system as the result of genetic, infectious, or environmental factors.

Immunopotentiation: Increasing the strength of future immune functions in any manner that is not antigen-specific.

Immunosuppression: Decreasing the strength of future immune functions in any manner that is not antigen-specific.

In vitro **fertilization:** Fertilization that takes place in glassware or plastic vessels instead of inside the body.

Inborn error of metabolism: One of several genetic conditions in which an important enzyme is missing or nonfunctional.

Inbreeding depression: An increase in the appearance of harmful recessive characteristics as the result of inbreeding.

Inclusive fitness: The total fitness of all individuals sharing one's genotype, including fractional amounts of the fitness of individuals sharing fractions of one's genotype.

Independent assortment, law of: Genes carried on different chromosomes segregate independently of one another; the separation of alleles for one trait has no influence on the

separation of alleles for traits carried on other chromosomes. Also called "Mendel's second law."

Indicator species: A bacterial or other species that serves as a convenient indication of a particular type of pollution.

Induction: Reasoning from specific instances to general principles, as in "these five animals have hearts, so all animals must have hearts."

Inflammation: A physiological response to cellular injury that includes capillary dilation, redness, heat, and immunological activity that stimulates healing and repair.

Informed consent: A voluntary agreement to submit to certain risks by a person who knows and understands those risks.

Instinct: Complex behavior that is innate and need not be learned.

Integrated pest management (IPM): An approach to the management of pest populations that emphasizes biological controls and frequent monitoring of pest populations.

Interbreeding: The mating of unrelated individuals or the exchange of genetic information between populations.

Intrinsic rate of natural increase (*r*): The constant of geometric increase, equal to the population increase during a specified time interval (usually a year) divided by the population size at the beginning of that time interval.

Inverse proportion: A situation in which one variable decreases while the other increases.

Inversion: The turning of part of a chromosome end-to-end.

Irrigation: Supplying of water to crops.

Isogamy: A condition in which gametes are all similar in size.

***K*-selection:** Natural selection that characterizes populations living at or near the carrying capacity (K) of their environments, emphasizing parental care and efficient exploitation of resources.

Karyotype: The chromosomal make-up of an individual.

Kilocalories (kcal): The amount of energy required to raise the temperature of 1000 grams of water 1 degree Celsius; equal to 1000 calories.

Kin selection: Selection that favors characteristics that decrease individual fitness but that are nevertheless favored because they increase inclusive fitness.

Koch's postulates: A set of test results that must be obtained to demonstrate that a particular species of microorganism is the cause of a particular infectious disease.

Krebs cycle: A series of biochemical reactions that break apart pyruvate and use the chemical bond energy to make ATP.

Lamarckism: A theory of evolution in which species evolve in a single ascending line, and in which the use and disuse of organs was thought to result in inherited adaptations to local environments.

Learning: The modification of behavior on the basis of experience.

Limiting amino acid: An amino acid present in small amounts that, when used up, prevents the further synthesis of proteins requiring that amino acid.

Limiting nutrient: Any nutrient whose amounts constrain the growth of an organism or population; supplying greater amounts of this nutrient therefore allows a population of organisms to increase or grow more vigorously.

Lineage: A succession of species in an ancestor-to-descendent sequence.

Linear dominance hierarchy: A social organization in which one individual is dominant to all others, a second individual to all others except the first, and so on.

Linkage: An exception to the law of independent assortment in which genes carried on the same pair of chromosomes tend to assort together, with the parental combinations of genes predominating.

Lipids: Nonpolar molecules formed primarily of carbon and hydrogen, occurring in cell membranes and also used as energy sources.

Locus: The location of a gene on a chromosome.

Logistic growth: Growth that begins exponentially but then levels off to a stable population size (K), acording to the equation

$$\frac{dN}{dT} = rN\left(\frac{K - N}{K}\right)$$

Low-density lipoproteins (LDLs): Proteins that carry lipids to tissues via the bloodstream; often referred to as "bad cholesterol."

Lymphatic circulation: An open circulatory system in vertebrate animals that gathers intracellular fluid and returns it to the blood along with cells of the immune system.

Lymphocytes: Cells of the immune system with specific receptors for antigens.

Malaria: A mosquito-borne disease caused by the single-celled parasite *Plasmodium falciparum* and related species.

Males: Individuals who produce small gametes (sperm).

Malignant: A tumor that has grown through the extracellular matrix.

Marker: An easily detected trait or chromosome pattern that can be used to detect the probable presence of a different gene located nearby or to determine the location of a closely linked gene.

Maternal effect gene: A gene that is transcribed in an egg prior to fertilization.

Mechanical digestion: Breaking food into smaller particles by physical means such as chewing and churning, exposing new surfaces for chemical digestion.

Meiosis: A form of cell division in which the chromosome number is reduced from the diploid to the haploid number. Compare *mitosis*.

Memory: The ability to recall past learning.

Messenger RNA (mRNA): A strand of RNA that leaves the nucleus following transcription and passes into the cytoplasm, where it functions in protein synthesis.

Metastasis: The ability of transformed cells to leave the original tumor, travel through the body, and adhere and form new tumors in other locations.

Mimicry: A situation in which one species of organisms derives benefit from its deceptive resemblance to another species.

Minerals: Inorganic (non-carbon-containing) atoms or molecules needed to regulate chemical reactions in the body.

Mitochondria: Organelles in eucaryotic cells that produce most of the energy-rich ATP that the cells use.

Mitosis: The usual form of cell division, in which the number of chromosomes does not change. Compare *meiosis*.

Model: A mathematical, pictorial, or physical representation of how something is presumed to work.

Monoculture: Growth of only one species in a particular place, as in a field planted with a single crop.

Monogamy: A mating system in which each adult forms a mating pair with only one member of the opposite sex.

Moral rules (morals): Rules governing everyday human conduct.

Mutagen: An agent that causes mutation in DNA.

Mutation: A heritable change in a DNA sequence or gene.

Mutualism: A type of symbiosis in which each of two species benefits the other.

Myelin sheath: A lipid-rich covering that surrounds and insulates many neurons.

Natural immunity: Immunological protection, not specific to any particular antigen, that is inborn, not acquired.

"Natural law" ethics: Ethics in which acts are judged right or wrong according to whether they occur in nature.

Natural selection: Naturally occurring consistent differences in the contributions of different genotypes to future generations.

Naturalistic sciences: Sciences based on the observation of naturally occurring events under conditions in which nature is manipulated as little as possible.

Negative reinforcement: The removal of an unpleasant stimulus that may result in learning.

Nerve: A bundle of axons outside the central nervous system.

Nerve impulse: An electrical excitation that travels along a nerve cell without decreasing in strength.

Neuroendocrine system: The nervous system and the endocrine system considered as an interactive whole.

Neuron: A nerve cell.

Neurotransmitter: Any chemical that transmits a nerve impulse from one cell to another.

Niche: The way of life of a species, or its role in a community.

Noncognitive stimuli: Stimuli received by the organism that are not processed by the brain.

Nonpolar: Having a molecular structure in which electric charges are evenly distributed (or nearly so); nonpolar substances are not stable in water because water is a polar solvent.

Normal science: Science that proceeds step-by-step within a paradigm.

Nucleic acids: DNA and RNA; compounds containing multiple phosphate groups, 5-carbon sugars, and nitrogen-containing bases.

Nucleotide: A compound formed from a nitrogen-containing base, a 5-carbon sugar, and a phosphate group; nucleic acids are made of many such nucleotides strung together.

Nucleus: The central part of an animal or plant or other eucaryotic cell, containing the chromosomes.

Obesity: A condition in which ideal body weight is exceeded by at least 20%.

Oncogene: A gene that leads to the transformation of a cell, which may then lead to cancer.

Opportunistic infection: Infection in a host with suppressed immunity, resulting from microorganisms that are normally present in the host's environment but that do not cause disease in a host with a normal immune system.

Organ: A group of tissues structurally and functionally working together.

Organelles: Specialized membrane-enclosed structures within cells, such as mitochondria, golgi, endoplasmic reticulum, etc.

Organizer: An embryonic tissue whose chemical secretions induce the differentiation of other cells.

Osmosis: Diffusion of water molecules across a semipermeable membrane in response to a concentration gradient of some other molecule or ion.

Osmotic pressure: Fluid pressure that exists as a result of differences in water concentration.

Oxidation: Removal of electrons from an atom or a molecule.

Paradigm: A coherent set of theories, beliefs, values, and vocabulary terms used to organize scientific research.

Parasympathetic nervous system: A division of the autonomic nervous system that brings about the relaxation response and that secretes acetylcholine as the final neurotransmitter.

Parental investment: The energy or resources that a parent invests in the production of offspring and the raising of offspring.

Passive immunity: Antigen-specific immunity acquired by one organism and then transferred to another organism in the form of antibodies or specific immune cells.

Pedigree: A chart showing inheritance of genetic traits within a family.

Peptides: Strings of amino acids too short to be functional proteins.

Perception: The interpretation that the brain gives to a particular stimulus.

Peripheral nervous system: The nervous system except for the brain and spinal cord.

Phenocopy: An altered phenotype that is not the usual product of an individual's genotype. Dyed hair is an example.

Phenotype: The visible characteristics or traits of an organism.

Phenylketonuria (PKU): An "inborn error of metabolism" in which phenylalanine (an amino acid) cannot be broken down and builds up to toxic levels, causing mental retardation if untreated.

Pheromones: Chemical signals by which organisms communicate with other members of their species.

Photosynthesis: A process by which plants and certain other organisms use energy captured from sunlight to build energy-rich organic compounds, especially carbohydrates.

Placebo: Something that lacks the ingredient thought to be effective, usually given as an experimental control for comparison with an active drug given to other subjects.

Placebo effect: Physiological response to a placebo, usually producing the response expected by the subject.

Plasmid: A bacterial DNA fragment that can separate from the main chromosome and later reattach at the point of separation.

Point mutation: A mutation resulting from a change in a single nucleotide.

Polar: Having a molecular structure in which electrons are shared unevenly, producing one part of the bond that has more negative charge than another part; water is polar, therefore other polar molecules are generally stable in contact with water.

Pollution: Contamination of an environment by substances present in undesirable quantities or locations.

Polyandry: An uncommon mating system in which each mating unit consists of one female and many males.

Polygyny: A mating system in which each mating unit consists of one male and many females.

Polymerase chain reaction: An artificial replication process in which many DNA copies are made from each DNA fragment.

Polymorphism: The persistence of several alleles in a population at levels too high to be explained by mutation alone.

Polyploidy: The presence of more than two complete sets of chromosomes, with chromosome numbers that are multiples of the haploid number ($3N$ or higher).

Population: A group of organisms capable of interbreeding among themselves and often sharing common descent as well; a group of individuals within a species living at a particular time and place.

Population control: All measures that limit or reduce the rate of population growth.

Population ecology: The study of populations and the forces that control them.

Population selection: In immunology, a process by which lymphocytes are allowed to develop if their predetermined antigen receptors do not bind to any self-antigen.

Positive reinforcement: A pleasant or pleasurable stimulus.

Potentiality: The range of possible futures for a cell's progeny.

Procaryotic: Containing a simple nuclear region that is never surrounded by a nuclear envelope, and a single chromosome (usually circular) containing nucleic acid only and no protein.

Procedural memory: Memory of how to do things, which does not require the hippocampus and is not necessarily conscious.

Progesterone: A hormone that maintains the uterine lining in its enlarged, blood-rich condition, ready for implantation of a zygote.

Promiscuity: A mating system in which no permanent mating units are formed and in which each adult of either sex mates with many individuals of the opposite sex.

Prospective study: An experimental design in which data are gathered on events as they happen.

Proteins: Molecules built of amino acids linked together in straight chains, which then fold upon themselves to produce complex shapes, functioning most often as enzymes or as structural materials in or around cells or their membranes.

Proto-oncogene: The normal gene from which an oncogene is derived; encodes products that regulate cell division.

Pseudoextinction: Extinction of a taxon by its evolution into something else, thus continuing to have descendents.

Psychoactive drug: Any chemical substance that alters consciousness, mood, or perception.

Psychoneuroimmunology: A theory that postulates that the mind and the body are a single entity interconnected through interactions of the nervous, endocrine, and immune systems.

Public health: Measures designed to improve the health of populations.

***r*-selection:** Natural selection that characterizes populations living far below the carrying capacity of their environments and that favors high rates of reproduction (high *r*) and maximum dispersal ability.

Race: A geographic subdivision of a species distinguished from other subdivisions by the frequencies of a number of genes; a genetically distinct group of populations possessing less genetic variability than the species as a whole. This concept is called the *population genetics race concept* and is distinguished from other, older race concepts by defining race as a characteristic that can only apply to populations, not to individuals. Important older meanings include the following:

Socially constructed race concept: A definition of an oppressed group and the individuals in that group by their oppressors, using whatever cultural or biological distinctions the oppressors wish to use.

Morphological (typological) race concept: A definition of each race by its physical characteristics, based on the assumption that each characteristic is unvarying and reflects an ideal type or form shared by all members of the group.

Racism: A belief that one race is superior to others and has the right to oppress others.

Recapitulation: A theory, no longer accepted, that the sequence of embryonic-through-adult stages in development is a brief summary (a recapitulation) of the sequence of ancestors and that embryos therefore resemble ancestral evolutionary stages.

Receptor: A protein or other molecule that binds to a drug or other chemical substance and responds to the binding by initiating some cellular activity.

Recessive: A trait that is not expressed in heterozygotes; an allele that only expresses its phenotype when no dominant allele of the same gene is present.

Reciprocal altruism: A system in which altruists benefit from the altruism of others.

Relaxation response: A process, mediated by the parasympathetic branch of the autonomic nervous system, in which the stress response is ended, blood pressure and breathing are reduced, and the threshhold of excitation of nerve cells becomes higher.

Replication: A process in which DNA is used as a template to make more DNA.

Reproductive isolating mechanism: Any biological mechanism that hinders the interbreeding of populations belonging to different species.

Reproductive isolation: The existence of biological barriers to interbreeding.

Reproductive strategy: A pattern of behavior and physiology related to reproduction.

Restriction enzyme: An enzyme that cuts nucleic acids by attacking certain DNA sequences only.

Restriction fragment: A DNA fragment created by breaking up DNA with a restriction enzyme.

Retrospective study: An experimental design in which data are collected about events that have already happened.

Reverse transcription: Transcription of complementary DNA from a template of RNA.

RFLP: A **R**estriction **F**ragment **L**ength **P**olymorphism, i.e., variation among individuals in the lengths of selected restriction fragments.

Rights: Any privilege to which individuals automatically have a just claim or to which they are entitled out of respect for their dignity and autonomy as individuals.

Risk: The probability of occurrence of a specified event or outcome.

RNA: Ribonucleic acid, a nucleic acid containing ribose sugar and usually existing in single-stranded form.

Rough-and-tumble play: Play in which large muscle groups are used in pushing, pulling, climbing, and mock fighting with other individuals.

Saturated fats: Lipids with no double bonds between their carbon atoms.

Savanna: A tropical grassland with few or no trees.

Scale of being: In past centuries, a linear arrangement of species in a hierarchy that was assumed to be divinely ordained and continuous (with no gaps).

Scaling: The study of the numerical relations of other variables to size.

Science: An endeavor in which falsifiable hypotheses are systematically tested.

Scientific revolution: The establishment of a new scientific paradigm, including the replacement of any earlier paradigms.

Second messenger: Molecules within the cytoplasm of a cell that carry information from membrane receptors to other locations in the cell.

Segregation, law of: When a heterozygous individual produces gametes, the different alleles separate so that some gametes receive one allele and some receive the other, but no gamete receives both.

Selection: See *Artificial selection, Natural selection,* and *Sexual selection.*

Sensation: Perception of a stimulus.

Sensitivity: The lowest amount of some substance that can be detected by a clinical or other test.

Sensitization: A form of learning in which an intense and often aversive stimulus increases subsequent responses to other stimuli.

Seroconversion: Development in a person or other host of an antibody specific for some microorganism to which they have been exposed, either through infection or vaccination.

Sex-linked: Carried on the X chromosome.

Sexual dimorphism: Differences in size and other physical traits between males and females.

Sexual reproduction: Reproduction in which recombination of genes occurs.

Sexual selection: Consistent differences in the contributions of different genotypes to future generations on the basis of their success in attracting a mate and reproducing.

Side effect: A drug effect other than the one for which the drug was therapeutically intended.

Slash-and-burn agriculture: A form of agriculture in which forests are first cleared by burning.

Social behavior: Any behavior that influences the behavior of other individuals of the same species.

Social organization: A set of behaviors that define a social group and the role of individuals within that group.

Sociobiology: The biological study of social groups and social behavior and their evolution.

Sodium-potassium pump: A group of membrane proteins that can actively transport sodium ions from the inside to the outside of a cell, such as a nerve cell, while actively transporting potassium ions in the opposite direction.

Somatic cells: In multicellular organisms, all cells except gametes. Somatic cells are usually diploid.

Speciation: The process by which a new species comes into being, especially by a single species splitting into two new species.

Species: Reproductively isolated groups of interbreeding natural populations.

Specific immunity: An acquired antigen-specific ability to react to a previously encountered antigen.

Specificity: The degree to which a test will detect only the molecule it is meant to detect and not detect other molecules.

Sperm: The male gamete, smaller and more motile than the egg in most species.

Stem cell: An undifferentiated cell in an adult organism that retains the ability to divide and differentiate but that is still under the regulatory influence of other cells.

"Sticky ends": Short single-stranded sequences at the ends of double-stranded DNA molecules or fragments, so named because complementary single-stranded sequences will stick to one another.

Strategy: A pattern of behavioral and structural characteristics in a species, as viewed by game theory.

Stress: (1) (also called *stress response*): A physiological response or state of heightened activity brought about by the sympathetic nervous system and maintained for a longer time by the immune and endocrine systems. (2) In mechanics, any force that tends to deform an object.

Subspecies: A geographical subdivision of a species, characterized by less genetic variation within the subspecies than in the species as a whole.

Succession: Orderly replacement of one group of species by another in a given location, the process continuing until a climax community is reached.

Surrogate pregnancy: Pregnancy in which a female is under contract to gestate a fetus on behalf of someone else.

Susceptibility: The likelihood that a person who is exposed to a microorganism will become infected with that organism.

Sustainable use: Use that can be maintained without time limits because only renewable resources are exploited at rates that do not exceed their rates of renewal.

Symbiosis: Any situation in which two species live together.

Sympathetic nervous system: A division of the autonomic nervous system that brings about the fight-or-flight response and that secretes epinephrine as the final neurotransmitter.

Synapse: A meeting of cells in which a nerve cell stimulates another cell by secreting a neurotransmitter; the postsynaptic cell must have a receptor to which the neurotransmitter binds.

Synergistic effect: A combination of two causes that lead to an effect greater than the sum of the effects that would have been produced by the two causes independently; for example, a combined effect in which two drugs together produce a greater physiological response than the sum of the effects of each drug given separately.

T lymphocyte: See *Cytotoxic (CD8) T lymphocytes* and *Helper (CD4) T lymphocytes*.

Taxon: A species or any other collective group of organisms.

Taxonomy: The study of how taxa are recognized and how classifications are made.

Territorial behavior: Defense of an area against intruders.

Theory: A coherent set of well-tested hypotheses that guide scientific research.

Threshhold: (1) A minimum level of a drug below which no physiological response can be detected. (2) The minimum level of a stimulus that is capable of producing an action potential.

Tissue: A group of similar cells and their extracellular products that are built together (integrated structurally) and that function together (integrated functionally).

Tolerance: (1) A condition in which a greater amount of a drug is required to produce the same physiological effect that a smaller amount produced originally. (2) In immunology, acquired unreactivity to a specific antigen after contact with that antigen.

Transcription: A process in which DNA is used to make RNA.

Transfer RNA (tRNA): A highly twisted RNA molecule that contains an anticodon; it attaches at one end to an amino acid that it inserts into a growing protein chain.

Transformation: The multistage process that a cell undergoes in changing from a normal cell to an unregulated, less-differentiated, immortal cell lacking contact inhibition and anchorage dependance. Also, in bacteria, a hereditary change caused by the incorpororation of DNA fragments from outside the cell.

Transgenic: Containing genes from another species.

Translation: A process in which amino acids are assembled into a polypeptide chain (part or all of a protein) in a sequence determined by codons in a messenger RNA molecule.

Transmission: The transfer of microorganisms from one individual to another; does not imply any particular route by which the transfer may occur.

Transpiration: Evaporation of water from the leaves of plants.

Tropical rainforest: A tropical biome (occurring between 25°S and 25°N latitude) characterized by tall trees and high rainfall.

True extinction: See *Extinction*.

Tumor: A solid mass of transformed cells that may also contain induced normal cells such as blood vessels.

Tumor initiator: Agents that begin the process of transformation by causing permanent changes in the DNA; mutagens and radiation are tumor initiators.

Tumor promoter: An agent that completes the process of cell transformation after the process is started by a tumor initiator; tumor promoters are not mutagenic by themselves but cause partially transformed cells to go into cell division.

Turgor: Fluid pressure that causes swelling and stiffness in plant cells and some other types of cells.

Typological species concept: An older definition of each species as a fixed and unvarying entity with its own ideal form or type. Also called *morphological species concept* because each species was described in terms of its visible form (morphology).

Typology: The belief in ideal (Platonic) types or forms.

Unsaturated fats: Lipids with one or more double bonds between their carbon atoms.

Utilitarian ethics: Ethics in which the rightness or wrongness of an act is judged according to its consequences.

Vaccine: Harmless (or less harmful) molecules from a disease-causing (or structurally very similar) microorganism, given in order to stimulate the immune system to protect an individual from future exposures to that microorganism.

Vascular plants: Plants containing tissues that efficiently conduct fluids from one part of the plant to another.

Verifiable: Capable of being proved true by experience.

Vestigial: Reduced to a nonfunctional size, but often showing resemblance to functional organs in related species.

Virulence: The ability of a microorganism to cause a disease.

Virus: A particle of nucleic acid (RNA or DNA) enclosed in a protein coat that cannot replicate itself but that can cause a cell to replicate it.

Vitamins: Carbon-containing molecules needed in small amounts to regulate chemical reactions in the body.

Voltage: A difference in electrical potential energy between two points.

Withdrawal: Physiological changes or unpleasant symptoms associated with the cessation of drug taking.

Zygote: The cell that results when a sperm fuses with an egg, doubling the number of chromosomes.

Credits

Chapter 1:
Fig. 1.1. ROSE IS ROSE reprinted by permission of UFS, Inc.

Chapter 2:
Quotation on p. 26 from Rollin, Bernard E., *Animal Rights and Human Morality* (Buffalo, NY: Prometheus Books). Copyright 1992 by Bernard E. Rollin. Used by permission of the publisher.

Chapter 3:
Fig. 3.1. Modified from Sinnott, Dunn, and Dobshansky, *Principles of Genetics*, 5th ed. (McGraw-Hill, 1958), p. 39, fig. 3.5.
Fig. 3.2. Modified from Sinnott, Dunn, and Dobzhansky, *Principles of Genetics*, 5th ed. (McGraw-Hill, 1958), p. 72, fig. 6.1.
Fig. 3.3. Modified from Sinnott, Dunn, and Dobzhansky, *Principles of Genetics*, 5th ed. (McGraw-Hill, 1958), p. 163, fig. 13.2.
Fig. 3.6. Modified from Pelczar, Chan, and Krieg, *Microbiology Concepts and Applications* (McGraw-Hill, 1993), pp. 44–45, figs. 1.22, 1.23, and 1.24.
Fig. 3.7. From Postlethwaite and Hopson, *The Nature of Life*, 2d ed. (McGraw-Hill, 1992), p. 198, fig. 9.11.
Fig. 3.8. Modified from Postlethwaite and Hopson, *The Nature of Life*, 2d ed. (McGraw-Hill, 1992), p. 210, fig. 10.3.
Fig. 3.9. Modified from Postlethwaite and Hopson, *The Nature of Life*, 2d ed. (McGraw-Hill, 1992), p. 209, fig. 10.2.

Fig. 3.10. Modified from Postlethwaite and Hopson, *The Nature of Life*, 2d ed. (McGraw-Hill, 1992), p. 211, fig. 10.4.
Fig. 3.11. From Postlethwaite and Hopson, *The Nature of Life*, 2d ed. (McGraw-Hill, 1992), p. 214, fig. 10.7.
Fig. 3.12. Photo in [F] by E. Kiselva and D. Fawcett/Visuals Unlimited; other portions modified from Postlethwaite and Hopson, *The Nature of Life*, 2d ed. (McGraw-Hill, 1992), pp. 215–216, figs. 10.8 and 10.10.
Fig. 3.13. Modified from Vander, Sherman, and Luciano, *Human Physiology*, 6th ed. (McGraw-Hill, 1994), p. 77, fig. 4-21.
Fig. 3.14C. From Sinnott, Dunn, and Dobshansky, *Principles of Genetics*, 5th ed. (McGraw-Hill, 1958), p. 60, fig. 5.1.
Fig. 3.16. Karyotypes courtesy of Laurent J. Beauregard, Genetics Laboratory, Eastern Maine Medical Center, Bangor ME.
Fig. 3.17. From Postlethwaite and Hopson, *The Nature of Life*, 2d ed. (McGraw-Hill, 1992), p. 247, fig. 12.5.
Fig. 3.18. Karyotypes courtesy of Laurent J. Beauregard, Genetics Laboratory, Eastern Maine Medical Center, Bangor ME. Photo of Klinefelter patient courtesy of March of Dimes Birth Defects Foundation. Photo of Turner patient courtesy of D. S. Borgaonkar, Ph.D., Medical Center of Delaware.
Fig. 3.20A. March of Dimes Birth Defects Foundation.

Fig. 3.20B. Laurent J. Beauregard, Genetics Laboratory, Eastern Maine Medical Center, Bangor ME.
Fig. 3.21. From Postlethwaite and Hopson, *The Nature of Life*, 2d ed. (McGraw-Hill, 1992), p. 261, fig. 12.18.
Box 3.1. Portions modified from Postlethwaite, Hopson, and Veres, *Biology! Bringing Science to Life* (McGraw-Hill, 1991), p. 137, fig. 8.3.
Box 3.2. Modified from Postlethwaite and Hopson, *The Nature of Life*, 2d ed. (McGraw-Hill, 1992), pp. 154, 146–147, 158–159, figs. 7.17, 7.10, 7.20.
Box 3.5. Excerpt from an address by I. King Jordan, "Ethical Issues in the Genetic Study of Deafness," *Annals N.Y. Acad. Sci.*, 630: 236–239.

Chapter 4:
Fig. 4.1. From *Introduction to Evolution*, 3d ed., by Paul A. Moody. Copyright 1953 by Harper & Row, Publishers, Inc. Copyright © 1962, 1970 by Paul Amos Moody. Reprinted by permission of Harper-Collins Publishers, Inc.
Fig. 4.2. From Lack, *Darwin's Finches* (Cambridge University Press, 1947), fig. 3. Reprinted with the permission of Cambridge University Press.
Fig. 4.3. From Postlethwaite and Hopson, *The Nature of Life*, 2d ed. (McGraw-Hill, 1992), p. 721, fig. 35.15.
Fig. 4.4A,B. Photos from the estate of E. B. Ford, courtesy of J. S. Haywood.
Fig. 4.5. From Postlethwaite and Hopson, *The Nature of Life*, 2d ed.

(McGraw-Hill, 1992), p. 682, fig. 33.21.

Fig. 4.6. From Colbert & Morales, *Evolution of the Vertebrates*, 4th ed. (Wiley-Liss, 1991), p. 186, fig. 14-2, as redrawn from G. Heilmann, *Origins of Birds* (D. Appleton & Co., 1927). Copyright © 1991 by John Wiley & Sons, Inc. Reprinted by permission of John Wiley & Sons., Inc.

Fig. 4.7. Modified from Hopson & Wessells, *Essentials of Biology* (McGraw-Hill, 1990), endpapers.

Fig. 4.9. From Postlethwaite and Hopson, *The Nature of Life*, 2d ed. (McGraw-Hill, 1992), p. 678, fig. 33.17.

Fig. 4.10. Modified from Hopson & Wessells, *Essentials of Biology* (McGraw-Hill, 1990), p. 737, fig. 40-5.

Fig. 4.12. From Hopson & Wessells, *Essentials of Biology* (McGraw-Hill, 1990), p. 855, fig. 46-11.

Fig. 4.13. From Hopson & Wessells, *Essentials of Biology* (McGraw-Hill, 1990), p. 857, fig. 46-13.

Fig. 4.14. Reprinted by permission of the publishers from *The Nariokotome Homo erectus Skeleton*, edited by Alan Walker and Richard Leakey, Cambridge, Mass.: Harvard University Press, Copyright © 1993 by Alan Walker and Richard Leakey.

Box 4.1. Modified from Moore, Lalicker, and Fisher, *Invertebrate Fossils* (McGraw-Hill, 1957), pp. 336, 338, 344–345, 364, 394, figs. 9-2, 9-3, 9-4, 9-5, 9-23, 9-47.

Box 4.2. In part from Hopson & Wessells, *Essentials of Biology* (McGraw-Hill, 1990), pp. 736, 734, figs. 40-3, 40-1.

Box 4.3. Portions modified from Postlethwaite and Hopson, *The Nature of Life*, 2d ed. (McGraw-Hill, 1992), pp. 58, 56, fig. 3.7, table 3.1, and from Pelczar, Chan, and Krieg, *Microbiology Concepts and Applications* (McGraw-Hill, 1993), p. 58, fig. 2.2.

Box 4.4. Modified from Hopson & Wessells, *Essentials of Biology* (McGraw-Hill, 1990), p. 337, fig. 19-10.

Chapter 5:

Fig. 5.2. From Buettner-Janusch, *Origins of Man* (John Wiley, 1966), pp. 499, 500, 501. Copyright © 1966 by John Wiley & Sons, Inc. Reprinted by permission of John Wiley & Sons, Inc.

Fig. 5.3. Modified from Carola, Harley, and Noback, *Human Anatomy and Physiology*, 2d ed. (McGraw-Hill, 1992), p. 571, fig. 18.11.

Fig. 5.4. From Stein and Rowe, *Physical Anthropology*, 5th ed. (McGraw-Hill, 1993), p. 190, fig. 8.9.

Fig. 5.6. Courtesy of Patricia Farnsworth, U.M.D.N.J.

Fig. 5.8. From Stein and Rowe, *Physical Anthropology*, 5th ed. (McGraw-Hill, 1993), pp. 126–127, fig. 6.6.

Fig. 5.9. From Stein and Rowe, *Physical Anthropology*, 5th ed. (McGraw-Hill, 1993), p. 177, fig. 8.2.

Box 5.2. Portions modified from Hopson & Wessells, *Essentials of Biology* (McGraw-Hill, 1990), p. 191, fig. 11-6.

Chapter 6:

Fig. 6.4B. Portion modified from Postlethwaite and Hopson, *The Nature of Life*, 2d ed. (McGraw-Hill, 1992), p. 704, fig. 34.20(a).

Box. 6.1. From Postlethwaite and Hopson, *The Nature of Life*, 2d ed. (McGraw-Hill, 1992), pp. 293, 296, figs. 14.5, 14.7.

Box. 6.2. From Postlethwaite and Hopson, *The Nature of Life*, 2d ed. (McGraw-Hill, 1992), p. 297, fig. 14.8.

Chapter 7:

Fig. 7.1. Modified from Postlethwaite, Hopson, and Veres, *Biology! Bringing Science to Life*, (McGraw-Hill, 1991), p. 614, drawings 1, 2, 3.

Fig. 7.2. Max and Bea Hunn/Visuals Unlimited.

Fig. 7.3A. Richard Gross.

Fig. 7.3B. John D. Cunningham/Visuals Unlimited.

Fig. 7.3C. Len Rue, Jr./Visuals Unlimited.

Fig. 7.3D. Courtesy of David Baker.

Fig. 7.5. From Hopson & Wessells, *Essentials of Biology* (McGraw-Hill, 1990), p. 840, fig. 45-9.

Fig. 7.6A. From Postlethwaite and Hopson, *The Nature of Life*, 2d ed. (McGraw-Hill, 1992), p. 789, fig. 38.25.

Fig. 7.6B. Courtesy of Paul W. Sherman, Cornell University.

Fig. 7.8A. David S. Addison/Visuals Unlimited.

Fig. 7.8B. Richard Gross.

Fig. 7.9A. Irene Vandermolen/Visuals Unlimited.

Fig. 7.9B. Leonard Lee Rue III/Visuals Unlimited.

Fig. 7.9C. James R. McCullagh/Visuals Unlimited.

Fig. 7.10. Harlow Primate Laboratory, University of Wisconsin.

Fig. 7.11. From Hinde, *Biological Basis of Human Social Behavior* (McGraw-Hill, 1974), pp. 294–295, figs. 18.1, 18.2.

Fig. 7.12A. Irven DeVore/Anthro-Photo #5045.

Fig. 7.12B. From Postlethwaite, Hopson, and Veres, *Biology! Bringing Science to Life*, (McGraw-Hill, 1991), p. 332, drawing 4.

Fig. 7.12C. From Hinde, *Biological Basis of Human Social Behaviour* (McGraw-Hill, 1974), p. 333, fig. 21.1.

Chapter 8:

Fig. 8.1. Portions redrawn from Vander, Sherman, and Luciano, *Human Physiology*, 6th ed. (McGraw-Hill, 1994), p. 562, fig. 8.1, and Carola, Harley, and Noback, *Human Anatomy and Physiology*, 2d ed. (McGraw-Hill, 1992), p. 778, fig. 8.1.

Fig. 8.3. Portions redrawn from Van Wynsberghe, Noback, and Carola, *Human Anatomy and Physiology*, 3d ed. (McGraw-Hill, 1995), p. 63, fig. 3.4.

Fig. 8.4. Portions modified from Postlethwaite, Hopson, and Veres, *Biology! Bringing Science to Life* (McGraw-Hill, 1991), p. 37, fig. 2.20.

Fig. 8.5. Portions modified from Postlethwaite, Hopson, and Veres, *Biology! Bringing Science to Life* (McGraw-Hill, 1991), p. 38, fig. 2.21.

Box 8.1. Portions redrawn from Carola, Harley, and Noback, *Human Anatomy and Physiology*, 2d ed. (McGraw-Hill, 1992), p. 58, fig. 3.2, and from Postlethwaite and Hopson, *The Nature of Life*, 2d ed. (McGraw-Hill, 1992), p. 44, fig. 2.23(a).

Box 8.3. In part from Postlethwaite, Hopson, and Veres, *Biology! Bringing Science to Life*, (McGraw-Hill, 1991), p. 100, fig. 5.16.

Box 8.4. Recipe for Chile Con Elote excerpted from *Laurel's Kitchen, A Handbook for Vegetarian Cookery and Nutrition*, copyright © 1976 by L. Robertson, Flinders, and Godfrey, with permission from Ten Speed Press, P.O. Box 7123, Berkeley, CA 94707. Recipe for Carrot and Onion Soup excerpted from *Diet for a Small Planet*, 10th anniversary edition (Ballantine Books, 1982), p. 301.

Chapter 9:

Fig. 9.1. From Pelczar, Chan, and Krieg, *Microbiology Concepts and Applications* (McGraw-Hill, 1993), p. 274, fig. 10.1.

Fig. 9.2. Modified from Postlethwaite and Hopson, *The Nature of Life*,

2d ed. (McGraw-Hill, 1992), pp. 382–383, figs. 18.5(b), 18.7.

Fig. 9.3. From Vander, Sherman, and Luciano, *Human Physiology*, 6th ed. (McGraw-Hill, 1994), p. 74, fig. 4-19.

Fig. 9.4. From Postlethwaite and Hopson, *The Nature of Life*, 2d ed. (McGraw-Hill, 1992), p. 151, fig. 7.13.

Fig. 9.6. From Postlethwaite and Hopson, *The Nature of Life*, 2d ed. (McGraw-Hill, 1992), p. 220, fig. 10.14.

Fig. 9.8. Modified from Hopson & Wessells, *Essentials of Biology* (McGraw-Hill, 1990), pp. 280, 289, figs. 16-12(b), 17-2.

Fig. 9.11. From Hopson & Wessells, *Essentials of Biology* (McGraw-Hill, 1990), p. 291, fig. 17-4.

Fig. 9.13. From Carola, Harley, and Noback, *Human Anatomy and Physiology*, 2d ed. (McGraw-Hill, 1992), p. 89, fig. 3.23.

Fig. 9.14. From Postlethwaite and Hopson, *The Nature of Life*, 2d ed. (McGraw-Hill, 1992), p. 286, fig. 13.22.

Fig. 9.15. From R. Doll and A. B. Hill, *British Medical Journal* 1: 1399–1410 (1964).

Table 9.1. From Hopson & Wessells, *Essentials of Biology* (McGraw-Hill, 1990), p. 292, table 17-2.

Chapter 10:

Fig. 10.1 From Tullar, *The Human Species* (McGraw-Hill, 1977), p. 18, fig. 1-8.

Fig. 10.2. Modified from Noback and Demarest, *The Human Nervous System*, 3d ed. (McGraw-Hill, 1981), p. 6, fig. 1-6.

Fig. 10.3A. Modified from Van Wynsberghe, Noback, and Carola, *Human Anatomy and Physiology*, 3d ed. (McGraw-Hill, 1995), p. 390, fig. 13.3.

Fig. 10.4A. Modified from Van Wynsberghe, Noback, and Carola, *Human Anatomy and Physiology*, 3d ed. (McGraw-Hill, 1995), p. 356, fig. 12.2.

Fig. 10.4B,C. From Vander, Sherman, and Luciano, *Human Physiology*, 6th ed. (McGraw-Hill, 1994), p. 182, fig. 8-3.

Fig. 10.5A. In part from Hopson & Wessells, *Essentials of Biology* (McGraw-Hill, 1990), p. 616, fig. 34-3(a).

Fig. 10.6. Portion redrawn from Hopson & Wessells, *Essentials of Biology* (McGraw-Hill, 1990), p. 618, fig. 34-6.

Fig. 10.7A. From Postlethwaite and Hopson, *The Nature of Life*, 2d ed.

(McGraw-Hill, 1992), p. 556, fig. 27.5.

Fig. 10.8. From Hopson & Wessells, *Essentials of Biology* (McGraw-Hill, 1990), p. 621, fig. 34-9.

Fig. 10.10A. From Carola, Harley, and Noback, *Human Anatomy and Physiology*, 2d ed. (McGraw-Hill, 1992), p. 420, fig. 14.17(b).

Fig. 10.10B. From Vander, Sherman, and Luciano, *Human Physiology*, 6th ed. (McGraw-Hill, 1994), p. 372, fig. 13-6.

Chapter 11:

Fig. 11.1. Modified from Carlson, *Patten's Foundations of Embryology*, 5th ed. (McGraw-Hill, 1988), pp. 279, 281, figs. 7-21(C), 7-22.

Box 11.1. Portions from Vander, Sherman, and Luciano, *Human Physiology*, 6th ed. (McGraw-Hill, 1994), pp. 475, 477, figs. 15-1, 15-3(B).

Box 11.2A,B. From Hopson & Wessells, *Essentials of Biology* (McGraw-Hill, 1990), p. 599, fig. 33-5(a,b).

Box 11.2C. Modified from Postlethwaite and Hopson, *The Nature of Life*, 2d ed. (McGraw-Hill, 1992), p. 523, fig. 25.10(a).

Table 11.3. From Segal, *Drugs and Behavior* (Gardner Press, 1988), tables 11-2 and 15-1, pp. 258–259, 369.

Chapter 12:

Fig. 12.1. From Van Wynsberghe, Noback, and Carola, *Human Anatomy and Physiology*, 3d ed. (McGraw-Hill, 1995), p. 711, fig. 22.1(A).

Fig. 12.3. From Van Wynsberghe, Noback, and Carola, *Human Anatomy and Physiology*, 3d ed. (McGraw-Hill, 1995), p. 740, fig. 23.10.

Fig. 12.4. Portion redrawn from Van Wynsberghe, Noback, and Carola, *Human Anatomy and Physiology*, 3d ed. (McGraw-Hill, 1995), pp. 66–67, fig. 3.5.

Fig. 12.6. From Hopson & Wessells, *Essentials of Biology* (McGraw-Hill, 1990), p. 548, fig. 30-14.

Fig. 12.7. Modified from Hopson & Wessells, *Essentials of Biology* (McGraw-Hill, 1990), p. 626, fig. 34-14.

Fig. 12.8. Modified from Carola, Harley, and Noback, *Human Anatomy and Physiology*, 2d ed. (McGraw-Hill, 1992), p. 534, fig. 17.13.

Fig. 12.9. Courtesy of David and Suzanne Felten, University of Rochester.

Chapter 13:

Fig. 13.2. From Van Wynsberghe, Noback, and Carola, *Human Anatomy and Physiology*, 3d ed. (McGraw-Hill, 1995), p. 752, unnumbered fig.

Fig. 13.3. Modified from Postlethwaite and Hopson, *The Nature of Life*, 2d ed. (McGraw-Hill, 1992), p. 477, fig. 22.14.

Fig. 13.5. In part from Postlethwaite and Hopson, *The Nature of Life*, 2d ed. (McGraw-Hill, 1992), p. 201, fig. 1(a).

Fig. 13.6. Modified from Pelczar, Chan, and Krieg, *Microbiology Concepts and Applications* (McGraw-Hill, 1993), p. 626, fig. 23.9.

Box 13.2. Photo courtesy of Mark Dixon and Robert Thomas.

Chapter 14:

Fig. 14.5. From Vincent, *Structural Biomaterials* (John Wiley, 1982), p. 175, fig. 6.33.

Fig. 14.6. Frog photo from Richard Gross.

Fig. 14.8. Kevin and Betty Collins/Visuals Unlimited.

Fig. 14.11. From Hopson & Wessells, *Essentials of Biology* (McGraw-Hill, 1990), p. 489, fig. 27-5.

Fig. 14.12. From Hopson & Wessells, *Essentials of Biology* (McGraw-Hill, 1990), p. 488, fig. 27-4.

Fig. 14.13. Richard Gross.

Fig. 14.14. Hank Andrews/Visuals Unlimited.

Fig. 14.17. Courtesy of Elin Haugen, Bigelow Laboratory.

Box 14.1. Redrawn from Postlethwaite and Hopson, *The Nature of Life*, 2d ed. (McGraw-Hill, 1992), p. 455, fig. 21.9, and from Hopson & Wessells, *Essentials of Biology* (McGraw-Hill, 1990), p. 519, fig. 29-5.

Box 14.2. In part from Hopson & Wessells, *Essentials of Biology* (McGraw-Hill, 1990), p. 486, fig. 27-2.

Chapter 15:

Fig. 15.2. From Hopson & Wessells, *Essentials of Biology* (McGraw-Hill, 1990), p. 139, fig. 8-6.

Fig. 15.3. Modified from Hopson & Wessells, *Essentials of Biology* (McGraw-Hill, 1990), pp. 138, 143, figs. 8-4, 8-9.

Fig. 15.4. From Hopson & Wessells, *Essentials of Biology* (McGraw-Hill, 1990), p. 485, fig. 27-1.

Fig. 15.6. From Postlethwaite and Hopson, *The Nature of Life*, 2d ed. (McGraw-Hill, 1992), p. 740, fig. 36.12.

Fig. 15.7. Root nodule photo by C. P. Vance/Visuals Unlimited; other portions from Hopson & Wessells, *Essentials of Biology* (McGraw-Hill, 1990), p. 494, fig. 27-8.

Fig. 15.8. From Tullar, *The Human Species* (McGraw-HIll, 1977), p. 320, fig. 11-7.

Fig. 15.9. From Judson and Kauffman, *Physical Geology*, 8th ed., © 1990, pp. 88–89, figs. 5.18, 5.20. Reprinted by permission of Prentice-Hall, Inc., Englewood Cliffs, N.J.

Fig. 15.10. From *Agronomy Journal*, vol. 44 (1952), p. 61, by permission of the American Society of Agronomy.

Box 15.1. Photos on p. 441 from Richard Gross. Drawings on p. 441 and photo on bottom of p. 442 from Slack, *Carnivorous Plants* (MIT Press, 1980), pp. 19, 59, 126; copyright 1980 by MIT Press. Top left photo on p. 442 from Howard A. Miller, Sr./Visuals Unlimited. Top right photo on p. 442 from Cabisco/Visuals Unlimited.

Chapter 16:

Fig. 16.1. Reprinted by permission of the publishers from *The Diversity of Life* by Edward O. Wilson, Cambridge, Mass.: The Belknap Press of Harvard University Press, Copyright © 1992 by Edward O. Wilson.

Fig. 16.2. Includes Moore, Lalicker, and Fisher, *Invertebrate Fossils* (McGraw-Hill, 1957), portions of figs. 3-4, 4-17, 4-29, 5-9, 6-20, 6-24, 6-36, 13-7, 13-20, 18-20, 18-29, 23-1.

Fig. 16.3. *Ginkgo* and *Limulus* from Palmer & Fowler, *Fieldbook of Natural History*, 2d ed. (McGraw-Hill, 1974), pp. 112, 433. *Lingula* from Hyman, *The Invertebrates, vol. 5* (McGraw-Hill, 1959), p. 519, fig. 183C. *Latimeria* from Weichert, *Anatomy of the Chordates* (McGraw-Hill, 1970), p. 27, fig. 2.18.

Fig. 16.4A. Photo by Don Prothero. Other portions from Moore, Lalicker, and Fisher, *Invertebrate Fossils* (McGraw-Hill, 1957), portions of figs. 9-40, 9-41.

Fig. 16.4B. Painting by Rudolph Zallinger, Courtesy of the Peabody Museum of Natural History, Yale University.

Figs. 16.5, 16.6. Reprinted by permission of the publishers from *The Diversity of Life* by Edward O. Wilson, Cambridge, Mass.: The Belknap Press of Harvard University Press, Copyright © 1992 by Edward O. Wilson.

Fig. 16.7. From Hopson & Wessells, *Essentials of Biology* (McGraw-Hill, 1990), p. 756, fig. 41-7.

Fig. 16.8. Photo courtesy of Sharon Kinsman. Other portions from Postlethwaite and Hopson, *The Nature of Life*, 2d ed. (McGraw-Hill, 1992), p. 756, fig. 37.8.

Fig. 16.9. Courtesy of Sharon Kinsman.

Fig. 16.10. Courtesy of Sharon Kinsman.

Fig. 16.12. Kjell B. Sandved/Visuals Unlimited.

Fig. 16.13A. G. Prance/Visuals Unlimited.

Fig. 16.13B. Courtesy of Sharon Kinsman.

Fig. 16.14. Modified from Hopson & Wessells, *Essentials of Biology* (McGraw-Hill, 1990), p. 755, fig. 41-6.

Fig. 16.15A. Leonard Lee Rue, III/Visuals Unlimited.

Fig. 16.16A. Richard Gross.

Fig. 16.16B. From D. W. Schindler et al. Long-term ecological stress: The effects of years of experimental acidification on a small lake, *Science*, 228: 1395–1401 (1985). Copyright 1985 by the AAAS.

Fig. 16.16C. John D. Cunningham/Visuals Unlimited.

Color figures:

A. R. Calentorce/Visuals Unlimited.

B. From Wickler, *Mimicry in Plants and Animals* (McGraw-Hill, 1968), p. 29, fig. 4b.

C. Courtesy of Sharon Kinsman.

D. Richard Thom/Visuals Unlimited.

E. G. Prance/Visuals Unlimited.

F. Courtesy of Sharon Kinsman.

G. Max and Bea Hunn/Visuals Unlimited.

H,I,J. Courtesy of Sharon Kinsman.

K. From Hopson & Wessells, *Essentials of Biology* (McGraw-Hill, 1990), p. 787, fig. 42-13; adapted from Wessells, *Biology* (Random House, 1988), p. 1125, fig. 45-21.

L. Hal Beral/Visuals Unlimited.

M,N. Michael Fogden.

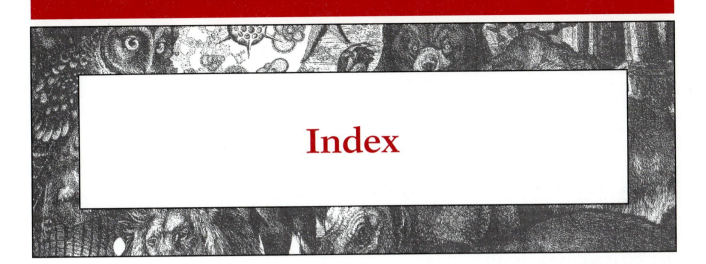

Index

The color figures will be found following page 490.

aberration, chromosomal, 55
ablation, 289
ABO blood groups, 140, 142–144
abortion, 183, 185, 504
absolute dating, 103
absorption, 221
    water, 225
abstinence, 177
abstraction, 306
abuse:
    drug, 329, 333
    substance, 329
Acanthocephala, 493
accessory pigments, 431
acetaldehyde dehydrogenase, 315
acetyl coenzyme A, 231
acetylcholine, 295–296, 352
acid:
    bog, 440
    nitric, 487
    rain, 487–488
    sulfuric, 487
acid-tolerant plants, 440
acquired characteristics, 90, 504
acquired immunodeficiency
  syndrome, 367–397
ACTH, 354–355
actin, 296, 405
Actinopodia, 491
action potential, 292, 294, 504
activation, 347
active transport, 224–225, 504
active traps, 440, 442
activity, drug, 309
acute effects, 329, 504
ADA, 80–81
adaptation, 90, 98, 149, 194, 196, 504
    physiological, 304

adaptive technologies, 409
ADD, 322
addiction, drug, 324–329
Addison's disease, 360
additive effect, 314, 504
additives, drug, 331
adenine, 46
adenosine deaminase, 80–81
ADHD, 322
adhesion, 254, 264, 406, 504
    capillary, 406
adipose cells, 229
adoption, 66
adoption studies, 66, 193
adrenocorticotrophic hormone, 354
adult cancers, 267
Africa, climate, 480–481
age distribution, 170–171
age pyramid, 170–171, 504
age structure, 170
agents of selection, 97
Agnatha, 494
agonist, 314, 504
agricultural pests, 449–452
agriculture, 439, 443–459
    and biodiversity, 462
    diversified, 244
    slash-and-burn, 481
    sustainable, 484
Agrobacterium, 455–456
AIDS, 367–397, 504
    and mosquitoes, 390
    transmission, 388
AIDS-related complex, 373, 381
air pollution, 486
    indoor, 486
alarm, 354
alarm signals, 196
albinism, 57–59
alcohol, 316, 323–325, 329–330, 332
    and cancer, 272

alcohol (continued)
    dehydrogenase, 315
    metabolism of, 315
alcoholism, 214
alfalfa, 452
algae, 426, 431, 491–492
alkaloids, 436
alkaptonuria, 57, 59
alleles, 38, 504
    blood group, 143
Allen's rule, 159, 504
allergens, 350
allergies, 349–350
allometry, 400–401, 504
altered perception, 323
alternative technologies, 399
altruism, 198, 504
    reciprocal, 200–202
Alvarez, L., 467
Alvarez, W., 467
Alzheimer's disease, 84, 299
Ames test, 278–279
amino acid, 53, 56, 230, 234, 318
ammonia, 436, 438, 445
ammonifying bacteria, 438
ammonoids, 468, 493
amniocentesis, 67, 504
amphetamines, 322
Amphibia, 494
amphibians, 487
amphioxus, 493
amylase:
    pancreatic, 221
    salivary, 219
anaerobic, 118, 425
analogy, 100, 504
anchorage dependence, 254, 264, 504
anemia, 152, 504
    hemolytic, 156
    sickle-cell, 152–155